ECOSYSTEM ECOLOGY

ECOSYSTEM ECOLOGY

Editor-in-Chief

SVEN ERIK JØRGENSEN

Copenhagen University,
Faculty of Pharmaceutical Sciences,
Institute A,
Section of Environmental Chemistry, Toxicology and Ecotoxicology,
University Park 2,
Copenhagen Ø, 2100,
Denmark

ELSEVIER

AMSTERDAM   BOSTON   HEIDELBERG   LONDON   NEW YORK   OXFORD
PARIS   SAN DIEGO   SAN FRANCISCO   SINGAPORE   SYDNEY   TOKYO

Elsevier B.V.
Radarweg 29, 1043 NX Amsterdam, The Netherlands

First edition 2009

British Library Cataloguing in Publication Data
A catalogue record for this book is available from the British Library

Library of Congress Catalog Number: 2009929111

ISBN: 978-0-444-53466-8

For information on all Elsevier publications
visit our website at books.elsevier.com

CONTENTS

ECOSYSTEMS AS SYSTEMS

ECOSYSTEM PROPERTIES

ECOSYSTEMS

LIST OF CONTRIBUTORS

W H Adey
Smithsonian Institution, Washington, DC, USA

C Alfsen
UNESCO, New York, NY, USA

T F H Allen
University of Wisconsin, Madison, WI, USA

S Allesina
University of Michigan, Ann Arbor, MI, USA

T R Anderson
National Oceanography Centre, Southampton, UK

O Andrén
TSBF-CIAT, Nairobi, Kenya

A Basset
Università del Salento – Lecce, Lecce, Italy

K M Bergen
University of Michigan, Ann Arbor, MI, USA

C L Bonin
University of Wisconsin, Madison, WI, USA

H Bossel
University of Kassel (retd.), Zierenberg, Germany

K A Brauman
Stanford University, Stanford, CA, USA

J M Briggs
Arizona State University, Tempe, AZ, USA

D E Burkepile
Georgia Institute of Technology, Atlanta, GA, USA

T V Callaghan
Royal Swedish Academy of Sciences Abisko Scientific Research Station, Abisko, Sweden

J L Casti
International Institute for Applied System Analysis, Laxenburg, Austria

L M Chu
The Chinese University of Hong Kong, Hong Kong SAR, People's Republic of China

E A Colburn
Harvard University, Petersham, MA, USA

J Colding
Royal Swedish Academy of Sciences, Stockholm, Sweden

S L Collins
University of New Mexico, Albuquerque, NM, USA

W H Conner
Baruch Institute of Coastal Ecology and Forest Science, Georgetown, SC, USA

K W Cummins
Humboldt State University, Arcata, CA, USA

W S Currie
University of Michigan, Ann Arbor, MI, USA

G C Daily
Stanford University, Stanford, CA, USA

R F Dame
Charleston, SC, USA

D L DeAngelis
University of Miami, Coral Gables, FL, USA

S Dudgeon
California State University, Northridge, CA, USA

T Elmqvist
Stockholm University, Stockholm, Sweden

B D Fath
Towson University, Towson, MD, USA and International Institute for Applied System Analysis, Laxenburg, Austria

J A D Fisher
University of Pennsylvania, Philadelphia, PA, USA

D G Green
Monash University, Clayton, VIC, Australia

N B Grimm
Arizona State University, Tempe, AZ, USA

R Harmsen
Queen's University, Kingston, ON, Canada

T K Harms
Arizona State University, Tempe, AZ, USA

G Harris
University of Tasmania, Hobart, TAS, Australia

M E Hay
Georgia Institute of Technology, Atlanta, GA, USA

R A Herendeen
University of Vermont, Burlington, VT, USA

C Holzapfel
Rutgers University, Newark, NJ, USA

F G Howarth
Bishop Museum, Honolulu, HI, USA

L B Hutley
Charles Darwin University, Darwin, NT, Australia

D M Johnson
USDA Forest Service, Corvallis, OR, USA

S E Jørgensen
Copenhagen University, Copenhagen, Denmark

W J Junk
Max Planck Institute for Limnology, Plön, Germany

P C Kangas
University of Maryland, College Park, MD, USA

T Kätterer
Department of Soil Sciences, Uppsala, Sweden

P Keddy
Southeastern Louisiana University, Hammond, LA, USA

J E Keeley
University of California, Los Angeles, CA, USA

A K Knapp
Colorado State University, Fort Collins, CO, USA

V Krivtsov
University of Edinburgh, Edinburgh, UK

C Körner
Botanisches Institut der Universität Basel, Basel, Switzerland

D J Larkin
University of Wisconsin, Madison, WI, USA

T G Leishman
Monash University, Clayton, VIC, Australia

B G Lockaby
Auburn University, Auburn, AL, USA

M I Lucas
National Oceanography Centre, Southampton, UK

F Médail
IMEP Aix-Marseille University, Aix-en-Provence, France

J M Melack
University of California, Santa Barbara, Santa Barbara, CA, USA

J Mitchell
Auburn University, Auburn, AL, USA

P Moreno-Casasola
Institute of Ecology AC, Xalapa, Mexico

F Müller
University of Kiel, Kiel, Germany

S N Nielsen
Danmarks Farmaceutiske Universitet, Copenhagen, Denmark

D W Orr
Oberlin College, Oberlin, OH, USA

M Pell
Swedish University of Agricultural Sciences, Uppsala, Sweden

P S Petraitis
University of Pennsylvania, Philadelphia, PA, USA

K Reinhardt
Wake Forest University, Winston-Salem, NC, USA

L Sabetta
Università del Salento – Lecce, Lecce, Italy

S Sadedin,
Monash University, Clayton, VIC, Australia

A K Salomon
University of California, Santa Barbara, Santa Barbara, CA, USA

U M Scharler
University of KwaZulu-Natal, Durban, South Africa

S A Setterfield
Charles Darwin University, Darwin, NT, Australia

W K Smith
Wake Forest University, Winston-Salem, NC, USA

M Soderstrom
Montreal, QC, Canada

R A Sponseller
Arizona State University, Tempe, AZ, USA

J Stanturf
Center for Forest Disturbance Science,
Athens, GA, USA

C Trettin
USDA, Forest Service, Charleston, SC, USA

R R Twilley
Louisiana State University, Baton Rouge, LA, USA

R E Ulanowicz
University of Maryland Center for Environmental Science,
Solomons, MD, USA

A Varty
University of Wisconsin, Madison, WI, USA

D H Vitt
Southern Illinois University, Carbondale, IL, USA

R B Waide
University of New Mexico, Albuquerque, NM, USA

K M Wantzen
University of Konstanz, Konstanz, Germany

M A Wilzbach
Humboldt State University, Arcata, CA, USA

A Wörman
The Royal Institute of Technology, Stockholm,
Sweden

J B Zedler
University of Wisconsin, Madison, WI, USA

D Zhang
Auburn University, Auburn, AL, USA

J-J Zhu
Institute of Applied Ecology, CAS, Shenyang, People's
Republic of China

PREFACE

S ystems ecology, also called ecosystem theory, offers today a complete theory about how ecosystems are working as systems. The theory will inevitably be improved in the coming years, when it hopefully will be used increasingly to explain ecological observations and to facilitate environmental management including the use of ecotechnology. The theory is, however, sufficiently developed today to be presented as a complete theory that offers a wide spectrum of applications. Only through a wider application of the theory – or let us call what we have today propositions of a theory – it will be possible to see the shortcomings of the present theory and propose improvement of the theory.

The book consists of three parts. The part Ecosystems as Systems emphasizes the system properties of ecosystems including the presentation of basic scientific propositions to a theory in the chapter Fundamental Laws in Ecology, while the part Ecosystem Properties gives a more comprehensive overview of the holistic properties of ecosystems, which of course – not surprisingly – are rooted in the system properties and covered by the propositions. The part Ecosystems gives an overview of different types of ecosystems, how they function due to their characteristic ecosystem properties, and how the scientific propositions can be applied to understand and illustrate their characteristic properties.

It is my hope that this book will be utilized intensively by ecologists and system ecologists to gain a deeper understanding of ecosystems and their function and to initiate the development of ecology toward a more theoretical science that can explain and predict reactions of ecosystems. By such a development, it will be possible to replace many measurements that are often expensive to perform with sound theoretical considerations.

The book is based on the presentation of

I. systems ecology as an ecological subdiscipline and
II. a very comprehensive overview of all types of ecosystems with many illustrations of their characteristic properties

in the recently published *Encyclopedia of Ecology*.

Due to an excellent work by the editor of the Ecosystem Section, Donald de Angelis, and the editor of the Systems Ecology Section, Brian Fath, in the *Encyclopedia of Ecology*, it has been possible to present a comprehensive and very informative overview of all types of ecosystems and an updated ecosystem theory. I would therefore like to thank Donald and all the authors of ecosystem entries and Brian Fath and all the authors of systems ecology entries for their contributions to the *Encyclopedia of Ecology*, which made it possible to produce this broad and up-to-date coverage of a very important subdiscipline in ecology.

Sven Erik Jørgensen
Copenhagen, May 2009

ECOSYSTEMS AS SYSTEMS

Introduction

S E Jørgensen, Copenhagen University, Copenhagen, Denmark

According to the definition by Tansley (1935), an ecosystem is an integrated system composed of interacting biotic and abiotic components. It is important in this definition that an ecosystem is a system, which implies that it has boundaries and that we can distinguish between the system and its environment – environment in principle means the rest of the world beyond the boundaries of the system. The components – biotic as well as abiotic – are interacting, which means that they are connected directly or indirectly. All systems that encompass interacting biotic and abiotic components may be considered as an ecosystem. A drop of polluted water may for instance be considered an ecosystem, because it contains microorganisms, organic matter, and inorganic salts and these components are interacting. Usually, our ecosystem research and management is interested in a larger area of nature characterized by its function and properties, for instance a lake, a forest, or a wetland. All these three examples of ecosystems have very characteristic functions and have several unique properties that are different from other types of ecosystems. The scale that is applied for the definition of an ecosystem is dependent on the function of the ecosystem and is determined by the addressed problem.

Because an ecosystem has interacting and connected biotic and abiotic components, it has system properties in the sense that the components work together to give the system emerging properties and make the system more than just the sum of the components. A living organism is much more than the cells and the organs that make up the organism. Similarly, a forest is more than just the trees – it is a cooperative working unit with emerging unique properties characteristic of a forest.

It is important to understand fully the function and the reactions of ecosystems in both ecological research and environmental management. The two basic questions in this context are

1. Which fundamental properties characterize ecosystems?
2. Is it possible to formulate basic scientific propositions that are able to explain the functions of ecosystems?

It is attempted to answer these two core questions in the parts Ecosystems as Systems and Ecosystem Properties of this book, while the part Ecosystems gives an overview of different types of ecosystems, how they function due to their characteristic ecosystem properties, and how the scientific propositions can be applied to understand and illustrate their characteristic properties. The part Ecosystems as Systems emphasizes the system properties of ecosystems and also presents basic scientific propositions, while the part Ecosystem Properties gives a more comprehensive overview of the holistic properties of ecosystems, which of course – not surprisingly – are rooted in the system properties.

The chapters Ecosystem Ecology, Ecological System Thinking, and Ecosystems in the part Ecosystems as Systems focus on the most fundamental system properties that are derived from the above-presented definition of ecosystems. The definition is repeated in all three chapters with slight modifications. The system properties presented in these three chapters may be summarized as follows:

1. Ecosystems cycle energy.
2. Ecosystems cycle matter.
3. Life and environment are connected, which implies that the environment of an ecosystem influences the ecosystem. This influence determines the prevailing conditions of the ecosystems, or expressed differently the external variables (also called forcing functions) determine the conditions for the internal variables (also called state variables) of an ecosystem. The wide spectrum of different ecosystems (the part Ecosystems gives an overview) is the result of an overwhelmingly large number of different conditions (combinations of external variables).
4. Ecosystems are whole systems and studies of ecosystem dynamics therefore require holistic views.

The human society is very dependent on the proper functioning of ecosystems, because humans are using a wide spectrum of services offered by the ecosystems. It is therefore important to understand the ecosystem properties on which these services are based. The chapter Ecosystem Services and partly the chapter Ecosystems present the ecosystem services, which may be classified into three groups:

- production services as we know them from agriculture, fishery, forestry, and so on;
- regulation services due to cycling, filtration, translocation, and stabilization processes;
- cultural services such as recreation, spiritual inspiration, and esthetic beauty.

The chapter Fundamental Laws in Ecology gives a brief summary of the ecosystem properties that are rooted in the system properties of ecosystems:

- Ecosystems are complex (many steadily varying interacting components).
- Ecosystems are open.
- Ecosystems are hierarchically organized.
- Ecosystems are self-organizing and self-regulated due to a very large number of feedback mechanisms.

These properties are discussed in more detail in the part Ecosystem Properties.

The chapter Fundamental Laws in Ecology proposes 10 fundamental laws of ecosystems that are consistent with the system properties presented in the other chapters of the part Ecosystems as Systems. The 10 propositions are able to explain ecosystem behavior and properties. The fundamental tentative laws presented in this chapter are furthermore able to explain many ecological observations and rules, which is a great advantage of having a good theory. By use of the theory, it is possible to conclude, without the need for observations, how an ecosystem will react to different impacts. It is therefore indeed possible to improve research plans and develop environmental management plans on the basis of theoretical considerations. The 10 propositions (tentative laws) can be shown to be rooted in five basic ecological system properties.

The part Ecosystem Properties gives more information on the basic properties of an ecosystem. The chapter Autocatalysis focuses on autocatalysis, which frequently increases the efficiencies and rates of ecological processes. The chapter Body Size Patterns discusses the body size pattern of ecosystems. The rate of biological processes such as growth, metabolism, mortality, generation time, and respiration is dependent on the size of the organisms. The spectrum of conditions in an ecosystem determines the spectrum of these fundamental ecological processes, which would allow the best utilization of the resources in ecosystems. It implies that the conditions also determine the body size pattern. Different ecosystems at different conditions may therefore have a different body size pattern, which therefore becomes a characteristic property of an ecosystem.

All ecosystems cycle the elements that are essential for the living matter, and thereby the growth and development of ecosystems can continue, because the essential elements are steadily recovered with a certain rate. The living matter needs about 22 different elements, of which the cycling of nitrogen, carbon, phosphorus, sulfur, silica, calcium, sodium, and magnesium is of utmost importance. The cycling is possible due to the ecological networks that are formed in all ecosystems. The network may be considered a 'map' of the connections of abiotic and biotic components. The network indicates the possibilities for interactions among the components of the ecosystem. Obviously, cycling is very important for ecosystems, because without cycling the growth and development of biological components would stop due to the lack of one or more essential elements. The chapter Cycling and Cycling Indices covers cycling and cycling indices, which quantify the network's possibilities to support the cycling processes.

The chapters Ecological Network Analysis, Ascendancy; Ecological Network Analysis, Energy Analysis; Ecological Network Analysis, Environ Analysis; and Indirect Effects in Ecology present different aspects of the ecological network. Network analysis, ENA (Ecological Network Analysis), uses network theory to study the interactions between organisms or populations within their environment. Ascendancy, which is covered in the chapter Ecological Network Analysis, Ascendancy, quantifies the efficiency of the networks on the basis of the actual flows. Development of an ecosystem will usually imply that the ascendancy is increasing. The chapter Ecological Network Analysis, Energy Analysis analyzes the ecological network by use of the energy flows, while the chapter Indirect Effects in Ecology uses the so-called environ analysis. Each object in the system has two 'environs', one receiving and one generating interactions in the system. It is by analyzing these flows that it is possible to deduce network properties such as network mutualism and network synergy. Cycling – the topic of the chapter Cycling and Cycling Indices – may of course also be considered a network property. The chapter Indirect Effects in Ecology focuses on perhaps the most important network property: the presence of a strong indirect effect that in many cases may even exceed the direct effect.

The chapter Emergent Properties deals with the topic of emergent properties – the ecosystem as an integrated system is more than the sum of the components. The emergent properties are the result of all the system properties. Due to the synergistic effect of the network, autocatalysis, cycling, self-regulation and self-organization, and so on, an ecosystem acquires a number of very useful, holistic properties as a system – properties that are often called emergent properties. Self-organization itself is perhaps the most clear example of an emergent property. The chapter Self-organization looks into the emergent property of self-organization and how it is rooted in complex adaptive ecosystems. This chapter discusses how the spatial patterns, persistence, stability, and ability to develop and evolve can be explained as a result of the self-organization. The differences between ecosystems at an early stage and mature ecosystems can also be explained by self-organization.

Ecosystems are very complex systems. They have a large number of components with a large diversity, hierarchical organization, and nonlinear behavior. The chapter Ecological Complexity presents various aspects of ecological complexity, while the chapter Hierarchy Theory in Ecology presents the application of hierarchy theory in ecology. The hierarchical organization makes it possible to overview the complexity. It is also possible to get a better overview of the complex behavior of ecosystems by

Table 1 The five basic properties that are rooted in the 10 tentative fundamental laws encompass all the system properties presented

Basic property	Derived system properties
1. Ecosystems are open	The forcing functions (external variables) determine the ecosystem conditions
2. Ecosystems have directionality	Ecosystems show autocatalysis
	Ecosystems grow and develop
	Ecosystems have the propensity to maximize exergy storage and power
	Ecosystems have a body size pattern
3. Ecosystems have connectivity	The biotic and abiotic components of an ecosystem are connected in a network
	The network gives the ecosystem mutualism and synergy
	The indirect effect is significant due to the network and may even exceed the direct effect
	Ecosystems are self-organizing and self-regulated
	Ecosystems cycle energy, matter, and information
4. Ecosystems have emergent hierarchies	Ecosystems are organized hierarchically
5. Ecosystems have complex dynamics	Ecosystems grow and develop by increasing the biomass, the network, and the level of information
	Ecosystems are adaptive systems
	Ecosystems grow and develop and can cope with disturbances by a propensity to increase the exergy storage and the power
	Ecosystems, particularly under natural conditions, often have a large diversity, which gives the ecosystems a wide spectrum of different buffer capacities
	Ecosystems have high buffer capacities as a result of the complex dynamics
	Ecosystems recover usually rapidly and effectively after disturbances

use of goal functions and orientors that are presented in the chapter Goal Functions and Orientors. They are able to quantify the development of ecosystems as a result of the complex dynamics of ecosystems. One of the most useful orientors is exergy, which is presented in the chapter Exergy. The complex dynamics of ecosystems determine how they are able to develop and cope with disturbances. The exergy or energy that can do work of ecosystems – we cannot calculate exergy for an ecosystem due to its enormous complexity but we can calculate exergy for a model of the ecosystem – will have the tendency to be as high as possible under the prevailing conditions. Disturbances may of course cause a reduction in the ecosystem exergy, but the organisms try to organize themselves by their network and interactions to get the best out of the situation – it means in the Darwinian sense most survival, which may be expressed by exergy, as it covers the product of biomass and information of the ecosystem.

The five fundamental properties (see chapter Fundamental Laws in Ecology) cover *all* the ecosystem properties that are presented in the parts Ecosystems as Systems and Ecosystem Properties. An overview of the five basic properties and the derived additional system properties can be obtained from **Table 1**. Some of the properties are derived from more than one of the five fundamental properties, but to simplify the overview the derived system properties are associated with one of the basic properties. Particularly, the basic property that ecosystems have connectivity, which means that they form a network, and have a complex dynamics has been used to derive several system properties that could also be derived partly from one of the four other basic properties.

The chapter Overview of Ecosystem Types, Their Forcing Functions, and Most Important Properties, which is the last chapter in the part Ecosystem Properties, gives an overview of the 39 different types of ecosystems that are presented in the part Ecosystems. For all the 39 ecosystem types, the most important forcing functions are indicated, that is, the forcing functions (impacts) that may be considered a threat to the ecosystem or the forcing functions that most frequently determine the ecosystem function. It is possible to classify the forcing functions of the 39 ecosystems into four groups. The most basic properties of the four ecosystem classes are presented. They are the result of the prevailing conditions that are determined by the forcing functions. The most important properties are those that need to be maintained for the ecosystem to be able to meet the threats or those that are particularly important for the maintenance of the ecosystem function in spite of the impact.

The part Ecosystems has 40 chapters covering 39 different types of ecosystems. Most of the Earth's ecosystems are covered by the 39 types of ecosystems. A few rare types of ecosystems are not included, but all ecosystems frequently represented in nature are included. The ecosystems that are not included will however have properties close to one or more of the 39 types covered.

See also: Autocatalysis; Body Size Patterns; Cycling and Cycling Indices; Ecological Complexity; Ecological Network Analysis, Ascendancy; Ecological Network Analysis, Energy Analysis; Ecological Network Analysis, Environ Analysis; Ecosystem Ecology; Ecosystem Services; Ecological System Thinking; Ecosystems; Emergent Properties; Exergy; Fundamental Laws in

Ecology; Goal Functions and Orientors; Hierarchy Theory in Ecology; Indirect Effects in Ecology; Overview of Ecosystem Types, Their Forcing Functions, and Most Important Properties; Self-Organization.

Further Reading

Jørgensen SE (2004) Information theory and energy. In: Cleveland CJ (ed.) *Encyclopedia of Energy*, vol. 3. pp. 439–449. San Diego, CA: Elsevier.
Jørgensen SE (2006) *Eco-Exergy as Sustainability*. 220pp. Southampton: WIT Press.

Jørgensen SE (2008b) *Evolutionary Essays. A Thermodynamic Interpretation of the Evolution,* 210pp.
Jørgensen SE (ed.) (2008a) *Encyclopedia of Ecology*, 5 vols. 4122pp, Amsterdam: Elsevier.
Jørgensen SE and Fath B (2007) *A New Ecology. Systems Perspectives.* 275pp. Amsterdam: Elsevier.
Jørgensen SE, Patten BC, and Straskraba M (2000) Ecosystems emerging: 4. growth. *Ecological Modelling* 126: 249–284.
Jørgensen SE and Svirezhev YM (2004) *Towards a Thermodynamic Theory for Ecological Systems.* 366pp. Amsterdam: Elsevier.
Ulanowicz R, Jørgensen SE, and Fath BD (2006) Exergy, information and aggradation: An ecosystem reconciliation. *Ecological Modelling* 198: 520–525.

Ecosystem Ecology

B D Fath, Towson University, Towson, MD, USA and International Institute for Applied System Analysis, Laxenburg, Austria

Introduction

History of the Ecosystem Concept

Defining an Ecosystem

Energy Flow in Ecosystems

Biogeochemical Cycles

Ecosystem Studies

Human Influence on Ecosystems

Summary

Further Reading

Introduction

Ecology is a broad and diverse field of study. One of the basic distinctions in ecology is between autecology and synecology, in which the former is considered the ecology of individual organisms and populations, mostly concerned with the biological organisms themselves; and the latter, the ecology of relationships among the organisms and populations, which is mostly concerned with communication of material, energy, and information of the entire system of components. In order to study an ecosystem, one must have knowledge of the individual parts; thus, it is dependent on fieldwork and experiments grounded in autecology, but the focus is much more on how these parts interact, relate to, and influence one another including the physical environmental resources on which life depends. Ecosystem ecology, therefore, is the implementation of synecology. In this manner, the dimensional units used in ecosystem studies are usually the amount of energy or matter moving through the system. This differs from population and community ecology studies in which the dimensional units are typically the number of individuals (**Table 1**). This simple dimensional difference has served as an unfortunate divide between research conducted at the different ecological scales. While ecosystem ecologists maintain that it is always possible to convert species numbers into biomass or nutrient mass, population and community ecologists often feel that too much unique biological detail is discarded by

abstracting to energetic or material units. The advantage of this abstraction, of course, is that energy and mass are conserved quantities, whereas number of individuals is not. Therefore, using conserved units it is possible to construct balance equations and input–output models. In fact, dimensionally, ecosystem ecology has more in common with organismal ecology in which the thermoregulation and physiology of a single organism is studied, which also often relies on energetic units. Indeed, all scales of ecological study have a role to contribute to general scientific understanding and have been developed to address a wide range of interesting and relevant questions regarding the natural world and the impact humans have on it.

History of the Ecosystem Concept

Systems concepts of the environment have long played a role in the development of ecology as a discipline, but these came to a head in the early twentieth century. During this period, the two dominant and competing ecological paradigms were the organismic (e.g., Clements) and individualistic (e.g., Gleason) views. The organismic approach held that communities and ecosystems were discernible objects that had an inherent and organized complexity resulting in a cybernetic and self-governing system, similar in ways to how an organism

Table 1 Typical dimensional units of study at different ecological scales

Ecological scale	Dimensions
Organismal ecology	dE/dt
Population ecology	dN/dt
Community ecology	dN/dt
Ecosystem ecology	dE/dt

dE/dt = change in energy over time; dN/dt = change in number over time.

regulates itself. The individualistic approach held that communities had observer-dependent boundaries and internal development was stochastic and individual. In this paradigm, the internal relations were synergistic, but not cybernetic since the individual parts functioned independently. The organismic ideas grew out of the functional understanding of whole systems such as lakes, and also out of the discussions involving how communities changed over time during succession. These ideas were influenced by philosophers of the day such as Jan Smuts. This was particularly true of German holists, such as the limnology group at the Kaiser-Wilhelm-Instituts in Plön led by Thienemann, and others such as Leick (plant ecology) and Friedrich (zoology). **Table 2** shows a summary of some of the main ecosystem and related concepts. This dialog between the holists and reductionists affected the main currents of ecological thought during this period, and it was in part resolved by the introduction of 'ecosystem', which is both physical in nature and also systemic.

The term ecosystem, which is ubiquitous today, both as scientific terminology and in common vernacular, grew out of this climate. It was first used by Arthur Tansley in 1935 in a seminal paper in the journal *Ecology*, entitled 'The use and abuse of vegetational concepts and terms'. In fact, his reason for coining the term 'ecosystem' was in response, as the title says, to a perceived abuse of community concepts by some such as Clements and Cowles. While Tansley himself brought a systems perspective, the community as organism metaphor bothered him to the extent that he wanted to provide a more scientific footing for the processes and interactions occurring during community development. Tansley

describes the ecosystem thus, "... the fundamental conception is... the whole system, including not only the organism-complex, but also the whole complex of physical factors forming what we call the environment of the biome – the habitat factors in the widest sense." The definition he proposed over 70 years ago sounds fresh today, since it has changed little if at all. The major tenets of this approach are the explicit inclusion of abiotic processes interacting with the biota – in this sense it is more along the Haeckelian lines of ecology than the Darwinian, with an additional emphasis on the system. The latter tied the field closely to the burgeoning disciplines of general system theory and systems analysis.

While the conceptual underpinning of the ecosystem was now established, the introduction of this term was theoretical, lacking guidance as to how it might be applied as a field of study. There were around this time several whole system energy budgets being developed, particularly for lake ecosystems by North American ecologists such as Forbes, Birge, and Juday in Wisconsin, and which were ideal test cases for the ecosystem concept. Building on this work, in 1942, Lindeman's study of Cedar Bog Lake also in Wisconsin was published, providing, for the first time, a clear application of the ecosystem concept. In addition to constructing the food cycle of the aquatic system, he developed a metric – now called the Lindeman efficiency – to assess the efficiency of energy movement from one trophic level to the next based on ecological feeding relations. His conceptual model of Cedar Bog Lake included passive flows to detritus, but these were not included in the trophic enumeration. Since then numerous additional studies have followed this same approach and it has been applied to many habitats such as terrestrial, aquatic, and urban ecosystems.

Defining an Ecosystem

An ecosystem, as a unit of study, must be a bounded system, yet the scale can range from a puddle, to a lake, to a watershed, to a biome. Indeed, ecosystem scale is defined more by the functioning of the system than by any checklist of constituent parts, and the scale of analysis

Table 2 Ecosystem and related concept

Year	Term	Author	Concept
1887	Microcosm	Forbes	Broadening of the biocoenosis concept
1914	Ecoid	Negri	Unholistic, based on Gleasonian ideas
1928	Ökologisches system	Woltereck	Still being used to avoid argument
1930	Holocoen	Friedrich	Holistic, biologistic
1935	Ecosystem	Tansley	Antiholistic, physicalist
1939	Biosystem	Thienemann	Stressing functional organization
1944	Geobiocönose	Sukacev	Geographic, landscape ecological
1944	Bioinert body	Vernadsky	Biogeochemical
1948	Biochore	Pallmann	Landscape ecological
1950	Landschaft	Troll	Holistic, 'Gestalt' viewing

Modified from Wiegleb G (2000) Lecture Notes on The History of Ecology and Nature Conservation.

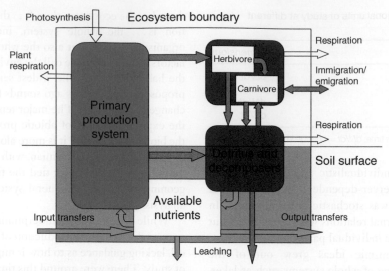

Figure 1 Conceptual diagram of a simplified ecosystem. Clear arrows, energy; dark arrows, biomass; blue arrows, water.

should be determined by the problem being addressed. Whereas individuals perish over time and even populations cannot survive indefinitely – none can fix their own energy and process their own wastes – every ecosystem contains the ecological community necessary for sustaining life: primary producers, consumers, and decomposers, and the physical environment for oikos (**Figure 1** shows a simple ecosystem model). It is this feature of ecosystems, that they are the basic unit for sustaining life over the long-term, which provides one of the main reasons for studying them for environmental management and conservation. The two main features of the ecosystem, energy flow and nutrient biogeochemical cycling, comprise the major areas of ecosystem ecology research.

Energy Flow in Ecosystems

The thermodynamic assessment of an ecosystem starts with the recognition that an ecosystem is an open system, in the sense of physics, such that it receives energy and matter input from outside its borders and transfers output back to this environment. Thus, every ecosystem must have a system boundary and must be embedded in an environment that provides low-entropy energy input and can receive high-entropy energy output. In addition to the external resource source–sink, there is another internal, within-system boundary environment with which each organism directly and indirectly interacts. Patten proposed the concept of these two environments, one external and mostly unknowable (other than the input–output interactions), and the second internal and measurable (i.e., external to the specific organismal component but within system boundary) as a systems approach to quantify indirect, yet within-system interactions. This approach – called environ analysis – relying on the

methodologies of input–output analysis has developed into a powerful analysis tool for understanding complex interactions and dependencies in ecological networks. For now though, let us concern ourselves more generally with what occurs within the ecosystem boundary.

Energy flow in ecosystems begins with the capture of solar radiation by photosynthetic processes in primary producers (eqn [1]). Note, there are also chemoautotrophs that capture energy in the absence of sunlight, but while biologically fascinating, contribute negligible energy flux to the overall global ecological energy balance

$$\text{Energy} + 6CO_2 + 6H_2O \rightarrow C_6H_{12}O_6 + 6O_2 \qquad [1]$$

The accumulated organic matter, first as simple sugars then combined with other elements to more complex molecules, represents the gross primary production in the system, some of which is released and used for the primary producers' growth and maintenance through respiration:

$$C_6H_{12}O_6 + 6O_2 \rightarrow 6CO_2 + 6H_2O + \text{Energy} \qquad [2]$$

The remainder, or net primary production, is available for the rest of the ecosystem consumers including decomposers. Secondary production refers to the energetic availability of the heterotrophic organisms, which accounts for the energy uptake by heterotrophs and the energy used for their maintenance. Overall ecosystem production is supported by the primary producers, whereas ecosystem respiration includes the metabolic activity of all the ecosystem biota (**Table 3**). In this manner, plants provide the essential base for all ecological food webs. Since it is often difficult to make direct measurements of ecological production, the change in biomass measures growth, which can be used as representative of production.

The captured energy moves through a reticulated network of interactions forming the complex dependency

Table 3 Ecosystem energetics defined by net and gross production

Net primary production = gross primary production − respiration (autotrophs)
Net secondary production = gross secondary production − respiration (heterotrophs)
Net ecosystem production = gross primary production − ecosystem respiration (autotrophs + heterotrophs)
Net production = biomass (now) − biomass (before)

patterns known as food webs. In a simplified food chain, and as first described by Lindeman, the trophic concept is used to assess the distance away from the original energy importation, but in reality the multiple feeding pathways found in ecological food webs make discrete trophic levels a convenient yet inaccurate simplification. Elton observed that one typically finds a decreasing number of organisms as one proceeds up the food chain from primary producers to herbivores, carnivores, and top carnivores – leading him to propose a pyramid of numbers. One can control for the individual variation in body size by considering the biomass at each trophic level rather than the number of individuals – resulting in a pyramid of biomass. The trophic pyramid is a thermodynamically satisfying view of interactions since according to the second law energy must be lost during each transformation step; in addition, energy is used at each level for the maintenance of that level. Under this paradigm, the trophic levels apparently cap out around five or six levels. Fractional trophic levels have been employed to account for organisms feeding at multiple levels, but even these do not usually account for the role of detritus and decomposition, which extend the feeding pathways to higher numbers. However, instead of linking detritus as a source compartment in the ecosystem conceptual model, the standard paradigm is to envision two parallel food webs one with primary producers as the base and the other with detritus as the base without any input from the rest of the web. If detritus were properly linked as both a source and sink in the ecosystem, then it would be clear that higher-order trophic levels are possible, if not common. The higher observed trophic levels observed in some studies are not in conflict with the laws of thermodynamics, but they show that ecosystems are more thorough at utilizing the energy within the system, mostly by decomposers, before it is lost as degraded, unavailable energy.

Energy resources flowing through the ecosystem are necessary to maintain all growth and development activities. Organisms follow a clear life-history pattern, and while the timescales differ depending on the species, early stage energy availability is generally used for growth, while later energy surplus is used for maintenance or reproduction. A similar pattern is visible in ecosystem-level growth and development. Net primary production is used to build biomass and physical structure of the ecosystem. The additional structure of photosynthetic material allows for the additional import of solar energy until saturation is reached at about 80% of the available solar radiation. At this point the overall growth of the ecosystem begins to level off because although gross primary production is high, the overall system supports more and more nonphotosynthetic biomass both in terms of nonphotosynthetic plant material and heterotrophs. When the average gross production is entirely utilized to support and maintain the existing structure, net production is zero and the system has reached a steady state regarding biomass growth. However, the ecosystem continues to develop both in terms of the network organization and in the information capacity. In addition to being a dynamic steady state, it does not persist indefinitely because disturbances afflict the system setting it back to earlier successional stages in which the growth and development processes begin anew, possibly with different results. In this manner, the disturbance acts according to Holling's creative destruction providing the system the opportunity to develop along a different pathway. Recent work on ecosystem growth and development has focused on the orientation of thermodynamic indicators such as energy throughflow, energy degradation, exergy storage, and specific entropy. These orientors provide good system-level indicators of development during succession or restoration of impaired ecosystems.

Biogeochemical Cycles

Another major focus of ecosystem ecology is understanding how the chemical elements necessary for life persist and translocate in pools and fluxes within the ecosphere. The biosphere actively interacts with the three abiotic spheres (hydrosphere, atmosphere, and lithosphere) to provide the available concentration of each for life. This action has a significant impact on the relative distribution of these elements. The simple sugar products of photosynthesis, $C_6H_{12}O_6$, are the base for organic matter, so carbon, hydrogen, and oxygen dominate the composition of life, and while oxygen is available in the lithosphere, and hydrogen in the hydrosphere, carbon is actually quite scarce in the environment, making the disproportionate amount of carbon in biomass a hallmark of life. In fact, there are about 20 elements used regularly in living organisms, of which nine called the macronutrients are the major constituents of organic matter: hydrogen, oxygen, carbon, nitrogen,

Table 4   Percentage atomic composition of the biosphere, hydrosphere, atmosphere, and lithosphere for first 10 elements

Biosphere		Hydrosphere		Atmosphere		Lithosphere	
H	49.8	H	65.4	N	78.3	O	62.5
O	24.9	O	33.0	O	21.0	Si	21.22
C	24.9	Cl	0.33	Ar	0.93	Al	6.47
N	0.073	Na	0.28	C	0.03	H	2.92
Ca	0.046	Mg	0.03	Ne	0.002	Na	2.64
K	0.033	S	0.02			Ca	1.94
Si	0.031	Ca	0.006			Fe	1.92
Mg	0.030	K	0.006			Mg	1.84
P	0.017	C	0.002			K	1.42

calcium, potassium, silicon, magnesium, and phosphorus. Some of these elements are readily available in the abiotic environment, in which case conservation through cycling of the elements is not paramount; however, those in scarce supply, such as nitrogen and phosphorus (**Table 4**), are reused many times before being released from the system. These biogeochemical cycles provide the foundation to understand how human modification leads to eutrophication (N and P cycles) and global climate change (C cycle). Therefore, much effort has been made to study and understand these cycles, particularly the carbon, nitrogen, and phosphorus cycles, details of which are addressed elsewhere in this encyclopedia.

Ecosystem Studies

The ecosystem perspective achieved footing in the ecological academic community since it was central to E. P. Odum's seminal textbook *Fundamentals of Ecology* first published in 1953. An early implementation of this approach at the institutional scale was attempted was in the International Biological Program (IBP), which was run from 1964 to 1974. The program had many successes in assessing and surveying the Earth's ecosystems, but faced the difficulty of compelling a top-down, holistic research paradigm on individual scientific endeavors. As a result of this conflict, the program did not deliver as much as had been hoped, but set the stage for the next generation of ecosystem-scale research. One feature of the IBP that did continue was the use of computer simulation modeling as a tool to understand the complex ecological interrelations. The journal *Ecological Modelling and Systems Ecology* started in 1975 continues as an active repository for mathematical and computer-based ecosystem research.

Subsequent to the IBP, the US National Science Foundation officially established the Long-Term Ecological Research Sites (LTER) in 1980 but research at several of the sites dates much earlier. Currently, there are 26 such sites ranging from the Coweeta Hydrological Lab in North Carolina, Hubbard Brook Ecosystem Study in New Hampshire, Sevilleta National Wildlife Refuge in

New Mexico, to the Baltimore Urban Ecosystem Study. These projects rely on a vast team of scientists to study the many interactions at this spatial scale. Still, the difficulty lies in putting together all the pieces into an integrated whole picture of the ecosystem.

Smaller-scale, individual-led ecological research is commonly conducted using microcosm and mesocosm experiments. A mesocosm experiment uses designed equipment or enclosures in which environmental factors can be controlled and manipulated to approximate natural conditions. The prevalence of this approach created a wealth of small-scale experimentation but at the expense of larger observational studies, which sparked a fierce debate in the 1990s between the 'field' versus 'bottle' approach. Indeed, the usefulness of microcosm experiments for ecosystem ecology was brought into question, but the resolution has been that a multiplicity of approaches is useful to address ecological questions.

Human Influence on Ecosystems

Humans have greatly altered and impacted the global biosphere. We recognize now the importance of maintaining functioning ecosystem services both out of our own necessity and for the obligation we have to the ecosphere. In 2000, the United Nations Secretary General called for a global ecological assessment, which was recently published as the Millennium Ecosystem Assessment (MEA) (www.mawed.org). The report compiled by over 1350 experts from 95 countries found that humans have changed ecosystems more rapidly and extensively over the last 50 years than in any comparable period of time in human history, resulting in a substantial and largely irreversible loss in the diversity of life on Earth (other highlights from the report are presented in **Table 5**). The MEA operated within a framework that identified four primary ecosystem services needed by humans: supporting (nutrient cycling, primary production, soil formation, etc.), provisioning (food, water, timber, fuel, etc.), regulating (climate, flood, disease, etc.), and cultural (esthetic, spiritual, educational,

Table 5 A few of the trends identified in the Millennium Ecosystem Assessment

50% of all the synthetic nitrogen fertilizer ever used has been used since 1985	20% of the world's coral reefs were lost and 20% degraded in the last several decades
60% of the increase in the atmospheric concentration of CO_2 since 1750 has taken place since 1959	35% of mangrove area has been lost in the last several decades
Approximately 60% of the ecosystem services evaluated are being degraded or used unsustainably	Withdrawals from rivers and lakes doubled since 1960

Table 6 Ecosystem Approach principles of the Convention on Biological Diversity

1 The objectives of land, water, and living resource management are a matter of societal choices
2 Management should be decentralized to the lowest appropriate level
3 Ecosystem managers should consider the effects (actual or potential) of their activities on adjacent and other ecosystems
4 Recognizing potential gains from management, there is usually a need to understand and manage the ecosystem in an economic context. Any such ecosystem-management program should
  (a) reduce those market distortions that adversely affect biological diversity;
  (b) align incentives to promote biodiversity conservation and sustainable use; and
  (c) internalize costs and benefits in the given ecosystem to the extent feasible
5 Conservation of ecosystem structure and functioning, in order to maintain ecosystem services, should be a priority target of the ecosystem approach
6 Ecosystem must be managed within the limits of their functioning
7 The ecosystem approach should be undertaken at the appropriate spatial and temporal scales
8 Recognizing the varying temporal scales and lag-effects that characterize ecosystem processes, objectives for ecosystem management should be set for the long term
9 Management must recognize the change is inevitable
10 The ecosystem approach should seek the appropriate balance between, and integration of, conservation and use of biological diversity
11 The ecosystem approach should consider all forms of relevant information, including scientific and indigenous and local knowledge, innovations, and practices.
12 The ecosystem approach should involve all relevant sectors of society and scientific disciplines

The 12 principles mentioned above are complementary and interlinked.

recreational, etc.). All have shown signs of stress and human pressures during the past century. One positive trend was the increase in food production (crops, livestock, and aquaculture), but this occurred with a concomitant loss of wild fisheries and food capture, along with a substantial increase in the resource inputs required to maintain the high agricultural production. While these observed changes to ecosystems have contributed to substantial net gain in human well-being and economic development, they have come at an increasing cost to the ecosystem health. The loss of this natural capital is typically not properly reflected in economic accounts.

Since the ecosystem provides the necessary functions for life, environmental management principles being devised and implemented today use the ecosystem concept as foundation. In particular, there have been several high-profile international efforts such as with the Convention on Biological Diversity (CBD), a treaty initiated in 1992 and signed by 150 government leaders with the expressed aim to protect and promote biological diversity and sustainable development. The 'ecosystem approach' adopted within this convention uses scientific methodologies regarding ecological interactions among organisms, their environment, and human activity to promote conservation, sustainability, and equity for managing natural resources. The approach deals with the complex socioecological–economic systems by promoting integrated assessment and adaptive management. The ecosystem approach of the CBD is outlined below in 12 principles (**Table 6**). Note particularly principles 5–8 that deal with ecosystem functioning, and taken in the context of the other principles assert how this ecological functioning provides opportunities and constraints for economic and social well-being. Research in ecosystem ecology today is directed toward improved understanding of key issues such as ecosystem services, resilience, spatial and functional scale, time lags, dynamics, and indirect effects.

Summary

Ecosystem ecology deals with the functioning at the system level of the ecological community with its abiotic environment, primarily in terms of the energy flow and nutrient cycling. Research in ecosystem ecology has given us a much better understanding of the processes and functions necessary to sustain life. The work in natural sciences has outpaced the ability of the social institutions to adapt and implement this knowledge. However, there is reason to be optimistic because the recent focus on the ecosystem approach in major international efforts recognizes that humans, with their cultural diversity, are an integral component of ecosystems.

See also: Ecological Network Analysis, Environ Analysis; Ecosystem Services; Ecosystems; Goal Functions and Orientors.

Further Reading

Chapin III FS, Matson PA, and Mooney HA (2002) *Principles of Terrestrial Ecosystem Ecology*. New York: Springer.

Fath BD, Jørgensen SE, Patten BC, and Straškraba M (2004) Ecosystem growth and development. *Biosystems* 77: 213–228.

Golley FB (1993) *A History of the Ecosystem Concept in Ecology*. New Haven: Yale University Press.

Likens GE, Borman FH, Johnson NM, Fisher DW, and Pierce RS (1970) Effects of forest cutting and herbicide treatment on nutrient budgets in the Hubbard Brook watershed-ecosystem. *Ecological Monographs* 20: 23–47.

Lindeman RL (1942) The trophic-dynamic aspect of ecology. *Ecology* 23: 399–418.

Odum EP (1969) The strategy of ecosystem development. *Science* 164: 262–270.

Odum HT (1957) Trophic structure and productivity of Silver Springs, Florida. *Ecological Monographs* 27: 55–112.

Patten BC (1978) Systems approach to the concept of environment. *Ohio Journal of Science* 78: 206–222.

Tansley AG (1935) The use and abuse of vegetational concepts and terms. *Ecology* 16: 284–307.

Weigert RG and Owen DF (1971) Trophic structure, available resources and population density in terrestrial versus aquatic ecosystems. *Journal of Theoretical Biology* 30: 69–81.

Wiegleb G (2000) Lecture Notes on 'The History of Ecology and Nature Conservation'. http://board.erm.tu-cottbus.de/index.php?id=5&no_cache=1&file=33&uid=14 (accessed May 2007).

Relevant Website

http://www.maweb.org – Millennium Ecosystem Assessment

Ecological Systems Thinking

D W Orr, Oberlin College, Oberlin, OH, USA

Introduction
Applied Systems Thinking
Environmental Education

Summary
Further Reading

Introduction

The greatest discovery of the past century had nothing to do with nuclear physics, or computer science, or genetic engineering. Rather it was the discovery of the essential connectedness of life and environment. The primary discipline of interrelatedness is ecology beginning with the work of Ernst Haeckel in the nineteenth century. The discovery of evolution extended the awareness of our connections to life in time and more extensively to the story of life on Earth. Fields such as ecology, general systems theory, systems dynamics, operations research, and chaos theory added details and theoretical depth, but with each advance in the precision and extent of knowledge the larger story remained the same. Living systems are linked in food webs and ecological processes into larger systems whether called the noosphere, biosphere, ecosphere, or Gaia. The boundaries between life forms and between what we take to be living and nonliving things shift and

sometimes morph into other forms and processes. In Earth systems, small changes can have large effects somewhere else and at some later time. Natural systems and the world made by humans are intertwined in more ways than we can possibly imagine. The result is less like a machine than it is like a web stretching across all life forms and back through time. The effects of human actions millennia ago still ripple forward, intersect with other changes sometimes amplifying, sometimes diminishing in intensity. Some human-wrought changes, such as deforestation and saline soils throughout much of the Middle East, are permanent as we measure time.

Nothing in the preceding paragraph is particularly new or controversial. But the idea of interrelatedness has yet to take hold of us in a deep way. We still live in thrall to a world created by Descartes, Bacon, Galileo, and their heirs who taught us to dissect, divide, parse, and analyze by reduction but not how to put things back together or see the world as systems and patterns. The results were intellectual power without perspective so that, in time, overspecialization became a kind of a cultural disease. There are many reasons why things do not change long after their deficiencies are apparent: the inertia of habit, economic inconvenience, the preservation of reputation, and intellectual laziness. But the most important barrier to change remains simply that science and the technology it spawned works and is a powerful presence in our daily lives. Automobiles, airplanes, the cornucopia evident in every supermarket, miracle cures, and the wonders of computers and communications are a constant reminder of the powers of a particular kind of science and a promise of things to come. That much of our technology also 'bites back' and incurs costs that we do not see is mostly lost on us. Many live in what has been called a 'consensus trance', believing that things will go well for us, which is to say that progress will continue indefinitely. Beneath such ideas is the faith that nature does not "set traps for unwary species," as biologist Robert Sinsheimer once put it or that progress itself is not a self-made trap.

There have always been skeptics, however. Toward the end of his life, H. G. Wells could see no grounds for hope. More recently, Joseph Tainter, Martin Rees, and Jared Diamond have expressed doubts about our longevity based in no small part on their views of scientific progress. Rees, for example, believes that our odds of making it to the year 2100 are no better than fifty-fifty. Diamond has cataloged the reasons why past societies have collapsed and they bear more than a passing resemblance to our present behavior. James Lovelock, coauthor of the Gaia hypothesis, believes that we are approaching a climate-tipping point somewhere between 400 and 500 ppm CO_2 in the atmosphere after which "nothing the nations of the world do will alter the outcome and the Earth will more irreversibly to a new hot state." In various ways, each of these attributes our vulnerability to the failure to see systems, patterns, and to exercise foresight. As a result, we stumble toward a time of severe climate destabilization, biotic impoverishment, and ecological surprises.

The failure of ecological knowledge to penetrate very deeply into the larger society and its decision-making systems ought to be a matter of grave concern. The early work of ecologists Howard and Eugene Odum on the productivity of salt marshes, for example, may have slowed but certainly did not stop the juggernaut of development that has severely damaged coastal ecosystems virtually everywhere. Similarly, we know a great deal about the services of natural systems and the impossibility of duplicating these by human means. Yet the drawdown of natural capital and the destruction of ecosystems are still trumped by narrow short-term concerns of profit and economic expansion. Sometimes the costs of ecological folly become starkly apparent as they did following hurricane Katrina in the fall of 2005 in which the damage done by a class III hurricane (at landfall) was amplified by the removal of mangroves and coastal forests that would otherwise have absorbed much of its energy and dampened the destructive effects. That, too, was known in many circles but did not have much effect on the policies that prevailed along the Gulf Coast, where oil extraction, commerce, and gambling ruled the day.

Public attitudes toward science are often undermined by poor education, inadequate public funding, and, sometimes, religious dogma. In the USA, evolution, once thought to be an established part of science, is hotly contested as just another 'theory' by advocates of 'intelligent design'. The scientific evidence about human-driven climate change is indisputable, but ignored or underestimated even when alternatives are economically advantageous. The results are evident in the considerable data describing ecological deterioration virtually everywhere and the failure to seize better alternatives as well. Law based on ecological knowledge and the hope that we might calibrate our public business with the way the world works as a physical system is under constant assault. Evidence about the health and ecological effects of toxins is downplayed. Public access to information about the release of toxics is restricted. The result is a significant gap between what is known about how the world works as a physical system and the public policy in every country. The cumulative result is that we are much more vulnerable to ecological ruin and extreme events than we might otherwise be.

What can be done with ecological knowledge? One answer is that ecology as a science ought to do what it has been doing, which is to say document the deterioration of ecosystems in ever finer detail. Ecology, the argument goes, is a science and its practitioners ought to maintain their credibility as scientists and not assume the role of advocates and risk losing their credibility even when they recognize folly disguised as public policy. If that is the future of the discipline it will, I think, flourish for a time while the human prospect withers.

There is, however, another perspective on the uses of ecology. Paul Sears in 1964 and later Paul Shepard and Daniel McKinley in 1969 once called the discipline "the subversive science." They proposed ecology as an integrative discipline, "a kind of vision across boundaries" and a "resistance movement" – an alternative to being "man fanatic." Ecology in their view "offers an essential factor . . . to all our engineering and social planning." In their perspective, the world needs to know what ecologists know and needs to take that knowledge seriously enough to transform the ways by which we provision ourselves with food, energy, materials, shelter, and livelihood. Ecology as a subversive science would be integrated with building, industry, agriculture, landscape management, economics, and governance. In short, the idea of interrelatedness would move from the pages of obscure scientific journals out to the main street, and into board rooms, editorial offices, courtrooms, legislatures, and classrooms.

It would progress from being just one more interesting but obsolete idea to become the design principles for a better world – the default setting for everyday behavior.

Applied Systems Thinking

In this regard, the news is guardedly optimistic. The art and science of high-performance building is growing. The result is a new generation of buildings that require a fraction of the energy of conventional buildings, use materials screened for environmental effects, minimize water consumption, and are landscaped to promote biological diversity, moderate microclimates, and grow foods. The best of these are highly efficient, powered substantially by sunlight and feature daylight, water recycling, and interior green spaces. They are a finer calibration between our five senses and the built environment and tend to promote higher user satisfaction and productivity. The costs of building green, as it turns out, are not necessarily higher than conventional buildings while having lower operating costs. The goal is to design buildings as whole systems, not as disjointed components. The green building movement is now a worldwide movement and is transforming the practice of architecture, landscape architecture, and engineering. It could, in time, transform the design of communities and cities as well.

Business, too, is beginning to go green. The best example of a well-run environmentally sensitive business is that of Interface, Inc., a global manufacturer of carpet tiles and raised flooring. In the mid-1990s, company founder and CEO, Ray Anderson, decided to transform the company to eliminate waste and carbon emissions. Interface launched a pioneering effort to develop carpet products that were returned to the company as a "product of service" not otherwise discarded in a landfill. Interface now leases carpet to its customers and takes it back to be remade into new products, thereby eliminating much of the petrochemical sources at one end and waste at the other. In the past decade, the company has eliminated 56% of its carbon emissions and is on track to becoming carbon neutral. The model for the company is consciously that of ecology all the way down to carpet products that mimic a forest floor. Interface is not alone. Other companies like Wal-Mart and DuPont are beginning to transform themselves as well. Some day, perhaps, all business will be powered by sunlight with materials cycles that mimic the circular flow of nutrients in ecosystems.

In agriculture, Wes Jackson, co-founder of the Land Institute, is pioneering the development of natural systems agriculture. The goal is to model agriculture on ecological systems such as forests and prairies. If successful, the end product will be agricultural polycultures of high yield perennials, long thought to be a biological impossibility. The early results, however, have confirmed Jackson's hypothesis that the two can be stitched together, thereby eliminating a great deal of fossil energy and soil erosion.

Materials science is a fourth area in which ecology is being taken seriously. Nature, as chemist Terry Collins has noted, uses only a relatively few ingredients while industrial chemistry uses virtually the entire periodic table, creating ecological havoc. The field of biomimicry has grown in response by studying how nature works in fine detail. Natural systems are a carnival of color, for instance, but nature does not use paints. To answer such questions, Janine Benyus, author of *Biomimicry*, is developing a database of the ways nature works to filter, reduce, recycle, color, purify, form, and join – all done without the use of toxics and fossil fuels and all of it biodegradable. The result could be a transformation of materials and industry that dramatically reduce pollution and energy use.

In these examples and elsewhere, the science of applied ecology has begun to seriously influence decisions and behavior and the evolution of architecture, engineering, materials science, agronomy, urban planning, and economics. The driving force is partly economic (to reduce the costs of unnecessary energy, materials and water use) and partly a matter of conviction (that it is wrong to leave a legacy of ruin behind us). While promising, such measures are necessary but insufficient. Ecological thinking, in one way or another, must become a more central part of global society and this is the task of education.

Environmental Education

The idea of specifically environmental education entered the public discourse in the late 1960s. Among the recommendations of the Stockholm Conference in 1972 was to "establish an international programme in environmental education." UNESCO and UNEP subsequently undertook to prepare curricular materials, establish priorities, develop pilot projects, and organize meetings. The result was a UN-sponsored Conference at Tbilisi, Georgia, in 1978 that produced a consensus statement including the words:

> Environmental education . . . should constitute a comprehensive lifelong education . . . it should prepare the individual for life through an understanding of the major problems of the contemporary world, and the provision of skills and attributes needed to play a productive role towards improving life and protecting the environment with due regard given to ethical values. By adopting a holistic approach, rooted in a broad interdisciplinary base, it recreates an overall perspective which acknowledges the fact that natural environment and manmade environment are profoundly interdependent . . . .

The Tbilisi Conference produced 41 recommendations spanning the needs for environmental education between developed and less-developed countries. In the subsequent decades, initiatives, including those spawned by Agenda 21 and discussions about the Earth Charter, have advanced the discussion of environmental education into a major part of the dialog about the role of education relative to the human prospect. There is no serious discussion about the transition to sustainability launched by the Brundtland Report in 1987 that does not include changing the goals and methods of education. From Tbilisi (1978), Talloires (1990), and subsequent international gatherings, a strong consensus about the importance of environment in higher education is clearly apparent.

Despite considerable progress, both conceptually and practically, there are serious differences about the goals and methods of environmental education that reflect and, in some ways, amplify larger disagreements about education. At the lowest level, there is a general consensus that the young ought to know something about how nature works as a physical system – the rudiments of biology and planetary science. There is considerably less agreement about how this should be incorporated into the standard curriculum or at what level. Most elementary schools include curricular components such as 'Project Learning Tree' or 'Wet and Wild' that introduce children to what was once called natural history along with some field experience and practical outdoor skills. But the later inclusion of values or discussion about the causes of environmental ills has often been controversial, especially when it has led to questions about conventional economic or political wisdom.

In important respects, all education is environmental education, that is, by what is included or excluded students are taught that they are part of or apart from ecological systems. The standard, discipline-centric curriculum may have contributed to a mindset that helped to create environmental problems by separating subjects into boxes and conceptually by separating people from nature. As a result, graduates are often ignorant of ecological relationships or why they are worthy of consideration. Not surprisingly, the first response to proposals for environmental education attempted to accommodate environmental issues and ecology into formal education as a kind of add-on. More radical critics proposed that formal education ought to be reformed along ecological lines, raising another and no less contentious issues. From either perspective, environmental mismanagement and the larger discussion of sustainability raise questions about the meaning of human mastery over nature, or more accurately as C. S. Lewis once put it: what does it mean for some men to control other men through the mastery of some parts of nature? What is the core knowledge of the environment that ought to be standard in an educational curriculum? At the heart of such questions are important differences about what it means to be human, what part of that definition ought to remain inviolable, and about the manipulation of natural systems through technological means such as genetic engineering. Is the problem, in other words, one in education or one of education?

What can be said with certainty is that public schooling and higher education have been underachievers in the task of inculcating essential knowledge about the environment. Public opinion surveys show high levels of support for environmental quality but little ecological knowledge. In the words of one typical survey, people have acquired a "substantial familiarity with environmental issues, but [have] a long way to go in developing a working environmental/energy knowledge." Much of what people know about the environment is derived from television in bits and pieces and not through direct experience with nature or through cultural transmission.

One particularly encouraging aspect is the development of environmental education in institutions of higher education. Stemming from innovations in the 1980s, a vibrant campus ecology movement has emerged in Europe, Australasia, and the USA, along with a wide discussion of sustainability of educational institutions. Beginning with the studies of college food, energy use, and pollution, the movement has grown in subsequent decades to a worldwide scale. Hundreds of colleges and universities globally have organized efforts to systematically reduce energy use, water consumption, and material flows. Campus sustainability and climate stability have come to the center of institutional planning, purchasing, and construction. Beginning in the late 1990s with the advent of means to promote and measure environmental performance of buildings, the construction of academic facilities is undergoing a rapid revolution. Green or high-performance building standards are increasingly regarded as necessary to reduce energy and maintenance costs as well as laboratories for research and education. Many of the problems of sustainability – ecological design, applications of solar energy, water purification, food production, ecological restoration, and landscape management – can be studied in buildings and adjacent landscapes at a scale that is both significant yet manageable. Given recent developments on many campuses, it is not inconceivable that educational institutions at all levels will one day become models of ecological design mirroring the larger solutions necessary to the transition to sustainability.

Summary

In the decades since the Stockholm Conference in 1972, environmental education has emerged as a significant component of education virtually everywhere in the world. It has, for the most part, flourished at all levels of education. There are magazines and journals such as *Sustainability in Higher Education*, professional associations, and regular conferences. It is not difficult to imagine all of this as the start of something like an ecological enlightenment emerging in the decades or centuries ahead. But no such thing is certain. If education is to be midwife to a deeper, broader, and sustainable transformation, it will have to surmount serious challenges.

Further Reading

Barlett P and Chase G (eds.) (2004) *Sustainability on Campus*. Cambridge, MA: MIT Press.

Benyus J (1998) *Biomimicry*. New York: William Morrow.

Bowers C (1993) *Education, Cultural Myths, and the Ecological Crisis*. Albany, NY: SUNY Press.

Bowers C (1995) *Educating for an Ecologically Sustainable Culture*. Albany, NY: SUNY Press.

Corcoran P and Wals A (eds.) (2004) *Higher Education and the Challenge of Sustainability*. Dordrecht, The Netherlands: Kluwer Academic.

Coyle K (2005) *Environmental Literacy in America*. Washington, DC: The National Environmental Education & Training Foundation.

Creighton S (1998) *Greening the Ivory Tower*. Cambridge, MA: MIT Press.

de Chardin T (1965) *The Phenomenon of Man*. New York: Harper Torchbooks.

Fischetti M (2001) Drowning New Orleans. *Scientific American* (October, 2001): 76–85.

Kuhn T (1963) *The Structure of Scientific Revolutions*. Chicago: University of Chicago Press.

Lovelock J (2006) *The Revenge of Gaia*. London: Penguin Books.

Lovelock J *The Gaia Hypothesis*. New York: Oxford University Press.

Lovins A (2005) *Winning the Oil Endgame*. Snowmass, CO: Rocky Mountain Institute.

Oakeshott M (1989) *The Voice of Liberal Learning*. New Haven, CT: Yale University Press.

Orr D (1992) *Ecological Literacy*. Albany, NY: Suny Press.

Orr D (1994) *Earth in Mind*. Washington, DC: Island Press.

Orr D (2006) *Design on the Edge*. Cambridge, MA: MIT Press.

O'Sullivan E (2005) *Millennium Ecosystem Assessment Report*, vols. 1–5. Washington, DC: Island Press.

Rees M (2003) *Our Final Hour*. New York: Basic Books.

Sears P (1964) Ecology – A subversive subject. *BioScience* 14(7): 11–13.

Shepard P and McKinley D (eds.) (1969) *The Subversive Science*. Boston: Houghton Mifflin.

Sinsheimer R (1978) The Presumptions of Science. *Daedalus* 107: 23–36.

Sobel D (1996) *Beyond Ecophobia*. Great Barrington, MA: The Orion Society.

Steffen W, Sanderson A, Jäger J, et al. (2004) *Global Change and the Earth System*. Berlin: Springer.

Tenner E (1996) *Why Things Bite Back: Technology and the Revenge of Unintended Consequences*. New York: Knopf.

Union of Concerned Scientists (1992) *World Scientists Warning to Humankind*. Boston: Union of Concerned Scientists.

US Department of Health, Education, and Welfare (1978) *Toward an Action Plan: A Report on the Tbilisi Conference on Environmental Education*. Washington, DC: US Government Printing Office.

Vernadsky V (1998) *The Biosphere*. New York: Springer.

Washburn J (2005) *University INC: The Corporate Corruption of Higher Education*. New York: Basic Books.

Wright R (2005) *A Short History of Progress*. New York: Carroll & Graf.

Wright T (2004) Evolution of sustainability declarations in higher education. In: Corcoran PB and Wals AEJ (eds.) *Higher Education and the Challenge of Sustainability*, pp. 7–19. Dordrecht, The Netherlands: Kluwer Academic.

Ecosystems

A K Salomon, University of California, Santa Barbara, Santa Barbara, CA, USA

What Is an Ecosystem?

Coined by A. G. Tansley in 1935, the term 'ecosystem' refers to an integrated system composed of a biotic community, its abiotic environment, and their dynamic interactions. A diversity of ecosystems exist through the world, from tropical mangroves to temperate alpine lakes, each with a unique set of components and dynamics (**Figure 1**). Ecosystems can be classified according to their components and physical context yet their classification is highly dependent on the spatial scale of scrutiny.

Typically, boundaries between ecosystems are diffuse. An 'ecotone' is a transition zone between two distinct ecosystems (i.e., the tundra–boreal forest ecotone).

History

Over 70 years ago, Sir Arthur Tansley (**Figure 2**) presented the notion that ecologists needed to consider 'the whole system', including both organisms and physical factors, and that these components could not be separated or viewed in isolation. By suggesting that ecosystems are

Figure 1   (a) Kelp forest, (b) subarctic alpine tundra, (c) tropical coastal sand dune, (d) tropical mangrove, (e) alpine lake, and (f) temperate coastal rain forest. Photos by Anne Salomon, Tim Storr, and Tim Langlois.

A. G. Tansley

Figure 2   Sir Arthur G. Tansley coined the term ecosystem in 1935. From *New Phytologist* 55: 145, 1956.

dynamic, interacting systems, Tansley's ecosystem concept transformed modern ecology. It led directly to considerations of energy flux through ecosystems and the pathbreaking, now classic work of R. L. Lindeman in 1942, one of the first formal investigations into the functioning of an ecosystem, in this case a senescent lake, Cedar Creek Bog, in Minnesota. Inspired by the work of C. Elton, Lindeman focused on the trophic (i.e., feeding) relationships within the lake, grouping together organisms of the lake according to their position in the food web. To study the cycling of nutrients and the efficiency of energy transfer among trophic levels over time, Lindeman considered the lake as an integrated system of biotic and abiotic components. He considered how the lake food web and processes driving nutrient flux affected the rate of succession of the whole lake ecosystem, a significant departure from traditional interpretations of succession.

By the late 1950s and early 1960s, system-wide energy fluxes were quantified in various ecosystems by E. P. Odum and J. M. Teal. In the late 1960s, Likens, Bormann, and others took an ecosystem approach to studying biogeochemical cycles by manipulating whole watersheds in the Hubbard Brook Experimental Forest to determine whether logging, burning, or pesticide and herbicide use had an appreciable effect on nutrient loss from the ecosystem. This research set an important precedent in demonstrating the value of conducting experiments at the scale of an entire ecosystem (see the section entitled 'Whole ecosystem experiments'), a significant advancement which continues to inform ecosystem studies today.

Ecosystem Components and Properties

Ecosystems can be thought of as energy transformers and nutrient processors composed of organisms within a food web that require continual input of energy to balance that lost during metabolism, growth, and reproduction. These organisms are either 'primary producers' (autotrophs), which derive their energy by using sunlight to convert inorganic carbon into organic carbon, or 'secondary producers' (heterotophs), which use organic carbon as their energy source. Organisms that perform similar types of ecosystem functions can be broadly categorized by their 'functional group'. For example, 'herbivores' are heterotophs that eat autotrophs, 'carnivores' are heterotophs that eat other heterotophs, while 'detritivores' are heterotophs that eat nonliving organic material (detritus) derived from either autotrophs or heterotrophs (**Figure** 3). Herbivores, carnivores, and dertitivores are collectively known as 'consumers'.

Classifying organisms according to their feeding relationships is the basis of defining an organism's 'trophic level'; the first trophic level includes autotrophs; the second trophic level includes herbivores and so on. Ecosystem components that make up a trophic level are quantified in terms of biomass (the weight or standing crop of organisms), while ecosystem dynamics, the flow of energy and materials among system components, are quantified in terms of rates.

Typically, ecologists quantifying ecosystem dynamics use carbon as their currency to describe material flow and energy to quantify energy flux. Material flow and energy flow differ in one important property, namely their ability to be recycled. Chemical materials within an ecosystem are recycled through an ecosystem's component. In contrast, energy moves through an ecosystem only once and is not recycled (**Figure** 3). Most energy is transformed to heat and ultimately lost from the system. Consequently, the continual input of new solar energy is what keeps an ecosystem operational.

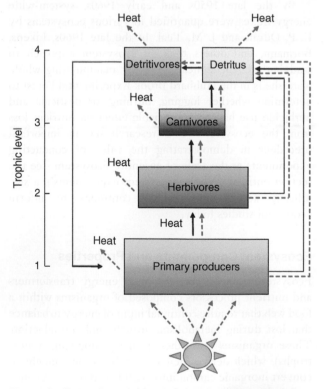

Figure 3 Energy flows and material cycles in an ecosystem. Materials move through the trophic levels and eventually cycle back to the primary producers via the decomposition of detritus by microorganisms. Energy, originating as solar energy, is transferred through the trophic levels via chemical energy and is lost via the radiation of heat at each step. Adapted from DeAngelis DL (1992) *Dynamics of Nutrient Cycling and Food Webs*. New York, NY: Chapman and Hall.

Solar energy is transformed into chemical energy by primary producers via photosynthesis, the process of converting inorganic carbon (CO_2) from the air into organic carbon ($C_6H_{12}O_2$) in the form of carbohydrates. Gross primary production is the energy or carbon fixed via photosynthesis over a specific period of time, while net primary production is the energy or carbon fixed in photosynthesis, minus energy or carbon which is lost via respiration, per unit time. Production by secondary producers is simply the amount of energy or material formed per unit term.

A careful distinction needs to be made between production rates and static estimates of standing crop biomass, particularly because the two need not be related. For example, two populations at equilibrium, in which input equals output, might have the same standing stock biomass but drastically different production rates because turnover rates can vary (**Figure** 4). For example, on surf swept shores from Alaska to California, two species of macroalgal primary producers grow in the low rocky intertidal zone of temperate coastal ecosystems (**Figure** 5). The ribbon kelp, *Alaria marginata*, is an annual alga with high growth rates, whereas sea cabbage, *Hedophyllum sessile*, is a perennial alga with comparatively lower growth rates. Although they differ greatly in their production rates, in mid-July, during the peak of the growing season, these two species can have almost equivalent stand crop biomasses.

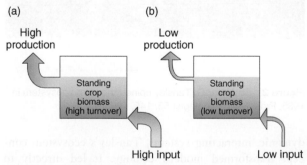

Figure 4 Standing crop biomass is not always correlated to production rates. Here, two hypothetical species with populations at equilibrium, where input equals output, have an equivalent standing crop biomass but differ in their turnover rates. Population (a) has high input, high production, and high turnover rates, whereas population (b) has low input, low production, and low turnover rates. In reality, populations are rarely at equilibrium so standing crop biomass fluctuates depending on input rates and the amount of production consumed by higher trophic levels. Adapted from Krebs C (2001) *Ecology: The Experimental Analysis of Distribution and Abundance*, 5th edn. San Francisco: Addison-Wesley Educational Publishers, Inc.

Figure 5 (a) In the low intertidal zone of temperate coastal ecosystems, (b) the ribbon kelp, *Alaria marginata*, is an annual alga with high growth rates, whereas (c) the sea cabbage kelp, *Hedophyllum sessile*, is a perennial alga with lower growth rates. During the peak of the growing season, these two species can have a similar stand crop biomass but differ greatly in their production rates because one is an annual and the other is a perennial. Photo by Anne Salomon and Mandy Lindeberg.

Ecosystem Efficiency

The efficiency of energy transfer within an ecosystem can be estimated as its 'trophic transfer efficiency', the fraction of production passing from one trophic level to the next. The energy not transferred is lost in respiration or to detritus. Knowing the trophic transfer efficiency of an ecosystem can allow researchers to estimate the primary production required to sustain a particular trophic level.

For example, in aquatic ecosystems, trophic transfer efficiency can vary anywhere between 2% and 24%, and average 10%. Assuming a trophic efficiency of 10%, researchers can estimate how much phytoplankton production is required to support a particular fishery. Consider the open ocean fishery for tuna, bonitos, and billfish. These are all top predators, operating at the fourth trophic level. According to world catch statistics recorded by the Food and Agriculture Organization, in 1990, 2 975 000 t of these predators were caught, equivalent to 0.1 g of carbon per m^2 of open ocean per year. To support this yield of tuna, bonitos, and billfish, researchers can calculate the production rates of the trophic levels below, assuming a trophic efficiency of 10% and equilibrium conditions. Essentially, to produce of $0.1\,gC\,m^{-2}\,yr^{-1}$ of harvested predators (tuna, bonitos, and billfish) requires $1\,gC\,m^{-2}\,yr^{-1}$ of pelagic fish to have been consumed by the top predators, $10\,gC\,m^{-2}\,yr^{-1}$ of zooplankton to be consumed by the pelagic fishes, and $100\,gC\,m^{-2}\,yr^{-1}$ of phytoplankton. Note that these values represent the production that is transferred up trophic levels. They do not represent the standing stock of biomass at each trophic level. Knowing the net primary production of the

photoplankton allows researchers to estimate the proportion of this production that is taken by the fishery.

It has been estimated that 8% of the world's aquatic primary production is required to sustain global fisheries. Considering continental shelf and upwelling areas specifically, these ecosystems provide one-fourth to one-third of the primary production required for fisheries. This high fraction leaves little margin for error in maintaining resilient ecosystems and sustainable fisheries.

Large-Scale Shifts in Ecosystems

A growing body of empirical evidence suggests that ecosystems may shift abruptly among alternative states. In fact, large-scale shifts in ecosystems have been observed in lakes, coral reefs, woodlands, desserts, and oceans. For example, a distinct shift occurred in the Pacific Ocean ecosystem around 1977 and 1989. Abrupt changes in the time series of fish catches, zooplankton abundance, oyster condition, and other marine ecosystem properties signified conspicuous shifts from one relatively stable condition to another (**Figure 6**). Also termed 'regime shifts', the implications of these abrupt transitions for fisheries and oceanic CO_2 uptake are profound, yet the mechanisms driving these shifts remain poorly understood. It appears that changes in oceanic circulation driven by weather patterns can be evoked as the dominant causes of this state shift. However, competition and predation are becoming increasingly recognized as important drivers of change altering

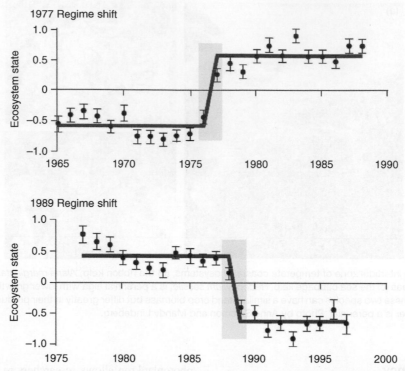

Figure 6 Distinct shifts in ecosystem states, also referred to as 'regime shifts', occurred in the Pacific Ocean ecosystem around 1977 and 1989. The ecosystem state index shown here was calculated based on the average of climatic and biological time series. From Scheffer M, Carpenter S, Foley JA, Folke C, and Walker B (2001) Catastrophic shifts in ecosystems. *Nature* 413: 591–596.

oceanic community dynamics. In fact, fisheries are well known to affect entire food webs and the trophic organization of ecosystems. Therefore, one could imagine that the sensitivity of a single keystone species to subtle environmental change could cause major shifts in community composition. Given this interplay between and within the biotic and abiotic components of an ecosystem, resolving the causes of regime shifts in oceanic ecosystems will likely require an understanding of the interactions between the effects of fisheries and the effects of physical climate change.

Studying Ecosystem Dynamics

Stable Isotopes

Important insights into ecosystem dynamics can be revealed through the use of naturally occurring 'stable isotopes'. These alternate forms of elements can reveal both the source of material flowing through an ecosystem and its consumer's trophic position. This is because different sources of organic matter can have unique isotopic signatures which are altered in a consistent manner as materials are transferred throughout an ecosystem, from trophic level to trophic level. Consequently, stable

isotopes provide powerful tools for estimating material flux and trophic positions.

The elements C, N, S, H, and O all have more than one isotope. For example, carbon has several isotopes, two of which are ^{13}C and ^{12}C. In nature, only 1% of carbon is ^{13}C. Isotopic composition is typically expressed in δ values, which are parts per thousand differences from a standard. For carbon,

$$\delta^{13}\text{C} = \left[\left(\frac{^{13}\text{C}/^{12}\text{C}_{\text{sample}}}{^{13}\text{C}/^{12}\text{C}_{\text{standard}}}\right) - 1\right] \times 10^3$$

Consequently, δ values express the ratio of heavy to light isotope in a sample. Increases in these values denote increases in the amount of the heavy isotope component. The standard reference material for carbon is PeeDee limestone, while the standard for nitrogen is nitrogen gas in the atmosphere. Natural variation in stable isotopic composition can be detected with great precision with a mass spectrometer.

Stable isotopes record two kinds of information. Process information is revealed by physical and chemical reactions which alter stable isotope ratios, while source information is revealed by the isotopic signatures of source materials. When organisms take up carbon and nitrogen, chemical reactions occur which discriminate among isotopes, thereby altering the ratio of heavy to light isotope. This is known as 'fractionation'. Although carbon fractionates very little (0.4‰, 1 SD = 1‰), the mean trophic fractionation of $\delta^{15}N$ is 3.4‰ (1 SD = 1‰), meaning that $\delta^{15}N$ increases on average by 3.4‰ with every trophic transfer. Because the $\delta^{15}N$ of a consumer is typically enriched by 3.4‰ relative to its diet, nitrogen isotopes can be used to estimate trophic position. Stable isotopes can provide a continuous measure of trophic position that integrates the assimilation of energy or material flow through all the different trophic pathways leading to an organism. In contrast, $\delta^{13}C$ can be used to evaluate the ultimate sources of carbon for an organism when the isotopic signatures of the sources are different.

Stable isotopes can track the fate of different sources of carbon through an ecosystem, because a consumer's isotopic signature reflects those of the key primary producers it consumes. For example, in both lake and coastal marine ecosystems, $\delta^{13}C$ is useful for differentiating between two major sources of available energy, benthic (nearshore) production from attached macroalgae, and pelagic (open water) production from phytoplankton. This is because macroalgae and macroalgal detritus (specifically kelp of the order Laminariales) is typically more enriched in $\delta^{13}C$ (less negative $\delta^{13}C$) relative to phytoplankton due to boundary layer effects. Researchers have exploited this difference to answer many important ecosystem-level questions. Below are two examples.

During the late 1970s and early 1980s, in the western Aleutian Islands of Alaska, where sea otters had recovered from overexploitation and suppressed their herbivorous urchin prey, productive kelp beds dominated. There, transplanted filter feeders, barnacles and mussels, grew up to 5 times faster compared to islands devoid of kelp where sea otters were scarce and urchin densities high. Stable isotope analysis revealed that the fast-growing filter feeders were enriched in carbon suggesting that macroalgae was the carbon source responsible for this magnification of secondary production.

In four Wisconsin lakes, experimental manipulations of fish communities and nutrient loading rates were conducted to test the interactive effects of food web structure and nutrient availability on lake productivity and carbon exchange with the atmosphere. The presence of top predators determined whether the experimentally enriched lakes operated as net sinks or net sources of atmospheric carbon. Specifically, the removal of piscivorous fishes caused an increase in planktivorous fishes, a decrease in large-bodied zooplankton grazers, and enhanced primary production, thereby increasing influx rates of atmospheric carbon into the lake. Atmospheric carbon was traced to upper trophic levels with $\delta^{13}C$. Here, naturally occurring stable isotopes and experimental manipulations conducted at the scale of whole ecosystems illustrated that top predators fundamentally alter biogeochemical processes that control a lake's ecosystem dynamics and interactions with the atmosphere.

Whole Ecosystem Experiments

Large-scale, whole ecosystem experiments have contributed considerably to our understanding of ecosystem dynamics. With its beginnings in wholesale watershed experiments in the 1960s, ecosystems are now being studied experimentally and analyzed as system of interacting species processing nutrients and energy within the context of changing abiotic conditions. This is particularly relevant these days given the effects of anthropogenic climate forcing and pollution in both terrestrial and oceanic ecosystems.

A classic series of whole-lake nutrient addition experiments conducted in northwestern Ontario by David Schindler and his research group illustrated the role of phosphorus in temperate lake eutrophication. To separate the effects of phosphorus and nitrate, the researchers split a lake with a curtain and fertilized one side with carbon and nitrogen and the other with phosphorus, carbon, and nitrogen. Within 2 months, a highly visible algal bloom had developed in the basin in which phosphorus had been added providing experimental evidence that phosphorus is the limiting nutrient for phytoplankton production in freshwater lakes. Certainly, algae may show signs of nitrogen or carbon limitation when phosphorus is added to a lake; however, other processes often compensate for these deficiencies. For instance, CO_2 is rarely limiting because physical factors such as water turbulence and gas exchange regulate its availiblity. Further, nitrogen can be fixed by blue-green algae. These species, which are favored when nitrogen is in short supply, increases the availability of nitrogen to algae, and the lake eventually returns to a state of phosphorus limitation. The practical significance of these results is that lake europhication can be prevented with management policies that control phosphorus input into lake and rivers.

Using Management Policies as Ecosystem Experiments

It has become increasingly common to use management policies as experiments and test their effects on ecosystem dynamics. An excellent example of this approach is the use of marine reserves to investigate the ecosystem-level consequences of fishing. Essentially, well-enforced marine reserves constitute large-scale human-exclusion experiments and provide controls by which to test the ecosystem effects of reducing consumer biomass via fishing at an ecologically relevant scale. Dramatic shifts in nearshore community structure have been documented in well-established and well-protected marine reserves in both Chile and New Zealand. In northeastern New Zealand's two oldest marine reserves, the Leigh Marine Reserve and Tawharanui Marine Park, previously fished predators, snapper (*Pagrus auratus*) and rock lobster

(*Jasus edwardsii*), have increased in abundance by 14- and 3.8-fold, respectively, compared to adjacent fished waters. Increased predation leading to reduced survivorship and cryptic behavior of their herbivorous prey, the sea urchin (*Evechinus chloroticus*), has allowed the macroalga (*Ecklonia radiata*) to increase significantly within the reserves, a trend that has been developing in the Leigh reserve for the past 25 years (**Figure 7**). Although this provides evidence that fishing can indirectly reduce ecosystem productivity, the trophic dynamics described above are context dependent and vary as a function of depth, wave exposure, and oceanographic circulation (**Figure 8**). For example, both in the presence and absence of fishing, urchin densities decline to nearly 0 individuals per m^2 below depths greater than 10 m due to unfavorable conditions for recruitment, despite the presence or absence of snapper and lobster, while at depths above 3 m, wave surge can preclude urchin grazing both inside and outside the reserves. Furthermore, where oceanic conditions hinder urchin recruitment, the effects of fishing on macroalgae become less clear-cut. These physical constraints highlight the importance of abiotic context on biotic interactions. Ultimately, one can gain a lot of information by using management policies as experiments.

Although policy experiments have played an important role in elucidating ecosystem dynamics, in many cases, it is politically intractable or logistically impossible to experiment with whole ecosystems. Under such circumstances, researchers have used alternative techniques to explore ecosystem dynamics. Models in ecology have a venerable tradition for both teaching and understanding complex processes. Ecosystem models are now being used to gain insight into the ecosystem-level consequences of management policies, from fisheries to carbon emissions. For more information on ecosystem models and using management policies as experiments, see the section entitled 'Social–ecological systems, Humans as key ecosystem components'.

Ecosystem Function and Biodiversity

Accelerating rates of species extinction have prompted researchers to formally investigate the role of biodiversity in providing, maintaining, and even promoting 'ecosystem function'. Typically, studies experimentally modify species diversity and examine how this influences the fluxes of energy and matter that are fundamental to all ecological processes. In many cases, studies are designed to document the effects of species richness on the efficiency by which communities produce biomass, although the effects of species diversity on other ecosystem functions such as decomposition rates, nutrient retention, and CO_2 uptake rates have also been examined.

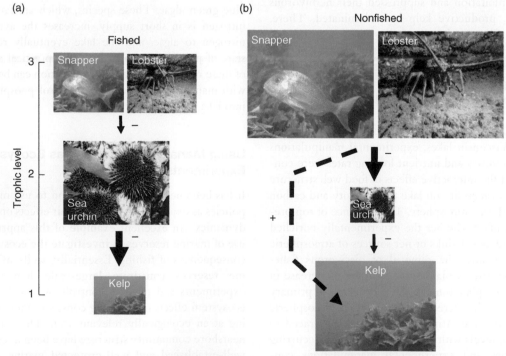

Figure 7   (a) In nearshore fished ecosystems in northeastern New Zealand, snapper and lobster densities have been reduced due to fishing pressure resulting in high sea urchin densities, urchin barrens, and reduced kelp production. (b) In marine reserves, where previously fished snapper and lobster have recovered, sea urchins that have not been consumed by these predators behave cryptically, hiding in crevices. Consequently, kelp forests of *Ecklonia radiata* dominate. Photos by Nick Shears, Hernando Acosta, and Timothy Langlois.

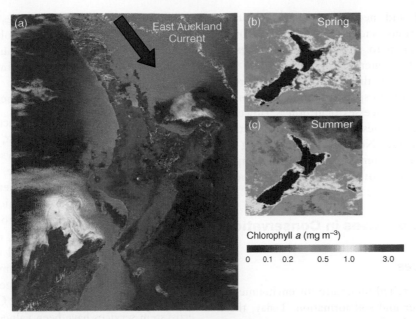

Figure 8 The effects of fishing on nearshore ecosystems are influenced locally by wave exposure and regionally by oceanographic circulation. (a) In northeastern New Zealand, ocean circulation patterns influence nutrient delivery and thus (b) spring and (c) summer pelagic primary production. Satellite images: SeaWiFs Project, Ocean Color Web.

Several seminal studies report a positive relationship between biodiversity and ecosystem function. Yet, the generality of the results, and the mechanisms driving them, have provoked considerable debate and several counterexamples exist.

At the crux of the debate lies a question with deep historical roots: do some species exert stronger control over ecosystem processes than others? Imagine two distinct positive relationships between biodiversity and ecosystem function (**Figure 9**). In type A communities, every single species contributes to the ecosystem function measured, even the rare species. By contrast, in type B communities, almost all of the ecosystem function measured can be provided by relatively few species, suggesting that many species are in fact redundant. Few empirical studies support type A relationships, rather, empirical evidence points to the prevalence of type B relationships. In fact, a recent meta-analysis of 111 such studies conducted in multiple ecosystems on numerous trophic groups found that the average effect of decreasing species richness is to decrease the biomass of the focal trophic group, leading to less complete depletion of resources used by that group. Further, the most species-rich polycultures performed no differently than the single most productive species used in the experiment. Consequently, these average effects of species diversity on ecosystem production are best explained by the loss of the most productive species from a diverse community. These results could be considered consistent with what has become known as the 'sampling effect'.

Critics argue that a positive relationship between species diversity and ecosystem function is a sampling artifact rather than a result of experimentally manipulated biodiversity *per se*. Such a 'sampling effect' can arise because communities comprising more species have a greater chance of being dominated by the most productive taxa. Yet, controversy surrounding the 'sampling effect' itself exists given the duality in its possible interpretation: is this a real biological mechanism that operates in nature or is it an experimental artifact of using random draws of species to assemble experimental communities? To add to the ecosystem function–biodiversity debate is the critical issue that many of these studies focus on a

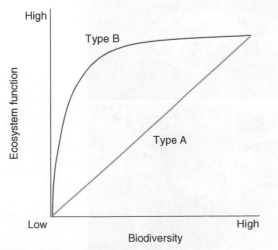

Figure 9 Type A communities: every single species contributes equally to ecosystem functioning. Type B communities: ecosystem function is provided by only a few species.

single trophic level and neglect or dismiss multiple trophic-level interactions, such as herbivory and other disturbances well known to alter ecosystem processes, calling into question the generality of these results.

Despite the controversy, these studies generally reinforce the notion that certain species exert much stronger control over ecological processes than others. However, identifying which species these are in advance of extinction remains a challenge. Nonetheless, identifying the mechanisms driving ecosystem functioning is an important conservation priority given that human well-being relies on a multitude of these functions.

Ecosystem Perspectives in Conservation Science

Ecosystem Services

Humans have always relied on nature for environmental assets like clean water and soil formation. Today, these assets are receiving global attention as 'ecosystem services', the conditions and processes by which natural ecosystems sustain and fulfill human life. Natural ecosystems perform a diversity of ecosystem services on which human civilization depends:

1. regulating services – purification of air and water, detoxification and decomposition of wastes, moderation of weather extremes, climate regulation, erosion control, flood control, mitigation of drought and floods, regulation of disease carrying organisms and agricultural pests;
2. provisioning services – provision of food, fuel, fiber, and freshwater;

3. supporting services – formation and preservation of soils, protection from ultraviolet rays, pollination of natural vegetation and agricultural crops, cycling of nutrients, seed dispersal, maintenance of biodiversity, primary production; and
4. cultural services – spiritual, esthetic, recreational.

Although critical to human existence, ecosystem services are often taken for granted or at best, greatly undervalued. This is ironic given that many ecosystem services are very difficult and expensive to duplicate, if they can be duplicated at all. Normally, ecosystem services are considered 'free' despite their obvious economic value. For example, over 100 000 species of animals provide free pollination services, including bats, bees, flies, moths, beetles, birds, and butterflies (**Figure 10**). Based on the estimate that one-third of human food comes from plants pollinated by wild pollinators, pollination has been valued at US$4–6 billion per year in the US alone. Globally, the world's ecosystem services have been valued at US$33 trillion a year, nearly twice as much as the gross national product of all of the world's countries.

The idea of paying for ecosystem services has been gaining momentum. Yet, because ecosystem services are typically not sold in markets, they usually lack a market value. Given the value of natural capital, nonmarket valuation approaches are being developed by economists and ecologists to account for ecosystem services in decision-making processes. The notion being that economic valuation gives decision makers a common currency to assess the relative importance of ecosystem processes and other forms of capital.

Figure 10 Pollination services, provided by bees, bats, butterflies, and birds to name a few, have been valued at US$4–6 billion per year in the US alone. Consider the global value of this important ecosystem service. Photos by Steve Gaines, Heather Tallis.

Yet, assigning value to ecosystem services is tricky and some analysts object to nonmarket valuation, because it is a strictly anthropogenic measure and does not account for nonhuman values and needs. Yet, in democratic countries, environmental policy outcomes are determined by the desires of the majority of citizens, and voting on a preferred policy alternative is ultimately an anthropogenic activity. A second objection to nonmarket valuation is a disagreement with pricing the natural world and dissatisfaction with the capitalistic premise that everything is thought of in terms of commodities and money. The point of valuation, however, is to frame choices and clarify the tradeoffs between alternative outcomes (i.e., draining a wetland may increase the supply of developable land for housing but does so at the cost of decreased habitat and potential water quality degradation). Finally, a third objection to nonmarket valuation stems from the uncertainty in identifying and quantifying all ecosystem services. Advocates argue that economic valuation need not cover all values and that progress is made by capturing values that are presently overlooked.

Despite the uncertainties, valuing ecosystem services can sometimes pay off. When New York City compared the coast of an artificial water filtration plant valued at US$6–8 billion, plus an annual operating cost of US$300 million, the city chose to restore the natural capital of the Catskill Mountains for this watershed's inherent water filtration services and for a fraction of the cost (US$660 million). Ultimately, the valuation of ecosystem services, even if flawed, may get ecosystem processes on the decision-making table and lead to more sustainable policies in light of ever-expanding human populations.

Ecosystem services are threatened by growth in the scale of human enterprise (population size, per-capita consumption rates) and a mismatch between short-term needs and long-term societal well-being. With a global population soon to number 9 billion people, ecosystem services are becoming so degraded, some regions in the world risk ecological collapse. Many human activities alter, disrupt, impair, or reengineer ecosystem services such as overfishing, deforestation, introduction of invasive species, destruction of wetlands, erosion of soils, runoff of pesticides, fertilizers, and animal wastes, pollution of land, water, and air resources. The consequences of degrading ecosystem services on human well-being were examined in the Millennium Ecosystem Assessment (MA) 2005, which concluded that well over half of the world's ecosystems services are being degraded or used unsustainably. The MA developed global ecological scenarios as a process to inform future policy options. These scenarios were based on a suite of models that were designed to forecast future change. The MA based its scenario analyses on ecosystem services. Specifically, scenarios were developed to anticipate responses of ecosystem services to alternative futures driven by different sets of policy decisions. Following the completion of this ambitious ecological study, there is now a growing movement to make the value of ecosystem services an integral part of current policy initiatives.

Social–Ecological Systems, Humans as Key Ecosystem Components

Humans are a major force in global change and drive ecosystem dynamics, from local environments to the entire biosphere. At the same time, human societies and global economies rely on ecosystem services. As such, human and natural systems can no longer be treated independently because natural and social systems are strongly linked. Accumulating evidence suggests that effective environmental management and conservation strategies must take an integrated approach, one that considers the interactions and feedbacks between and within social, economic, and ecological systems. As a result, the concept of coupled 'social–ecological systems' has become an emerging focus in environmental and social science and ecosystem management. Social–ecological systems are considered as evolving, integrated systems that typically behave in nonlinear ways. The concept of resilience – the capacity to buffer change – has been increasingly used as an approach for understanding the dynamics of social–ecological systems. Two useful tools for building resilience in social–ecological systems are structured scenario modeling and active adaptive management.

Models of linked social–ecological systems have been developed to inform management conflicts over water quality, fisheries, and rangelands. These models represent ecosystems coupled to socioeconomic drivers and are explored with stakeholders to probe the management decision-making processes. Alternative scenarios force participants to be absolutely explicit about their assumptions and biases, thereby improving communication between stakeholders and exposing the ecological consequences of various management policies.

Adaptive management is an approach where management policies themselves are deliberately used as experimental treatments. As information is gained, policies are modified accordingly. This approach helps isolate anthropogenic effects from sources of natural variation and, most importantly, considers the consequences of a human perturbation on the whole ecosystem. In contrast, basic research on various parts of an ecosystem leads to the challenge of assembling all the data into a practical framework. Yet, biotic and abiotic ecosystem components are not additive, they interact. Due to these interactions, the dynamics of an ecosystem cannot be extrapolated from the simple addition of an ecosystem's components. Adaptive management examines the response of the system as a whole rather than a sum of its parts. Furthermore, this approach involves adaptive learning and adaptive institutions that acknowledge uncertainties and can respond to nonlinearities. In sum, structured scenario modeling and policy experimentation are tools that can be used to examine the resilience of social–ecological systems to alternative management policies and conservation strategies.

Ecosystem-Based Management

Recognizing the need to sustain the integrity and resilience of social–ecological systems has led to calls for 'ecosystem-based management', a management approach that considers all ecosystem components, including humans and the physical environment. With the overall goal of sustaining ecosystem structure and function, this management approach:

- focuses on key ecosystem processes and their responses to perturbations;
- integrates ecological, social, and economic goals and recognizes humans as key components of the ecosystem;
- defines management based on ecological boundaries rather than political ones;
- addresses the complexity of natural processes and social systems by identifying and confronting uncertainty;
- uses adaptive management where policies are used as experiments and are modified as information is gained;
- engages multiple stakeholders in a collaborative process to identify problems, understand the mechanisms driving them, and create and test solutions; and
- considers the interactions among ecosystems (terrestrial, freshwater, and marine).

Ecosystem-based management is driven by explicit goals, executed by policies and protocols, and made adaptable by using policies as experiments, monitoring their outcomes and altering them as knowledge is gained.

Traditionally, management practices have focused on maximizing short-term yield and economic gain over long-term sustainability. These practices were driven by inadequate information on ecosystem dynamics, ignorance of the space and timescales on which ecosystem processes operate, and a prevailing public perception that immediate economic and social value outweighed the risk of alternative management. Seeking to overcome these obstacles, ecosystem-based management relies on research at all levels of ecological organization, explicitly recognizes the dynamic character of ecosystems, acknowledge that ecological processes operate over a wide range of temporal and spatial scales and are context dependent, and presupposes that our current knowledge of ecosystem function is provisional and subject to change. Ultimately, ecosystem-based management recognizes the importance of human needs while addressing the reality that the capacity of our world to meet those needs in perpetuity has limits and depends on the functioning of resilient ecosystems.

See also: Ecosystem Ecology.

Further Reading

Cardinale BJ, Srivastava DS, Duffy JE, *et al.* (2006) Effects of biodiversity on the functioning of trophic groups and ecosystems. *Nature* 443: 989–992.

Daily GC (ed.) (1997) *Nature's Services: Societal Dependence on Natural Ecosystems*. Washington, DC: Island Press.

DeAngelis DL (1992) *Dynamics of Nutrient Cycling and Food Webs*. New York, NY: Chapman and Hall.

Krebs C (2001) *Ecology: The Experimental Analysis of Distribution and Abundance,* 5th edn. San Francisco: Addison-Wesley Educational Publishers, Inc.

Millennium Ecosystem Assessment (2005) *Ecosystems and Human Well-Being: Synthesis*. Washington, DC: Island Press.

Pauly D and Christensen V (1995) Primary production required to sustain global fisheries. *Nature* 374: 255–257.

Scheffer M, Carpenter S, Foley JA, Folke C, and Walker B (2001) Catastrophic shifts in ecosystems. *Nature* 413: 591–596.

Ecosystem Services

K A Brauman and G C Daily, Stanford University, Stanford, CA, USA

Introduction	Capturing the Value of Ecosystem Services
Defining Ecosystem Services	Conclusions
Examples of Ecosystem Services	Further Reading

Introduction

The world's ecosystems yield a flow of essential services that sustain and fulfill human life, from seafood and timber production to soil renewal and personal inspiration. Although many societies have developed the technological capacity to engineer replacements for some services, such as water purification and flood control, no society can fully replace the range and scale of benefits that ecosystems supply. Thus, ecosystems are capital assets, worthy of at least the level of attention and investment given to other forms of capital. Yet, relative to physical, financial, human, and social capital, ecosystem capital is poorly understood, scarcely monitored, and, in many cases, undergoing rapid degradation and depletion.

Recognition of ecosystem services dates back at least to Plato. This recognition of human dependence on ecosystems, in the past and today, is often triggered by their disruption and loss. Direct enjoyment of services, such as the extraction of timber, fish, and freshwater, can reduce the quantity and quality produced. The provision of ecosystem services can also be affected indirectly and inadvertently. Deforestation, for instance, has exposed the critical role of forests in the hydrological cycle – mitigating flooding and reducing erosion. Release of toxic substances has uncovered the nature and value of physical and chemical processes, governed in part by microorganisms that disperse and break down hazardous materials. Thinning of the stratospheric ozone layer has sharpened awareness of the value of its service in screening out harmful ultraviolet radiation.

Defining Ecosystem Services

Simply put, ecosystem services are the conditions and processes through which ecosystems, and the biodiversity that makes them up, sustain and fulfill human life. Ecosystem services are tightly interrelated, making their classification somewhat arbitrary. The Millennium Ecosystem Assessment (MA) – the formal international effort to elevate awareness and understanding of societal dependence on ecosystems – has suggested four categories.

First, 'provisioning services' provide goods such as food, freshwater, timber, and fiber for direct human use; these are a familiar part of the economy. Second, and much less widely appreciated, 'regulating services' maintain a world in which it is biophysically possible for people to live, providing such benefits as water purification, pollination of crops, flood control, and climate stabilization. Third, 'cultural services' make the world a place in which people want to live; they include recreation as well as esthetic, intellectual, and spiritual inspiration. Fourth, 'supporting services' create the backdrop for the conditions and processes on which society depends more directly. All of these services are provided by complex chemical, physical, and biological cycles, powered by the sun, and operate at scales ranging from smaller than the period at the end of this sentence to as large as the entire biosphere (**Table 1**).

Table 1 A classification of ecosystem services. Examples of ecosystem services and how they can be categorized

Provisioning services: Production of...
Food
   Seafood, agricultural crops, livestock, spices
Pharmaceuticals
   Medicinal products, precursors to synthetic pharmaceuticals
Durable materials
   Natural fiber, timber
Energy
   Biomass fuels, low-sediment water for hydropower
Industrial products
   Waxes, oils, fragrances, dyes, latex, rubber
Genetic resources
   Intermediate goods that enhance the production of other goods

Regulating services: Generation of...
Cycling and filtration processes
   Detoxification and decomposition of wastes
   Generation and renewal of soil fertility
   Purification of air and water
Translocation processes
   Dispersal of seeds to sustain tree and other plant cover
   Pollination of crops and other plants
Stabilizing processes
   Coastal and river channel stability
   Control of the majority of potential pest species
   Carbon sequestration
   Partial stabilization of climate
   Protection from disasters:
      regulation of hydrological cycle (mitigation of floods and droughts)
      moderation of weather extremes (such as of temperature and wind)

Cultural services: Provision of...
Esthetic beauty, serenity
Recreational opportunities
Cultural, intellectual, and spiritual inspiration

Supporting services: Preservation of...
Processes underlying services in the classes above
Options
   maintenance of the ecological components and systems needed for future supply of the goods and services above and others awaiting discovery

Tradeoffs in Managing the Flows of Ecosystem Services

Biophysical constraints on human activities, such as limited supplies of energy, land, and water, typically manifest themselves as tradeoffs between different uses. Thus, managing ecosystem services involves difficult ethical and political decisions about which services to develop and how to do so. At local scales, allocation of limited resources to alternative activities typically involves a zero-sum game, illustrated by the widespread redirection of water from agriculture to urban and industrial purposes. At global scales, different groups of people compete for use of Earth's open-access resources and waste sinks, such as the atmosphere's capacity to absorb CO_2 and other greenhouse gases without inducing climate change.

Making informed decisions about how to use ecosystem goods and services hinges on understanding these tradeoffs: knowing the joint products – the suite and level of services – that ecosystems can provide. For example, an ecosystem managed exclusively for agriculture may yield a greater return on agricultural products than one managed for multiple services, but understanding that diversified management may produce greater overall returns could influence management decisions (**Figure 1**).

Provision of biodiversity is one supporting service that has historically been discounted when managing for other ecosystem services. Biodiversity, however, can provide irreplaceable benefits. Genetic diversity, for example, allows for both the survival and evolution of the ecosystems we depend on for myriad benefits. Recent research indicates that diverse systems are more resilient, and therefore provide ecosystem services more reliably in the long term, than monocultures. While under optimal conditions managing for a single species may provide superior timber supplies or nutrient sequestration, given natural and human-caused variability in temperature, rainfall, and other environmental factors, managing for a diverse system will more consistently provide services in an uncertain world.

Examples of Ecosystem Services

Ecosystem services can be explored by focusing either on a single service that may be provided by various ecosystems or by looking at a single ecosystem that may provide a variety of services. Here we illustrate both approaches, considering first pollination services provided by bees then the suite of services provided by wetlands and forests. We highlight the differences in scale of delivery of

services, from local to global, and explore the tradeoffs inherent in their management.

Pollination Services Provided by Bees

Pollination, the movement of genetic material in the form of pollen grains, is a key step in the development of most food crops. Even crops that do not rely on insect pollination – wind pollinated or self-pollinated crops – are sometimes more productive when visited by an insect pollinator. Bees are a particularly important group of insect pollinators, responsible for pollinating 60–70% of the world's total flowering plant species, including nearly 900 food crops worldwide, such as apples, avocados, cucumbers, and squash. These crops comprise 15–30% of the world's food production, and bees are credited with $4.2 billion in annual crop productivity in California alone. Bees are especially important pollen vectors in part because physical adaptations, such as hairs designed to pick up pollen, and behavioral adaptations, such as fidelity to a single species of plant on each pollen-gathering trip, ensure good pollen transport and cross-pollination.

In the US, most major agricultural enterprises that rely on bee pollination import managed bees, almost always the European honeybee *Apis mellifera*. The available stock of managed honeybees has declined dramatically, however, dropping by over 50% in the last 50 years, while demand for pollination services has increased in many areas. This decline in managed bee populations has many causes, including increased pesticide use, disease in the hives, and downsizing of stocks that have hybridized with Africanized bees, introducing traits that make managed bees more aggressive and thus a liability to the farmer.

The contribution of native, wild bees to agricultural pollination was ignored, and assumed to be negligible, until the early 2000s. Since then, research has shown that native bees serve an important role in pollination, picking up slack when managed bee pollination is

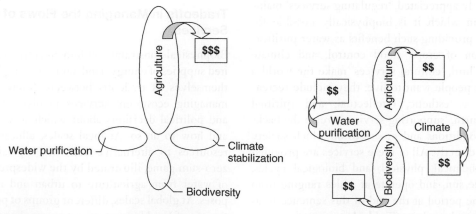

Figure 1   Joint products of ecosystems. Many ecosystems are currently managed to exploit only one service. Managing for multiple services can increase ecosystem benefits.

insufficient and enhancing crop production in general. Farms with generous native bee habitat nearby may be able to fully or partially replace pollination by managed bees. In some cases, native bees are more efficient pollinators than European honeybees. The variety of wild bees, with distinct physical and behavioral traits, allows them, as a group, to pollinate a wide variety of flowering plants. Tomatoes, for example, have pollen that is accessible only by vibrating the flower, which bumble bees and some other native bees can, while honeybees cannot. Though tomatoes are self-pollinating and do not require an insect vector, native bees promote cross-pollination, which, for example, significantly increases the fruit set and size of Sungold cherry tomatoes.

The contributions of native bees to crop production are usually undocumented and underestimated, and they are always unpaid, at least directly. Though hives of managed honeybees must be rented or maintained, wild bees pollinate at no cost to the farmer. Populations of native bees are under great threat, however, by land management practices that promote the use of pesticides and the loss, fragmentation, and degradation of habitat. Protecting native bees without protecting the ecosystems in which they live is impossible. Native habitat, unlike agricultural monocropping, provides the year-round supply of blooming plants that wild bees require for sustenance. Native habitat also provides nesting areas; most wild bees are solitary, laying a single egg in a nest cavity dug into the ground or into dead wood, not forming social hives. In order to reap the benefits of native pollinators, food resources and nesting habitat must be available within a short distance of crops, possibly as hedgerows, in ditches, or around water ponds. A study of wild bee pollination of coffee in Costa Rica showed that farms closer to tropical forest remnants were visited by many more species of wild bees than those further away. Had the far sites been adequately pollinated, coffee yield would have been increased by nearly 20% and misshapen coffee beans reduced by 27%. A lower-bound estimate of the pollination services from these patches is US $62 000 per year (in the early 2000s).

The diversity of the native bee population is one of its strengths. Many species of bees participate in pollination, and the abundance of different species varies year by year. This diversity allows the native pollinator community to be both resistant, maintaining functionality in the face of environmental upheaval, and resilient, able to reestablish itself in the wake of a destructive event. When the population of *Apis* declined dramatically in the second year of the Costa Rica study, sites close to forest fragments showed minimal loss of pollination while pollinator visits dropped by nearly 50% further away. Thus, as well as enhancing pollination services in conjunction with managed bees, native bee populations provide important insurance against

the possibility that managed bee populations could fail because of disease, hybridization, or other causes.

Services Provided by Wetlands

Areas inundated by fresh, brackish, and salt water are all considered wetlands; among many wetland types are fens and bogs, tidal marshes, riparian zones, and lakeshores. Wetlands, which cover less than 9% of the Earth's surface, can be extremely productive and many are disproportionately large providers of ecosystem services. Three of the key services that wetlands provide are flood mitigation, water purification, and biodiversity support.

In the upper part of a watershed, many wetlands store water that flows overland toward rivers and streams. They can release this water into the main channel slowly, reducing and delaying flood peaks. Downstream, wetlands can absorb and reduce peak flood levels, providing area into which flood waters can spread, dissipating flood energy by slowing water movement, and removing flood water through transpiration and infiltration.

The same physical characteristics of wetlands that slow and absorb overland flow related to flooding can also provide a mechanism for storing and detoxifying urban and agricultural wastewater before it discharges directly into a main channel. Wetlands filter out various nutrients, other pollutants, and sediment: they support anaerobic bacteria that denitrify waste; the plants take up and store nutrients; and by slowing and redirecting water flow, wetlands enhance sedimentation – the accreting sediments can effectively bury pollutants. While many wetlands can purify water very economically, their effectiveness depends on many factors, including rate of inflow, amount of sediment and organics in the wastewater, residence time of wastewater in the wetland, and total surface area.

A wide variety of animals rely on wetlands for survival. Plant species that deliver flood abatement and water purification can also support biodiversity, providing varied food and shelter. A riparian wetland, for example, might provide food plants and underground burrows for muskrats; seeds, food plants, and nest-building materials for ducks; and food and shelter for fish and invertebrates.

Wetlands provide a variety of other services as well. Major products associated with wetlands are peat, timber, and mulch. Regulating services in addition to flood mitigation and water purification include waste detoxification, carbon storage, and control of pests and diseases. Wetlands provide many cultural services as well, particularly recreation services such as bird watching, boating, and hunting. Wetlands also provide key supporting services, such as soil formation and buffering freshwater aquifers from saltwater intrusion.

Worldwide, wetlands are estimated to provide many billions of dollars in services each year. They are recognized by the international treaty, the Ramsar Convention on Wetlands, and regulated by domestic law in many countries. Nonetheless, they have historically undergone widespread losses in favor of other land uses; worldwide, 50% of wetlands are estimated to have been lost since 1900.

While the services provided by wetlands are widely recognized, simultaneously maximizing multiple services may not be possible. In some cases this is related to location: upland watersheds may be very important for flood control but may be too far upstream to have an impact on water purification. In other cases one service may thrive to the detriment of another: a wetland that is absorbing a heavy nutrient load may be overtaken by a single, aggressive plant species and thus fail to be an effective reservoir for biodiversity. Finally, it can be costly to measure function and hence difficult to judge how effectively a wetland is performing a given service or how to manage for that particular service.

Services Provided by Forests

Forests provide a wide array of services, such as timber production, climate stabilization, provision of water quantity and quality, and cultural benefits, such as recreation. Some management options increase the supply of several services, but often one service is enhanced to the detriment of others.

Forests are often managed for provisioning services, particularly for timber. But even within the category of provisioning services, management options differ. If a forest is considered exclusively a supplier of timber, managers will encourage the growth of only certain kinds of trees, possibly nonnative fast-growing trees, and will cultivate them so that they grow in a uniform way, typically straight and tall. When the trees are deemed mature, they will be cut down, often all at once. By contrast, if a forest is regarded as a supplier of diverse benefits, it may be managed to nurture a wide array of valued species that would not be available in the monocrop forest described above.

Forests also have both short-term and medium-term impacts on climate. Temperature regulation happens in forests when the canopy shades the ground and when dark-colored foliage absorbs heat. Forests can in certain circumstances also influence precipitation – in cloud forests, for example, trees and epiphytes intercept and condense water directly from the air, and that water runs down trunks to plants and soil below. On a longer timescale, forests play a role in carbon cycling and sequestration; when forest plants, bacteria, and algae respire, they take CO_2 out of the atmosphere. Plants, soils, and the animals that eat them in forests, grasslands, and other terrestrial ecosystems store ~2000 billion tons of carbon worldwide, about half the amount of carbon stored in the ocean and nearly three times that stored in the atmosphere. However, if these ecosystems are burned or

destroyed, as happens when timber is harvested, the carbon they are sequestering is released to the atmosphere. Although most organic compounds do return to the atmosphere as CO_2 when living organisms die and decompose, in a functioning forest ecosystem some is buried and sequestered. About 25% of the human-caused increase in CO_2 concentration in the atmosphere during the past 20 years resulted from land-use change, primarily deforestation.

Forests in a watershed, on the hillslopes that drain into a river, influence the water quality in that river. In part this is because higher-intensity uses, such as agriculture input pollutants like nutrients and pesticides into a system while forests do not. Forests themselves also reduce sediment and nutrient runoff. Clearing trees can have an impact as soon as the next rainy season on sediment and nutrient loads in streams, as demonstrated in the classic Hubbard Brook experiment. In some cases, water users have invested in forests to keep their water supplies clean. New York City recently invested US$ 250 million to acquire and protect land in the Catskills watershed that supplies water to the city. By working with landowners to reduce pesticide and fertilizer application and to plant buffer strips along waterways, New York City reduced potential contamination of its drinking water. In conjunction with related conservation investments amounting to ~US$ 1.5 billion, the city thereby obviated the need to build a filtration plant projected to cost between US$ 6 and US$ 8 billion.

Forests can also play an important role regulating the timing and quantity of runoff. The economic value of forests in the watershed of the Yangtze River above Three Gorges Dam, in western Hubei Province, Central China, was quantified in a study published in 2000. Here, the Gexhouba Hydroelectric Power Plant, the largest hydro-facility in China, producing 15.7 billion kW annually, requires a narrow range of flows on the Yangtze in order to run at full power. If the water level is too high, then water must be released through the sluice gates, causing the water level below the dam to rise, reducing the amount of power that can be produced; at very high flows, turbines are drowned and cannot work at all. If the water is too low, then generators cannot run at full power.

The goal of the hydroelectric facility's managers is for the river to have flow depths that vary as little as possible, as this has been shown to be much more important for power generation than the total flow. Upstream forests damp fluctuations in stream flow by reducing runoff in wet periods through canopy interception, leaf litter absorption, and soil and groundwater storage; increased infiltration provides base flow in dry periods through groundwater discharge. Though water flow regulation is a function of vegetation, soil type, and slope, which occur in a heterogeneous mix through the watershed, forests and even shrubs with all types of soils and slopes consistently provided better water regulation than grasses, orchards, and crop agricultural fields. This study estimated the value of electricity produced by the hydro-facility due to water regulation by the forest at over US$ 600 000 per year (in the early 2000s), or about 2.2 times the income derived from forest product services in this area. Because trees lose water to the atmosphere through transpiration, however, the total water available downstream was decreased by the forest.

Different management regimes will yield different suites of services. Some services can never be coproduced; other services will almost always be produced in tandem, though often to differing degrees. For the hypothetical forest illustrated in **Figure 2**, cattle and timber cannot be produced on the same parcel of land – conversion to pasture optimizes livestock but reduces timber output dramatically. Under timber maximization, once trees are harvested they are not available for climate or hydrologic regulation, though before harvest those services will be produced, as well as some habitat and hiking trails. Carbon sequestration, hydropower, recreation, and preservation of biodiversity tend to be coproduced, but there are tradeoffs in their optimal supply. Maximizing biodiversity, for example, produces all four to their fullest extent but allows for no timber supply. Bringing selective logging back into the management regime reduces supply of the other services somewhat; maximizing timber yield reduces them much more dramatically.

Tradeoffs between services are also tradeoffs between consumers, such as local recreationalists, regional users of hydropower, and global beneficiaries of carbon sequestration and biodiversity conservation. These tradeoffs underscore the importance of valuation, making explicit who benefits from ecosystem services and who pays for them. Conceiving of ecosystem functions as services and assigning a monetary value to them provides a tool for decision-makers to weigh different management options.

Capturing the Value of Ecosystem Services

Despite their obvious importance to human well-being, people tend to think of ecosystems as being economically productive in narrow terms, often assigning value only to the production of conventional commodities or to real estate development. Provision of ecosystem services is only rarely considered in cost–benefit analyses, preparation of environmental impact statements, or other assessments of alternative paths of development. There is no shortage of markets for ecosystem goods (such as clean water and water-melons), but the services underpinning these goods (such as water purification and bee pollination) often have no monetary value. This is in part because ecosystem services are generally public goods, free to any user, and therefore difficult to value. Because people mostly do not pay for them, it can be difficult to discern what the supply, demand, and willingness to pay for services actually are. As a result, there are no direct price mechanisms to signal the scarcity or degradation of these public goods before they fail.

While for some goods and services price reflects value or importance, when ecosystem services are assigned monetary value they tend to be priced much lower than their importance suggests. This is true in part because when supply is much larger than demand, prices are low, no matter how necessary the good. The pricing of diamonds and water is illustrative. Lost in the desert, a traveler would happily trade all the diamonds in the world for a single cup of water; back in the marketplace, our traveler would find that diamonds are many, many times more costly than water. Water is inexpensive or free because, like many ecosystem goods and services, it tends to be far more abundant than the volume demanded by people; when ecosystems are functioning well, even more is available.

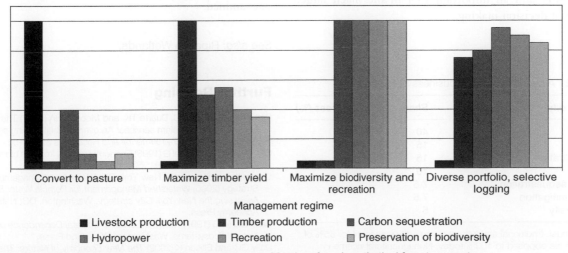

Figure 2 Tradeoffs associated with alternative management objectives for a hypothetical forest ecosystem.

Ecosystem services are also often undervalued because prices are based on current supplies and demands, so the amount we are willing to pay for continued nutrient retention in a wetland may be low today even if we can predict that nutrient-laden runoff from increased agriculture will threaten a downstream fishery tomorrow. Further, prices are based on marginal utility – for example, the amount someone would be willing to pay for the carbon stored in one more tree in a forest. If that forest is clear-cut, we lose all of the carbon storage and, since the loss of each tree changes the value of the next, we cannot account for the whole loss using the price of the first tree.

Precise valuation of ecosystem services is often not required to provide appropriate economic incentives for protecting the ecosystems that supply them. Incentives need only make it more economically appealing to a landowner to maintain hedgerows as habitat for native pollinators than to cultivate every last square meter of a field, for instance, or make it pay to preserve a wetland rather than filling it to build houses. A farm, as illustrated below, might generate enough income from nonagricultural commodities to alter its land management regime (**Table 2**). Incentives to protect and maintain ecosystems can be provided by the government, privately through markets, or through hybrid institutions such as cap-and-trade systems supported by government policy.

A variety of tools for valuing ecosystem services and creating incentives for their conservation are currently being developed, including capital markets such as the Chicago Climate Exchange, wetland mitigation banks, and outright payments, often involving private–public partnerships, for services, as is occurring in Australia, Costa Rica, and Mexico. These market-based approaches provide a much better indication of value than early, more theoretical attempts to quantify the value of ecosystem services. While valuation is not necessarily a solution or end in itself, it is a powerful way of organizing information and an important tool in the much larger process of decision making.

Conclusions

Because ecosystem services explicitly invoke human beneficiaries, basic scientific understanding of the ecosystem processes producing goods and services is meaningful only in the context of economic valuation and institutional structures. There is still much to learn on many fronts. Important questions include: Which ecosystems supply which services? What levels and types of ecosystem protection are required to sustain service supply? Can we develop robust methodologies for the valuation of ecosystems? Even if clear answers are absent to all of these questions, numerous and diverse efforts are now underway worldwide to protect vital ecosystem services, often using innovative economic incentives.

Explicitly identifying and valuing the goods and services provided by ecosystems has two obvious benefits. First, understanding the role of ecosystem services powerfully justifies habitat preservation and biodiversity conservation as vital, though often overlooked, policy objectives. While a wetland surely provides existence and option values to some people, the benefits provided by the wetland's nutrient retention and flood mitigation services are both universal and undeniable. Tastes may differ over beauty, but they are in firm accord over the high costs of polluted water and flooded homes. Second, if given the opportunity, natural systems can in many cases quite literally pay their own way. Market mechanisms and institutions that can capture and maximize service values can effectively promote environmental protection at the local, regional, national, and international levels. In some cases, however, protection of ecosystem services will not justify conservation of natural habitats. In other cases, the services will be largely irrelevant to environmental protection efforts. While a focus on ecosystem services provides great potential to promote environmental protection, its practical implications remain largely unexamined.

See also: Riparian Wetlands.

Table 2   A hypothetical farm business in 15 years

Commodity	Share of farm business (%)
Wheat	40
Wool	15
Water filtration	15
Timber	10
Carbon sequestration	7.5
Salinity mitigation	7.5
Biodiversity	5

In this model, traditional agricultural commodities account for 55% of revenues, as opposed to 100% today. Nonagricultural income is supplied by a mature market for ecosystem goods and services.

Further Reading

Brauman KA, Daily GC, Duarte TK, and Mooney HA (2007) The nature and value of ecosystem services: An overview highlighting services. *Annual Review of Environmental and Resources* 32: 67–98.

Chichilnisky G and Heal G (1998) Economic returns from the biosphere – Commentary. *Nature* 391: 629–630.

Committee to Review the New York City Watershed Management Strategy (2000) *Watershed Management for Potable Water Supply: Assessing the New York City strategy*. Washington, DC: National Academy Press.

Daily GC (ed.) (1997) *Nature's Services: Societal Dependence on Natural Ecosystems*. Washington, DC: Island Press.

Daily GC and Ellison K (2002) *The New Economy of Nature: The Quest to Make Conservation Profitable*. Washington, DC: Island Press.

Daily GC, Söderqvist T, Aniyar S, *et al.* (2000) The value of nature and the nature of value. *Science* 289: 395–396.

Findlay SEG, Kiviat E, Nieder WC, and Blain BA (2002) Functional assessment of a reference wetland set as a tool for science, management and restoration. *Aquatic Sciences* 64: 107–117.

Guo Z (2000) An assessment of ecosystem services: Water flow regulation and hydroelectric power production. *Ecological Applications* 10: 925–936.

Heal G (2000) *Nature and the Marketplace: Capturing the Value of Ecosystem Services*, Washington, DC: Island Press.

Heal G, Daily GC, and Salzman J (2001) Protecting natural capital through ecosystem service districts. *Stanford Environmental Law Journal* 20: 333–364.

Kremen C, Williams NM, and Thorp RW (2002) Crop pollination from native bees at risk from agricultural intensification. *Proceedings of the*

National Academy of Sciences of the United States of America 99: 16812–16816.

Millennium Ecosystem Assessment (2005) *Ecosystems and Human Well-being: Current State and Trends: Findings of the Condition and Trends Working Group*. Washington, DC: Island Press.

Postel SL and Thompson BH (2005) Watershed protection: Capturing the benefits of nature's water supply services. *Natural Resources Forum* 29: 98–108.

Ricketts TH, Daily GC, Ehrlich PR, and Michener C (2004) Economic value of tropical forest to coffee production. *Proceedings of the National Academy of Sciences of the United States of America* 101: 12579–12582.

Zedler JB and Kercher S (2005) Wetland resources: Status, trends, ecosystem services, and restorability. *Annual Review of Environment and Resources* 30: 39–74.

Fundamental Laws in Ecology

S E Jørgensen, Copenhagen University, Copenhagen, Denmark

The Need for Fundamental Laws

Humans have always strived toward finding a structure or a pattern in their observations – to develop a theory. Science does not make sense without theory. Without theory, our observations become only a beautiful collection of impressions without explanation or application to solve problems of human interest. The alternative to scientific theory is to observe everything which is not possible. A well-developed theory can be used to make predictions.

Our scientific knowledge has to be coherent in order to apply the underlying theory and explain our observations. Ecology has only partially been able to condense the systematic collection of observations and knowledge about ecosystems into testable laws and principles. During the last few decades systems ecologists have developed hypotheses that together with basic laws from biochemistry and thermodynamics are proposed as a first attempt to formulate fundamental laws in ecology. The inherent complexity of ecosystems means that it is necessary to break from the long reductionistic scientific tradition to a new holistic ecological approach. Reductionistic science has had a continuous chain of successes since Descartes and Newton. Lately, however,

there is an increasing understanding for the need of knowledge syntheses to a more holistic image. Today this search for a holistic understanding of complex systems is considered one of the greatest scientific challenges of the twenty-first century by many scientists.

Several important contributions to systems ecology have attempted to capture the features and characteristics of ecosystems, their processes, and their dynamics. The different theories and approaches look inconsistent at first glance, but when examined more closely, their complementarity becomes evident . This commonality and consensus regarding ecosystem dynamics was asserted by Jørgensen in the first edition of *Integration of Ecosystem Theories: A Pattern* (1992), and later editions (2nd edn. 1997 and 3rd edn. 2002) have only enhanced the perception that the theories form a pattern and that they are highly consistent. It is clear from recent meetings and discussions that today we have a general ecosystem theory which is rooted in a consensus of the pattern of ecosystem dynamics. The ecosystem theory presented here combines the work of several scientists, and provides a foundation for further progress in systems ecology, ecosystem theory, and ecology. Furthermore, it may be feasible to use a few fundamental laws to derive other laws to explain most observations. We do not know yet to what extent this is possible in ecology, but at least we propose a promising direction for a useful, comprehensive ecosystem theory. Only by the application of the theory can we assess how and where the theory needs improvements.

Systems Ecology in the Jet Stream of Scientific Development

Seven general scientific theories have changed our perception of nature radically during the last 100 years: general and special relativity, quantum theory, quantum complementarity, Gödel's theorem, chaos theory, and theory for far-from-thermodynamic equilibrium systems. With these seven theories, we understand today that nature is much more complex than we thought 100 years ago, but we also have tools to understand this complexity better, which has entailed that we have a general ecosystem theory today.

The speed of light is the absolute upper limit for any transmission of matter, energy, and information according to the special relativity theory. This has given a completely new meaning to the concept of locality. It has also in systems ecology brought another meaning of network: links among components that share a locality and of the hierarchical organization: networks of smaller and smaller localities that are linked together on the next level of the hierarchy. Relativity theory also gives us a clear understanding of the lack of absolute measures, which was the governing scientific perception before the twentieth century. When we use ecological indicators to assess ecosystem health, we can only apply them relatively to other (similar) ecosystems; and, when we use thermodynamic calculations of ecosystems we know that we cannot get the absolute value but only an index or relative value because ecosystems are too complex to allow us to include all the components in our calculations. Quantum theory and later chaos theory upended the deterministic world picture: we cannot determine the future in all detail, even if we know all details of the present conditions. The world is ontically open. In the nuclear world, uncertainty is due to our inevitable impact on nuclear particles, while in ecology the uncertainty is due to the enormous complexity. Ecosystems are middle number systems. The number of components in such systems is many orders of magnitude smaller than the number of atoms in a room but too many to be countable. Further complicating the situation is that while the atoms are represented by a few different types all ecosystem components are different even among organisms of the same species. A room may contain 10^{28} components but they are represented by only 10 or 20 different types of molecules with exactly the same properties. An ecosystem contains in the order of 10^{15}–10^{20} different components

all with different individual properties and interaction potentials. It would be impossible to observe all components and even more impossible to observe all the possible interactions among these 10^{15}–10^{20} different components. Such complexity leads to a nondeterministic picture in ecology. In accordance with quantum complementarity, light can only be described by an interpretation as both waves and particles (photons). An ecosystem is much more complex than light. Therefore, a full (holistic) description of an ecosystem will also, not surprisingly, require two or more complementary descriptions. Various descriptions suggest ecosystems as dissipative, self-organizing systems that follow a dynamic to increase energy, emergy, ascendency (see Ecological Network Analysis, Ascendency), or eco-exergy which are not in conflict, because they cover different aspects of the ecosystem. All descriptions help to understand ecosystem dynamics, but some may be more applicable for addressing specific ecosystem questions.

Gödel's theorem that there are no complete theories – they are all based on some assumptions – is of course also valid for ecological theories. We shall not expect a complete theory based on no assumptions and which can be used in all contexts.

Newtonian Physics is based on the reversibility of all processes. Prigogine's new interpretation of the second law of thermodynamics has shown that time has an arrow. All processes are irreversible and evolution is rooted in this irreversibility. Einstein's special relativity theory, which provides the speed of light as an upper speed making it impossible to change the light signals which give information about a previous event, also supports the principle of irreversibility. We cannot change the past but only the future. With the enormous complexity of ecosystems it also implies that the same conditions will never be repeated . Ecosystems are always confronted in space and time with new challenges, which explains the enormous diversity that characterizes the biosphere. Clearly, systems ecology has not developed in a vacuum, but has been largely influenced by the general scientific development during the last 100 years. A summary of a general ecosystem theory is presented here. The current proposed theory consists of ten laws.

Systems Ecology: Ten Tentative Fundamental Laws – An Attempt to Formulate an Ecosystem Theory

A tentative ecosystem theory consisting of eight basic laws has previously been presented, but it seems to be an advantage to split one of the laws into three due to some recent results, which are presented below with a few comments.

1. *All ecosystems are open systems embedded in an environment from which they receive energy (matter) input and discharge energy (matter) output.* From a thermodynamic viewpoint, this principle is a prerequisite for ecological processes. If ecosystems could be isolated, then they would be at thermodynamic equilibrium without life and without gradients. This law is rooted in Prigogine's use of thermodynamics far-from-thermodynamic equilibrium. The openness explains, according to Prigogine, why the system can be maintained far-from-thermodynamic equilibrium without violating the second law of thermodynamics.

2. *Ecosystems have many levels of organization and operate hierarchically.* This principle is used again and again when ecosystems are described: atoms, molecules, cells, organs, organisms, populations, communities, ecosystems, and the ecosphere. The law is based on the differences in scale of local interactions. The distance between components becomes essential because it takes time for events and signals to propagate. Ecological complexity makes it necessary to distinguish between different levels with different local interactions.

3. *Thermodynamically, carbon-based life has a viability domain determined between about 250 and 350 K.* It is within this temperature range that there is a good balance between the opposing ordering and disordering processes: decomposition of organic matter and building of biochemically important compounds. At lower temperatures the process rates are too slow and at higher temperatures the enzymes catalyzing the biochemical formation processes decompose too rapidly. At 0 K there is no disorder, but no order (structure) can be created. At increasing temperatures, the order (structure)-creating processes increase, but the cost of maintaining the structure in the face of disordering processes also increases.

4. *Mass, including biomass, and energy are conserved.* This principle is used again and again in ecology and particularly in ecological modeling.

5. *The carbon-based life on Earth has a characteristic basic biochemistry which all organisms share.* It implies that many similar biochemical compounds can be found in all living organisms. They have largely the same elementary composition, which can be represented using around 25 elements. This principle allows one to identify stoichiometric relations in ecology.

6. *No ecological entity exists in isolation but is connected to others.* The theoretical minimum unit for any ecosystem is two populations, one that fixes energy and another that decomposes and cycles waste, but in reality viable ecosystems are complex networks of interacting populations. This reinforces the openness principle at the scale of the individual component. The network interactions provide the environmental niche in which each component acts. This network has a synergistic effect on the components: the ecosystem is more than the sum of the parts.

7. *All ecosystem processes are irreversible (this is probably the most useful way to express the second law of thermodynamics in ecology).* Living organisms need energy to maintain, grow, and develop. This energy is lost as heat to the environment, and cannot be recovered again as usable energy for the organism. Evolution can only be understood in the light of the irreversibility principle rooted in the second law of thermodynamics. Evolution is a step-wise development based on previously achieved solutions to survive in a changing and dynamic world. Due to the structural and genetic encapsulation of these solutions, evolution has produced more and more complex solutions. Eco-exergy expressed by Kullbach's measure of information (see Exergy) is one way to measure this development.

8. *Biological processes use captured energy (input) to move further from thermodynamic equilibrium and maintain a state of low-entropy and high-exergy relative to its surrounding and to thermodynamic equilibrium.* This is just another way of expressing that ecosystems can grow. It has been shown that eco-exergy of an ecosystem corresponds to the amount of energy that is needed to break down the system.

9. *After the initial capture of energy across a boundary, ecosystem growth and development is possible by (1) an increase of the physical structure (biomass), (2) an increase of the network (more cycling), or (3) an increase of information embodied in the system.* All three growth and development forms imply that the system is moving away from thermodynamic equilibrium and all three are associated with an increase of (1) the eco-exergy stored in the ecosystem, (2) the energy flow in the system (power), and (3) the ascendency. When cycling increases, the eco-exergy storage capacity, the energy-use efficiency, and space–time differentiation all increase. When the information increases, the feedback control becomes more effective, the animal gets bigger, which implies that the specific respiration decreases, and there is a tendency to replace r-strategist with K-strategists. Notice that the first growth form corresponds to the eco-exergy of organic matter, $18.7 \, \text{kJ} \, \text{g}^{-1}$, while the increase of the network plus the increase of the information correspond to the eco-exergy calculated as $(\beta-1)c$ (see Exergy). Notice also that the three growth and development forms are in accordance with EP Odum's trends of ecosystem development (**Table 1**). A typical growth and development sequence is present as follows (**Figure 1**): increased biomass (form 1) has a positive feedback allowing even more additional solar energy capture, until a limit of around 75% of the available solar energy is reached. Thereafter the ecosystem continues to grow and develop by increasing network interactions (form 2) and improving energy efficiencies (form 3).

Table 1 Differences between initial stage and mature stage according to Odum (1959 and 1969) are indicated with reference to the three growth forms

Growth form	Properties	Early stages	Late or mature stage
1 (biomass)	Production/respiration	>> 1 << 1	Close to 1
	Production/biomass	High	Low
	Respiration/biomass	High	Low
	Yield (relative)	High	Low
	Total biomass	Small	Large
	Inorganic nutrients	Extra biotic	Intra biotic
2 (network)	Patterns	Poorly organized	Well organized
	Niche specialization	Broad	Narrow
	Life cycles	Simple	Complex
	Mineral cycles	Open	Closed
	Nutrient exchange rate	Rapid	Slow
	Life span	Short	Long
	Ecological network	Simple	Complex
	Stability	Poor	Good
	Ecological buffer capacity	Low	High
3 (information)	Size of organisms	Small	Large
	Diversity, ecological	Low	High
	Diversity, biological	Low	High
	Internal symbiosis	Undeveloped	Developed
	Stability (resistance to external perturbations)	Poor	Good
	Ecological buffer capacity	Low	High
	Feedback control	Poor	Good
	Growth form	Rapid growth	Feedback controlled growth
	Types	r-strategists	K-strategists

Reproduced by permission of Elsevier.

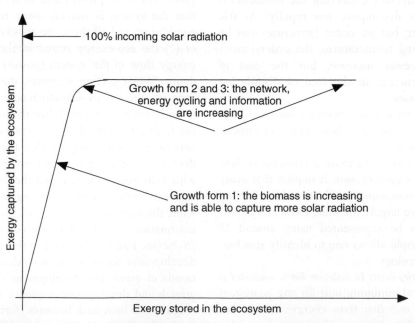

Figure 1 The development of an ecosystem is illustrated by plotting exergy captured from the inflowing solar radiation toward the exergy stored in the ecosystem. Growth form 1 is dominant in the first phase of the development from an early-stage ecosystem to a mature ecosystem. By increasing the biomass the percentage of solar radiation captured increases up to about 80% corresponding to what is physically possible. Growth forms 2 and 3 are dominant in the intermediate phase and when the ecosystem is in a mature stage. Thereby more exergy is stored without increasing the exergy needed for maintenance. The system becomes in other words more effective in the use of the solar radiation according to Prigogine's minimum-entropy principle. The exergy stored is increased for all three growth forms. Reproduced by permission of Elsevier.

10. *An ecosystem receiving solar radiation will attempt to maximize eco-exergy storage or maximize power such that if more than one possibility is offered, then in the long-run the one which moves the system furthest from thermodynamic equilibrium will be selected.* The eco-exergy storage and energy flow increase during all three growth and development forms – see above. When an ecosystem evolves it can apply all three forms in a continuous Darwinian selection process. The nested space–time differentiation in organisms optimizes thermodynamic efficiency as expressed in the tenth law, because it allows the organism to simultaneously exploit equilibrium and nonequilibrium energy transfer with minimum dissipation.

A special issue of *Ecological Modelling* (vol. 158) was devoted to use the proposed ecosystem theory to explain ecological observations that were unexplained in the ecological literature.

Other Ecosystem Theories

The ten fundamental laws presented above have been formulated in a slightly different manner in the scientific literature and other systems ecologists may emphasize other aspects. For instance, H. T. Odum could emphasize maximum power more than eco-exergy storage; but, as pointed out, they are two perspectives of the same basic dynamic. Such perspectives show that a complex system should be described by several different viewpoints according to the complementarity theory. However, this attempt to provide fundamental laws does not mean that there are no other candidates in the literature. For example, the allometric principles are fundamental principles in ecology. Emergent properties are also sometimes considered sufficiently general to be considered as a fundamental principle. Other ecologists still withhold that fundamental laws exist, preferring to focus on descriptions of fundamental properties and processes. Therefore, the discussion about which laws should be considered the fundamental laws in ecology and systems ecology is still open.

Summary

Advancement in our understanding of ecosystem theories has led to a tentative consensus of the principle laws in ecology as outlined above. A priority now is to gain wider application of the theory and to promote, in general, ecology as a theoretical science. As such the following synthesis was recently put forth:

I. Ecosystems are physically and ontically open, meaning that they can exchange mass, energy, and information with the surroundings and that it is not possible to make exact predictions on their development due to their enormous complexity.

II. Ecosystems have directionality.

III. Ecosystems have connectivity.

IV. Ecosystems have emergent hierarchies.

V. Ecosystems have a complex dynamics (growth and disturbances).

Steps toward a wider application of a theoretical explanation of ecological observations will reinforce the fundamentals of ecology. The experience from physics shows that a theory advances through wider use, because every application will either support the theory or improve it by demonstrating where it fails. In conclusion, a tentative ecosystem theory is proposed which has broad explanatory power today, but will improve with more experience providing an even stronger theoretical basis for ecology.

See also: Exergy; Hierarchy Theory in Ecology.

Further Reading

Elsasser WM (1975) *The Chief Abstraction of Biology.* Amsterdam: North-Holland.

Fath B, Jørgensen SE, Patten BC, and Strakraba M (2004) Ecosystem, growth and development. *BioSystems* 77: 213–228.

Fath BD, Patten BC, and Choi JS (2001) Complementarity of ecological goal functions. *Journal of Theoretical Biology* 208(4): 493–506.

Ho MW and Ulanowicz R (2005) Sustainable systems as organisms? *BioSystems* 82: 39–51.

Jørgensen SE (1990) Ecosystem theory, ecological buffer capacity, uncertainty and complexity. *Ecological Modelling* 52: 125–133.

Jørgensen SE (1995) The growth rate of zooplankton at the edge of chaos: Ecological models. *Journal of Theoretical Biology* 175: 13–21.

Jørgensen SE (2002) *Integration of Ecosystem Theories: A Pattern*, 3rd edn. Dordrecht, The Netherlands: Kluwer Academic Publishing Company (1st edn. 1992, 2nd edn. 1997).

Jørgensen SE and Fath B (2004) Application of thermodynamic principles in ecology. *Ecological Complexity* 1: 267–280.

Jørgensen SE, Fath BD, Bastianoni S, *et al.* (2007) *Systems Ecology: A New Perspective*, 275pp. Amsterdam: Elsevier.

Jørgensen SE, Patten BC, and Strakraba M (2000) Ecosystems emerging: 4. growth. *Ecological Modelling* 126: 249–284.

Jørgensen SE and Svirezhev YM (2004) *Towards a Thermodynamic Theory for Ecological Systems*, 366pp. Oxford: Elsevier.

Margalef RA (1968) *Perspectives in Ecological Theory.* Chicago, IL: Chicago University Press.

Margalef RA (1995) Information theory and complex ecology. In: Patten BC and Jørgensen SE (eds.) *Complex Ecology*, pp. 40–50. Princeton, NJ: Prentice-Hall.

Margalef RA (1997) *Our Biosphere*, 178pp. Nordbunte, Oldendorf, Germany: Ecology Institute.

Margalef RA (2001) Exosomatic structures and captive energies relevant in succession and evolution. In: Jørgensen SE (ed.) *Thermodynamics and Ecological Modelling*, pp. 117–132. Boco Raton, FL: CRC Press.

Morowitz HJ (1968) *Energy Flow in Biology. Biological Organisation as a Problem in Thermal Physics*, 179pp. New York: Academic Press (see also the review by H.T. Odum in *Science* 164: 683–684).

Odum EP (1959) *Fundamentals of Ecology*, 2nd edn. Philadelphia, PA: W.B. Saunders.

Odum HT (1983) *System Ecology*, 510pp. New York: Wiley-Interscience.

Odum HT (1996) *Environmental Accounting – Emergy and Decision Making*, 370pp. New York: Wiley.

Odum HT (1998) Self-organization, transformity, and information. *Science* 242: 1132–1139.

Patten BC (1991) Network ecology: Indirect determination of the life environment relationship in ecosystems. In: Higashi M and Burns TP (eds.) *Theoretical Studies of Ecosystems: The Network Perspective*, pp. 288–351. Cambridge: Cambridge University Press.

Schrødinger E (1994) *What is Life?* Cambridge: Cambridge University Press.

Svirezhev YM (2001) Thermodynamics and theory of stability. In: Jørgensen SE (ed.) *Thermodynamics and Ecological Modelling*, pp. 117–132. Boco Raton, FL: CRC Press.

Ulanowicz RE (1986) *Growth and Development. Ecosystems Phenomenology*, 204pp. New York: Springer.

Ulanowicz RE (1997) *Ecology, The Ascendent Perspective*. New York: Columbia University Press.

ECOSYSTEM PROPERTIES

Autocatalysis

R E Ulanowicz, University of Maryland Center for Environmental Science, Solomons, MD, USA

Introduction

In chemistry the term catalysis means the speeding up of a chemical reaction. It follows that autocatalysis then means "the catalysis of a chemical reaction by one of the products of the reaction." For example, oxalic acid oxidizes purple permanganate. When a few crystals of $MnSO_4$ are added to a mixture of the chemicals, the conversion to $Mn(II)$ is sped up. If no $MnSO_4$ is added, then the reaction will gradually speed up of itself, because $Mn(II)$ is gradually being created by the reaction, and this product autocatalyzes the reaction itself. Autocatalysis in chemistry is usually considered to occur among relatively simple, fixed, and inflexible reactants. As such it is commonly regarded as a subclass of general mechanisms.

Autocatalysis in Ecology

In systems ecology, autocatalysis is regarded as a generalized form of mutualism, that is, an association between organisms of two different species in which each member benefits. In systems ecology focus remains more on processes and less on objects. Hence, an autocatalytic configuration of two or more ecological processes is one in which the processes can be arrayed in a closed cycle, wherein each process in the cycle facilitates the next. Without loss of generality, one may focus on a serial, circular conjunction of three processes – A, B, and C (**Figure 1**). Thus, any increase in the rate of process A is likely to induce a corresponding increase in process B, which in turn elicits an increase in process C, and whence back to A.

A didactic example of autocatalysis in ecology is the community that builds around the aquatic macrophyte, *Utricularia* (commonly called Bladderwort). All members of the genus *Utricularia* are carnivorous plants. Scattered along its feather-like stems and leaves are small bladders, called utricles (**Figure 2a**). Each utricle has a few hair- like triggers at its terminal end, which, when touched by a feeding microheterotroph, opens the end of the bladder, and the animal is sucked into the utricle by a negative osmotic pressure that the plant had maintained inside

the bladder. This feeding upon microheterotrophs helps the *Utricularia* to grow and increase its surface area (process A). In nature the surface of *Utricularia* plants is always host to a film of diatomaceous algal growth known as periphyton, so that more surface area encourages the growth of more periphyton (process B). More periphyton in its turn means more food to support the growth of any number of species of small microheterotrophs (process C). The autocatalytic cycle is closed when it is noted that a greater density of microheterotrophs provides more resources for the *Utricularia* to grow (process A again) by capturing and absorbing more abundant zooplankton (**Figure 2b**).

Unlike in chemistry, the actors in ecology are more complex, malleable entities with capabilities to undergo small, incremental alterations. Such malleability substantially enhances the repertoires of autocatalysis and enables it to exhibit some very nonmechanical behaviors. This is especially the case when autocatalysis involves processes that can change in stochastic and nonpredictable ways. An important characteristic of causal cycles (e.g., autocatalysis) is that when random events impinge upon them, they usually yield nonrandom results. This is the consequence of the first and foremost attribute of autocatalysis – its generation of selection pressure.

To see how autocatalysis generates selection, one begins by considering a small spontaneous change in process B. If that change either makes B more sensitive to A or a more effective catalyst of C, then the transition will receive enhanced stimulus from A. In the *Utricularia* example, diatoms that have a higher P/B ratio and are more palatable to microheterotrophs would be favored as members of the periphyton community. Conversely, if the change in B makes it either less sensitive to the effects of A or a weaker catalyst of C, then that perturbation will likely receive diminished support from A. Hence, the response of this causal circuit is decidedly not symmetric, and out of this asymmetry emerges a direction. This direction is not imparted or cued by any externality; its action resides wholly within the system. As one might expect from a causal circuit, the resulting directionality is in

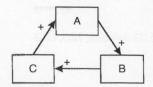

Figure 1 Schematic of a hypothetical three-component autocatalytic cycle.

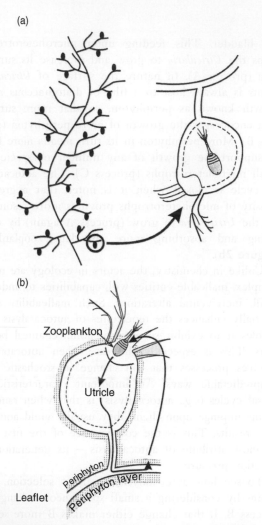

Figure 2 (a) Sketch of a typical 'leaf' of *Utricularia floridana*, with detail of the interior of a utricle containing a captured invertebrate. (b) Schematic of the autocatalytic loop in the *Utricularia* system. Macrophyte provides necessary surface upon which periphyton (speckled area) can grow. Zooplankton consumes periphyton, and is itself trapped in bladder and absorbed in turn by the *Utricularia*.

part tautologous, that is, autocatalytic systems respond to random events over time in such a way as to increase their degree of autocatalysis. It should be emphasized that this directionality, by virtue of its internal and transient nature, should not be conflated with teleology. There is no externally determined or preexisting goal toward which the system strives. Direction arises purely out of the immediate response by the internal system to a novel, random event impacting one of the autocatalytic members.

Centripetality and Agency

A second important and related directionality emerges out of autocatalysis – that of centripetality. To see this one notes in particular that any change in B is likely to involve a change in the amounts of material and energy that are required to sustain process B. As a corollary to selection pressure one immediately recognizes the tendency to reward and support any changes that serve to bring ever more resources into B. Because this condition pertains to any and all members of the causal circuit, any autocatalytic cycle becomes the epicenter of a centripetal flow of resources toward which as many resources as possible will converge (**Figure 3**). That is, an autocatalytic loop defines itself as the focus of centripetal flows. A didactic example of such centripetality is coral reef communities, which by their considerable synergistic activities draw a richness of nutrients out of a desert-like and relatively inactive surrounding sea.

The centripetality generated by autocatalysis is a much-neglected and essential attribute of the life process. For example, evolutionary narratives are replete with explicit or implicit references to such actions as striving or struggling, but the origin of such directional behaviors almost always remains unmentioned. Such actions are simply postulated. Centripetality, however, appears to be at the very roots of such behaviors. To see this, one only needs to recognize that it is centripetality that gives rise to the much vaunted competition, which is the crux of evolutionary theory. For centripetality guarantees that, whenever two or more autocatalytic loops

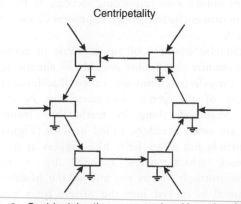

Figure 3 Centripetal action as engendered by autocatalysis.

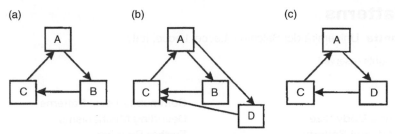

Figure 4 (a) Original configuration. (b) Competition between component B and a new component D, which is either more sensitive to catalysis by A or a better catalyst of C. (c) B is replaced by D, and the loop section A–B–C by that of A–D–C.

exist in the same system and draw from the same pool of finite resources, competition among the foci necessarily ensues. In particular, whenever two loops share pathway segments in common, the result of this competition is likely to be the exclusion or radical diminution of one of the nonoverlapping sections. For example, should a new element D happen to appear and to connect with A and C in parallel to their connections with B (**Figure 4**), then if D is more sensitive to A and/or a better catalyst of C, the ensuing dynamics should favor D over B to the extent that B will either fade into the background or disappear altogether. That is, the selection pressure and centripetality generated by complex autocatalysis (a macroscopic ensemble) is capable of shaping and replacing its own elements. Perhaps the instances that spring most quickly to mind here involve the evolution of obligately mutualistic pollinators, such as yuccas and yucca moths, which coevolve with the yucca so as to displace other pollinators.

One notes in passing that the same tendency to replace B with D could as readily replace a defective or destroyed B with another similar component B', that is, autocatalysis lies behind the ability of living systems to repair themselves.

It becomes obvious that the autocatalytic system is no longer acting merely at the behest of externalities, but it is actively drawing ever more resources unto itself. In fact, the tendency of centripetality to transform as much as possible into itself lies at the very crux of evolutionary drive; for absent such striving, there would be no competition at the next level up.

Furthermore, one perceives autocatalytic action as the agency behind one of a pair of agonistic tendencies that together account for the patterns of life-forms and

functions. On the one hand is the stochastic, entropic tendency to fall apart, which at the same time generates new diversities of form and behavior. Arrayed against the inevitable centrifugal drift toward disorder is the autocatalytic selection and centripetal pull toward greater activity and tighter organization. Opposing thrusts though they are, the continued development of life would be impossible without the actions of both.

Finally, the focal position that autocatalytic configurations of processes occupy in the phenomenon of life is aptly illustrated by considering what differs between a living organism (say a deer) and the same entity immediately upon death. The mass of the deer remains the same, as does its overall form, chemical constitution, embodied energy, and genomic configuration. What the live deer had that the dead deer no longer possesses is simply its configuration of autocatalytic processes.

See also: Ecological Network Analysis, Ascendency; Ecological Network Analysis, Energy Analysis.

Further Reading

Eigen M and Schuster P (1979) *The Hypercycle: A Principle of Natural Self-Organization*. Berlin: Springer.

Kauffman SA (1995) *At Home in the Universe: The Search for Laws of Self-Organization and Complexity*. New York: Oxford University Press.

Ulanowicz RE (1995) *Utricularia's* secret: The advantages of positive feedback in oligotrophic environments. *Ecological Modelling* 79: 49–57.

Ulanowicz RE (1997) *Ecology, the Ascendent Perspective*. New York: Columbia University Press.

Body-Size Patterns

A Basset and L Sabetta, Università del Salento – Lecce, Lecce, Italy

Background	Community Level Patterns
The Problem of Measuring Body Size	Decoding Mechanisms
Population and Species-Level Patterns	Further Reading

Background

Life in the biosphere shows an impressive variety of individual shapes and body sizes. From the smallest microorganism (approx. 10^{-13} g) to the largest mammal ($>10^8$ g), living things cover more than 21 orders of magnitude of body size. The largest living organisms are actually plants (giant sequoia, *Sequoiadendron giganteum* (Lindl.) Buchholz), but since most of their bodies are actually dead bark tissues, their living biomass is lower than that of the largest mammals. Given this impressive variability of sizes, consistent body-size patterns, so common at every scale of observation as to be considered universal, can be detected.

The first body-size patterns to be emphasized were that there are many small and few large individuals and species in the biosphere. The range of body sizes from the smallest to the largest individuals may vary substantially, when moving from marine to brackish water to freshwater and terrestrial ecosystems, as well as from tropical to polar ecosystems or from lowlands to highlands, but the pattern of many small and few large individuals still holds. This simple and universal observation was reported by Charles Elton in the first half of the twentieth century in his pivotal book *Animal Ecology*. This pattern can be explained by means of simple, 'taxon-free', energy-related arguments: since small individuals require less energy per unit of time than large individuals for their maintenance and activity, a fixed productivity will support, at equilibrium, a higher density of small than large individuals. This explanation is actually an oversimplification of the real world; there are at least two other components that need to be taken into account in order to decode the body-size-abundance patterns into a deterministic mechanism of community organization: a phylogenetic and evolutionary component, determining the actual diversity of species and body sizes at continental and global scales; and an interaction component, selecting the body sizes and the species best suited to withstand the locally occurring abiotic conditions and structural habitat architecture (abiotic niche filtering), and determining

trophic links and competitive ranking among co-occurring species differing in body size. However, the simple, energy-related, 'taxon-free' explanation emphasizes the ecological relevance of body-size patterns, which are conceptually independent of species-specific resource requirements and species composition.

Body-size patterns include population or species-level patterns, such as the body-size range pattern, and community, landscape, or continental-level patterns. The latter include variations with individual body size in number and biomass of individuals, number of species, population densities, and energy used by populations. The body-size ratio between co-occurring species pairs, known as the Hutchinson ratio, also shows deterministic and consistent patterns of variation within the community-level body-size patterns.

The Problem of Measuring Body Size

Measurements of body size include linear dimensions (e.g., body length, body width, and the length or width of some morphological attribute of individuals), body surface area and biovolume, and weight (e.g., wet weight, dry weight, and body mass as ash-free dry weight). The energy content of biomass, measured in energy units, can also be used as a measurement of body size.

Body-size patterns are derived from individual biomass data, although the original data may have been obtained as the body length, morphological attribute length or size, body wet weight, body volume, or cell volume of unicellular individuals. Some conversion is required, because individual biomass cannot always be measured, although body size in general is an easily measurable characteristic of individuals. Indeed, in many cases it is necessary to avoid the destructive analysis required to measure individual biomass, and in other cases individuals are simply too small.

Indirect measurements of individual biomass, where lengths or biovolumes are converted to weights, may make the body-size patterns weaker or harder to detect, depending on the dimensions measured, on the precision of the allometric relationship used with respect to the specific set of data, on the precision of biovolume detection, and on the adequateness of the conversion equations. As regards the weight-per-length allometry, the comparability of the seasonal period, climatic conditions, sex ratio and the reproductive status of individuals, and resource availability all have to be taken into account as major sources of variation. As regards biovolume, the complexity of individual or cell shape, taxon-specific weight per unit of biovolume, and the type of weight unit used (C, biomass) all have to be taken into account in order to minimize the bias introduced by using indirect measurements and conversion factors.

Population and Species-Level Patterns

Range-Size Patterns

The range of a species is its natural area of geographic distribution. Considering the overall range of species and body sizes occurring in the biosphere, there does not seem to be any simple and deterministic relationship between body size and species range size: very large species, such as some cetaceans, and very small species, such as many microorganisms, can have very wide natural ranges. However, within much more restricted taxonomic groups, small-bodied species tend to have smaller minimum geographic ranges than large-bodied species. The interspecific relationships of body size to geographic range size commonly exhibit an approximately triangular form, where species of all body sizes may have large geographic ranges while the minimum range size of a species tends to increase with body size.

The relationship between body size and home-range size (i.e., the minimum space needed by an individual to successfully complete its life cycle) can help to account for patterns of natural range size. Since home-range size (H) scales with individual body size (BS) according to an allometric equation ($H = a\text{BS}^b$), in which the slope (b) is significantly larger than 1, large-bodied species may require a larger total geographic range than small species in order to maintain minimum viable population sizes in all local areas. This results in the triangular relationship between body size and range size, because there is not necessarily an upper limit on the range size of small-bodied species.

The dependence of a species' fundamental niche space and dispersal ability on body size may also help to explain range-size patterns, since species of large body size are potentially able to maintain homeostasis in a wider range of conditions and to successfully colonize a larger proportion of their potential range than small-bodied species.

These mechanistic explanations of the relationship between range size and body size are not mutually exclusive and may be reinforcing.

Community Level Patterns

Body Size–Abundance Distributions

Body size–abundance distributions describe the variation of some measurements of individual abundance with individual body mass. The measurements of abundance used are number and biomass of individuals of each population within a guild or a community, number or biomass of individuals in successful populations within a species range, at the regional, continental, and global scale, number or biomass of individuals within a community and number or biomass of individuals in base 2 logarithmic body-size classes. Whatever criteria for grouping individuals are selected, within populations, communities or size classes, at the guild, community, landscape, continental or global scale, as number or biomass, a negative relationship between individual density and body size is generally observed. However, the shape and coefficient of these relationships, the mechanisms involved and the ecological significance vary according to the criteria selected, and each single body-size pattern provides different information, contributing to a better understanding of the role of individual body size in structuring and organizing ecological communities.

The selection of either species populations or body-size classes as a grouping criterion creates two main categories of size–abundance distributions: 'taxonomically based and nontaxonomically based'. The latter are commonly referred to as 'size spectra'. Studies of terrestrial ecosystems have preferentially used body size–abundance distributions based on the taxonomic grouping of individuals into populations and communities, whereas studies of aquatic ecosystems have preferentially used body size–abundance distributions as 'taxon-free' patterns, grouping individuals into logarithmic body-size classes independently of their taxonomy.

Taxonomically based size–abundance distributions

On average, population densities (PDs) scale with individual body size (BS) according to the allometric equation

$$\text{PD} = a_1\text{BS}^{b_1}$$

where b_1 is typically lower than 0 and a_1 is the specific density. a_1 expresses the combined action of factors such as average energy transfer efficiency, average energy availability, and temperature-driven shifts in the metabolic rates of the populations in question.

Broadly speaking, taxonomically based size–abundance distribution derives from the notion that since the energy requirement of individuals (Met) increases with individual body size according to a well-known allometric equation

$$Met = a_2 BS^{b_2}$$

where b_2 has been consistently found to be close to 0.75, the number of individuals of each population supported by the available resources must decrease with average individual body size. Assuming that resource availability is homogeneous across species and body sizes, the slope of the body size–abundance distribution (b_1) is expected to be −0.75.

The processes underlying body size–abundance distributions, and hence their information content and ecological meaning, depend on whether they account for density values and average body sizes of species on a regional, continental, or global scale (hereafter, 'global-scale size–abundance distributions'), for density and average individual body size of co-occurring populations within guilds or communities (hereafter, 'local-scale size–abundance distributions'), or for average population densities and individual body sizes of entire guilds or communities along ecological, climatic, or biogeographic gradients (hereafter, 'cross-community size–abundance distributions').

Global-scale size–abundance distributions are among the most extensively studied. They cover regional, continental, and global scales, and the broadest range of taxonomic variation, with a bias toward birds and mammals, for which more extensive databases of population densities and body sizes are available at every spatial scale. Data used to compile global-scale size–abundance distributions typically describe densities of successful populations within the species' geographic range, which may be close to the maximum carrying capacity. Most commonly, populations included in the global-scale size–abundance distributions do not coexist, and affect each other through vertical or horizontal interactions. For large compilations of population densities, population density generally scales very closely with body size, with a slope near the value of −0.75. The close agreement between the slope observed for global-scale size–abundance distributions and that expected on the grounds of simple energetic arguments confirms that at the continental and global scales, availability of resources or energy is not correlated with species body size. The homogeneity of resource or energy availability across species body sizes is an interesting, but far from straightforward, aspect of global size–abundance distributions. It implies that the advantage for large species arising from their wider niches (and thus greater availability of resources) with respect to small species, is counter-balanced by the presence of other body-size-related factors which compensate.

These include the resource density perceived by individuals and the individuals' exploitation efficiency, both of which decrease with increasing body size. Intercepts of global size–abundance distributions express the average energy-use efficiency of the group of populations considered. Compilations of global size-abundance distributions for ectothermic and endothermic species show different intercept (a_1) values, the former having less negative intercepts than the latter due to the cost of being homoeothermic. Similarly compilations of size-abundance distributions of herbivores have higher a_1 values than those obtained for carnivores, reflecting the overall efficiency of energy transfer in food webs.

When size–abundance distributions are compiled at the local level, where the body size and abundance of each species (N) is measured at the same location, body size generally explains only a small part of the variation in population abundance, and the regression slope is much higher than the expected −0.75. The observed deviations from the expected slope in local size–abundance distributions are suggestive of size biases in resource acquisition that could be driven by size asymmetry in competition. An alternative hypothesis to explain the deviation of local size–abundance distributions from global ones is that the former typically examines a smaller range of sizes than the latter. Observing a smaller portion of the overall relationship accentuates the noise in the local sample. This could explain why local size–abundance distributions in aquatic environments, covering a larger spectrum of body sizes than terrestrial ones, are also generally stronger. In fact, at the local scale, triangular-shaped size–abundance distributions are much more commonly observed than simple allometric relationships. Triangular distributions have three major attributes: an 'upper bound', a 'lower bound', and a dispersion of points in the size–abundance space (**Figure 1a**). The 'upper bound' of the triangular-shaped size–abundance distributions is determined by the body-size scaling of the dominant species' population densities. The 'upper bound' has been used as a proxy of the complete local size–abundance distribution, under the assumption that the ecological role of rare and occasional species, being weak, is unclear. The procedure may be useful for applied purposes but since most species are rare the assumption is not generally acceptable. The body-size dependency of the minimum viable population may explain an expected 'lower bound', which is difficult to measure because of the problems with correctly quantifying the rarity of populations. The density of points between these two bounds is determined mainly by regional processes and horizontal and vertical partitioning rules. The ecological information carried by the intercepts of the size–abundance distributions is of lower value at local than at global scale because whenever slopes are different, as often

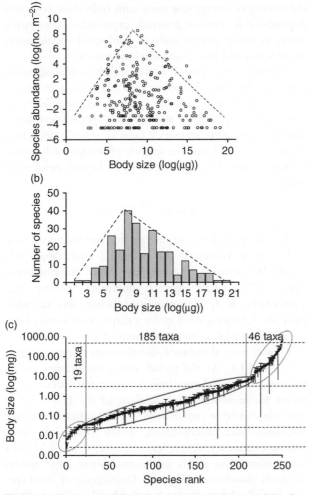

Figure 1 Body-size patterns of macroinvertebrate guilds of transitional water ecosystems in the Mediterranean and Black Sea Eco-regions. Both local size–abundance distributions (a) and body size–species distributions (b) are triangular shaped. The 'upper bounds' of the triangular distributions are reported. The graph (c) emphasizes that species of transitional water macroinvertebrates are clumped around the mode of the body size–species distribution, with 74% of the species being grouped in 2 out of the 5 order of magnitudes occurring between the size of the smallest and the largest species.

occurs when comparing local size–abundance distributions, comparisons between intercepts are not possible.

Classifying all the individuals in a population into guilds or communities and averaging their mass, we may then describe every guild and community with two simple parameters: mean organism size (BS) and total community abundance (N_{tot}). The scaling of total community abundance with mean organism size leads to cross-community size–abundance distributions. Cross-community size–abundance distributions were first studied in self-thinning plant and sessile communities, where, as organisms grow, there is space for fewer and fewer individuals, determining a negative relationship between

BS and N_{tot}. In general cross-community size–abundance distributions tend to be well described by allometric equations, whose slopes tend to be similar to the inverse of the scaling exponent of metabolic rates with individual body size. A similarity between observed and expected slopes has been also detected in guilds and communities which are not regulated by self-thinning rules, such as bird and phytoplankton guilds. However, since much fewer data are available for cross-community size–abundance distributions than for global and local size–abundance distributions, the underlying mechanisms remain to be determined.

Nontaxonomically based size–abundance distributions

In large aquatic ecosystems, early studies of body size–abundance distributions focused on energy transfer (i.e., how information on productivity and energy transfer may be gained from body-size data, which can be collected relatively easily). In accordance with this objective, they dealt with particles rather than with species, dividing particles suspended in the water column into logarithmic-base 2 size classes, irrespective of species and including nonliving organic particles. Thus the n_i particles in the ith body-size class of average mass BS_i may represent more than one species, and every species can occur in more than one class. Nontaxonomic size–abundance distributions (hereafter referred to as size spectra) have been quantified for many different guilds and communities, including plankton, benthos and fish guilds, woodland and forest plant guilds, as well as marine, freshwater, and terrestrial ecosystems; however, a large proportion of the ecological literature addressing size spectra deal with the pelagic marine environment.

According to the classification reported for taxonomically based body size–abundance distributions, almost all size spectra are local, being determined at the guild or community scale. Size spectra can be compiled with two different types of data, that is, biomass and number of individuals. Both biomass-size spectra and number-size spectra can cover different body-size ranges, describing either entire communities or single guilds.

Regarding biomass-size spectra, the amount of biomass has been shown both empirically and theoretically to be constant when plankton individuals are organized into logarithmic size classes. As a result of this equal partitioning of biomass, the slope of a straight line fitted to plankton biomass-size spectra is expected to be 0; this relationship is known as the 'linear biomass hypothesis', which has strong experimental support in aquatic pelagic environments, particularly when a large spectrum of sizes and trophic levels are considered. Often the data is subjected to a normalization procedure, which consists of dividing the biomass in each size class by the width of the size class. In normalized biomass-size spectra, biomass in each size class decreases isometrically with the average class size, the slope being close to −1. The linear biomass hypothesis implies that in pelagic systems, the number of individuals within logarithmically increasing size classes declines linearly with average body size. The slope of the allometric equation tends to be close to −1; when number-size spectra are normalized, the expected slope is equal to −2. Nevertheless, within pelagic-size spectra, a series of dome-like distributions are typically detected, corresponding mainly to different functional guilds within which there is a poor fit with linear statistical regressions.

'Dome-like' distributions and gaps in number- and biomass-size spectra occur not only between but also within functional groups, such as phytoplankton and zooplankton, even when they are not attributable to incomplete censuses of species or to systematic underestimation of intraspecific size variation. Dome-like patterns of biomass distribution have been observed both in freshwater and marine ecosystems, as well as in macro-zoobenthos and fish. Therefore, by restricting the range of body size considered and addressing specific functional groups, size spectra tend to have a shape similar to the triangular shape of local size–abundance distribution. Most commonly, the maximum number and biomass of individuals, either partitioned into species or irrespective of species, occur at some small but intermediate body size, rather than at the smallest size.

Two kinds of scaling in the relationship between body size and abundance within size spectra may be recognized. A unique and primary slope reflecting the size dependency of metabolism ('metabolic scaling'), and a collection of secondary slopes which represent the scaling of numerical or biomass abundance with body size within groups of organisms having similar production efficiencies ('ecological scaling'). Size-dependent coexistence relationships are likely to be representative of the secondary slopes, leading to a dominance of large cells/species, and slopes that are less negative than predicted by the 'linear biomass hypothesis'. Ecological scaling can also produce dome-like patterns in size spectra within the size range of each functional group.

Body Size–Energy Use Distributions

The body-size dependence of both metabolic rates and population densities makes it possible to evaluate populations' rates of energy use and how they scale with individual body size. Indeed, the rate at which energy flows through a population (E) can be evaluated as the product of individual metabolism (Met) and population density (PD), as follows:

$$E = \text{Met} \times \text{PD} = a_2 \text{BS}^{b_2} \times a_1 \text{BS}^{b_1} = (a_2 \times a_1) \text{BS}^{(b_2 + b_1)}$$

Since b_2 has been found to be consistently close to 0.75, the scaling of energy-use rates with individual body size depends on b_1, which is generally expected to be negative, since, at every spatial scale of ecological organization, many small and few large individuals occur.

Assuming that resource availability is homogeneous across species and that species do not limit each other's resource availability and have optimized the efficiencies of resource exploitation and use, then population densities are expected to scale with individual body size with a slope (b_1) of −0.75, and the amount of energy each species uses per unit of area is expected to be independent of body size:

$$E = (a_2 \times a_1) \text{BS}^{(0.75 - 0.75)} = a_3 \text{BS}^0$$

The independence of energy use per unit area from body size is known as the energetic equivalence rule (EER). Whenever b_1 is consistently lower, more negative, than −0.75, small species dominate energy use. Conversely, if b_1 is consistently larger, less negative, than −0.75, large species make a disproportionately large use of the available energy per unit of area.

Global size–abundance distributions seem to agree with the EEF. At the global scale, the energy use of the most successful populations within the species range seems to be actually independent of the body size of individuals within populations. On the other hand, local size–abundance distributions, which commonly show scaling exponents higher, less negative, than −0.75 show that within local guilds and communities large species normally dominate energy use. Dominance of small species has also been detected at the local scale usually in relation to some degree of stress. Therefore, the shape and slope of local size–abundance distribution, and consequently the body-size scaling of energy use, can have practical applications in ecology.

Body-Size–Species Distributions

Understanding biodiversity is a major goal of ecology. Since many small and few large species occurs in the biosphere, at every scale, from the community to the continental and global level, describing and understanding the scaling of biodiversity patterns with individual body size is also a key topic.

Basically, whenever organisms perceive a two-dimensional (2D) habitat, they sample habitats on a grid proportional to the reciprocal of the square of their linear dimension (L). Therefore, the likelihood of niches being opened up to species specializing in particular resources and habitat patches is proportional to $\approx L^{-2}$ or to $\text{BS}^{-0.67}$. Consequently, the number of species (S) is expected to decrease with individual body size according to $\approx L^{-2}$. Whenever organisms perceive a 3D habitat, the species number is expected to be proportional to $\approx L^{-3}$ or $\approx BS^{-1}$. Considering that the linear dimension of individuals represents the 'ruler' (L) they use to sample the habitat and that habitats are rarely completely homogeneous at every scale of perception, individuals perceive the habitat to be fractal. The perceived 2D habitat scale is $\approx L^{-2D}$, where D is the fractal dimension of the habitat as well as of the resources. The fractal dimension is a habitat property that in many field studies has been found to be close to 1.5. This would mean that in 2D habitats the number of species (S) is expected to be between L^{-D} and L^{-2D}, that is, between $L^{-1.5}$ and $L^{-3.0}$, where L^{-2D} is analogous to the 'upper bound' of size–abundance distributions. In 3D habitats the number of species (S) is expected to be proportional to L^{-3D}, that is, to $L^{-4.5}$. Assuming $D \approx 1.5$, a tenfold decrease in individual size determines a threefold increase in the perceived length of each habitat edge, a tenfold increase in apparent habitat surface and a maximum tenfold increase in S.

Available data on both the full range of taxa and particular groups of species consistently show that the species–size distributions are humped, with the mode in some small but intermediate size class (**Figure 1b**). The underestimation of the number of existing small species may be an explanation of humped distributions covering the whole scale of size from the smallest to the largest species. Underestimation of small species is less likely to explain humped distributions observed within restricted taxonomic groups, such as invertebrates, birds, and mammals. Within restricted taxonomic groups, it seems likely that an optimal body size exists, where species and individuals perform optimally and tend to be clumped (**Figure 1c**). An optimal body size of between 100 g and 1 kg has been proposed for mammals and an optimal body size of 33 g has been proposed for birds. Two hypotheses have been proposed to explain the size dependency of species performance at every scale: the energy conversion hypothesis, addressing optimal size according to the size dependency of the efficiency of energy conversion into offspring; and the energy control hypothesis, addressing optimal body size according to the species' performance in monopolising resources.

Body-Size Ratios

Coexisting species of potential competitors commonly differ in body size. Using consumer body size as a proxy of resource size, this difference may explain competitive coexistence; in his famous paper 'Homage to Santa Rosalia: or why are there so many kind of animals' Hutchinson proposed that in order to coexist species must be spaced in size with a ratio between their linear dimensions of at least 1.28 (2.0–2.26 in biomass), which is commonly referred to as the 'Hutchinson ratio'. Patterns of body-size spacing between coexisting species pairs

consistent with the 'Hutchinson ratio' have been observed for many groups of animals, including, birds, desert rodents, and lizards. The 'Hutchinson ratio' corresponds to a limiting similarity threshold; therefore, the average size ratio between species is expected to vary with resource limitation. Actually, the average size ratio between species pairs decreases with increasing richness and with decreasing guild trophic level.

That co-occurring species within guilds tend to have different body size, with an average size ratio close to the expected 2–2.26, is a very general observation in ecology. However, the ecological relevance of the 'Hutchinson ratio' has been questioned, mainly due to two key critical observations, apparently in contrast with the interpretation that size ratios between species correspond to a low-enough niche overlap between species pairs to allow interspecific coexistence: (1) many nonliving things in nature as well as many objects built by humans, from nails to musical instruments, are scaled in size according to the 'Hutchinson ratio'; and (2) size spacing between species pairs does not always seem to be related to niche spacing. The latter is the most critical issue. However, a functional link between the body-size ratios of coexisting species and competitive coexistence conditions may also be derived independently of any niche spacing. Body-size-mediated coexistence between species differing in size may result from simple energetic constraints on individual space use regardless of any *a priori* resource partitioning: that is, size ratios between species may critically affect species coexistence even if niche spacing is not detected.

Decoding Mechanisms

At their most simple, body-size patterns depend on phylogenetic and evolutionary constraints, on energetic constraints, and on interactions with the habitat structure and with co-occurring species.

As regards phylogenetic and evolutionary constraints, body-size patterns are in some way dependent on existing biodiversity and its evolutionary basis. Each taxon performs best under a fixed set of conditions and has bioengineering constraints on its performance. For example, insects cannot be too large and birds cannot be too small; therefore, although both of them can take advantage of a 3D space, the complete spectrum from insects to birds has 'dome-like' distributions which incorporate the bioengineering constraints of the two groups of species.

As regards energy constraints, metabolic theory gives a general explanation of body-size patterns in terms of energy and temperature constraints on metabolism and the intrinsic properties of energy partitioning. Metabolic theory sets the theoretical expectations of body-size patterns, under the assumption that they are basically driven by simple energy constraints

As regards interactions, clearly populations interact with their environment and with co-occurring species. 'Textural habitat architecture' and 'body-size-mediated coexistence' hypotheses have been proposed to explain the abiotic and biotic components of interaction regarding its influence on the observed body-size patterns.

Further Reading

Brown JH, Gillooly JF, Allen AP, Savage VM, and West GB (2004) Towards a metabolic theory of ecology. *Ecology* 85: 1771–1789.

Brown JH and West GB (2000) *Scaling in Biology*. Oxford: Oxford University Press.

Elton C (1927) *Animal Ecology*. London: Sidgwick and Jackson.

Gaston K (2003) *The Structure and Dynamics of Geographic Ranges*. Oxford: Oxford University Press.

Holling CS (1992) Cross-scale morphology, geometry and dynamics of ecosystems. *Ecological Monographs* 62: 447–502.

Hutchinson GE (1959) Homage to Santa Rosalia, or why are there so many kinds of animals? *American Naturalist* 93: 145–159.

Lawton JH (1990) Species richness and population abundance of animal assemblages. Patterns in body size: Abundance space. *Philosophical Transactions of the Royal Society of London, Series B* 330: 283–291.

May RM (1986) The search for patterns in the balance of nature: Advances and retreats. *Ecology* 67: 1115–1126.

Peters RH (1983) *The Ecological Implications of Body Size*. Cambridge, UK: Cambridge University Press.

Sheldon RW, Prakas A, and Sutcliffe WH, Jr. (1972) The size distribution of particles in the ocean. *Limnology and Oceanography* 17: 327–340.

White EP, Ernest SKM, Kerkhoff AJ, and Enquist BJ (2007) Relationship between body size and abundance in ecology. *Trends in Ecology and Evolution* 22: 323–330.

Cycling and Cycling Indices

S Allesina, University of Michigan, Ann Arbor, MI, USA

Introduction

Given the finite amount of chemical compounds in the biosphere, it is inevitable that the same material will be utilized repeatedly by different organisms. This phenomenon is addressed as 'recycling' or simply 'cycling' of energy and matter. Familiar examples of recycling of nutrients involve the so-called 'detritus chain', which decomposes organic matter that is unusable for some organism to its basic compounds that can be recycled into the grazing chain.

This article provides an overview of cycles and cycling indices in ecosystems ecology. Depending on the way of modeling ecosystems, cycles assume different meanings. In what follows, a general definition of cycles, taken from graph theory will be introduced. Then the concept of cycling is applied to (1) food webs (description of who eats whom in the ecosystem) and (2) ecological networks (weighted, mass-balanced versions of food webs). Simple ways of computing cycling indices and removing cycles will be provided.

Definition of Cycle

A very common way of describing ecosystems is by means of graphs. Graphs are constituted by nodes (representing species or functional groups of species) connected by arrows (or edges, arcs, links, representing relationships between species).

The simplest way of sketching ecosystems using graphs is the food web representation. In this way of drawing species relations, edges connect prey to their predators (see **Figure 1**).

This food web representation can be associated with a matrix that expresses the relationships between species. This is the so-called 'adjacency matrix', A. If the row species is a food source of the column species then the corresponding coefficient will be 1. More generally, this relation is a consumer–resource relation, as nodes can represent nutrient pools, etc. Elsewhere, the coefficients will be 0. The food web in **Figure 1** can therefore be represented by this adjacency matrix:

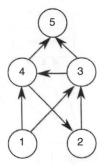

Figure 1 Example of food web containing five species and seven feeding relations (arrows, edges).

$$A = \begin{pmatrix} 0 & 0 & 1 & 1 & 0 \\ 0 & 0 & 1 & 0 & 0 \\ 0 & 0 & 0 & 1 & 1 \\ 0 & 1 & 0 & 0 & 1 \\ 0 & 0 & 0 & 0 & 0 \end{pmatrix}$$

The adjacency matrix represents direct interactions between species. These direct interactions, however, yield chains of indirect interactions. These will be sequences of nodes and edges that are called 'paths'. We can discriminate between different kinds of paths:

1. Open paths connect two different nodes. They can be subdivided into 'simple paths', containing no repeated nodes (e.g., $A \rightarrow B \rightarrow C$, **Figure 2a**) and 'compound paths', which contain repeated nodes (e.g., $A \rightarrow B \rightarrow C \rightarrow B \rightarrow D$, **Figure 2b**).

2. Closed paths start and end at the same node. Also closed paths can be divided into 'simple cycles', containing no repeated nodes except the initial one (e.g., $A \rightarrow B \rightarrow C \rightarrow A$, **Figure 2c**) and 'compound cycles', representing repeated cycles (e.g., $A \rightarrow B \rightarrow A \rightarrow B \rightarrow A$, **Figure 2d**, where double arrows mean that the cycle is traversed twice).

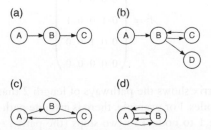

Figure 2 Classification of pathways in (a) simple paths (open pathways start and end at different nodes); (b) compound paths (open pathways start and end at different nodes, contain repeated nodes); (c) simple cycles (closed pathways start and end at the same node); and (d) compound cycles (same cycle traversed more than once).

All kinds of paths, other than simple paths, contain at least one cycle. For example, the graph in **Figure 1** contains just the simple cycle $2 \rightarrow 3 \rightarrow 4 \rightarrow 2$. A graph containing no cycles is said to be acyclic. Every pathway can be classified according to its length that is given by the number of nodes involved.

Cycles in Food Webs

Cycles in food webs can be divided into two main classes: feeding cycles and nonfeeding cycles. The former involve species and their feeding relations (e.g., species A eats species B; species B eats species A); cannibalism is a simple kind of feeding cycle. The latter are typical of food webs that comprise detritus compartments and nutrient pools: organic matter is recycled in the system via mineralization, creating a huge number of detritus-mediated cycles.

Feeding cycles are rare in published food webs. This is mainly due to the fact that the resolution of food webs is usually at the species/group of species level. The number of feeding cycles becomes more significant when age-structured populations are considered, especially in aquatic food webs. Nonfeeding cycles, on the other hand, are extremely abundant in published networks, being several billion cycles for highly resolved ecosystem models.

Structure of Cycles in Ecological Networks: Strongly Connected Components

Two nodes A and B are said to belong to the same strongly connected component (SCC) if they are reachable from each other, that is to say if we can find a path going from A to B and a path coming back from B to A.

If A and B belong to the same SCC, then they are connected by cycles. A graph can be divided into its SCCs, considering every node that is not involved in cycles as an SCC by itself. **Figure 3a** represents the Baltic Sea ecosystem. One can individuate 6 SCCs: 4 of them are composed by a single node, while 2 of them comprise more than 1 node (**Figure 3b**).

If we compact every SCC into a single node, we produce an acyclic graph (**Figure 3c**). Further analysis shows how one component contains just pelagic species and the other one just benthic. Acyclic graphs can be ordered so that all edges point in the same direction (from bottom to top in **Figure 3c**) using a procedure known as 'topological sort' (or partial ordering). Acyclic graphs are therefore intrinsically hierarchical. In this case, the flows find a sink in the benthic compartment, while the pelagic compartment acts as a bridge between

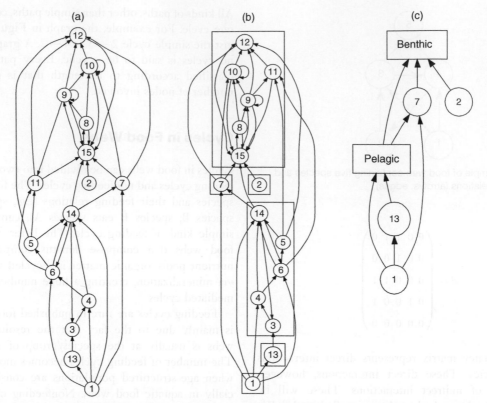

Figure 3 Schematic representation of Baltic Sea ecosystem (a). The boxes define different strongly connected components. Condensing each box into a single node yields an acyclic graph (b). This graph can be sorted so that all arrows point in the same direction, showing the underlying straight flow between compartments (c).

the primary producer 1 and the benthic compartment. The same structure was found for other aquatic networks as well. Note that this feature depends drastically on the presence/absence of resuspension of nutrients. If this is negligible, then the network presents several SCCs. When remineralization is strong, however, the process joins the benthic and pelagic components, thus forming a giant SCC.

Quantifying Cycled Fraction: Finn's Cycling Index

Ecological networks are food webs where the edges are quantified and represent exchanges of nutrients (usually grams of carbon per m^2 per year, but also nitrogen or phosphorous) or energy. Moreover, inputs to the system and outputs from the system are explicitly represented by flows involving 'special compartments' (i.e., nodes that act as a source (imports) or sink (exports and respirations) for the system). Besides the graph representation, a system can be described using the so-called flow matrix T, where each coefficient t_{ij}

describes the flow of energy–matter from the row-compartment (i) to the column compartment (j). An example of network and its matrix representation is given in **Figure 4**.

In order to show the computation of the Finn's cycling index, it is necessary to introduce the concept of power of adjacency matrices. Take as an example the matrix introduced in the first section. If we square it, we obtain

$$A^2 = \begin{pmatrix} 0 & 1 & 0 & 1 & 2 \\ 0 & 0 & 0 & 1 & 1 \\ 0 & 1 & 0 & 0 & 1 \\ 0 & 0 & 1 & 0 & 0 \\ 0 & 0 & 0 & 0 & 0 \end{pmatrix}$$

This matrix shows the pathways of length 2 that connect to two nodes. For example, there is just one path connecting node 1 to node 4 in two steps (the path $1 \rightarrow 3 \rightarrow 4$), while there are two pathways connecting 1 to 5 ($1 \rightarrow 4 \rightarrow 5$ and $1 \rightarrow 3 \rightarrow 5$).

In the same way, if we multiply this matrix with the adjacency matrix we get A^3, which describes all the pathways connecting two nodes in three steps; A^4 will

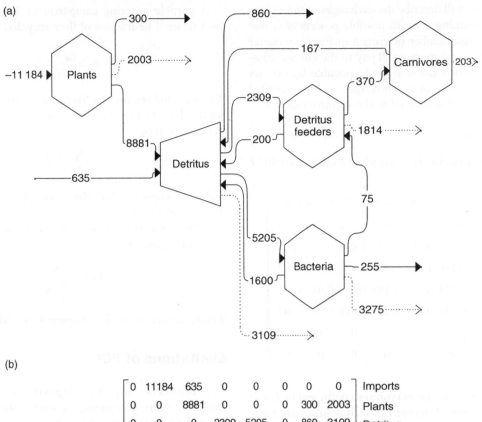

Figure 4 Schematic representation of cone spring ecosystem (a). There are two imports (to Plants and Detritus), three exports (from Plants, Detritus, and Bacteria) and five Dissipations (dashed arrows). The network can be associated with a matrix of transfers (b). The first row represents imports, the last two columns stand for exports and dissipations, and the internal 5 × 5 part depicts intercompartment flows.

similarly contain all the pathways of length 4, and so forth. The power A^x will contain all the pathways of length x. If the food web contains no cycles, then for some $x < n$ (where n is the number of species) the matrix will contain just zeros. If the food web contains cycles, on the other hand, the powers never converge to 0. The pathways enumerated in these matrices belong to all the different types that we illustrated in the first section. Now we can see how these considerations apply to quantified networks.

Dividing each coefficient t_{ij} for the row sum produces the coefficients g_{ij} (of matrix G), which describe the fraction of flow leaving each compartment:

$$g_{ij} = \frac{t_{ij}}{\sum_k t_{ik}}$$

For example, the G matrix for the network in **Figure 4** would be

$$G = \begin{pmatrix} 0 & 0.946 & 0.054 & 0 & 0 & 0 & 0 & 0 \\ 0 & 0 & 0.794 & 0 & 0 & 0 & 0.027 & 0.179 \\ 0 & 0 & 0 & 0.201 & 0.453 & 0 & 0.075 & 0.271 \\ 0 & 0 & 0.084 & 0 & 0 & 0.155 & 0 & 0.761 \\ 0 & 0 & 0.307 & 0.014 & 0 & 0 & 0.049 & 0.629 \\ 0 & 0 & 0.451 & 0 & 0 & 0 & 0 & 0.549 \\ 0 & 0 & 0 & 0 & 0 & 0 & 0 & 0 \\ 0 & 0 & 0 & 0 & 0 & 0 & 0 & 0 \end{pmatrix}$$

Multiplying G by itself, one obtains the fraction of flow leaving the row compartment and reaching the column compartment in two steps (i.e., passing by an intermediate compartment). G^3 will describe the exchanges in three steps, and so forth. Summing over all possible powers of G, one obtains the average number of visits a quantum of matter leaving the row compartment will pay to the column compartment. This computation is made possible by the fact that the power series of G converges to the so-called Leontief matrix L. G^0 is defined as the identity matrix I:

$$I + G + G^2 + G^3 + G^4 + \cdots = [I-G]^{-1} = L$$

The Leontief matrix for the network in Figure 4 would be

$$L = \begin{pmatrix} 1 & 0.946 & 0.946 & 0.202 & 0.440 & 0.031 & 0.120 & 0.880 \\ 0 & 1 & 0.958 & 0.199 & 0.434 & 0.031 & 0120 & 0.880 \\ 0 & 0 & 1.207 & 0.251 & 0.547 & 0.039 & 0.117 & 0.883 \\ 0 & 0 & 0.186 & 1.039 & 0.084 & 0.161 & 0.018 & 0.982 \\ 0 & 0 & 0.374 & 0.092 & 1.169 & 0.014 & 0.085 & 0.915 \\ 0 & 0 & 0.545 & 0.113 & 0.247 & 1.018 & 0.053 & 0.947 \\ 0 & 0 & 0 & 0 & 0 & 0 & 1 & 0 \\ 0 & 0 & 0 & 0 & 0 & 0 & 0 & 1 \end{pmatrix}$$

In an acyclic network, the maximum coefficient of L will be 1 (i.e., a quantum of matter can visit another compartment maximum once). This is because a particle of matter leaving a compartment will never be recycled to the same compartment again. This is not true when cycles are present. In fact, when matter cycles in the network, a particle can be recycled into the same compartment many times, raising the maximum value of the coefficients of the Leontief matrix. Therefore, the Leontief matrix of an acyclic network would contain unitary coefficients on the diagonal for all compartments (a particle starting at any compartment will never come back). Consequently, a simple way of estimating the cycled fraction would be to see how much these coefficients deviate from 1. This is at the heart of the so-called 'Finn's cycling index' (FCI). There are various formulations for this index, but here we present the simplest one, adapted from the one developed in 1980 by J. T. Finn; the reader is referred to the 'Further reading' section for a complete account of the possible variations. The following computation is valid for steady-state network only, that is, for networks where the input to any node equals the output from the same node.

We will call T_k the sum of all flows entering the compartment k:

$$T_k = \sum_i t_{ik}$$

For example, in Figure 4 the sum of the flows to the 'Plants' compartment T_1 would be 11 184.

A particle entering compartment k will be recycled $l_{ij}-1$ times. The fraction of flow recycled is therefore

$$R_k = \frac{l_{kk}-1}{l_{kk}}_k$$

The recycled fraction for 'Bacteria' (fourth compartment) would be $(1.018-1)/1.018 = 0.0172$. The total flow cycled C will be

$$C = \sum_k R_k T_k$$

which, computed for the example, will result in 2777.23 units recycled.

The total fraction of recycled flow for the whole system will therefore be

$$\text{FCI} = \frac{C}{\sum_{ij} t_{ij}}$$

which, for the network in **Figure 4**, would be 0.0654.

Limitations of FCI

FCI considers only the diagonal coefficients of the Leontief matrix, accounting therefore only for paths starting and ending at the same node.

Using the notation introduced above, we see that FCI accounts for simple cycles and compound cycles, but does not consider the contribution of compound paths, as they never appear on the diagonal. Compound paths, however, contain cycles that should be included in the definition of cycling index. Unfortunately, there is no simple linear algebra technique that can account both for cycles and compound paths, and counting all the pathways in an ecological network is computationally very intense.

As an example of the limitation of the FCI, we see that in **Figure 4** the pathway Plants → Detritus → Detritus feeders → Detritus → Bacteria will not contribute to any diagonal coefficient, even if it contains a cycle. Because each quantum of matter can be recycled into the same compartment many times, it will also move around compound paths many times. This may result in off-diagonal coefficients in the Leontief matrix that are greater than 1, stressing the need for counting compound paths in the cycling process.

Number of Cycles in Food Webs

In order to quantify the abundance of simple cycles in food webs, one should know the maximum possible number of simple cycles. The maximum number of simple cycles will be associated with a completely connected food web, that is, a food web whose adjacency matrix contains just 1s.

In order to count the maximum number of simple cycles, we start from the ones with maximum length (in graph theory they are called Hamiltonian cycles). In a completely connected food web composed of n species, the number of simple cycles of level (i.e., length) n is $(n-1)!$. This simple formula can be explained combinatorically using permutations: we can see a cycle of level n as a permutation of the n labels of the nodes: for example, ABCD will represent the cycle $A \to B \to C \to D \to A$. Now, the number of permutations of n elements is $n!$. We note, however, that every cycle gives rise to n possible sequences (e.g., ABCD, BCDA, CDAB, and DABC represent the same cycle of length 4). Therefore, the total number of simple cycles of maximum length is $n!/n = (n-1)!$.

This is an enormous number, as soon as n becomes large. For example, in a 100 species food web, we can find almost 10^{155} simple cycles of level n.

Now that we know the total number of simple cycles of level n in a completely connected food web, we can easily derive the number of simple cycles of level $(n-1)$. For each subgraph containing $(n-1)$ species we will have $(n-1)!/(n-1) = (n-2)!$ simple cycles of length $(n-1)$. The number of possible subgraphs containing $(n-1)$ species is given by the binomial coefficient

$$\binom{n}{n-1}$$

Therefore, the total number of simple cycles of level $(n-1)$ in a completely connected food web composed of n species is $n(n-2)!$.

Similarly, we can define the total number of simple cycles of length k in a completely connected food web of n species as

$$C(k,n) = (k-1)! \binom{n}{k}$$

Table 1 represents the number of cycles of level k (column) for a completely connected food web of n species (rows).

The total number of cycles is therefore given by the following formula:

$$\mathrm{Tot}_{\mathrm{Cycles}} = \sum_{k=1}^{n} C(k,n) = \sum_{k=1}^{n} (k-1)! \binom{n}{k}$$

The first 10 values are represented in **Table 1**. Note that this sequence is defined, in combinatorics, as 'logarithmic numbers'.

Finding Cycles in Ecological Networks

Finding cycles in graphs is a computationally difficult task. Nevertheless, published ecosystems contain a few hundred nodes at most, and the low connectance (fraction of realized connections) displayed by these systems ensures that the number of simple cycles is much lower than the theoretical case illustrated above, where all possible cycles are present.

The idea behind most algorithms for cycle search is simple: one should construct a path inside the network until the same node is found twice. In this case the path is either a cycle (the initial and final nodes do coincide) or a compound path (initial and final nodes are different).

Of the various possible ways of searching the cycles, backtracking-based ones, such as 'depth first search' (DFS) are surely the easiest to implement.

Removing Cycles in Ecological Networks

We have stated above that it is possible to enumerate all the cycles in a food web. In an ecological network, however, each cycle will also possess a 'weight', given by the amount of flow passing through the cycle.

Some network analysis applications (e.g., the so-called 'Lindeman spine') require an acyclic network as an input. The removal of the cycles therefore becomes an important topic for network analysis.

The current procedure requires the removal of cycles according to their 'nexus'. Two cycles are in the same

Table 1 Number of simple cycles of length k (column) in a completely connected food web formed by n species (rows)

n	|_ k _| 1	2	3	4	5	6	7	8	9	10	Total
1	1										1
2	2	1									3
3	3	3	2								8
4	4	6	8	6							24
5	5	10	20	30	24						89
6	6	15	40	90	144	120					415
7	7	21	70	210	504	840	720				2 372
8	8	28	112	420	1344	3 360	5 760	5 040			16 072
9	9	36	168	756	3024	10 080	25 920	45 360	40 320		125 673
10	10	45	240	1260	6048	25 200	86 400	226 800	403 200	362 880	1 112 083

nexus if they share the same weak arc, defined as the smallest flow in the cycle. Cycles are then removed dividing the flow constituting the weak arc among all the cycles sharing the same nexus. The resulting amounts are then subtracted from each edge of the cycles. This process results in the removal of the weak arc. The procedure is then repeated until the resulting network is acyclic.

A nice by-product of the procedure is the creation of a network composed of all the cycles in the original network. This is usually referred to as 'aggregated cycles' network in ecological literature. This network will receive no input, produce no output, and will be balanced (i.e., incoming flows equal outgoing ones) for all nodes. If the resulting aggregated cycle network is composed of several subgraphs, each subgraph is a strongly connected component.

Note that while some applications require acyclic networks, most of them are actually based on the fact that empirical networks contain millions of cycles. As explained in the next section, in fact, cycles are among the most important features of ecosystems.

Ecological Applications of Cycle Analysis

The recycling of energy–matter is an important process that occurs in every ecosystem. Cycling is believed to be a buffering mechanism that allows ecosystems to face shortage of nutrient inflows. This process, however, has been neglected in many theoretical models, which concentrated on communities rather than ecosystems, and which usually comprised just a few species due to constraints of modeling techniques. Food web ecologists always had an ambivalent attitude toward cycling. For example, the first collection of food webs published (which contained poorly resolved food webs with just a few nodes) showed that cycles are very rare. This was justified by the fact that cycles are likely to destabilize a system, because they introduce positive feedbacks. This result was, however, challenged by the discovery of many cycles in larger food webs, and the role of cannibalism in age-structured population dynamics. In recent times, the importance of cycles in food webs has been reconsidered, thanks to the switch of focus from local stability dynamics toward a more comprehensive approach to ecosystems persistence and nonlinear dynamics. Moreover, a greater attention has been devoted to the microbial loop, which, in some aquatic ecosystems, receives more than 50% of the primary production, remineralizes it and feeds it back to higher trophic levels.

Ecosystem oriented modeling, on the other hand, included cycles as the very foundation of the discipline. The first clear reference to the importance of cycling in ecological network comes from the work of Lindeman who, in his seminal paper in 1942, described food webs as cycling material and energy. Odum then included the amount of recycling as one of the 24 criteria for evaluating if an ecosystem is 'mature' (i.e., developed).

The request for a quantification of cycling was then answered by the FCI illustrated above. Modified versions of the FCI, including biomass storage, utilizing the so-called 'total dependency and contribution matrices' were published, increasing the possibilities for modelers and therefore the number of applications of such indices to empirical studies.

Recently, it was pointed out how all these calculations ignore some cycling that involves just off-diagonal terms in the Leontief matrix. Unfortunately, in order to compute the exact amount of cycling in an ecosystem one should utilize a computationally intensive method, which is therefore unfit to be applied to large ecosystem networks. Fortunately, studies conducted on many small networks showed that the total amount of cycling and the FCI seem linearly related, with the total cycling being around 1.14 times the FCI.

The relation between cycling and maturity of ecosystems was challenged by the work of Ulanowicz. He showed how cycling could be inversely related to the developmental status of an ecosystem, and how perturbations could be reflected into a higher cycling index. These considerations suggest that cycling could be seen as a homeostatic response to stress: impacts on ecosystems free nutrients from the higher trophic levels; this freed matter is then recycled into the system by microorganisms, generating cycles at the lower trophic levels. In this view, responding to stress ecosystem would show a decrease in cycle length and an increase in total cycling. It is therefore important to know the distribution of cycle lengths together with the total amount of cycling in the ecosystem when one wants to assess the ecosystem status and maturity. Ulanowicz also presented important insights on cycling as autocatalytic processes. The cycling feature of ecosystems is at the basis of the views of several authors on ecosystem function and dynamics, such as, for example, the work of Patten and colleagues.

Another aspect of cycling is represented by the compartmentalization into SCCs. Although ecosystems comprise myriad interactions, they still can be divided into a few subsystems that are connected by linear chains of energy transfers. In several aquatic food webs, SCC analysis shows a subdivision into pelagic and benthic components of the ecosystem. This result is, however, dependent on the way the ecosystem is modeled, with particular emphasis on the importance of including several detritus compartments.

Summarizing, cycling is an important aspect of ecosystem dynamics. Although cycles seem to be rare in published community food webs and models, their number is very large when detritus compartments are considered. Moreover, it is important to stress that the role of the so-called microbial loop, neglected in studies that concentrate on larger organisms, can dramatically change the cycling performance of the system. These considerations lead ecosystem ecologists to the formulation of the amount of cycling in ecosystem networks. The FCI, even though it is a biased count of the cycling in ecosystems, has found wide application in ecosystem studies. The problem of measuring the exact amount of cycling in an ecosystem is still an open problem, as it could be possible to ameliorate the algorithms for finding and removing cycles. Finally, the network building process is likely to determine the outcome in terms of cycling. It would therefore be important to have shared rules for network building that would result in the comparability between different networks and ecosystems.

See also: Autocatalysis; Ecological Network Analysis, Ascendency.

Further Reading

Allesina S, Bodini A, and Bondavalli C (2005) Ecological subsystems via graph theory: The role of strongly connected components. *Oikos* 110: 164–176.

Allesina S and Ulanowicz RE (2004) Cycling in ecological networks: Finn's index revisited. *Computational Biology and Chemistry* 28: 227–233.

De Angelis DL (1992) *Dynamics of Nutrient Cycling and Food Webs*, 270pp. London: Chapman and Hall.

Finn JT (1976) Measures of ecosystem structure and functions derived from analysis of flows. *Journal of Theoretical Biology* 56: 363–380.

Finn JT (1980) Flow analysis of models of the Hubbard Brook ecosystem. *Ecology* 61: 562–571.

Patten BC (1985) Energy cycling in the ecosystem. *Ecological Modelling* 28: 1–71.

Patten BC and Higashi M (1984) Modified cycling index for ecological applications. *Ecological Modelling* 25: 69–83.

Ulanowicz RE (1983) Identifying the structure of cycling in ecosystems. *Mathematical Biosciences* 65: 219–237.

Ulanowicz RE (1986) *Growth and Development: Ecosystems Phenomenology*. New York: Springer.

Ulanowicz RE (2004) Quantitative methods for ecological network analysis. *Computational Biology and Chemistry* 28: 321–339.

Ecological Network Analysis, Ascendency

U M Scharler, University of KwaZulu-Natal, Durban, South Africa

Introduction
Principle of Ascendency

Ascendency Applications
Further Reading

Introduction

In the search for a nonmechanistical explanation of ecosystem behavior and development, Ulanowicz developed the theory of ascendency. Direct cause and effect mechanisms, as known from the Newtonian world, are believed not to be sufficient to describe, or predict, the behavior of ecosystems. Such mechanisms are inherently reversible and are not seen to be sufficient to explain the behavior of single components (e.g., species) within the context of the ecosystem. Ecosystems are believed to behave and evolve in a nonmechanistic fashion. The theory of ascendency tries to capture this nonmechanistic behavior in a single index, indicative of ecosystem state and development, and of ecosystem health.

Principle of Ascendency

Ascendency

Conditional probabilities and ecosystem complexity

In a mechanistic world, the probabilities of events following specific causes can be calculated by joint probabilities $p(a_i, b_j)$. These describe an absolute probability of an effect occurring in response to a cause. In ecosystems, it is believed that no such direct, mechanistic cause and effect behavior exists due to the interaction with other elements which in turn influence the patterns of cause and effects between pairs. Instead of the absolute probability, Popper introduces the term propensity, which describes a bias that events might (not will) happen. Popper therefore calls for a measure of such relative or conditional probabilities. Conditional probabilities are denoted by $p(a_i|b_j)$, and are calculated by dividing the absolute probabilities $p(a_i, b_j)$ by the marginal probability $p(a_i)$, or the sum of all probable effects of one cause (**Tables 1–3**). The conditional probability thus describes cause and effect in the context of other absolute probabilities, considering that a cause might have more than one effect. This eliminates the pitfall of disregarding the influence of other interactions on the one in question. It is, of course, possible to calculate the conditional probability of a mechanical cause–effect pair, that is, the case of having one cause and one effect. This turns out to be 1, or in other words, there is certainty that the effect in question will follow the cause in question.

Since ecosystems are open, not all causes can be accounted for. Some of them might originate outside the

Table 1 Frequencies of joint occurrences of events (total = 60)

	b_1	b_2	b_3	b_4
a_1	4	5	7	9
a_2	2	4	2	1
a_3	6	7	9	4

Table 2 Joint probabilities ($p(a_i, b_j)$) and their column/row sums or marginal probabilities ($p(a_i)$, $p(b_j)$)

		b_1	b_2	b_3	b_4	$p(a_i)$
	a_1	0.07	0.08	0.12	0.15	0.42
$p(a_i,b_j)$	a_2	0.03	0.07	0.03	0.02	0.15
	a_3	0.10	0.12	0.15	0.07	0.43
	$p(b_j)$	0.20	0.27	0.30	0.23	1.00

Values are obtained by dividing the number of occurrences (**Table 1**) by the total number of observations (60).

Table 3 Conditional probabilities

		b_1	b_2	b_3	b_4
	a_1	0.33	0.31	0.39	0.64
$p(a_i|b_j)$	a_2	0.17	0.25	0.11	0.07
	a_3	0.50	0.44	0.50	0.29

Values are obtained by dividing the values in the joint probability matrix by the column sums ($p(a_j)$) (see **Table 2**).

system. Therefore, an open ecosystem can never evolve toward a mechanistic behavior of cause and effect. Ulanowicz states that autocatalysis, or indirect mutualism, is an important cause in ecosystem growth and development. Autocatalysis is apparent when members of a feeding loop positively enhance the following member of the loop, which eventually leads back to a positive enhancement of the starting member. Autocatalytic loops exert a selection pressure on its members in that a member of the loop might be replaced with a new constituent who has a more positive effect. Autocatalytic loops exhibit a centripetality, which enables them to attract more resources (available energy). These are reasons for the growth and development of, or increase of order in, ecosystems.

To quantify growth and development, ecosystems are portrayed as networks of material or energy exchanges. These networks of feeding transfers are believed to adequately describe an ecosystem. It is assumed that other significant aspects of ecological systems, such as behavioral aspects, are in one form or another imprinted on the amount of energy transferred, through their effect on population size and predator avoidance.

Ascendency describes both growth and development. Growth of the ecosystem is measured as any increase in total system throughput (TST), which is the sum of all exchanges within the ecosystem and between the system in question and its outside (imports, exports, respirations). Total system throughput can rise either by increasing the extent of the system (more species, or by extending ecosystem borders) or by an increased activity of the system (e.g., during phytoplankton blooms).

Ecosystem development is quantified from the same networks of material exchanges with the help of information theory. In autocatalytic loops, the trend for transferring material is as follows: those linkages which are most rewarding to the loop will transfer more material than those which are not (compartments have in general more than one outgoing link and can thus have pathways to compartments outside the loop). The latter are not necessarily discarded, but transfer only a small amount of material. If a quantum of material sits in a compartment in an autocatalytic loop, then it is therefore more likely to be transferred along a route with high material transfer than along a route with low material transfer. The probability that a quantum of material flows along the highly frequented routes is, therefore, higher compared to a network where all routes transfer the same amount of material. Conversely, the probability that a quantum of material flows along the less-frequented routes will be lower compared to a network where all routes transfer the same amount of material. Such a change in probability can be quantified with the help of information theory. Information is defined as the agent that causes a change in probability. Ulanowicz uses the term information to describe 'the effects of that which imparts order and pattern to the system'.

In the calculation of information, the starting point is to quantify ecosystem complexity. The complexity of a system is mirrored in the system configuration (amount of links and distribution of transfers along those links). According to Boltzmann, the potential of each configuration contributing to systems complexity, s, can be calculated as the negative logarithm of the probability that the event (the system configuration) will occur ($s = -k \log p$, where k is a constant of proportionality, i.e., a scaling factor). If a system configuration (the event) will occur always ($p = 1$), then the contribution to complexity is diminished ($\log(1) = 0$, uncertainty is at its lowest) and the system behavior is simple (i.e., it always behaves the same way). If a system configuration (or event) occurs only rarely, then there is a large potential for complexity (i.e., it can behave in many different ways, uncertainty is high). Behavior of a truly complex system is unique each time it functions (uncertainty is highest).

To calculate how much a rare configuration contributes to system complexity, it is weighted by the (low) frequency of its occurrence. The potential contributions (or events) are averaged by the configurations of the system by weighting each s_i by its corresponding p_i (Shannon's formula: $H = -K \sum_i p_i \log p_i$). In other words, each potential contribution of occurrence is weighted by its corresponding probability that it will occur, which is summed over all system configurations. A high H corresponds to high uncertainty, complexity, and diversity.

Average mutual information

The above discussion serves to illustrate how events can contribute to the complexity of a system. Next to consider is whether these events contribute to an ordered pattern in the system, or whether they contribute to random behavior. If all events are equiprobable, then the average uncertainty about what event will happen next is the highest. This hypothetical situation can serve as a starting point to calculate how much less uncertainty there is under circumstances where not all events are equiprobable. The decrease in uncertainty from a situation of equiprobability to any other is called information. From an ecosystem perspective, a situation of equiprobability is one where material flows in equal amounts along all pathways (**Figure 1a**). One that is not equiprobable is where

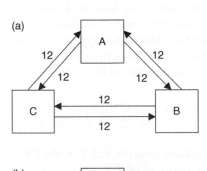

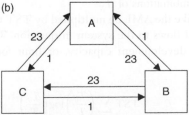

Figure 1 (a) Hypothetical unconstrained network: low AMI. (b) Hypothetical constrained network: higher AMI.

more material flows along some pathways, and less material along others (**Figure 1b**). Thus, the most indeterminate network is one where all compartments are connected with each other and where, in proportion to the compartmental throughput, equal amounts of material flow along the ingoing and outgoing pathways. Quantifying the information which is gained by transferring material along more- and less-frequented routes thus gives a clue about the unevenness of material flowing along pathways.

The change in probability from a situation where a quantum of material flows along an equiprobable pathway and along a pathway which is not equiprobable is calculated using conditional probabilities. To start with, the uncertainty that an event occurs is

$$H = -K \log p(a_i) \qquad [1]$$

and the uncertainty that an event occurs provided certain information (b_j) is available is

$$H = -K \log p(a_i|b_j) \qquad [2]$$

The information then is the *a priori* uncertainty minus the uncertainty if b_j is known or

$$I = -K \log p(a_i) - \left[-k \log p(a_i|b_j) \right] \qquad [3a]$$

or

$$I = K \log p(a_i|b_j) - K \log p(a_i) \qquad [3b]$$

or

$$I = K \log \left[p(a_i|b_j)/p(a_i) \right] \qquad [3c]$$

I is not positive for all pairs of occurrences. The sum of all I's which have been weighted by the corresponding joint probability turns out always to be positive, however. The joint probability of each occurrence serves, as in Shannon's formula, as a weighting for the frequency of occurrence of each event (i.e., each co-occurrence of a_i and b_j). The result is called the average mutual information (AMI) or

$$\text{AMI} = K \sum_i \sum_j p(a_i, b_j) \log \left[p(a_i|b_j)/p(a_i) \right] \qquad [4]$$

AMI is the amount of uncertainty reduced by knowing b_j. Results are in units of K.

As in the hypothetical example above, the *a priori* uncertainty about where a quantum of material flows in ecological networks is given by Shannon's formula. The additional information (b_j) to calculate the conditional probability is the knowledge of the outputs from each compartment in the flow network a time step earlier.

Since, from an ecological network point of view, joint and conditional probabilities refer to transfers of material from compartment i to compartment j, the above formula can be rewritten as

$$\text{AMI} = K \sum_{i,j} \left(\frac{T_{ij}}{T_{..}}\right) \log\left(\frac{T_{ij} T_{..}}{T_{i.} T_{.j}}\right) \qquad [5]$$

where the joint probability of a quantum of material $(p(a_i,b_j))$ flowing from species i to species j can be denoted as $T_{ij}/T_{..}$, remembering that the events in an events table are material flows in a system. $T_{..}$ is the total system throughput, or the sum over all combinations of T_{ij}. The summation among all rows of the matrix is denoted by the first dot, while the second dot stands for summation among columns.

The conditional probability

$$p(a_i|b_j) = p(a_i, b_j)/p(a_i)$$

can be rewritten as $T_{ij}/T_{i.}$ and the marginal probability (sum of all probable outcomes, $p(a_i)$) as $T_{.j}/T_{..}$.

To summarize, the AMI describes the information gained by knowing the outputs from each compartment in the flow network a time step earlier (b_j) in addition to the *a priori* situation describing the flow of a quantum of energy or material between two compartments (a_i). The uncertainty of where a quantum flows is calculated through Shannon's index of flow diversity. The uncertainty of where a quantum of material will flow by knowing b_j is calculated by the conditional probability.

Ascendency

The scalar constant, k, has been retained throughout all calculations. To be able to combine growth and development into one single index, k is substituted by the 'total system throughput' or TST in order to scale the AMI to the size of the system in question. The resulting index is called ascendency and is denoted by

$$A = \text{TST} \sum_{i,j} \left(\frac{T_{ij}}{T_{..}}\right) \log\left(\frac{T_{ij} T_{..}}{T_{i.} T_{.j}}\right) \qquad [6a]$$

or

$$A = \sum_{i,j} T_{ij} \log\left(\frac{T_{ij} T_{..}}{T_{i.} T_{.j}}\right) \qquad [6b]$$

Besides indirect mutualism there are a number of influences that can change the ascendency of a system. These influences are thought to not have any favored direction of change, whereas indirect mutualism is believed to drive development toward increased

ascendency. Mutualism is furthermore not a result of events elsewhere in the system's hierarchy but can arise at any level. Therefore it is theorized that in the absence of overwhelming external disturbances, the ascendency of a system has a propensity to increase, that is, both activity (TST) and structure (AMI) increase. The theoretical behavior of mutual information conforms to most of the 24 ecosystem properties originally put forward by Odum to characterize mature ecosystems.

Ascendency is limited by any constraints on the increase in either TST or AMI. Limits to TST are set by the finite imports from outside system boundaries and by the second law of thermodynamics, which requires that a portion of the compartmental throughput be lost as dissipation. Therefore the TST cannot increase indefinitely via recycling. The limits to the AMI, or system development, are set by the flow structure. It limits the extent to which the flows can be organized without a change to the structure itself. Further limits to the AMI in real networks are discussed in the section titled 'Overhead'.

In theory, ascendency is higher when pathways are fewer in numbers (more specialization) and more articulated (few pathways transport most of the material). The highest theoretical value of ascendency is achieved when all players in the system have one input and one output only, and are thus joined in one big single loop. This configuration mirrors highest specialization, and in this case AMI = H (diversity of flows, see below). This situation cannot be achieved in real systems, due to reasons discussed in the section titled 'Overhead'.

Development Capacity

As mentioned above, the limit to development is set by Shannon's diversity index pertaining to the material transfers or flows. MacArthur applied Shannon's diversity index to the material flows in an ecosystem to arrive at a measure for the diversity of flows, H:

$$H = -k \sum_{i,j} \left(\frac{T_{ij}}{T_{..}}\right) \log\left(\frac{T_{ij}}{T_{..}}\right) \qquad [7]$$

where k is a scalar constant, and $T_{..}$ is the TST, or the sum over all combinations of T_{ij}.

H can, like the AMI, be multiplied by TST to scale the diversity of flows to the system in question. TST $\times$ H is called the development capacity, or limit for development, C:

$$C = -\text{TST} \sum_{i,j} \left(\frac{T_{ij}}{T_{..}}\right) \log\left(\frac{T_{ij}}{T_{..}}\right) \qquad [8a]$$

or

$$C = -\sum_{i,j} T_{ij} \log\left(\frac{T_{ij}}{T_{..}}\right) \qquad [8b]$$

The development capacity is limited by two factors, namely TST and the number of compartments. The limits to TST are the same as in the case of ascendency. If a certain amount of TST is split between too many compartments, then some compartments will end up with a very small throughput. These are, in turn, prone to extinction should the system undergo disturbances. This process is believed to reduce the number of compartments and thus the number of flows. More stable systems are thus believed to show a higher C compared to systems undergoing frequent perturbations.

The initial complexity, H, consists of two elements. One is the AMI, describing the information gained by reducing the uncertainty in flow probability. It is an index of the organized part of the system. The other is the residual uncertainty, or H_c (also called conditional diversity). Thus, $H = \text{AMI} + H_c$.

H_c or Overhead

The residual uncertainty H_c, when scaled by TST is also called the overhead. The overhead represents the unorganized, inefficient, and indeterminate part of the flow structure and is considered an insurance for the system. Should the system become overly organized (high ascendency), it will also be prone to perturbations. The overhead is split into four components: overhead due to imports, exports, respiration, and internal pathways.

The combined overhead is denoted by

$$H_c = -k \sum_{i,j} \left(\frac{T_{ij}}{T_{..}}\right) \log\left(\frac{T_{ij}^2}{T_{i.}\,T_{.j}}\right) \qquad [9]$$

Scaling H_c to the system by replacing k with TST yields

$$\Phi = -\sum_{i,j} T_{ij} \log\left(\frac{T_{ij}^2}{T_{i.}\,T_{.j}}\right) \qquad [10]$$

The relationship between C, A, and Φ so becomes $C = A. + \Phi$.

Imports

The overhead due to imports is dependent on the number of pathways originating outside the system, and on the magnitude of the material transferred along those pathways. If all sustenance is equally distributed among all import pathways, then the contribution to the overhead will be maximal. It will decrease when some pathways import more and others less. It will also decrease if the overall magnitudes of the imports decrease. If there is

only one import path, then the overhead due to imports is minimal and equals zero. From a systems point of view it is regarded as counterproductive to minimize the magnitude of the import, or to import only via one pathway. The insurance lies in being able to receive imports via several pathways in case one is lost. In the case of increased recycling within the system, the imports will occupy a smaller and smaller part of the TST. In this case, the development capacity will rise faster than the overhead on imports.

If the imports enter the system via fewer pathways or compartments, then the ascendency will increase at the expense of the overhead. Systems are expected to progress toward fewer import pathways. The number of such pathways can be changed should those links be disrupted and others become necessary. Overall it is expected that systems in a more stable environment rely on fewer import pathways compared to perturbed systems.

The formula for the overhead on imports is as follows:

$$\Phi_I = -\sum_{j=1}^{n} T_{0j} \log\left(\frac{T_{0j}^2}{T_{0.}\,T_{.j}}\right) \qquad [11]$$

where imports are assumed to originate in the fictitious compartment 0.

Exports

Similar to the overhead on imports, the overhead on exports depends on the amount of exporting pathways leaving the system and the amount transferred along those pathways. The overhead due to export diminishes whenever there are fewer export pathways, lower magnitude of transfers, or an uneven distribution of amounts transferred along the pathways. An increase in exports becomes beneficial to the system whenever there is positive feedback via another system. The overhead on exports is denoted by

$$\Phi_E = -\sum_{i=1}^{n} T_{i,n+1} \log\left(\frac{T_{i,n+1}^2}{T_{i.}\,T_{.,n+1}}\right) \qquad [12]$$

where exports are assumed to flow into a fictitious compartment $n+1$.

Respiration

Again, the overhead regarding the dissipations depends on the magnitude lost to the environment, on the number of pathways, and the distribution of the magnitude transferred. Losses through dissipation are required by the second law of thermodynamics and are necessary to maintain metabolisms. The overhead on dissipation is

$$\Phi_D = -\sum_{i=1}^{n} T_{i,n+2} \log\left(\frac{T_{i,n+2}^2}{T_{i.}\,T_{.,n+2}}\right) \qquad [13]$$

where respiration is assumed to flow into a fictitious compartment $n+2$.

Redundancy

The fourth part of the overhead is that of internal transfers and represents the extent of pathway redundancy. There are disadvantages to the system in maintaining redundant, or parallel pathways. For one, there can be an increase in dissipations, whenever transfers occur not only along the most efficient route, but also along leakier pathways. Also, the resource transferred along different parallel pathways might not always end up at the right time at the consumer.

An obvious advantage of parallel pathways is the insurance of having more than one route of transfer in case of disturbances of other routes. Redundancy is denoted by

$$R = -\sum_{i=1}^{n}\sum_{j=1}^{n} T_{ij}\log\left(\frac{T_{ij}^2}{T_{i.}T_{.j}}\right) \qquad [14]$$

Biomass Inclusive Ascendency

The above indices were calculated on the trophic flows between compartments. It is also possible to calculate a systems ascendency that embraces the connection between biomass stocks and the trophic flows. This biomass inclusive ascendency can be used as a theoretical basis to derive element limitations for compartments, to identify limiting nutrient linkages, and to quantify the successional trend to include larger species with slower turnover times.

Above, AMI was calculated as the difference between two flow probabilities, the unconstrained or *a priori* joint probability, and the constrained or *a posteriori* conditional probability. AMI can also be calculated between a biomass (unconstrained or joint) probability and the resulting flow (constrained or conditional) probability, thereby calculating a relationship between biomass and flows. From the principal of mass action, the joint probability of whether a quantum of biomass leaves compartment i (B_i/B) and enters compartment j (B_j/B) is B_iB_j/B^2. This expression constitutes the unconstrained joint probability that a quantum flows from i to j. No constraining assumptions are made about this exchange, with the exception of the magnitudes of the stocks. The corresponding constrained distribution is taken as the conditional probability of the actual flow from i to j or T_{ij}/T. This constraint is an addition to the probability calculated from the stocks only, and therefore, structure and function are tied together. The information gained is calculated as follows:

$$I_B = -K\log\left(\frac{B_iB_j}{B^2}\right) - \left[-K\log\left(\frac{T_{ij}}{T_{..}}\right)\right] \qquad [15a]$$

or

$$I_B = k\log\left(\frac{T_{ij}}{T_{..}}\right) - K\log\left(\frac{B_iB_j}{B^2}\right) \qquad [15b]$$

or

$$I_B = K\log\left(\frac{T_{ij}B^2}{T_{..}B_iB_j}\right) \qquad [15c]$$

Summing over all realized combinations of i and j and weighted by the joint probability of occurrence, one arrives at the biomass inclusive AMI, AMI_B:

$$AMI_B = K\sum_{i,j}\frac{T_{ij}}{T_{..}}\log\left(\frac{T_{ij}B^2}{T_{..}B_iB_j}\right) \qquad [16]$$

which is also called the Kullback–Leibler information. Scaling by the total system throughput gives the biomass inclusive ascendency, A_B:

$$A_B = TST\sum_{i,j}\frac{T_{ij}}{T_{..}}\log\left(\frac{T_{ij}B^2}{T_{..}B_iB_j}\right) \qquad [17a]$$

or

$$A_B = \sum_{i,j} T_{ij}\log\left(\frac{T_{ij}B^2}{T_{..}B_iB_j}\right) \qquad [17b]$$

A_B is sensitive to changes in biomass and can thus show the sensitivity of the whole system to changes in stock of a particular compartment.

The above term can be split into the following terms:

$$A_B = \sum_{i,j} T_{ij}\log\left(\frac{T_{ij}T_{..}}{T_{i.}T_{.j}}\right) + \sum_{i} T_{i.}\log\left(\frac{T_{i.}B}{T_{..}B_i}\right) + \sum_{j} T_{.j}\log\left(\frac{T_{.j}B}{T_{..}B_j}\right) \qquad [18]$$

The first term is exactly the same as in the above definition of the flow ascendency. Therefore, also the biomass inclusive ascendency rises with an increased number of compartments, increased specialization of flows, and increased throughput. The second and third terms become zero whenever the proportional flow through each compartment is the same as its proportion of the biomass. Only in this case would A_B equal A. In all other cases, A_B will exceed A.

Limiting elements in compartments and limiting flows

If one is interested in calculating a compartment's contribution to the ascendency of a particular element k (e.g., C, N, P, S, ...) during a certain time step l, then one has to substitute into above equation the element and the time step:

$$A_B = \sum_{i,j,k,l} T_{ijkl} \log\left(\frac{T_{ijkl}B^2}{T_{..}B_{ikl}B_{jkl}}\right) \qquad [19]$$

where T_{ijkl} is the flow from i to j of element k during time step l.

To show how the ascendency responds to turnover times of various elements, the differential of A_B regarding compartment p is given as

$$\frac{\partial A_B}{\partial B_{pk}} = 2\left(\frac{T_{..}}{B_{.}} - \frac{1}{2}\frac{T_{.pk} + T_{p.k}}{B_{pk}}\right) \qquad [20]$$

Here the relative contributions of all elements investigated to the system's ascendency can be calculated. Results will show that the system is most sensitive to the element with the slowest turnover rate. The element with the slowest turnover rate is also the element which enters the compartment in its least relative proportion. The last statement accords with Liebig's law of the minimum for which ascendency provides a theoretical basis. The same results could have been obtained by comparing elemental turnover rates of all compartments. However, ascendency provides yet another level of information, namely it identifies which source provides the limiting flow of the controlling element. To calculate this, the sensitivities of the individual biomasses can be expanded to include the sensitivities of the individual flows from source r to predator p. The following equation calculates the contributions of each flow:

$$\frac{\partial A_B}{\partial T_{rp}} = \log\left(\frac{T_{rp}B^2}{T_{..}B_r B_p}\right) \qquad [21]$$

The limiting source of the controlling element is the one which is depleted fastest in relation to its available stock, that is, the one with the highest (T_{rp}/B_r). Knowing the sensitivity of the flow for each element and compartment, it is thus possible to pinpoint nutrient limitations and the limiting flows for each compartment in the food web. In ecosystems, not all species are limited by the same nutrient. For instance when primary producers are limited by nitrogen, it does not necessarily mean that the entire food web is limited by nitrogen.

Ascendency Applications

Principles of ascendency, as they have been shown here, have been applied to compare similar ecosystems (e.g., estuaries), or the same ecosystems over a period of time including the response of systems to disturbances. Examples of such applications are the description of spatial and temporal change of ascendency in marine microbial systems. They revealed that ascendency is strongly related to the functionality of the microbenthic loop. Important parameters determining the

value of ascendency were the decomposition activity and the capacity for resource exploitation. Ascendency was found to be a useful indicator for the health assessment of marine benthic ecosystems over space and time.

Ascendency has also been applied to establish ecosystem responses to eutrophication and other anthropogenic system alterations of carbohydrates, proteins, lipids, and carbon biopolymers in various parts of the globe. Whereas ascendency is, in general, believed to rise with eutrophication due to an increase in TST, this is not always the case. Depending on the extent and frequency of the eutrophication event, it might disturb the system to an extent where ascendency reflects a decrease in ecosystem stability through a decrease in AMI and TST. Another case of system perturbation was described for pesticide-perturbed microcosms, using an index called 'scope for change in ascendency' (SfCA). SfCA is an analogy to scope for growth of an organism and is the balance of the ascendency of individual compartment inputs and outputs. SfCA was hypothesized to decrease in the presence of a disturbance and was ultimately found to be a useful indicator for the short-term assessment of perturbations in herbicide-treated microcosms.

Ascendency has also been used to assess the whole ecosystem impacts of severe freshwater abstractions from an estuarine catchment. The interdecadal comparison between light and severe freshwater abstraction and the consequential reduction in sustained and pulsing freshwater inflow into the Kromme estuary revealed a decrease in ascendency under the present, freshwater-starved condition. The spatial comparison with other, similar, estuaries that do not have such severe freshwater abstractions in the catchment shows a higher ascendency in estuaries with higher freshwater inflow that ensures sustained renewal of the nutrient pool to fuel primary production.

Since ascendency is very often influenced by a change in the magnitude of TST, the organization of a system is frequently reported as a ratio of ascendency/development capacity (A/C), which cancels out the influence of TST. Also the AMI is used as an unscaled index in a comparative way. In general, it is advised to take the behavior of other indicators of ecosystem health (e.g., exergy) into account in combination with ascendency to arrive at a representative assessment of ecosystem state. Ascendency has been shown to vary with the degree of aggregation of the network. In general, ascendency decreases in highly aggregated networks, even if the TST is the same. The type of aggregation, that is, which compartments are aggregated, also significantly affects the value of ascendency. This is equally true for the aggregation of living and nonliving components of the network.

The biomass inclusive version of ascendency and the sensitivities of the individual flows were determined for the Chesapeake Bay system to identify the limiting nutrient in the ecosystem and bottlenecks in carbon, nitrogen, and phosphorus transfers. The comparison over four seasons revealed that, in general, the primary producers were nitrogen limited, which was in concordance with previous studies on these groups. However, the nitrogen limitation on the primary producer level was not propagated throughout the entire web, but all nekton was found to be phosphorus limited. The type of nutrient limitation changed over the course of the year for a few primary producers and invertebrates, but not for the nekton. It is important to note that nutrient limitations in a trophic flow network are not determined by the type of limitation of the primary producer, since the various organisms have different stoichiometric requirements.

See also: Autocatalysis; Emergent Properties; Goal Functions and Orientors; Indirect Effects in Ecology.

Further Reading

Baird D and Heymans JJ (1996) Assessment of the ecosystem changes in response to freshwater inflow of the Kromme River estuary, St. Francis Bay, South Africa: A network analysis approach. *Water SA* 22(4): 307–318.

Fabiano M, Vassallo P, Vezzulli L, Salvo VS, and Marques JC (2004) Temporal and spatial changes of exergy and ascendency in different benthic marine ecosystems. *Energy* 29: 1697–1712.

Genoni GP (1992) Short-term effect of a toxicant on scope for change in ascendency in a microcosm community. *Ecotoxicology and Environmental Safety* 24: 179–191.

MacArthur R (1955) Fluctuations of animal populations, and a measure of community stability. *Ecology* 36(3): 533–536.

Morris JT, Christian RR, and Ulanowicz RE (2005) Analysis of size and complexity of randomly constructed food webs by information theoretic metrics. In: Belgrano A, Scharler UM, Dunne JA, and Ulanowicz RE (eds.) *Aquatic Food Webs,* vol. 7, pp. 73–85. New York: Oxford University Press.

Odum EP (1969) The strategy of ecosystem development. *Science* 164: 262–270.

Patrício J, Ulanowicz RE, Pardal M, and Marques J (2006) Ascendency as ecological indicator for environmental quality assessment at the ecosystem level: A case study. *Hydrobiologia* 555: 19–30.

Popper KR (1982) *A World of Propensities*, 51pp. Bristol: Thoemmes.

Rutledge RW, Basore BL, and Mulholland R (1976) Ecological stability: An information theory viewpoint. *Journal of Theoretical Biology* 57: 355–371.

Scharler UM and Baird D (2005) A comparison of selected ecosystem attributes of three South African estuaries with different freshwater inflow regimes, using network analysis. *Journal of Marine Systems* 56(3–4): 283–308.

Tobor-Kaplon MA, Holtkamp R, Scharler UM, Bloem J, and de Ruiter PC (2007) Evaluation of information indices as indicators of environmental stress in terrestrial. *Ecological Modelling* 208: 80–90.

Ulanowicz RE (1986) *Growth and Development: Ecosystems Phenomenology*. New York: Springer.

Ulanowicz RE (1997) *Ecology, The Ascendent Perspective*. New York: Columbia University Press.

Ulanowicz RE (2004) Quantitative methods for ecological network analysis. *Computational Biology and Chemistry* 28: 321–339.

Ulanowicz RE and Abarca-Arenas LG (1997) An informational synthesis of ecosystem structure and function. *Ecological Modelling* 95: 1–10.

Ulanowicz RE and Baird D (1999) Nutrient controls on ecosystem dynamics: The Chesapeake mesohaline community. *Journal of Marine Systems* 19: 159–172.

Relevant Websites

http://www.dsa.unipr.it – Dipartimento di Scienze Ambientali.

http://www.ecopath.org – Ecopath with Ecosim.

http://www.cbl.umces.edu – Ecosystem Network Analysis.

http://www.glerl.noaa.gov – National Oceanic and Atmospheric Administration, Great Lakes Environmental Research Laboratory.

Ecological Network Analysis, Energy Analysis

R A Herendeen, University of Vermont, Burlington, VT, USA

Introduction
Level of Analysis
Steady-State Analysis: Energy and Nutrient Intensities
Steady-State Analysis: Other Indicators

Indicators in Dynamic Systems
Applications
Further Reading

Introduction

Ecologists have long told us that all flesh is grass, which in turn is sunlight. Thanks in large part to the oil embargo in 1973, we appreciate that bread is not just sunlight, but also

oil. And economists have shown that demand for a shirt produces a demand for steel. These are all examples of indirect effects. The techniques used to quantify them span systems ecology, engineering, and economics – a compelling example of cross-disciplinary fertilization.

Understanding them yields insights in diverse applications, from bioaccumulation of pollutants in ecosystems to labor demand in economies.

In principle, one could discern all aspects of indirectness from the full diagram of flows between compartments in a system. In practice, we often desire, or accept, summary variables or indicators which are specific to a particular application and convey the concept more concisely, though often with a loss of details.

In this article, such indicators are discussed, often using explicit calculations applied to a simple, idealized two-compartment ecosystem. The indicators are energy and nutrient intensities, trophic position (TP), path length (PL), and residence time. Besides application to steady state, the concept is also extended to dynamic ecosystems such as those responding to perturbations. Finally, calculating energy intensity of goods and services in economic systems is discussed. The latter is a crucial step in determining the energy cost of living in a consumer society, and has specific application in analyzing consequences of an energy tax.

Level of Analysis

In a multicompartment system, interactions can be analyzed at three levels of aggregation/detail:

1. *Single compartment* (*isolated*). This addresses direct effects (e.g., an eagle eats mice but no grass). Traditional population biology often works at this level, which includes no indirectness at all.
2. *Single compartment in-system.* This addresses direct plus indirect effects (e.g., by eating mice which eat grass, the eagle is consuming embodied grass, which is embodied sunlight).
3. *Whole-system.* This addresses system-wide processes (e.g., to what extent is the entire system recycling phosphorus vs. leaking it immediately?).

This article concentrates on level 2. The indicators calculated are the property of a single compartment explicitly connected to other compartments in an ecosystem. Level 3 is beyond the scope of this article.

Level 2 analysis is the basis for most of the energy analysis started in the early 1970s. It led to then-surprising results such as these:

1. Only *c.* 60% of the energy to own and operate a car is the fuel in the tank. Around 15% is required to produce the car, and *c.* 25% is for parts, maintenance, insurance, registration, parking, etc.
2. Only *c.* 10% of the energy to make the car is consumed at the assembly plant. The remainder is consumed at the steel mill, glass works, iron mine, rubber plantation, etc.

3. Switching from throwaway to returnable beverage bottles saves energy and increases jobs.

A more recent example is that suburban living ('sprawl') is only *c.* 10% more energy-intensive than urban ('compact') living. A biological example is the trophic cascade, exemplified by consequences of recent wolf reintroduction into Yellowstone National Park. Adding wolves has suppressed elk activity, resulting in increased regeneration of browse vegetation.

Steady-State Analysis: Energy and Nutrient Intensities

The bookkeeping of energy analysis can be used to allocate many other kinds of indirectness. Starting with energy, we extend the method to other entities. To illustrate this, a hypothetical two-compartment system at steady state is used (**Figure 1**). This is complex enough to allow feedback (recycling), yet simple enough to allow using standard algebra. All that is done here can be couched in matrix notation, and shorthand is useful for systems with many compartments, but algebra is more transparent.

Figure 1 shows the flow of something, say energy, in a two-compartment system containing producers (e.g., green plants) and consumers (e.g., herbivores). (See **Table 1** for definitions of terms.) The input to producers comes from outside the system (the Sun), and the export from consumers leaves the system. There is a variable amount of feedback from consumers to producers; it is possible for some plants to eat animals. In all diagrams of this type we decide which flows convey the direct and indirect influences we deem important; our judgement is required. For example, the standard energy intensity concept is that the energy losses (e.g., low-temperature heat) are assumed to be embodied in the remaining flows of high-quality metabolizable biomass. This would give a modified diagram (**Figure 2**).

The remaining flows thus convey the input, which we wish to account for. The missing losses are now implicitly embodied in the flows that remain, and the energy intensities carry this formally. The parallelism and difference

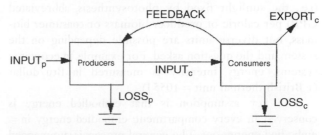

Figure 1 Energy flows (cal/day) in a two-compartment system at steady state. INPUT$_p$ is gross primary production of biomass by photosynthesis.

Table 1 Definitions of terms

Symbol	Definition	Units[a]
Δt_j	Time step	day
E_j	Energy input to compartment j	cal/day
ε_j	Energy intensity of compartment j's output	cal/cal
ECOL	Energy cost of living for a household	Btu/yr
$\langle\varepsilon\rangle$	Household average energy intensity ($=$ECOL/Y)	Btu/$
ε_{impj}	Energy intensity of imported material functionally identical to that produced by compartment j	cal/cal
EXPORT$_j$	Export from compartment j	cal/day
FEEDBACK$_j$	Flow from consumers to producers	cal/day
gloof	Generic term for system input which is allocated by method used in this paper	?/day
GPP	Gross primary production, the energy fixed by plants	cal/day
IMPORT$_j$	Import of material functionally identical to that produced by compartment j	cal/day
INPUT$_j$	Input to compartment j	cal/day
LOSS$_j$	Loss from compartment j	cal/day
N_j	Nutrient input to compartment j	g/day
η_j	Nutrient intensity of compartment j's output	g/cal
η_{impj}	Nutrient intensity of import of material identical to that produced by compartment j	g/cal
OUTPUT$_j$	Output of compartment j	cal/day
PL$_j$	Path length of compartment j	dimensionless
S_j	Stock of compartment j	cal
TP$_j$	Trophic position of compartment j	dimensionless
t_j	Isolated-compartment residence time for compartment j	day
τ_j	In-system residence time for compartment j	day
X_{ij}	Flow from compartment i to compartment j	cal/day
X_j	OUTPUT$_j$	cal/day
Y_i	Annual household expenditure for consumption category i	$/yr
Y	Sum of all annual household expenditures	$/yr
Z	FEEDBACK	cal/day

[a]Grams, calories, and days are arbitrarily chosen as the units of mass, energy, and time for the examples in this article. Other units, as appropriate, are used for the applications.

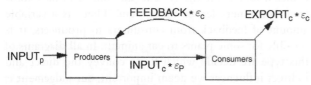

Figure 2 Embodied energy flows (cal GPP/day) in a two-compartment system. The energy intensities ε (cal GPP/cal) convert energy flows (cal/day) of **Figure 1** to embodied energy flows (cal GPP/day).

between **Figures 1** and **2** crystallize the entire import of indirectness presented in this article. The intensities are thus the conceptual, dimensional, and numerical bridge between system input and flows. In this system, the units of intensities will be calories of gross primary production (i.e., the sunlight fixed by photosynthesis, abbreviated GPP) per calorie of producer biomass or consumer biomass, but diverse units are possible depending on the system and the question asked. For example, in economic systems, energy intensity is measured in Btu/dollar (1 British thermal unit $=$ 1055 J).

The key assumption is that embodied energy is conserved in every compartment: embodied energy in $=$ embodied energy out. The general process is summarized in **Figures 3a** and **3b**. Because the method can be used to allocate many things besides energy, the generic term

'gloof' is used in **Figure 3a**. **Figure 3b** shows the balance equation for energy. **Figures 3a** and **3b** allow for imports to inject embodied energy into the system. An example is Howard Odum's study of Silver Spring, Florida, where tourists fed bread to fish whose normal food was plants growing in the spring. Another example of imported embodied energy is America's importing of clothes made in China.

The energy intensities are obtained by solving the (linear) equations implied by **Figure 3b**:

$$\sum_{i=1}^{n} \varepsilon_i X_{ij} + \varepsilon_{impj}\text{IMPORT}_j + E_j = \varepsilon_j X_j \qquad [1]$$

The sum is over all within-system inputs. In **Figure 3b** E is the energy from the Earth, or at least external to the system. The implicit energy intensity of energy itself is 1.0. Imported materials already have an energy intensity because of their production elsewhere. This is discussed in more detail in the section on dynamic indicators. Let us apply this to a specific set of flows shown in **Figure 4a**. For conciseness, feedback will be denoted by Z.

Equation [1] gives one equation for each compartment:

$$100 + Z\varepsilon_c = 10\varepsilon_p$$
$$10\varepsilon_p = (5 + Z)\varepsilon_c \qquad [2]$$

(a)

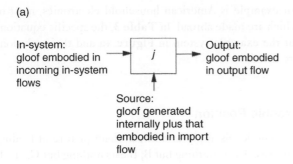

In-system: gloof embodied in incoming in-system flows → j → Output: gloof embodied in output flow

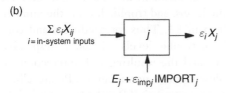

Source: gloof generated internally plus that embodied in import flow

(b)

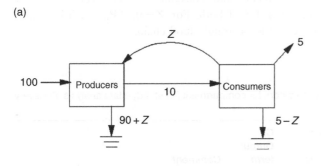

$\Sigma\ \varepsilon_i X_{ij}$
$i =$ in-system inputs → j → $\varepsilon_i X_j$

$E_j + \varepsilon_{\text{imp}j} \text{IMPORT}_j$

Figure 3 (a) Generic scheme for allocating the embodied generic system input gloof for steady-state system. Gloof may be energy, nutrient, or even time (for calculating residence time). In economic applications, gloof may be money, labor, or pollution assimilation. For each compartment j, embodied gloof in = embodied gloof out, by assumption. (b) Embodied energy balance equation for compartment j, which follows from (a). X_{ij} are in-system inputs to j; X_j is j's total output; E_j is the (direct) energy input to j.

(a)

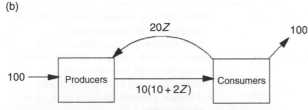

Z

100 → Producers → 10 → Consumers → 5

90 + Z 5 − Z

(b)

20Z 100

100 → Producers → Consumers → 100

10(10 + 2Z)

Figure 4 (a) Explicit energy flows (cal/day) for a two-compartment system. Feedback, Z, can vary between 0 and 5 cal/day. Example is arbitrary, but the output/input ratio of 0.1 for producers is appropriate for many real systems. (b) Embodied energy flows (cal GPP/day) in system shown in (a) as a function of feedback, Z. Both compartments are in embodied energy balance, as is the entire system.

The solution is $\varepsilon_p = 10 + 2Z$, $\varepsilon_c = 20$, both expressed in cal GPP/cal biomass. Substituting these energy intensities in **Figure 3b** gives the embodied energy flows shown in **Figure 4b**. The entire system is in embodied energy

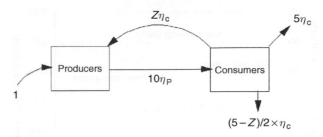

$Z\eta_c$ $5\eta_c$

Producers → 10η_P → Consumers

1

$(5 - Z)/2 \times \eta_c$

Figure 5 Embodied nutrient flows (g/day) in system of **Figure 4a**. η (g/cal) are the nutrient intensities.

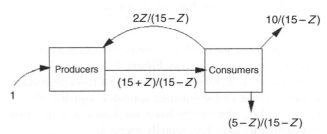

$2Z/(15 - Z)$ $10/(15 - Z)$

Producers → $(15 + Z)/(15 - Z)$ → Consumers

1

$(5 - Z)/(15 - Z)$

Figure 6 Embodied nutrient flows (g/day) in system shown in **Figure 4a** as a function of feedback, Z. Both compartments are in embodied nutrient balance, as is the entire system.

balance, a consequence of the assumption that each individual compartment is in balance.

Now suppose we want to track the flow of nutrient. Nutrient intensities (η) will be expressed in grams of nutrient/cal of biomass. Here we will (arbitrarily) assume that nutrient is taken up only by producers and passed on without loss to consumers, but that some nutrient is leaked from consumers during metabolism. Specifically, assume that the nutrient intensity of the consumer metabolic loss is half that of the consumer export flow. This is shown in **Figure 5**.

The balance equations are then:

$$1 + Z\eta_c = 10\eta_p$$

$$10\eta_p = 5\eta_c + Z\eta_c + \frac{(5 - Z)}{2}\eta_c \qquad [3]$$

which are solved to yield $\eta_p = (1/10)(15 + Z)/(15 - Z)$, $\eta_c = 2/(15 - Z)$. Using these intensities yields **Figure 6** for embodied nutrient flow.

The system is in embodied nutrient balance, but the flows have a surprising feature: for $Z > 0$, the nutrient flow from producers to consumer (an internal flow) exceeds 1 g/day, the system's input flow. Critics have called this apparent contradiction a damning flaw of the method, but actually it is to be expected. Feedback speeds up a system's flows: more molecules pass a given point per unit time. Because here embodied nutrient is actual nutrient, the effect could be measured experimentally. This validates the method generally.

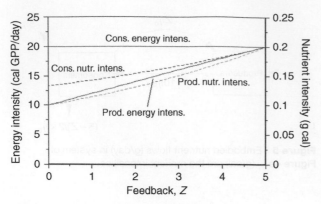

Figure 7 Energy and nutrient intensities vs. Z.

Energy intensities and nutrient intensities as a function of feedback, Z, are shown in **Figure 7**.

As feedback increases (and hence loss from consumers decreases), the two intensities approach equality. When $Z = 5$ cal/day, consumers have no losses and the two compartments have functionally merged.

Steady-State Analysis: Other Indicators

Below are discussed three other indicators: TP, PL, and residence time. In **Table 2**, the equations for each are listed. The possibility of imports is explicitly allowed for.

An example is American household electronics, most of which are made abroad. In **Table 3**, the specific equations for the example system in **Figure 4a** and their solutions are listed.

Trophic Position

Trophic levels apply to a linear chain picture of feeding patterns: A eats nothing but B; B eats nothing but C, etc. If there are n compartments in the chain, then there are n integral trophic levels, and trophic level is the number of steps from the Sun $+ 1$. Thus for producers and consumers in a chain, trophic levels $= 1$ and 2, respectively. With omnivory and the resulting web interactions, this view breaks down unless nonintegral TPs are allowed. Simply put, a compartment's TP is the (energy) weighted average of the TPs of each of its inputs plus 1. Caution: trophic interactions are always expressed in energy flows, so here one must use energy flows only, not nutrients or other flows. (There is also a dual approach, which results in an infinite series of integral trophic levels, which is not covered here.)

The standard convention of setting $TP_{Sun} = 0$ is used. From **Table 3**, the TPs are $1 + 2Z/100$ and $2 + 2Z/100$ for producers and consumers, respectively. Feedback increases TP of both. For $Z = 0$, $TP_p = 1$, $TP_c = 2$, as expected for a straight food chain.

Table 2 Explicit forms of the input and outputs in calculating several indicators for compartment j in steady-state analysis. For every indicator, the equation to solve is Col A + Col B = Col C

Indicator	A. In-system inputs term	B. Source term: Internal or imported inputs	C. Output term	Comment
Energy intensity (ε)	$\sum\limits_{i=\text{in-system inputs}} \varepsilon_i X_{ij}$	$\varepsilon_{impj} * \text{IMPORT}_j + E_j$	$\varepsilon_j X_j$	Flows can have different units for different compartments. Intensities will correspondingly have different units
Nutrient intensity (η)	$\sum\limits_{i=\text{in-system inputs}} \eta_i X_{ij}$	$\eta_{impj} * \text{IMPORT}_j + N_j$	$\eta_j X_j$	Flows can have different units for different compartments. Intensities will correspondingly have different units
Trophic position (TP)	$\dfrac{\sum\limits_{i=\text{in-system inputs}} TP_i X_{ij} + TP_{impj} \text{IMPORT}_j}{\sum\limits_{i=\text{in-system inputs}} X_{ij} + \text{IMPORT}_j + E_j}$	1	TP_j	For trophic position, all flows must be in energy terms
Path length (PL)	$\dfrac{\sum\limits_{i=\text{in-system inputs}} (PL_i + 1) X_{ij}}{\sum\limits_{i=\text{in-system inputs}} X_{ij} + \text{IMPORT}_j + E_j}$	None	PL_j	Path length is almost the same as trophic position. Flows need not be energy but must have the same units for every compartment
Residence time (τ)	$\dfrac{\sum\limits_{i=\text{in-system inputs}} \tau_i X_{ij}}{\sum\limits_{i=\text{in-system inputs}} X_{ij} + \text{IMPORT}_j + E_j}$	t_j	τ_j	t_j is the isolated compartment residence time for compartment j ($=$ stock/throughflow), assumed to be constant. Flows need not be energy but must have the same units for every compartment

X_{ij}, flow of i to j; E_j, energy flow to j; N_j, nutrient flow to j.
See Table 1 for additional definitions.

Table 3 Explicit balance equations and solutions for five indicators

Indicator	Refer to figure	Equations	Solutions	Units
Energy intensity (ε)	4a	$100 + Z\varepsilon_c = 10\varepsilon_p$ $10\varepsilon_p = (5 + Z)\varepsilon_c$	$\varepsilon_p = 10 + 2Z$ $\varepsilon_c = 20$	cal GPP/cal
Nutrient intensity (η)	5	$1 + Z\eta_c = 10\eta_p$ $10\eta_p = 5\eta_c + Z\eta_c + \dfrac{(5-Z)}{2}\eta_c$	$\eta_p = \dfrac{1}{10}\dfrac{15+Z}{15-Z}$ $\eta_c = \dfrac{2}{15-Z}$	g nutrient/cal
Trophic position (TP)	4a	$\dfrac{TP_c Z}{100 + Z} + 1 = TP_p$ $\dfrac{TP_p 10}{10} + 1 = TP_c$	$TP_p = 1 + \dfrac{2Z}{100}$ $TP_c = 2 + \dfrac{2Z}{100}$	
Path length (PL)	4a	$(PL_c + 1)\dfrac{Z}{100 + Z} = PL_p$ $PL_p + 1 = PL_c$	$PL_p = \dfrac{2Z}{100}$ $PL_c = 1 + \dfrac{2Z}{100}$	
Residence time (τ)	4a	$\tau_c \dfrac{Z}{100 + Z} + t_p = \tau_p$ $\tau_p + t_c = \tau_c$	$\tau_p = \left(1 + \dfrac{Z}{100}\right)t_p + \dfrac{Z}{100}t_c$ $\tau_c = \left(1 + \dfrac{Z}{100}\right)(t_p + t_c)$	day

Path Length

This can be expressed in two ways:

- *Looking backward in time.* A molecule is just now leaving compartment j. How many intercompartment transits has it made between its entering the system and now?
- *Looking forward in time.* A molecule is just now leaving compartment j. How many intercompartment transits will it make on average before exiting the system?

For a steady-state system, these are equally easy to calculate. For a dynamic system, the backward-looking PL is preferable because it can be calculated without knowing the future. Therefore we calculate only the backward-looking PL. In words, PL is the weighted sum of the quantity $(PL + 1)$ for each input. Imports do not figure in PL, which is based upon internal flows only. The input flows need not be energy, but they must all be in the same units so that the weighted average can be calculated.

PL is almost the same as TP. For a system with only sunlight as energy input, $TP_i = PL_i + 1$. If there are other system energy inputs such as imported feed, the difference between the two is more significant. As shown in **Table 3**, $PL = 2Z/100$ and $1 + 2Z/100$ for producers and consumers, respectively. For $Z = 0$, $PL_p = 0$ because there are no in-system inputs to producers.

Residence Time (τ)

As with PL, this can be expressed looking either backward or forward in time, but here only the former is treated: a molecule is just now leaving compartment j. How long has it been in the system? It is assumed that we know the isolated compartment residence times t_i, typically defined as the ratio of stock to throughflow. (Unlike all the indicators so far discussed, this requires that we know the stocks at steady state.) The system residence time is a function of these isolated-compartment residence times and the degree of connectedness of the compartments. In words, residence time for compartment j is the weighted average of the residence time of each input $+ t_j$. From **Table 3**, the residence times are $(1 + Z/100)t_p + (Z/100)t_c$ and $(1 + Z/100)(t_p + t_c)$ for producers and consumers, respectively. Without feedback, the residence time for producers, τ_p, is just t_p because the only input is from outside the system. Consumer residence time, τ_c, is $t_p + t_c$, because a molecule leaving consumers has passed exactly once through producers and consumers. TP and residence time are graphed versus Z in **Figure 8**.

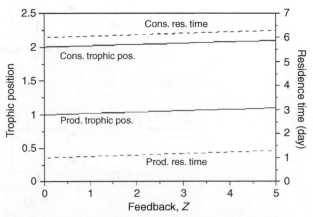

Figure 8 Trophic position and residence time vs. Z. The isolated compartment residence times are 1 and 5 days for producers and consumers, respectively.

Indicators in Dynamic Systems

Calculating Dynamic Indicators

Most of energy analysis and systems ecology stressing indirectness has assumed a steady state in which flows and stocks are constant over time. Yet real systems are almost always dynamic. All the indicators addressed in this article can have a dynamic interpretation, as long as we use the back-looking form. Any dynamic analysis must be explicit about stocks, flows, and time steps. The elements of the dynamic view are shown in **Figure 9**.

Figure 9 summarizes the assumption that in a time step Δt, the gloof embodied in the inflows and in the stock is distributed over the final stock and outflows. At the end of the time step, there has been the mathematical equivalent of perfect mixing, so that the energy intensity

of stock and output are the same. **Figures 10a** and **10b** illustrate this in detail for energy intensity and residence time.

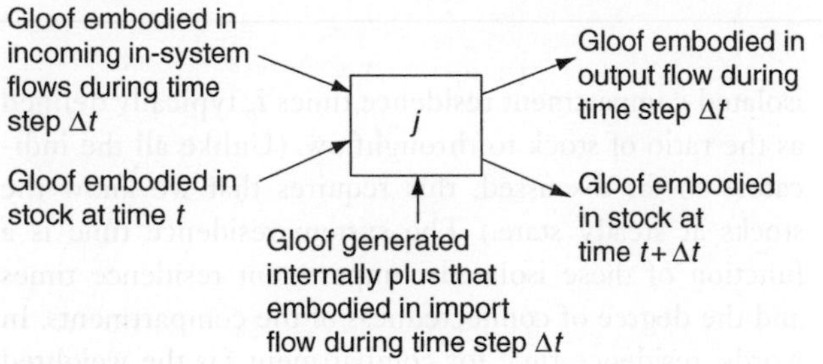

The flows are multiplied by the time step Δt for dimensional commensurateness with the stocks. Output can include a change in stock (inventory change in economic terminology), so that the new stock is the old stock plus this change. **Figure 10a** shows that energy intensity at time $t + \Delta t$ is a function of the flows at time $t + \Delta t$, and the stocks and energy intensities at time t. If one knows the initial energy intensities and stocks, and one has a dynamic model to specify stocks and flows over time, one can use the equation implied by **Figure 10a** to calculate dynamic energy intensities:

$$\sum_{i=\text{in-system inputs}} \varepsilon_i^{t+\Delta t} X_{ij}^{t+\Delta t} \Delta t + \varepsilon_j^t S_j^t + \varepsilon_{\text{imp}j}^{t+\Delta t} \text{IMPORT}_j^{t+\Delta t} \Delta t$$

$$+ E_j^{t+\Delta t} \Delta t = \varepsilon_j^{t+\Delta t} X_j^{t+\Delta t} \Delta t$$

$$+ \varepsilon_j^{t+\Delta t} S_j^t \qquad [4]$$

Similarly, from **Figure 10b**, one obtains for dynamic residence time:

$$\sum_{i=\text{in-system inputs}} \tau_i^{t+\Delta t} X_{ij}^{t+\Delta t} \Delta t + \tau_j^t S_j^t + S_j^t \Delta t$$

$$= \tau_j^{t+\Delta t} \left(\sum_{i=\text{in-system inputs}} X_{ij}^{t+\Delta t} \Delta t + \text{IMPORT}_j + E_j \right) + \tau_j^{t+\Delta t} S_j^t$$

$$[5]$$

Figures 10a and **10b** also demonstrate how intensity (of anything) is injected into a system as a source term

Gloof embodied in incoming in-system flows during time step Δt

Gloof embodied in stock at time t

j

Gloof generated internally plus that embodied in import flow during time step Δt

Gloof embodied in output flow during time step Δt

Gloof embodied in stock at time $t+\Delta t$

Figure 9 Generic scheme for allocating the embodied generic system input gloof for the dynamic system. In the underlying dynamics, stock changes over time as $S_{t+\Delta t} = S_t + (\text{OUTPUT}_{t+\Delta t} - \sum \text{INPUTS}_{t+\Delta t}) \Delta t$.

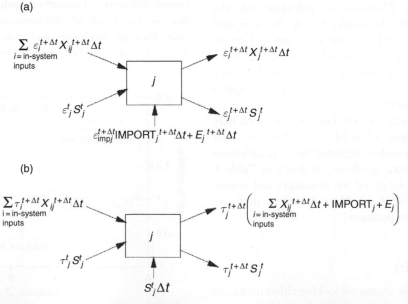

(a)

$$\sum_{\substack{i=\text{in-system} \\ \text{inputs}}} \varepsilon_i^{t+\Delta t} X_{ij}^{t+\Delta t} \Delta t$$

$$\varepsilon_j^t S_j^t$$

j

$$\varepsilon_{\text{imp}j}^{t+\Delta t} \text{IMPORT}_j^{t+\Delta t} \Delta t + E_j^{t+\Delta t} \Delta t$$

$$\varepsilon_j^{t+\Delta t} X_j^{t+\Delta t} \Delta t$$

$$\varepsilon_j^{t+\Delta t} S_j^t$$

(b)

$$\sum_{\substack{i=\text{in-system} \\ \text{inputs}}} \tau_i^{t+\Delta t} X_{ij}^{t+\Delta t} \Delta t$$

$$\tau_j^t S_j^t$$

j

$$S_j^t \Delta t$$

$$\tau_j^{t+\Delta t} \left(\sum_{\substack{i=\text{in-system} \\ \text{inputs}}} X_{ij}^{t+\Delta t} \Delta t + \text{IMPORT}_j + E_j \right)$$

$$\tau_j^{t+\Delta t} S_j^t$$

Figure 10 (a) Scheme for calculating dynamic energy intensity. The source term is energy itself plus energy embodied in imports. The energy intensity of imports of type j, $\varepsilon_{\text{imp}j}$, is specified exogenously. (b) Scheme for calculating dynamic residence time. The source term is the aging of the existing stock in the time period Δt.

and then allocated by internal flows. For energy intensity, the source is the embodied energy in imports of similar entities (competitive imports in economic terminology) plus imports of different entities, here just energy itself.

For residence time, the source term is just the aging of the stock; external inputs do not contribute to residence time by definition. Similar comments apply to TP and PL.

Simulations of Dynamic Indicators

Figure 11 shows a two-compartment dynamic model system. Initially the system is at steady state with no feedback ($Z = 0$), but feedback is switched on ($Z = 3$) at time = 20 days, and then off again ($Z = 0$) at 500 days. The details of the underlying model are not important here; it incorporates a nonlinear ratio-dependent feeding response by consumers to abundance of producers, and vice versa when feedback is on. Producer output depends on light level, which is assumed constant, and producer biomass. Simulation is performed using the modeling software STELLA.

Figures 12a–12d show dynamic behavior of four of the indicators calculated here: energy intensity, TP, PL, and residence time. On all graphs, the stock of producers and consumers is shown as well. Immediately after the onset of nonzero feedback, producer stock increases as more material now enters that compartment and consumer stock drops. But then consumer stock increases in response to increased producer stock, and both stocks asymptotically increase. This is reasonable, because along with increased feedback comes decreased loss, as shown in **Figure 11**.

Similar to the steady-state calculations, energy intensity, TP, PL, and residence time all increase for both producers and consumers. However, the values are not given exactly by the static equations for $Z = 3 \, \text{cal d}^{-1}$. This is because in the dynamic model, all flows and stocks

change when feedback changes, while in the static model used in the previous two sections all flows except feedback are assumed to remain constant.

Applications

Ecological Example: Four-Compartment Food Web

Figure 13 shows steady-state energy flows and stocks in a bog in Russia. The analysts disaggregated this ecosystem into four compartments: plants, animals, decomposers, and detritus. Detritus consists of undifferentiated dead material and therefore has no metabolic losses. All other compartments contribute to detritus. Additionally, animals and decomposers also eat detritus, resulting in two feedback flows and a web structure. All indicators can be calculated from the equations given in **Table 2**. The results are given in **Table 4**. Because plants have only a solar input, their energy intensity is quite low and TP $= 1$. For the other compartments, however, the energy intensities are higher and the TPs are high. Decomposers have a TP $= 4.9$, higher than the value of 4 which one would expect for a food chain instead of this web. Decomposers come out on top in both energy intensity and TP. Because this system has only one input, PL is just TP-1.

Table 4 shows that residence times are affected dramatically by web structure. Isolated compartment residence times ($=$stock/throughflow) are long for plants and detritus, and short for animals and decomposers. The longest, detritus, is 760 times the shortest, animals. In contrast, in-system residence times differ by only a factor of 4. Both animals and decomposers, which in isolation would be fast, have large input flows from detritus (which in isolation is slow). The consequence is that all three are comparably slow. This is one aspect of the notion that detritus links tend to slow down the response of ecosystems to perturbations.

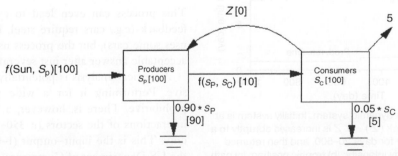

Figure 11 Model for dynamic simulation. Figures in square brackets are initial steady-state values, before feedback is started. Figures within boxes are stocks; others are flows.

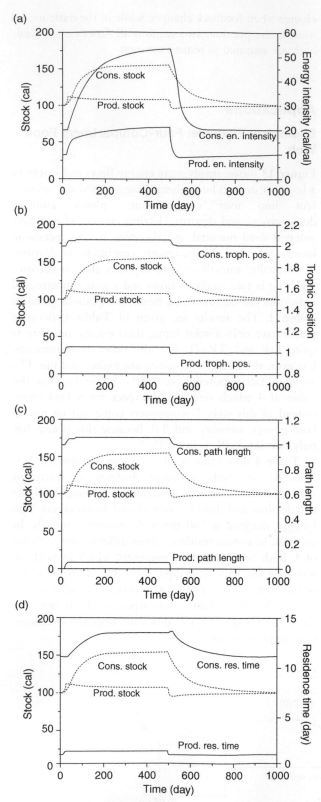

(a)

(b)

(c)

(d)

Figure 12   Indicators in dynamic system. Initially system is at steady state with feedback (Z) = 0. Z is increased abruptly to a steady value of 3 cal/day for days 20–500, and then returned abruptly to zero. (a) energy intensity; (b) trophic position; (c) path length; (d) residence time.

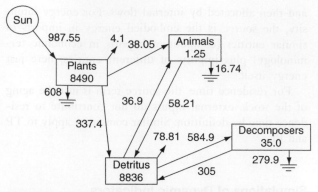

Figure 13   Energy flows (g fixed carbon per m² per year) in a bog in Russia. Detritus is undifferentiated dead material, and therefore has no metabolic loss. Numbers in compartments are stocks (g fixed carbon per m²). From Logofet DO and Alexandrov GA (1984) Modelling of matter cycle in a mesotrophic bog ecosystem. Part 1: Linear analysis of carbon environs. *Ecological Modelling* 21: 247–258.

Energy/Economic Example: Energy Intensity of Consumer Goods and Services, Energy Cost of Living

This topic is included to emphasize and illustrate the breadth of applicability of energy analysis and the analogies between ecological and economic systems. The question is how much energy is required to support, directly and indirectly, human household consumption patterns. The approach is in two steps: (1) determine how much energy is needed, directly and indirectly, to produce a product and (2) determine how much of it a household consumes.

Consider a loaf of bread. The energy to grow the ingredients, make the bread, and transport and market it can be determined by a detailed vertical analysis (also called process analysis), in which one sums:

1. the energy used in the supermarket;
2. the energy consumed in the bakery;
3. the energy consumed at the flour mill;
4. the energy used on the farm;
5. the energy for transport at every link; and so on.

This process can even lead to cycles in systems with feedback (e.g., cars require steel, but the steel industry uses some cars), but the process usually converges to an acceptable answer after just several steps.

A vertical analysis is potentially accurate, but expensive. Performing it for a wide range of products is prohibitive. There is, however, a large database on the interactions of the sectors (c. 350–500) of the US economy. This is the input–output (I–O) table published by the US Department of Commerce. Many other countries have similar I–O tables. With a number of fairly

Table 4 Energy intensities, trophic positions, path lengths, and residence times for the Russian bog food web of **Figure 13**

Compartment	Energy intensity, ε (cal GPP/cal)	Trophic position (TP)	Path length (PL)	Isolated-compartment residence time, t (years)	In-system residence time, τ (years)
Plants	2.60	1.00	0.00	8.60	8.60
Animals	9.56	3.43	2.43	0.017	20.5
Detritus	12.4	3.90	2.90	12.6	32.7
Decomposers	23.8	4.90	3.90	0.060	32.8

stringent assumptions, this table can be combined with direct energy use data for each sector to produce energy intensities using the equation implied by **Figure 3b**. One such assumption is necessitated by the fact that the units in I–O tables are monetary units per year, so one must accept dollars as an appropriate allocator of embodied energy. Because in the American energy industry, energy is usually measured in Btu, the energy intensity of goods and services is then expressed in Btu/$. I–O-based determination of energy intensities has been performed for c. 35 years. Under further assumptions, the intensities can be used to evaluate the energy impact of different expenditure patterns. Doing this for a household yields the so-called energy cost of living.

The I–O data are available, but gathering the associated direct energy data and performing the computation is tedious, though today's computers make it increasingly easier. Solving 500 simultaneous equations of the form in eqn [1] is done by inverting a 500-rank matrix.

Once we have the energy intensities, we need details on how households spend their money over the range of consumer product categories, also known as their market basket. This information is collected by the US Bureau of Labor Statistics. Putting the two together yields the energy cost of living (ECOL):

$$\text{ECOL} = \sum_{i=\text{all expenditure categories}} \varepsilon_i Y_i \qquad [6]$$

where Y_i is the household's annual expenditure for expenditure category i. Applying eqn [6] allows one to analyze the effect of overall spending and the mix in the market basket. The latter will be significant only if the energy intensities are different for different expenditure categories.

Table 5 shows I–O-based energy intensities determined by Carnegie Mellon University for 1997, and updated and aggregated by the author into 15 categories covering all household expenditures. The intensities are indeed different, especially energy itself and service industries such as health care.

Figure 14 shows the result of transforming of a household market basket (in dollars/yr) to its energy impact (in

Table 5 Energy intensities for household consumption categories (Btu/$, 2003 technology, 2003 dollars)

1. Residential fuel, electricity	139 300
2. Vehicle fuel	94 300
3. Vehicle purchase, maintenance	5 400
4. Food	6 100
5. Alcohol, tobacco	3 700
6. Apparel	6 500
7. Communication, entertainment	4 000
8. Health, personal care	2 400
9. Reading, education	3 000
10. Insurance, pension	1 600
11. Contributions	3 800
12. Public transportation	21 200
13. Asset gain	4 700
14. Miscellaneous	4 200
15. Housing	5 100
Direct energy ((1) + (2))	118 100
Nonenergy (sum of (3)–(15))	4 700
All personal consumption	11 100
Energy/GDP	8 900
Sprawl ((1) + (2) + (3) + (15))	15 700
Nonsprawl (sum of (4)–(14))	4 300
Auto and related ((2) + (3))	21 200

Shaded categories indicate an intensity greater than the energy/GDP ratio. 'Sprawl' contains housing and auto ownership and operation.
Source: Author's calculations based on Carnegie Mellon University data.

Btu/yr) using eqn [6] and intensities such as those shown in **Table 5**. In **Figure 14a**, we see that of the average household's expenditures of $49 300 in 1973, only 6.4% was for direct energy (residential fuel and electricity and auto fuel). After conversion to energy requirements (**Figure 14b**), this portion was 63% of the total impact of 604 million Btu. The total is roughly the energy equivalent of 100 barrels of oil. **Figures 15a** and **15b** show the energy pie for the lowest expenditure decile ($11 500/yr, 241 million Btu/yr) and highest decile ($140 200/yr, 1233 million Btu/yr). The direct fraction is largest, 79%, for the lowest decile, and lowest, 47%, for the highest decile.

Figure 16 shows a statistical fit to energy versus expenditures for a representative sample of several thousand American households. It confirms that because the mix changes, energy is not a linear function of total

(a)

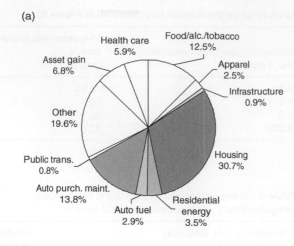

(b)

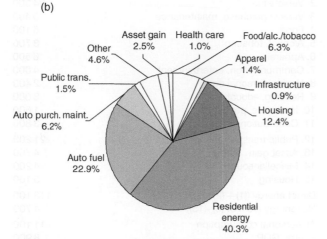

Figure 14   For the average American household in 2003:
(a) expenditures, which total $49 300; (b) energy impacts, which
totaled 604 million Btu. Source: Unpublished calculations by
R. Shammin, R. A. Herendeen, M. Hanson, and E. Wilson.

(a)

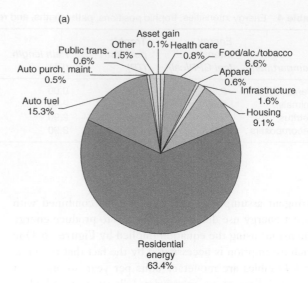

(b)

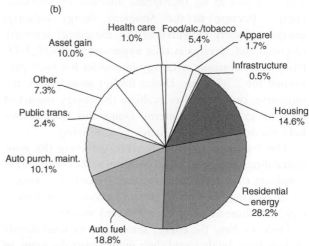

Figure 15   2003 household energy impacts for (a) lowest
income decile ($11 500; 241 million Btu), (b) highest income decile
($140 200; 1233 million Btu). Source: Unpublished calculations by
R. Shammin, R. A. Herendeen, M. Hanson, and E. Wilson.

expenditures, but rather bends down and away from a
straight line through the origin. The reason is that direct
energy (auto fuel and residential fuel and electricity)
tends to level out as household expenditures increase.
Expenditures increase for other products, but these
tend to be less energy intensive. For developed countries,
the shape of **Figure 16** seems robust: studies of
Norway, the Netherlands, and Australia have found a
similar result.

Energy/Economic Example: Regressive Effects of an Energy Tax

Concern with global warming, energy security, and pollution strongly implies that fossil energy is too cheap to
compensate for its drawbacks. Energy taxes of various
sorts have been proposed to stimulate more efficient use
and to fund alternatives. In the debate, equity issues
quickly surface. Because direct energy is a larger fraction

of the total for less affluent households, regressive impacts
of an energy price increase would be expected if one
ignored the indirect portion.

However, because the total energy curve in **Figure 16**
bends down, some regressiveness is still expected. To
compensate, one could design an income tax rebate to
even out the impacts over income classes. **Figure 16** is the
key, as follows.

Suppose that fossil energy at the wellhead or mine
mouth is taxed at rate of p dollars per Btu. Assume that
economic sectors maintain their patterns of using inputs
to produce inputs, that is to say, technology is constant.
Assume further that each sector can successfully pass on
its increased costs to the consumers of its output. Then a
household's market basket, if unchanged, will now cost an
additional amount $= p \times$ ECOL. The fractional increase

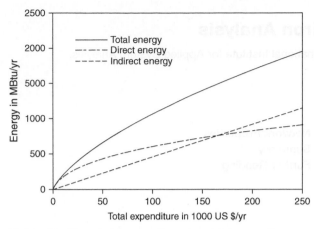

Figure 16 Household energy impact vs. total expenditures. Direct energy is auto and residential fuel and electricity. Total energy = direct energy plus energy impact of all other purchases. Source: Unpublished calculations by R. Shammin, R. A. Herendeen, M. Hanson, and E. Wilson.

is this quantity divided by the total market basket's original cost, denoted by Y. then

$$\text{Fract.incr.in mkt. basket cost} = \frac{p * \text{ECOL}}{Y} = p\langle\varepsilon\rangle \quad [7]$$

where $\langle\varepsilon\rangle$ is the average energy intensity of the market basket. The average is just energy/expenditure at the appropriate point on **Figure 16**; it is the slope of a straight line connecting the origin and the point.

As an example, consider a tax of $0.50 per gallon of gasoline equivalent. This is about $4/million Btu, or $24/barrel of oil. The latter is about 25% of the world crude oil price as of 7 November 2007.

Using **Figures 14–16**, we perform the calculation in **Table 6**. The increase in market basket price ranges from 3.5% for the highest expenditure decile to 8.9% for the lowest. For full equity, income tax rebates, or other measures, could address this differential.

Needless to say, a proper calculation is much more involved than this one, but the idea of indirectness will pervade it.

See also: Cycling and Cycling Indices; Ecological Network Analysis, Ascendency; Ecological Network Analysis, Environ Analysis; Indirect Effects in Ecology.

Table 6 Consequences of a $4 per million Btu mine mouth fossil energy tax, based on assumptions in text

	Expenditure level		
	Lowest decile	Average	Highest decile
Market basket expenditure (thousand $/yr)	11.5	49.3	140.2
Total energy (million Btu/yr)	241	604	1233
$\langle\varepsilon\rangle$ (thousand Btu/$)	21.0	12.3	8.8
Market basket price increase ($/yr)	964	2416	4932
Market basket price increase (%)	8.4	4.9	3.5

Further Reading

Bullard C, Penner P, and Pilati D (1978) Net energy analysis: Handbook for combining process and input–output analysis. *Resources and Energy* 1: 267–313.

Burns T (1989) Lindeman's contradiction and the trophic structure of ecosystems. *Ecology* 70: 1355–1362.

Fath BD and Patten BC (1999) Network synergism: Emergence of positive relations in ecological models. *Ecological Modelling* 107: 127–143.

Finn JT (1976) Measures of ecosystem structure and function derived from analysis of flows. *Journal of Theoretical Biology* 56: 115–124.

Hannon B (1973) The structure of ecosystems. *Journal of Theoretical Biology* 41: 535–546.

Herendeen R (1989) Energy intensity, residence time, exergy, and ascendency in dynamic ecosystems. *Ecological Modelling* 48: 19–44.

Herendeen R and Fazel F (1984) Distributional aspects of an energy conserving tax and rebate. *Resources and Energy* 6: 277–304.

Herendeen R, Ford C, and Hannon B (1981) Energy cost of living, 1972–1973. *Energy* 6: 1433–1450.

Lenzen M, Wier M, Cohen C, et al. (2006) A comparative multivariate analysis of household energy requirements in Australia, Brazil, Denmark, India, and Japan. *Energy* 31: 181–207.

Logofet DO and Alexandrov GA (1984) Modelling of matter cycle in a mesotrophic bog ecosystem. Part 1: Linear analysis of carbon environs. *Ecological Modelling* 21: 247–258.

Odum HT (1996) *Environmental Accounting*. New York: Wiley.

Ulanowicz R (1986) *Growth and Development: Ecosystems Phenomenology*. New York: Springer.

Relevant Website

http://www.eiolca.net – Economic Input–Output Life Cycle Assessment, Carnegie Mellon University.

Ecological Network Analysis, Environ Analysis

B D Fath, Towson University, Towson, MD, USA and International Institute for Applied System Analysis, Laxenburg, Austria

Introduction

Environ Analysis is in a more general class of methods called ecological network analysis (ENA) which uses network theory to study the interactions between organisms or populations within their environment. Bernard Patten was the originator of the environ analysis approach in the late 1970s, and he, along with his colleagues, has expanded the analysis to reveal many insightful, holistic properties of ecosystem organization. ENA follows along the synecology perspective introduced by E. P. Odum which is concerned with interrelations of material, energy, and information among system components.

ENA starts with the assumption that a system can be represented as a network of nodes (compartments, vertices, components, storages, objects, etc.) and the connections between them (links, arcs, flows, etc.). In ecological systems, the connections are based on the flow of energy, matter, or nutrients between the system compartments. If such a flow exists, then there is a direct transaction between the two connected compartments. These direct transactions give rise to both direct and indirect relations between all the objects in the system. Network analysis provides a systems-oriented perspective because it uncovers patterns and relations among all the objects in a system. Therefore, showing how system components are tied to a larger web of interactions.

Theoretical Development of Environ Analysis

Patten was motivated to develop environ analysis to answer the question, "What is environment?". In order to study environment as a formal object, a system boundary is a necessary condition to avoid the issue of infinite indirectness, because in principle one could trace the environment of each object out in space and back in time to the big-bang origins. The inclusion of a boundary is, in fact, one of the three foundational principles in his seminal paper introducing the environ theory concept. The necessary boundary demarcates two environments, the unbound external environment, which indeed includes all space–time objects in the universe, and the second internal, contained environment of interest. This quantifiable, internal environment for each system object is termed 'environ', and is the focus of environ analysis. An object's environ stops at the system boundary, but as ecosystems are open systems, they require exchanges across the boundary into and out of the system environs. Therefore, input and output boundary flows are necessary to maintain the system's far-from-equilibrium organization. Objects and connections that reside wholly in the external environment are not germane to the analysis.

Another foundational principle of environ analysis theory is that each object in the system itself has two 'environs' one receiving and one generating interactions in the system. In other words, an object's input environ includes those flows from within the system boundary leading to the object, and an output environ, those flows emanating from the object back to the other system objects before exiting the system boundary. This alters the perception of a system component from internal–external to receiving–generating. Thus, the object, while distinct in time and space, is more clearly embedded in and responsive to the couplings with other objects within the network. This shifts the focus from the objects themselves to the relations they maintain; or from parts to processes (or what Ilya Prigogine called from 'being' to 'becoming').

The third foundational principle is that individual environs (and the flow carried within each one) are unique such that the system comprises the set union of all environs, which in turn partitions the system level of organization. This partitioning allows one to classify environ flow into what have been called different modes: (1) boundary input; (2) first passage flow received by an object from other objects in the system (i.e., not boundary flow), which has not cycled; (3) cycled flow, which returns to a compartment before leaving the system; (4) dissipative flow that it has left the focal object not to return, but does not directly cross a system boundary (i.e., it flows to another within system object); and (5) boundary outflow. The modes have been used to understand better the general role of cycling and the flow contributions from each object to the other, which has had application in showing a complementarity of several of the holistic, thermodynamic-based ecological indicators.

Data Requirements and Community Assembly Rules

Network environ analysis could be referred to as a holistic/reductionistic approach. It is holistic because it considers simultaneously the whole influence of all system objects, yet it is reductionistic in that the fine details of all object transactions are entailed in the analysis. In other words, it is the opposite of a black box model. The network data requirements are considerable, which include the complete flow–storage quantities for each identified link and node (note flow and storage are interchangeable as determined by the turnover rate). Data can be acquired from empirical observations, literature estimates, model simulation results, or balancing procedures, when all but a few are unknown. This difficulty in obtaining data has resulted in a dearth of available complete network data sets. Due to this lack of requisite data for fully quantified food webs, researchers have developed community assembly rules that are heuristics to construct ecological food webs. Assembly rules are in general a set of rules that will generate a connectance matrix for a number of species (N). Common assembly rules that have been developed are random or constant connectance, cascade, niche, modified niche, and cyber-ecosystem, each with its own assumptions and limitations. In all but the last case, the assembly rules construct only the structural food web topology. The cyber-ecosystem methodology also includes a procedure for quantifying the flows along each link. It uses a metastructure of six functional groups: producer (P), herbivore (H), carnivore (C), omnivore (O), detritus (D), and detrital feeders (F), within which random connections link species based on these definitional constraints. Flows are assigned based on realistic thermodynamic constraints.

Methods and Sample Network

To demonstrate basic environ analysis, it is best to proceed with an example. Consider the network in **Figure 1**, which has five compartments or nodes (x_i, for $i = 1$–5).

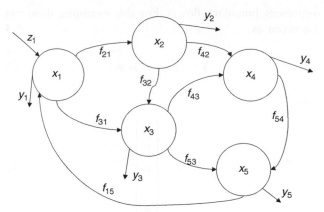

Figure 1   Sample network with five compartments used to demonstrate environ analysis notation and methodology.

Compartments are connected by transaction of the energy–matter substance flowing between them. These pairwise couplings are the basis for the internal network structure. A structural connectance matrix, or adjacency matrix, **A**, is a binary representation of the connections such that $a_{ij} = 1$ if there is a connection from j to i, and a 0 otherwise (eqn [1]):

$$\mathbf{A} = \begin{bmatrix} 0 & 0 & 0 & 0 & 1 \\ 1 & 0 & 0 & 0 & 0 \\ 1 & 1 & 0 & 0 & 0 \\ 0 & 1 & 1 & 0 & 0 \\ 0 & 0 & 1 & 1 & 0 \end{bmatrix} \quad [1]$$

Storage and flows must have consistent units (although it is possible to consider multiunit networks). Typically, units for storages are given in amount of energy or biomass per given area or volume (e.g., $g\,m^{-2}$), and units for flows are the same but as a rate (e.g., $g\,m^{-2}\,d^{-1}$). The intercompartmental flows for **Figure 1** are given in the following flow matrix, **F**:

$$\mathbf{F} = \begin{bmatrix} 0 & 0 & 0 & 0 & f_{15} \\ f_{21} & 0 & 0 & 0 & 0 \\ f_{31} & f_{32} & 0 & 0 & 0 \\ 0 & f_{42} & f_{43} & 0 & 0 \\ 0 & 0 & f_{53} & f_{54} & 0 \end{bmatrix} \quad [2]$$

Note that the orientation of flow from j to i is used because that makes the direction of ecological relation from i to j. For example, if i preys on j, the flow of energy is from j to i. All compartments experience dissipative flow losses (y_i, for $i = 1$–5), and here the first compartment receives external flow input, z_1 (arrows not starting or ending on another compartment represent boundary flows). For this example, these can be given as

$$\mathbf{y} = [y_1 \ y_2 \ y_3 \ y_4 \ y_5] \qquad [3]$$

and

$$\mathbf{z} = \begin{bmatrix} z_1 \\ 0 \\ 0 \\ 0 \\ 0 \end{bmatrix} \qquad [4]$$

Total throughflow of each compartment is an important variable, which is the sum of flows into, $T_i^{\text{in}} = z_i + \sum_j^n f_{ij}$, or out of, $T_i^{\text{out}} = y_i + \sum_j^n f_{ji}$ the ith compartment. At steady state, compartmental inflows and outflows are equal such that $dx_i/dt = 0$, and therefore, incoming and outgoing throughflows are also equal: $T_i^{\text{in}} = T_i^{\text{out}} = T_i$. In vector notation, compartmental throughflows are given by

$$\mathbf{T} = \begin{bmatrix} T_1 \\ T_2 \\ T_3 \\ T_4 \\ T_5 \end{bmatrix} \qquad [5]$$

This basic information regarding the storages, flows, and boundary flows provides all the necessary information to conduct environ analysis. Environ analysis has been classified into a structural analysis, dealing only with the network topology, and three functional analyses (flow, storage, and utility) – which requires the numerical values for flow and storage in the network (**Table 1**).

The technical aspects of environ analysis are explained in detail elsewhere, so rather than repeat those here, the remainder of the article highlights some of the important results from environ analysis. But first, one issue that must be covered is the way in which network analysis identifies and quantifies indirect pathways and flow contributions.

Indirectness originates from transfers or interactions that occur nondirectly, and are mediated by other within-system compartments. These transfers could travel two, three, four, or many links before reaching the target destination. For example, the flow analysis starts with the calculation of the nondimensional flow intensity matrix, $\mathbf{G}$, where $g_{ij} = f_{ij}/T_j$. The generalized G matrix corresponding to **Figure 1** would look as follows:

$$\mathbf{G} = \begin{bmatrix} 0 & 0 & 0 & 0 & g_{15} \\ g_{21} & 0 & 0 & 0 & 0 \\ g_{31} & g_{32} & 0 & 0 & 0 \\ 0 & g_{42} & g_{43} & 0 & 0 \\ 0 & 0 & g_{53} & g_{54} & 0 \end{bmatrix} \qquad [6]$$

These values represent the fraction of flow along each link normalized by the total throughflow at the donating compartment. These elements give the direct, measurable flow intensities (or probabilities) between any two nodes j to i. To identify the flow intensities along indirect paths (e.g., $j \to k \to i$), one need only consider the matrix $\mathbf{G}$ raised to the power equal to the path length in question. For example, $\mathbf{G}^2$ gives the flow intensities along all paths of length 2, $\mathbf{G}^3$ along all paths of length 3, etc. This well-known matrix algebra result is the primary tool to uncover system indirectness. In fact, it turns out that due to the way in which the $\mathbf{G}$ matrix is constructed, all elements in $\mathbf{G}^m$ go to zero as $m \to \infty$. Therefore, it is possible to sum the terms of $\mathbf{G}^m$ to acquire an 'integral' flow matrix (called $\mathbf{N}$), which gives the flow contribution from all path lengths:

$$\mathbf{N} = \mathbf{G}^0 + \mathbf{G}^1 + \mathbf{G}^2 + \mathbf{G}^3 + \cdots = \sum_{m=0}^{\infty} \mathbf{G}^m = (\mathbf{I} - \mathbf{G})^{-1} \qquad [7]$$

where $\mathbf{G}^0 = \mathbf{I}$, the identity matrix, $\mathbf{G}^1$ the direct flows, and $\mathbf{G}^m$ for $m > 1$ are all the indirect flows' intensities. Note, that the elements of $\mathbf{G}$ and $\mathbf{N}$ are nondimensional; to retrieve back the actual throughflows, one need only multiply the integral matrix by the input vector: $\mathbf{T} = \mathbf{Nz}$. In other words, $\mathbf{N}$ redistributes the input, $\mathbf{z}$, throughout each compartment to recover the total flow through that compartment. Similarly, one could acquire any of the direct or indirect flows by multiplying $\mathbf{G}^m \mathbf{z}$ for any m.

Table 1 Basic methodologies for network environ analysis

Structural analysis	Functional analyses

Path analysis
Enumerates pathways in a network (connectance, cyclicity, etc.)
	Flow analysis: $g_{ij} = f_{ij}/T_j$
	    Identifies flow intensities along indirect pathways
	Storage analysis: $c_{ij} = f_{ij}/x_j$
	    Identifies storage intensities along indirect pathways
	Utility analysis: $d_{ij} = (f_{ij} - f_{ji})/T_i$
	    Identifies utility intensities along indirect pathways

A similar argument is made to develop integral storage and utility matrices:

storage: $\mathbf{Q} = \mathbf{P}^0 + \mathbf{P}^1 + \mathbf{P}^2 + \mathbf{P}^3 + \cdots \sum_{m=0}^{\infty} \mathbf{P}^m = (\mathbf{I} - \mathbf{P})^{-1}$

$$[8]$$

utility: $\mathbf{U} = \mathbf{D}^0 + \mathbf{D}^1 + \mathbf{D}^2 + \mathbf{D}^3 + \cdots \sum_{m=0}^{\infty} \mathbf{D}^m = (\mathbf{I} - \mathbf{D})^{-1}$

$$[9]$$

where $p_{ij} = (f_{ij}/x_j)\Delta t$, and $d_{ij} = (f_{ij} - f_{ji})/T_i$.

Network Properties

Patten has developed a series of 'ecological network properties' which summarize the results of environ analysis. The properties have been used to assess the current state of ecosystem networks and to compare the state of different networks. Furthermore, while interpreting some of the properties as ecological goal functions, it has been possible to identify the structural or parametric configurations that positively affect the network property values as a way to detect or anticipate network changes. For example, certain network alterations, such as increased cycling, lead to greater total system energy throughflow and energy storage, so one could expect that if possible ecological networks are evolving or adapting to such configurations. This leads to a new area of research on evolving networks. In this section, a brief overview is given for four of these properties: dominance of indirect effects (or nonlocality), network homogenization, network mutualism, and environs.

Dominance of Indirect Effects

This property compares the contribution of flow along indirect pathways with those along direct ones. Indirect effects are any that require an intermediary node to mediate the transfer and can be of any length. The strength of indirectness has been measured in a ratio of the sum of the indirect flow intensities divided by the direct flow intensities:

$$\frac{\sum_{i,j=1}^{n} (n_{ij} - g_{ij} - \delta_{ij})}{\sum_{i,j=1}^{n} g_{ij}} \quad [10]$$

where δ_{ij}, the Kronecker delta, is 1 if and only if $i = j$ and is 0 otherwise. When the ratio is greater than 1, then dominance of indirect effects is said to occur. Analysis of many different models has shown that this ratio is often greater than 1, revealing the nonintuitive result that indirect effects have greater contribution than direct effects. Thus, each compartment influences each other, often significantly, by many indirect, nonobvious pathways.

The implications of this important result are clear in that each compartment is embedded in and dependent on the rest of the network for its situation, thus calling for a true systems approach to understand such things as feedback and distributed control in the network.

Network Homogenization

The homogenization property yields a comparison of resource distribution between the direct and integral flow intensity matrices. Due to the contribution of indirect pathways, it was observed that flow in the integral matrix was more evenly distributed than that in the direct matrix. A statistical comparison of resources distribution can be made by calculating the coefficient of variation of each of the two matrices. For example, the coefficient of variation of the direct flow intensity matrix $\mathbf{G}$ is given by

$$\mathrm{CV}(\mathbf{G}) = \frac{\sum_{j=1}^{n} \sum_{i=1}^{n} (\bar{g}_{ij} - g_{ij})^2}{(n-1)\bar{g}} \quad [11]$$

Network homogenization occurs when the coefficient of variation of $\mathbf{N}$ is less than the coefficient of variation of $\mathbf{G}$ because this says that the network flow is more evenly distributed in the integral matrix. The test statistic employed here looks at whether or not the ratio $\mathrm{CV}(\mathbf{G})/\mathrm{CV}(\mathbf{N})$ exceeds 1. The interpretation again is clear that the view of flow in ecosystems is not as discrete as it appears because in fact the material is well mixed (i.e., homogenized) and has traveled through and continues to travel through many, if not, most parts of the system.

Network Mutualism

Turning now to the utility analysis, the net flow, utility matrix, $\mathbf{D}$, can be used to determine quantitatively and qualitatively the relations between any two components in the network such as predation, mutualism, or competition. Entries in the direct utility matrix, $\mathbf{D}$, or integral utility matrix, $\mathbf{U}$, can be positive or negative ($-1 \leq d_{ij}$, $u_{ij} < 1$). The elements of $\mathbf{D}$ represent the direct relation between that (i, j) pairing; for the example in **Figure 1**, this produces the following:

$$\mathbf{D} = \begin{bmatrix} 0 & -\dfrac{f_{21}}{T_2} & -\dfrac{f_{31}}{T_3} & 0 & \dfrac{f_{15}}{T_5} \\ \dfrac{f_{21}}{T_2} & 0 & -\dfrac{f_{32}}{T_3} & -\dfrac{f_{42}}{T_4} & 0 \\ \dfrac{f_{31}}{T_3} & \dfrac{f_{32}}{T_3} & 0 & -\dfrac{f_{43}}{T_4} & -\dfrac{f_{53}}{T_5} \\ 0 & \dfrac{f_{42}}{T_4} & \dfrac{f_{43}}{T_4} & 0 & -\dfrac{f_{54}}{T_5} \\ -\dfrac{f_{15}}{T_5} & 0 & \dfrac{f_{53}}{T_5} & \dfrac{f_{54}}{T_5} & 0 \end{bmatrix} \quad [12]$$

Table 2 Direct and integral relations in sample network from **Figure 1**

Direct	Integral
$(sd_{21}, sd_{12}) = (+, -) \rightarrow$ exploitation	$(su_{21}, su_{12}) = (+, -) \rightarrow$ exploitation
$(sd_{31}, sd_{13}) = (+, -) \rightarrow$ exploitation	$(su_{31}, su_{13}) = (+, -) \rightarrow$ exploitation
$(sd_{41}, sd_{14}) = (0, 0) \rightarrow$ neutralism	$(su_{41}, su_{14}) = (+, +) \rightarrow$ mutualism
$(sd_{51}, sd_{15}) = (-, +) \rightarrow$ exploited	$(su_{51}, su_{15}) = (+, +) \rightarrow$ mutualism
$(sd_{32}, sd_{23}) = (+, -) \rightarrow$ exploitation	$(su_{32}, su_{23}) = (-, -) \rightarrow$ competition
$(sd_{42}, sd_{24}) = (+, -) \rightarrow$ exploitation	$(su_{42}, su_{24}) = (+, -) \rightarrow$ exploitation
$(sd_{52}, sd_{25}) = (0, 0) \rightarrow$ neutralism	$(su_{52}, su_{25}) = (+, +) \rightarrow$ mutualism
$(sd_{43}, sd_{34}) = (+, -) \rightarrow$ exploitation	$(su_{43}, su_{34}) = (+, -) \rightarrow$ exploitation
$(sd_{53}, sd_{35}) = (+, -) \rightarrow$ exploitation	$(su_{53}, su_{35}) = (+, -) \rightarrow$ exploitation
$(sd_{54}, sd_{45}) = (+, -) \rightarrow$ exploitation	$(su_{54}, su_{45}) = (+, -) \rightarrow$ exploitation

The direct matrix $\mathbf{D}$, being zero-sum, always has the same number of positive and negative signs:

$$sgn(\mathbf{D}) = \begin{bmatrix} 0 & - & - & 0 & + \\ + & 0 & - & - & 0 \\ + & + & 0 & - & - \\ 0 & + & + & 0 & - \\ - & 0 & + & + & 0 \end{bmatrix} \quad [13]$$

The elements of $\mathbf{U}$ provide the integral, system-determined relations. Continuing the example, and now including flow values derived from 10% transfer efficiency along each link ($g_{ij} = 0.10$, if $a_{ij} = 1$, and $g_{ij} = 0$ otherwise), we get the following integral relations between compartments:

$$sgn(\mathbf{U}) = \begin{bmatrix} + & - & - & + & + \\ + & + & - & - & + \\ + & - & + & - & - \\ + & + & + & + & - \\ + & + & + & + & + \end{bmatrix} \quad [14]$$

Unlike, the direct relations, this is not zero-sum. Instead, we see that there are 17 positive signs (including the diagonal) and 8 negative signs. If there are a greater number of positive signs than negative signs in the integral utility matrix, then network mutualism is said to occur. Network analysis demonstrates the positive mutualistic relations in the system. Specifically, here, we can identify two cases of indirect mutualism, seven of exploitation, and one of competition (**Table 2**).

Environ Analysis

The last property mentioned here is the signature property, the quantitative environ, both in the input and output orientation. Since each compartment has two distinct environs, there are in fact $2n$ environs in total. The output environ, $\mathbf{E}$, for the ith node is calculated as:

$$\mathbf{E} = (\mathbf{G} - \mathbf{I})\hat{N}_i \quad [15]$$

where $\hat{N}_i$ is the diagonalized matrix of the ith column of $\mathbf{N}$. When assembled, the result is the output-oriented flow from each compartment to each other compartment in the system and across the system boundary. Input environs are calculated as

$$\mathbf{E}' = \hat{N}_i'(\mathbf{G}' - \mathbf{I}) \quad [16]$$

where $g_{ij}' = f_{ij}/T_i$, and $\mathbf{N}' = (\mathbf{I} - \mathbf{G}')^{-1}$. These results comprise the foundation of network environ analysis since they allow for the quantification of all within-system interactions, both direct and indirect, on a compartment-by-compartment basis.

Summary

A practical objective of ENA in general, and environ analysis in particular, is to trace material and energy flow–storage through the complex network of system interactions. The network environ approach has been a fruitful way of holistically investigating ecological systems. In particular, a series of 'network properties' such as indirect effects ratio, homogenization, and mutualism have been observed using this analysis, which consider the role of each entity embedded in a larger system.

See also: Cycling and Cycling Indices; Ecological Network Analysis, Ascendency; Ecological Network Analysis, Energy Analysis; Emergent Properties; Indirect Effects in Ecology.

Further Reading

Dame RF and Patten BC (1981) Analysis of energy flows in an intertidal oyster reef. *Marine Ecology Progress Series* 5: 115–124.
Fath BD (2007) Community-level relations and network mutualism. *Ecological Modelling* 208: 56–67.

Fath BD and Patten BC (1998) Network synergism: Emergence of positive relations in ecological systems. *Ecological Modelling* 107: 127–143.

Fath BD and Patten BC (1999) Review of the foundations of network environ analysis. *Ecosystems* 2: 167–179.

Fath BD, Jørgensen SE, Patten BC, and Straškraba M (2004) Ecosystem growth and development. *Biosystems* 77: 213–228.

Gattie DK, Schramski JR, Borrett SR, *et al.* (2006) Indirect effects and distributed control in ecosystems: Network environ analysis of a seven compartment model of nitrogen flow in the Neuse River Estuary, North Carolina, USA - Steady state analysis. *Ecological Modelling* 194(1–3): 162–177.

Halnes G, Fath BD, and Liljenström H (2007) The modified niche model: Including a detritus compartment in simple structural food web models. *Ecological Modelling* 208: 9–16.

Higashi M and Patten BC (1989) Dominance of indirect causality in ecosystems. *American Naturalist* 133: 288–302.

Jørgensen SE, Fath BD, Bastianoni S, *et al.* (2007) *Systems Ecology: A New Perspective*. Amsterdam: Elsevier.

Patten BC (1978) Systems approach to the concept of environment. *Ohio Journal of Science* 78: 206–222.

Patten BC (1981) Environs: The superniches of ecosystems. *American Zoologist* 21: 845–852.

Patten BC (1982) Environs: Relativistic elementary particles or ecology. *American Naturalist* 119: 179–219.

Patten BC (1991) Network ecology: Indirect determination of the life–environment relationship in ecosystems. In: Higashi M and Burns TP (eds.) *Theoretical Ecosystem Ecology: The Network Perspective*, pp. 288–315. London: Cambridge University Press.

Whipple SJ and Patten BC (1993) The problem of nontrophic processes in trophic ecology: Towards a network unfolding solution. *Journal of Theoretical Biology* 163: 393–411.

Indirect Effects in Ecology

V Krivtsov, University of Edinburgh, Edinburgh, UK

Introduction

Interrelations among ecosystem components and processes can be subdivided into direct (i.e., those which are restricted to the direct effect of one component/process on another, and are attributable to an explicit direct transaction of energy and/or matter between the components in question) and indirect (i.e., those that do not comply with the above restriction). The history of natural sciences is inseparable from the gradually increasing awareness and understanding of indirect effects. By nineteenth century the significance of indirect interactions was well realized, and was (sometimes implicitly) accounted for in the classic studies of Darwin, Dokuchaiev, Gumboldt, Engels, and many other scientists. In the twentieth century, however, appreciation of indirect effects in nature received considerable acceleration, predominantly due to the accumulating interdisciplinary knowledge of natural ecosystems, the development of appropriate mathematical techniques, and the urgent necessity to resolve the growing problems of environmental damage, resulting, ironically, from the uncurbed expansion of the human population backed by the advances of the technological progress. It should also be noted that the boost of the growing appreciation of indirect effects in twentieth century was partly initiated by Vernadsky's fundamental theories about the 'biosphere', the 'noösphere', and interrelations between biota and geochemical cycling. Popularization of these views 50 years later (e.g., by Lovelock's Gaia theory) stimulated investigations of indirect effects even further.

Basics

There have been many definitions of direct and indirect effects. Information on indirect interactions is scattered in the literature, and may appear under various terms. For example, among ecological phenomena which may (depending on the exact definition) be regarded as indirect effects are exploitative and apparent competition, facilitation, mutualism, cascading effects, tri-trophic-level interactions, higher-order interactions, interaction modification, nonadditive effects, etc.

First of all, it is important to distinguish between direct and indirect effects. Usually, the interactions between two components not involving direct transfer of energy and/ or matter are viewed as indirect, while those that involve an explicit direct transaction are viewed as direct. The literature is inconsistent on the definitions of indirect effects, and one way to clarify the problem is to stress the difference between a transaction and a relation. A simple transaction between two ecosystem components is always direct since it is the transfer of matter and/or energy, whereas a relation is the qualitative type of interaction. Relations include predation, mutualism, competition, commensalism, ammensalism, etc. Hence a direct relationship is the one which is based on a direct (i.e., unmediated by another ecosystem component) transaction only. For example, the classic predation (not to be mistaken with, for example, keystone predation, indirect predation, etc.) is direct, and so is the nutrient uptake by plants, algae, and bacteria, whereas mutualism and competition are always indirect, as they result from the combination of a number of simple transactions. It is worth pointing out that the observed patterns of interrelations between ecosystem components (e.g., correlation between abundance indices) frequently result from a combination of direct and indirect effects, as each component is involved in a large number of pathways. Furthermore, if a direct relationship between two ecosystem components (say A and B) is modified by a third ecosystem component, attribute, or forcing function (the two latter notions will include, for example, such modifiers as sunlight, temperature, pH, external and internal concentrations of alternative nutrients) then the indirect relationship between the modifying agent and the first two components (i.e., A and B) becomes superimposed upon the direct relationship between the components A and B. Consequently, the observed pattern of interrelation (e.g., correlation between the abundance data) between A and B will in this case result from the combination of direct and indirect effects.

Examples of factors known to modify the strength of density-mediated indirect interactions include differences in the specific growth rates (important, for example, for apparent competition), density dependence of the transmitting compartment, and the possibility of stochastic physical disruption. On the other hand, issues important in determining the manifested strength of the behavior-mediated indirect interactions involve ability of a focal species to detect changes in factors which matter for energetic costs and benefits of its behavior, sensitivity of its optimum behavior to these costs and benefits, and available behavioral options.

For density-mediated effects, presence and strength of indirect interactions can be determined by analyzing partial derivatives of the abundance of a species on the abundances of other (not immediately connected) species. However, indirect interaction may involve ecologically important changes other than changes in abundance, for example, demographic changes in the population structure, changes in the genotypic composition, and changes in behavior (e.g., searching rates, antipredator behaviors), morphology, biochemistry (e.g., nutrient content, toxin concentration), or physiology.

Most Commonly Studied Indirect Effects

Among a plethora of possible indirect effects, there are five that have been studied most commonly. Their essence is depicted in **Figure 1** and is briefly explained below.

Interspecific competition

Interspecific competition (also called exploitative competition) takes place whenever two (or several) species compete for the same resource. In **Figure 1a**, an increase in Component 1 will lead to the increased consumption of the shared resource (Component 2), and consequently to the decrease in a competitor (Component 3). Examples of this include, for example, two predators sharing the same prey, or two microbial species whose growth is limited by the availability of the same nutrient.

Apparent competition

Apparent competition occurs when two species have a common predator. In **Figure 1b** an abundant population of species 1 sustains a high-density population of predator 2, who, in turn, may limit the population of another prey species 3. From practical point of view, it is worth noting here that this situation sometimes happens as an unwanted result in biocontrol, when a biocontrol agent (species 2), specifically introduced to control a target (species 1), may increase the risk of a nontarget's (species 3) extinction.

Trophic cascades

Trophic cascades involve propagation of the effect along a vertical trophic chain consisting of three or more components connected by grazing or predation. In **Figure 1c**, an increase/decrease in Component 4 will lead to the decrease/increase in Component 3, increase/decrease in Component 2, and decrease/increase in Component 1. These effects are particularly well studied in aquatic food chains (see examples below), but have also been studied in terrestrial systems.

It is worth pointing out, however, that the structure of real ecosystems hardly ever fits tidily into the concepts of simple trophic levels (e.g., omnivory is widespread in nature), and trophic cascades, therefore, are often complicated by the interlinks within and among trophic levels (e.g., in terrestrial ecosystems insectivorous birds prey on predatory, herbivorous, and parasitoid insects, and the resulting effect of birds on the

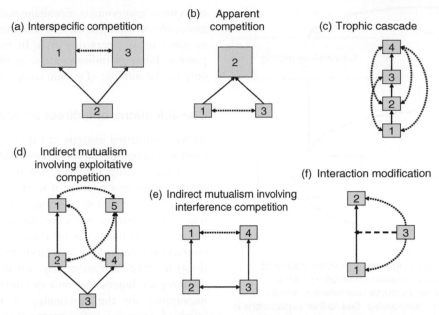

Figure 1 Diagrams of the most commonly studied indirect affects. Direct effects are shown using solid lines, while indirect effects (only the effects relevant to the accompanying discussion are illustrated) using dotted lines. Interaction modification is illustrated using a dashed line. Numbers in the compartments are used solely for labeling to distinguish between different compartments, and do not relate to any kind of hierarchy. Likewise, the box sizes do not bear any relevance to the sizes or significance of the compartments drawn, and the relative size of the arrows relates neither to the effect's strength no to the preferential directionality. See further explanations in the text. (a) Interspecific competition; (b) apparent competition; (c) trophic cascade; (d) indirect mutualism involving exploitative competition; (e) indirect mutualism involving interference competition; (f) interaction modification. Modified from Wootton JT (1994) The nature and consequences of indirect effects in ecological communities. *Annual Review of Ecology and Systematics* 25: 443–466.

primary producers and their damage by herbivory may, therefore, depend on the specific species and the conditions involved). In particular, proper consideration of detritus contributions to the energy flows may prove the 'trophic cascade' simplification unsuitable, as the detritus compartment often has direct links to a number of trophic levels.

Indirect mutualism and commensalism

Indirect mutualism and commensalism involve a consumer–resource interaction coupled with either exploitative (**Figure 1d**) or interference (**Figure 1e**) competition. For instance, starfish and snails reduce the abundance of mussels, a dominant space occupier, and increase the abundance of inferior sessile species. The presence of grazers on oyster farms in Australia increases oyster recruitment by removing algae, who otherwise preempt the available spaces. In **Figure 1d**, an increase in species 1 should lead to a decrease in species 2 and an increase in species 3. The latter positive effect would propagate up the right branch of the diagram, increasing the abundances of species 4 and 5. This situation arises when, for example, planktivorous fish preferentially feeding on large zooplankton indirectly increase the abundance of small zooplankton. Cases involving interference competition are well known from, for example, the intertidal environment,

where birds increase the abundance of acorn barnacles by consuming limpets that otherwise dislodge the young barnacles off the rock.

Interaction modification

Interaction modification occurs when the relationship between a species pair is modified by a third species (**Figure 1f**). Examples include positive effects of macroalgae on zooplankton through interference with the hunting potential of fish and changing of a chemical's bioavailability due to the activity of a species, when the chemical in question is important for the functioning of another species (e.g., acids produced by one microbial population may increase bioavailability of compounds that are bound or unaccessible for another microbial population).

It is worth pointing out that 'interaction modification' is often, and quite rightly, considered as a principally different type of indirect effect. By coupling interaction modifications with other types of relationships (e.g., trophic), one may arrive at possibilities of numerous (including very complex) relationships. One of the more simple of such combinations may be exemplified (**Figure 2**) with an indirect effect of grazers and certain agricultural practices on the population density of foxes (*Vulpes vulpes*) and the rodent *Marmota bobac* in Eastern Europe (V. Takarsky, personal communication): lower

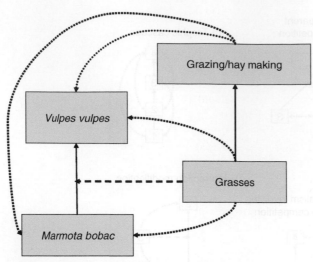

Figure 2 Diagram illustrating a positive indirect effect of grazing on *Marmota bobac* population resulting from a combination of consumer–resource relationships with an interaction-modification relationship. See further explanations in the text.

grazing rates lead to a denser and taller grass cover, enabling more successful hunting of predators. Conversely, higher grazing rates lead to a lower grass cover, thus enhancing the detection of predators by the rodents. As a result, increase in grazing may have an indirect positive effect on the *Marmota bobac* population, and an indirect negative effect on the population of foxes.

It should also be noted that some of the known examples of ammensalism and commensalism do actually fit in the description either of a simple interaction modification or interaction modification coupled with a number of tropic relationships. For instance, the bioavailability example described above has been quoted by Atlas and Bartha as an example of commensalism. If, however, the chemical in question is not nutritional, but harmful for the second species, then the relationship fits the criteria for ammensalism. In a similar vein, protocooperative and mutualistic relationships are easily envisaged from certain combinations of interaction modifications and tropic relationships.

It is worth pointing out that although the indirect relationships listed above are mainly studied in relation to pairs of biological species, they are applicable to a wider range of system components. It should also be noted that many more types of indirect effects are easily envisaged from various possible combinations between interacting compartments, and quite a few have indeed been observed in nature. For example, Menge distinguished 83 subtypes of indirect effects. However, an attempt to exemplify every possible type of indirect effects would be outside the scope of this article. The readers could easily construct, for example, many further types of indirect effects combining the most commonly studied ones depicted in **Figure 1**. In a real world,

ecosystem components simultaneously take part in a multitude of interactions, and it is therefore appropriate to name it an interaction web. In fact, the number of possible kinds of indirect effects is likely to be limited only by the number of system components considered.

Classifications of Indirect Effects

Although detailed analysis of various possible classifications would be outside the scope of this publication, it is worth mentioning, however, that indirect effects can be characterized in a number of ways, related, for example, to the characteristics of exerting, receiving, and transmitting compartments, presence/absence of a lag phase before the manifestation of a response, strength of the interaction (particularly in relation to the direct interactions) and its directionality (e.g., whether it is isotropic or anisotropic), dependence on a specific ecosystem context, importance for the functioning of the compartments involved, importance for structural (e.g., successional or evolutionary) changes in the populations involved and the whole biological community, and significance for overall ecosystem functioning. In the author's view, the different ways to characterize indirect interactions are not contradictory, but rather complementary, and may conveniently contribute to the toolbox for comparative ecosystem analysis.

Indirect Effects

All the relations not restricted to the effects of a direct transaction of matter and energy between the adjacent ecosystem components are treated as indirect. Hence, for the purpose of the forgoing sections, all types of indirect interactions mentioned above will be considered as indirect effects. However, the distinction between directly and indirectly mediated effects will be made where deemed appropriate. The terms 'relationship' and 'interaction' will be used interchangeably. Furthermore, although it is realized that for the purpose of quantitative assessment the distinction between the terms 'effect' and 'interaction' may be helpful, no such distinction has been made in this article, as in many studies addressing indirect effects these terms are used interchangeably.

The definition of indirect effects given above is very encompassing, and will include some of the effects which may fall into the category of 'direct' under a different definition. For example, it is useful to account for the distinction between those effects that are directly and indirectly mediated, since the latter ones are particularly difficult to observe, especially if the cause and effect are substantially separated in time.

The directly mediated effects have previously been regarded as direct (i.e., as regards to the properties of their propagation). Here, however, the directly mediated effects will be treated as indirect, and the definition of indirect effects will, therefore, include such effects as trophic cascades, top-down and bottom-up controls, etc. The classification of indirect effects into directly and indirectly mediated is applicable to a wide range of environmental processes and bears certain similarities with the distinction between 'interaction chains' and 'interaction modifications' earlier recognized for purely biotic relationships.

Examples of Occurrence and Importance of Indirect Effects

Indirect Effects in Terrestrial Environment

Arguably, the awareness of natural scientists as regards indirect effects in the terrestrial environment can be traced back at least to the end of nineteenth century, when the school of thought founded by Dokuchaiev had developed a theory that soil was a product of complex interactions between climate and geological and biological components of the terrestrial landscape. To date, the importance of indirect interactions in the terrestrial environment is well recognized. Indirect effects in terrestrial ecosystems relate, for instance, to the dependence of plant nutrient supply on mineralization of nutrients by soil biota, and to the propagation of these effects through the food chain. Soil fauna may help to disperse microorganisms crucial for plant functioning and biogeochemical cycling, and physically modify the habitat, thus changing environmental conditions for all the biological community. Plants, in turn, modify the habitat for other organisms, for example, by producing litter, providing shade, shelter, etc. All in all, indirect effects in the terrestrial environment are widespread; below are just a few examples of their recent studies.

A number of studies conducted in the terrestrial environment (this includes both field experiments and soil microcosms) adopted experimental approach focusing on the density-manipulation experiments followed by analysis of the results obtained using parametric (e.g., ANOVA, Tukey's HSD) and nonparametric (e.g., Kruskal–Wallis and Mann–Whitney U-tests) statistical tests. For instance, Miller used exclusion experiments to elucidate direct and indirect species interactions in a field plant community. Experimental results were analyzed by parametric and nonparametric techniques, which yielded interesting information on the ecological characteristics of the species involved. Particularly, it was established that species with a large competitive ability due to direct effects generally had almost as large indirect effects, so that the two effects almost cancelled each other.

A number of terrestrial studies used various mathematical methods to investigate indirect interactions. In particular, a good insight into specific indirect effects was gained using simulation modeling to interpret monitoring or experimental results. For example, Hunt and co-authors found that the increase in net N mineralization with precipitation is a consequence of not only the direct effect of moisture supply on decomposition, but also an indirect effect of changes in substrate supply and quality. de Ruiter and co-authors studied nitrogen mineralization conducted at a wheat field. The impact of microfaunal functional groups on N mineralization was evaluated by calculating the impact of group deletion. The results showed that the effect of the removal of a group may exceed the direct contribution of this group to N mineralization rather considerably, with amoebae and bacterivorous nematodes having values of 18% and 28%, and 5% and 12% for, respectively, direct contribution toward and impact of deletion upon overall N mineralization. Influence of the transitions of soil microorganisms between dormant and active stages was studied by Blagodatsky and co-authors. Such transitions were shown to be important for biogeochemical cycling and the rate of organic matter decomposition.

A combination of a detailed monitoring program, and statistical and simulation modeling has been used in a study of ecological patterns in the Heron Wood Reserve, located at the Dawyck Botanic Garden in Scotland. The suite of statistical techniques included ANOVA, ANCOVA, correlation analysis, CCA, factor analysis, and stepwise regression modeling. The study revealed a number of indirect effects resulting from a complex multivariate interplay among ecosystem components. For example, the results suggested that both direct negative and indirect positive effects of the microarthropod community on specific fungal groups appeared to take place. The relatively high local abundances of the dominant collembolan *Folsomia* might have caused local declines in ectomycorrhizal fungi, reflected, in turn, in the increase in pH (Whist this work was in press, Dr. Peter Shaw has checked identification of the dominant Folsomia species (previously referred to as *F. candida*) from the Dawyck ecosystem study, and has shown that it appears to fit the description of *F. inoculata*.) However, for those samples where the dominant *Folsomia* were less abundant, overcompensatory fungal growth due to grazing by mites and other collembola was implicated. Complex effects were also shown for bacteria, nematodes, protozoa, plants, and soil properties.

Indirect Effects in Aquatic Systems

Awareness of indirect interactions in aquatic environment has rather a considerably long history, and clearly presented examples can be found in works (among others) of, for example, Mortimer, Hutchinson, and Reynolds. In particular, in an earlier review by Abrams it was even suggested that most studies specifically addressing behavior-mediated indirect effects tend to be conducted in freshwater ecosystems, while many of the early demonstrations of density-mediated indirect effects were done in community studies in marine habitats. Likewise, much of the knowledge related to indirect ecological interactions has been contributed through the development and applications of the methods of simulation modeling and network analysis in relation to aquatic environment. Consequently, simulation models capable of demonstrating indirect interactions in aquatic biogeocenoses (e.g., the Lake 2 model of J. Solomonsen) are widely used for teaching in the educational establishments across the world.

Recent studies of indirect effects in aquatic environment variously involved a combination of the empirical approach and an application of statistical techniques, methods of network analysis, simulation modeling using 'What if' scenarios, and sensitivity analysis. One of the perhaps most frequently addressed examples of indirect effects in aquatic environment relate to trophic cascades, which involve propagation of the effect along a vertical trophic chain consisting of three or more components connected by grazing or predation. For instance, as was recently investigated by Daskalov, a decrease in the top predator's population in the Black Sea due to overfishing resulted in a 'trophic casade', leading to an increase in the abundance of planktivorous fish, a decline in zooplankton biomass, and an increase in phytoplankton crop.

The previously made statements regarding the abiotic components (see above) can be emphasized with examples related to the importance of detritus. For instance, Carrer and Opitz found that in the Lagoon of Venice about half of the food of nectonic benthic feeders and nectonic necton feeders passed through detritus at least once, while there was no direct transfer of such food according to the diet matrix. Whipple provided an analysis of the extended path and flow structure for the well-documented oyster reef model. Few simple paths and large number of compound paths were counted. The study provided structural evidence for feedback control in ecosystems, and illustrated importance of nonliving compartments (in this case, detritus) for the ecosystem's functioning. Even for the model with a low cycling index (i.e., 11%) multiple cyclic passage paths provided a considerable (22%) flow contribution. Therefore, it was envisaged that for ecosystems with higher cycling indexes the patterns observed should be even more pronounced.

Another noteworthy illustration of indirect effects in aquatic ecosystems relates to the interdependency of biogeochemical cycles. For example, Dippner concluded that indirect effect of the silicate reduction in coastal waters causes an increased flagellate bloom, due to a high availability of riverborne nutrient loads. In a study of lake Suwa (Japan), Naito and co-authors have shown that the physiological parameters of the diatom *Melosira* were the important sources of the cyanobacterium *Microcystis'* production variability. These results agree well with our

work on Rostherne Mere and suggest that the underlying mechanism might be a common inverse relationship between spring diatom and summer cyanobacterial blooms resulting from the fact that the biogeochemical cycles of Si and P in the aquatic environment are coupled via the dynamics of primary producers (i.e., increased concentrations of Si in spring lead to an increase in a spring diatom bloom, and an increase in the removal of P, N, and microelements from the water column with easily sedimenting biomass at the end of the bloom; consequently, this may lead to a decrease in the summer cyanobacterial development).

Role of Abiotic Components

Although the importance of abiotic ecosystem components is commonly recognized, most of the ecological studies (including those addressing the indirect effects) tend to study in detail only relationships among biota. The restriction of the integrative synthesis to species interaction only cuts off a plethora of useful environmental studies related, for example, to issues of global climate change. It should be noted, however, that the science of ecosystem dynamics is highly interdisciplinary, and the information relevant to the present discussion can, therefore, be found not only in ecology and biology, but also virtually in any section of natural and environmental sciences, with geography, palaeontology, geoecology, and climatology comprising the most obvious candidates.

In ecology, it is widely recognized that species interaction can be mediated by a nonliving resource, and that a species can potentially exert a selective force on another species through nontrophic interactions. It should also be noted that in nature many species are very well adapted to modify their community and habitat (e.g., beavers by changing the habitat's hydrological regime, humans by initiating dramatic changes in global climate and geochemical fluxes, earthworms by increasing aeration and redistributing organic matter in soil, etc.). Changes in physical characteristics of a habitat caused by the activity of so-called 'ecosystem engineers' may be regarded as an extreme case of such nontrophic interactions. Often, however, even if abiotic components are considered in terms of detrital pathways and/or nutrient cycling, the effects studied in detail are mostly confined to trophic interactions only. Furthermore, many indirect interactions occur between different stages of ecosystem development and are therefore easily overlooked and understudied. In ecological literature these interactions are sometimes called 'historical effects', 'priority effects', or 'indirect delayed regulations'. Consideration of these effects is particularly important for the correct understanding of an overall ecosystem functioning. Hence, if one abstracts from the labels given to different branches of science, the importance of abiotic ecosystem components and physical environment for ecosystem dynamics and evolutionary development becomes increasingly obvious.

Indirect Effects of Global Relevance

Indirect relationships important on the global or subglobal scale are often separated from their cause spatially and/or temporally. For example, the dramatic increase in volcanic activity (possibly caused by the impact of an asteroid) at the end of the Mesozoic era is thought to have led to the extinction of dinosaurs, which arguably stimulated the eventual evolution of mammals (including humans). The increased production and use of fertilizers in the 1950s led to the increased phosphate inputs, eutrophication, and decrease in water quality in many lakes, ponds, and reservoirs during the subsequent decades. The increased consumption of fossil fuels in the twentieth century led to the increased emissions of carbon dioxide, which were eventually followed by global warming and an apparent increase in the frequency of natural disasters. This climate change was probably accelerated by the depletion of the planet's ozone layer due to the CFC (chlorofluorocarbon)-containing deodorants and refrigerants.

It should be noted that indirect relationships are not related just to the activities of humanity, but have been important throughout the history of our planet. For example, a gradual development of the modern atmosphere was largely due to the activity of cyanobacteria, which were among the first organisms to produce oxygen as a by-product of their metabolism. The indirect implications of the atmospheric oxygen enrichment were far reaching, and led not only to profound global biological and geochemical changes, but also ultimately enabled the development of *Homo sapiens* and its current civilization.

Last century, the line of thought started by Vernadsky has eventually led to the creation of a new integrative branch of natural sciences, sometimes referred to as 'global ecology'. Essentially, 'global ecology' encompasses methods and scope of virtually all other environmental disciplines, and is predominantly concerned with the dynamics (including past and future) of the global ecosystem – the biosphere. As an example, it is worth mentioning the now classic climatological research carried out by Budiko and co-workers, which led to the creation of a half-empirical model of the thermal regime of the atmosphere. This model was subsequently used to simulate past and future dynamics of the atmosphere, and changes between glaciation and interglacial periods. Furthermore, the results obtained aided interpretation of human evolution, and led to further research aiming to counteract possible global change, for example, by injecting certain substances into the stratosphere, and direct and indirect consequences to which such manipulations may lead.

Currently, global climate change (principally related to the increased concentrations of greenhouse gases) is still one of the most discussed topics in ecology and environmental sciences in general. While the detailed review and the lively controversy of the discussions related to this topic is outside the scope of this publication, it is worth pointing out that the absolute majority of studies dealing with it also inevitably deal with indirect effects (although the exact term is often not mentioned).

Indirect Effects and Industrial Ecology

This article would be incomplete without mentioning of studies and methods used in 'industrial ecology'. Industrial ecology is based on the analogy between natural and industrial ecosystems, and aims to facilitate the development of industrial recycling and cascading cooperative systems by minimizing the energy consumption, generation of wastes, emissions, and input of raw materials. Complex interplay among system components has been taken into account in a large number of waste management and industrial ecology studies. Consequently, throughout the second half of the last and the beginning of the present century, some substantial progress has been made in various aspects of industrial ecology, and in particular in understanding and accounting for indirect effects.

One of the commonly used methods of industrial ecology is 'life cycle assessment' (LCA). It studies the environmental aspects and potential impacts throughout a product's life (commonly referred to as cradle to grave approach), from raw material acquisition through production, use, and disposal, and the same methodological framework allows analysis of the impacts associated with physical products (e.g., cars, trains, electronic equipment), and services such as waste management and energy systems. Similar to LCA, but usually with considerably narrower system boundaries, are methods of energy analysis, including, for example, energy footprinting (which, effectively, constitutes calculations of how much energy is spent and saved/recovered in all the processes included within the chosen system boundary) and net energy analysis (which in addition to the detailed energy budgeting involves calculation of indicators such as incremental energy ratio and absolute energy ratio). For example, on the basis of the energy budget estimates for case studies from the UK and Switzerland it has been argued that increasing recycling rates for plastic and glass would improve the energy budget of waste management programmes, and, therefore, benefit the corresponding industrial ecosystems. Further modifications of the energy analysis methods make fruitful use of emergy and exergy budgets.

Another method popular in 'industrial ecology' is 'ecological footprinting'. Basically, the method estimates the area necessary to support (i.e., in terms of, for example, production of food, energy, processing of wastes) current, past, or probable future functioning of particular geographical (often administrative, for example, countries, counties, towns) units. Despite numerous logistical problems of interconversions, system boundary definitions, and coefficient estimates, application of this method is very useful and illustrative. For example (as illustrated by Herendeen), out of all Western industrialized countries, only the ecofootprints of Australia and Canada appear to fit inside their borders (the rest of the 'developed' countries appear to live on the expense of other territories).

Evolutionary Role of Indirect Effects

It has been postulated by a number of authors, and has been proved mathematically by Fath and Patten, that indirect effects often promote coexistence and the role of indirect effects should, in general, increase in the course of evolution. For example, in grassland communities containing *Rumex* spp., insect herbivory (by *Gastrophysa viridula*) appears to be a cost inherent in the development of plants' resistance to pathogenic fungi (*Uromyces rumicus*). Another example relates to the fact that infection of plants with endophytic fungi often enhances plants' competitive abilities via deterring grazers by production of toxic compounds (as a result, some plants might have coevolved together with their endophytes, for example, coupled evolution of *Festuca* and *Acremonium* spp.).

It should be noted, that indirect effects are important for the evolution of not only natural, but also industrial ecosystems. Traditionally, human society has developed without the necessary due respect to the rules and processes governing the stability of its environment. However, by analogy with natural ecosystems (i.e., as regards recycling and cascading networks) industrial ecosystems should aim to facilitate the development of recycling and cascading cooperative systems by minimizing the energy consumption, generation of wastes, emissions, and input of raw materials.

Approaches and Techniques Used to Detect and Measure Indirect Effects

Detection and measurements of indirect effects are often far from straightforward, and are mostly based on the intuition, common sense, and prior knowledge of any particular system. Abrams and co-authors described two major approaches adopted in ecological studies, namely theoretical and experimental. They stated that in practice, the theoretical and empirical approaches may be regarded as endpoints of a methodological continuum. Recently, however, we have argued that the methodological continuum to study indirect interactions is best represented by a triangle, with observational, experimental, and theoretical nodes.

Within the theoretical approach, observations (and/or carefully considered experimental data) are used together with theoretical considerations to construct a model capable of investigating interactions among the components incorporated in the model structure. This model is subsequently used to examine indirect effects between the components. There are a number of drawbacks of this approach, for example, difficulties related to obtaining sufficient details about the components represented in the models, unavoidable uncertainty as regards fluxes, parameters, initial values, etc. This uncertainty may mask the significance of the relationships studied, including indirect effects. Furthermore, as it is impossible to reproduce all the complexity of a real ecosystem, any model is a simplification of reality. Therefore, some of the potentially important interactions may be lost just by defining the model structure, while the importance of the others may be considerably altered.

Within the experimental approach, densities of individual species are manipulated (e.g., by total removal) in microcosms or experimental plots, and statistical analysis (e.g., ANOVA, ANCOVA) are subsequently applied to estimate the magnitude of indirect effects of manipulations on densities of other species. It has been argued that this approach is best applied using a factorial design, where the densities of a number of components (e.g., species or trophic groups) are changed both alone and in combination. If implemented properly, this approach leads to a straightforward estimation of net effects. However, there is always a danger that some of the indirect interactions have not manifested owing to unavoidable time constraints of any experiment. Also, partitioning of the registered net effects may be subject to speculation. Experiments are often costly and by definition are limited by their design and the hypotheses tested. The simplicity of the experimental design may mask the significance of the relationships studied for trait-mediated effects; measurements of population abundances may need to be supplemented by behavioral observations, and/or biochemical, physiological, genetic, and other analyses. Furthermore, there is always a big question mark how applicable are the results obtained to the processes happening in the real world.

Among mathematical methods which have been used in studies of indirect effects in natural ecosystems are statistical methods (e.g., regression and correlation analysis, PCA, factor analysis, CCA, ANCOVA, ANOVA), simulation modeling (e.g., using 'what-if scenarios', sensitivity and elasticity analysis), and methods of network analysis. In particular, indirect interactions have often been analyzed using methods of network analysis. For example, Fath and Patten used methods of network analysis to show that, in the ecosystem context, direct transactions between organisms produce integral effects more positive than a simple sum of direct effects. This was in line with the view that mutualism is an implicit consequence of indirect interactions and ecosystem organization, and that the contribution of positive relationships should increase along the course of evolution and ecological succession.

It should be noted that all the methods so far applied to investigations of indirect effects have both advantages and limitations. Many of these have been previously addressed and no attempt to discuss the benefits and disadvantages of the techniques used to investigate indirect interactions has been done in this article. Neither was it intended to address any controversy and related discussion resulting from specific applications (and/or implications of such applications) of any particular method. It should be noted, however, that the methodological framework of 'comparative theoretical ecosystem analysis' (CTEA) (see below) suggests that the mathematical techniques may be best used in concert, thus allowing a detailed complementary insight into complex patterns of mechanisms underpinning dynamics of natural ecosystems.

Problems and Implications for Environmental Management

There are many problems associated with studies of indirect effects. Here we list the most general ones, in the author's view, resulting from the very nature of such relationships, and the complexity of natural environment. We also emphasize the potential of using indirect effects in environmental management and caution as regards their misuse and careful consideration.

Complexity and Uncertainty

Although the characteristics of indirect effects are fairly readily established in a controlled laboratory experiment involving a very limited number (typically <5) of interacting components, in the natural environment the complexity of interactions renders their characterization, in particular in practical terms as regards the outcome for any specific ecosystem, rather extremely challenging. Using the inverse of 100 community matrices (elements of such an inverse matrix specify how the change in density of a particular species affects the density of the other species, with species pairs determined by the element's position in the matrix) obtained by randomly changing the species-specific parameters, Yodzis has shown that a large number of predictions for overall interactions between separate species were directionally undetermined, that is, positive in some cases and negative in others. Owing to such an uncertainty, some theoreticians have even questioned whether credible predictions of the overall outcome of indirect interactions could be made at all. As a consequence, there are numerous cases where the application of

seemingly appropriate environmental management methods, and in particular biomanipulation, resulted in failure.

Separation in Time

Another problem related, in particular, to the detection of indirect effects, is the fact that they often occur after a considerable time lag. Unfortunately, research grants are typically no longer than 3 years (often less). Hence it is likely that in many studies the potential for indirect effects has been considerably underestimated (if not totally overlooked).

Not only are cause and effect often separated in time, but they also may occur at different stages of the system's succession. This, however, may be turned to human advantage, and used as a complementary measure of environmental control. For example, it has been shown that in freshwater lakes and reservoirs biogeochemical cycle of P may be regulated by alterations in the biogeochemical cycle of Si, and stimulation of the spring diatom growth may help to alleviate nasty cyanobacterial blooms in summer. Clearly, due to inherent problems with ecosystem complexity and uncertainty any application of environmental management measures utilizing time lags of indirect effects should be done with extreme caution.

Separation in Space

Geographical separation between the cause and the effect is another inherent problem of indirect effects. For economic and societal reasons, it is particularly well documented in studies and practical applications of pollution and biological control, but numerous examples are readily available from virtually any areas of ecology and environmental sciences. As regards pollution damage, the well-known examples relate to coastal and benthic effects of oil spills, consequences of lake acidification in Scandinavia due to the transboundary transfer of sulphur dioxide since the beginning of industrial revolution, bioaccumulation of radionuclides in the food webs of, for example, British pastures following the Chernobyl fallout, etc.

In ecology, there is a growing awareness of placing the dynamics of a local biological community in the broader landscape and biogeocoenological context. For example, in biocontrol applications, failure to anticipate indirect effects resulting from the community openness and exchange between localized subpopulations has led to a number of costly failures and unexpected outcomes. Following introduction, migrations of biocontrol agents were shown to be capable of suppressing nontarget species in spatially removed areas. Certain traits of the agent were shown to be related to elevated risks of nontarget species extinction, for example, the risk of extinction of a nontarget species increases with the decrease of its growth rate and the increase of the agent's attack rate compared to the target species. Hence the threat of extinction appears to be particularly acute for a nontarget species occupying habitat fragments co-mingled with the habitats occupied by target species, and when the target species is relatively productive and the agent is only moderately efficient at limiting the target species numbers.

Defining System Boundaries

Closely related to the 'separation' problem is the problem of drawing the system's boundaries. By narrowing the boundaries of the system considered, many of indirect effects may be reduced to the effects attributable to the variation in external forcing. This is often convenient, in particular, if the aim is to study responses in a controlled environment. However, the scope for investigations of indirect effects in a system with narrow system boundaries inevitably becomes limited; by widening the boundaries and including more components the potential for indirect effects is greatly enhanced, but as the system's complexity increases, so does the associated uncertainty of any interpretations!

Implications for Environmental Management

Qualitative and quantitative account of indirect effects is (albeit often implicitly) becoming a common part of environmental management, and is indispensable for successful application of, for example, landscape engineering, biomanipulation, biogeochemical manipulation, strategic environmental assessment (SEA), and environmental impact assessment (EIA). In particular, mathematical methods can be helpful in this respect. For example, Ortiz and Wolff used Ecopath with Ecosim software to study benthic communities in Chile. They found that a simulated harvest of the clam *Mulinia* generated a complex interplay involving direct and indirect effects, and drastically changed the properties of the whole system.

McClanahan and Sala used a simulation model of the Mediterranean infralittoral rocky bottom to study possible effects of various management options. Running a number of 'what-if' scenarios they concluded that many of potential changes are likely to be indirect effects caused by changes in trophic composition. For example, if invertivorous fish were removed as part of a management scenario, sea urchins would reduce algal abundance and primary production, leading to competitive exclusion of herbivorous fish. Although similar interactions were known from tropical seas, these results were not anticipated by previous field studies in the Mediterranean.

In some cases appreciation and reliance on indirect effects may form the basis of environmental management measures. To date, there are numerous relevant examples, including, for example, biocontrol, bio- and biogeochemical manipulation or application of chemicals to reduce sediment P release for subsequent control of blue-green

algae, infection of grass cultivars by endophytic fungi in turf industry, etc. It should be reiterated, however, that a sound knowledge of the system's natural history is absolutely indispensable for an application of any environmental control measure. The problems related to the detection and investigations of indirect effects (with the major ones listed above) are likely to provide challenging, and often unexpected complications, frequently after a considerable time period. Hence, a combination of empirical and theoretical work should precede any practical steps, and any desk study should be backed up by a thorough monitoring plus (where necessary) experimental program.

Current and Further Directions

Further investigations of indirect effects are important both for enhancing our understanding and therefore improving management of specific ecosystems, and for general development of ecology. Due to the increasing pace of technological progress, collection of monitoring data using automated techniques, and in particular remote sensing, is becoming increasingly easy. Combined with rapidly increasing computing power and progressive development of mathematical methods, this may provide the necessary basis for a dramatic acceleration of investigations of indirect effects, in particular the ones manifesting on a global level, and in cases when the geographical separation is an issue. With progressive accumulation of long-term data sets, it is also likely that the effects occurring after a time lag will become more readily discernable. To maximize the benefits (i.e., for investigations of indirect effects) from the technological development, however, it should be supplemented by contemporary advances in the methodology of their investigations.

It has been argued that analysis of indirect relationships in ecosystems may be greatly enhanced by the application of a specialized methodological framework, called CTEA. CTEA is aimed at bringing together separate lines of current investigations, hence combining them in an integrative approach (see the section titled 'Further reading'). Further development and systematic application of CTEA is vital for improving the accuracy of ecological forecasting, and has, therefore, potential societal benefits related to issues of EIA and sustainable development. It is suggested that further developments should pay much attention to similarities and differences of the indirect effects revealed in various types of ecosystems, or at different stages of the ecosystem development, and that the characteristics (e.g., magnitude, sign, etc.) of indirect interactions should be increasingly used for describing differences in ecosystem state, structure, and overall functioning. For example, analysis of specific ecosystems may benefit from answering the following (to name but a few) questions:

- What types of indirect effects are important for the overall functioning of an ecosystem under investigation?
- How does the importance of indirect effects compare with the importance of direct effects?
- Is the pattern of indirect effects relatively constant, or subject to (system specific) seasonal and longer-term changes?
- How does the pattern of indirect interactions change due to pollution, disturbance, and various management practices?
- Do indirect interactions predominant in an ecosystem help to stabilize this ecosystem?
- What is the relative contribution of indirect interactions to resistance, resilience, and facilitation of successional changes?
- How have the indirect effects changed during the evolution of a particular ecosystem, and what was their contribution toward the driving forces of this evolution?

Further Reading

Abrams PA, Menge BA, Mittelbach GG, Spiller DA, and Yodzis P (1996) The role of indirect effects in food webs. In: Polis G and Winemillor K (eds.) *Food Webs: Integration of Patterns and Dynamics*, pp. 371–395. New York: Chapman and Hall (also see other papers in this book).

Budiko MI (1977) *Global'naya Ekologiya (Global Ecology)*. Moscow: Misl'(in Russian).

Fath BD and Patten BC (1998) Network synergism: Emergence of positive relations in ecological systems. *Ecological Modelling* 107: 127–143.

Fath BD and Patten BC (1999) Review of the foundations of network environ analysis. *Ecosystems* 2: 167–179.

Fleeger JW, Carman KR, and Nisbet RM (2003) Indirect effects of contaminants in aquatic ecosystems. *Science of the Total Environment* 317: 207–233.

Herendeen RA (1998) *Ecological Numeracy. Quantitative Analysis of Environmental Issues*. Toronto: Wiley.

Kawanabe H, Cohen JE, and Iwasaki K (1993) *Mutualism and Community Organisation: Behavioral, Thoretical and Food Web Approaches*. Oxford: Oxford University Press.

Korhonen J (2001) Industrial ecosystems – Some conditions for success. *International Journal of Sustainable Development and World Ecology* 8: 29–39.

Krivtsov V (2004) Investigations of indirect relationships in ecology and environmental sciences: A review and the implications for comparative theoretical ecosystem analysis. *Ecological Modelling* 174(1–2): 37–54.

Menge BA (1995) Indirect effects in marine rocky intertidal interaction webs – Patterns and importance. *Ecological Monographs* 65: 21–74.

Miller TE and Travis J (1996) The evolutionary role of indirect effects in communities. *Ecology* 77: 1329–1335.

Patten BC, Bosserman RW, Finn JT, and Gale WG (1976) Propagation of cause in ecosystems. In: Patten BC (ed.) *Systems Analysis and Simulation in Ecology*, pp. 457–579. New York: Academic Press.

Schooner TW (1983) Field experiments on interspecific competition. *American Naturalist* 122: 240–285.

Strauss SY (1991) Indirect effects in community ecology – Their definition, study, and importance. *Trends in Ecology and Evolution* 6: 206–210.

Wardle DA (2002) *Communities and Ecosystems. Linking the Aboveground and Belowground Components*. Princeton: Princeton University Press.

Wootton JT (1994) The nature and consequences of indirect effects in ecological communities. *Annual Review of Ecology and Systematics* 25: 443–466.

Wootton JT (2002) Indirect effects in complex ecosystems: Recent progress and future challenges. *Journal of Sea Research* 48: 157–172.

Emergent Properties

F Müller, University of Kiel, Kiel, Germany

S N Nielsen, Danmarks Farmaceutiske Universitet, Copenhagen, Denmark

Introduction
The History of the Concept
Emergence and Hierarchy
How Emergence Emerges

Classification of Emergent Properties
Quantifying Emergence
Further Reading

Introduction

Many biologists will recognize the statement that "the whole is more than the sum of the parts" as a commonly used (but hardly understood) phrase. This formulation refers to the idea that there are systems which possess additional qualities or quantities, beyond easily measurable or predictable physical parameters. The resulting properties have been described on many levels of the biological hierarchy, from simple physical systems, like laser beams, to the organization of the whole biosphere within the Gaia concept of J. Lovelock. The emergent property at one level is in general finding its causality at the subsystem components and the interaction between them. For example, an organized form of cell functions, stemming from self-organized transformations of cellular compounds, known as hypercycling may be considered as an emergent entity; at the physiological levels we are, for example, dealing with the mating behaviors of organisms as results of hormone interactions. Similarly, motion, feelings, or intelligent behavior occur as a consequence of special couplings of neurons. In addition, the patterns in the development of ecosystems may not be predictable from knowledge of organisms alone. Therefore, emergent properties are not unusual phenomena, but simply consequences of hierarchical organizations.

The concept of emergence found its way into ecology through the proposal of E. P. Odum, who suggested that the study of emergence should lead to a 'new integrative discipline'. This idea was due to the fact that studies of complex systems had shown that the investigations of the details alone were not adequate in predicting ecosystem function and behavior. Neither were they sufficient to explain a more advanced pattern like behavior and performance of ecosystems, for example, during the succession from young systems toward more mature states.

The History of the Concept

The concept of emergent properties originates in the nineteenth century, finding the primary roots back in Kantian philosophy. The term was coined by G. H. Lewis as far back as 1875. A common definition from that time states that "emergence is the denomination of something new which could not be predicted from the elements constituting the preceding condition."

Throughout the last century, several scientists have addressed the concept from a more philosophical point of view, resulting in the appearance of different descriptions and explanations. The definitions have, in general, been referring to subjective arguments, such as surprise, unexpectancy, thus being clearly observer dependent. This has strongly influenced the present approaches, and this comprehension has often been connected with a flavor of mysticism. Thus, the seriousness of the concept has often been underestimated.

During the last decades, the use of the term emergent properties has found widespread use in biological sciences, especially, because it is clearly connected with the growing implementation of the system approach in ecology. The need for a holistic concept was due to the failure of the traditional reductionistic research strategies to explain the properties of ecosystems by the knowledge of the behavior and the properties of the ecosystem constituents alone. Ecosystems are highly complex middle-numbered systems dominated by nonlinear relationships between their constituents. In such systems, things are

bound to happen that are not easy to predict from the basic knowledge of the system, no matter how extensive this knowledge is.

Highly relevant to biology and ecology is the question when an emergent property appears. This leads to the distinction of 'primary' and 'secondary' emergence, primary emergence being the first time an emergent property appears. To be conserved the property can be reproduced again and again but in this case it is nominated as a secondary emergence. Recent approaches to emergence have come up with three further notions of emergence: 'computational emergence', 'thermodynamic emergence', and 'emergence relative to a model'. The computational emergence deals with the patterns produced by different computer programs, for example, cellular automata systems developing complex distributions out of simple rules from game theory. Thermodynamic emergence covers the establishment of highly complex, self-organized structures and their relations to the nonlinear, far from equilibrium thermodynamics. Emergence relative to a model defines emergence as the deviation of the actual behavior of a physical system in comparison with an observer's model of it.

Summarizing these historical notions of emergence, the following features can be stated:

- Emergent properties are properties of a system which are not possessed by component subsystems alone.
- The properties emerge as a consequence of the interactions within the system.
- Two fundamental types of interactions are found that may be characterized as intra- and inter-connectedness, that is, connections within and between levels, including controls. This point does not consider the direction of the intra-level interactions. Emergence is based on both, upward and downward causation.
- The historically emerged properties are considered 'new' with reference to their primary appearance.
- These new properties appear at one level of a system and are not immediately deducible from observation of the levels or units of which the system consists.

Emergence and Hierarchy

Emergence has been described at many levels of the biological hierarchy. As argued above, the reason for emergence is to be found in the hierarchical organization of the system and the quantitative and qualitative characters of the 'linkages' within the structure. As biological structures are often complex, this makes it hard to determine the actual cause of emergence.

Hierarchy theory (see Hierarchy Theory in Ecology) states that middle-numbered systems – such as ecosystems – can be comprehended if they are investigated on different levels of integration. Broad scale levels can be assigned to high spatial extents, and low typical frequencies, filtering the signals from lower levels. These scale levels are spatially smaller and their typical frequencies are higher. They are not able to filter constraints from the higher levels, but their potentials and interactions are building up the material basis and the coordination functions of the higher level. Emergent properties are created by both types of nonlinear interactions. Therefore, the properties of specific levels can be termed 'hierarchical emergent properties'. Of course, the interesting question is how these properties emerge. Some examples might be helpful to illuminate this question.

Prebiological Emergence

Several examples of emergent properties can be found in physical and chemical sciences. They form an important prerequisite for protobiology and evolutionary processes. Within the area of physics some examples are nearly classical: Water (e.g., its wetness), which is a simple molecule with a rather complex behavior, that is unpredictable from knowledge about oxygen and hydrogen alone, has often been used to demonstrate emergent properties. Similarly, the sense of colors by the eyes is not predictable by knowing a certain wavelength of light.

Two famous chemical examples related to selforganized behavior of systems may also be mentioned, the Bénard cells and the Bhelusov–Zabotinsky (BZ) reaction. In the case of Bénard cells, during specific conditions, hexagonal, convective cells (the emerging structures) form a fluid when a thermal gradient is imposed on the experimental setup containment. In the BZ-reaction a special ratio of chemicals causes a mixture to perform a pulsing pattern in colors with a period of about one minute. The structure of these physicochemical processes, gradients resulting in convective cells, pulsing patterns, together with other observations like the occurrence of Turing structures in chemical fluids, spontaneous formation of lipid coacervates, might be crucial to our understanding of the emergence of life.

Protobiological Emergence

The appearance of the earliest life forms has often been referred to as a primary emergence. Although many of the properties occurring during this phase of evolution have been repeated, over and over again their appearance still qualifies them as emergent properties. As examples, the emergence of life, emergence of animals, or the emergence of bird feathers from reptile scales can be mentioned to characterize situations of primary emergence.

Many examples found in the literature deal with the formation of the earliest cells. Biochemical cycles, the organization and exchange of information by DNA or RNA and the compartmentalization of material within membranes are but a few examples. Molecular complementarity, defined as "nonrandom, reversible coupling of the components of a system," has been argued to be a widespread mechanism in biological systems and important for the understanding of the processes lying behind emergent properties. The seemingly (self)organization of molecules observed in prebiotic systems, such as Turing structures and autocatalytic hypercycles (see Autocatalysis), can be seen as emergent properties already at a very low level of organization.

Emergence in Biological Systems

Emergent properties really come into play when biological systems reach higher levels of complexity. This becomes evident already when cells or groups of cells communicate with each other as in the case of hormones and natural neural networks. Organs are composed of cells, their individual functions are important only to the organism as a whole. A heart, kidneys or lungs, are vital but their function is not existent when they are on their own. Organisms interacting as populations or societies provide properties which cannot be explained by properties of the individual organisms alone. They all go together in what we consider as ecosystems and thus are a part of the biosphere.

Cellular level. In regarding the outcome of interacting cells many studies have been concentrating on the organization of neuronal systems, which result in unexpected properties like the ability to move, to sense, to be intelligent, and to emote. The sensory systems, being connected to visual, auditory, or other communicative processes are all playing a major role in how successful living organisms are in performing specific life strategies. Reliable senses, and responding the right way to the received stimuli are crucial to the existence of many life forms, in processes like finding food, knowing when and where to escape, or creating bonds with other members of the species, for example, during reproduction.

Neural networks, like in our brains, consisting of a huge number of interconnected neurons, are so complex that unforeseen patterns in responses are bound to occur and have also been reported to exist. During the evolution of the brain, emergent properties, together with new cell types, local and large circuits have added up to the increasing complexity of brain function. Motor control, the control and coordination of motor activity are taken care of by our brain passing on signals to the limbs or organs involved.

Organ level. Numerous cells, often during morphogenesis differentiated in certain, specialized directions, form organs, take up a particular task of the organism, like for instance liver cells secreting enzymes, or kidney cells filtering and cleaning the coelom. Although the formal 'layout' for this functionality is existent in the genetic material of all cells, the eventual determination occurs during the development of the organism and the actual function of the organs may be viewed as emergent. The brain as an organ may serve as an example of this emergence: Here differentiated cells, with highly specialized physiological properties, go together and create activity patterns that are far more complex than expected from knowing the physiology of neural cells alone. The whole becomes more than the sum of the parts.

Organism level. Complex behavior occurs among the individual organisms that cannot be determined exclusively by internal factors. The sending, reception and interpretation of signals from interagent organisms, the relationship(s) to the outside, and thus semiotics play an important role, creating patterns impossible to foresee if only the subsystems are known. For example, in trees, the formation of branches and leaf mosaics have been studied in a number of recent investigations with modeling approaches as well as the allocation of resources between above- and belowground biomass and the related physiological mechanisms. A modeling study of this problem indicates a 'complex integrated growth pattern' which may only be understood as an emergent property as it is claimed to have no direct or indirect mechanistic basis related to subcellular activities. In a similar manner it was shown that whole-plant behavior is an emergent property arising from a rule-based model of the system. Communications between individuals, that is, their social interactions within a population, are important to the function of the organism as a whole and are indistinguishable from the emergence of ethological features. Stressing the importance of communication, may lead to an interpretation of the communicative process as an emergent interpretation of signs, which is described within the discipline of semiotics.

Population level. Populations are composed of individual organisms, interacting in various ways, differing in quantity and quality, throughout the biological system. The interactions may vary in character according to the complexity. At the one end of the spectrum, we find the single cell organisms interacting mostly on a material basis (matter fluxes). At the other end, there are colonial organisms forming complex societies, where brains, senses, memory, and thus informational interaction become dominating. Emergent properties as a result of individual level behavior and interactions in populations of social insects have been argued in several studies. For instance, the distribution of food to larvae of the fire ant has been argued as emerging from interactions between individuals, workers, and larvae. Cellular automata models were used to study the short time oscillations in ant colonies. The nonlinear dependencies describing the relationships between, and the movement of, individuals

explain this behavior. The resulting oscillations were found to be emergent properties of the colony.

Ecosystem level. Ecosystems are inherently complex as they are composed of an embedded hierarchy of all the previously mentioned subsystems in close interaction with abiotic factors. Emergence is to be expected, but surprisingly few reports exist at this level, before all analyzing microcosms, forest ecosystems, predator–prey relationships, food webs, and the organization of aquatic communities.

Ecosystem behavior is often analyzed through modeling studies. The relation to emergent properties becomes clear when looking at recent efforts of structural dynamic modeling, where the changes in ecosystem composition and structure over time are analyzed. Another example is the work of B. C. Patten on the propagation of matter–energy through the ecosystem network, leading to the discovery of the importance of 'indirect effects, quantitative and qualitative utilities' of the system, results that are highly surprising and unexpected, and as such are emergent properties (see Ecological Network Analysis, Environ Analysis). Both examples link to higher-level information expressions such as ascendency, different kinds of entropy or information derived descriptors like exergy (see Exergy).

The ability of the ecosystem to perform with systematic directional changes in some macroscopic characters, not predictable from knowledge about the single ecosystem members alone, has been discussed since 1967 on the basis of the 24 principles of ecosystem development during succession in the second edition of E. P. Odum's *Fundamentals of Ecology.* Many other factors, known as indicators, orientors, or goal functions have been presented since then (**Table 1**).

How Emergence Emerges

The concept of emergent properties refers very clearly to, and must be seen in tight connection with, at least two other concepts often occurring in literature on modern ecosystem theory, the concepts of hierarchy and self-organization. In connection with hierarchy, the emergent properties are seen as outcomes of ecosystem organization where supersystems are formed with subsystems as constituents and where the properties are observable at the supersystem level only. Here the emergent property is an outcome of a certain way of organization. To exemplify this point, we might look at the following hierarchical features:

1. Individual level: individual nutrition budgets – foraging strategies.
2. Population level: species nutrition efficiencies – intraspecific food competition.
3. Ecosystem level: nutrient cycling – food webs.

Table 1 Some ecosystem orientors

Immature state	*Mature state*
Properties of the dominating species	
Rapid growth	Slow growth
R-selection	K-selection
Quantitative growth	Qualitative development
Small size	Large size
Short life spans	Long life spans
Broad niches	Narrow niches
Properties of production	
Small biomass	Large biomass
High P/B ration	Low P/B ratio
Low respiration	High respiration
Small gross production	Medium gross production
Properties of nutrient flows and cycles	
Simple, rapid, and leaky	Complex, slow, and closed cycles
Small storage	Large storage
Extrabiotic	Intrabiotic nutrient distribution scheme
Small amounts of detritus	Large amounts of detritus
Rapid nutrient exchange	Slow nutrient exchange
Short residence times	Long residence times
Minor chemical heterogeneity	High chemical heterogeneity
Loose network articulation	High network articulation
Low diversity of flows	High diversity of flows
Undeveloped symbiosis	Developed symbiosis
Properties of the community	
Low diversity	High diversity
Poor feedback control	Developed feedback control
Poor spatial patterns	Developed spatial patterns
Thermodynamic and integrative system properties	
Poor hierarchical structure	Developed hierarchical structure
Close to equilibrium	Far from equilibrium
Low exergy storage	High exergy storage
Small total entropy production	High total entropy production
High specific entropy production	Small specific entropy production
Small level of information	High level of information
Small internal redundancy	High internal redundancy
Small path lengths	High path lengths
Low ascendency	High ascendency
Poor indirect effects	Developed indirect effects
Small respiration and evapo-transpiration	High respiration and evapo-transpiration
Small energy demand for maintenance	High energy demand for maintenance

The features, that are optimized throughout natural successions, provide several characteristics of emergent properties: They are only observable at the ecosystem level (which is the typical and the lowest logical level to describe, e.g., cycling phenomena), and they are based on self-organized processes. They can not be explained on the basis of knowledge of the parts alone, and the emergence-creating processual linkages between the subsystems are nonlinear processes. From the hierarchy-based viewpoint also the additive features (e.g., size, biomass, life spans) can be categorized as emergent properties because their extensions are dependent on the scale of observation and because they also are based on internal system interrelations.

4. Landscape level: lateral nutrient transfers – food webs including large scale predators.

On the other hand, the ability of biological systems to arrange themselves in a special manner, for example, in a hierarchical way, is in itself a property which emerges as a consequence of the properties of its constituents, but the organization and the function for sure cannot always be foreseen. Thus, the capability of self-organization can be seen as an emergent property itself (**Figure 1**).

The existence of emergent properties is based on the system's organization (built up by structures and functions) whereby the interrelations (energy, matter, water and information flows, communications) play an important role. Some conditions of the system's state add up to the increased chances that emergent properties will appear. For example, instabilities seem to be important conditions that support emerging processes, especially referring to evolutionary emergence. Stable periods may lead to the emergence of new structures through bifurcations. As systems move toward the state of minimum

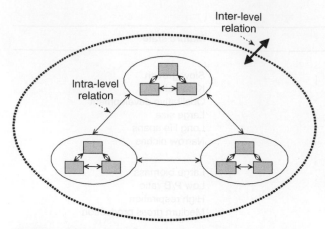

Figure 1 Biological entities are often organized in a hierarchical manner, whereby the emergent properties of a certain level are based on the interrelations between the lower levels, while both are constrained from the highest level linkages.

dissipation they are, at the same time, moving toward bifurcation points with possibilities of further evolution to occur. Similarly broken symmetries, complementarity has been proposed as a global mechanism.

Classification of Emergent Properties

From the presentation of the concepts above it can be seen that emergence and emergent properties will not easily find a clear, consistent and unifying definition for covering all the cases described. The widespread and

'loose' use of the concepts over a vast range of areas at first glimpse simply shows confusion. However, it is possible to establish some typology of the areas where, and the ways in which, the concepts have been used, following **Figure 2**.

First, emergent properties might appear through evolution of the systems, primary emergence, hereafter only being repeated. This characteristic may be called 'evolutionary emergence'. As structures are integrated, new organizational forms, as previously mentioned often hierarchical, occur ('hierarchical emergence').

Taking the view that emergent properties do exist and that the reductionistic approach to science will not (dis)-solve the problem so it eventually disappears may allow us to establish a schematic relationship between the various categories of ecosystem properties.

One major line follows a direction of research problems, the search for the unexplained and not understood. This lies close to using emergent properties as research strategies, while the extreme leads to the reductionist approach. This is more or less the situation at the second line, where properties are 'collective' and additive, that is, that the properties are the sum of the whole, and may be explained at subsystem level, provided sufficient knowledge exists. At the other end, the attitude that only holistic studies will lead to increased understanding might be taken.

Along the third line, we find the core of emergence, and following the above points the respective features may be divided in an evolutionary line and in a hierarchal line. Here emergence is basically represented as a function of time and space. The evolutionary process was

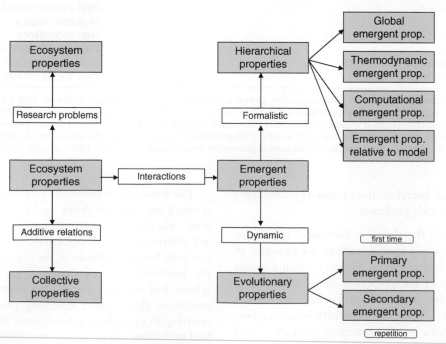

Figure 2 An attempt to form a typology of emergent properties.

described above and deals with primary and secondary emergence. The organizational, hierarchical line includes four areas described in the previous sections: 'global emergent properties' as a function of local rules and local interactions, 'thermodynamic emergent properties' dealing with emergence as a consequence of mainly the second thermodynamic law, the emergence of (dissipative) structures as a result of thermodynamic gradients. 'Computational emergence' is also based on global patterns emerging form local rules. As mentioned above, emergent properties also appear as a result of models being used to analyze the problem, which is called 'emergent property relative to a model'.

Quantifying Emergence

Several authors argue that in any attempts to formalize or quantify the concepts, true emergent properties should be observer independent. This does not necessarily mean that emergent properties should be observation-independent. Observations undertaken by different methods result in differences in acquired knowledge. This means that emergent properties can be defined as the differences in knowledge gained by the observation of a system by two different methods. This is partly reflected by the computational emergence.

It is this observer dependency that leaves a way open for the quantification of emergent properties. Emergent properties could then be expressed in a semiquantitative way by the use of an 'index' derived of Kullback's measure of information (**Figure 3**). This involves moving the normal reference frame in information theory assuming the *a priori* knowledge of the system to be zero, which is not necessarily the case.

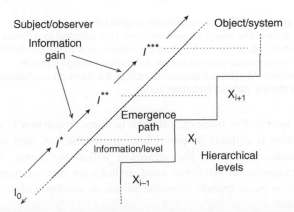

Figure 3 Quantification of emergence, based on Kullback's measure of information, might be carried out from quantifying the difference between actual observed, *a posteriori*, behavior or composition of a system and what may be predicted from *a priori* knowledge about subsystems. The analysis may be carried out at various levels of hierarchy, differing in emergence value.

Rather in ecology, we do possess some knowledge about the system and what we usually refer to are the deviations in what we observe in the systems or models of systems compared with our expectations built on previous knowledge. The way of quantifying emergence has to be built on the use of computers and models. If our knowledge gained hitherto is synthesized and treated in a computer model (from traditional ecological science) is p^*, and the outcome of an experiment or observations of a system differs by p^{**} the emergent properties can be calculated by the following:

$$\text{Emergence} = \sum p^{**} \ln \frac{p^{**}}{p^*}$$

which correlates emergence to the concept of exergy. Emergence now is a consequence of information gained between observations.

The question is if emergence in this manner will, at the end, dissolve itself and disappear as knowledge increases, which refers to the above debate of reductionism versus holism.

Many of the concepts used to characterize ecosystems are based on various numerical treatments of data observed in the ecosystem. Since the concepts are immediately deducible (calculable) from certain knowledge about the components of the ecosystem, for example, numbers, species, biomass etc. such concepts cannot be coined as emergent property but rather as a 'collective' property of the system. An interesting corresponding analog in this context are the macroscopic properties from thermodynamics such as entropy and parallels in formulation of formulas. Reductionism cannot win the debate since it will be impossible to achieve enough knowledge. If not for anything else, then for thermodynamic reasons, since the achievement of more and more detailed knowledge becomes more and more expensive in terms of not only energy but also dissipation.

Meanwhile, what strikes is that such a traditional, vertical organization of systems is not mandatory in order to produce emergent behavior. Vertical, here, refers to levels being either higher or lower in the hierarchy. Rather only parts are needed, of which none have actual regulatory functions and therefore should be evaluated or ranked higher than the other(s). Emergent properties can occur also in horizontally organized systems, emergence appearing alone as a consequence of interactions at the same level. The study of these intra-level relationships and their consequences to the higher levels in the hierarchy may be important to investigate in the future.

See also: Autocatalysis; Ecological Network Analysis, Environ Analysis; Exergy; Hierarchy Theory in Ecology.

Further Reading

Bhalla US and Iyengar R (1999) Emergent properties of networks of biological signalling pathways. *Science* 283: 381–387.

Breckling B, Müller F, Reuter H, Hölker F, and Fränzle O (2005) Emergent properties in individual-based ecological models – introducing case studies in an ecosystem research context. *Ecological Modelling* 186: 376–388.

Cariani P (1992) Emergence and artificial life. In: Langton G, Taylor C, Farmer JD, and Rasmussen S (eds.) *Artificial life II*, pp. 775–797. Redwood City: Addison-Wesley.

Conrad M and Rizki MM (1989) The artificial worlds approach to emergent evolution. *BioSystems* 23: 247–260.

Emmeche C, Køppe S, and Stjernfelt F (1993) Emergence and the Ontology of Levels. *In Search of the Unexplainable. Arbejdspapir.*

Afdeling for Litteraturvidenskab. Copenhagen: University of Copenhagen.

Morgan CL (1923) *Emergent Evolution. Williams and Norgate.*

Nielsen SN and Müller F (2000) Emergent properties of ecosystems. In: Joergensen SE and Müller F (eds.) *Handbook of Ecosystem Theories and Management*, pp. 195–216. Boca Raton, FL: Lewis Publishers.

Salt GW (1979) A comment on the use of the term emergent properties. *American Naturalist* 113(1): 145–148.

Wicken JS (1986) Evolution and emergence. A structuralist perspective. *Rivista di Biologia/Biology Forum* 79(1): 51–73.

Wieglieb G and Bröring U (1996) The position of epistemological emergentism in ecology. *Senckenbergiana maritima* 27(3/6): 179–193.

Self-Organization

D G Green, S Sadedin, and T G Leishman, Monash University, Clayton, VIC, Australia

Introduction	Self-Organization in an Ecological Setting
Historical Comments	Practical Considerations
Theories of Self-Organization	Further Reading

Introduction

Self-organization is the appearance of order and pattern in a system by internal processes, rather than through external constraints or forces. Plant distributions provide examples of both constraints and self-organization. On a mountainside, for instance, cold acts as an external constraint on the ecosystem by limiting the altitude at which a plant species can grow. Simultaneously, competition for growing sites and resources leads to self-organization within the community by truncating the range of altitudes where plant species do grow. Self-organization can also be seen among individuals within a population (e.g., within an ant colony or a flock of birds) and within individuals (e.g., among cells during development) (**Figure 1**).

A growing understanding of ways in which internal processes contribute to ecological organization has provided new perspectives on many phenomena familiar from traditional ecology. Self-organization usually involves interactions between components of a system, and is often closely identified with complexity. Also associated with self-organization is the idea of emergence: that is, features of the system emerge out of interactions, as captured by the popular saying, "the whole is greater than the sum of its parts." It is necessary to distinguish between emergent features and other global properties of

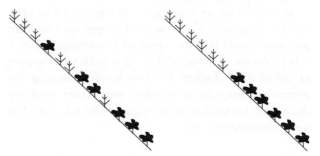

Figure 1 Effect of competition on plant distributions on a gradient. The two plant species shown are adapted to different conditions, which are here found at either end of the slope. At left, there is no competition, so the distributions merge into one another. At right, competition truncates the distributions, leading to sharply defined altitudinal zones.

a system. For instance, although biomass production in a forest is a global property, it is simply the sum total of production by all the organisms within the forest. A stampede, on the other hand, is behavior that emerges when panic spreads from one animal to another within a herd.

Semantic and philosophical issues sometimes lead to confusion about self-organization. Self-organizing systems are usually open systems, that is, they share information, energy, or materials with their surroundings. However this does not necessarily mean that the external environment controls or determines the way they organize. A growing plant, for instance, absorbs water, light, and nutrient from its environment, but its shape and form are determined largely by its genes.

Also, in considering self-organization, it is important to clearly identify the system concerned, and in particular, what is external and what is internal? This issue arises in the difference between a community and an ecosystem. For a community, which consists of the biota of an area, the effect of (say) soil is an external constraint. However, for the corresponding ecosystem, which would include soils, the interactions between plants, microorganisms, and soil formation are internal processes. Defining the physical limits of an ecosystem poses similar problems. A lake, for instance, is not a closed ecosystem. Among other things, water birds come and go, removing some organisms and introducing others.

Historical Comments

Self-organization as a widespread phenomenon first came to the attention of researchers during the mid-twentieth century. The interest in self-organization comes from many different fields of study. The biologist Ludwig von Bertalanffy drew attention to the role of internal interactions and processes in creating organization within biological systems. His 'general systems theory' drew heavily on analogies to highlight the existence of common processes in superficially different systems. Meanwhile, W. Ross Ashby and Norbert Wiener explored self-organization from the perspective of communications and feedback in the control of systems. Ashby introduced the term self-organizing in 1947. Wiener coined the term cybernetics to refer to the interplay of control systems and information. In the 1950s, systems ecologist H. T. Odum collaborated with engineer Richard Pinkerton to develop the principle of maximum power, which states that systems self-organize to maximize energy transformation.

During the 1970s and 1980s, increasing computing power made it possible to use simulation to explore the consequences of complex networks of interactions. By the last two decades of the twentieth century, the nature and implications of biological self-organization were increasingly being explored as a part of the complexity theory. The new field of Artificial life (Alife), initiated by pioneers such as Chris Langton, Pauline Hogeweg, and Bruce Hesper, has produced a series of seminal models that demonstrate self-organization in a variety of ecological and evolutionary contexts. Around the same time, H. T. Odum introduced the systems concept of 'emergy' to represent the total energy used in developing a process.

By the 1990s, researchers were looking for broad-based theories of self-organization. John Holland stressed the role of adaptation in self-organization. He suggested that seven basic elements are involved in the emergence of order in complex adaptive systems. These include four properties – aggregation, nonlinearity, flows, and diversity – and three mechanisms – tagging, internal models, and building blocks. In contrast, Stuart Kauffman's work on autocatalytic sets within Boolean networks emphasizes ways in which self-organization may structure biological systems independent of selection. Likewise, embryologist Brian Goodwin suggested that to understand macroevolution, we require a theory of morphogenesis which takes account of physical, spatial, and temporal dynamics in addition to selection. The work of James Kay provided an interpretation of life from a thermodynamic perspective, arguing that self-organizing systems maximize the dissipation of gradients in nature. In particular, Kay argues that over time, ecosystems evolve to dissipate energy more efficiently by becoming increasingly complex and diverse.

Theories of Self-Organization

Thermodynamic Basis

In physical terms, the phenomenon of self-organization appears at first sight to be ruled out by the second law of thermodynamics, which states that in any closed system, entropy increases with time. In this sense, living systems seem to fly in the face of thermodynamics by accumulating order. However, self-organizing systems need not be closed. Open systems, including living things, share energy and information with the outside environment. In the late 1960s, Ilya Prigogine introduced the idea of dissipative systems to explain how this happens. He defined dissipative systems to be open systems that are far from equilibrium. Dissipative systems have no tendency to smooth out irregularities and to become homogeneous. Instead, they allow irregularities to grow and spread. Physical examples include crystal formation. Biological systems, including cells, organisms, and ecosystems, are all examples.

The Network Model

An important source of self-organization is provided by the interactions and relationships between the objects that comprise a complex system. Patterns of such relationships are captured by the network model of complexity.

Networks capture the essence of interactions and relationships, which is a fundamental source of complexity. A graph is defined to be a set of nodes (objects) joined by edges (relationships) and a network is a graph in which the nodes and/or edges have values associated with them. In a food web, for instance, the populations form nodes, and the interactions between them (e.g., predation) form the edges. In a landscape, spatial processes and relationships create many networks. For instance, the nodes might be individual plants and the corresponding edges would be any processes that create relationships between them, such as dispersal or overshading. In an animal social group, the nodes would be individuals and the edges would be relationships such as kinship or dominance.

Nodes that are joined by an edge are called neighbors. The degree of a node is the number of immediate neighbors that it has. A path is a sequence of edges in which the end node of one edge is the start node of the next edge, for example, the sequence of edges A–B, B–C, C–D, D–E forms a path from node A to node E. A cycle is a path that ends where it starts, for example, A–B, B–C, C–A. A network is called connected if, for any pair of nodes, there is always some path joining them (otherwise it is disconnected). The diameter of a network is the maximum separation between any pair of nodes. Clusters are highly connected sets of nodes.

The importance of networks stems from their universal nature. Network structure is present wherever a system can be seen to be composed of objects (nodes) and relationships (edges). Less obvious is that networks are also implicit in the behavior of systems. In this respect, the nodes are states of the system (e.g., species composition) and the edges are transitions from one state to another.

Sometimes, network structure plays a more important part in determining the behavior of a system than the nature of the individual components. In dynamic systems, for instance, cycles are associated with feedback loops. In disconnected networks, the nodes form small, isolated components, whereas in connected networks, they are influenced by interactions with their neighbors. Self-organization in a network can occur in two ways: by the addition or removal of nodes or edges, or by changes in the values associated with the nodes and edges.

Several kinds of network patterns are common and convey important properties.

- A random network is a network in which the nodes are connected at random. In a random network of n nodes, the degrees of the nodes approximate a Poisson distribution, and the average length (L) of a path between any two nodes is given by $L = \log(n)/\log(d)$, where d is the average degree.
- A regular network is a network with a consistent pattern of connections, such as a lattice or cycle.
- Small worlds fall between random networks and regular networks. They are typically highly clustered, but with low diameter. A common scenario is a system dominated by short-range connections, but in which some long-range connections are also present.
- A tree is a connected network that contains no cycles. A hierarchy is a tree that has a defined root node. For instance, the descendents of a particular individual animal (the root of the tree) form a hierarchy determined by birth. Trees and hierarchies are closely associated with the idea of encapsulation.
- A scale-free network is a connected network in which the degrees of the nodes follow an inverse power law. That is, some nodes are highly connected, but most have few (usually just one) connections.

Encapsulation

Encapsulation is the process by which a set of distinct objects combine to act as a single unit. Individual fish, for example, form a school by aligning their movements with their neighbors. Because smaller objects usually merge into larger wholes, encapsulation is often linked to questions of scale. Encapsulation is closely associated with the idea of emergence. The whole emerges when individuals become subsumed within a group in relation to the outside world. There are many examples in ecology. Ecosystems are communities of interacting organisms; populations are groups of interbreeding organisms; and schools, flocks, and herds are groups of animals moving in coordinated fashion. In all of these cases, the individuals may not be permanently bound to the group, unlike cells within the human body. Cellular slime molds present an intermediate case in which cells sometimes act independently but at other times aggregate to form a multicellular individual.

Various ecological theories are based on the assumption that encapsulation plays an important role in ecosystem structure and function. The concept of ecosystem compartments implies that a community is formed of distinct groups (compartments) consisting of mutually interacting species, but the interactions between the groups are limited.

Connectivity and Criticality

Criticality is a phenomenon in which a system exhibits sudden phase changes. Examples include water freezing, crystallization, and epidemic processes. Associated with every critical phenomenon is an order parameter, and the phase change occurs when the order parameter reaches a critical value. For example, water freezes, when its temperature falls to $0\,°C$. A wildfire spreads when fuel moisture falls below a critical level (else it dies out).

Changes in the connectivity of a network have important consequences and often underlie critical phenomena. When a network is formed by adding edges at random to a set of N nodes, a connectivity avalanche occurs when the number of edges is approximately $N/2$. This avalanche is characterized by the formation of a connected subnet, called a unique giant component (UGC), which contains most of the nodes in the full network. The formation of the UGC marks a phase change in which the network shifts rapidly from being disconnected to connected.

Any system that can be identified with nodes and edges forms a network, so the connectivity avalanche occurs in many settings and is the usual mechanism underlying critical phase changes.

The connectivity avalanche has several important implications. For interacting systems, it means that the group behaves either as disconnected individuals, or as a connected whole. Either global properties emerge, or they do not: there is usually very little intermediate behavior. Landscape connectivity provides an important ecological example of critical phase change.

Phase changes in connectivity also underlie criticality in system behavior. The degree of connectivity between states of a system determines the richness of its behavior. Studies based on automata theory show that if connectivity is too low, systems become static or locked in narrow cycles. If connectivity is too high, systems behave chaotically. The transition between these two phases is a critical region, popularly known as the 'edge of chaos'. It has been observed that automata whose state spaces lie in this critical region exhibit the most interesting behavior. This observation led researchers such as James Crutchfield, Christopher Langton, and Stuart Kauffman to suggest that automata need to reside in the critical region to perform universal computation. More speculative is their suggestion that the edge of chaos is an essential requirement for evolvability (the ability to evolve) in complex systems, including living things. Others have proposed that living systems exploit chaos as a source of novelty, and that they evolve to lie near the edge of chaos. These ideas are closely related to self-organized criticality (SOC).

Self-Organized Criticality

SOC is a phenomenon wherein a system maintains itself in a critical or near-critical state. A classic example is the pattern of collapses in a growing pile of sand. Because information theory suggests that systems in critical states are most amenable to information processing and complexity, self-organized criticality has been proposed as a component of collective behavior in ant colonies, societies, ecosystems, and large-scale evolution. SOC is characterized by events whose size and frequency distributions follow an inverse power law. However, it is often difficult to distinguish genuine cases of SOC from simple cause and effect processes that exhibit similar distributions.

For example, ecosystems might tend toward critical states through the following mechanism. If new species or mutations appear in an ecosystem occasionally, then as the variation in the ecosystem increases over time, so does the probability of forming destabilizing positive feedback loops. Such destabilizing interactions could initiate avalanches of extinctions, and the probable size of such avalanches would be related to the preexisting connectivity of the system. In this way, mutation, migration, and extinction could keep the system near the critical region, as the addition of new variation drives the ecosystem out of subcriticality, while extinction avalanches prevent supercriticality. Proponents of this idea point to extinction events, whose distribution follows an inverse power law, as supporting evidence. However, other explanations of this pattern, such as cometary impacts, are also plausible.

Feedback

Feedback is a process in which outputs from a system affect the inputs. Predator–prey systems are examples of negative feedback. For instance, any increase in the size of a predator population means that more prey are caten, so the prey population decreases, which in turn leads to a decrease in the predator population. Reproduction is an example of positive feedback: births increase population size, which in turn increases the rate of reproduction, which leads to yet more births. Feedback loops arise when a sequence of interactions form a closed loop, for example, A–B–C–A. Feedback loops play an important role in food webs and ecosystem stability. Time delays in the response within a feedback loop often lead to cyclic behavior (e.g., in predator–prey systems).

Both positive and negative feedback are important in self-organization. By dampening changes, negative feedback acts as a stabilizing force. It is one of the principal mechanisms of homeostasis, the maintenance of dynamic equilibrium by internal regulation. In contrast, positive feedback magnifies minor deviations. An example is competitive exclusion: any small decrease in size of a competing population is likely to lead to further decreases, until it dies out (**Figure 2**).

Stigmergy

Stigmergy is a form of self-organization that occurs when parts of a system communicate by modifying their environment. Many examples of stigmergy occur in the organization of eusocial insect colonies. For example, in ant colonies, objects such as food, larvae, and corpses are often stored in discrete larders, nurseries, and cemeteries. Models show that this civic order can emerge through interactions between the ants and their environment. In the model, ants pick up objects at random, and may drop them when they encounter similar objects. Over time, this process creates piles of similar objects. Positive feedback causes larger piles to grow at the expense of smaller ones (**Figure 3**).

Synchronization

Synchrony can alter system-level behavior by enhancing or dampening nonlinearities. For example, when predator and

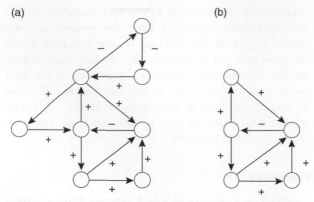

Figure 2 The role of feedback in self-organization of a food web. In this diagram, circles represent populations and arrows indicate the influence (positive or negative) of one population on another. In a food web, circular chains of interaction between populations form feedback loops, as in the example shown here. (a) The initial food web contains both positive and negative feedback loops. Internal dynamics within the positive feedback loops leads to the local extinction of several populations. (b) The resulting food web contains only negative feedback loops, which stabilize the community.

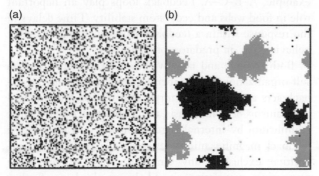

Figure 3 The emergence of order by stigmergy and feedback in an ant colony. Given a random scatter of objects (a), ants sort objects by picking them up and dropping them again when they find a similar object. This process creates piles, which grow over time. Large piles grow at the expense of smaller ones until only a few large piles remain (b).

prey populations are tightly coupled to one another, a stable, negative feedback relationship can result where an increase in prey causes increased predators and a subsequent decrease in prey. In this case, the ecological interaction acts like a thermostat regulating population size. However, if the two populations respond at different rates, oscillations or even chaotic behavior can occur instead. A classic example of such oscillations occurs in the interaction between populations of hares and lynxes in the Arctic Circle.

Synchronized breeding behavior is common and includes mass flowering in plants, mass breeding in birds, and mass spawning among marine animals such as corals and squid. In these cases, synchrony is usually achieved by individuals responding to a common environmental cue, such as a change in temperature or day length. Synchronized breeding conveys distinct advantages such as maximal exploitation of resources and satiation of predators.

Different species often have co-adapted simultaneous seasonal behavior, such as birds that breed when butterflies emerge. However, both the environmental cues, and the physiological response, may differ among these co-adapted species. For example, great tits time their egg laying by photoperiod. Winter moths are an important food source during the breeding season, and they develop more quickly at higher temperatures. As a result, recent warm springs in Europe caused by climate change have disrupted the synchronization between these species, reducing food availability for nesting great tits and potentially destabilizing populations.

In other cases, synchronous behavior arises through social contagion, where individuals imitate others. The dynamics of such behavior are similar to those seen in epidemiology. Social contagion can lead to coordinated group behavior such as flocking, as well as disparate phenomena such as synchronized flashing in fireflies, and 'fashions' in mate choice among birds and fish. The emergence of synchronous behavior in these cases is highly sensitive to the structure of social networks. Synchrony is easily achieved when networks are highly connected (i.e., individuals can perceive a large number of other individuals, or some individuals have very large influence). However, in loosely connected networks, social contagion can result in asynchronous waves or chaos.

Complex Adaptive Systems

Complex adaptive systems (CASs) consist of diverse, locally interacting components that are subject to selection. Examples include learning brains, developing individuals, economies, ecosystems, and the biosphere. In such systems, hierarchical organization, continual novelty and adaptation, and nonequilibrium dynamics are known to emerge. As a result, the behavior of a CAS is characterized by nonlinearity, historical contingency, thresholds, and multiple basins of attraction. A key question in current CAS research has been the relationship between resilience and criticality. Some authors suggest that a CAS will generally evolve toward self-organized criticality. By being maintained near the edge of chaos, such systems might maximize information processing. In this way, criticality might enhance the ability of CASs to adapt to changing environments and efficiently utilize resources, making systems become more resilient over time.

Artificial Life

The field of Alife uses simulation models to understand biological organization by abstracting crucial features and examining living systems 'as they could be'.

One of the most widespread representations used in Alife models is the cellular automaton (CA). This is a grid of cells in which each cell has a state (some property of interest) and is programmed to behave in identical fashion. Each cell has a neighborhood (usually the cells immediately adjacent to it) and the states of it neighbors affect changes in a cell's state. The most famous example is the Game of Life, in which each cell is either 'alive' or 'dead' at any time. Despite its extreme simplicity, the game showed that large numbers of interactions governed by simple rules lead to the emergence of order within a system. Cellular automata have been used to model many biological and ecological systems. In models of fires, epidemics, and other spatial processes, each cell represents a fixed area of the landscape and the cell states represent features of interest (e.g., susceptible, infected, or immune organisms in an epidemic model).

Other prominent ALife models include Tom Ray's Tierra model, which demonstrated adaptation within self-reproducing automata. Craig Reynolds' boids model demonstrated that flocking behavior emerges from simple interactions between individuals. James Lovelock's Daisyworld model showed the potential for biotic feedback and adaptation to stabilize the biosphere.

Self-Organization in an Ecological Setting

Social Groups

Relationships between individuals create several kinds of organizations within groups of animals.

Coordination between moving animals leads the formation of groups. Examples include swarms of insects, flocks of birds, schools of fish, and herds of mammals. Coordinated group movements, even in very large groups, can be achieved by individuals obeying simple rules, such as 'keep close, but not too close, to your neighbors' and 'head in the same general direction as your neighbors.'

Several mechanisms that channel aggressive behavior create social organization. In social animals, dominance hierarchies reduce the potential costs of conflict over mates and food. Adominance hierarchy emerges when interactions between individuals result in physiological and behavioral changes: for example, 'winning' a contest may elevate testosterone, causing increased dominance behavior, and evoking submissive behavior from individuals who have been less successful in the past. In this way, coherent transitive hierarchies can emerge even when all individuals were initially equal. Similarly, territoriality reduces the costs of conflict over resources by partitioning a landscape among a population. Territoriality often generates spatial patterns, such as regular distances between nests in seabird colonies. In this case, the distance between nests is defined by the maximum area that a sitting bird can defend without abandoning her nest. More complex coordinated group behaviors can emerge when individuals take on different tasks and roles within groups. For example, within ant and termite colonies, individuals can develop into a variety of castes, each with distinct roles such as foraging, nest defense, and nursing young. In honeybees, individuals take on different roles at different life stages.

In some cases, upper limits exist on the size that social groups can attain and depend on interactions between the animals. In apes, for instance, where social bonds are maintained by grooming, troop sizes tend to be 30–60 individuals. Larger troops tend to fragment. Among humans, social groups are usually much larger. The anthropologist Robin Dunbar argues that this is a consequence of speech providing more efficient social bonding than grooming, leading to a natural group size of 100–150 individuals.

In most cases, group size may be the outcome of several interacting ecological and social factors. For example, although lions hunt cooperatively, prides and hunting groups are usually larger than is optimal for hunting efficiency. Lionesses cooperate to defend cubs against infanticidal males by forming crèches. In addition, hunters are vulnerable to attack by larger groups, and territories are more effectively defended by larger prides.

The origin of cooperation among groups of cells and organisms can also be examined from the perspective of self-organization. The paradox of the evolution of cooperation is that (by definition) selfish individuals outcompete altruists, and therefore in a population of self-replicators, a selfish mutant should always spread at the expense of altruists. Nonetheless, altruism does occur among humans and cooperative behavior is often seen among animals. Such cooperative behavior can self-organize when the network structure that governs interactions among individuals results in the same individuals encountering one another repeatedly (e.g., when individuals are fixed in space, so that their only interactions are with their neighbors), or when their reproductive fate is very closely tied to that of others (as is the case for cells within a multicellular organism). Experimentally, the evolution of cooperation has been induced in bacterial populations by production of adhesive, causing individual cells to clump together. Cooperation can also evolve in marginal environments, where the evolutionary impact of competition between individuals is outweighed by the need to survive. Experimental studies of bacteria in marginal environments show that complex spatial patterns and signaling behaviors can emerge as a result of this selection. In theoretical models, the inclusion of policing behavior (punishing nonconformists) can also enforce high levels of cooperation even when interactions occur at random in large societies.

Persistence and Stability in Ecosystems

One of the most puzzling topics in systems ecology is how ecosystems emerge that are at once complex and stable. Field studies suggest that the most complex (diverse) ecosystems are also the most stable. However, this observation runs counter to expectation from systems theory. It shows that the more components a dynamic system has, the more likely it is that a destabilizing interaction (such as a positive feedback loop) will cause it to collapse and lose species. Consequently, systems theory suggests that simpler ecosystems should be more stable than complex ones. The paradox implies that the complex, stable ecosystems seen in nature are not random assemblages. Self-organization in this case involves removal of destabilizing positive feedback loops.

Communities versus Assemblages

The question of how important self-organization is in ecosystems has long been debated in ecology. Are ecosystems communities of co-adapted species, or are they simply random assemblages? Some early theorists, such as Clements, believed that the groups of species found together were specialized for living together, whereas others, such as Gleason, stressed the importance of chance and individuals.

The idea of succession concerns the patterns and processes involved in community change, especially after disturbance. A form of self-organization often associated with succession is facilitation. That is, plants and animals present in an area can alter the local environment, thereby facilitating the appearance of populations that replace them. After a fire, for example, a forest will regenerate with herbs and shrubs growing back almost immediately. The first trees to reappear will be 'pioneer' (disturbance) species, which disperse well, grow fast, and can tolerate open, exposed conditions. These trees create shade and leaf litter, which favor slow-growing, shade-tolerant trees.

Recent theoretical work (such as Hubbell's neutral theory of biodiversity and biogeography) emphasizes the role of chance and spatial dynamics in generating ecological patterns. In these models, self-organization is trivial because all individuals and species are effectively identical, and species abundances are driven by random birth, migration, and death processes. Both neutral and self-organizing models have been successful in explaining real relative abundance and species–area curves.

Food Webs

Species interactions lead to the flow of material within an ecosystem. For animals the most common processes are eating, respiration, excretion, and egestion. For plants, they are root uptake of water and nutrients, respiration,

and photosynthesis. The outputs of material from one organism often become inputs to other. This focus on 'what eats what' led Elton to identify several patterns, notably the food chain and the food web, the food cycle, the ecological niche, and the pyramid of numbers.

Self-organization in ecosystems is evident in the structure in food webs, networks that describe trophic interactions among species. Within food webs, specific patterns of interaction may be prevalent. These patterns, termed ecological motifs, are thought to represent especially stable interactions. The concept of keystone species supposes that certain species play a crucial role in maintaining the integrity and stability of an ecosystem.

Analysis of food webs suggests that a small-world structure is common. That is, most species interact with only a small number of other species, but the connectivity of the web as a whole is maintained by a few species that interact with a large number of others. This observation provides a theoretical basis for the idea of keystone species. Functionally, small world networks are thought to be robust to random loss of nodes (e.g., species), but vulnerable to attacks that target their highly connected nodes (e.g., keystone species).

Spatial Patterns and Processes

Spatial processes lead to the formation of distribution patterns. Seed dispersal, for instance, often produces concentrations of seedlings around parent plants and leads to the formation of clumped distributions. When local dispersal is combined with patchy disturbance, such as fire, the result is a distribution composed of patches. When combined with environmental gradients, such as soil moisture, local dispersal can produce zone patterns, with different species dominating different areas (**Figure 4**).

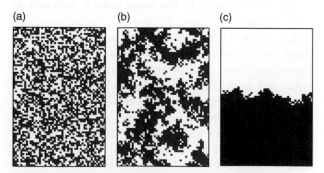

(a) (b) (c)

Figure 4 Emergence of spatial patterns from dispersal. This CA model shows the hypothetical distributions of two plant populations that result in three different scenarios. (a) Global dispersal, in which seeds can spread anywhere, results in random distributions of plants. (b) Dispersal from local seed sources leads to clumped distributions. (c) The combination of local dispersal and environmental gradients (from top to bottom) creates vegetation zones.

Fragmentation is one of the most important consequences of landscape connectivity. When the density of (randomly located) objects in a landscape falls below a critical density, they are mostly isolated individuals. When the density exceeds the critical threshold, they become connected. The density at which the critical threshold occurs depends on the size of the neighborhood of the objects. There are many cases where landscape connectivity plays an important role. Epidemic processes require a critical density of resources to spread. Instances include disease outbreaks (susceptible individuals), fire spread (fuel), and invasions of exotic plants (suitable sites). Populations become fragmented if individuals cannot interact with one another. For instance, in wet years the water bodies of central Australia are essentially connected for water birds, which can fly from one body to another almost anywhere in the continent. In dry years, however, many water bodies shrink or dry up and become too widely separated for birds to migrate between them (**Figure 5**).

Self-Organization in the Biosphere

Arguably the most ambitious ecological theory based on self-organization is the Gaia hypothesis, which postulates that the biosphere itself evolves to a homeostatic state. Lovelock suggested the Daisyworld model as an illustration of how this process might occur. On the hypothetical Daisyworld, black and white daisies compete for space. Although both kinds of daisies grow best at the same temperature, black daisies absorb more heat than white daisies. When the Sun shines more brightly, heating the planet, white daisies spread, and the planet cools again. When the Sun dims, the black daisies spread, warming the planet. In this way, competitive interactions between daisies provide a homeostatic mechanism for the planet as a whole.

The idea behind Gaia is that ecosystems will survive and spread more effectively if they promote the abiotic conditions required for their own persistence. If so, ecosystems might gradually evolve to be increasingly robust,

and if this happened on a global scale, then the biosphere itself might behave as a self-regulating system. However, evidence for Gaian processes in real ecosystems remains tenuous and their theoretical plausibility is disputed.

Evolution

Self-organization may play a prominent role in evolution, especially in the context of landscapes, which regulate interactions between individuals. One consequence is the evolution of cooperation in marginal and viscous habitat networks, whereas randomly interacting populations are more dominated by intraspecific competition and therefore more likely to behave selfishly.

Landscape structure influences genetic diversity and speciation. In connected landscapes, genes flow freely throughout a species and speciation is inhibited. However, in fragmented landscapes, a species breaks into isolated subpopulations. Fragmentation increases the risk of inbreeding and loss of genetic diversity in these subpopulations. Divergence between population fragments may also underlie adaptive radiations, in which many novel species suddenly emerge simultaneously.

As species adapt to their environment, they are often faced by tradeoffs in allocating resources for different purposes. These tradeoffs can lead to the evolution of distinct morphs within a species, or to speciation. For example, many mangrove species face a conflict between salt tolerance and competitive ability. Mangroves grow in estuaries, where salinity varies along the gradient between land and sea. Mangroves growing landward will be under strong selection for competitive ability, while those growing closer to the sea require better salt tolerance. The tradeoff, combined with local seed dispersal, can generate discrete banding patterns in the distribution of mangrove species, where each species is displaced by a more salt-tolerant one closer to the sea.

Contingency also plays a large part in the organization of spatial distributions. Spatial dominance occurs when a particular species is overwhelmingly abundant in a local environment. In this situation, the species can resist invasion, even by a superior competitor, because its propagules are much more numerous locally than those of any other population. For the same reason, a mutation that enables a species to exploit a novel environment may result in it permanently excluding potential competitors from that environment, even after they have evolved similar adaptations.

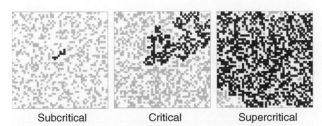

Subcritical   Critical   Supercritical

Figure 5 Critical phase changes in connectivity within a fragmented landscape. In this CA model, grid cells represent sites in a landscape. Gray and black cells represent vegetation and white cells have no cover. The black cells show examples of patches of vegetation sites that are connected, for example, by spread of a fire ignited in the center of the grid. Notice that only a small increase in the density of covered sites makes the difference between subcritical and supercritical.

Practical Considerations

The insights provided by theories of self-organization have many practical implications, both for ecology and for conservation. The sharp end of the conservation debate often hinges on the question of which areas and which sites to conserve. If ecosystems consist of random collections of species, then one site in a landscape is as good as another. All that matters is to preserve representative populations of each species. However, if the ecosystems consist of self-organized communities, in which the species are adapted to depend on one another for survival, then whole communities need to be conserved.

Closely related to the above issue is that the tendency for randomly constructed food webs to be unstable raises questions about the long-term viability of artificially created communities in which translocated species are introduced into new areas. Self-organization is evident even in artificial ecosystems. In biosphere 2, for instance, a closed, experimental environment designed to emulate natural ecosystems, the environment was found to favor species that collect more energy and internal processes led to unexpected problems, such as runaway depletion of oxygen levels.

The need to understand self-organization is important when considering altered ecosystems. For instance, it is usually not possible to carry out experiments to determine the long-term effects of current ecological management practices such as translocation of populations, controlled burning or allocation of reserves and wilderness areas. This problem makes simulation modeling a potentially crucial tool of ecological theory and practice. New methods of field observation are also appearing. For instance, the need to understand landscape fragmentation has led to studies of connectivity in landscapes, both field based, and using data from remote-sensing and geographic information.

See also: Autocatalysis; Ecological Complexity; Emergent Properties; Hierarchy Theory in Ecology.

Further Reading

Ball P (1999) *The Self-Made Tapestry: Pattern Formation in Nature.* Oxford: Oxford University Press.

Camazine S, Deneubourg J-L, Franks NR, *et al.* (2003) *Self-Organization in Biological Systems.* Princeton: Princeton University Press.

Green DG, Klomp NI, Rimmington GR, and Sadedin S (2006) *Complexity in Landscape Ecology.* Amsterdam: Springer.

Holland JH (1996) *Hidden Order: How Adaptation Builds Complexity.* New York: Addison-Wesley.

Levin SA (1998) Ecosystems and the biosphere as complex adaptive systems. *Ecosystems* 1(5): 431–436.

Patten BC, Fath BD, and Choi JS (2002) Complex adaptive hierarchical systems – Background. In: Costanza R and Jørgensen SE (eds.) *Understanding and Solving Environmental Problems in the 21st Century*, pp. 41–94. London: Elsevier.

Prigogine I (1980) *From Being to Becoming.* New York: Freeman (ISBN 0-7167-1107-9).

Rohani P, Lewis TJ, Gruenbaum D, and Ruxton GD (1997) Spatial self-organization in ecology: Pretty patterns or robust reality? *Trends in Ecology and Evolution* 12(8): 70–74.

Solé RV and Levin S (2002) Preface to special issue: The biosphere as a complex adaptive system. *Philosophical Transactions of the Royal Society of London B* 357: 617–618.

Watts DJ and Strogatz SH (1998) Collective dynamics of 'small-world' networks. *Nature* 393(6684): 440–442.

Ecological Complexity

J L Casti, International Institute for Applied System Analysis, Laxenburg, Austria

B D Fath, Towson University, Towson, MD, USA and International Institute for Applied System Analysis, Laxenburg, Austria

Complexity as a Systems Concept	'Would-Be' Worlds
Surprise-Generating Mechanisms	Conclusions
Emergent Phenomena	Further Reading
Ecological Complexity	

Complexity as a Systems Concept

In everyday parlance, the term 'complex' is generally taken to mean a person or thing composed of many interacting components whose behavior and/or structure are difficult to understand. The behavior of national economies, the human brain, and a rain forest ecosystem are all good illustrations for complex systems.

These examples show that there is nothing new about complex systems. But what is new is that for perhaps the first time in history, we have the knowledge – and the tools – to study such systems in a controlled, repeatable, scientific fashion. So there is reason to believe that this newfound capability will eventually lead to a viable theory of such systems.

Prior to the recent arrival of cheap and powerful computing capabilities, we were hampered in our ability to study a complex system like a road-traffic network, a national economy, or a supermarket chain because it was simply too expensive, impractical, too time consuming – or too dangerous – to tinker with the system as a whole. Instead, we were limited to biting off bits and pieces of such processes that could be looked at in a laboratory or in some other controlled setting. But with today's computers we can actually build complete silicon surrogates of these systems, and use these 'would-be worlds' as laboratories within which to look at the workings – and behaviors – of the complex systems of everyday life.

In coming to terms with complexity as a systems concept, we first have to realize that complexity is an inherently subjective concept; what is complex depends upon how you look. When we speak of something being complex, what we are really doing is making use of everyday language to express a feeling or impression that we dignify with the label 'complex'. But the meaning of something depends not only on the language in which it is expressed (i.e., the code), the medium of transmission, and the message, but also on the context. In short, meaning is bound up with the whole process of communication and does not reside in just one or another aspect of it. As a result, the complexity of a political structure, an ecosystem, or an immune system cannot be regarded as simply a property of that system taken in isolation. Rather, whatever complexity such systems have is a joint property of the system and its interaction with another system, most often an observer and/or controller.

So just as with truth, beauty, good, and evil, complexity resides as much in the eye of the beholder as it does in the structure and behavior of a system itself. This is not to say that there do not exist 'objective' ways to characterize some aspects of a system's complexity. After all, an amoeba is just plain simpler than an elephant by whatever notion of complexity you happen to believe in. The main point, though, is that these objective measures only arise as special cases of the two-way measures, cases in which the interaction between the system and the observer is much weaker in one direction than in the other.

A second key point is that common usage of the term 'complex' is informal. The word is typically employed as a name for something that seems counterintuitive, unpredictable, or just plain hard to understand. So if it is a genuine 'science' of complex systems we are after and not just anecdotal accounts based on vague personal opinions, we are going to have to translate some of these informal notions about the complex and the commonplace into a more formal, stylized language, one in which intuition and meaning can be more or less faithfully captured in symbols and syntax. The problem is that an integral part of transforming complexity (or anything else) into a science involves making that which is fuzzy precise, not the other way around, an exercise we might more compactly express as 'formalizing the informal'.

To bring home this point a bit more forcefully, let us consider some of the properties associated with 'simple' systems by way of inching our way to a feeling for what is involved with the complex. Generally speaking, simple systems exhibit the following characteristics.

Predictable behavior. There are no surprises in simple systems; simple systems give rise to behaviors that are easy to deduce if we know the inputs (decisions) acting upon the system and the environment. If we drop a stone, it falls; if we stretch a spring and let it go, it oscillates in a fixed pattern; if we put money into a fixed-interest bank account, it grows to a predictable sum in accordance with an easily understood and computable rule. Such predictable and intuitively well understood behavior is one of the principal characteristics of simple systems.

Complex processes, on the other hand, generate counterintuitive, seemingly acausal behavior that is full of surprises. Lower taxes and interest rates lead to higher unemployment; low-cost housing projects give rise to slums worse than those the 'better' housing replaced; the construction of new freeways results in unprecedented traffic jams and increased commuting times. For many people, such unpredictable, seemingly capricious, behavior is the defining feature of a complex system.

Few interactions and feedback/feedforward loops. Simple systems generally involve a small number of components, with self-interactions dominating the linkages among the variables. For example, primitive barter economies, in which only a small number of goods (food, tools, weapons, clothing) are traded, seem much simpler and easier to understand than the developed economies of industrialized nations, in which the pathways between raw material inputs and finished consumer goods follow labyrinthine routes involving large numbers of interactions between various intermediate products, labor, and capital inputs.

In addition to having only a few variables, simple systems generally consist of very few feedback/feedforward loops. Loops of this sort enable the system to restructure, or at least modify, the interaction pattern among its variables, thereby opening up the possibility for a wider range of behaviors. To illustrate, consider a large organization that is characterized by variables like employment stability, substitution of capital for human labor, and level of individual action and responsibility (individuality). Increased substitution of work by capital decreases the individuality in the organization, which in turn may reduce employment stability. Such a feedback loop exacerbates any internal stresses initially present in the system, leading possibly to a collapse of the entire organization. This type of collapsing loop is especially dangerous for social structures, as it threatens their ability to absorb shocks, which seems to be a common feature of complex social phenomena.

Centralized decision making. In simple systems, power is generally concentrated in one or at the most a few decision makers. Political dictatorships, privately owned corporations, and the Roman Catholic Church are good examples of this sort of system. These systems are simple because there is very little interaction, if any, between the lines of command. Moreover, the effect of the central authority's decision upon the system is usually rather easy to trace.

By way of contrast, complex systems exhibit a diffusion of real authority. Such systems seem to have a nominal supreme decision maker, but in actuality the power is spread over a decentralized structure. The actions of a number of units then combine to generate the actual system behavior. Typical examples of these kinds of systems include democratic governments, labor unions, and universities. Such systems tend to be somewhat more resilient and stable than centralized structures because they are more forgiving of mistakes by any one decision maker and are more able to absorb unexpected environmental fluctuations.

Decomposable. Typically, a simple system involves weak interactions among its various components. So if we sever some of these connections, the system behaves more or less as before. Relocating American Indians to reservations produced no major effects on the dominant social structure in New Mexico and Arizona, for example, since, for various cultural reasons, the Indians were only weakly coupled to the dominant local social fabric in the first place. Thus, the simple social interaction pattern present could be further decomposed and studied as two independent processes – the Indians and the settlers.

Complex processes, on the other hand, are irreducible. Neglecting any part of the process or severing any of the connections linking its parts usually destroys essential aspects of the system's behavior or structure. You just cannot start slicing up systems of this complexity into subsystems without suffering an irretrievable loss of the very information that makes these systems a 'system'.

Surprise-Generating Mechanisms

The vast majority of counterintuitive behaviors shown by complex systems are attributable to some combination of the following five sources: paradoxes/self-reference, instability, incomputability, connectivity, and emergence. With some justification, we can think of these sources of complexity as 'surprise-generating mechanisms', whose quite different natures each lead to their own characteristic type of surprise. Let us take a quick look at each of these mechanisms before turning to a more detailed consideration of how they act to create complex behavior.

Paradox. Paradoxes arise from false assumptions about a system leading to inconsistencies between its observed behavior and our 'expectations' of that behavior. Sometimes these situations occur in simple logical or linguistic situations, such as the famous 'liar paradox' ("This sentence is false."). In other situations, the paradox comes from the peculiarities of the human visual system, as with the impossible staircase shown in **Figure 1**, or simply from the way in which the parts of a system are put together, like the developing economy discussed in the preceding section.

Instability. Everyday intuition has generally been honed on systems whose behavior is stable with regard to small disturbances, for the obvious reason that unstable systems tend not to survive long enough for us to develop good intuitions about them. Nevertheless, the systems of both nature and humans often display pathologically sensitive behavior to small disturbances, as for example, when stock markets crash in response to seemingly minor economic news about interest rates, corporate mergers, or bank failures. Such behaviors occur often enough that they deserve a starring role in our taxonomy of surprise.

Incomputability. The kinds of behaviors seen in models of complex systems are the end result of following a set of rules. This is because these models are embodied in computer programs, which in turn are necessarily just a set of rules telling the machine what bits in its memory array to turn on or off at any given stage of the calculation. By definition, this means that any behavior seen in such worlds is the outcome of following the rules encoded in the program. Although computing machines are *de facto* rule-following devices, there is no *a priori* reason to believe that any of the processes of nature and humans are necessarily rule based. If incomputable processes do exist in nature – for example, the breaking of waves on a beach or the movement of air masses in the atmosphere – then we could never see these processes manifest themselves in the surrogate worlds of their models. We

(a)   (b)

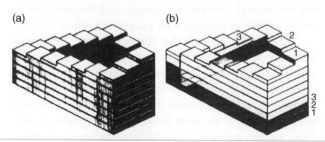

Figure 1   The impossible staircase.

may well see processes that are close approximations to these incomputable ones, just as we can approximate an irrational number as closely as we wish by a rational number. However, we will never see the real thing in our computers, if indeed such incomputable quantities exist outside the pristine world of mathematics.

Connectivity. What makes a system a system and not simply a collection of elements is the connections and interactions among the individual components of the system, as well as the effect these linkages have on the behavior of the components. For example, it is the inter-relationship between biota and abiota that makes an ecosystem. Each component taken separately would not suffice. The two must interact for sustainable life to take place. Complexity and surprise often reside in these connections.

Emergence. A surprise-generating mechanism depen-dent on connectivity for its very existence is the phenomenon of emergence. This refers to the way the interactions among system components generate unex-pected global system properties not present in any of the subsystems taken individually. A good example is water, whose distinguishing characteristics are its natural form as a liquid and its nonflammability, both of which are totally different from the properties of its component gases, hydrogen and oxygen.

The difference between complexity arising from emergence and that coming only from connection pat-terns lies in the nature of the interactions among the various component pieces of the system. For emergence, attention is not simply on whether there is some kind of interaction between the components, but also on the specific nature of that interaction. For instance, connec-tivity alone would not enable one to distinguish between ordinary tap water involving an interaction between hydrogen and oxygen molecules and heavy water (deu-terium), which involves interaction between the same components albeit with an extra neutron thrown in to the mix. Emergence would make this distinction. In prac-tice it is often difficult (and unnecessary) to differentiate between connectivity and emergence, and they are fre-quently treated as synonymous surprise-generating procedures. A good example of emergence in action is the organizational structure of an ant colony.

Like human societies, ant colonies achieve things that no individual ant could accomplish on its own. Nests are erected and maintained, chambers and tunnels are exca-vated, and territories are defended. All these activities are carried on by individual ants acting in accord with simple, local information; there is no master ant overseeing the entire colony and broadcasting instructions to the individual workers. Somehow each individual ant processes the partial information available to it in order to decide which of the many possible functional roles it should play in the colony.

Recent work on harvester ants has shed considerable light on the process by which an ant colony assesses its current needs and assigns a certain number of members to perform a given task. These studies identify four distinct tasks an adult harvester ant worker can perform outside the nest: foraging, patrolling, nest maintenance, and mid-den work (building and sorting the colony's refuse pile). So it is these different tasks that define the components of the system we call an ant colony, and it is the interaction among ants performing these tasks that gives rise to emergent phenomena in the colony.

One of the most notable interactions is between forager ants and maintenance workers. When nest main-tenance work is increased by piling some toothpicks near the opening of the nest, the number of foragers decreased. Apparently, under these environmental conditions, the ants engaged in task switching, with the local decision made by each individual ant determining much of the coordinated behavior of the entire colony. Task allocation depends on two kinds of decisions made by individual ants. First, there is the decision about which task to per-form, followed by the decision of whether to be active in this task. As already noted, these decisions are based solely on local information; there is no central decision maker keeping track of the big picture.

Figure 2 gives a summary of the task-switching roles in the harvester ant colony, showing that once an ant becomes a forager it never switches back to other tasks outside the nest. When a large cleaning chore arises on the surface of the nest, new nest-maintenance workers are recruited from ants working inside the nest, not from workers performing tasks on the outside. When there is a disturbance like an intrusion by foreign ants, nest-main-tenance workers will switch tasks to become patrollers. Finally, once an ant is allocated a task outside the nest, it never returns to chores on the inside.

The ant colony example shows how interactions among the various types of ants can give rise to patterns of global work allocation in the colony, patterns that could not be predicted or that could not even arise in any single ant. These patterns are emergent phenomena due solely to the types of interactions among the different tasks.

Table 1 gives a summary of the surprise-generating mechanisms just outlined.

Emergent Phenomena

Complex systems produce surprising behavior; in fact, they produce behavioral patterns and properties that just cannot be predicted from knowledge of their parts taken in isolation. These so-called 'emergent properties' are probably the single most distinguishing feature of com-plex systems. An example of this phenomenon occurs

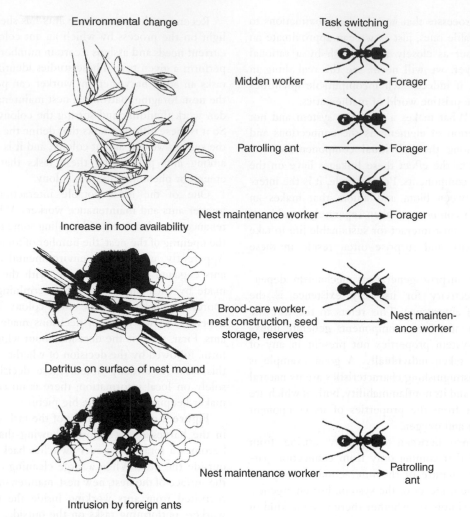

Figure 2 Task switching in a harvester ant colony.

Table 1 The main surprise-generating mechanisms

Mechanism	Surprise effect
Paradoxes	Inconsistent phenomena
Instability	Large effects from small changes
Incomputability	Behavior transcends rules
Connectivity	Behavior cannot be decomposed into parts
Emergence	Self-organizing patterns

when one considers a collection of independent random quantities, such as the heights of all the people in New York City. Even though the individual numbers in this set are highly variable, the distribution of this set of numbers will form the familiar bell-shaped curve of elementary statistics. This characteristic bell-shaped structure can be thought of as 'emerging' from the interaction of the component elements. Not a single one of the individual heights can correspond to the normal probability distribution, since such a distribution implies a population. Yet when they are all put into interaction by adding and

forming their average, the 'central limit theorem' of probability theory tells us that this average and the dispersion around it must obey the bell-shaped distribution.

Ecological Complexity

Complexity research has discovered that many systems display common structural and behavioral/dynamical characteristics (**Table 2**). The interplay of these complex system characteristics entails systems to exhibit properties such as surprise, emergence, and power law scaling. Ecological complexity is the observation that ecological systems exhibit many of the same properties as physical complex systems, and thus an active research program has arisen over the analysis of ecological data to see to what extent ecosystems share these common properties with other complex systems.

Ecosystems are composed of a large number of highly diverse components interacting with self-stabilizing and self-promoting feedback to produce emergent patterns. As

Table 2 Some characteristics of complex systems

Structural characteristics	Behavioral/dynamical characteristics
Large number of components	Nonlinear
Large number of components	Chaotic
High diversity of components and connections	Catastrophic
Asymmetries	Self-organization
Strong interactions	Multiple steady states
Hierarchic organization	Adaptive

such, ecological systems have been described as complex, adaptive, hierarchical systems (CAHS) or self-organized, hierarchical open (SOHO) systems. Unlike with complex physical systems, openness is a property that is required of all ecological systems. This is because ecological systems are self-perpetuating through means of capturing energy, doing useful work (biochemical reactions, growth, and maintenance) to persist at least momentarily at a highly organized state far from thermodynamic equilibrium – this is the metabolic process. A second ecological-defining feature is that organisms are able to replicate themselves such that the system outlives the constituent parts – this is the reproductive process. Therefore, it is often said of an ecosystem that, 'the whole is more than the sum of the parts'.

Two of the most pressing issues regarding ecological complexity are the need to develop appropriate measures to quantify the structural and behavioral complexity of ecosystems, and to identify the underlying processes that generate this complexity, through theory, analysis, modeling, and field studies.

A new journal, *Ecological Complexity* (Elsevier), is one forum for this research since 2004. The journal considers papers dealing with biocomplexity related to the environment with an emphasis on interdisciplinary and integrated natural and social systems science.

Topics typically found in the journal include:

- all aspects of biocomplexity in the environment and theoretical ecology;
- ecosystems and biospheres as complex adaptive systems;
- self-organization of spatially extended ecosystems;
- emergent properties and structures of complex ecosystems;
- ecological pattern formation in space and time;
- the role of biophysical constraints and evolutionary attractors on species assemblages;
- ecological scaling (scale invariance, scale covariance, and across-scale dynamics), allometry, and hierarchy theory;
- ecological topology and networks;
- studies toward an ecology of complex systems;
- complex systems approaches for the study of dynamic human–environment interactions;

- using knowledge of nonlinear phenomena to better guide policy development for adaptation strategies and mitigation to environmental change; and
- new tools and methods for studying ecological complexity.

The emphasis on integrated natural and social systems addresses the growing interest to understand the role of human influence on the environment. There is a recent awareness of the need to alter this influence in some fashion both for ourselves and our environment. New tools and approaches incorporating self-organization, emergence, and co-adaptation are needed to improve our ability to manage and restore natural systems. These new approaches to ecosystem management must also account for the natural dynamics and integrate concepts of sustainable development. Advances in ecological complexity science are essential in successfully navigating this transition.

'Would-Be' Worlds

In the past few years, a number of electronic worlds have been created by researchers associated with the Santa Fe Institute and elsewhere to study the properties of complex, adaptive systems. The authors cite just three such worlds here as prototypical examples of how to use the computer as a kind of information laboratory to investigate such systems.

Tierra. This world, created by naturalist Tom Ray, is populated by binary strings that serve as electronic surrogates for genetic material. As time unfolds, these strings compete with each other for resources, with which they create copies of themselves. New strings are also created by computational counterparts of the real-world processes of mutation and crossover. Over the course of time, the world of Tierra displays many of the features associated with evolutionary processes seen in the natural world, and hence can be used as a way of experimenting with such processes – without having to wait millions of years to bring the experiment to a conclusion. But it is important to keep in mind that Tierra is not designed to mimic any particular real-world biological process; rather, it is a laboratory within which to study neo-Darwinian evolution, in general.

TRANSIMS. For the past 3 years, a team of researchers at the Los Alamos National Laboratory headed by Chris Barrett has built an electronic counterpart of the city of Albuquerque, New Mexico, inside their computers. The purpose of this world, which is called TRANSIMS, is to provide a testbed for studying the flow of road traffic in an urban area of nearly half a million people. In contrast to Tierra, TRANSIMS is explicitly designed to mirror the real world of Albuquerque as faithfully as possible, or at least to mirror those aspects of the city that are relevant for road-traffic flow. Thus, the simulation contains the entire road traffic network from freeways to back alleys, together with information about where people live and work, as well as demographic information about incomes, children, type of cars, and so forth. So here we have a would-be world whose goal is to indeed duplicate as closely as possible a specific real-world situation.

Sugarscape. Somewhere between Tierra and TRANSIMS is the would-be world called Sugarscape, which was created by Joshua Epstein and Rob Axtell of The Brookings Institution in Washington, DC. This world is designed as a tool to study processes of cultural and economic evolution. On the one hand, the assumptions about how individuals behave and the spectrum of possible actions at their disposal is a vast simplification of the possibilities open to real people as they go through everyday life. On the other hand, Sugarscape makes fairly realistic assumptions about the things that motivate people to act in the way they do, as well as about how they go about trying to attain their goals. What is of considerable interest is the rich variety of behaviors that emerge from simple rules for individual action, and the uncanny resemblance these emergent behaviors have to what is actually seen in real life.

In order to conduct the kinds of repeatable, controlled experiments that natural scientists take for granted when trying to understand and create theories of physical and engineering systems, Epstein and Axtell decided to 'grow' a social order from scratch by creating an ever-changing environment and a set of agents who interact with each other and the environment in accordance with simple rules of survival. An entire social structure – trade, economy, culture – then evolves from the interactions of the agents. As Epstein remarks about social problems, "You don't solve it, you evolve it." Epstein and Axtell call their laboratory in which societies evolve the 'CompuTerrarium'. Here is how it works.

The interacting agents are each graphically represented by a single colored dot on the landscape they inhabit, which is called the Sugarscape. Every location in the landscape contains time-varying concentrations of a food resource, called sugar. Each individual has a unique set of characteristics; some are fixed like sex, visual range for food detection, and metabolic rate, whereas others are variable like health, marital status, and wealth. The

behavior of these agents is determined by a set of extremely simple rules that constitute nothing more than common-sense rules for survival and reproduction. A typical set of rules might be:

1. Find the nearest location containing sugar. Go there, eat as much as necessary to maintain your metabolism, and save the rest.
2. Breed if you have accumulated enough energy and other resources.
3. Maintain your current cultural identity (set of characteristics) unless you see that you are surrounded by many agents of different types ('tribes'). If you are, change your characteristics and/or preferences to fit in with your neighbors.

With even such primitive rules as this, strange and wondrous things begin to happen. A typical scenario is shown in **Figure 3**, where we see the sugar marked by yellow dots on the Sugarscape. The agents are initially distributed randomly on the landscape, red dots being agents that have a good ability to see food at a distance, blue dots representing more myopic agents. It is reasonable to expect that if no other considerations enter, natural selection would tend to favor good vision over time. Indeed this is the case, as seen by the center panel in **Figure 3**, showing a preponderance of red agents in the population. However, if the experiment is run again, giving agents the possibility of passing wealth on to their offspring in the form of sugar, we find that inheritance has a pronounced effect on survival. This is shown in the third panel of the figure, in which many more agents having poor vision are able to survive by making use of sugar willed to them by their parents.

Although this simple example is useful in illustrating the workings of the CompuTerrarium, it hardly suggests a revolution in our way of thinking about and studying social structures. For that we need to add a lot more whistles and bells to the system. Epstein and Axtell have done exactly this. When they add seasons so that sugar concentrations change periodically over time, the agents begin to migrate. When a second resource, spice, is introduced, a primitive economy emerges as a result of a new elementary rule: "Look around for a neighbor having a commodity you need. Bargain with that neighbor until you reach a mutually satisfactory price. Trade at that price." **Figure 4** shows the effect of this type of trading

| Initial condition | No inheritance | With inheritance |

Figure 3 Evolution on the Sugarscape.

① Agents forage for 'sugar and spice'   ② If trade is allowed, they flourish   ③ Without trade, many starve

Figure 4 The effects of trade in the Sugarscape.

economy. In the first part of the figure, agents are simply foraging independently for both sugar and spice. In the middle panel we see the effect of beginning trade; now lots of agents flourish. Finally, the third panel shows the effect of turning off the trade. Without trade being allowed, many of the agents cannot survive.

There is certainly much more that can be said about the social laboratory constructed by Epstein and Axtell. Issues involving the emergence of cultural groups, combat, institutional structures, and the like can all be introduced to study myriad questions of interest to social scientists. The interested reader will certainly want to consult the monograph by Epstein and Axtell that details these and many other matters. It is cited in the references for this article. Here we must content ourselves with simply noting that the CompuTerrarium offers a platform to study society from the bottom up. With this view, we can explore social behavior that is dynamic, evolutionary, and locally simple. What could be better than to have a laboratory like this in which to do such experiments?

The main point in presenting these discussions of Tierra, TRANSIMS, and Sugarscape is to emphasize two points: (1) we need different types of would-be worlds to study different sorts of questions, and (2) each of these worlds has the capability of serving as a laboratory within which to test hypotheses about the phenomena they can represent. And, of course, it is this latter property that encourages the view that such computational universes will play the same role for the creation of theories of complex systems that chemistry labs and particle accelerators have played in the creation of scientific theories of simple systems. Gleick has given a fuller account of the technical, philosophical, and theoretical problems surrounding the construction and use of these silicon worlds.

Conclusions

The key components in each and every complex, adaptive system and a decent mathematical formalism to describe and analyze them would go a long way toward the creation of a viable theory of such processes. These key components are given as follows.

A medium-sized number of agents. In contrast to simple systems – like superpower conflicts, which tend to involve a small number of interacting agents – or large systems – like galaxies or containers of gas, which have a large enough collection of agents that we can use statistical means to study them – complex systems involve what we might call a medium-sized number of agents. Just like Goldilocks's porridge, which was not too hot and not too cold, complex systems have a number of agents that are not too small and not too big, but just right to create interesting patterns of behavior.

Intelligent and adaptive agents. Not only are there a medium-sized number of agents, these agents are intelligent and adaptive. This means that they make decisions on the basis of rules, and that they are ready to modify the rules they use on the basis of new information that becomes available. Moreover, the agents are able to generate new rules that have never before been used, rather than being hemmed in by having to choose from a set of preselected rules for action. This means that an ecology of rules emerges, one that continues to evolve during the course of the process.

Local information. In the real world of complex systems, no agent knows what 'all' the other agents are doing. At most, each person gets information from a relatively small subset of the set of all agents, and processes this 'local' information to come to a decision as to how he or she will act. In the Sugarscape, for instance, what the traders adjacent to a given individual in the market are doing constitutes the local information that the individual has available to help decide what to do next.

So these are the components of all complex, adaptive systems like the Sugarscape, TRANSIMS, or Tierra situations – a medium-sized number of intelligent, adaptive agents interacting on the basis of local information. At present, there appears to be no known mathematical structures within which we can comfortably accommodate a description of 'any' of these worlds. This suggests a situation completely analogous to that faced by gamblers in the seventeenth century, who sought a rational way to divide the stakes in a game of dice when the game had to be terminated prematurely (probably by the appearance of the police or, perhaps, the gamblers' wives). The description and analysis of that very definite real-world problem led Fermat and Pascal to the creation of a mathematical formalism we now call probability theory. At present, complex-system theory still awaits its Pascal and Fermat. The mathematical concepts and methods currently available were developed, by and large, to describe systems composed of material objects like planets and atoms. It is the development of a proper theory of complex systems that will be the capstone of the transition from the material to the informational.

See also: Emergent Properties; Hierarchy Theory in Ecology; Self-Organization.

Further Reading

Casti J (1992) *Reality Rules: Picturing the World in Mathematics. I – The Fundamentals, II – The Frontier.* New York: Wiley (paperback edition, 1997).

Casti J (1994) *Complexification.* New York: HarperCollins.

Casti J (1997) *Would-Be Worlds.* New York: Wiley.

Epstein J and Axtell R (1996) *Growing Artificial Societies.* Cambridge, MA: MIT Press.

Gleick J (1987) *Chaos.* New York: Viking.

Jackson E (1990) *Perspectives of Nonlinear Dynamics.* Cambridge: Cambridge University Press, vols. 1 and 2.

Mandelbrot B (1982) *The Fractal Geometry of Nature.* San Francisco, CA: W.H. Freeman.

Nicolis J (1991) *Chaos and Information Processing.* Singapore: World Scientific.

Peitgen H-O, Jürgens DH, and Saupe D (1992) *Fractals for the Classroom,* Parts 1 and 2. New York: Springer.

Ray T (1991) An approach to the synthesis of life. In: Langton CG, Taylor C, Farmer JD, and Rasmussen S (eds.) *Artificial Life II, SFI Studies in the Science of Complexity,* vol. X, pp. 371–408. Reading, MA: Addison-Wesley.

Schroeder M (1991) *Fractals, Chaos, Power Laws.* New York: W.H. Freeman.

Stewart I (1989) *Does God Play Dice?* Oxford: Basil Blackwell.

Hierarchy Theory in Ecology

T F H Allen, University of Wisconsin, Madison, WI, USA

Need for Hierarchy Theory	History of the Field
Hierarchy and Hypothesis	Scale and Type
Hierarchical Levels	Further Reading

Need for Hierarchy Theory

In ecology we need a body of theory to address relationships that are consequences of changing levels of analysis, which call for altered definitions. For instance, the contemporary fracas over definitions of plant competition could benefit from recognizing differences in the scope and type of the framework used by the respective partisans. With distinctions between levels of analysis made clear, each school may test their respective hypotheses in peace aware of which theories actually compete and when they merely address some other level of discourse. The contentious literature surrounding overcompensation of plants in response to losses to grazing was significantly a matter of pulse versus press consumption in relation to different timescales for assessing recovery. The Clements/Gleason debate over the proper definition of plant community might not have lasted the body of the twentieth century had hierarchy theory been available at

the onset of hostilities when Nichols attacked Gleason's paper at the 1926 International Congress of Plant Sciences. Hierarchy theory's focus on level of analysis offers such clarification. It lays subtle distinctions bare, so that definitions work for the ecologist instead of ecologists working for their definitions.

If big, slow things were always on top, such that hierarchical levels were only a matter of scale, the problem would reduce to a straightforward technical scaling issue. Not to underestimate the challenges of scaling in engineering, but that technical setting does not need something as grand as a theory to deal with hierarchies. But in ecology, scaling is complicated by higher ecological levels giving lower levels meaning. In ecology, the move upscale to be more inclusive often changes significance more than it invokes a change of size, and so we do indeed need a special body of theory to deal with difference of quality, not just quantity. Differences in scale quickly become large enough to cause qualitative change in perception, which forces a change in the level of analysis. Thus scale is soon embroiled in values, judgment, and arbitrary choices, not just as an inconvenience, but as a necessity for proper understanding. While scaling as an engineering technicality actively ignores such messy issues, hierarchy theory explicitly includes value-based decisions of the observer in creating hierarchies.

To control for observer values, technical measurement and analysis in science keeps its criteria constant across the local discourse. But large discoveries precisely amount to the recognition of a change in value. New scientific ideas indicate a specific change in the preanalytic stage, before deciding what might be relevant data. In the terms of Russell and Whitehead (made accessible by Gregory Bateson), new scientific ideas amount to the definition of new logical types. Hierarchy theory's central activity is recognizing logical type. Logical types are tied to some new level of inclusivity, a new hierarchical level with its own meaning. Notice how left and right sides are possessed by organisms at their own level of existence. Meanwhile, the notions of up and down refer to a larger discourse that includes an environment, which is shared by many organisms. As a result, a mirror switches the image left and right, but with no switch in up versus down. The larger scope invoked by the idea of up introduces a new logical type, even though left and right may often be simply at right angles to up and down. If left and right contrasted with up and down can be problematic, ecosystems are a nightmare. While exquisitely holding criteria constant in formal scientific calibration will help, it is insufficient for large discoveries, which turn on recognizing when a new type is necessary to solve some puzzle.

Ecology in particular invites many logical types because its hierarchies are so rich. A new type invokes new aggregation criteria, which come explicitly from observer decisions. Consider the difference between a community conception of vegetation as opposed to the process-functional conception that prevails in ecosystem modeling. A forest can be considered as a collection of trees on a tract of land. Alternatively those same tree trunks may be aggregated as a separate class from the leaves (**Figure 1**). If leaves in a forest are the production system independent of species, then the boles are part of the carbon storage function. This assignment has the peculiar effect of unifying the tree trunks with soil carbon in a single carbon storage compartment. A community focus aggregates trees set in an environment of soil and atmosphere. Meanwhile a flux-process conception splits the trees into at least two parts, one of which aggregates with the soil. But the soil was part of the environment in the community conception. Thus the same pieces of soil and plant biomass are aggregated into different higher units, depending on the type of system that is recognized as being in the foreground by the observer. Note how forests under either conception may be called forest ecosystems, suggesting that one use of hierarchy theory is to untangle alternative meanings in commonplace ecological terminology. The difference between a process-focused ecosystem and a community is a change in logical type.

Hierarchy and Hypothesis

Hierarchy theory is a body of thought that relates chosen levels of analysis to defined levels of organization, all in

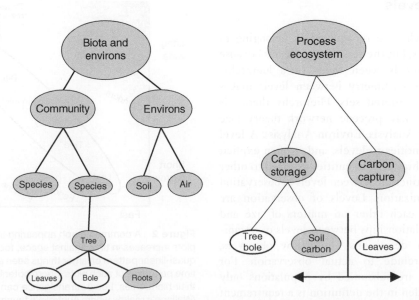

Figure 1 A community conception leads to an expected situating of whole trees in an environment. But a process-functional ecosystem conception of that same forest can lead to a hierarchy where tree boles are separated from the leaves, and then united with soil elements in a carbon storage compartment. Each respective hierarchy takes its form from the purpose for which it is intended.

the context of scaling. It advises the scientist of subtle but crucial distinctions that follow from observing and making analytical decisions. A significant part of hierarchy theory is observation in relation to conceptions of order in complex systems that would otherwise invite confusion. It is a metatheory that guides the generation, fine-tuning and testing of other bodies of thought, themselves more easily recognizable as theories in the conventional sense. Some theories in ecology are associated with answering questions taken as given. Such theory is validated in testing hypotheses. Other ecological theories may fine-tune questions, perhaps clarifying what is meant by competition, so that worthwhile hypotheses may be generated. By contrast, hierarchy theory applies in the preanalytical stage, when the questions are being framed rather than when clarified or answered. In the preanalytical stage, the boundaries of things are established, and structures are assigned to types or classes. With the discourse laid out unambiguously by hierarchy theory, other theories may come into play, testing explicit hypotheses with measurement and models. Thus hierarchy theory does not have its own hypotheses *per se*, but rather opens the way for subsequent testing of specific hypotheses. Like multivariate description in ecology, hierarchy theory focuses on hypothesis generation and clarification. From a small number of first principles, it highlights what would be otherwise taken for granted and then forgotten in the muddle that ensues. Hierarchy theory is explicit, as it positions the tacit next to the focal. Its precision is in thought and choosing definitions, more than action in quantitative experimentation.

Hierarchical Levels

Entities in a hierarchy are recognized as belonging to levels. Levels are sets, but the sets become levels because of robust asymmetry between them in a hierarchy. Mathematically, the asymmetry between levels makes hierarchies partially ordered sets. Hierarchy theory is the set theory that may precede network theory (see Ecological Network Analysis, Environ Analysis). A level of analysis assigns entities to levels, and is often explicit about their relationship to other entities assigned to other levels. There is a distinction between levels of observation and levels of organization. Levels of observation are ordered relative to each other on matters of size and scale. Meanwhile relationships between levels of organization follow from definitions chosen by observers, sometimes as a prelude to actual observation. For instance, organisms are subsumed by populations only by a definition. Hidden in the definition is a requirement for equivalence between population members. Meanwhile, a host and its parasite, while both organisms, are generally not assigned to the same population, in part

because of inequivalent size. Host versus parasite is the basis of a hierarchy employing levels subtly different from those in the population/organism distinction. Hierarchy theory places entities in levels, taking care to be explicit about the definitions that lead to those levels, and the criteria that create order and linkage.

Scale versus definition has potential for generating different sorts of hierarchies. Some hierarchies focus on size and containment, while others are control hierarchies where upper level entities simply control lower levels. A Watt governor may be placed at a higher level in a control hierarchy, while being smaller than the whole steam engine it controls. Whether it is a scalar or a control hierarchy depends on the use for which the hierarchical conception is intended, something for which the observer must take responsibility. Time against space plots are popular in landscape ecology. But such hierarchies can miss out on the interesting situations where large space maps onto short time spans, or long time spans map onto small places (**Figure 2**). The globe is large enough to be the context of continental movement over hundreds of millions of years, but at the same time the rotation of the globe is also responsible for diurnal phenomena, at the fast end of ecological happenings. Surfaces arise when narrow space applies to large differences in time constants (strong temporal connection within, but weak connections across surfaces). Ecotones would be a case in point, because there is rapid exchange and fast process inside the abutting ecosystems or communities areas, while the exchanges across the narrow ecotone may be remarkably slow. Thus ecotones are spatially small, while

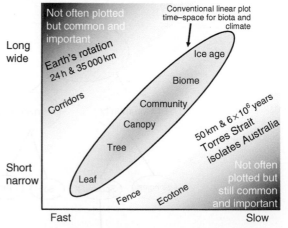

Figure 2 A common graph appearing in landscape ecology plots increases in time against space, focusing on the quasi-linear pattern of larger things seen as behaving over longer time periods. But such plots ignore potential control systems and their hierarchies. Local intransigence can control large entities (Wallace's realms where Australia's fauna is isolated from Asia by the narrow Torres Strait, separating millions of years of evolution). Barriers and surfaces occur at the lower right, while communication channels and corridors appear upper left.

representing slow exchanges that might cause the ecotone to move in a process of gradual encroachment. Conversely, in a communication channel, small differences in time constants apply along the long connection. Corridors would be an example here, where there is rapid movement along the extended length of the corridor. In ecology, these special places, such as ecotones and corridors, are at least as interesting as situations where time and space widen in concert. Complexity in hierarchies arises from the challenge of mapping between levels, as scale and definition entwine.

History of the Field

Hierarchy theory has its roots in economics and business administration of the 1960s, suggesting that the world appears nearly decomposable. We can decompose wholes into parts, but only to a degree, in that parts communicate and leak onto each other. Complete decomposability would deny upper-level structures' existence. Completely decomposed, the parts would not to be able to communicate with each other in making the larger whole. Parts have strong connections within, but weak connections between, and those weak connections may be precisely what links hierarchical levels (**Figure 3**).

While some practitioners in subsequent studies have sought real hierarchies in an external world, much of the early literature of business administration hierarchies is agnostic about the ultimate reality of hierarchical structure. In this spirit, hierarchy theory in social organizations

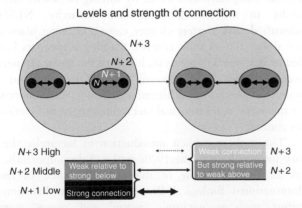

Levels and strength of connection

Figure 3 In nested hierarchies, the bonds that unite members of the lowest level N are strong, as they make entities of level $N+1$. The bonds that create level $N+2$ are weaker. Nevertheless, these weaker bonds appear as the strong bonds making the entities at level $N+2$ when seen from level $N+3$. If the two largest units in the figure are molecules, then the entities at $N+1$ are atoms. Breaking their atomic bonds releases huge amounts of atomic energy as subatomic particles, N, are freed. Atomic bonds are stronger and release more energy when broken, compared to breaking the chemical bonds that make molecules, the $N+2$ entities.

operates largely in the realm of epistemology, as a theory of observation and analysis. The discourse generally takes the position that hierarchies appear somewhere between the material world and human understanding. If there are complex material systems that are not hierarchic, we might expect to have great difficulty in observing or understanding them. There appear to be points of passage of information up hierarchies, where details are explicitly lost. A military command is a favorite hierarchic example of a human organization. There, details of how an individual soldier observed local enemy concentrations fall away as the intelligence passes up the command. Setting the detail aside allows the top brass to make sweeping decisions, without being encumbered by a blizzard of local happenings. Not only must the general in command let go of details of grains so as to get a handle on the wide extent, but so too must the observers of hierarchical structure. To understand what a general is doing, the observer of a command structure needs to integrate away the details inside the army. By the late 1960s, the notion of hierarchy had moved beyond administration systems, and was being taken up across a range of disciplines.

In the following decade, hierarchy theorists from physics addressed hierarchical complexity after Heisenberg, invoking dualities, uncertainty, and complementarity between dynamics versus structure. Important developments have turned on the tension across the dilemmas presented by dual structures, as in the holon, a generalized entity in a hierarchy. Holons have been equated with the concept of system, with the advantage that holon does not appear in common parlance. The holon can therefore escape the reification and slovenly usage in the vernacular where the model is mistaken for the materiality. Conceptual developments suggest that what is inside a given holon is chosen by an observer. This emphasizes that holons are abstractions more than material objects; a point forgotten when 'system' is used for 'holon'. In the holon, the tension is between system and subsystem. But the subsystem is a system in its own right, thus offering some sort of dual existence that invites contradiction.

The concept of the holon takes the whole to be a surface that integrates the parts to give a unified signal to the rest of the universe. At the same time, the holon is the surface that integrates the external environment for the parts to experience. In ecology, the environment falls away at the level of the holon when viewed from the perspective of the parts. A forest raises humidity and lowers temperature, thus allowing survival of some of its parts, tree seedlings that are its future. The parts are protected. Conversely, viewed from the context of the hot, dry environment surrounding the forest, the contributions of each tree to the water vapor inside the forest are lost in a more general flux of water from the canopy. Thus, loss of information occurs with movement both up and down the hierarchy. While the environment is too large to catch the details of the working of the parts, the parts themselves cannot span wide enough to see large, slow differences in the larger context. In all this, we see again the tension embodied in hierarchical discussions between scale, organization, and uncertainty in observation.

Earlier, five general principles for ordering ecological hierarchies were recognized:

(1) As to frequency of behavior, higher-level holons operate at a lower frequency, taking longer to exhibit returns in behavior than holons at lower levels.

(2) Higher levels in a hierarchy constrain lower levels by displaying intransigent constancy. Deans constrain faculty by not changing the budget, except once a year.

(3) Higher levels in a hierarchy will be contextual to lower levels. The environment would be seen as operating at a higher level.

(4) With regard to bond strength, higher-level holons are held together by weaker forces than those that integrate lower-level holons (e.g., chemical vs. nuclear bonds) (**Figure 3**).

(5) As to containment, if higher-level holons consist of lower holons, which they contain, then the hierarchy is said to be nested. Not all the criteria apply to all hierarchies, but all five principles may apply simultaneously.

The distinction between nested and non-nested hierarchies matters (**Figure 3**). In nested systems, upper-level entities contain and consist of lower-level entities. In non-nested hierarchies, containment is not a criterion, but principles (1)–(3) can still apply. In nested hierarchies, containment applies even if aggregation criteria between levels change type. Western medicine generally uses nested hierarchies for the human condition. Thus organelles may be aggregated into cells by biochemical interaction. Meanwhile, nesting of organs inside the whole body may invoke fluid mechanics as a principal on which parts make the whole person. When whole humans nest inside groups, relationships may be in epidemiological terms. In Western medicine, there are regular changes in aggregation criteria from biochemical, through fluid dynamic, to epidemiological. Despite inconstant criteria for linking levels, the nesting keeps such hierarchies straight. But in non-nested hierarchies, such as food chains or pecking orders, the top dog neither contains nor consists of the subordinate individuals. Because there is no nesting to maintain order, non-nested hierarchies embody only one specific rule for moving between levels. As a result, the criteria for moving up a food chain must be consistently 'is eaten by', or conversely going down it is 'eats'. In this way, the hierarchy is consistent top to bottom. Because of their robustness to changes in aggregation criteria, nested hierarchies are

particularly useful for exploration before firm criteria connecting levels have been established. Concomitantly, non-nested hierarchies are for mature ideas, where focused sets of relationships are organized and abstracted in a control system.

In thermodynamic studies of ecological emergence, nested hierarchies are essential, because otherwise the bookkeeping of energy flow between the system and its environment would not sum. In complexity theory, self-organized emergence is a matter of thermodynamic gradients being applied to material systems that are pushed away from equilibrium. Thus, nested hierarchies apply when self-organization is invoked, when holons emerge at a new level without any plan. Planned systems often yield to a non-nested conception. A surprising and important new turn in applied ecology of human management systems links non-nested human socioeconomic hierarchies to nested thermodynamic hierarchies. The whole system is embodied in energy flow and control through the twinned social and biogeochemical hierarchies.

These thermodynamic approaches develop self-organizing holarchic open systems (SOHOs), using the term holarchy for nested hierarchies. The word holarchy appears in part to sidestep the political unacceptability of hegemonic hierarchical control. Using the SOHO approach, the full power of hierarchy theory in solving real time problems has been developed by Waltner Toews and colleagues at NESH, a Canadian centered, complex systems group. They solved some critical problems in Peru, Kenya, and Nepal. For instance, a Kathmandu sewer had children playing around slaughterhouse waste. By linking the social hierarchy to the ecological process hierarchy, NESH identified that a street cleaner caste was being blamed for things out of their control. Blaming scapegoats had led to inaction and paralysis, but once the street cleaners were no longer held responsible, the SOHO thermodynamic methodology achieved significant rehabilitation as the social and ecological hierarchies began to function in concert.

The earliest explicit introduction of hierarchy theory into ecology in the 1970s spoke of decomposability as an issue in some of the biomes studied in the International Biological Program (IBP). At that time, terms, such as 'environ', 'creon', and 'genon' were coined as extensions of the concept of holon. Environ addresses the environment acting as an integrated whole for its residents (see Ecological Network Analysis, Environ Analysis). The inward direction toward the holon pertains to the creon, whereas the outward direction pertains to the genon generating new things and experiences for the environment and its residents. Holon remains the central concept. The first fully integrated treatments of hierarchies in ecology turned on epistemological implications of scale and dynamics. Following shortly, evolutionary ideas focused on the structural elements in hierarchies, in a more ontological spirit. The structural elements were cast as a triadic view of holons, where the level above and the level below, as well as the level of the holon in between, are all required for an adequate treatment. Recently, two more crucial levels were added: the level above the context keeps the context of the holon stable, while the level below the parts provides stability for the material of which the parts are made.

Scale and Type

Scale problems invited hierarchy theory into the discipline. Ecologists have long been aware of scale, investigating the properties of quadrats in obtaining estimates of vegetation in the 1950s. Then change in variance across quadrat size was used to measure aggregation of plants on the ground. Hierarchy theory remains associated with scale today. The observation protocol brings attention to a universe of a certain extent, while making a second distinction, the finest grain at which observation units are distinguished from one another. Grain and extent together characterize the scalar level in question in many ecological hierarchies. Grain and extent are connected. Wider extents require coarser grains, if the mass of data are to be remembered, analyzed, and understood. Modern computational power has widened the gap between grain and extent, where remotely sensed areas are captured in billions of pixels. Even so, explicitly linking items in the grain across the extent becomes difficult, and generally impossible as the extent widens by much.

In contrast to linking across scales, it is possible to unify ecology across types of ecological system that correspond to the main subdisciplines of ecology: organism; population; community; ecosystem; landscape; biome; and biosphere. These types for ecological subdisciplines are explicitly not scale based, and so are not required to be assigned to level in the order given in the previous sentence. When scale is parsed away from type, the various approaches to ecology achieve a sharper depth of focus, offering clear relief between types of investigation. The subdisciplines of ecology are not scalar levels. If they are levels at all, they are type-based levels of organization, with the different types related to one another by asymmetric relationships made explicit in the definitions. As a separate issue, a typed level of organization itself contains scale-based hierarchies, as in fractal landscapes. In that scaled universe, the ecosystem modeling strategy may apply across a range of sizes, where local processes are part of more global processes. Communities too

may be variously inclusive of species across narrow or wider areas. Under the organism criterion, examples are found from redwood trees to mites. An ecological hierarchy may change the scale and type at the same time, but it is fraught with conceptual danger. Indeed, hierarchy theory is often invoked to clean up the mess in the aftermath of scale and type being mixed together. There is no prohibition changing both together, but only so long as the relationships at each new level are explicit. This matters because most descriptions of ecological material precisely do change type across widening scalar levels, although most of them do not follow the textbook ordering from organism to biosphere. For instance, in a forest community, a rotting tree trunk may be considered an ecosystem, whose upper surface is landscape, on which grows a community of bryophytes.

The copious variety of materials, entities, and sizes in ecology invites hierarchy theory into ecology. Indeed, it is in ecology that hierarchy theory has been used most often to significant effect, as in the NESH studies mentioned above. Hierarchy theory can capture a rich set of scaled examples across a mixture of types. Ecology is a multiple-scaled labyrinth of types. Hierarchy theory is the ball of string that we can trail behind, so that ecological scientists do not get lost.

See also: Ecological Network Analysis, Environ Analysis.

Further Reading

Ahl V and Allen TFH (1996) *Hierarchy Theory, A Vision Vocabulary and Epistemology*. New York: University of Columbia Press.
Allen TFH and Hoekstra TW (1992) *Toward a Unified Ecology*. New York: University of Columbia Press.
Allen TFH, O'Neill RV, and Hoekstra TW (1984) Interlevel relations in ecological research and management: Some working principles from hierarchy theory. *General Technical Report R.M.110*. Fort Collins: USDA Forest Service (republished in 1987 in *Journal of Applied Systems Analysis* 14: 63–79).
Allen TFH and Starr TB (1982) *Hierarchy: Perspectives for Ecological Complexity*. Chicago: University of Chicago Press.
Kay J, Regier H, Boyle M, and Francis G (1999) An ecosystem approach for sustainability: Addressing the challenge of complexity. *Futures* 31: 721–742.
Koestler A (1967) *The Ghost in the Machine*. Chicago: Gateway.
O'Neill RV, DeAngelis D, Waide J, and Allen TFH (1986) *Monographs in Population Biology 23: A Hierarchical Concept of Ecosystems*. Princeton: Princeton University Press.
Overton WS (1975) Decomposability: A unifying concept? In: Levin S (ed.) *Proceedings of the SIAM–SIMS Conference on Ecosystems Analysis and Prediction*, pp. 297–299. Philadelphia: Society for Industrial and Applied Mathematics.
Patten BC (1978) Systems approach to the concept of environment. *Ohio Journal of Science* 78: 206–222.
Pattee HH (ed.) (1973) *Hierarchy Theory: The Challenge of Complex Systems*. New York: Braziller.
Salthe SN (1985) *Evolving Hierarchical Systems*. New York: Columbia University Press.
Simon HA (1962) The architecture of complexity. *Proceedings of the American Philosophical Society* 106: 467–482.

Waltner-Toews D, Kay JJ, Neudoerffer C, and Gitau T (2003) Perspective changes everything: Managing ecosystems from the inside out. *Frontiers in Ecology and the Environment* 1(1): 23–30.

Webster JR (1979) Hierarchical organization of ecosytems. In: Halfon E (ed.) *Theoretic Systems Ecology*, pp. 119–131. New York: Academic Press.

Whyte LL, Wilson AG, and Wilson D (1969) *Hierarchical Structures*. New York: Elsevier.

Relevant Websites

http://www.nesh.ca – James Kay Web Page, Network for Ecosystem Sustainability and Health (NESH).

http://www.nbi.ku.dk – Stanley N. Salthe Web Page, Center for the Philosophy of Nature and Science Studies (CPNSS), Niels Bohr Institute.

Goal Functions and Orientors

H Bossel, University of Kassel (retd.), Zierenberg, Germany

Introduction
System Concepts
System Orientation in a Complex Environment

Simulation of the Evolution of System Orientation
Further Reading

Introduction

The global ecosystem is made up of an ensemble of interacting local and regional ecosystems, each composed of biotic and abiotic subsystems. The evolution of these systems is constrained by physical and system laws and by the basic properties of their environment, including the constraints of exergy (energy that can be usefully transformed into work), material, and information flows. Sustainability (persistence) of a system in its environment therefore requires respecting these constraints. Conversely, the very fact of its persistence demonstrates that a system has successfully adapted to its operating conditions. Evolution has forced it to respect physical and system laws and the basic properties of its environment. To an observer, the system's behavior appears to be guided by a particular attractor state, or by attention to a number of orientors.

System Concepts

System Organization

'System' is anything that is composed of system elements connected in a characteristic system structure (**Figure 1**). This configuration of system elements allows it to perform specific functions in its environment. These functions can be interpreted as serving a distinct system purpose. The system boundary is permeable for inputs from, and outputs to, the environment. It defines the system's identity and autonomy.

When we talk about a viable system, we mean that this system is able to survive, be healthy, and develop in its particular environment. In other words, system viability has something to do with both the system and its properties, and with the environment and its properties. And since a system usually adapts to its environment in a process of coevolution, we can expect that the properties of the system's environment will be reflected in the properties of the system; for example, the form of a fish and its mode of motion reflect the laws of fluid dynamics of its aquatic environment.

Systems are termed complex if they have an internal structure of many – qualitatively different – processes, subsystems, interconnections, and interactions. Besides assuring their own viability, the individual systems that are part of a complex total system specialize in certain functions that contribute to the viability of the total system. Viability of subsystems and the total system requires that subsystem functions and interactions are organized efficiently (or at least effectively). In the evolution of complex systems, two organizing principles in particular have established themselves: hierarchy and subsidiarity. They can be found in all successful complex systems: biological, ecological, social, political, technological.

Hierarchical organization means a nesting of subsystems and responsibilities within the total system. Each subsystem has a certain degree of autonomy for specific actions, and is responsible for performing certain tasks contributing to the viability of the total system. For example, body cells are relatively autonomous subsystems, but contribute specific functions to the operation

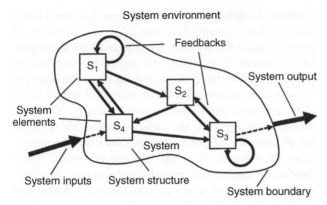

Figure 1 System notation.

of particular body organs, which in turn contribute to the viability of an organism.

Subsidiarity means that each subsystem is given the responsibility and the means for keeping its own house in order, within the range of its own abilities and potential. Only if conditions occur that cannot be handled by the subsystem would the suprasystem step in and help. The principles of hierarchical organization and subsidiarity require that each subsystem has a certain measure of autonomy. In its particular environment, each subsystem must be viable. The total system can only be viable if each of the subsystems supporting it is viable. Each subsystem reflects the properties of its individual environment; its behavior is informed (oriented) by that environment.

Note that this way of looking at complex systems is recursive. If necessary, we can apply the same system/subsystem dichotomy of viable systems again at other organizational levels. For example, a person is a subsystem of a family; a family is a subsystem of a community; a community is a subsystem of a state; a state is a subsystem of a nation, etc.

It is not enough to be concerned with the viability of individual systems. There are no isolated systems in the real world; all systems depend in one way or another on other systems. Hence their viability, and ultimately the viability of the total system are also preconditions for sustainable development. This means that a holistic system view must be adopted.

Evolution of Systems, and Emergence of Orientors and Goal Functions

The adaptation of a system to its environment is reflected in its structure, including its nonmaterial, cognitive structure. This system structure determines its behavior, and hence the adaptive response to its particular environment. System structures of material systems are dissipative; they require exergy and material flows for their construction, maintenance, renewal, and reproduction.

The dissipative structures of the global ecosystem are constructed and maintained by a finite rate of exergy input (mostly solar energy) and a finite stock of materials. The global ecosystem is therefore forced to recycle all of its essential material resources. The development of local ecosystems is constrained by the local rate of exergy flux (solar radiation input) and by the local rate of material recycling (weathering rate, absorption rate, decomposition rate, etc.) that it produces.

Evolution favors those species or (biotic) subsystems of the ecosystem that have learned to use available resources more efficiently and effectively than their competitors. This learning is embedded in their genetic code, and it is manifest in the dissipative structures they construct. Both will increase in complexity as a species evolves. At the ecosystem level, species evolution will cause increasingly better use of (exergy and material) resources. Species as well as ecosystems as a whole therefore tend to progress toward more complex dissipative structure producing more complex behavior.

Interacting species in a common ecosystem coevolve in the direction of increasing fitness of each individual species. Evolution of ecosystems therefore proceeds in the direction (arrow of time) of specialization, speciation, synergy, complexification, diversity, maximum throughflow of exergy, and more efficient use of material resources. This development becomes manifest in the corresponding emergent properties: exergy degradation, recycling, minimization of output, efficiency of internal flows, homeostasis and adaptation, diversity, heterogeneousness, hierarchy and selectivity, organization, minimization of maintenance costs, storage of available resources. These properties can be viewed as orientors, propensities, or attractors guiding system evolution and development. They are not limited to ecosystems; they are a general feature of living systems, including human organizations. When quantified and used in models, we refer to them as goal functions.

In particular, ecosystems will therefore build up in the course of their development as much dissipative structure as can be supported by the available exergy gradient. Available opportunities will eventually be found out by the processes of evolution, and will then be utilized. The ability to respond successfully to environmental challenges can be 'interpreted' as intelligent behavior, although it is strictly the result of nonteleological evolutionary development.

System Orientation in a Complex Environment

Basic concepts can be introduced by visualizing a simple animal with limited vision in a simple environment. The animal requires exergy for self-organization, motion, harvesting food, and maintenance. The environment provides food in certain locations, usually associated with obstacles that must be avoided since they have an exergy cost.

In a stable environment where sufficient (regenerating) food is distributed in a completely regular pattern, evolutionary adaptation would eventually lead to optimization of an animal's movements in a regular grazing pattern, with a single objective, optimum exergy uptake and use. The regular grazing pattern reflects the complete certainty of the next step, which the animal learns by accumulating and internalizing experience in a cognitive structure aiding its limited vision.

In more complex and diverse environments the animal, because of its limited vision, may not know for several steps which situation it will encounter next. It will therefore have to develop decision rules that have greater generality and are applicable to (and will be reinforced by) different motion sequences with different outcomes. In addition to the requirement of harvesting and using exergy resources effectively and efficiently, another objective is now implicitly added, to secure food under the constraint of incomplete information, that is, a security objective. Note that this is an emergent property that is not explicit in the reward system (which still rewards only food uptake). Failure to heed this implicit security objective will reduce food uptake and may endanger survival. On the other hand, the pressure to play it safe will occasionally mean giving up relatively certain reward. With other words, efficiency is traded for more security, and both are now prominent normative orientations (goals, values, interests) incorporated in the cognitive structure.

Orientation theory deals in a more general way with the emergence of behavioral objectives (orientors) in self-organizing systems in general environments. The proposition is that if a system is to survive in a given environment – characterized by a specific normal environmental state, sparse resources, variety, unreliability, change, and the presence of other systems – it must be able to physically exist in (be compatible with) this environment, effectively harvest necessary resources, freely respond to environmental variety, protect itself from unpredictable threats, adapt to changes in the environment, and interact productively with other systems. These essential orientations emerge in the course of the system's evolution in its environment.

Properties of Environments

There is obviously an immense variety of system environments, just as there is an immense variety of systems. But all of these environments have some common general properties. These properties will be reflected in systems. These reflections, or basic orientors, orient not just structure and function of systems, but also their behavior in the environment. The term orientor is used to denote (explicit or implicit) normative concepts that direct behavior and development of systems in general. In the social context, values and norms, objectives and goals are important orientors. Ecosystems and organisms tend toward certain attractor states whose specific characteristics can be viewed as orientors. Orientors exist at different levels of concreteness within an orientor hierarchy. The most fundamental orientors, the basic orientors, are identical for all complex adaptive systems. Orientors are dimensions of concern; they are not specific goals. Their satisfaction can be determined by observation of corresponding indicators, which can also be used to define goal functions for model studies.

In addition to the physical constraints of exergy and material flows, ecosystem and species development is determined by the 'general properties of the environment':

1. *Normal environmental state*. The actual environmental state can vary around this state in a certain range.
2. *Scarce resources*. Resources (exergy, matter, information) required for a system's survival are not immediately available when and where needed.
3. *Variety*. Many qualitatively very different processes and patterns occur in the environment constantly or intermittently.
4. *Reliability*. The normal environmental state fluctuates in random ways, and the fluctuations may occasionally take it far from the normal state.
5. *Change*. In the course of time, the normal environmental state may gradually or abruptly change to a permanently different normal environmental state.
6. *Other systems*. The behavior of other systems changes the environment of a given system.

Basic Orientors

If evolution enforces fitness of (natural) systems, then persistent systems must reflect the properties of their environment in their structure. More generally, the basic properties of the environment require corresponding basic system features. Since the basic environmental properties are independent of each other, a similar set of independent system features must exist, and it must find expression in the concrete features of the system structure.

There is a one-to-one relationship between the properties of the environment and the 'basic orientors of systems' (**Figure 2**):

1. *Existence*. Attention to existential conditions is necessary to insure the basic compatibility and immediate survival of the system in the normal environmental state.
2. *Effectiveness*. In its efforts to secure scarce resources (exergy, matter, information) from, and to exert influence on its environment, the system should on balance be effective.

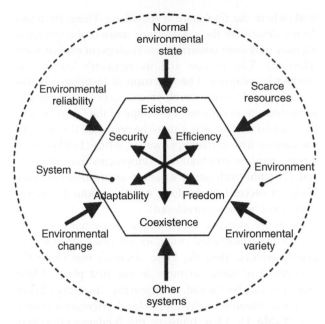

Figure 2 A tentative typology of emergent properties.

3. *Freedom of action.* The system must have the ability to cope in various ways with the challenges posed by environmental variety.
4. *Security.* The system must have the ability to protect itself from the detrimental effects of variable, fluctuating, unpredictable, and unreliable environmental conditions.
5. *Adaptability.* The system should be able to change its parameters and/or structure in order to generate more appropriate responses to challenges posed by changing environmental conditions.
6. *Coexistence.* The system must modify its behavior to account for behavior and interests (orientors) of other systems.

Obviously, the system equipped to secure better overall orientor satisfaction will have better fitness, and will therefore have a better chance for long-term survival and sustainability. In persistent systems or species, these orientors will be found as emergent objectives (or system interests).

Properties of Orientors

Each of the basic orientors stands for a unique requirement. Attention (conscious or unconscious) must therefore be paid to each of them, and the compensation of deficits of one orientor by over-fulfillment of other orientors is not possible. Fitness forces a multicriteria response, and comprehensive (conscious or unconscious) assessments of system behavior and development must also be multicriteria assessments.

In the assessment and orientation of system behavior, we deal with a two-phase assessment process where each phase is different from the other.

Phase 1. First, a certain minimum satisfaction must be guaranteed separately for each of the basic orientors. A deficit in even one of the basic orientors threatens long-term survival. The system will have to focus its attention on this deficit.

Phase 2. Only if the required minimum satisfaction of all basic orientors is guaranteed is it permissible to try to raise system satisfaction by improving satisfaction of individual orientors further.

Adequate satisfaction of each of the basic orientors requires, on a lower level, system- and environment-specific satisfaction of thermodynamic, structural, functional, ecophysiological, and system orientors. Network analysis suggests complementarity of different formulations of extremal principles as orientors describing ecosystem development.

Characteristic differences in the behavior of otherwise very similar systems (animals, humans, political, or cultural groups) can often be explained by differences in the relative importance attached to different basic orientors (i.e., emphasis on freedom, or security, or effectiveness, or adaptability) in phase 2 (i.e., after minimum requirements for all basic orientors have been satisfied in phase 1).

The basic orientor proposition has three important implications:

1. If a system evolves in a normal environment, then that environment forces it to implicitly or explicitly ensure minimum and balanced satisfaction of each of the basic orientors (and of lower-level orientors contributing to this satisfaction).
2. If a system has successfully evolved in a normal environment, its behavior will exhibit balanced satisfaction of each of the basic orientors.
3. If a system is to be designed for a normal environment, proper and balanced attention must be paid to satisfaction of each of the basic orientors.

The third implication has particular relevance for the creation of programs, institutions, and organizations in the sociopolitical sphere, among other things. Note that for a specific system in a specific environment, each orientor will have a specific meaning. For example, security of a nation is a multifacetted objective set with very different content from the security of an individual particular organism. However, the systems theoretical background for satisfaction of the security orientor is the same in both cases.

Orientors as Implicit Attractors

Better orientor satisfaction (better fitness) for more participants in a system requires more dissipative structure, which requires more exergy throughput as well as exergy accumulation. Since the exergy flow of ecosystems is limited (capture of solar radiation by photoproduction), increasingly better utilization is to be expected in the course of system development. This saturates at maximum exergy flow utilization for the ecosystem as a whole. Ecosystems as a whole therefore move in the direction of using all available exergy gradients. For organisms in the ecosystem, this implies development tendencies (orientors, propensities, attractors) toward specialization (using previously unused gradients), more complex structure (greater use efficiency), larger individuals (less maintenance exergy required per biomass unit), mutualism, etc. For species development, this translates into a principle of maximum exergy use efficiency. On the basis of these principles, prediction of development trends in ecosystems is possible.

The selection for better fitness in evolutionary processes favors systems (organisms) with better coping ability. Aspects of the behavioral spectrum of a system that improve coping ability (basic orientors) can be understood as implicit goals or attractors: existence, security, effectiveness, freedom, adaptability, coexistence. In the developmental stage of ecosystems, emphasis is on the basic orientors: existence, effectiveness, and freedom; in the mature stage it shifts to security, adaptability, and coexistence (see **Table 1**, where orientor concepts have been linked to E. P. Odum's classical model of ecological succession).

The existence of these implicit goals does not imply teleologic or teleonomic development toward a given goal (where the final state is specified). These attractors do not determine the exact future states of the system at all; they only pose constraints on choices (or evolutionary selection). The process and its rules are known, the product is unknown. The spectrum of (qualitatively different) possible future development paths and sustainable states remains enormous. The shape of the future, and of the systems that shape it, cannot be predicted this way. All one can say with certainty, however, is that (1) all possible futures must be continuous developments from the past, and (2) paths with better orientor satisfaction are more likely to succeed in the long run (if options to change paths have not been foreclosed).

In many systems, in particular ecosystems, specific attractors or functional orientors are often more immediately obvious than the basic orientors that cause the emergence of these orientors in the first place. These orientors can be viewed as appearing on a level below the basic orientors in the hierarchical orientation system (see **Table 1**). They translate the fundamental system needs expressed in the basic orientors into concrete attractor states linking system response to environmental properties. In models and ecosystem analyses, measures of ecosystem integrity can be based on corresponding ecosystem goal functions. Ecosystem attractor states emerge as general ecosystem properties in the coevolution of ecosystem and environment. They can be viewed as ecosystem-specific responses to the need to satisfy the basic orientors. Major ecosystem orientors are optimization of use of solar radiation, material, and energy flow intensities (networks); matter and energy cycling (cycling index); storage capacity (biomass accumulation); nutrient conservation, respiration, and transpiration; diversity (organization); hierarchy (signal filtering).

The emergence of basic orientors in response to the general properties of environments can be deduced from general systems theory, but supporting empirical evidence and related theoretical concepts can also be found in such fields as psychology, sociology, and the study of artificial life.

Table 1 Orientor concepts in the context of ecological succession

	Developmental stage	Mature stage
	Basic orientor emphasis	
	Existence	Coexistence
	Freedom	Security
	Effectiveness	Adaptability
Ecosystem orientor	*Orientor emphasis (goal function)*	
Growth and change	High	Low
Life cycle	Short, simple	Long, complex
Biomass	Low	High
Energy conservation	Low	High
Nutrient conservation	Low	High
Nutrient recycling	Low	High
Specialization	Low	High
Diversity	Low	High
Organization	Low	High
Symbiosis	Low	High
Stability; feedback control	Low	High
Structure	Linear, simple	Network, complex
Information	Low	High
Entropy	High	Low

Orientor Guidance in System Development, Control, Adaptation, and Evolution

Environmental influences partially determine system behavior. The magnitude of their effect on behavior depends on the influence structure of the system.

Sometimes systems can be controlled by controlling the inputs from their environment. However, the feedbacks in the system itself are usually more important for system control and adaptation of behavior to environmental conditions. Feedback means that the system state influences itself. Behavior-changing internal feedbacks are possible on several hierarchical levels in complex systems with different typical response characteristics and time constants (typical response times). These possibilities are also shown in **Figure 3**.

Response time	Level	Response
Immediate	Process	Cause–effect
Short	Feedback	Control
Medium	Adaptation	Parameter change
Long	Self-organization	Structural change
Very long	Evolution	Change of identity
Always	Basic orientors	Maintaining integrity

The simplest type of system response is the cause–effect relationship. It occurs at once as in, for example, stimulus–response reflex. It is the only type of system behavior which can legitimately be described by relating the output directly to the input. Unfortunately, it is often assumed that the same simple relationship is also applicable to other types of system response (such as the following), and this erroneous assumption often leads to fundamental mistakes.

On the next higher level we find responses which are generated by feedback in the system, involving at least one state variable or delay – such as an empty stomach causing hunger and the search for food. Control processes belong to this category. The response time is short, and influence structure and system parameters remain invariant.

On the next higher level we find processes of adaptation. In this case the system maintains its basic influence structure, but parameters are adjusted to adapt to the situation, possibly changing the response characteristics in the process. For example, a tree may adapt to the gradual lowering of the groundwater level by growing its roots to greater depth. This constitutes a parameter change (root length and root surface). The fundamental system structure of a tree, in particular, the function of the roots, has not changed in this case.

On the next higher level we find processes of self-organization in response to environmental challenges. This means structural change in the system. Processes of this kind have a longer response time and can only be conducted by systems having the capability for self-organization. Adult organisms or technical systems rarely or never belong to this category; on the other hand, this characteristic is often found in the development of organisms, social systems, organizations, and ecosystems.

A system may also change its identity in the course of an evolutionary process. This means that its functional characteristics, and hence its system purpose, change with time. Adaptations of this kind take place as a result of reproduction and evolution of living organisms. It is characteristic of this process that the system change coincides with a possibly drastic shift in system identity (change of goal function and of system purpose). An evolutionary example is the development of flying animals (birds) from water-dwelling reptiles.

All of these system responses to challenges from the environment in essence constitute attempts to maintain system integrity (possibly over many generations and over a long time period) even if it means changing system identity, that is, system purpose. From this observation it can be deduced that a system must orient its development with respect to certain basic criteria (basic orientors) to assure its long-term existence and development in an often hostile environment. This orientation may be implicit (forced upon the system) or explicit (actively pursued by the system). It does not require conscious decision or

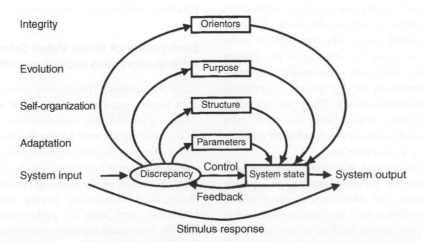

Figure 3 System response can be caused by different processes with very different time constants: stimulus–response, feedback control, adaptation, self-organization, evolution, maintaining system integrity.

even cognitive ability, although resulting action may appear to an observer as intelligent or even goal- or value-oriented behavior.

Simulation of the Evolution of System Orientation

Animats and Genetic Algorithms for Orientation

Orientation theory is not just a conceptual framework for understanding system evolution and behavior under the exergy availability constraint. It also allows quantitative and comparative analysis of system performance under different environmental conditions.

Genetic algorithms are models of biological adaptive processes that are being widely and successfully applied to a wide spectrum of adaptation and optimization problems. In particular, these algorithms have been used to simulate learning and adaptation of artificial animals (animats) in simulated environments containing food and obstacles. They can be used to demonstrate the emergence of basic orientors in self-organizing systems having to cope with complex environments.

The animat model incorporates essential features of a simple animal in a diverse environment. Being an open system, an animal depends on a flow of exergy from the environment. In the course of its (species) evolution, it has to learn to associate certain signals from the environment with reward or pain and to either seek or avoid their respective sources (exergy gain or exergy loss). This learning phase (of populations) will eventually lead to the establishment of cognitive structure and behavioral rules which are approximately optimal in the particular environment (with respect to maximization of reward, minimization of pain, and securing survival). These behavioral rules incorporate knowledge which enables intelligent behavior.

The animat is designed to simulate this process. It can pick up sensory signals from its environment (containing food and obstacles), and classify them with available rules to determine an appropriate action (direction of movement). After a successful move, the strength of rules leading up to it is increased by sharing in the reward (i.e., exergy gain). New rules are occasionally generated by either random creation, or by genetic operations (crossing-over and recombination). They are added to the existing rule set, and compete with the other rules for reward. Unsuccessful rules are not reinforced and lose strength and influence in the rule set.

The training process consists of placing the animat at a random empty location in an environment with specific environmental properties, and allowing it to move around searching for food. A collision with an obstacle causes a loss of exergy and throws the animat back to its previous position. Rules leading to success are rewarded. A genetic event of rule generation may occur with a prescribed probability.

Random rules are created in unknown situations. The process is repeated for a large number of steps (typically 10 000). Eventually, a set of behavioral rules develops which allows optimal behavior under the given set of conditions.

Note that this optimal behavior has not been defined in terms of an objective function guiding the evolution of the set of behavioral rules. The rule set develops solely from the reinforcement of rules which lead to food or avoid collisions. An explicit exergy balance accounts for all exergy losses associated with movement, collisions with obstacles, and rule generation, and exergy gains due to uptake of food. The development of the rule set is then driven by the requirement to optimize exergy pickup in the given environment (with specific resource availability), while allowing for environmental variety, variability, and change specific for that environment. Neglect of these properties is penalized by lack of fitness, and threat to survival, and causes disappearance of deficient rules. Other criteria besides efficiency will therefore be reflected in the set of behavioral rules. Since these were not expressly introduced, we must recognize them as emergent value orientations or objective functions.

The animat experiment contains all components necessary for a study in the basic orientor framework. Animat fitness depends on the ability to maintain a positive exergy balance in the long term. This exergy balance is therefore at the core of the orientor satisfaction assessment. At each step, exergy uptake (by food consumption) and exergy losses (by collisions with obstacles, motion, and learning of rules) are recorded and used to compute the momentary exergy balance. Attention to all orientors is mandatory to ensure a positive exergy balance even under adverse environmental conditions.

Quantitative measures must be defined for characterizing the different properties of the environments used in the animat experiments. Animat performance in different environments is compared by using measures of orientor satisfaction. These have to be defined using relevant parameters of animat performance.

Emergence of Basic Value Orientations, Anticipation, and Individual Differences

Since the animat's training depends on a number of random factors, each animat develops a different cognitive system (classifier set and decision rules), even though final performance may be similar. In order to show general tendencies despite these individual differences, mean values over large populations were obtained. These dealt with (1) results of the training process in two (otherwise identical) environments having different variety and variability, and with (2) performance of animats after transfer from their training environment to environments challenging them with more variety, or variability, or change.

One remarkable result from these experiments is that individuals achieve comparable performance in a given training environment with very different cognitive systems, and in particular with different orientor emphasis. While this may not provide any particular advantage in the training environment, it may provide distinct fitness advantages if the animat is moved to a different environment. Three particular types of individuals stand out: generalists (type F) stressing freedom of action, specialists (type E) focusing on effectiveness, and cautious type (type S) emphasizing security. **Figure 4** shows the different orientor stars for these three types.

The ability to develop a cognitive system reflecting its environment makes the animat a suitable vehicle for investigating goal function emergence and value orientation. Genetic algorithms are very effective processes that seem to capture the essentials of real processes found in the evolution of organisms and ecosystems. In the animat, they very effectively build up a cognitive model (or goal function) that enables anticipatory behavior; since rewards flow back to earlier rules leading to later pay-off, the activation of the initial rules in a pay-off chain means that the system suspects possible pay-off and anticipates the near future, that is, it has an internal model of the results of its actions under the given circumstances.

In the animat experiments (and similarly, in real life), implicit and (more or less) balanced multidimensional attention to the basic orientors emerges from the simple one-dimensional mechanism of rewarding success in the given environment. Thus, in the course of its evolutionary development in interaction with its environment, the system evolves a complex multidimensional behavioral objective function from the very unspecific requirement of fitness. Conversely, this also means that balanced attention to the emergent basic orientors is necessary for

system viability and survival – they would not have emerged unless important for the viability of the system.

Balanced attention still leaves room for individual differences in the relative emphasis given to the different orientors. Individuals belonging to the populations used in the animat experiments evolve significant differences in value emphasis (e.g., specialist, generalist, cautious type). These individual variations, while not significantly reducing performance in the standard training environment, provide comparative advantage and enhanced fitness when resource availability, variety, or reliability of the environment change. They also result in distinctly different behavioral styles. However, pathological behavior will follow if orientor attention becomes unbalanced (e.g., dominant emphasis on one orientor).

Training of animats in different environments, the performance of animat individuals in environments that differ from their training environments, and the simulation of adaptive learning in a changing environment, lead to some general conclusions that are in full agreement with everyday observations and general systems knowledge:

- Generalists have a better survival chance than others if moved to an environment of greater variety.
- Cautious types have a better survival chance than others if moved to a less reliable environment.
- Training in more unreliable and/or more diverse environments increases satisfaction of the security and/or freedom of action orientors at the cost of the effectiveness orientor.
- Training in an uncertain environment teaches caution and improves fitness in a different environment.
- Learning caution (better satisfaction of the security orientor) takes time and decreases effectiveness, but increases overall fitness.
- Investment in learning (exergy cost of learning in the animat) pays off in better fitness; the learning investment is (usually) much smaller than the pay-off gain.

Animat individuals not only develop behavior that can be interpreted as intelligent, they also develop a complex goal function (balanced attention to basic orientors), or value orientation. Serious attention to basic values (basic orientors: existence, effectiveness, freedom, security, adaptability, coexistence) is therefore an objective requirement emerging in, and characterizing self-organizing systems. These basic values are not subjective human inventions; they are objective consequences of the process of self-organization in response to normal environmental properties.

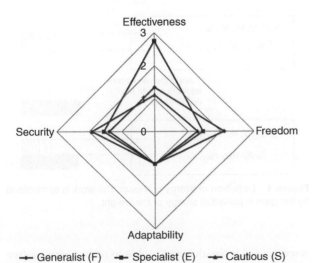

Figure 4 Generalist (F) Specialist (E) Cautious (S)

Figure 4 In an identical training environment, different lifestyles may evolve. Generalists stress freedom of action, specialists focus on effectiveness, while cautious types emphasize security.

See also: Ecological Network Analysis, Ascendency; Ecological Network Analysis, Environ Analysis; Exergy; Fundamental Laws in Ecology.

Further Reading

Ashby WR (1962) Principles of the self-organizing system. In: von Foerster H and Zopf GW (eds.) *Principles of Self-Organization*, pp. 255–278. New York: Pergamon.

Bossel H (1977) Orientors of nonroutine behavior. In: Bossel H (ed.) *Concepts and Tools of Computer-Assisted Policy Analysis*, pp. 227–265. Basel: Birkhäuser.

Bossel H (1999) *Indicators for Sustainable Development: Theory, Method, Applications*. Winnipeg: IISD International Institute for Sustainable Development.

Bossel H (2001) Exergy and the emergence of multidimensional system orientation. In: Jørgensen SE (ed.) *Thermodynamics and Ecological Modelling*, pp. 193–209. Boca Raton, FL: Lewis.

Fath BD, Patten BC, and Choi JS (2001) Complementarity of ecological goal functions. *Journal of Theoretical Biology* 208(4): 493–506.

Holland JH (1992) *Adaptation in Natural and Artificial Systems*. Cambridge, MA: MIT Press.

Jantsch E (1980) *Self-Organizing Universe: Scientific and Human Implications of the Emerging Paradigm of Evolution*. New York: Pergamon.

Jørgensen SE (2001) A tentative fourth law of thermodynamics. In: Jorgensen SE (ed.) *Thermodynamics and Ecological Modelling*, pp. 305–347. Boca Raton, FL: Lewis.

Krebs F and Bossel H (1997) Emergent value orientation in self-organization of an animat. *Ecological Modelling* 96: 143–164.

Mayr E (1974) Teleological and teleonomic: A new analysis. *Boston Studies in the Philosophy of Science* 14: 91–117.

Mayr E (2001) *What Evolution Is*. New York: Basic Books.

Miller JG (1978) *Living Systems*. New York: McGraw-Hill.

Odum EP (1969) The strategy of ecosystem development. *Science* 164: 262–270.

Müller F and Leupelt M (eds.) (1998) *Eco Targets, Goal Functions, and Orientors*. Berlin/Heidelberg/New York: Springer.

Wilson SW (1985) Knowledge growth in an artificial animal. In: Grefenstette JJ (ed.) *Proceedings of the First International Conference on Genetic Algorithms and Their Applications*, pp. 16–23. Pittsburgh PA and San Mateo: Lawrence Earlbaum and: Morgan Kaufmann.

Exergy

S E Jørgensen, Copenhagen University, Copenhagen, Denmark

Definition: Exergy
Definition: Eco-Exergy
Exergy and Information
Exergy and the Dissipative Structure
How to Calculate Eco-Exergy of Organic Matter and Organisms?
Why Do Living Systems Have Such a High Level of Eco-Exergy?

Losses and Gains of Eco-Exergy by Human Activities Included Pollution
Formulation of a Thermodynamic Hypothesis for Ecosystems
Support to the Maximum Eco-Exergy Hypothesis
Further Reading

Definition: Exergy

Energy is defined as the amount of work (= entropy-free energy) a system can perform when it is brought into thermodynamic equilibrium with its environment (**Figure 1**). The considered system is characterized by the extensive state variables S, U, V, N_1, N_2, N_3, ..., where S is the entropy, U is the energy, V is the volume, and N_1, N_2, N_3, ... are moles of various chemical compounds, and by the intensive state variables, T, p, μ_{c_1}, μ_{c_2}, μ_{c_3}, ..., where T is the temperature, p the pressure, and μ symbolizes the chemical potential of the components 1,2,3,.... The system is coupled to a reservoir or reference state by a shaft, together forming a closed system. The reservoir (the environment) is characterized by the intensive state variables T_o, p_o, μ_{oc_1}, μ_{oc_2}, μ_{oc_3}, ..., and as the system is small compared with the reservoir the intensive state variables of the reservoir will not be changed by interactions between the system and the reservoir. The system develops toward equilibrium with the

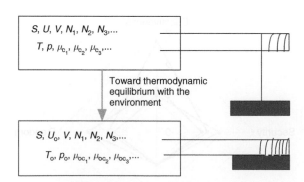

Figure 1 Definition of exergy is shown. The work is symbolized by the gain in potential energy of the weight.

reservoir and is simultaneously able to release entropy-free energy to the reservoir. During this process the volume of the system is constant as the entropy-free energy (i.e., work energy) must be transferred through the shaft only.

According to the definition of exergy, Ex, we have

$$Ex = \Delta U = U - U_o \qquad [1]$$

As

$$U = TS - pV + \sum_c \mu_c N_i \qquad [2]$$

(when we only consider the three energy forms: heat, spatial energy (displacement work), and chemical energy; see any textbook in thermodynamics), and correspondingly

$$U_o = T_o S - p_o V + \sum_c \mu_{c_o} N_i \qquad [3]$$

we get the following expression for exergy, when in this case kinetic energy, potential energy, electrical energy, radiation energy, and magnetic energy are excluded we have

$$Ex = S(T - T_o) - V(p - p_o) + \sum_c (\mu_c - \mu_{c_o}) N_i \qquad [4]$$

The total transfer of entropy-free energy in this case is the exergy of the system. It is seen from this definition that exergy is dependent on the state of the total system (= system + reservoir) and not dependent entirely on the state of the system. Exergy is therefore not a state variable.

This definition of exergy is used in engineering to express the efficiency of power plants. The energy efficiency of power plants is of course 100%, according to the first law of thermodynamics, while the interesting efficiency is the exergy efficiency: how much of the chemical energy (exergy) in the applied fossil fuel if fossil fuel is the energy source is converted to useful work (exergy)? What is not converted to exergy in form of electricity is lost as heat to the environment at the temperature of the environment – it contains therefore no work potential.

Notice that the exergy of the system is dependent on the intensive state variables of the reservoir. Notice that exergy is not conserved – only if entropy-free energy is transferred, which implies that the transfer is reversible. All processes in reality are, however, irreversible, which means that exergy is lost (and entropy is produced). Loss of exergy and production of entropy are two different descriptions of the same reality, namely, that all processes are irreversible, and we unfortunately always have some loss of energy forms which can do work to energy forms which cannot do work (heat at the temperature of the environment). So, the formulation of the second law of thermodynamic by use of exergy is 'all real processes are irreversible which implies that exergy inevitably is lost'. 'Exergy is not conserved', while energy of course is conserved by all processes according to the first law of thermodynamics.

The efficiency of concern is the ratio of useful energy (work) to total energy which always is less than 100% for real processes, which always are irreversible. This efficiency expresses that a part of the energy cannot be utilized as work and that all processes are irreversible because exergy is lost by all energy transfer processes as heat to the environment.

Exergy efficiency, defined as work performed divided by the total exergy available, is also of interest, particularly in technology. It expresses how much of the work capacity we are able to utilize.

All transfers of energy imply that exergy is lost because energy is transformed to heat at the temperature of the environment. It has therefore been of interest to set up for all environmental systems an exergy balance in addition to an energy balance. Our concern is exergy loss because it means that 'first class energy' which can do work is converted to 'second class energy' (heat at the temperature of the environment) which cannot do work. So, the particular properties of heat and temperature are a measure of the movement of molecules, given limitations in our possibilities to utilize energy to do work. Due to these limitations, we have to distinguish between exergy which can do work and anergy which cannot do work, and all real processes imply inevitably a loss of exergy as anergy (see also the next section).

Exergy or rather the loss of exergy as heat, which means production of entropy, seems more useful to apply than entropy to describe the irreversibility of real processes. It has the same unit as energy and is an energy form, while the definition of entropy is more difficult to relate to concepts associated to our usual description of reality. In addition entropy is not clearly defined for 'far from thermodynamic equilibrium systems', particularly for living systems. Moreover, it should be mentioned that the self-organizing abilities of systems depend strongly on the temperature. Exergy takes the temperature into consideration as the definition shows, while entropy does not. It implies that exergy at 0 K is 0 and at minimum. Negative entropy is not expressing the ability of the system to do work (we may call it 'the creativity' of the system as creativity requires work), but exergy becomes a good measure of 'the creativity', which is increasing proportional with the temperature. Furthermore, exergy facilitates the differentiation between low-entropy energy and high-entropy energy, as exergy is entropy-free energy.

Information contains exergy. Boltzmann showed that the free energy of information (it means exergy) that we actually possess (in contrast to the information we need to describe the system) is $kT \ln I$, where I is the information we have about the state of the system, for instance, that the configuration is 1 out of W possible ones and k is Boltzmann's constant $= 1.3803 \times 10^{-23}$ J/(molecules deg). It implies that one bit of information has the exergy equal to $kT\ln 2$. Transformation of information from one system to another is often almost an entropy-free energy transfer. If the two systems have different temperatures, then the entropy lost by one system is not equal to the entropy gained by the other system, while the exergy lost by the first system is equal to the exergy transferred and equal to the exergy gained by the other system, provided that the transformation is not accompanied by any loss of exergy. Also, in this case, it is obviously more convenient to apply exergy than entropy.

Definition: Eco-Exergy

In ecology, technological exergy is not so useful because the reference state, the environment, would be the adjacent ecosystem and we would like to find an expression that can measure how developed an ecosystem is, that is, how far it is from thermodynamic equilibrium. For a reservoir or reference state, it is therefore advantageous in ecology to select the same system but at thermodynamic equilibrium, that is, that all components are inorganic and at the highest oxidation state, if sufficient oxygen is present (nitrogen as nitrate, sulfur as sulfate, etc.). The reference state will in this case correspond to the ecosystem without life forms and with all chemical energy utilized or as an 'inorganic soup'. Usually, it implies that we also consider $T = T_0$, and $p = p_0$, which means that the exergy becomes equal to the difference of Gibb's free energy of the system and the same system at thermodynamic equilibrium, or the chemical energy content included the thermodynamic information (see below) of the system. Gibb's free energy is defined according to the following equation:

$$dG = dE + p\,dV - S\,dT$$

where dV is the change in volume and dS is the change in entropy. T and p are the temperature and pressure, respectively. The exergy becomes by this definition clearly a measure of how far the ecosystem is from thermodynamic equilibrium, that is, how much (complex) organization the ecosystem has build up in the form of organisms, complex biochemical compounds, and complex ecological network. Here, we use the available work, that is, the exergy, as a measure of the distance from thermodynamic equilibrium.

This description of exergy development in an ecosystem makes it pertinent to assess the exergy of ecosystems. It is not possible to measure exergy directly – but it is possible to compute it by eqn [4]. **Figure 2** illustrates the definition of 'eco-exergy'. As the chemical energy embodied in the organic components and the biological structure contributes far most to the exergy content of

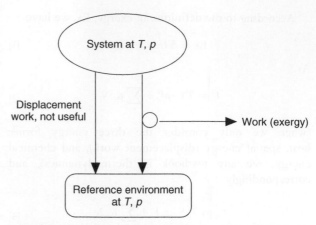

Figure 2 The exergy content of the system is calculated in the text for the system relatively to a reference environment of the same system at the same temperature and pressure, as an inorganic soup with no life, biological structure, information, or organic molecules.

the system, there seem to be no reason to assume a (minor) temperature and pressure difference between the system and the reference environment. Under these circumstances we can calculate the exergy content of the system as coming entirely from the chemical energy:

$$\mathrm{Ex} = \sum_c (\mu_c - \mu_{c_0}) N_i \qquad [5]$$

This represent the nonflow chemical eco-exergy. The difference in chemical potential $(\mu_c - \mu_{c_0})$ between the eco-system and the same system at thermodynamic equilibrium determines the eco-exergy. This difference is determined by the concentrations of the considered components in the system and in the reference state (thermodynamic equilibrium), as it is the case for all chemical processes.

We can measure the concentrations in the eco-system, but the concentrations in the reference state (thermodynamic equilibrium) can only be based on the usual use of chemical equilibrium constants. If we have the process

Component A $\leftrightarrow$ inorganic decomposition products

It has a chemical equilibrium constant, K:

$$K = [\text{inorganic decomposition products}]/[\text{Component A}] \qquad [6]$$

The concentration of component A at thermodynamic equilibrium is difficult to find, but we can find the concentration of component A at thermodynamic equilibrium from the probability of forming A from the inorganic components.

We find by these calculations the exergy of the system compared with the same system at the same temperature and pressure but in form of an inorganic soup without any life, biological structure, information, or organic molecules. As $(\mu_c - \mu_{c_0})$ can be found from the definition of the chemical potential replacing activities by concentrations, we get the following expressions for eco-exergy:

$$\text{Ex} = RT \sum_{i=0}^{i=n} c_i \ln c_i / c_{i,o} \qquad [7]$$

where R is the gas constant ($8.314\,\text{J\,K}^{-1}\,\text{mol}^{-1} = 0.08207\,\text{l\,atm.\,K}^{-1}\,\text{mol}^{-1}$), T is the temperature of the environment (and the system; see **Figure 2**), while c_i is the concentration of the ith component expressed in a suitable unit, for example, for phytoplankton in a lake c_i could be expressed as $\text{mg\,l}^{-1}$ or as $\text{mg\,l}^{-1}$ of a focal nutrient. $c_{i,o}$ is the concentration of the ith component at thermodynamic equilibrium and n is the number of components. $c_{i,o}$ is of course a very small concentration (except for $i = 0$, which is considered to cover the inorganic compounds), corresponding to a very low probability of forming complex organic compounds spontaneously in an inorganic soup at thermodynamic equilibrium. $c_{i,o}$ is even lower for the various organisms because the probability of forming the organisms is very low with their embodied information, here represented by the genetic code.

By using this particular exergy based on the same system at the thermodynamic, chemical equilibrium as reference, the eco-exergy depends only on the chemical potential of the numerous biochemical components that are characteristic for life. It is consistent with Boltzmann's statement that life is a struggle for free energy. Eco-exergy has a definition close to the free energy, but unlike free energy, eco-exergy is not a state variable. It will depend on the reference state that will vary from ecosystem to ecosystem. Furthermore, it is difficult to apply the classic state variables in thermodynamics far from thermodynamic chemical equilibrium. Classic thermodynamics presumes that the system is close to equilibrium, which makes it possible to show that for instance free energy is a state variable that gives the same result independent on the pathway. We want to use eco-exergy far from thermodynamic equilibrium and can therefore not use free energy in this context.

As we know that ecosystems due to the throughflow of energy have the tendency to move away from thermodynamic equilibrium losing entropy or gaining exergy and information, we can at this stage formulate the following proposition of relevance for ecosystems: 'ecosystems attempt to develop toward a higher level of exergy'.

Exergy and Information

Information means 'acquired knowledge'. The thermodynamic concept of exergy is closely related to information. A high local concentration of a chemical compound, for

instance, with a biochemical function that is rare elsewhere, carries exergy and information. On more complex levels, information may still be strongly related to exergy but in more indirect ways. Information is also a convenient measure of physical structure. A certain structure is chosen out of all possible structures and defined within certain tolerance margins.

It is possible to distinguish between the exergy of information and the exergy of biomass. p_i defined as c_i / A, where

$$A = \sum_{i=1}^{n} c_i \qquad [8]$$

is the total amount of matter in the system, is introduced as new variable in eqn [7]:

$$\text{Ex} = ART \sum_{i=1}^{n} p_i \ln (p_i / p_{i_o}) + A \ln A / A_o \qquad [9]$$

As $A \approx A_o$, exergy becomes a product of the total biomass A (multiplied by RT) and Kullback's measure:

$$K = \sum_{i=1}^{n} p_i \ln (p_i / p_{i_o}) \qquad [10]$$

where p_i and p_{i_o} are probability distributions, *a posteriori* and *a priori* to an observation of the molecular detail of the system. It means that K expresses the amount of information that is gained as a result of the observations.

If we observe a system, which consists of two connected chambers (see **Figure 3**), then we expect the molecules to be equally distributed in the two chambers, that is, $p_1 = p_2$ is equal to $1/2$. If, on the other hand, we observe that all the molecules are in one chamber, then we get $p_1 = 1$ and $p_2 = 0$. Let us presume that the chamber to the left contains 1 mol of a pure ideal gas, while the chamber to the right is empty. If we open the valve between the two chambers, the loss of eco-exergy (and also of technological exergy) for the system would be $RT \ln 2$ in accordance with eqns [7], [9], and [10]. We could utilize at least a part of the exergy by installation of a small propeller in the valve. The system will by this process increase its entropy $R \ln 2$. This is in accordance with eqn [8].

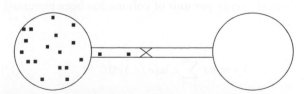

Figure 3 The left chamber contains 1 mole of a pure ideal gas, while the right chamber is empty. If we open the valve, the system will loose eco-exergy (or technological exergy) = $RT \ln 2$, which we could utilize by installation of a propeller in the valve. The entropy of the system will simultaneously increase by $R \ln 2$.

Exergy and the Dissipative Structure

As an ecosystem is nonisolated, the entropy changes during a time interval, dt, can be decomposed into the entropy flux due to exchanges with the environment, and the entropy production due to the irreversible processes inside the system such as diffusion, heat conduction, and chemical reactions. It can also be expressed by use of exergy:

$$Ex/dt = d^e Ex/dt + d^i Ex/dt \qquad [11]$$

where $d^e Ex/dt$ represents the exergy input to the system and $d^i Ex/dt$ is the exergy consumed (is negative) by the system for maintenance, etc. e is used to indicate an external source and i to indicate the internal exergy change.

Equation [11] shows among other things that systems can only maintain a nonequilibrium steady state by compensating the internal exergy consumption with a positive exergy influx ($d^e Ex/dt > 0$). Such an influx induces order into the system. In ecosystems, the ultimate exergy influx comes from solar radiation, and the order induced is, for example, biochemical molecular order. If $d^e Ex > -d^i Ex$ (the exergy consumption in the system), the system has surplus exergy input, which may be utilized to construct further order in the system, or as Prigogine calls it, dissipative structure. The system will thereby move further away from thermodynamic equilibrium. Evolution shows that this situation has been valid for the ecosphere on a long-term basis. In spring and summer, ecosystems are in the typical situation that $d^e Ex$ exceeds $-d^i Ex$. If $d^e Ex < -d^i Ex$, the system cannot maintain the order already achieved, but will move closer to the thermodynamic equilibrium, that is, it will lose order. This may be the situation for ecosystems during fall and winter or due to environmental disturbances.

How to Calculate Eco-Exergy of Organic Matter and Organisms?

The following expression for what we could call the ecological exergy per unit of volume has been presented; see eqn [7]:

$$Ex = RT \sum_{i=0}^{i=n} c_i \ln(c_i/c_{i_o}) \left[ML^{-1}T^{-2}\right] \qquad [12]$$

where R is the gas constant, T is the temperature of the environment, c_i is the concentration of the ith component expressed in a suitable unit, and c_{i_o} is the concentration of the ith component at thermodynamic equilibrium and n is

the number of components. c_{i_o} is very low for living component because the probability that living components are formed at thermodynamic equilibrium is very low. It implies that living components get a high eco-exergy. c_{i_o} is not zero for organisms, but will correspond to a very low probability of forming complex organic compounds spontaneously in an inorganic soup at thermodynamic equilibrium. c_{i_o} on the other hand is high for inorganic components, and although c_{i_o} still is low for detritus, it is much higher than for living component.

The exergy of structurally complicated material can be estimated based on the elementary composition. This has, however, the disadvantage that a higher organism and a microorganism with the same elementary composition will get the same exergy which is in complete disagreement with the lower probability to form a more complex organism, that is, the lower concentration of c_{i_o} in the equation. The composition will not account for the contribution of Kullbach's measure of information, which is often the major part of the eco-exergy, as it is shown below.

The problem related to the assessment of c_{i_o} has been discussed and a possible solution proposed. For dead organic matter, detritus, which is given the index 1, it can be found from classical thermodynamics.

For the biological components, $2, 3, 4, \ldots, N$, the probability, p_{i_o}, consists at least of the probability of producing the organic matter (detritus), that is, p_{1_o}, and the probability, $p_{i,a}$, to find the correct composition of the enzymes determining the biochemical processes in the organisms. Living organisms use 20 different amino acids and each gene determines in average the sequence of about 700 amino acids. $p_{i,a}$ can be found from the number of permutations among which the characteristic amino acid sequence for the considered organism has been selected. It means that

$$p_{i,a} = a - N_{g_i} \qquad [13]$$

where a is the number of possible amino acids $= 20$, N is the number of amino acids determined by one gene $= 700$, and g_i is the number of non-nonsense genes. The following two equations are available:

$$p_{i_o} = p_{1_o} p_{i,a} = p_{1_o} a - Ng \approx p_{1_o} \cdot 20 - 700g \qquad [14]$$

The exergy contribution of the ith component can be found by combining eqns [12] and [14]:

$$
\begin{aligned}
Ex &= RT \, c_i \ln c_i/(p_{1_o} a - Ng c_{0_o}) = (\mu_1 - \mu_{1_o})c_i - c_i \ln p_{i,a} \\
&= (\mu_1 - \mu_{1_o})c_i - c_i \ln (a - Ng_i) \\
&= 18.7 c_i + 700(\ln 20) \, c_i \, g_i \left[ML^{-1}T^{-2}\right]
\end{aligned}
\qquad [15]
$$

The total eco-exergy can be found by summing up the contributions originated from all components. The contribution by inorganic matter can be neglected as the contributions by detritus and even to a higher extent from the biological components are much higher due to an extremely low concentration of these components in the reference system (thermodynamic equilibrium for the system). The contribution by detritus, dead organic matter, is $18.7\,kJ\,g^{-1}$ times the concentration (in gram per unit of volume) corresponding to the composition of detritus, namely lipids, carbohydrates, and proteins mainly, while the eco-exergy of living organisms with approximations consists of

$$Ex_{1chem} = 18.7\,kJ\,g^{-1} \text{ times the concentration } c_i$$

$$(\text{gram per unit of volume})$$

and

$$Ex_{ibio} = RT(700\ln 20)\,c_i\,g_i = RT2100\,g_i\,c_i \qquad [16]$$

$R = 8.314\,J\,mole^{-1}$ and if we presume a molecular weight of an average 105 for the enzymes, we obtain the following equation for Ex_{ibio} at 300 K:

$$Ex_{ibio} = 0.0529\,g_i\,c_i \qquad [17]$$

where the concentration now is expressed in g per unit of volume and the exergy in kilojoules per unit of volume.

For the entire system the eco-exergy, Ex-total = exergy-chemical + exergy-biological can be found as

$$\text{Ex-total} = 18.7\sum_{i=1}^{N} c_i - 0.0529\sum_{i=1}^{N} c_i\,g_i\,[ML^{-1}T^{-2}] \qquad [18]$$

where g for detritus $(i=1)$ of course is 0. **Table 1** shows the weighting factor, β, which is introduced to be able to cover the exergy for various organisms in the unit detritus equivalent or chemical exergy equivalent:

$$\text{Ex-total} = \sum_{i=1}^{N} \beta_i c_i (\text{as detritus equivalent}) \qquad [19]$$

The calculation of eco-exergy accounts for the chemical energy in the organic matter as well as for the (minimum) genetic information embodied in the living organisms. The latter contribution is measured by the extremely small probability to form the living components, for instance algae, zooplankton, fish, mammals, etc., spontaneously from inorganic matter. Weighting factors defined as the exergy content relatively to detritus (see **Table 1**) may be considered quality factors reflecting how developed the various groups of organisms are and to what extent they contribute to the exergy due to their content of information which is reflected in the computation. The β-values in **Table 1** are found on basis of latest

knowledge of the genome size and the complexity of different organisms. A β-value of 2.0 means that the eco-exergy embodied in the organic matter and the information are equal. As the β-values in **Table 1** are much bigger than 2.0 (except for virus, where the β-value is 1.01) the information eco-exergy is the most significant part of the eco-exergy of organisms.

The eco-exergy due to the 'fuel' value of organic matter (chemical energy) is about $18.7\,kJ\,g^{-1}$ (compare with coal: about $30\,kJ\,g^{-1}$ and crude oil: $42\,kJ\,g^{-1}$). It can be transferred to other energy forms for instance mechanical work directly, and be measured by bomb calorimetry, which requires destruction of the sample (organism), however. The information eco-exergy $= (\beta - 1)\,c$ is taken care of by the control and function of the many biochemical processes. The ability of the living system to do work is contingent upon its functioning as a living dissipative system. Without information eco-exergy, the organic matter could only be used as fuel similar to fossil fuel. But due to the information eco-exergy, organisms are able to make a network of the sophisticated biochemical processes that characterize life. The eco-exergy (of which the major part is embodied in the information) is a measure of the organization. This is the intimate relationship between energy and organization that Schrödinger was struggling to find.

As calculated here, eco-exergy is a result of evolution and of what Elsasser calls re-creativity to emphasize that the information is copied and copied again and again in a long chain of copies where only minor changes are introduced for each new copy. The energy required for the copying process is very small, but it has of course required a lot of energy to come to the 'mother' copy through the evolution for instance from prokaryotes to human cells.

Kullback's measure of information covers the gain in information when the distribution is changed from p_{ion} to p_l. Note that K is a specific measure (per unit of matter). Expressed by the Kullbach's measure of information, we get the following equation for eco-exergy:

$$\text{Ex-organism} = cRTK \qquad [20]$$

β is therefore RTK.

The total eco-exergy of an ecosystem cannot be calculated exactly, as we cannot measure the concentrations of all the components or determine all possible contributions to exergy in an ecosystem. If we calculate the exergy of a fox for instance, then the above shown calculations will only give the contributions coming from the biomass and the information embodied in the genes, but what is the contribution from blood pressure, sexual hormones, network interactions, etc.? These properties are at least partially covered by the genes but is that the entire story? We can calculate the contributions from the dominant components, for instance by the use of a model or

Table 1 Eco-exergy of living organisms

Early organisms	Plants	Animals
Detritus	1.00	
Virus	1.01	
Minimal cell	5.8	
Bacteria	8.5	
Archaea	13.8	
Protists (Algae)	20	
Yeast	17.8	
	33	Mesozoa, Placozoa
	39	Protozoa, amoebe
	43	Phasmida (stick insects)
Fungi, molds	61	
	76	Nemertina
	91	Cnidaria (corals, sea anemones, jelly fish)
Rhodophyta	92	
	97	Gastroticha
Prolifera, sponges	98	
	109	Brachiopoda
	120	Plathyhalminthes (flatworms)
	133	Nematoda (round worms)
	133	Annelida (leeches)
	143	Gnathostomulida
Mustard weed	143	
	165	Kinorhyncha
Seedless vascular plants	158	
	163	Rotifera (wheel animals)
	164	Entoprocta
Moss	174	
	167	Insecta (beetles, flies, bees, wasps, bugs, ants)
	191	Coleodiea (Sea squirt)
	221	Lepidoptera (butterflies)
	232	Crustaceans
	246	Chordata
Rice	275	
Gymnosperms (incl. Pinus)	314	
	310	Molluska, bivalvia, gastropodea
	322	Mosquito
Flowering plants	393	
	499	Fish
	688	Amphibia
	833	Reptilia
	980	Aves (birds)
	2127	Mammalia
	2138	Monkeys
	2145	Anthropoid apes
	2173	*Homo sapiens*

β-values = exergy content relatively to the exergy of detritus (Jørgensen et al.).

measurements that cover the most essential components for a focal problem. The 'difference' in eco-exergy by 'comparing' two different possible structures (species composition) is decisive here. Moreover, eco-exergy computations give always only relative values, as the eco-exergy is calculated relatively to the reference system.

Eco-exergy calculated using the above equations has some clear shortcomings:

1. We have made some although minor approximations in the equations presented above.

2. We do not know the genes in all details for all organisms.

3. We calculate only in principle the eco-exergy embodied in the proteins (enzymes), while there are other components of importance for the life processes. These components are contributing less to the exergy than the enzymes and the information embodied in the enzymes control the formation of these other components, for instance hormones. It can however not be excluded that these components will contribute to the total exergy of the system. The life processes are of course considered indirectly as the enzymes determine the life processes.

4. We do not include the eco-exergy of the ecological network. If we calculate the exergy of models, the network will always be relatively simple and the contribution coming from the information content of the network is considerably less than the exergy contribution from the organisms. The real ecological network may contribute much more to the total exergy. When network models are compared it may also be relevant to compare exergy of different networks.

5. We will always use a simplification of the ecosystem, for instance by a model or a diagram or similar. This implies that we only calculate the exergy contributions of the components included in the simplified image of the ecosystem. The real ecosystem will inevitably contain more components which are not included in our calculations.

It is therefore proposed to consider the eco-exergy found by these calculations as a 'relative minimum eco-exergy index' to indicate that there are other contributions to the total exergy of an ecosystem, although they may be of minor importance. In most cases, however, a relative index is sufficient to understand the reactions of ecosystems because the absolute exergy content is irrelevant for the reactions. In most cases, the change in eco-exergy is of importance to understand ecological responses.

The weighting factors presented in **Table 1** have been applied successfully to calculate eco-exergy applied as an indicator to assess ecosystem health and in several structurally dynamic models to express the model goal function, and furthermore in many illustrations of the maximum eco-exergy principle, that is presented below. Structural dynamic models are able to take a shift of species composition into account: which combinations of properties are able to offer most survival? Further information about structural dynamic models is given in Structural Dynamic Models. The relatively good results in applying the weighting factors in this context, in spite of the uncertainty of their assessment, seems only to be explicable by the robustness of the application of the factors in modeling and other quantifications. The differences between the factors of the microorganism, the vertebrates, and invertebrates are so clear that it does not matter if the uncertainty of the factors is very high – the results are influenced slightly.

On the other hand, from a theoretical point of view it would be an important progress to get better weighting factors but also because it would enable us to model the competition between species which are closely related.

The key to find better β-values maybe the proteomes (the total compositions of the proteins that as enzymes determine the life processes). Our knowledge about the composition of proteomes in various organisms is, however, more limited than for the number of the genes.

Why Do Living Systems Have Such a High Level of Eco-Exergy?

A frog of 20 g will have an eco-exergy content of $20 \times 18.7 \times 688\,\text{kJ} \approx 257\,\text{MJ}$, while a dead frog will have only an exergy content of $374\,\text{kJ}$, although they have the same chemical composition, at least a few seconds after the frog has died. The difference is rooted in the information or rather the difference in the useful information. The dead frog has the information a few seconds after its death (the amino acid composition of the proteins has not yet been decomposed), but the difference between a live frog and a dead one is the ability to utilize the enormous information stored in the genes and the proteomes of the frog.

The amount of information stored in a frog is really surprisingly high. The number of amino acids placed in the right sequence is about 200 000 000 and for each of these 200 000 000 amino acids there are 20 possibilities. This information is again repeated in billions of cells that are cooperating to make up the frog. This enormous amount of information is able to allow reproduction and is transferred from generation to generation which implies that the evolution can continue because what is already a favorable combination of properties is conserved through the genes.

The information in living organisms applies 'nanotechnology' in the sense that the weight of 200 000 000 amino acids is for an average amino acid molecular weight of $125\,\text{g moles}^{-1}\ 2.5 \times 10^{10}\,\text{g/A} = 4 \times 10^{-14}\,\text{g}$, where A is Avogadro's number ($A = 6.2 \times 10^{23}$). A book with the same amount of information would weigh several hundreds of kilograms.

Because of the very high number of amino acids, about 200 000 000, it is not surprising that there will always be a minor difference from frog to frog in the amino acid sequence. It may be a result of mutations or of a minor mistake in the copying process. These variations are important because they give possibilities to 'test' which amino acid sequence gives the best result with respect to survival and growth. The best – representing the most favorable combination of properties – will offer the highest probability of survival and give the most growth and the corresponding genes will therefore prevail. Survival and growth mean more exergy, resulting in a bigger distance from thermodynamic equilibrium. Exergy could therefore be used as a thermodynamic function which could be used to quantify Darwin's theory. In this context, it is interesting that it has been demonstrated that eco-exergy also represents the amount of energy needed to tear down the system. It means that the more exergy the system possesses the more difficult it becomes to degrade the system and the higher is therefore the probability of survival. Consequently, eco-exergy can be applied as a measure of sustainability. The crucial question is therefore: do we hand over the Earth to our children and grandchildren with the same distance from the thermodynamic equilibrium, that is, the same exergy, as we received it from our ancestors?

Losses and Gains of Eco-Exergy by Human Activities Included Pollution

When contaminants, for instance heavy metals, are widely dispersed, eco-exergy is lost. When leaded gasoline was used to obtain a higher octane number, on the order of 400 000 t of lead were dispersed annually around the globe. Lead was even found in the ice pack of Greenland! A typical concentration in lead ore is about 5% or $0.05 \, kg \, kg^{-1}$ ore, while a typical concentration in the environment after the dispersion is $1 \, \mu g \, kg^{-1}$ soil. If we presume 300 K, the annual eco-exergy lost can be found by eqn [7] as

$$Ex \, lost = (8.314 \times 0.300 \times 4 \times 10^{11}/207)$$
$$\ln(0.05/10^{-9}) \approx 85\,000 \; GJ \, yr^{-1} \quad [21]$$

where 207 is the atomic weight of lead. The consumption of lead has decreased due to shifts to other additives in the gasoline in most countries. This loss of eco-exergy by the use of lead as additive to gasoline is therefore today only estimated to be around $40\,000 \, GJ \, yr^{-1}$.

'The loss of exergy due to dispersion of resources in general' can be calculated parallel to the application of eqn [7] as shown in eqn [21]. In addition to lead (only the dispersion of lead as gasoline additive is considered in this context), the loss of exergy by dispersion of other nonrenewable resources is shown in **Table 2**.

The 'loss of eco-exergy due to the consumption of fossil fuel' is found by addition of the chemical free energy (the work capacity) of the fossil fuel and the loss of eco-exergy due to the dispersion of the gases resulting from the chemical processes. The exergy loss due to dispersion of the components of fossil fuel is found by the following calculations: if we consider 1 g of coal that contains 1% of sulfur and 99% of carbon (coal contains

Table 2 Loss of eco-exergy due to dispersion of nonrenewable mineral resources

Element	$GJ \, yr^{-1}$
Chromium	32 000
Nickel	15 000
Zinc	80 000
Copper	18 000
Mercury (included fossil fuel)	27 000
Lead (today)	40 000

Calculations based upon principles shown in eqn [21].

also ash, but let us not consider it in our calculations), the exergy loss due to the dispersion can be determined by the following calculations:

$$0.01(8.314 \times 0.300/32) \ln 0.01/50 \times 10^{-9})$$
$$+ 0.99(8.314 \times 0.300/12) \ln 0.99/4 \times 10^{-4})$$
$$= 1617 \, J \approx 1.6 \, kJ$$

where 50×10^{-9} and 4×10^{-4} represents concentrations (expressed as ratios, i.e., no units) of sulfur dioxide-S and carbon dioxide-C in a typical town atmosphere. The chemical exergy content of 1 g coal is about 32 kJ. The loss of exergy by dispersion is therefore only 5% of the loss directly of chemical exergy by burning coal. As all our calculations will have a higher uncertainty than 5%, and the quality of coal may vary more than 5%, it seems acceptable not to include the dispersion exergy loss by use of fossil fuel or as alternative to multiply all exergy losses due to our consumption of fossil fuel by a factor of 1.05 to compensate approximately for the exergy loss due to the dispersion of the formed gases in the atmosphere.

The deterioration of ecosystems. By the use of eqn [20] it is possible to find the eco-exergy of an ecosystem or rather of the ecosystem corresponding to our model of the ecosystem. Consequently, the loss of eco-exergy due to deterioration and pollution of ecosystems can be found by calculation of the eco-exergy before and after deterioration. The difference will directly yield the loss.

The use of renewable resources. The formation of renewable resources are found separately by multiplication of the annual consumption of the various resources by the exergy content of each renewable resource. If, for instance, the annual fishery in the North Sea has the last many years been in the order of 100 000 t, which implies that the eco-exergy of the North Sea has been reduced $1011 \times 499 \times 18.7 \, kJ = 9.3 \times 1017 \, J$, then 499 is the β-value for fish (**Table 1**). Sustainability requires that the growth of fish biomass compensate for this loss of eco-exergy. It has, unfortunately, for a couple of decades in many marine ecosystems, including the North Sea, not been the case due to over fishing.

Dispersion of waste. This to a certain extent can be calculated parallel to eqn [21]. This is often named the external costs of our activities including the industrial and agricultural activities, but it is actually as the other four points just a question about the loss of eco-exergy. It is not surprising that the cost of treating waste is increasing as the environmental agencies require a more and more complete elimination of these exergy losses. Or expressed differently: we are coming closer and closer to the carrying capacity of the Earth for man-made production. In this context it should not be forgotten that also the treatment of waste costs eco-exergy.

Consumption of nonrenewable fuel, including fossil fuel and nuclear fuel. The annual loss of eco-exergy is found by multiplication of the exergy content and the annual consumption.

We calculated above the loss of eco-exergy by dispersion of waste due to consumption of nonrenewable resources. This is, however, not the entire story, as energy is required in producing various materials from ore and raw source materials. The energy requirement when the material is produced from scrape is included. As it can be seen, the energy requirement is less when scrap if used for the production. Reuse and recycling gives therefore double benefits: we save to draw on the limited resources and we save energy. Notice, particularly for aluminum, that the energy requirement by the use of scrap instead of ore is considerable. As energy consumption explains one of our major losses of eco-exergy, the latter benefit is of great importance.

Formulation of a Thermodynamic Hypothesis for Ecosystems

If an (open, nonequilibrium) ecosystem receives a boundary flow of energy from its environment, it will use what it can of this energy, the free-energy or the exergy content, to do work. The work will generate internal flows, leading to storage and cycling of matter, energy, and information, which move the system further from equilibrium. Self-organizing processes get started. This is reflected in decreased internal entropy and increased internal organization.

The open question of this section is which of many possible pathways will an ecosystem take in realizing its three forms of growth? The answer given is that an ecosystem will change in directions that most consistently create additional capacity and opportunity to achieve increasing deviation from thermodynamic ground, that is, the exergy stored in the ecosystem will increase. Abundant and diverse living biomass represents abundant and diverse departure from thermodynamic equilibrium, and both are captured in this parameter. If multiple growth pathways are offered from a given starting state, those producing greatest exergy storage will tend to be selected, for these in turn require greatest energy dissipation to establish and maintain, consistent with the second law. Energy storage by itself is not sufficient, but it is the increase in specific exergy, that is increased exergy/energy ratio, that reflects improved usability, and this represents the increasing capacity to do the work required for living systems to continuously evolve new adaptive 'technologies' to meet their changing environments.

These considerations lead to a thermodynamic hypothesis which is able to explain the growth and development of ecosystems and the reactions of ecosystems to perturbations: "If a system receives an input of exergy, it will utilize this exergy after the maintenance of the system far from thermodynamic has been covered to move the system further from thermodynamic equilibrium If more than one pathway to depart from equilibrium is offered, the one yielding the most gradients, and most exergy storage (dEx/dt is maximum) under the prevailing conditions, to achieve the most ordered structure furthest from equilibrium, will tend to be selected."

Just as it is not possible to prove the first three laws of thermodynamics by deductive methods, so can the above hypothesis only be 'proven' inductively. In the next section we do examine a number of concrete cases which contribute in a general way to the weight of evidence in favor.

This tentative law may be considered a translation of Darwin's theory into thermodynamics. Exergy measures survival: the biomass and the network, information, and organization that imply that the resources are used in the best possible way to gain most survival. The question is which of the possible combinations of properties by the entire spectrum of organisms in an ecosystem will be able to store most exergy (obtain most survival)? The organisms with the properties that make it possible to gain most survival (exergy) will win in accordance to Darwin and in accordance to the tentative fourth law of thermodynamics. Notice that the resources are always limited relatively to the number of possible offsprings. Therefore there will always be a competition about the resources – and this competition explains together with the huge variation of properties even by the same species, that an evolution has taken place.

Support to the Maximum Eco-Exergy Hypothesis

Eight supporting arguments for the hypothesis presented above. More evidence has been provided; but the eight supporting evidences presented here give a good idea of the theoretical support for the hypothesis.

1. The exergy-storage hypothesis might be taken as a generalized version of 'Le Chatelier's Principle.' Biomass synthesis can be expressed as a chemical reaction:

Energy + nutrients = molecules with more free energy

(exergy) and organization

+ dissipated energy

According to Le Chatelier's Principle, if energy is put into a reaction system at equilibrium the system will shift its equilibrium composition in a way to counteract the change. This means that more molecules with more free energy and organization will be formed. If more pathways are offered, those giving the most relief from the disturbance (using most of the inflowing energy) by using the most energy, and forming the most molecules with the most free energy, will be the ones followed in restoring equilibrium.

2. The sequence of organic matter oxidation takes place in the following order: by oxygen, by nitrate, by manganese dioxide, by iron (III), by sulfate, and by carbon dioxide. This means that oxygen, if present, will always outcompete nitrate which will outcompete manganese dioxide, etc. The amount of exergy stored as a result of an oxidation process is measured by the available $kJ\,mole^{-1}$ electrons which determine the number of adenosine tri-phosphate molecules (ATPs) formed. ATP represents an exergy storage of $42\,kJ\,mole^{-1}$. Usable energy as exergy in ATPs decreases in the same sequence as indicated above. This is as expected if the exergy-storage hypothesis was valid (**Table 3**). If more oxidizing agents are offered to the system, the one giving the resulting system the highest storage of free energy will be selected.

3. Numerous experiments have been performed to imitate the formation of organic matter in the primeval atmosphere on Earth 4 billion years ago. Energy from various sources were sent through a gas mixture of carbon dioxide, ammonia, and methane. Analyses showed that a wide spectrum of compounds, including several amino acids contributing to protein synthesis, is formed under these circumstances. There are obviously many pathways to utilize the energy sent through simple gas mixtures, but mainly those forming compounds with rather large free energies (high exergy storage, released when the compounds are oxidized again to carbon dioxide, ammonia, and methane) will form an appreciable part of the mixture.

4. There are three biochemical pathways for photosynthesis: (1) the C3 or Calvin–Benson cycle, (2) the C4 pathway, and (3) the Crassulacean acid metabolism (CAM) pathway. The latter is least efficient in terms of the amount of plant biomass formed per unit of energy received. Plants using the CAM pathway are, however, able to survive in harsh, arid environments that would be inhospitable to C3 and C4 plants. CAM photosynthesis will generally switch to C3 as soon as sufficient water

becomes available. The CAM pathways yield the highest biomass production, reflecting exergy storage, under arid conditions, while the other two give highest net production (exergy storage) under other conditions. While it is true that one gram of plant biomass produced by each of the three pathways has different free energies, in a general way improved biomass production by any of the pathways can be taken to be in a direction that is consistent, under the conditions, with the exergy-storage hypothesis.

5. Givnish and Vermelj observed that leaves optimize their size (thus mass) for the conditions. This may be interpreted as meaning that they maximize their free-energy content. The larger the leaves the higher their respiration and evapotranspiration, but the more solar radiation they can capture. Deciduous forests in moist climates have a leaf area index (LAI) of about 6%. Such an index can be predicted from the hypothesis of highest possible leaf size, resulting from the tradeoff between having leaves of a given size versus maintaining leaves of a given size. Size of leaves in a given environment depends on the solar radiation and humidity regime, and while, for example, sun and shade leaves on the same plant would not have equal exergy contents, in a general way leaf size and LAI relationships are consistent with the hypothesis of maximum exergy storage.

6. The general relationship between animal body weight, W, and population density, D, is $D = A/W$, where A is a constant. Highest packing of biomass depends only on the aggregate mass, not the size of individual organisms. This means that it is biomass rather than population size that is maximized in an ecosystem, as density (number per unit area) is inversely proportional to the weight of the organisms. Of course the relationship is complex. A given mass of mice would not contain the same exergy or number of individuals as an equivalent weight of elephants. Also, genome differences (example 1) and other factors would figure in. Later we will discuss exergy dissipation as an alternative objective function proposed for thermodynamic systems. If this were maximized rather than storage, then biomass packing would follow the relationship $D = A/W^{0.65-0.75}$. As this is not the case, biomass packing and the free energy associated with this will lend general support for the exergy-storage hypothesis.

Table 3 Yields of kJ and ATPs per mole of electrons, corresponding to 0.25 moles of CH_2O oxidized

Reaction	kJ/(mol e⁻)	ATPs/(mol e⁻)
$CH_2O + O_2 \leftrightarrow CO_2 + H_2O$	125	2.98
$CH_2O + 0.8\,NO_3^- + 0.8\,H^+ \leftrightarrow CO_2 + 0.4\,N_2 + 1.4\,H_2O$	119	2.83
$CH_2O + 2\,MnO_2 + H^+ \leftrightarrow CO_2 + 2\,Mn^{2+} + 3\,H_2O$	85	2.02
$CH_2O + 4\,FeOOH + 8\,H^+ \leftrightarrow CO_2 + 7\,H_2O + Fe^{2+}$	27	0.64
$CH_2O + 0.5\,SO_4^{2-} + 0.5\,H^+ \leftrightarrow CO_2 + 0.5\,HS^- + H_2O$	26	0.62
$CH_2O + 0.5\,CO_2 \leftrightarrow CO_2 + 0.5\,CH_4$	23	0.55

The released energy is available to build ATP for various oxidation processes of organic matter at pH = 7.0 and 25 °C.

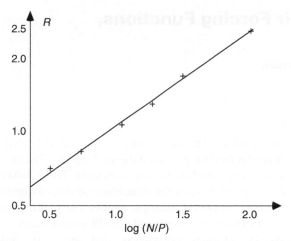

Figure 4 Log–log plot of the turnover rate ratio of nitrogen to phosphorus, R, at maximum exergy versus the logarithm of the nitrogen/phosphorus ratio, log N/P. The plot is consistent with Vollenweider (1975).

7. If a resource (for instance, a limiting nutrient for plant growth) is abundant, it will typically recycle faster. This is a little strange, because a rapid recycling is not needed when a resource is nonlimiting. A modeling study indicated that free-energy storage increases when an abundant resource recycles faster. **Figure 4** shows such results for a lake eutrophication model. The ratio, R, of nitrogen (N) to phosphorus (P) cycling which gives the highest exergy is plotted versus log (N/P). The plot in **Figure 4** is also consistent with empirical results. Of course, one cannot 'inductively test' anything with a model, but the indications and correspondence with data do tend to support in a general way the exergy-storage hypothesis.

8. Dynamic models whose structure changes over time are based on nonstationary or time-varying differential or difference equations. We will refer to these as 'structurally dynamic models'. A number of such models, mainly of aquatic systems, have been investigated to see how structural changes are reflected in free-energy changes. The latter were computed as exergy indexes.

Time-varying parameters were selected iteratively to give the highest exergy index values in a given situation at each time step. Changes in parameters, and thus system

structure, not only reflect changes in external boundary conditions, but also mean that such changes are necessary for the ongoing maximization of exergy. For all models investigated along these lines, the changes obtained were in accordance with actual observations (see references). These studies therefore affirm, in a general way, that systems adapt structurally to maximize their content of exergy. It is noteworthy that Coffaro et al., in their structural dynamic model of the Lagoon of Venice, did not calibrate the model describing the spatial pattern of various macrophyte species such as *Ulva* and *Zostera*, but used exergy-index optimization to estimate parameters determining the spatial distribution of these species. They found good accordance between observations and model, as was able by this method 'without' calibration to explain more than 90% of the observed spatial distribution of various species of *Zostera* and *Ulva*.

See also: Fundamental Laws in Ecology.

Further Reading

Fath B, Jørgensen SE, Patten BC, and Strakraba M (2004) Ecosystem growth and development. *BioSystem* 77: 213–228.

Jørgensen SE (2002) *Integration of Ecosystem Theories: A Pattern*, 3rd edn., 432pp. Dordrecht, The Netherlands: Kluwer Academic Publishing Company (1st edn. 1992, 2nd edn. 1997).

Jørgensen SE and Fath B (2004) Application of thermodynamic principles in ecology. *Ecological Complexity* 1: 267–280.

Jørgensen SE and Svirezhev YM (2004) *Towards a Thermodynamic Theory for Ecological Systems*, 366pp. Oxford: Elsevier.

Jørgensen SE, Patten BC, and Strakraba M (2000) *Ecosystems Emerging: 4. Growth. Ecological Modelling* 126: 249–284.

Jørgensen SE, Ladegaard N, Debeljak M, and Marques JC (2005) Calculations of exergy for organisms. *Ecological Modelling* 185: 165–176.

Morowitz HJ (1968) *Energy Flow in Biology. Biological Organisation as a Problem in Thermal Physics*, 179pp. New York: Academic Press. (See also the review by Odum HT (1969) *Science* 164: 683–84).

Schrödinger E (1944) *What is Life?* Cambridge: Cambridge University Press.

Svirezhev YM (2001) Thermodynamics and theory of stability. In: Jørgensen SE (ed.) *Thermodynamics and Ecological Modelling*, pp. 117–132. Boca Raton, FL: CRC Press, LLC.

Ulanowicz RE (1986) *Growth and Development. Ecosystems Phenomenology*, 204pp. New York, Berlin, Heidelberg, Tokyo: Springer.

Overview of Ecosystem Types, Their Forcing Functions, and Most Important Properties

S E Jørgensen, Copenhagen University, Copenhagen, Denmark

The 39 ecosystems that are described in the part Ecosystems may be classified into four groups according to their forcing functions.

Class I consists of ecosystems that are completely or almost completely managed by man. This class encompasses agriculture systems, biological waste water systems, botanical gardens, green houses, microcosms and mesocosms, landfills, forest plantations, urban systems, and wind shelterbelts. The properties of these nine types of ecosystems are to a large extent determined by the management strategy and plan. The properties of agriculture systems are for instance very much dependent on the management plan: is the system an industrialized agriculture system as it is known in the industrialized world, an integrated agriculture system as it is used in China and to a lesser extent in Europe, or is the agriculture based on the principles of organic farming. The last type of agriculture is increasingly practiced in industrialized countries, although the percentage of organic farming in countries where it is practiced mostly is only slightly more than 10%.

Class II comprises ecosystems that are highly dependent on human activities, which may cause pollution and deteriorate water quality, soil quality, or the quality of other important ecological properties. In other words, class II ecosystems are usually strongly affected by the impacts of pollution. Ecosystems that belong to this class are estuaries, floodplains, freshwater lakes, freshwater marshes, lagoons, mangrove wetlands, Mediterranean ecosystems, riparian wetlands, rivers and streams, salt marshes, and temporary waters. Discharge of pollutants in these ecosystems is usually – but not necessarily always – the most important forcing function.

Class III encompasses ecosystems where pollution may play a role while climate change may also be an important forcing function. For a few of the ecosystem types in this class, the conservation of the ecosystems and their function may be very important globally. It is the case for the rain forests and the temperate forests, which are ecosystems of importance for the global ecological balance, the global carbon cycle, and the climate. These two types of ecosystems may not be threatened by pollution but it is important that the areas occupied by forests – rain forests as well as temperate forests – are not reduced because of the significant role of these ecosystems in the global ecological cycling of carbon and nutrients. There is of course a gradual transition from class II to class III, usually dependent on how important the discharge of pollutants is as a forcing function for the ecosystem. The following ecosystems belong to this class: boreal forests, chaparrals, peatlands, savannas, swamps, steppes and prairies, rain forests, temperate forests, and upwelling ecosystems.

Class IV includes ecosystems that are very little influenced by man and it has been possible up to now to maintain them in almost entirely natural conditions. Some of these ecosystems have extreme conditions; for example, caves have no or almost no light and polar terrestrial ecosystems are characterized by relatively low temperatures. This class includes the following ecosystems: alpine ecosystem, alpine forest, caves, desert streams, dunes, polar terrestrial ecosystems, rocky intertidal ecosystems, saline and soda lakes, and tundra.

For the four classes of ecosystems, the properties of the ecosystems determine whether the ecosystems can remain in ecologically healthy conditions in spite of the forcing functions. In **Table 1**, the four ecosystem types and their characteristic forcing functions are listed together with the ecosystem properties that are most important for the maintenance of a good ecosystem health. The ecosystem properties mentioned in the table are the properties that must be kept at a stable level to ensure that the ecosystem health is not deteriorated. It is therefore the ecosystem properties that should be currently recorded to follow the development of the ecosystems as a result of the forcing functions.

The classification of the 39 ecosystem types is as follows:

Class I: agriculture systems, biological waste water systems, botanical gardens, green houses, microcosms and mesocosms, landfills, forest plantations, urban systems, and wind shelterbelts.

Class II: estuaries, floodplains, freshwater lakes, freshwater marshes, lagoons, mangrove wetlands, Mediterranean ecosystems, riparian wetlands, rivers and streams, salt marshes, and temporary waters.

Class III: boreal forests, chaparrals, coral reefs, peatlands, savannas, swamps, steppes and prairies, rain forests, temperate forests, and upwelling ecosystems.

Class IV: alpine ecosystem, alpine forest, caves, desert streams, deserts, dunes, polar terrestrial ecosystems, rocky intertidal ecosystems, saline and soda lakes, and tundra.

Table 1 Ecosystem classes according to their forcing functions with specification of the most important ecosystem properties

Ecosystem class	Forcing functions	Important properties
I[1]	Almost completely managed by man	Determined by the management
II[2]	Discharge of pollutants (and climate)	Buffer capacity, diversity, adaptation
III[3]	Discharge of pollutants, climate changes	Buffer capacity, ability to adapt and recover
IV	Almost only natural forcing functions	Buffer capacity, diversity, health of the ecological network. Often vulnerable due to extreme conditions

Further Reading

Jørgensen SE (2004) Information theory and energy. In: Cleveland CJ (ed.) *Encyclopedia of Energy*, vol. 3. pp. 439–449. San Diego, CA: Elsevier.

Jørgensen SE (2006) *Eco-Exergy as Sustainability*. 220pp. Southampton: WIT Press.

Jørgensen SE (2008b) *Evolutionary Essays. A Thermodynamic Interpretation of the Evolution,* 210pp.

Jørgensen SE (ed.) (2008a) *Encyclopedia of Ecology*, 5 vols. 4122pp. Amsterdam: Elsevier.

Jørgensen SE and Fath B (2007) *A New Ecology. Systems Perspectives.* 275pp. Amsterdam: Elsevier.

Jørgensen SE, Patten BC, and Straskraba M (2000) Ecosystems emerging: 4. growth. *Ecological Modelling* 126: 249–284.

Jørgensen SE and Svirezhev YM (2004) *Towards a Thermodynamic Theory for Ecological Systems*. 366pp. Amsterdam: Elsevier.

Ulanowicz R, Jørgensen SE, and Fath BD (2006) Exergy, information and aggradation: An ecosystem reconciliation. *Ecological Modelling* 198: 520–525.

ECOSYSTEMS

ECOSYSTEMS

Agriculture Systems

O Andrén, TSBF-CIAT, Nairobi, Kenya

T Kätterer, Department of Soil Sciences, Uppsala, Sweden

Introduction

The Agroecosystem

Further Reading

Introduction

An agricultural ecosystem is an ecosystem managed with a purpose. This purpose usually is to produce crops or animal products. Agricultural ecosystems are designed by humans, and current agroecosystems are products of a long chain of experimental work. These experiments have been performed by individual farmers as well as research institutions, and when results were positive for the purpose, the methods have been adopted.

The purpose has, however, changed with time. In highly productive regions, for example, Western Europe, the emphasis has changed from maximum productivity to environmental considerations, such as reduction of nutrient losses to groundwater and maintaining an open landscape with high biodiversity, etc. In less-productive regions, where resources such as water or fertilizers are scarce and production is too low to properly feed the farmer, environmental considerations have low priority. This is a major global problem, since this leads to land degradation and even lower production, etc. in a downward spiral.

Agroecosystems are conceptually fairly similar to managed forests and grasslands, and whether extensively cattle-grazed natural grasslands should be included under the category of agroecosystems is a matter of choice in the individual case. Arable land is defined as land that is soil cultivated regularly, but also here the boundaries are not sharp (seminatural grasslands, permanent crops, etc.). At the other end, agroecosystems border horticultural systems, that is, vegetable cropping. Alternatively, horticulture can be viewed as a subset of agriculture. Production of cabbage in a field can be considered as agriculture, but hydroponic (soil-less) production of tomatoes in a greenhouse under artificial light can perhaps not be included. However, in many respects even an artificial ecosystem such as this can be considered as an agricultural ecosystem. It is designed for production of a crop and is just managed to a higher extent than an arable field.

According to FAO statistics for 2002, agricultural ecosystems comprise almost 40% (5 Gha) of the total land area of the Earth. About 11% of the total land area is arable land (cultivated with crops), and approximately 27% of the total land area is under permanent pasture, grazed by cattle, goats, sheep, camels, etc. Clearly, we are actively managing a considerable part of our planet for agricultural purposes, and to this one can add other similar systems, such as intensively managed forest systems (planted and harvested, sometimes fertilized), etc.

Ecological research performed in agricultural systems has many advantages compared with research in most natural ecosystems. For example, there are a number of long-term field experiments running, although originally designed for, for example, crop production response to fertilizer dose, that can give us a 30-year integration of what has happened, for example, to organisms in the soil under different conditions. Further, agricultural fields are 'homogenized', that is, trees, larger stones, etc. are removed and regular soil cultivation evens out differences in topsoil properties over time. However, even after many years of cultivation, a fairly high variability in soil properties remain, which is the incentive for 'precision farming', where soil and crop properties are measured at high resolution (m^2), and management is based on these measurements. For ecological research, this is an opportunity, since any given hectare will yield numerous observation points, each helping us to answer questions such as: Why does this particular location yield more wheat, or why is more water present at that location?

Another advantage is that agricultural crops often have a short lifespan and a small size, compared with, for example, forest trees. Often, an experiment can be started when the soil is bare, and a single crop can be followed from sowing, through harvest, and finally when the stubble is plowed down at the end of the growing season. This life cycle can take a century for a tree in a northern forest, which, to add insult to injury, also may contain several other plant species. Therefore it is not surprising that a considerable part of modern ecological theory (predator–prey interactions, general soil ecology, above- and below-ground plant growth dynamics, organic matter decomposition, nutrient mineralization, etc.) is based on work performed in agricultural land, and that the reluctance of ecologists to work in agricultural systems that was obvious 30 years ago seems to have vanished.

The Agroecosystem

Figure 1 is an attempt to summarize the characteristics of an agroecosystem as compared to most natural ecosystems. Note that this comparison is between a typical natural ecosystem and a typical, high-production agroecosystem.

Abiotic Constraints

Just like natural ecosystems, agroecosystems are constrained by climate and soil properties – maize does not grow in Northern Sweden. However, climate can be modified, that is, in dry climates one can irrigate (with surface- or groundwater), and soil properties can be modified through, for example, liming, organic matter amendments, and fertilization. Too high water tables can be lowered through ditching or tile draining.

Nutrients

Highly productive agroecosystems need high inputs of plant nutrients (nitrogen, phosphorus, potassium, and other elements) to replenish the nutrients removed with the exported products. These inputs can be delivered either as commercial fertilizer, recirculated sewage sludge and ash from garbage burning, or manure from cattle, pigs, poultry, etc. All sources have their advantages and disadvantages. Commercial fertilizers are well defined, low in pollutants such as heavy metals (although exceptions exist), hygienically safe, and are concentrated, easy to transport, and rapidly available to the plant when applied in the field. However, production and long-range transport of fertilizers is energy consuming, and the concentrated product increases the risk for too high

doses, leading to environmental pollution. An even greater problem is that a large part of the farmers of the world cannot afford to buy enough fertilizer to maintain soil fertility and obtain good yields. In all, world N fertilizer production in 2001 was slightly less than 90 Mt, very unevenly distributed. In sub-Saharan Africa, only 1.1 kg fertilizer nitrogen is used per person and year, whereas in China the corresponding value is 22 kg.

In theory, recirculation of nutrients from waste of the exported products seems to be ecologically sound. In practice, there are a number of problems. First, sewage sludge mainly consists of water, which either must be removed (requires energy) or transported, which is expensive and impractical. Second, sewage sludge contains harmful bacteria, human parasites, etc. and has to undergo hygienic treatment. Third, and most severe, is the problem with contaminants, such as heavy metals and organic toxins. Therefore recirculation of sewage sludge and garbage incineration ash is strictly regulated in most countries. In this perspective, replacement of nutrients using newly produced fertilizer can be a better solution from an environmental viewpoint.

Naturally, animal manure produced on the farm should be and is recycled to soil as much as possible. Compared with fertilizers, manure has the advantage of containing organic matter, which improves soil structure. On the other hand, manure contains mostly water (expensive storage and transportation, heavy machinery needed for spreading), and it will lose nitrogen through ammonia emission, both at storage and spreading.

Crops, Varieties, and Cropping Systems

The vegetation found in an agroecosystem is usually divided into crop and weeds, where weeds are unwanted

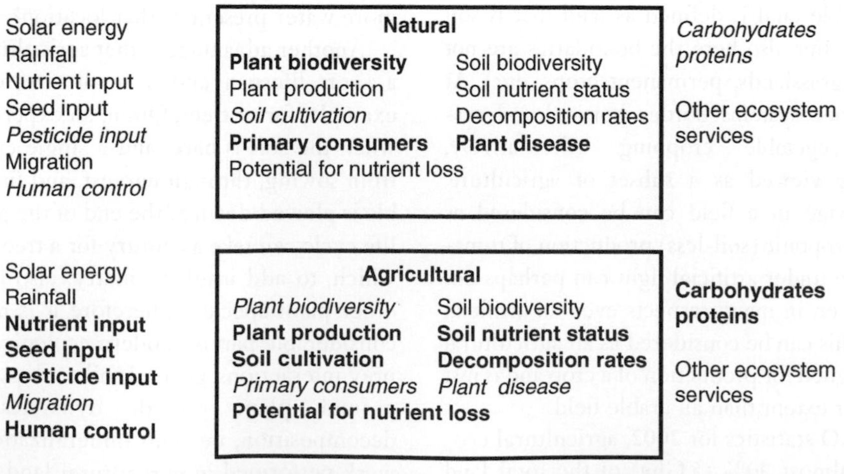

Figure 1   Similarities and differences between typical natural and high-production agricultural ecosystems. Inputs of energy, mass, and control (left), comparison of selected ecosystem properties (center), outputs (right). Note that cattle, etc. are not included as primary consumers here. (**Bold** = markedly higher value than in the other ecosystem type. *Italic* = very low.)

trespassers, which traditionally have been regarded only as negatives. More recently, this view has been modified, and weeds, particularly weedy border zones can be accepted to some extent, as biodiversity enhancers and refuges, for example, for beetles.

The crops used today are products of many years (in some cases millennia) of plant breeding, and properties selected for are usually productivity, product quality, pest resistance, etc. This directed selection, in recent years augmented by direct manipulation of DNA, is one of the main differences between agro- and natural ecosystems. Crop species and varieties are being redistributed all over the world; maize, a staple food in Africa, comes from Central America, common West European and North American cereals such as wheat come from the Middle East, etc. This breeding and distribution of improved crops, together with improved cultivation/fertilization techniques probably is the main reason for the global success of the human species (three billion in 1960, probably nine billion in 2050). For example, world grain production was 631 Mt in 1950, and in 2000 it had increased to 1840 Mt.

Herbicides, Pesticides, and Fungicides

To reach the goal of high production of crops of good quality, weeds (unwanted plants), pests (unwanted animals), as well as fungal, bacterial, and viral diseases must be kept in check. A monoculture crop is vulnerable to attacks, since one (or a pair) of the pests that enter a field will have a high concentration of food with no transport stretches in between. Potential predators may be absent, since they may need a litter layer on the ground for reproduction, which does not exist in the field, etc. Repeated monocultures may build up specialized pests, such as plant parasitic nematodes. Crop rotations (switching crops from year to year according to a predetermined pattern) can successfully deal with many pests and diseases, and careful soil cultivation can reduce weed problems. Intercropping (growing two or more crops together, such as barley/clover) may also help.

However, most fields will benefit from occasional chemical (or biological) pesticide/herbicide treatment. These types of agrochemicals have a somewhat dubious reputation among laymen and perhaps also ecologists (DDT, Agent Orange, mercury, etc.). Three things should be kept in mind, though. First, the substances and formulations used today are thoroughly tested before approval, and their side effects and the fate of their decomposition products are well known. Second, chemical warfare is common in natural systems – all successful plant species present today have at least some chemical defense against microorganisms and pests. Third, which alternatives do we have? A failed crop in a well-fertilized field will lead to high risks for nutrient losses to the environment. A failed crop in poorer conditions may lead to starvation for the farmer and her family.

Alternative methods, such as increased cultivation, hand weeding, or biological pest reduction by introduction of predators all have their advantages and disadvantages, but there is no 'silver bullet' available. In summary, an integrated approach with a combination of methods is the solution, and modern agriculture has moved and is moving in this direction. Of course, for commercial reasons it can be profitable to cultivate, for example, 'organic' crops (without fertilizer or pesticides) to obtain a higher price, but from an ecological or environmental viewpoint this approach is not necessarily better.

Agriculture can thus be classified according to the use of agrochemicals, for example, biodynamic, organic, integrated, and industrialized farming. Biodynamic farming forbids the use of conventional agrochemicals and replaces them with exotic homemade concoctions, and organic farming a priori forbids conventional agrochemicals. None of these farming systems is firmly based on scientific evidence; instead they are based on a green view of nature that leads to the banning of certain chemicals.

Integrated and industrial farming can also be called 'conventional', where economic, legal, and environmental constraints limit the end goal, maximum productivity, and profitability. The main difference between the latter two is that integrated is more environmentally concerned (reduced pesticide use, use of biological pest reduction methods, etc.), and industrialized is more leaning to maximum production with whatever means available, with a minimum of environmental concerns. It should be noted that 'conventional' and particularly 'industrialized' are somewhat derogatory terms, mainly used by those negative to these approaches.

Migration

Natural ecosystems, for example, East African savannas, can be subjected to major migrations of large herbivores that annually move long distances, following the seasonal changes in rainfall and consequential grass growth. Most natural ecosystems are less subjected to migrations, but, for example, in Northern forests at least migratory birds occur seasonally.

In agroecosystems, migration is usually kept to a minimum. Measures are taken to keep large or small grazers out from the cropped field. In some regions, wild grazers are exterminated (or close to extinction – Western European agricultural regions) and in other regions crop fields are guarded or fenced. However, migration is a component in animal husbandry; cattle is often shifted between pastures, which are given time to recover. Nomadic herding of cattle (Sami people, Masai) is similar to the savanna migrations mentioned above; the cattle and herdsmen follow the annual cycles in grazing opportunities.

Biodiversity

In a cereal monoculture, plant biodiversity is extremely low – if weed control is successful there may be only one species present, a highly specialized and genetically homogeneous wheat variety. This is not common in natural ecosystems, although it can occur in extreme environments. As mentioned above, this means that a pest can have a field day if it can reproduce in the field (or migrate into the field at a large scale).

However, agricultural monocultures still are common and continue to produce good yields. There are several reasons for this. First, there is no simple relation between biodiversity, productivity, or ecosystem stability. A plant monoculture that is well adapted, grows under good conditions, and has a reasonable resistance to pests and diseases can survive and produce well. This is exactly what a highly productive agricultural field is – a well-adapted monoculture. The crop variety has been selected for high production under a number of years with different weather (and on different soils) in a region. A variety that would demand intensive treatment with herbicides, pesticides, and fungicides will not be economical and will be rejected.

Second, the low plant diversity reduces animal diversity in the stand, but perhaps less than one would expect. In a cereal monoculture stand, there can be hundreds of species of insects, mites, springtails, snails, slugs, etc. In the soil under a monoculture the biodiversity is almost always extremely high, though usually lower than in natural systems. Thousands, perhaps millions of bacterial species, tens to hundreds of species of earthworms, enchytraeids, soil insects, springtails, mites, spiders, millipedes, flagellates, amoebae, blue-green algae, etc. can be found. There are no consistent indications that soil functions such as organic matter decomposition is hampered by a low biodiversity under monocultures – a given plant residue will decompose at the same rate under a monoculture as under mixed plants, if soil temperature and moisture are the same.

Third, the last line of defense is the crop protection measures that the farmer takes. For example, in several countries there is a sophisticated monitoring and prediction system for aphid outbreaks. Aphids suck the sap from the crop leaves, but they are also vectors for crop diseases. Therefore their hibernating stages are enumerated, weather is monitored, and if the conditions are 'right' the farmers are recommended to spray the fields with an insecticide (or a more specific aphicide) with dose x at date y. In less technically developed regions, experience and skill is a substitute for the model projections, but the principles are the same. It should also be mentioned that in spite of these defenses, pest insects, pathogens, and weeds still reduce worldwide crop yields considerably, and there is a great potential for improvements.

Other Ecosystem Services

The main ecosystem service from agricultural systems is simply to 'feed the world'. This simple fact is easily forgotten in the richer parts of the world. However, even in Europe, which for centuries has been thoroughly under agriculture, there are other ecosystem services that are appreciated. In the forest-dominated northern Europe, agriculture actually contributes to biodiversity and landscape diversity. Without agriculture, the forest would cover all land area – the only open areas at lower altitudes would be the lakes and rivers (and the newly clear-cut forest areas, rapidly covered by shrubs). The European rural landscape in general, that is so refreshing for the city-dweller, is an agricultural product.

In other areas of the world, where the agricultural land is not sufficient to properly feed the population, other ecosystem services become relatively less important. However, if agricultural productivity can be increased, some agricultural land can be returned to savanna, forest, or other natural or seminatural states – which would be another type of service from the agroecosystem.

Since the agroecosystem is managed, and more or less sophisticated machinery and management skills are in place, it can easily be converted according to new demands from the society. If the quality requirements are met, agricultural fields can be used for recycling organic waste and ashes, and even for drawing nutrients out of sewage water. Conversion to energy crops is not too difficult (grasses, sugar beet, willow, sugarcane, etc.). Another demand from society, to sequester carbon in the soil to reduce CO_2 in the atmosphere, has recently received much attention. Increasing soil carbon content usually has beneficial effects for soil structure, water-holding capacity and general fertility, and C sequestration, perhaps even with direct payments per ton C sequestered to the farmer, is a new potential service.

The Intelligent Choices

As mentioned in the introduction, an agroecosystem has a purpose. It is designed to obtain certain goals, and the state of the system at any given point in time is a consequence of an array of intelligent choices by the farmer, complementing the border conditions set up by weather and soils, etc. The following decision matrix (**Figure 2**) illustrates how decisions made by a maize farmer in sub-Saharan Africa can be supported by basic science knowledge. Note that the chemical analyses are not necessary for every farmer and decision. Instead, typical values for the different organic

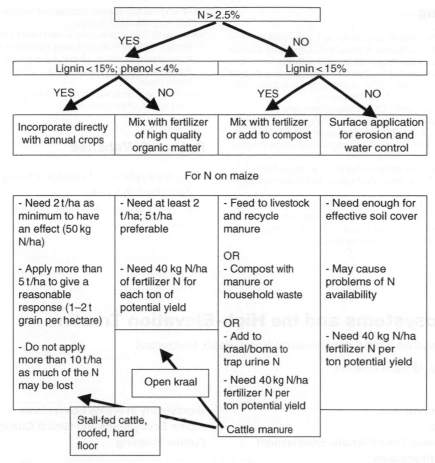

For N on maize

Figure 2 Example of farmer's decisions regarding N management for a maize crop in sub-Saharan Africa, using a decision support system for organic N management depending on resource quality, expressed as N, lignin, and soluble polyphenol content. General decision matrix (top), more detailed for N economy in a maize cropping system (bottom). Modified from Vanlauwe B, Sanginga N, Giller K, and Merckx R (2004) Management of nitrogen fertilizer in maize-based systems in subhumid areas of sub-Saharan Africa. In: Mosier AR, Syers JK, and Freney JR (eds.) *Agriculture and the Nitrogen Cycle*. 124p. SCOPE 65. Washington Island Press.

resources are estimated, and the individual farmer uses the rule of the thumb based on these estimates.

In the upper part of the **Figure 2**, the general decision matrix is shown. Let us assume that we have leaves from a tree, which we know have a low N content and less than 15% of lignin. Then we should mix the leaves with fertilizer or add to compost. Now, in the lower part of **Figure 2** we can see that if we look in more detail at the N economy of a maize system, we have other options – maybe add the low N material to the cattle corral (kraal/boma) to trap urine N or feed to livestock to produce higher quality organic inputs. Organic resources belonging to the third column from the left could be fed to livestock and the manure thus produced could belong to the first or the second organic resource class, depending on the management of that manure.

All over the world, farmers make these kinds of choices, based not only on biophysical knowledge and constraints, but also on economic and sociopolitical opportunities and constraints. An agroecosystem is not

only controlled by farmers, but also by the society the farmer operates in. Subsidies can make growing products that have no market an intelligent choice for the farmer; lack of money can make fertilization impossible, even if it would be profitable in the long run, or real or imaginary environmental concerns from the society can force a farmer to, for example, abandon fertilizer use, cereal cropping, or pig farming.

Summing up, the agroecosystem, although limited by climatic constraints, is a product of decisions made by generations of farmers, supported by advice from agronomists and extension workers – all within a societal context of values, traditions, and legislation. In fact, the present and future agroecosystems are at least equally dependent on the societal context as on the climate and soil. However, the organisms involved are, as in any ecosystem, products of millions of years of evolution, and crop and animal breeding has only contributed with small, although important changes to the germplasm.

Further Reading

Andrén O, Lindberg T, Paustian K, and Rosswall T (eds.) (1990) *Ecological Bulletins 40: Ecology of Arable Land - Organisms, Carbon and Nitrogen Cycling.* Copenhagen: Ecological Bulletins.

Brussaard L (1994) An appraisal of the Dutch program on soil ecology of arable farming systems (1985–1992). *Agriculture, Ecosystems and Environment* 51(1–2): 1–6 and following papers.

Clements D and Shrestha A (eds.) (2004) New Dimensions in Agroecology, 553p. Binghamton: The Hawort Press, Inc.

Eijsackers H and Quispel A (eds.) (1988) *Ecological Bulletins 39: Ecological Implications of Contemporary Agriculture.* Copenhagen: Ecological Bulletins.

Kirchmann H (1994) Biological dynamic farming – an occult form of alternative agriculture? *Journal of Agricultural and Environmental Ethics* 7: 173–187.

Mosier AR, Syers JK, and Freney JR (eds.) (2004) *Agriculture and the Nitrogen Cycle*, 124p. SCOPE 65 Washington: Island Press.

New TR (2005) *Invertebrate Conservation and Agricultural Ecosystems*, 354p. Cambridge: Cambridge University Press.

Newman EI (2000) *Applied Ecology and Environmental Management*, 2nd edn. Blackwell Science.

Vanlauwe B, Sanginga N, Giller K, and Merckx R (2004) Management of nitrogen fertilizer in maize-based systems in subhumid areas of sub-Saharan Africa. In: Mosier AR, Syers JK, and Freney JR (eds.) *Agriculture and the Nitrogen Cycle.* 124p. SCOPE 65. Washington Island Press.

Woomer PL and Swift MJ (1994) *The Biological Management of Tropical Soil Fertility.* Chichester: Wiley.

Relevant Websites

http://www.cgiar.org – Consultancy Group on International Agricultural Research.

http://www.fao.org – Food and Agriculture Organization of the United Nations.

Alpine Ecosystems and the High-Elevation Treeline

C Körner, Botanisches Institut der Universität Basel, Basel, Switzerland

Definitions and Boundaries	Biodiversity in Alpine Ecosystems
The Alpine Treeline	Alpine Ecosystems and Global Change
Alpine Plants Engineer Their Climatic Environment	Further Reading
Alpine Ecosystem Processes	

Definitions and Boundaries

Ecosystems above the upper climatic limit of trees are termed 'alpine'. Scientifically, the alpine life zone is an altitudinal belt defined by climatic boundaries (**Figure 1**) and the term 'alpine' does not refer to the European Alps, but refers to treeless high-elevation biota worldwide (mostly grassland and shrubland). 'Alpine' supposedly roots in the pre-Indogermanic word *alpo* for steep slopes, still used today in the Basque language. By contrast, in common language, 'alpine' is often used for places anywhere in mountainous terrain, irrespective of altitude (e.g., alpine village, even alpine cities). If a city were truly alpine it would have to be above the climatic treeline, but no such city does exist worldwide. Hence, a distinction must be made between the scientific, biogeographic meaning of alpine (the issue of this text) and common (often touristic) jargon.

The upper limit of the alpine life zone or alpine belt is reached where flowering plants have their high altitude limit. This is often close to the snow line (the altitude at which snow can persist year-round), but commonly a few scattered flowering plants also grow above the snow line, in favorable, equator-facing, and sheltered places. The uppermost part of the alpine belt, where closed ground cover by vegetation is missing, is often termed 'nival', referring to sparse vegetation in rock and scree fields. The highest place on Earth where flowering plants have been found is in the Central Himalayas at 6200–6350 m above sea level.

Depending on latitude, the climatic treeline and hence the lower limit of the alpine belt can be anywhere between close to sea level in subpolar regions (>70° N, >55° S) and close to 5000 m in subtropical continental climates (trees >3 m at 4800 m in Bolivia and at 4700 m in Tibet). In the cool temperate zone (45–50° N), the alpine belt may start anywhere between 1200 and 3500 m (in the European Alps at 2000 m, the Colorado Rocky Mountains at 3400 m); that is, it is lower under strong oceanic influence and higher in the inner parts of continents. The common natural treeline altitude near the equator is 3600–4000 m. The altitudinal width of the alpine belt above treeline is roughly 1000 m. It covers *c.* 3.5% of the globe's terrestrial area, if cold and hot deserts are disregarded (Antarctica, Greenland, Sahara, etc.).

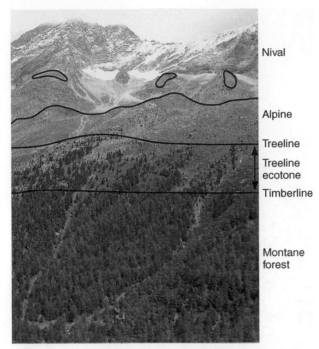

Figure 1 The altitudinal belts of mountain ecosystems. With increasing altitude these belts become fragmented and topography (exposure) plays an increasing role. (Example from the Swiss Central Alps with *Pinus cembra* forming the treeline at 2350 m.)

Given this convention on the two boundaries of the alpine belt, it is important to note that these boundaries are not sharp lines, but are centered across gradients which change from place to place and depend on topography and region. Usually these boundaries are obvious from great distance (an airplane), but hard to depict on the ground, hence depend on scale.

The Alpine Treeline

Since, by definition, the alpine belt is naturally treeless, the mechanisms by which trees are restricted from growing beyond a certain altitude are key to any understanding of alpine ecosystems. The so-called treeline marks the upper limit of the life-form 'tree' irrespective of the tree species involved (see Alpine Forest). Generally, species which form treelines are *Pinus, Picea, Abies, Juniperus,* and *Larix* among conifers, and *Betula, Alnus, Erica, Polylepis, Sorbus, Eucalyptus,* and others among non-coniferous families. Because tree occurrence does not stop abruptly, and trees gradually get smaller and finally become crippled, any definition of 'a line' is a convention. The forest line or timberline represents the edge of the closed upper montane forest (note, 'montane' is the biogeographic term for the next lower belt, not to be confused with 'mountain'), the zone of gradual forest opening near the treeline is often termed treeline parkland, and the uppermost position where tree species can survive as

small saplings or shrubs among other low-stature vegetation is called the tree species line, with the 'treeline' holding a middle ground, used for the line connecting the uppermost patches of trees >3 m. The whole transition zone from montane forest to alpine heathland is termed treeline ecotone, across which alpine vegetation gains space yielded by the thinning forest. The altitudinal range of the treeline ecotone may be 20–200 m, often <50 m.

Where moisture is permitting tree growth at these altitudes (a minimum of 250–300 mm of precipitation per year), the position of the natural climatic treeline matches with a mean growing season temperature of $6.6 \pm 0.8\,^\circ$C worldwide. The duration of the growing season may vary from 10 weeks at high latitude to a full year in the tropics and its onset and end are defined by a weekly mean air temperature of $0\,^\circ$C (corresponding to *c.* $3\,^\circ$C in 10 cm soil depth, where most roots occur). This isotherm sets the lower climatic threshold for alpine vegetation, which can be close to $5\,^\circ$C in dry subtropical mountains and $7.5\,^\circ$C in cool temperate mountains, which is a surprisingly narrow range, given the great difference in season length across latitudes.

It is very important not to confuse this climatic (physiological) limit of trees with a multitude of other natural or anthropogenic causes for the local absence of trees such as fire, avalanches, logging and pasturing, loose or missing substrate, waterlogging, or the regional lack of cold-adapted tree species (as is the case for instance in Hawaii or New Zealand). In the last case, the treeline observed is a specific tree species line, not representative of the climatic limit of the life-form tree, as can easily be demonstrated by the success of introduced tree species which grow well at much higher altitudes in such regions. Open 'alpine-looking' grassland and shrubland may thus occur several hundreds of meters below the climatic treeline; among the most famous of these are the Andean Páramo grasslands with their spectacular giant rosette plants (**Figure 2**).

Alpine Plants Engineer Their Climatic Environment

Why is there lush alpine vegetation but trees cannot grow? Are alpine plants physiologically superior, able to cope with those low temperatures which otherwise are harming trees? There is good evidence that thermal constraints for growth, that is, building new tissue, are the same for alpine plants, cold-adapted trees, and winter crops (winter rape and winter wheat), all being completely halted when tissue temperatures drop below $5\,^\circ$C, and growth is close to zero at 6–7 $^\circ$C. In contrast, all these species reach 30–50% of maximum rates of photosynthesis at these same temperatures; thus the provision of raw

Figure 2 Fire and grazing (both naturally and under human influence) can replace the montane forest, leading to 'alpine-looking' vegetation below the climatic treeline. Here is an example of the Ecuadorian páramos at 3600 m altitude, *c.* 400–500 m below the potential climatic treeline (Páramos El Angel). Giant rosettes of the genus *Espeletia* are the prominent feature of this landscape, with similar vegetation also found in African highlands.

material for growth (sugar) cannot be decisive. Neither are there critical differences in freezing resistance between alpine plants compared to trees. Hence, at tissue level, there is no physiological reason why alpine grasses, herbs, and shrubs should grow at a given low temperature and trees should not.

There are two reasons for alpine plant success above treeline:

1. By low stature and dense stand structure, alpine plants restrict aerodynamic exchange with the atmosphere, which causes heat to accumulate during periods with solar radiation and permit plants to operate at comparatively warm temperatures, much unlike those experienced by upright, ventilated trees. The life-form 'tree' does not permit any escape from the gradually declining ambient temperatures, whereas alpine plants engineer their microclimate and aircondition their meristems close to the ground so that they can build new tissue at otherwise cold air temperatures above the plant canopy (**Figure 3**).

2. By developmental flexibility and morphological adaptation, alpine plants are able to make use of short favorable weather conditions, they sprout rapidly, produce only a few, mostly short-lived leaves (*c.* 60 days), and have their meristems positioned very close to the ground, in the case of many grasses, sedges, or rosette plants, often 1–2 cm below ground, where the solar-heated soil provides a thermally buffered environment. In contrast trees operate at longer leaf duration (mostly >120 days, in evergreen treeline conifers 4–12 years) and leaves take longer to mature, and their aboveground meristems are fully exposed to the cold air temperatures.

Figure 3 Trees are coupled to air temperature and thus, appear 'cool' on this infrared thermograph taken at 10 a.m. on a bright midsummer morning in the Swiss Alps near Arolla. Alpine grassland and shrub heath accumulate heat by decoupling from atmospheric conditions (low stature, dense structures). So the treeline can clearly be depicted as a thermal boundary driven by plant architecture.

The transition from trees to alpine vegetation is thus dictated by plant architecture and not by tissue-specific inferiority of trees compared to alpine plants. This close coupling of trees to atmospheric conditions also explains the surprisingly uniform leveling of treelines across mountain valleys which reminds one of the level of a water reservoir. In contrast, the climate in alpine vegetation varies with compactness and height of the leaf canopy and exposure to the sun. A sun-exposed, sheltered microhabitat at 3000 m of altitude may be warmer than a shaded microhabitat at 1800 m. Altitude *per se*, or data from a conventional climate station, thus, tell us little about the climate actually experienced by alpine plants. It had long been known that mutual sheltering among alpine plants or leaves/tillers within a plant is very beneficial ('facilitation'), and removing this shelter effect by opening the plant canopy can be disastrous.

Alpine plants are small by design (genetic dwarfs); they are not forced into small stature by the alpine climate directly, though evolution had selected such morphotypes. What seems like a stressful environment is not really stressful for those well adapted. However, there is some additional modulative, direct effect on size by low temperature. Alpine plants that survive in low-altitude rock gardens indeed grow taller than their relatives in the wild. But plants grown in such rock gardens are commonly of montane origin, because most typical alpine plants fade at such high, low-altitude temperatures, possibly because of overshooting mitochondrial respiration.

Alpine Ecosystem Processes

Almost everything gets slower when it gets cold, but slow production of biomass and slow recycling of dead biomass (litter) go hand in hand, so that the carbon and nutrient cycles remain in balance. Recycling of organic debris is responsible for most of the steady-state nutrient provision and thus controls vigor of growth. When mineral nutrients are added, all alpine vegetation tested had shown immediate growth stimulation, but this holds for most of the world's biota and is not specific to alpine ecosystems. On the other hand, nutrient addition had been shown to make alpine plants more susceptible to stress (softer tissue, reduced winter dormancy) and pathogen impact (e.g., fungal infections) and causes nitrophilous grasses and herbs to overgrow the best-adapted slow-growing alpine specialist species.

It comes as a rather surprising observation that alpine plant productivity – at least in the temperate zone – is only low when expressed as an annual rate of biomass accumulation, but is not low at all when expressed per unit of growing season length. In a 2-month alpine season in the temperate zone alpine belt, the biomass production (above-plus below-ground) accumulates to $c.$ $400\,g\,m^{-2}$ (range $200-600\,g\,m^{-2}$). A northern deciduous hardwood forest produces $1200\,g\,m^{-2}$ in 6 months and a humid tropical forest $2400\,g\,m^{-2}$ in a 12-month season, all arriving at $c.$ $200\,g\,m^{-2}$ per month. Time constraints of growth are thus the major causes of reduced annual production in closed alpine grass- and shrubland and not physiological limitations in what seems to a human hiker like a rather hostile environment. Acclimation to lowtemperature, perfect plant architecture, and developmental adjustments can equilibrate these constraints on a unit of time basis. It makes little sense to relate productivity to a 12-month period when 9–10 months show no plant activity because of freezing conditions and/or snow cover.

Similar to carbon and nutrient relations, alpine ecosystem's water relations are largely controlled by seasonality. During the growing season in the humid temperate zone, daily water consumption during bright weather hardly differs across altitude ($c.$ $3.5-4$ mm evapotranspiration). However, because of the short snow-free season at such latitudes, annual evapotransiration may be only $250-300$ mm compared to $600-700$ mm at low altitude, hence runoff is much higher in alpine altitudes. Given that precipitation often increases with altitude in the temperate zone (a doubling across 2000 m of altitude is not uncommon), annual runoff may be 3–5 times higher in the alpine belt, with major implications for erosion in steep slopes.

In many tropical and subtropical mountains, moisture availability drops rapidly above the condensation cloud layer at 2000–3000 m altitude, causing the alpine belt to receive very little water, often not more than 200–400 mm per year (e.g., the high Andes, Tenerife, East African volcanoes). The resulting sparse vegetation is often termed alpine semidesert, but because of wide spacing of plants and very little ground cover, those plants which are found in this semiarid alpine landscape were found to be surprisingly well supplied with water even at the end of the dry season (**Figure 4**). As a rule of thumb, alpine plants are thus better supplied with moisture (even in dry alpine climates) than comparable low-altitude vegetation. True physiologically effective water stress is quite rare in the alpine belt, but moisture shortage in the top soil may restrict nutrient availablity periodically, which restricts growth.

Biodiversity in Alpine Ecosystems

For plants and animals to become 'alpine' they must pass through a selective filter represented by the harsh climatic conditions above treeline. It comes as another surprise that alpine ecosystems are very rich in organismic taxa. It was estimated that the $c.$ 3.5% of global land area that can be ascribed to the alpine belt hosts $c.$ 4% of

Figure 4 High-altitude semideserts (near Sajama, Bolivia, 4200 m) are often dominated by sparse tussock grasses, shrubs, and minor herbs in the intertussock space, all together preventing soil erosion, while being used for grazing. The wide spacing mitigates drought stress in an otherwise dry environment.

all species of flowering plants. In other words, alpine ecosystems are on average similarly rich or even richer in plant species than average low-altitude ecosystems. This is even more surprising if one accounts for the fact that the available land area above treeline shrinks rapidly with altitude (on average a halving of the area in each successive 170 m belt of altitude). A common explanation for this high species richness is the archipelago nature of high mountains (a fragmentation into climatic 'islands'), the high habitat diversity as it results from gravitational forces (topographic diversity, also termed geodiversity), and the small size of alpine plants, which partly compensates for the altitudinal loss of land area . The altitudinal trends for animal diversity are similar to plants, but some animal taxa decline in diversity with altitude more rapidly (e.g., beetles, earthworms, butterflies) than others (e.g., vertebrates, birds). Often animal diversity peaks at mid-altitudes (close to the treeline ecotone) and then declines.

The four major life-forms of flowering plants in the alpine belt are graminoids (grasses, mostly forming tussocks, sedges, etc.), rosette-forming herbs, dwarf shrubs, and cushion plants (**Figure 5**). In most parts of the world, bryophytes and lichens (a symbiosis between algae and fungi) contribute an increasing fraction of biodiversity as altitude increases. Each of these life-forms can be subdivided into several subcategories, mostly represented by different forms of clonal growth. Clonal (vegetative) spreading is dominant in all mountains of the world and it secures long-term space occupancy by a 'genet' (a single genetical individual) in a rather unpredictable environment. Because of the topography-driven habitat diversity, rather contrasting morphotypes and physiotypes may be found in close proximity, as for instance

succulent (water storing) plants such as alpine cactus or some leaf-succulent Crassulaceae (*Sedum* sp., *Echeveria* sp.) next to wetland or snowbed plants.

Alpine ecosystems are known for their colorful flowers, and it was often thought that this may be a selected-for trait, because it facilitates pollinator visitation. There is also morphological evidence that alpine plants invest relatively more in flowering, given that plant size (and biomass per individual) declines by nearly tenfold from the lowland to the alpine belt, whereas the size of flowers hardly changes. Futhermore, flower duration increases and so does pollinator visiting duration, and there is no indication that there is a shortage in alpine pollinators. The net outcome is a surprisingly high genetic diversity in what seems like highly fragmented and isolated habitats. Despite the successful reproductive system at the flower-pollinator scale and well-adapted (fast) seed maturation, the real bottleneck is seedling establishment (the risk to survive the first summer and winter), which explains why most alpine plants also propagate clonally.

Overall, mountain biodiversity (the montane belt, the treeline ecotone, and the alpine belt) is a small-scale analog of global biodiversity, because of the compression of large climatic gradients over very short distances. Across a vertical gradient from 1200 to 4200 m in the Tropics one may find a flora and fauna with a preference for climates otherwise only found across several thousand kilometers of latitudinal distance. This is why mountains are ideal places for biodiversity conservation as long as the protected mountain system is large and has migration corridors to prevent biota from becoming trapped in ever-narrowing land area should climatic warming induce altitudinal upward shifts of life zones.

Figure 5   The four major life-forms of flowering plants in alpine ecosystems: cushion plants (*Azorella compacta, Silene exscapa*), herbs (small: *Chrysanthemum alpinum*, tall: *Gentiana puncata*), dwarf shrubs (*Loiseleuria procumbens, Salix herbacea*), and tussock-forming graminoids (*Carex curvula*, diverse tall grass tussock).

Alpine Ecosystems and Global Change

'Global change' includes changes in atmospheric chemistry (CO_2, CH_4, N_xO_y), the climatic consequences of these changes, and the manifold direct influences of humans on landscapes. All three global change complexes affect alpine biota, either directly or indirectly.

Elevated atmospheric carbondioxide (CO_2) concentrations affect plant photosynthesis directly, although late-successional alpine grassland in the Alps was found to be carbon saturated at ambient CO_2 concentrations of the early 1990s. The effect of doubling CO_2 concentrations over four consecutive seasons on net productivity was zero. However, not all species within that sedge-grass-herb community responded identically, hence there is a possibility of gradual shifts in species compositition in the long run, with some species getting suppressed and others gaining.

In contrast, even very moderate additions of soluble nitrogen fertilizer at rates of those received today by mountain forelands in Central Europe with rains ($40–50\,kg\,N\,ha^{-1}a^{-1}$) doubled biomass in only 2 years. Even $25\,kg\,ha^{-1}a^{-1}$ had immediate effects on biomass (+27%), again favoring some species more than others. Atmospheric nitrogen deposition is thus far more important for alpine ecosystems than elevated CO_2. Just for comparison, in intense agriculture, cereals are fertilized with $>200\,kg\,N\,ha^{-1}\,a^{-1}$.

Consequences of climatic change for alpine ecosystems are hard to predict because of the interplay of climatic warming with precipitation. A warmer atmosphere can carry more moisture; hence increasing precipitation had been predicted for temperate mountain areas. Greater snowpack can shorten the growing season at otherwise higher temperatures. While the temperate zone has seen more late winter snow in recent years, the uppermost reaches of higher plants seem to have profited from climatic warming over the twentieth century. Several authors documented a clear enrichment of summit floras, accelerated in recent decades.

Treeline trees respond to warmer climates by faster growth, but whether and how fast this would cause the treelines of the world to advance upward depends on tree establishment, which is a slow process. Hence treelines always lagged behind climatic warming during the Holocene by centuries, as evidenced by pollen records. Current trends are largely showing an infilling of gaps in the treeline ecotone, but upward trends await larger-scale confirmation. Eventually any persistent warming will induce upward migration of all biota. By contrast, recent climatic warming has caused the tropical upper montane/alpine climate on Kilimanjaro to become drier, facilitating devastating fires, which depressed the montane forest by

several hundred meters with a downslope advance (expansion) of alpine vegetation following.

Land use is still the most important factor for changes in alpine ecosystems. Around the globe, alpine vegetation is used for herding or uncontrolled grazing by livestock. Much of the treeline ecotone has been converted into pasture land, with both overutilization and erosion (mainly in developing countries) and abandonment of many centuries old, high-elevation cultural landscapes (mainly industrialized countries) causing problems. The question is not whether there should be pasturing, but how it should be done. Sustainable grazing requires shepherding and observation of traditional practices, which largely prevent soil damage and erosion. Traditional alpine land use has a several thousand years history and was optimized for maintaining an intact landscape for future generations as opposed to land-hungry newcomers faced with the need of feeding a family today, rather than thinking of sustained livelihood in a given area. All other forms of land use (except mining), as dramatic their negative effects at certain places may look, are less important, because their impact is rather local (e.g., tourism, road projects). Agriculture is by far the most significant factor in terms of affected land area.

Mismanagement of alpine ecosystems has severe consequences (e.g., soil destruction, sediment loading of rivers) not only for the local population, but for people living in large mountain forelands, which depend on steady supplies of clean water from high-altitude catchments. Almost 50% of mankind consumes mountain resources, largely water and hydrolectric energy, hence there is an often overlooked teleconnection between alpine ecosystems and highly populated lowlands. Highland poverty is thus affecting the conditions and the economic value of catchments, which goes far beyond the actual agricultural benefits. This insight should lead to better linkages between lowland and highland communities and also include economic benefit sharing with those that perform sustainable land care in alpine ecosystem.

See also: Alpine Forest.

Further Reading

Akhalkatsi M and Wagner J (1996) Reproductive phenology and seed development of *Gentianella caucasea* in different habitats in the Central Caucasus. *Flora* 191: 161–168.

Bahn M and Körner C (2003) Recent increases in summit flora caused by warming in the Alps. In: Nagy L, Grabherr G, Körner C, and Thompson DBA (eds.) *Ecological Studies 167: Alpine Biodiversity in Europe*, pp. 437–441. Berlin: Springer.

Barthlott W, Lauer W, and Placke A (1996) Global distribution of species diversity in vascular plants: Towards a world map of phytodiversity. *Erdkunde* 50: 317–327.

Billings WD (1988) Alpine vegetation. In: Barbour MG and Billings WD (eds.) *North American Terrestrial Vegetation*, pp. 392–420. Cambridge: Cambridge University Press.

Billings WD and Mooney HA (1968) The ecology of arctic and alpine plants. *Biological Reviews* 43: 481–529.

Bowman WD and Seastedt TR (eds.) (2001) *Structure and Function of an Alpine Ecosystem – Niwot Ridge, Colorado*. Oxford: Oxford University Press.

Callaway RM, Brooker RW, Choler P, *et al.* (2002) Positive interactions among alpine plants increase with stress. *Nature* 417: 844–848.

Chapin FSIII and Körner C (eds.) (1995) *Arctic and Alpine Biodiversity: Patterns, Causes and Ecosystem Consequences. Ecological Studies 113*. Berlin: Springer.

Dahl E (1951) On the relation between summer temperature and the distribution of alpine vascular plants in the lowlands of Fennoscandia. *Oikos* 3: 22–52.

Fabbro T and Körner C (2004) Altitudinal differences in flower traits and reproductive allocation. *Flora* 199: 70–81.

Grabherr G and Pauli MGH (1994) Climate effects on mountain plants. *Nature* 369: 448.

Hemp A (2005) Climate change-driven forest fires marginalize the impact of ice cap wasting on Kilimanjaro. *Global Change Biology* 11: 1013–1023.

Hiltbrunner E and Körner C (2004) Sheep grazing in the high alpine under global change. In: Lüscher A, Jeangros B, Kessler W, *et al.* (eds.) *Land Use Systems in Grassland Dominated Regions*, pp. 305–307. Zurich: VDF.

Kalin Arroyo MT, Primack R, and Armesto J (1982) Community studies in pollination ecology in the high temperate Andes of central Chile. Part I: Pollination mechanisms and altitudinal variation. *American Journal of Botany* 69: 82–97.

Körner C and Larcher W (1988) Plant life in cold climates. In: Long SF and Woodward FI (eds.) *Symposium of the Society of Experimental Biology 42: Plants and Temperature*, pp. 25–57. Cambridge: The Company of Biology Ltd.

Körner C (2003) *Alpine Plant Life*, 2nd edn. Berlin: Springer.

Körner C (2004) Mountain biodiversity, its causes and function. *AMBIO* 13: 11–17.

Körner C (2006) Significance of temperature in plant life. In: Morison JIL and Morecroft MD (eds.) *Plant Growth and Climate Change*, pp. 48–69. Oxford: Blackwell.

Körner C and Paulsen J (2004) A world-wide study of high altitude treeline temperatures. *Journal of Biogeography* 31: 713–732.

Mark AF, Dickinson KJM, and Hofstede RGM (2000) Alpine vegetation, plant distribution, life forms, and environments in a perhumid New Zealand region: Oceanic and tropical high mountain affinities. *Arctic Antarctic and Alpine Research* 32: 240–254.

Messerli B and Ives JD (eds.) (1997) *Mountains of the World: A Global Priority*. New York: Parthenon.

Meyer E and Thaler K (1995) Animal diversity at high altitudes in the Austrian Central Alps. In: Chapin FS, III, and Körner C (eds.) *Ecological Studies 113: Arctic and Alpine Biodiversity: Patterns, Causes and Ecosystem Consequences*, pp. 97–108. Berlin: Springer.

Miehe G (1989) Vegetation patterns on Mount Everest as influenced by monsoon and föhn. *Vegetatio* 79: 21–32.

Nagy L, Grabherr G, Körner C, and Thompson DBA (2003) *Ecological Studies 167: Alpine Biodiversity in Europe*. Berlin: Springer.

Pluess AR and Stöcklin J (2004) Population genetic diversity of the clonal plant *Geum reptans* (Rosaceae) in the Swiss Alps. *American Journal of Botany* 91: 2013–2021.

Rahbek C (1995) The elevational gradient of species richness: A uniform pattern? *Ecography* 18: 200–205.

Sakai A and Larcher W (1987) *Ecological Studies 62: Frost Survival of Plants. Responses and Adaptation to Freezing Stress*. Berlin: Springer.

Spehn EM, Liberman M, and Körner C (2006) *Land Use Change and Mountain Biodiversity*. Boca Raton, FL: CRC Press.

Till-Bottraud J and Gaudeul M (2002) Intraspecific genetic diversity in alpine plants. In: Körner C and Spehn E (eds.) *Mountain Biodiversity: A Global Assessment*, pp. 23–34. New York: Parthenon.

Yoshida T (2006) *Geobotany of the Himalaya*. Tokyo: The Society of Himalayan Botany.

Alpine Forest

W K Smith, Wake Forest University, Winston-Salem, NC, USA

D M Johnson, USDA Forest Service, Corvallis, OR, USA

K Reinhardt, Wake Forest University, Winston-Salem, NC, USA

Introduction
Alpine Forest Biogeography
The Abiotic Environment
Altitude versus Microclimate
The Treeline Ecotone – Tree Distortion,
   Clustering, and Spacing

Mechanisms of Treeline Formation
Summary
Further Reading

Introduction

The forest of the alpine zone occurs near mountain tops and forms a transition zone between the subalpine forest below and the alpine zone above (**Figure 1**). Whether this zone of overlap represents a definable, stable community with its own inherent structure and stability is open for debate.

Observations of the spatial patterns of the tree species do insinuate some successional character, although the long-term encroachment of the subalpine forest into the alpine zone, or vice versa, is a slow process that is detectable only after centuries of change, at least. Although the alpine forest has a well-defined, characteristic vegetation pattern that contrasts with the subalpine forest and alpine zones, animal

Figure 1 Alpine forest landscape (~3200 m altitude) in the treeline ecotone of the Snowy Range, Medicine Bow Mountains, southeastern Wyoming (USA). Alternating snow glades (long-lasting snow pack) and ribbon forest are characteristic of this alpine forest, along with the potentially extreme distortion of individual tree structure and form (see **Figure 2**). Prevailing winds are from the right in this photo.

species are often viewed as community members of either or both. This boundary ecotone between two contiguous communities is often referred to as the upper (or cold) treeline (or timberline) ecotone where the treeline limit is reached. This limit is defined as the highest occurrence of a tree species in any form, or for a tree species that has a certain minimum tree stature (e.g., greater than 2 m vertical height). The latter definition is necessary because this upper limit of tree occurrence is often composed of disfigured (flagged branching) and stunted (krummholz mat) tree forms that

are more shrub-like than tree-like in appearance (**Figure 2**). This upper (cold) treeline ecotone can vary in altitude and width according to latitude and proximity to maritime influences, as well as the degree of slope and azimuth at a given location. In addition, plant demographics such as tree size, age, spacing, and clustering among individual trees, plus the structural distortion and disfigurement of individual trees further from the timberline, can vary dramatically. Regardless of the latitude or altitude of mountain areas, excessive steepness of the slope and, thus, poorly developed soils, will prevent tree establishment and result in sharp boundaries between the timberline and alpine community. Above the timberline, individual trees or patches occur sporadically associated with less wind-exposed microsites where aeolian soil and snow accumulate. These characteristics of the alpine forest landscape can also vary according to the proximity to oceans or other large bodies of water (e.g., 'lake effect' weather patterns). In general, greater latitudes result in a decrease in the altitude at which alpine forest is found, as does a closer proximity to oceans or other large bodies of water. In contrast, the dryer continental mountains tend to have timberlines and treelines at the highest altitude for a given latitude.

Alpine Forest Biogeography

Most of the ecological research focusing on the alpine forest has involved vegetation studies, although many animal species use this zone seasonally, especially later in summer when lower elevations have dried from the longer summer. This area is a prolonged green zone

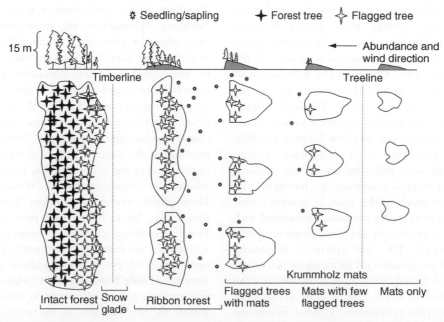

Figure 2 Schematic representation of **Figure 1** showing the relative size and spacing of individual tree forms and tree clusters making up a typical alpine forest within the treeline ecotone of a dry, continental mountain range. See text for further explanation.

where food for herbivores, especially, is still in abundance compared to lower elevations where most annual plants have completed their life cycle, and the perennial species have undergone a seasonal senescence due to accumulating summer drought. Alpine forest is found on all continents except Antarctica, as well as several oceanic islands. The mountain regions of the Western Hemisphere form large, N–S cordilleras that connect polar regions to the subtropics. For example, the Cascades, Rocky, and Sierra Nevada Mountains of the western US extend from the most northern boreal forest to southern Mexico where very high volcanic mountain ranges occur, while the high ranges of the Andes connect the full latitudinal extent of South America along its western seaboard. In contrast, the high ranges of the Alps of Central and Southern Europe, as well as the Himalayas of the Eurasia, are formed along an E–W axis and are much more discontinuous between the boreal and subtropical latitudes. Further south, high mountains of southern and eastern Africa represent much more isolated ranges compared to the more continuous cordilleras of the Western Hemisphere. In the Southern Hemisphere where there is much less land mass, alpine forests are less extensive and found in only a few mountain regions that tend to be close to coastlines and, thus, have a strong maritime influence (e.g., Andes, Australian Alps, New Guinea, and New Zealand).

The question of why treelines across the globe occur at specific altitudinal limits, and no higher, has been a focus of research and discussion for over a century and a half. Although it is well known that the altitude of upper treelines have been strongly influenced by anthropogenic causes (e.g., grazing and fires), the primary focus of these studies has been on identifying the abiotic factors that are most limiting to the growth and survival of trees. However, there is also evidence that certain seed-dispersing bird species (e.g., Clark's nutcracker and the gray jay) may play a crucial role in the distribution of certain species in the high-altitude treeline (e.g., limber and whitebark pine of the western US). The high-altitude environment involves particularly extreme values of cold temperature, high wind, high and low (clouds) sunlight levels, low air humidity, high long-wave energy exchange, and rapid mass diffusion due to low ambient pressure. On wetter tropical mountaintops, forests may be cloud-immersed for much of the year. In general terms, the temperature lapse rate (dry adiabatic) associated with altitude generates a maximum decrease in air temperature of approximately 1°C per 100 m of increasing altitude. Thus, this environmental factor alone is a dominant environmental factor influencing differences in the alpine forest located at dryer continental versus more moist coastal mountain ecosystems. Coastal mountains experience much lesser lapse rates (<0.3 °C per 100 m) because substantial condensation of moisture with greater

altitude transfers heat to the thinning atmosphere. In addition to this extreme abiotic environment, the total length of the growth season is severely curtailed (often <90 days), and even growth during summer is severely limited for short, but often frequent, time periods that occur periodically during the entire summer growth period. Because of these factors alone, adaptation and survival of alpine forest species is most often perceived as driven by abiotic pressures.

The distribution and species composition of alpine forests on a global scale vary strongly according to both latitude and longitude. In general, the Western Hemisphere of North and South America has an N–S and S–N cordillera that extends from the boreal forests of northern Canada (Canadian Rocky Mountains) all the way across North and Central America to the southernmost portions of South America (southern Andes). In addition, this long expanse of alpine forest may be strongly influenced by nearby oceanic influences. In contrast, the major mountain ranges of Europe and Asia have a more E–W distribution and in the case of the Asian provinces, are much further away from strong oceanic influences. There are also high mountains with treelines located on volcanic islands in all the major oceans. The mountains of New Guinea, southeastern Australia, Tasmania, and New Zealand are examples of island-like alpine forests with strong oceanic influence and that also extend from the subtropics to the extreme south temperate zone of the Southern Hemisphere. There are only a few Antarctic treelines that occur on small islands relatively close by. The distribution of different treeline tree species on a global scale and according to plant type and latitude is summarized in **Table 1** and **Figure 3**. Treelines of the northernmost temperate zones of the Northern Hemisphere are dominated in the eastern hemisphere by white birch at the highest latitudes of the Scandinavian, Ural, and eastern Siberia ranges, followed by the Scotch pine and European spruce treelines of central Norway and Sweden. In the more interior, continental areas of central Europe, the Swiss mountain pine (*Pinus mugo*) forms the alpine forest, while European larch (*Larix decidua*) and stone pine (*Pinus cembra*) form the treelines of the central Alps. In the maritime mountains of the western and southern Alps, European beech (*Fagus silvatica*) is dominant. In the Western and Northern Hemisphere, evergreen conifers dominate the alpine forest (e.g., larch, bristlecone pine, subalpine fir, and Engelmann spruce), while deciduous conifers and broadleaf species are found less frequently, along with the rare occurrence of evergreen broadleaf species in South America (**Table 1** and **Figure 4**). Many of these distribution patterns appear influenced by differences not only in abiotic factors, but also in historical factors related to dispersal mechanisms and historical factors related to continental drift over geological timescales.

Table 1 Biogeography of alpine forest areas worldwide

Latitude	Mountain range	Altitude (m)	Climate	Life form	Dominant tree genera
Western Hemisphere					
50–60° N	Northern Rockies, USA/Canada	2600–2900	Continental	DN, EN	*Abies, Picea, Pinus, Larix*
55–58° N	Scottish Highlands, UK	600–800	Oceanic	DB	*Pinus, Juniperus, Betula*
45–50° N	Northern Appalachians, USA	1500	Continental	EN	*Abies, Picea*
30–60° N	Pacific Coast Mtns, USA, Canada	To 3300	Oceanic	EN	*Tsuga, Pinus, Abies, Picea, Chamaecyparis*
40° N	Middle Rockies, USA	2900–3300	Continental	EN	*Abies, Picea, Pinus*
37° N	Sierra Nevadas, Spain	1950	Continental	DB, EN	*Quercus, Pinus*
35–40° N	Sierra Nevadas, USA	3000–3500	Oceanic	EN	*Pinus, Tsuga, Juniperus*
30–35° N	Southern Rockies, USA	3300–3800	Continental	EN	*Abies, Picea, Pinus*
18–25° N	Sierra Madres, Mexico	4000	Oceanic	EN	*Pinus*
9–11° N	Talamancas, Costa Rica	3000	Oceanic	DB	*Quercus*
10° N–20° S	Tropical Andes, Costa Rica–Peru	3150–4700	Oceanic	DB	*Polylepis*
23° S–50° S	South America, Temperate Andes	1100–1500	Oceanic	EB, DB	*Podocarpus, Nothofagus*
Eastern Hemisphere					
60–70° N	Skanderna, Scandinavia	700–900	Continental	DB	*Betula*
50° N	Altai, Mongolia	2000	Continental	EN, DN	*Pinus, Larix*
44° N	Tien Shan, Asia	3000	Continental	EN	*Picea*
46–43° N	Alps, Europe	1600–2300	Continental	EN, DN, DB	*Abies, Picea*
43° N	Caucasus, Georgia	2200	Continental	DB, EB, EN	*Betula, Rhododendron, Pinus*
38° N	Pamir, Tajikistan, Asia	3000	Continental	EN	*Picea*
35–36° N	Japanese Alps/Fuji, Japan	1950–2400	Oceanic	EN, DN	*Abies, Larix, Pinus, Tsuga*
31° N	Great Atlas, Northern Africa	2850		EN	*Cedrus, Juniperus*
28° N	Himalayas, Asia	3800–4500	Continental	DB, EB, DN, EN	*Betula, Rhododendron, Picea, Larix, Juniperus, Tsuga*
2° S	Maoke Mtns	3000–3600	Oceanic	EN	*Podocarpus*
3° S	East Africa Highlands, Africa	4050		EB	*Erica*
36° S	Australian Alps, Australia	1800–1950		EB	*Eucalyptus*
42° S	Tasmania	1200–1260		EN, EB, DB	*Athrotaxis, Eucalyptus, Nothofagus*
43° S	Southern Alps, New Zealand	1200–1500		DB	*Nothofagus*

DN, deciduous needleleaf; EN, evergreen needleleaf; DB, deciduous broadleaf; EB, evergreen broadleaf.

The Abiotic Environment

Aboveground

Abiotic factors have traditionally been viewed as dominating the ecology of high altitudes, including the alpine forest. Sunlight, temperature, water, and gas-phase nutrients (e.g., CO_2 and O_2) can vary substantially with altitude, regional climate, and orographics (e.g., maritime vs. continental mountain ranges). In addition, many factors influencing leaf energy balance and temperature may also vary with elevation, including solar and long-wave radiation, wind, and ambient humidity. Probably, the best-known abiotic change with increasing elevation is the decline in air temperature in response to lower ambient pressure. Ambient pressure decreases by over 20% at 2 km and over 50% at 6 km, leading to a maximum, dry adiabatic lapse potential of 1.0°C/100 m. Simulated dry (8.0°C km^{-1}) versus wet (3.0°C km^{-1}) lapse conditions resulted in a more rapid decline in air temperature with altitude for both winter and summer temperatures. Also,

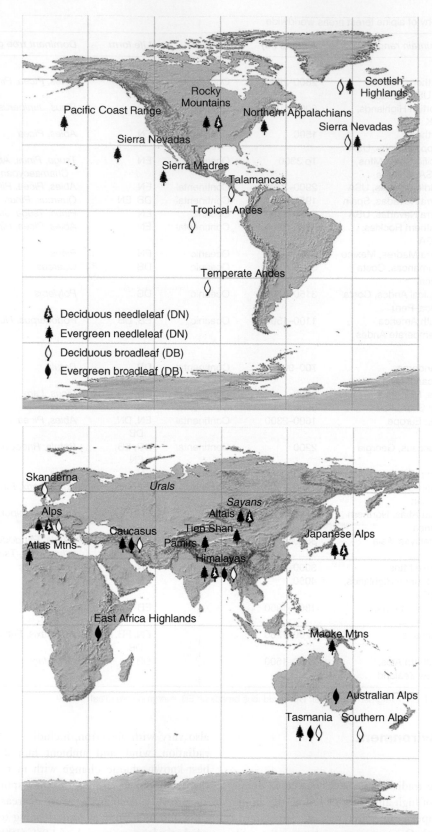

Figure 3 Global biogeography of alpine forest areas. *Italicized* mountain ranges have no corresponding information in **Table 1**.

Figure 4 Individual alpine forest tree in the high-altitude treeline (3306 m) at a wind-exposed site. At this high-altitude limit of tree growth, extreme distortion in tree structure results in the classic krummholz mat growth form showing the presence of flagged branches on the downwind edge. The prevailing winds are from the right in this photo.

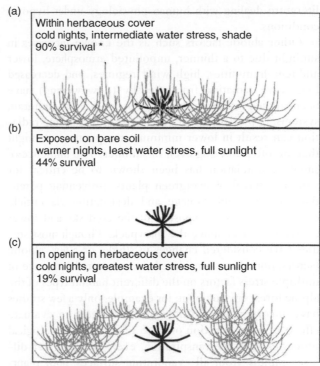

(a)
Within herbaceous cover
cold nights, intermediate water stress, shade
90% survival *

(b)
Exposed, on bare soil
warmer nights, least water stress, full sunlight
44% survival

(c)
In opening in herbaceous cover
cold nights, greatest water stress, full sunlight
19% survival

Figure 5 Microsite alteration experiment showing effects of facilitation vs. competition on survival of new (first-year) seedlings of *Picea engelmannii* Parry ex. Engelm. (Engelmann spruce) in an alpine treeline ecotone, southeastern Wyoming. Greatest survival (90%) occurred for seedlings growing in vegetative ground cover that resulted in low sky exposure and incident sunlight the following morning, intermediate water stress, and relatively cold nights. Removing all vegetation well away from a seedling reduced competition for soil water (higher xylem water potentials), but increased sky exposure, resulting in significantly lower survival (44%). The highest mortality occurred when only proximal vegetation was removed to increase sky exposure, while maintaining boundary layer effects, lower minimum needle temperatures, and competition for water (as validated by higher water potential values). Higher photosynthetic carbon gain due to less low-temperature photoinhibition of photosynthesis was also associated with greater survival. Thus, facilitated reduction in sky exposure (day and night) appeared to have a greater influence on photosynthesis and survival, compared to low temperatures or competition for water with neighbors, although all three stress factors had significant impact. From Germino MJ, Smith WK, and Resor C (2002) Conifer seedling distribution and survival in an alpine-treeline ecotone. *Plant Ecology* 162: 157–168.

dry lapse conditions in summer generated similarly cold air temperatures at higher elevations (>4 km) that were very near values computed for wet lapse conditions during winter (**Figure 5b**). Similar dry and wet lapse rates of 7.5 and 5.5°C km^{-1}, respectively, have been used previously to evaluate transpiration potential for plants growing on mountains of temperate and tropical zones.

Another fundamental change in abiotic factors of increasing altitude is the unique and colligative property of decreasing atmospheric pressure and, thus, the partial pressures of gas-phase molecules such as CO_2 and O_2. In contrast, the amount of water vapor in the air at saturation is dependent only on temperature and, thus, strongly influenced by the lapse rate in air temperature described above. Because ambient CO_2 concentration can have a strong, direct influence on plant photosynthesis via the leaf-to-air concentration gradient (driving force for diffusion), it has often been assumed to be a limiting factor for carbon gain and growth at high elevation. For plants, where the diffusion process is the primary mode of gas exchange, a lower ambient CO_2 concentration with altitude could result in a corresponding decrease in the leaf-to-air gradient, assuming a constant CO_2 concentration inside the leaf. For this reason, mountain ecosystems have been considered as natural field models for evaluating the effects of natural differences in atmospheric CO_2 concentrations. However, because molecular diffusion is more rapid at lower ambient pressure, a substantial compensatory effect on CO_2 uptake potential occurs with greater elevation. In agreement with this physiochemical property, little evidence has been found supporting the idea that lower partial pressures result in diffusion limitations at higher altitudes, at least for systems depending on the diffusion process for physiological gas exchange. Although quantitative

evaluations showing these compensating effects on photosynthetic CO_2 uptake exist in the literature, there are few comprehensive studies incorporating all of the potentially important factors influencing diffusional gas exchange at higher altitudes. Similar concerns for animal O_2 uptake at high altitude form a vast literature, although animals, depend primarily on bulk supply mechanisms for enhancing gas exchange. However, diffusion effects on animal ecophysiology at high elevations (e.g., eggs, burrowing and subnivian animals) are not well studied, except for a large literature dealing with human physiology under hypoxic conditions.

Other abiotic factors such as the known increases in sunlight due to a thinner, unpolluted atmosphere, lower ambient humidities, high wind regimes, and decreased long-wave radiation from the sky (downwelling) have been studied less thoroughly, and for only a few mountain systems. In particular, the decrease in downwelling radiation can result in lower minimum temperatures at night that are often freezing even in summer. The influence of snow accumulation has been shown to be critical for winter survival of evergreen plants, preventing potentially lethal wind damage and desiccation via cuticle abrasion, as well as exposure to the cold sky and lower air temperatures above the snowpack. Though most studies have considered changes in single, or a few, abiotic factors, none have considered the concerted influence of multiple stress factors on the different habitat types of the alpine forest environment; for example, only a few studies have incorporated multiple abiotic factors to evaluate effects of high elevation on such important physiological processes as evapotranspiration, even though water diffuses rapidly from all evaporating surfaces, both plants and animals, compared to sea level.

Belowground

The soil environment of the alpine forest, as for many other communities and ecosystems, is strongly dependent on the prevailing moisture regime, including the seasonal timing and physical nature of the precipitation (rain vs. snow). Although the winter season can result in snow even in tropical mountains, there can also be important impacts due to the occurrence of dry seasons, sometimes twice per year. Regardless, rainfall of a melting snowpack results in quite different impacts on soil nutrients based on the accompanying temperature regimes. Warmer periods with rainfall will result in the release of soil nutrients, while colder periods with snow accumulation will involve dormant periods for surface soil organisms and, thus, decomposition and nutrient release. In addition, the growth activities of the important mycorrhizal fungi of plant root systems are strongly limited by soil temperatures well above freezing (up to 7°C). Because more tropical alpine forests receive the majority of precipitation as snow during a relatively brief winter season, or wet season, snowmelt often occurs quickly and is followed by a rapid drying of surface soils where roots are found. Thus, plants must take up soil nutrients during a very abbreviated time period which can be limited by persistently cold soils that lag behind air temperatures on a daily and seasonal basis. Soils' freeze and thaw cycles at high altitudes can also create distinct patterns in the microtopography of the soil surface (e.g., polyhedrons), providing important

microsites for seedling establishment and small-scale differences in plant distribution patterns.

Altitude versus Microclimate

An important message concerning changes in abiotic factors with elevation is the realization that decreasing ambient pressure is the only physical property unaffected by microclimatic effects. All others (e.g., temperature, sunlight, wind, long-wave radiation, water and nutrient relations) can be strongly influenced by topography, microsite, and plant form at any altitude. Natural variability in these factors can substantially lower, or raise, the effective altitude of a microsite at any given altitude. In the Northern Hemisphere, south-facing, wind-sheltered microsites can effectively match conditions at an altitude thousands of meters lower, while similar north-facing microsites, sheltered from sun but not the cold night sky, could generate increases in effective altitude. Even smaller microsites around a fallen tree stem or an exposed boulder can result in effectively different altitudes based on differences in sunlight exposure and temperatures. Additionally, changes in leaf orientation of the plant can create different levels of sky and wind exposure, two primary factors influencing microclimate at any altitude. Leaf and plant aggregation (close spacing) and height patterns can also influence microclimate due to the potentially strong boundary layer effects on temperature and ambient gas concentrations. Thus, microclimate effects can significantly impact fundamental gas exchange processes at any altitude, with the exception of ambient pressure effects on molecular diffusion. In contrast to the potential effects of microsite and plant form on effective altitude, individual plants cannot escape the ambient pressure of their respective altitudes (only negligible changes in ambient pressure due to weather fronts). Thus, lower ambient pressure and more rapid molecular diffusion are the only immutable abiotic factors associated with increasing altitude, one that is not dependent on microsite/microclimate effects.

The Treeline Ecotone – Tree Distortion, Clustering, and Spacing

On a geographic scale, the appearance of an alpine forest community can vary considerably, depending primarily on latitude and the distance from oceanic influences. Alpine forests occur at higher altitudes at lower latitudes, but with stronger maritime impacts. However, similar variation at a particular site is also associated with the steepness of the site, along with sun and wind exposure. Typically, the altitude of the treeline falls as these site-specific factors increase. A typical alpine forest and treeline ecotone community of a dry, continental mountain range is shown in **Figures 1** and **2**, both of which represent the most extreme changes in tree form and landscape found in alpine treeline ecotones (alpine forest). As one progresses from the treeline toward the intact subalpine forest, individual trees occur as small (<2 m in width and a meter in height) krummholz mats, larger mats with flagged trees at their leeward edge, still larger tree islands (>10 m diameter) with more intensely flagged trees near the windward side, and, finally, ribbon forest alternating with snow glades (>10 m across) just prior to edge of the intact subalpine forest (although some flagging at tree tops is still noticeable). The density of these various structures also increases closer to the forest edge, along with the occurrence of young seedlings and saplings.

Mechanisms of Treeline Formation

Investigators have been interested for over a century in the question of why trees do not occur above certain altitudes. Ecological studies have shown that the occurrence of both timberlines and treelines decline steeply and almost linearly in altitude as latitude increases between about 30° N and S latitude and over 60° N and S latitude. This linear relationship results in an estimated change in timberline altitudes of ~100 m per degree of latitude. However, between about 30° on each side of the equator, there is a relatively constant, maximum altitude of occurrence that is near 3.5–4.0 km. Little information exists concerning the difference in altitude between the timberline and treeline, or the width of the treeline ecotone (alpine forest) as related to geography or any specific environmental factor. Although most of these studies have associated this altitude of occurrence to the colder temperature regimes at higher latitudes, the actual ecophysiological mechanisms are still being debated, and may involve a large number of abiotic and biotic factors. In addition, major changes in tree habit occur within this life zone, including dramatic alterations in plant height and crown features such as branching pattern. This change in growth form becomes more dramatic as distance from the forest edge (timberline) increases toward the ultimate treeline limit (**Figures 1** and **2**). Across this ecotone, the full-tree stature of a typical forest tree becomes twisted and distorted, forming ultimately a small, shrub-like habit commonly referred to as the 'krummholz' mat at treeline. During this transition, trees also become more and more flagged in appearance, with stems occurring only on the downwind side of trunks and main stems (**Figure 3**).

Because temperature data have been mostly available for the longest period of time and for most locations worldwide, a host of studies have attempted to correlate measured temperature regimes with the highest altitude of tree occurrence. Within these myriad studies, the occurrence of minimum temperatures and the amount and physical nature of the prevailing snowfall has been a central focus. For example, more continental (noncoastal) mountain ranges of both hemispheres have dryer, colder climates characterized by 'powder snow' conditions. This type of snow is strongly influenced by wind-driven snow that can generate strong abrasive forces due the sharp-edged, crystalline nature of these snow particles. These systems also have distinct snow accumulation patterns across the landscape that are the result of the strong turbulent and eddy flow characteristics. Moreover, snow burial and avoidance of excessive exposure to wind and colder temperatures may be critical for the winter survival of both plants and animals in this alpine-forest belt. In contrast, coastal ranges with lower altitudes of treeline also have higher air humidity levels, snowfall of high water content, and low abrasive power of softer ice crystals that are relatively uncoupled from the influence of wind patterns. This wetter, heavier snow can accumulate on exposed branches, creating severe mechanical forces that can bend, break, and distort stems due to snow and ice loading above the snowpack surface and freeze–thaw compression forces beneath the snow surface. Snow accumulation in the dryer continental alpine forest is much more dependent on drift mechanics and eddy flow dynamics (e.g., burial of krummholz mats), while the wetter snows of more coastal systems generate a more uniform depth and homogeneous distribution pattern across the treeline ecotone. For the dryer powder snow of the continental mountain tops, severe abrasive properties can lead to abrasion of leaf cuticles, removal of paint from highway traffic placards, and the common windburn suffered by skiers on windy days and powder-like snow. Thus, these differences in the physics of snow particles and spatial distribution dynamics also play a major role in the distortion and disfiguration effects on individual trees of the alpine forest (stunting, flagging, and krummholz tree forms), as well as the spatial patterns of tree spacing. These differences in the basic physical make-up of snow have not been considered systematically in terms of their influence on the vegetation patterns and distortions in growth form observed for individual trees within different alpine forests (appearance of krummholz and flagged growth forms). These effects for more maritime versus continental mountain ecosystems need further elucidation, in particular, the impact on the altitude at which trees can no longer regenerate.

The ecophysiological mechanisms regulating the upper elevational limits of treelines across the globe have been contemplated by plant ecologists, biogeographers, and biometeorologists for over a century. A recent review concluded that the elevation limits of the upper treelines on a global scale is the result of (1) the inability of alpine plants to metabolically process the carbon gained from daytime photosynthesis because of cold-temperature limitations (e.g., respiratory limitations), and (2) the large size of conifer trees which prevents adequate warming of the soil due to soil surface shading by the closed, overstory canopy. Thus, low soil temperatures due to self-shading was proposed as a major abiotic determinant of the elevational limits of upper forest treelines. However, other studies have provided evidence of strong limitations to resource acquisition at high altitudes, specifically the photosynthetic uptake of CO_2 by alpine forest trees. Many other investigators have also questioned conclusions (1) and (2) above.

Despite the longstanding interest in the environmental and physiological mechanisms generating observed altitudinal patterns in the formation of alpine forests and their respective treelines, virtually all of this research has focused on the ecophysiological effects measured for adult trees, even though they may show distortions in form and greatly diminished stature, for example, krummholz mats and stunted, flagged trees. Very little research has focused on the establishment of new seedlings away from the forest edge into the treeline ecotone. Yet, it is this life stage within the treeline ecotone that appears critical for migration to a higher altitude and formation of new subalpine forest. The formation of new subalpine forest at higher elevation is dependent on seedling regeneration into the ecotone, whereas the migration of the forest timberline to a lower altitude would require both the mortality of older trees and the successful seedling regeneration at the new altitude of occurrence. However, any mortality of the overstory trees could also introduce an important impact – a decrease in the ecological facilitation of seedling establishment. Likewise, a lack of establishing seedlings in the forest understory at the forest edge, in combination with the death of the overstory trees, would result in a lowering of the timberline and, most likely, the treeline as well. An important component of this process is the ecological facilitation of new seedling survival and growth that results from a more mature forest structure (**Figure 2**). In other words, the development of trees with forest-like stature (no flagging or krummholz distortion) requires the formation of an intact forest and the resulting amelioration of a host of extreme abiotic factors outside the forest. Thus, the altitudinal movement of timberline and treeline boundaries begins with new seedling establishment, either below of above the existing timberline that will act, ultimately, to facilitate further seedling establishment and the gradual development of new subalpine forest either above or below the altitude of the existing timberline. For example, the mechanisms involved in the migration of a timberline/treeline to a higher altitude must initially depend upon new seedling establishment above the existing timberline, into the treeline ecotone. Moreover, greater seedling/sapling abundance must follow to provide the ultimate facilitation required for continued growth to full

forest-tree stature and, thus, the formation of new subalpine forest at higher altitudes. At high elevation, this migration of timberline is possible only with the protective, mutual facilitation provided by neighboring trees and surroundings, similar to that found within intact subalpine forest. Thus, growth to forest-tree stature without structural distortion may require, to some degree, 'the forest before the tree'. In the Rocky Mountains of southeastern Wyoming (USA), the establishment of new tree seedlings into a treeline ecotone appears also to involve considerable microsite facilitation (**Figure 5**) by either inanimate objects (e.g., rocks, fallen logs, microtopography due to freezing and thawing of the soil surface), or by intra- and interspecific spatial associations generating ecological facilitation of microsites. Structural self-facilitation (e.g., cotyledon orientation and primary needle clustering, krummholz mats) may also act to enhance the growth and survival at all structural scales from the seedling to mature trees (**Table 2**). Increased seedling establishment and abundance is followed subsequently by even greater facilitation, which leads to even greater seedling establishment and sapling growth, and so on (**Table 3**). Thus, increased seedling/sapling abundance will lead to the same 'sheltering' effect that is necessary for the formation of the forest 'outposts', or islands, known to be important shelters for improved seedling establishment. In addition, the ultimate development into a forest tree (nondistorted growth form) is analogous functionally to the biophysical 'escape' of vertical stems from the surface boundary layer of a krummholz mat (**Figure 3**). Subsequently, continued facilitation of the sapling stage, approaching a similar level as found within the intact subalpine forest at lower elevation, is required before an establishing sapling can reach the stature of a subalpine forest tree.

Table 2 Factors identified as important for explaining the altitudinal occurrence of alpine forest and its maximum altitude of occurrence as an alpine treeline ecotone

---

1. *Seedling/sapling establishment* – seed germination, growth, and survival
2. *Mechanical damage* – wind abrasion of needle cuticles, apical bud damage, snow loading, and frost heaving cause tissue and whole-tree mortality
3. *Physiological tissue damage* – low temperature and desiccation limits growth and survival
4. *Annual carbon balance* – photosynthetic carbon gain minus respiratory demands is less than that needed for successful growth and reproduction
5. *Biosynthesis and growth limitation[a]* – greater cold temperature limitation to growth processes than to photosynthetic carbon gain

---

[a]Cold soil temperature due to the large size of conifer trees and consequential soil shading have been hypothesized as a primary environmental factor limiting the processing of assimilated carbon and, thus, maximum altitude of alpine treelines.

Table 3 The importance of ecological facilitation for seedling establishment, growth, and survival in the alpine forest

Source
  Biotic: Inanimate (rocks, dead wood, microtopography)
  Abiotic: Plant structure (clustering), intraspecific and interspecific facilitation of microsites

Benefits
  Winter
    Snow burial – prevents ice crystal abrasion and desiccation; warmer and less extreme diurnal temperature differences; no excessive sunlight exposure
    Clustering at the shoot-to-landscape scale – increased snow deposition and burial
    Flagging – prevents damage from snow loading and rime ice accumulation
  Summer
    Less sky exposure
      Day: Less sunlight and cooler temperatures
      Night: Higher minimum temperatures and less LTP; less dew and frost accumulation
    Less wind exposure – warmer needles in sun

Possible adaptive tradeoffs
  Less sun sky exposure due to burial and mutual shading
    Day: Less sunlight for photosynthesis and lower temperatures
  Less wind exposure
    Day: Warmer temperatures and greater transpiration
    Night: Colder minimum temperatures and greater LTP

Inanimate, intraspecific, interspecific, and structural facilitation can all generate protective snow burial, as well as amelioration of subsequent growth limitation factors within and just above associated ground cover. LTP represents low-temperature photoinhibition of photosynthetic carbon gain.

Summary

The alpine forest represents a transitional zone separating the alpine tundra and subalpine forest communities. This treeline ecotone is also the highest altitude at which trees are found to occur, although the exact environmental factors and mechanisms limiting this occurrence are just beginning to be unraveled. These treelines are composed of evergreen conifer species most often, although deciduous conifers and broadleaf species also occur, as well as evergreen broadleaves at lower latitudes. There is also a strong correlation between higher latitudes and a lower treeline altitude, as well as with more continental versus maritime mountains. Ecological facilitation of seedling microsites by inanimate structures and microtopography, along with intra- and interspecific facilitation, is a fundamental property of timberline migration up or down the mountain and, thus, the formation of new subalpine forests at a different altitude. This facilitation of microsites

involves environmental parameters such as avoidance of wind exposure, wind/snow abrasion, and exposure to sunlight and the cold nighttime sky. In addition, the ability to survive in exposed microsites appears coupled to developmental capabilities for forming krummholz and flagged forms that enable wind protection, including adequate snow collection and burial to prevent damage from the abiotic environment. As new seedling and sapling cover increases, facilitation of growth processes by microclimate amelioration leads to the ultimate growth of trees to a forest-tree stature, culminating in the protective environment of a new subalpine forest.

See also: Alpine Ecosystems and the High-Elevation Treeline; Boreal Forest.

Further Reading

Arno SF and Hammerly RP (1990) *Timberline: Mountain and Arctic Forest Frontiers*. Seattle, WA: The Mountaineers.
Callaway RM (1995) Positive interactions among plants. *Botanical Review* 61: 306–349.
Choler P, Michalet R, and Callaway RM (2001) Facilitation and competition on gradients in alpine plant communities. *Ecology* 82: 3295–3308.
Germino MJ, Smith WK, and Resor C (2002) Conifer seedling distribution and survival in an alpine-treeline ecotone. *Plant Ecology* 162: 157–168.
Grace J, Berniger F, and Nagy L (2002) Impacts of climate change on the treeline. *Annals of Botany* 90: 537–544.
Holtmeier FK (1994) Ecological aspects of climatically-caused timberline fluctuations: Review and outlook. In: Beniston M (ed.) *Mountain Environments in Changing Climates*, pp. 223–233. London: Routledge.
Innes JL (1991) High altitude and high latitude tree growth in relation to past, present and future climate change. *Holocene* 1: 168–173.
Jobbagy EG and Jackson RB (2000) Global controls of forest line elevation in the Northern and Southern hemispheres. *Global Ecology and Biogeography* 9: 253–268.
Körner C (1998) A re-assessment of high elevation treeline positions and their explanation. *Oecologia* 115: 445–459.
Smith WK, Germino MJ, Hancock TE, and Johnson DM (2003) Another perspective on the altitudinal limits of alpine timberline. *Tree Physiology* 23: 1101–1112.
Smith WK and Knapp AK (1985) Montane forests. In: Chabot BF and Mooney HA (eds.) *The Physiological Ecology of North American Plant Communities*, pp. 95–126. London: Chapman and Hall.
Stevens GC and Fox JF (1991) The cause of treeline. *Annual Review of Ecology and Systematics* 22: 177–191.
Sveinbjornsson B (2000) North American and European treelines: External forces and internal processes controlling position. *AMBIO* 29: 388–395.
Tranquillini W (1979) *Physiological Ecology of the Alpine Timberline*. New York: Springer.
Walter H (1973) *Vegetation of the Earth in Relation to Climate and Ecophysiological Conditions*. London: English University Press.
Wardle P (1974) Alpine timberlines. In: Ivey JD and Barry R (eds.) *Artic and Alpine Environment*, pp. 371–402. London: Meuthuen Publishers.

Biological Wastewater Treatment Systems

M Pell, Swedish University of Agricultural Sciences, Uppsala, Sweden

A Wörman, The Royal Institute of Technology, Stockholm, Sweden

Introduction
Life and Nutrient Transformation Processes
Biological Wastewater Treatment Systems

Perspective on Biological Wastewater Treatment
Further Reading

Introduction

Eutrophication of water courses, lakes, and marine environments is a major issue in most parts of the world. Looking back 150 years the urban situation in the emerging industrial part of the world led to the introduction of water-based systems for conveying and discharge of sewage. At first the wastewater was disposed into nearby watercourses and lakes. As the populations grow, this was not a sustainable solution – the natural wetlands became overloaded as evident from the odors. This untenable situation led to the development of more active treatment systems like shallow ponds and sand filters. In 1914 the activated sludge technique was introduced by Arden and Lockett, a technique that still probably is the most common technique for wastewater treatment (WWT) in the industrial part of world. In the 1960s eutrophication became evident due to the high amounts of plant nutrients discharged from sewage treatment plants. The first and maybe the simplest solution was to remove phosphorus by chemical precipitation. The European Commission and national authorities have gradually over the latest couple of decades sharpened the treatment demands, especially with regard to nitrogen, in order to avoid further eutrophication in the sea. Hence, WWT today probably is more focused on removing phosphorus and nitrogen than pathogens. It is still argued whether phosphorus or nitrogen is limiting for the eutrophication process, that is, should either one or both of these elements be eliminated.

Simply put, biological WWT can be defined as a natural process in which organisms assist in environmental cleanup simply through their own life-sustaining activities. By studying the organisms in natural ecosystems the biologists have explored their function and capacity to degrade organic matter and transform nutrients. Such information has then been used by engineers to design effective WWT systems, that is, the biological processes have been concentrated into well-regulated units. In addition, knowledge of geochemistry, hydrology, etc., is essential component of a successful system for treating polluted waters.

Hence, globally, WWT probably is the most common biotechnological process.

Though the same biological processes are the basis for most WWT systems, the number of technological solutions for achieving the goal probably is innumerable. The numbers of techniques are as many as there are sanitary engineers. However, the techniques may be categorized as follows: (1) soil filters and wetlands – terrestrial ecosystems working as natural filters; natural water courses, lakes, and wetlands; soils receiving irrigated wastewater; constructed wetlands and ponds; soil or sand absorption systems; and trickling filters; and (2) treatment plants – rotating biological contactors; fluidized beds; and activated sludge systems including sequencing batch reactors (SBRs). This array of techniques describes the systems on a scale from natural ecosystems at one end to high-technology solutions at the other end. In the choice of WWT system to be used many factors have to be considered like influent water characteristics, desirable effluent water quality, costs for building and maintenance, and population density and dimensioning.

In this article we have chosen first to give a general background on the microbial cell and biological processes important in all WWT and, second, to focus on the importance of understanding the interaction between hydraulic performance and microbial processes to achieve effective nitrogen removal, and third, to outline the function of two common systems: the constructed wetland, requiring in-depth knowledge on hydraulic properties, and the activated sludge process, relying on advanced control and optimization. Finally, we give some perspectives on the future development of biological WWT systems and their use.

Life and Nutrient Transformation Processes

The Cell

The cell is the smallest independent unit in all living organisms. The cell can also form an individual organism itself. Such organisms are referred to as microorganisms as they are not visible to the naked eye. Examination of the internal structure of the microbial cells reveals two structural types: the prokaryote (Bacteria or Archea) and the eukaryote (Eukarya) (**Table 1**). The previous group includes the bacteria while the latter contain protozoa, fungi, algae, plants, and animals. Prokaryotic cells have a very simple structure. They lack a membrane-enclosed nucleus and they are very small, typically being from less than 1 μm up to several micrometers. Eukaryotic cells are generally larger and structurally more complex. They contain a membrane-enclosed nucleus, and several membrane-enclosed organelles specialized in performing various cell tasks. The morphological differences between the two cell types have profound effect on their capacities to absorb and transform nutrients and energy. The prokaryotes have a large surface in relation to their volume meaning short

transportation distances within the cell not hindered by complex membrane systems. Their potential to transform and take up nutrients as well as to grow is very high; hence, they can be said to be tailor-made for high metabolic rates. Some bacteria may under optimal conditions multiply by binary division every 20 min. This will result in a rapid exponential increase in cells.

For its growth the cell needs energy, carbon, and macronutrients like nitrogen and phosphorus, and several elements in minor amounts. In addition, an adequate environment is needed, with oxygen, water, temperature, and pH being the most important regulators. Most microorganisms are heterotrophs and organotrophs meaning that they derive their energy and carbon, respectively, from organic molecules (**Table 2**). Other energy options available are inorganic chemicals (lithotrophs) and light

Table 1 Cell types and some typical characteristics

Characteristic	Prokaryotic		Eukaryotic
	Bacteria	Archaea	Eukarya
Morphology and genetic			
Cell size	Small, mostly 0.5–5 μm	Small, mostly 0.5–5 μm	Larger, mostly 5–100 μm
Cell wall components	Peptidoglucane	Protein, pseudopeptido-glucane	Absent, or cellulose or chitin
Cell membrane lipids	Ester-linked	Ether-linked	Ester-linked
Membrane-enveloped organelles	Absent	Absent	Mitochondrion, chloroplast, endoplasmatic reticulum, Golgi apparatus
DNA	One chromosome, circular, naked	One chromosome, circular, naked	Several chromosomes, straight, enveloped
Plasmids	Yes	Yes	Rare
Biochemistry and physiology			
Methane production	No	Yes	No
Nitrification	Yes	No	No
Denitrification	Yes	Yes	No
Nitrogen fixation	Yes	Yes	No
Chlorophyll-based photosynthesis	Yes	Yes	Yes
Fermentation end products	Diverse	Diverse	Lactate or ethanol

Table 2 Characterization of chemotrophic organisms according to their need of carbon and energy

Type	Carbon source	Examples of primary electron donors	Examples of terminal electron acceptors
Energy metabolism			
Lithotrophs	–	NH_3, NO_2^-, H_2S, S^0, Fe_2^+, H_2	Respiration: O_2, NO_3^-, NO_2^-, S^0, SO_4^{2-}, CO_2
Organotrophs	–	Organic	Respiration: O_2, NO_3^-, NO_2^-, SO_4^{2-}, Fe^{3+}, CO_2, organic; fermentation: organic
Carbon metabolism			
Autotrophs	CO_2	–	–
Heterotrophs	Organic	–	–

–, not relevant to this term.

(phototrophs). It is not uncommon that bacteria, like plants, can use carbon dioxide as the carbon source (autotrophs). Though the most common trait of living is organo-heterotrophic, virtually all combinations above of energy and carbon derivation exist.

Classical taxonomy of microbes is based on phenotypic characters like shape and size, and their relation to oxygen, as well as way of utilizing the carbon and energy source. Two classical shapes of bacteria are the rod and coccus, but filamentous and appendaged forms are also common. In addition to the shape, production of different enzymes is an important parameter in grouping and identifying bacteria. Recent developments within the nucleic acid-based molecular biology have provided invaluable tools in the systematic of life by genotypic characters. By comparing nucleotide sequences of not known organisms with the emerging database of sequence information, unknown organisms can be identified and/or classified.

The Microbial Community

Aggregated microbial communities called flocs or biofilms are the backbone of most WWT processes (**Figures 1a** and **1b**). The source of microorganisms is soil and sewage coming in with influent wastewater. In the WWT system the organisms are subjected to high selective pressure. Those tolerating the new environment will develop and even thrive to form the basis for an effective WWT process. In any system organic molecules due to their chemical/energetic properties will accumulate at

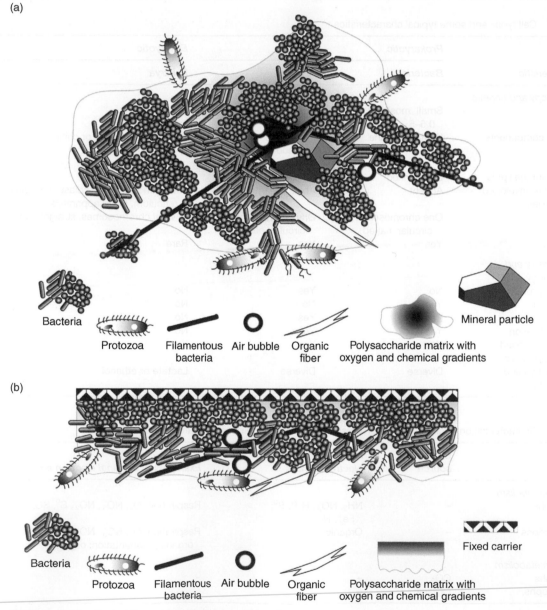

(a)

Bacteria   Protozoa   Filamentous bacteria   Air bubble   Organic fiber   Polysaccharide matrix with oxygen and chemical gradients   Mineral particle

(b)

Bacteria   Protozoa   Filamentous bacteria   Air bubble   Organic fiber   Polysaccharide matrix with oxygen and chemical gradients   Fixed carrier

Figure 1 Structure of (a) activated floc and (b) biofilm on solid surface.

interfaces (gas/liquid or liquid/solid). Hence, these niches will be the first to be colonized and microorganisms with features for keeping the community tightly together, for example, production of extracellular polysaccharides acting as glue, will dominate. The microbial community so formed will consist of a web of different species of bacteria, protozoa, and metazoa. Though present, fungi, algae, and virus probably play a less important role. The communites can be observed as sludge flocks or biofilms. Another advantage of living in dense communities is that environmental gradients, for example, of oxygen and substrate, are formed, allowing many types of organisms to share the space. From the WWT point of view the cooperation of micoorganisms will result in an effective degradation and mineralization of organic matter.

Investigation of activated sludge flocs and biofilms concerns the following issues: (1) morphology, that is, size and shape; (2) composition, that is, internal structure; (3) identification of microbial species; and (4) spatial arrangement of microorganisms. Traditionally, the detection of bacteria in wastewater is restricted to the ability to culture them. However, it has become evident that most organisms are unculturable which is the reason for our limited knowledge of the microbial actors in WWT processes. Recent advances in molecular techniques have supplied the means for examination community structure and detecting specific organisms in complex ecosystems without cultivation. Most techniques are based on nucleic acid fingerprinting after amplification by the polymerase chain reaction (PCR) of extracted DNA or RNA. Examples of techniques used are amplification of ribosomal DNA restriction analysis (ARDRA), denaturing gradient gel electrophoresis (DGGE), and terminal-restriction fragment length polymorphism (T-RFLP).

Microarray technology seems to be promising in capturing the taxonomical or functional structure of complex ecosystems. In this technique a vast number of oligonucleotide probes of known genes can be attached (spotted) to the surface of a glass slide. Extracted DNA or RNA from an unknown sample is then applied to the microarray plate. After hybridization the presence of target organisms will appear as radiant or fluorescent spots. Moreover, the intensity reflects the concentration of the sequence. By constructing a DNA microarray containing probes targeting the 16S rRNA of several groups of nitrifying bacteria the presence of *Nitrosomonas* spp. has been detected without need for PCR amplification prior to analysis. However, the technique failed to detect *Nitrospira* and *Nitrobacter*, but its future potential was clearly demonstrated. Fluorescence *in situ* hybridization (FISH) is an effective technique to detect specific bacteria in complex microbial communities. By use of confocal laser scanning microscopy (CLSM) FISH images of nitrifying bacteria in biofilms of domestic wastewater have been analyzed. Where the C/N ratio of the substrate was high, heterotrophic bacteria occupied the

outer part of the biofilm while ammonium oxidizing bacteria were distributed in the inner part. As the C/N ratio gradually decreased, the nitrifying bacteria began to colonize the outer layer.

The use of the molecular approaches discussed above has drastically widened our knowledge on bacterial diversity in WWT systems. Until 2002 more than 750 16S rRNA gene sequences derived from wastewater had been analyzed and sequences affiliated to the Beta-, Alpha-, and Gammaproteobacteria as well as the Bacteroidetes and the Actinobacteria were most frequently retrieved. Many new, previously unrecognized, bacteria have been detected, and many more, without doubt, await identification. Although some of the newly identified organisms can be attributed to the flocculation process as well as the biological nitrogen and phosphorus removal processes, most of them possess unknown functions. Not until it is fully understood can the potential of the biological component of the WWT system be fully utilized.

Microbial Carbon and Phosphorus Processes

Respiration

Respiration is probably the process most closely associated with life and in WWT systems it is attributed to a wide range of microorganisms such as bacteria and protozoa. Respiration is the aerobic or anaerobic energy-yielding process where reduced organic or inorganic compounds in the cell serve as primary electron donors and imported oxidized compounds serve as terminal electron acceptors (**Figure 2**). During respiration the energy-containing compound descends a redox ladder commonly consisting of the glycolysis, citric acid cycle (CAC), and finally the electron transport chain. The ultimate aim is to convert energy into proton gradients and ATP. During the metabolic pathway various intermediate organic molecules are withdrawn to enter the anabolic route, that is, building blocks incorporated into new cell material. Roughly, in actively growing heterotrophic cells, 50% of the substrate carbon will form new cells while the other 50% will be released as mineralized carbon dioxide (CO_2). In a less strict sense respiration can be defined as the uptake of oxygen while at the same time

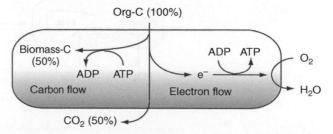

Figure 2 Carbon and electron flow in aerobic respiration. Box represents the microbial cell.

carbon dioxide is released. However, in the ecosystem, CO_2 is also formed by other processes such as fermentation and abiotic processes, for example, CO_2 release from carbonate. In addition, several types of anaerobic respiration can take place where, for example, nitrate or sulfate are used by microorganisms as electron acceptors; hence, O_2 is then not consumed as in aerobic respiration.

Precipitation and cellular uptake of phosphorus

Removal of phosphorus from the wastewater stream is a common strategy to control eutrophication. The idea is to limit this element in the ecosystem and hence starve the organisms to avoid growth and increase in biomass. In all cases, phosphorus is removed by converting the phosphorus ion into a solid fraction.

Chemical orthophosphate (PO_4^{3-}) removal uses the property of metal ions like Al^{3+}, Ca^{2+}, Fe^{2+}, Fe^{3+}, or Mg^{2+} to effectively react with phosphorus and form stables precipitates, under specific sets of pH. These ions may be naturally present in some soils and, hence, the phosphate will be adsorbed to surfaces. Alternatively, chemicals containing these ions can be added to the WWT system and the precipitate formed mechanically removed after having settled. Not only phosphorus is affected by the chemical addition, the pH may also change and the content of organic matter in the water may be reduced. Both these events will affect the microbial activity in the system.

An alternative to chemical precipitation is to employ plants, macrophytes, microalgae, or bacteria, or combinations of these, to concentrate the phosphorus. All cells need phosphorus and the uptake of this element is part of the natural metabolism. Phosphorus is an essential component of nucleic acids and phospholipids are located in the various cell membrane systems. In addition, the pH of the cell is regulated by a phosphate buffer system. Therefore, phosphorus is needed in high quantities and the cell normally constitutes 1–3% phosphorus per gram dry matter. To achieve real removal of phosphorus the produced biomass must be harvested.

In the activated sludge process under certain conditions, it may be possible to enhance the storage capacity of highly energy rich polyphosphate by the bacterial biomass. Under anaerobic conditions in the WWT reactor principally acetate, but also other volatile fatty acids (i.e., fermentation products) are taken up and incorporated in biopolymers like poly β-hydroxyalkanoate (PHA) or glycogen (**Figure 3**). In the anaerobic stage the level of polyphosphate in the cell decreases while at the same time soluble phosphate is released. When conditions are changed to aerobic and carbon-poor, the stored reserve of PHA is used as an energy and carbon source for uptake of even larger amounts of phosphorus than previously released to the system. The concentration of phosphorus in polyphosphate-accumulating (PAO) bacteria can then be increased up to >15%. In the end of a successful process the buoyant density of the sludge should have increased. The polyphosphate forms dense granules that can be stained and easily observed under the microscope. The ecological mechanisms selecting for polyphosphate-accumulating organisms are not clearly understood. Originally strains of *Acinetobacter* were thought to be the key players in the process. The role of *Acinetobacter* has been argued against as recent molecular biology-based tools for identification of bacteria have demonstrated that other bacteria, for example, *Rhodocyclus* spp., *Dechloromonas* spp., and *Tetrasphaera* spp., may dominate the polyphosphate-accumulating community.

Nitrogen Transformation Processes

In microbial ecosystems nitrogen is of special interest, as it can exist in several oxidation levels ranging from ammonium/ammonia (−III) to nitrate (+V). Moreover, the transitions of nitrogen, both oxidation and reduction, are mediated mostly by microorganisms and, in particular, bacteria. When transformed, the nitrogen compounds may serve as building blocks in the cell, as energy sources, or as a way of dumping electrons.

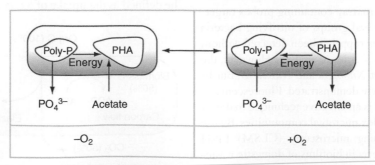

Figure 3 Release and uptake of phosphorus by polyphosphate-accumulating bacteria under varying oxygen status. Shaded boxes represent bacterial cells, Poly-P is polyphosphate, and PHA is poly β-hydroxyalkanoate.

Mineralization and immobilization

Virtually all microorganisms can mineralize and immobilize nitrogen, and the processes are more or less independent of oxygen. Proteins and nucleic acids, being the two dominating macromolecules in the cell, contain nitrogen as an essential component. Thus, most organic matter contains at least some nitrogen. By predation or after cell death and lysis the nitrogen-containing molecules will be released (**Figure 4a**). However, due to their molecular size, they cannot be directly taken up and immobilized by new bacteria. Growing bacteria exudes so-called exoenzymes that attack and degrade the macromolecules into smaller portions: amino acids and ammonia that can be transported through the cell membrane. The fate of the nitrogen part will depend on the nitrogen and carbon status of both the cells and the environment. In a carbon-rich environment with high ratios of carbon to nitrogen (>20) all nitrogen will be assimilated, that is, immobilized in the cell. If the ratio is low (<10), nitrogen will be mineralized and released to the environment. The release of ammonium might also lead to an increase in pH due to the alkaline properties of ammonia (NH_3).

Nitrification

In lithotrophic nitrification, ammonia is stepwise oxidized first via hydroxylamine (NH_2OH) to nitrite (NO_2^-) and then further to the end product nitrate (NO_3^-) (**Figures 4b** and **4c**). The two steps are carried out by two groups of specialists within the bacterial family Nitrobacteriaceae. Earlier *Nitrosomonas europaea* and *Nitrobacter* spp. were thought to be the ammonium and nitrite oxidizers in WWT systems. Molecular tools for analysis of the *amoA* gene present in all ammonia oxidizers have lately revealed a wide variety of different proteobacterial ammonia oxidizers (AOBs, ammonia-oxidizing bacteria) in nitrifying WWT processes. In addition, it seems that *Nitrospira*-like microorganisms and not *Nitrobacter* spp. are the dominating nitrite oxidizers. Through oxidation of the mineral nitrogen the bacteria derive energy for growth, that is, fixation of carbon dioxide into their biomass. In addition, most nitrifiers are strict aerobes, that is, they are completely dependent on oxygen in their respiration. Being both lithotrophic and autotrophic, they have a complex cell machinery including an extensive system of internal membranes, leading to both slow growth and sensitivity to environmental disturbances. One consequence of disturbance on the AOBs can be the formation of nitric or nitrous oxide that will be emitted to the atmosphere. In addition, AOBs produce protons which will lower the pH somewhat.

Denitrification

Denitrification is the anaerobic respiration process in which nitrogenous oxides, principally nitrate and nitrite, are used as terminal electron acceptors and, hence, reduced into the gaseous products nitric oxide (NO), nitrous oxide

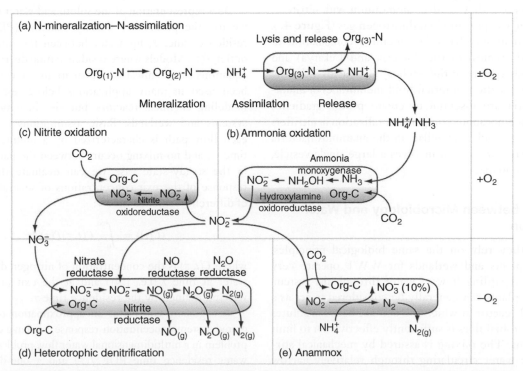

Figure 4 Five microbial nitrogen transformation processes leading to nitrogen removal in the ecosystem. For each process its relation to the environmental oxygen status is given. Shaded boxes represent microbial cells.

(N₂O), and dinitrogen (N₂) (**Figure 4d**). The process is controlled by factors such as pH, supply of organic carbon and mineral nitrogen, and aeration. Normally, dinitrogen dominates the end product but under conditions not optimal for the organisms nitrous oxide can constitute a considerable fraction. Nitrous oxide contributes to the global warming as well as to the depletion of the ozone layer in the stratosphere. Denitrifying capacity is represented within most taxonomical and physiological groups of bacteria. Denitrifiers are facultative anaerobes, that is, they prefer oxygen as terminal electron acceptors in their respiration but upon depletion of oxygen they rapidly switch into the use of a nitrogenous oxide. Denitrifiers, of which most are organotrophs and heterotrophs, are known to use only the easily available fraction of the organic matter. It should also be mentioned that some bacteria like *Paracoccus denitrificans* have been found to denitrify under aerobic conditions. In addition, the AOB *Nitrosomonas europea* may, under certain conditions, denitrify aerobically, resulting in dinitrogen gas and nitrite as end products. Conventional nitrogen removal techniques are based on a combination of autotrophic nitrification and denitrification.

Anaerobic *ammonium oxidation*

The nitrogen pathways in bacteria are more complex than earlier thought. It was recently discovered not only that some ammonium oxidation bacteria can denitrify, but also that lithoautotrophic bacteria belonging to the phylum Planctomycetes can perform anaerobic ammonium oxidation (anammox); they oxidize ammonium with nitrite as the electron acceptor to yield dinitrogen gas (**Figure 4e**). Lipids of anammox bacteria contain a combination of ester-linked (typical of the Bacteria and Eukarya) and ether-linked (typical of the Archea) fatty acids. Like the autotrophic nitrifiers, internal lipid membranes of anammox bacteria are essential to create proton gradients. While the membranous compartmentalization in nitrifiers is arranged as stacked lamellae, in the anammox bacteria the compartmentalization involves a large single vesicle, called anammoxozome.

Coupling between Microbiology and Water Circulation

Although they rely on the same biological principles, treatment plants and wetlands for WWT operate very differently regarding flow of water through the system. Treatment plants can generally be considered to act as a well-mixed reactor in which the contact between solutes and the microbial flocs is sufficiently effective not to limit the reactions. The mixing is assured by mechanical stirring in the water circulating through relatively narrow and long stretched basins. Because wetland systems rely on natural mixing of the water, it is essential to have an

appropriate planning of the wetland shape, bottom bathymetry, and placement of vegetation – factors that control the water circulation on its way through the wetland.

A canalized flow through the center of a wetland pond with vegetation on the side along the flow channel implies a separation of the main flowing water from the host environment for biofilms, that is, the vegetation stems. Introducing deep zones in the bottom or vegetation zones transverse to the flow direction can counteract this. Both measures even out differences in pressure head of the water flow, which should provide a more uniform flow and better utilization of the entire wetland volume. Generally, it is also better to place the in- and outlet in such a way that the water flow is as stretched out as far as possible. This can be arranged also by introducing hanging separation walls or dams.

Reaction Kinetics in Biological Treatment Systems

Coupling of water circulation and microbiology

The reactions of phosphorus and nitrogen in treatment systems involve kinetic processes related both to the timescale (on the order of seconds) for circulation of solutes and water as well as the kinetics of the biological processes and sorption. This section gives a general theoretical basis by which we can describe the interactions of processes in a manner that principally links the scientific basis of the individual mechanisms with the gross system response.

As a representation of the solute and water circulation we use the probability density function (PDF) of the residence time, τ, for water between the inlet and the outlet, $f(\tau)$. Models where residence time distribution and equations for chemical transformations are coupled have been used in many applications before, especially for modeling chemical reactors, but also in natural stream systems and wetlands. Basic assumptions include that each flow path is characterized by a unique residence time, τ, and no mixing occurs between the paths. Hence, in the steady-state case we can evaluate the average response of the exit concentrations of several pathways of different residence time τ:

$$\langle C(t) \rangle = \int_0^\infty C(t,\tau) f(\tau)\, \mathrm{d}\tau \qquad [1]$$

where $C(t,\tau)$ is the concentration of nitrogen dissolved in water at time t (i.e., [N]) and at the exit point of the wetland defined by the residence time τ.

Recognizing eqn [1] as an approximation of the integrated system concentration response, we have divided the problem in a multidimensional water flow problem to define water residence time PDF $f(\tau)$ and a one-dimensional problem of the nitrogen concentration response along each flow path $C(t,\tau)$. For a well-mixed reactor of a wastewater treatment plant (WWTP, see below), the all-water parcels stay equal time expressed by the 'nominal' water residence (detention) time, that is, the ratio between flow volume V and discharge Q. Mathematically, this means that $f(\tau) = \delta(\tau - V/Q)$, where δ is the Dirac delta function. Hence, eqn [1] yields $\langle C(t) \rangle = C(t, \tau - V/Q)$, suggesting that the average response is given by a single value of the response concentration curve associated with a residence time $\tau = V/Q$. Specifically, for a steady-state reactor with constant input and external constraints, we have $\langle C \rangle = C(\tau - V/Q)$.

Because of the complicated mixing conditions of treatment wetlands, the nominal residence time is often a poor approximation. Open water flow in two dimensions basically follows the so-called Saint-Venant equations, that is, the depth-averaged form of the momentum equation. However, generally scientists and engineers neglect inertia effects in the analysis of wetland flows, which yields a significantly simpler mathematical statement for the flow problem. A main problem is to relate the friction losses to, for example, the distribution of vegetation, in particular, since this controls the flow pattern and solute mixing in wetlands. In addition to a passive advection with the flow, solutes undergo mixing such as dispersion and exchange with stagnant zones in vegetation and bottom sediments. This leads to a calculation procedure, demonstrated in **Figure 5**, that results in a physically-mathematically based estimation of the flow residence times.

As a simpler alternative to represent the flow residence times utilizes the fact that the residence time distribution for treatment wetlands has been found to generally follow that of an idealized system consisting of M tanks in a series with the same residence time in each tank:

$$f(\tau) = \frac{(M/\langle \tau \rangle)^M}{(M-1)!} \tau^{M-1} e^{-\tau M/\langle \tau \rangle} \qquad [2]$$

where $\langle \tau \rangle$ is the expected value of the residence time. The lack of physical basis of the model implies that the number of compartments M needs to be determined explicitly from comparison of the model prediction and results of tracer injections in the wetland such as those shown in **Figure 5d**. Generally, it has been found that the M value falls around 3. A main advantage of the functional form is that it can easily be utilized in eqn [1].

Coupled enzyme kinetics and bacterial growth

Both nitrogen- and phosphorus-removal processes described earlier involve several reaction steps and controlling factors that are nested in a complex manner. This is why the treatment process often is described using system analysis and automatic control. A strong characteristic in most biologically catalyzed reactions, however, is the fundamental limitation on the process of the substances carbon, phosphorus, and nitrogen. For both traditional treatment plants and treatment wetlands, the total nitrogen in water is often assumed to follow a reduction with a type of Michaelis–Menten enzyme kinetics according to

$$\frac{dC(\tau)}{dt} = -\frac{qX}{K + C(\tau)} C(\tau) \qquad [3]$$

in which q is the specific bacterial activity, X is the number of bacteria, and K is a saturation coefficient or a critical

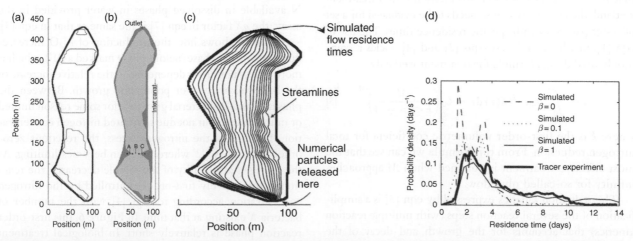

Figure 5 Figures exemplifying the procedure with which a physically based flow model can be used to derive the water residence time distribution in a treatment wetland pond. (a, b) Measured bathymetry in the pond as well as an estimated vegetation distribution. This information is used as input for the flow modeling that consists of two steps, one in which the water surface elevation and flow velocities are calculated (not illustrated here), and one in which numerical particles are released at the pond inlet to determine the streamlines and flow residence times (c). The simulated flow residence times can be compared with flow residence times determined from tracer experiments (d). Partly reprinted after Kjellin J, Wörman A, Johansson H, and Lindahl A (2007) Controlling factors for water residence time and flow patterns in Ekeby treatment wetland, Sweden. *Advances in Water Research* 30(4): 838–850.

concentration representing conditions when the bacteria are in a transition between limiting and nonlimiting state. The original Michaelis–Menten equation assumes that qX is constant, whereas here we assume that the number of bacteria can vary with time. For a constant number of bacteria, K can be seen as a constant saturation concentration. Equation [3] approaches the zero-order form $dC/dt = -qX$ for nonlimiting conditions (i.e., $C \gg K$) under which the nitrogen concentration is not important for the reaction. Only the amount of enzymes (number of bacteria) controls the reaction. Under limiting conditions (i.e., $C \ll K$), which is the main interest for treatment wetlands, eqn [3] approaches the first-order form $dC/dt = -(qXC)/K$ that reflects a control on the reaction of the nitrogen concentration. The ratio $(qX)/K$ is the denitrification rate coefficient.

If there is steady state in reaction-controlling factors, like the number of bacteria, and we have nitrogen-limiting conditions, the nitrogen concentration decays with residence time as

$$C(\tau) = C_0 e^{-k\tau} \qquad [4]$$

where C_0 is the initial concentration.

The activated sludge process in a WTP is run under carbon excess. By controlling the removal of the activated sludge, the concentration of bacteria is kept on a fairly constant level. Further, the circulation of wastewater ensures approximately a constant residence time that balances the nitrogen-reducing processes between nitrogen-limiting and -nonlimiting conditions. In treatment wetlands with carbon excess due to sufficient carbon production in plants, we can normally assume nitrogen-limited conditions.

In the activated sludge there is just a single residence time defined by the ratio between reactor volume and discharge, that is, $\tau = V/Q$ (see section 'Respiration'). In a treatment wetland, the average response needs to be evaluated for a set of water pathways with specific residence times. The use of eqn [1], in which we insert eqns [2] and [4], yields a commonly used design formula for treatment wetlands:

$$C_{\text{out}} = \int_0^\infty C_0 e^{-k\tau} f(\tau)\, d\tau = C_0 \left(\frac{M}{k \langle \tau \rangle + M} \right)^M \qquad [5]$$

where k is the first-order volumetric coefficient for total nitrogen reduction. From this formula we can see that the most effective treatment is obtained when M approaches infinity for so-called plug flow.

The reduction kinetics expressed by eqn [3] is a simplification of the several reaction steps (with multiple reaction kinetics) that accounts for the growth and decay of the microbial communities. In 1949 Monod proposed that the bacterial growth rate unlimited by environmental factors other than the substrate, such as concentration [N], follows

$$\frac{dX}{dt} = \mu_{\max} \frac{S}{S + K_s} X \qquad [6]$$

in which S is substrate concentration (carbon concentration), K_s is the substrate saturation coefficient, and $\mu_{\max}$ is the maximum population specific growth rate constant. The linear, first-order differential equation for bacterial number means that the bacteria grows exponentially with time as long as the growth process is unlimited. A change of the nitrogen load to a biological treatment system, thus, leads to a change in the bacterial community and in the denitrification rate coefficient. Sometimes one can see variants of the Monod growth rate formulation involving the nitrogen concentration in another factor similar to the factor in which substrate concentration is included in eqn [6].

Measurements of denitrification rates

Laboratory experimental techniques for determination of potential denitrification activity (PDA) are based on inhibiting the final denitrification step in which N_2O transforms to nitrogen gas. The PDA assay is prepared with an excess of carbon (e.g., glucose) and nitrate sources as well as inhibiting acetylene (C_2H_2). The preparation of the experiment implies, however, a growth of the bacterial population according to eqn [6] as long as this growth is not limited by the availability of nutrients. For relatively high nitrogen concentration in the substrate the production in terms of $[N_2O\text{-}N]$ is not limited by the solute concentrations, but only by bacterial growth. The corresponding zero-order reaction can be written as

$$\frac{d[N_2O\text{-}N]}{dt} = qX \qquad [7]$$

where $[N_2O\text{-}N]$ denotes the nitrogen concentration in the form of N_2O. Since the rate of mass production of $N_2O\text{-}N$ is assumed to be the same as the rate of mass reduction of total N available in dissolved phases in water provided in the assay, the qX factor in eqn [7] is the same as that in eqn [3].

Figure 6 shows how the production of N_2O increases during an initial phase (before phase marked as '1'), which is more or less nonlinear depending on the relative increase of the bacterial activity or population growth. Between the phases '1' and '2', bacterial growth is for some reason limited or much slower, but not due to limited nitrogen concentration. Because of the nitrogen excess, the reactions zero-order up to phase '2' where nitrogen becomes limiting. As the nitrogen availability of the sample decreases, the reaction is successively first-order-controlled by the nitrogen concentration according to eqn [4] with the number of bacteria X existing at that time. In **Figure 6**, this first-order reaction phase is relatively short. In biological treatment systems, the bacteriological composition reaches such an equilibrium in the bacteria populations after a short while due to limitation of carbon and/or nitrogen and follows the second phase of the experimental results shown in the figure. The PDA associated with the treatment system is defined from the initial slope of the N_2O production versus

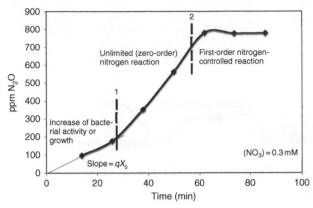

Figure 6 Example of laboratory measurement of kinetics in N_2O production with initially unlimited conditions in nitrate and carbon involving a first-order growth of the bacterial population and enzyme kinetics. The initial phase – before dashed line marked with '1' – is controlled by increased bacterial activity or growth. Thereafter, the number of bacteria is constant due to the linear increase of N_2O production. As nitrogen becomes limiting, the reaction is succeeded by nitrogen deficiency and a first-order production rate. This result was obtained from a sediment sample taken in Ekeby wetland, Eskilstuna, Sweden.

time relationship, since the initial amount of bacteria X_0 in the sample is believed to represent the state of the treatment system in which the sample was taken.

At the transition to nitrogen-limited conditions, the product qX is the time rate of reaction of nitrogen in units mass per unit time and volume, so the ratio qX/K is the denitrification rate coefficient, where K is a limiting nitrogen concentration at the transition.

A batch reactor undergoes all stages represented in **Figure 6**, whereas the continuous flow system is generally kept at steady state (see section entitled 'Continuous flow systems and SBRs'). In treatment wetlands with sufficient carbon supply from moldering vegetation, denitrification is nitrogen limited and first-order controlled by the nitrogen concentration. The denitrification rate coefficient is given by qX_0/K.

Biological Wastewater Treatment Systems

Treatment Wetlands

Classification of wetlands

There are several types of natural wetlands such as swamps, fens, bogs, marshland, and tidal freshwater areas. Swamps and marshes have open flowing water and are distinguished in terms of vegetation, soil type, and wild life. Mires such as fens and bogs are mainly subsurface wetlands with little open water. Bogs are isolated hydrological units that receive water only through precipitation, whereas fens have through flowing water. As water goes through such wetland areas it undergoes

great chemical transformation. Both nutrients and elements like heavy metals that attach chemically (sorb) on solid surfaces are effectively removed such that water reaches a status corresponding to 'natural' water quality.

Constructed wetlands or treatment wetlands are usually built where natural wetland conditions can be found and are, therefore, to some extent modifications of a natural system. By introducing dams and canals, however, it is possible to provide proper water depth for carbon providing vegetation species, like common reed (*Phragmites*), and a separation of oxygen conditions. In some cases, wetlands can be built in clay strata or artificially sealed using clay even if there is no natural groundwater reaching the ground surface. Hence, leakage through infiltration is an essential problem that needs to be accounted for in the design.

Constructed wetlands are commonly divided into subsurface flow (SSF) and surface flow or free water surface (FWS) wetlands. Both types are used for treating domestic, municipal, and industrial wastewater. In particular, these systems can be useful for treating landfill leachates, agricultural runoff, and wastewater from minor communities. SSF wetlands are commonly used as a polishing step after conventional treatment plants for municipal wastewater. SSF systems are often favored in minor communities due to the soil cover of possibly contagious wastewater. Subsurface systems require separation of solid material in the wastewater before solute fractions are led into a sand filter or other soil layer in which phosphorus is removed through sorption to the particulate matrix and nitrogen to denitrification supported by soil bacteria (**Figure 7**).

Because of the large discharge capacity surface flow wetlands are usually preferred as polishing step for municipal wastewater. Phosphorus is generally effectively removed in the treatment plant, whereas nitrogen treatment requires longer detention times that are provided in the wetland.

Both FWS and SSF wetlands used for treating municipal and industrial wastewater are designed with an area of *c.* 5–10 m^2 per person equivalent.

Functionality of FWS wetlands

Vegetation is important in surface flow wetlands to provide carbon for supporting denitrification, to offer host environment for biofilms that grow on stems, and to cause friction losses for flow water, which can be utilized to provide a beneficial flow pattern. Submersed vegetation also controls the oxygen level in the water.

Vegetation in a recently established wetland changes with time and this leads to a relatively long period (years) to approach equilibrium in the wetland ecosystem and the interacting treatment processes. In cold climates, the effectiveness of treatment wetlands also varies over the year, but there is a notable effect even during the winter.

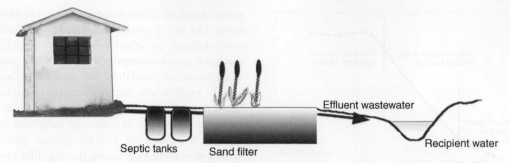

Figure 7 Typical subsurface treatment system for a single household including septic tanks and downstream sand filter.

The shores of wetland ponds in cold climates can be populated by reed sweet grass (*Glyceria maxima* L.), common reed (*Phragmites australis* L.), and cattail (*Typha latifolia* L.). Examples of submersed vegetation include slender waterweed (*Elodea nuttallii* L.), sago pondweed (*Potamogeton pectinatus* L.), coontail (*Ceratophyllum demersum* L.), and spiked watermilfoil (*Myriophyllum spicatum* L.). Coontail forms dense layers of vegetation that can be considered to be a porous medium for the water flow with a large inner surface available for biofilms.

A main role of treatment wetlands that are constructed as a polishing step after a conventional treatment plant is to remove nitrogen through denitrification in biofilms. Generally, the ordinary treatment process has included oxygenation of the water, which transforms most nitrogen fractions, like ammonium and nitrite, to nitrate before it enters the wetland. Biofilms grow both on vegetation stems and in the bottom sediments. Therefore, an important factor is the exchange rate for solute substances between flowing channels in the wetland with bottom sediments and vegetation zones. The potential for denitrifications in the host environments for biofilms, like bottom sediments, is usually significantly higher than actual rates on the scale of the entire system, because of the difficulty to arrange an effective contact between water and biofilms.

Functionality of SSF wetlands

The advantage of SSF wetlands is that the water is present below the ground surface, which decreases odors and the risk for public exposure for possible contagious bacteria. The construction usually includes a sand and/or gravel bed with supporting emergent vegetation such as cattail (*Typha*) and reeds (*Phragmites*). The systems are designed with aspect ratio ($L{:}W$) of about 15:1 and a flow velocity in the order of centimeters to decimeters per day.

A typical design layout for a single household is shown in **Figure 7**. A first step usually involves separation of coarse fractions of the wastewater in deposition basins, or, such as in this case, in septic tanks. This produces wastewater that can percolate and flow through the sand filter

without rapidly clogging the pores of the filter and end its lifespan too fast.

The active processes include mechanical filtering of particulate (organic) matter in the porous material, sorption of phosphorus and heavy metals to the solid matrix, and nitrogen decomposing reactions caused by nitrifying and denitrifying bacteria in the upper soil layer. Good performance is commonly reported for the removal of biological oxygen demand (BOD), total suspended solids (TSS), phosphorus, and nitrogen.

WWTPs – The Activated Sludge Process

General

Generally, WWT systems containing compartmentalized reactors (basins or tanks) for their performance often are termed a WWTP. In addition, the flow of wastewater through such systems is thoroughly controlled and optimized. The WWTP may consist of a mechanical, chemical, and biological step. In the mechanical step, heavy solid particles are allowed to settle at the bottom and light material floating on the water surface is removed. In the chemical step metal salts are added to precipitate phosphorus. Phosphorus removal can be performed at different stages in the treatment process: prior to, simultaneous with, or after the biological step and are hence called preprecipitation, coprecipitation, or postprecipitation, respectively. The biological step can be performed according to either of two basic principles. The reactor may contain solid surfaces to support bacterial growth and the development of a biofilm (trickling filters, rotating biological contactors, and fluidized beds) (**Figure 1b**). The other approach is to allow bacterial growth in the water body supported by natural occurring suspended solids (activated sludge process) (**Figure 1a**).

The activated sludge process can be designed as either a continuous flow system or as SBRs. Both systems normally include an aerated biological nutrient removal step followed by settlement of produced sludge. The difference between the systems is that in the continuous process these processes take place in two different reactors, whereas in the SBR process they occur sequentially in the same reactor.

Continuous flow systems and SBRs

In the conventional continuous flow system primary treated wastewater is conveyed into an aerated basin (**Figure 8a**). The feed of wastewater and supply of compressed air can be done in many ways, from being introduced at one end leading to gradients of oxygen and substrate throughout the basin to being introduced at several points giving a more homogeneous environment. Moreover, during its way through the basin, the water may be led through more or less open compartments. In a completely mixed process, the water is also circulated within the basin. The effluent is led to a clarifier to allow particles to settle before the clear phase leaves the process. The settled excess sludge containing a viable biomass is removed and treated separately; however, some is recycled to reinoculate the process. This procedure will ascertain stable function of the unit. The whole concept is not unlike the continuous culturing of microorganisms in the laboratory or in many industrial processes.

The operation of one or more SBRs in a series consists of a sequence of fill-and-draw cycles. Each cycle typically consists of a number of separate operational phases of fill, react, settle, draw, and idle. Hence, after the react phase, that is, growth phase, the produced biomass is allowed to settle and the clear treated supernatant can be removed. The process resembles that of batch-culturing of bacteria in the laboratory.

The biological process

In the biological step of the activated sludge process, suitable mixing of the water is necessary to allow suspended solids, air, nutrients, and microorganisms to make intimate contact. Based on the mixing regimes, plug flow or completely mixed systems can be differentiated. Large volumes of air are blown into the reactor tanks from beneath to achieve effective mixing and support the aerobic microorganisms with sufficient oxygen for respiration. The concentration of dissolved oxygen (DO) should be kept at approximately 2 mg l^{-1}. As described above, the

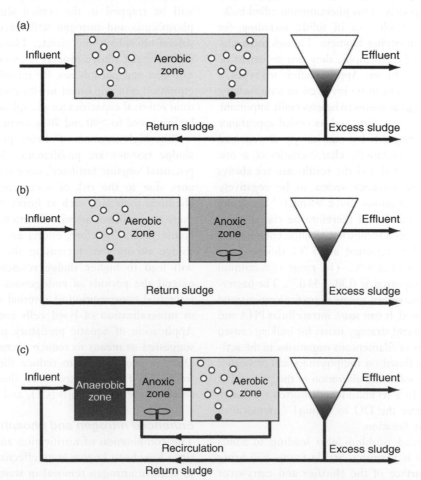

Figure 8 Examples of conventional activated sludge processes (a) without and (b) with nitrogen removal capacity, and (c) with capacity of both biological nitrogen removal and enhanced biological phosphorus removal.

phase transitions in the system will support an optimal environment for microbial growth. During the react phase, three-dimensional aggregates of highly active microbial communities, called flocs, are formed (**Figure 1a**). Flocs typically are 100–500 μm in diameter. New microscopic techniques such as epifluorescence and CLSM in combination with image analysis have been used to analyze the aggregates of activated sludge. Generally, four main structures can be discriminated in the flocs: (1) active and inactive microbial cells, mainly bacteria, protozoa, and metazoa; (2) extracellular polymeric substances like carbohydrates and proteins; (3) inorganic particles (sand); and (4) water. From a technical point of view, the sludge properties are essential. The forming of dense flocs with good properties for settlement will provide good operational conditions. Filamentous bacteria will always be present in a healthy process which operates normally, and which shows no symptoms of problems with bulking or foaming. Several sludge indices have been suggested to describe and characterize the sludge property.

Occasionally the proportion of filamentous bacteria will increase causing flocs of loose structure that settle only slowly and compact poorly. This phenomenon called bulking leads to uncontrollable loss of solids, including, for example, the active nitrifying biomass. Though most filamentous bacteria are heterotrophs, they are shown to be extremely difficult to cultivate. Approximately ten types of filamentous microbes seem to be involved in most bulking events. *Microthrix parvicella* seems to be especially important. The bacterium is long and thin, and its coiled appearance makes it easy to distinguish by microscopy of activated sludge samples. Only metabolic characteristics of a few isolates have been reported and the results are not always concordant. *Microthrix parvicella* seems to be negatively affected by DO concentrations above $>6 \, mg \, l^{-1}$ but grows well at $\sim0.4 \, mg \, l^{-1}$ and should therefore be considered a microaerophile. It prefers a somewhat alkaline environment and optimum growth is reported at 25 °C, though some growth was still observed at 8 °C. The range of maximum growth rates (μ_{max}) reported is 0.38–1.44 d^{-1}. The bacterium cannot utilize glucose but seems to prefer long-chained fatty acids like oleic acid. It can store intracellular PHA and lipids. No reliable control strategy exists for bulking caused by increased amounts of filamentous organisms in the activated sludge process. Based on the physiological properties of the bacterium, the following alteration of the process has been suggested to reduce its abundance: shorten the sludge retention time, increase the DO to $>2 \, mg \, l^{-1}$, removal of high lipid contents by flotation.

Another widespread problem also leading to solids separation problems is foaming. Stable foams will bring the sludge to the surface of the clarifier and carry-over of solids from the clarifier. The foam most often consists of a dense matrix of filamentous bacteria and air bubbles.

Foaming may have several causes. *Microthrix parvicella* seems to be more hydrophobic than most other bacteria in the activated sludge process and are frequently associated with foaming problems. Another group of bacteria identified in activated sludge foams is mycolic acid-producing actinomycetes. The most commonly methods for controlling foaming are the same as those for controlling bulking problems. However, the magnitude of the problem has forced the development of both physical and chemical short-term measures to control these situations.

Nutrient removal capacity

A properly controlled activated sludge process can remove very effectively the content of organic carbon, and mineralize and nitrify nitrogen. Typically, the chemical oxygen demand (COD) and BOD removal capacities for municipal wastewater are higher than 85 and 95%, respectively. The reduction of carbon is due to aerobic respiration losses, removal of settled sludge produced by biomass growth, as well as flocculation of dissolved and particulate organic matter. In addition, some 20–30% each of influent phosphorus and nitrogen will be trapped in the settled sludge; however, most phosphorus and nitrogen will leave the system as dissolved phosphate and nitrate. Thus, the basic design of the activated sludge process is less effective in reducing nitrogen and phosphorus. By introducing chemical precipitation and combined nitrification–denitrification the total removal capacities for phosphorus and nitrogen may be improved to >90 and 70%, respectively.

The high amounts of sludge produced by activated sludge systems are problematic. Although sludge is a potential 'organic fertilizer', since it is rich in plant nutrients, due to the risk of occurrence of pathogens and chemical toxicants, such as heavy metals in the sludge, there are problems associated with recycling the sludge to arable land. Therefore, efforts are made to reduce the sludge production. Increasing the periods of aeration will lead to higher sludge residence time which will extend the periods of endogenous metabolism, that is, microbial consumption of internal cell material as well as mineralization of lysed cells and particulate matter. Application of aquatic predatory oligichaetes has been suggested as means to reduce excess sludge production. One common means to reduce the amounts of sludge from WWTPs is to treat the sludge in an anaerobic reactor to produce biogas (CO_2 and CH_4).

Enhanced nitrogen and phosphorus reduction

The combination of nitrification and denitrification has since long been known as an effective biological solution to achieve nitrogen removal in wastewater. The obvious way to arrange suitable environments for the two groups of bacteria is to connect an aerobic compartment or zone prior to an anoxic in a so-called post-denitrification process (**Figure 8b**). However, since most organic matter is consumed in the aerobic zone, this setup may experience low effects due to lack of easy available energy to the denitrifiers. A more effective solution can be to place the anoxic zone prior to the oxic zone and circulate the water between the two zones. In this design, called pre-denitrification, the denitrifiers will meet both anoxic conditions and fresh organic material from the influent. Another solution is to support the denitrification with an external organic energy source. Effective denitrification has been reported with, for example, acetate, ethanol, and methanol. The response to acetate and ethanol is immediate as these molecules are part of the normal metabolic pathways of organotrophic bacteria. For effective denitrification with methanol a long period of adaptation is needed, typically several months. Only a few slow-growing specialists, for example, *Hyphomicrobium* sp., can use one-carbon compounds (CH_3OH) and the metabolic pathways are complex.

Recent developments in biological nitrogen removal techniques in combination with the discovery of novel bacteria have resulted in some new methods. By combining partial nitrification with the anammox process some nitrogen removal techniques have been set up that may consume lower resources (**Figures 4b** and **4e**). In the partial nitrification process a shortcut is taken by preventing the oxidation of nitrite to nitrate by nitrite-oxidizing bacteria. Instead the nitrite is removed directly by heterotrophic denitrification. In the single-reactor system for high ammonium removal over nitrite (SHARON), incomplete nitrification is achieved by use of the slower growth rate of nitrite oxidizers than ammonium oxidizers at higher temperatures ($>26\,^{\circ}C$). By applying higher hydraulic retention times, the nitrite oxidizer will be washed out. The nitrite thus accumulated can be removed by the anammox process in a succeeding reactor. In the anammox process nitrite is oxidized with ammonia as the electron donor. In the partial nitrification process, half the ammonium is converted into nitrite. One advantage with the process is that no extra organic energy is needed for the denitrification step. Another variation is to let nitrifiers oxidize ammonia to nitrate in a single reactor and consume oxygen to create the anoxic conditions needed by the anammox bacteria. This process is called CANON, the acronym for 'completely autotrophic nitrogen removal over nitrite'.

As both biological nitrogen removal and enhanced biological phosphorus removal need alternating cycles of aerobic and anoxic conditions, it seems logical to combine the two processes in the same WWTP. However, this is not as easy as it seems to be. In addition to alternating anoxic and aerobic regimes, the anoxic zone must be maintained completely anaerobic to provide fermentation end products like fatty acids to select for PAO bacteria. The level of nitrate in the anaerobic zone must be low; otherwise the heterotrophic denitrifiers will consume the organic molecules needed by the PAO bacteria. In the so-called three-stage PHOREDOX process, influent water is fed to an anaerobic reactor, and then conveyed to an anoxic reactor also fed with recycled activated sludge from the last aerobic reactor (**Figure 8c**). In this way less nitrate is returned with sludge from the clarifier to the head of the system. Thus, both phosphorus and nitrogen removal are accomplished by this design.

Regulation and simulation models

The activated sludge process does not only involve complex elements but also the influent wastewater characteristics vary temporarily. This emphasizes the need for thorough control and optimization to maintain and fine-tune the process performance. To describe the actual WWTP, a general model including the ensemble of an activated sludge model, hydraulic model, oxygen-transfer model, and sedimentation tank model can be used. The activated sludge model describes the biological reactions occurring in the process by a set of differential equations. In addition to use in control and optimization, a WWTP model can be used to simulate different scenarios for learning or to evaluate new alternatives for design.

Strengths and weaknesses of WWTP

In its basic design the activated sludge process has a high capacity to biologically oxidize carbon and nitrogen. In addition, this is achieved in comparable small units, that is, less space is needed, which most often is a prerequisite for WWT in urban areas. By modifying the design also high amounts of nitrogen and phosphorus can be removed by biological processes. The SBR process is both a stable and flexible activated sludge process. The biomass cannot be washed out and the possibility to handle shifts in organic and hydraulic loads is good. In addition, less equipment and operator attention are needed to maintain the SBR process.

WWT by the activated sludge process must be regarded as a highly technological process, that is, much knowledge and experience are needed to operate a system based on this technique. In the process design of activated sludge processes, much focus has been put into efficiency in nutrient removal. Although generally pathogens are acceptably removed, most WWTPs are not designed for treating pathogenic microorganisms. Moreover, the environmental selective pressure on the microbial communities probably leads to highly specialized ecosystems. Consequently, the treatment process may be sensitive to disturbances due to environmental variations such as sewage load and composition as well as influent toxicants. The costs for maintenance and care are high. The nitrogen removed from the system is left as gaseous emissions instead of using such a valuable plant nutrient in crop production. In addition, the plant-nutrient-rich sludge may contain heavy metals as well as anthropogenic organic pollutants that may pose a risk to the ecosystem and must therefore most often be deposited or possibly incinerated.

Finally, the activated sludge process most likely is a WWTP technique that will also prevail in the foreseeable future. Process designs are continuously evolving to meet the demands of upcoming wastewater types, improved performance, and less resource consumption.

Perspective on Biological Wastewater Treatment

Originally, organized WWT was introduced for sanitation reasons. Today, in the industrialized world, WWTPs and arable land contribute with a substantial proportion to the anthropogenic nitrogen load to the marine recipients, which severely enhances eutrophication of aquatic environments. Most natural ecosystems are controlled by a deficiency in macronutrients like phosphorus and nitrogen, which means that eutrophic level often directly controls ecosystem responses. This interplay stresses the importance that WWT systems are adapted to natural biogeochemical cycles and are aligned with a vision of a durable society.

An important question is to what extent wastewater, for example, municipal wastewaters and sewage sludge, should be considered a waste or valuable resource and recycled as plant nutrients in crop and in energy production. Key constraints for the growing global population are due to food and energy. Today, both extraction of phosphorus and production of mineral nitrogen fertilizers consume extensive resources of fossil fuels. Hence, one important future aim must be to create a sustainable loop of plant nutrients through food production and refinement, urban consumption, waste handling, and back to arable land. To achieve this, the effluent wastewater stream must contain as much phosphorus and nitrogen as possible in addition to minimal amounts of organic and inorganic toxicants.

Such global aims have to be linked with the ability to treat a growing amount of wastewater. Not only is it important to select specific solutions for specific treatment situations, but it will also be essential to be able to optimize treatment with account to the broad scientific basis involving both water dynamics and biological processes. The coupled scientific basis is essential for an in-depth understanding of the key microbiological processes involved in nitrogen removal and for optimizing biological treatment systems.

Another future perspective is the contribution of treatment wetlands to maintain biological diversity in

the ecosystem as well as to create easy accessible recreational and educational meetings between urban citizens and the ecosystem. Most importantly, this would create awareness of the waste stream as a resource and probably encourage the citizen to contribute to this idea.

Further Reading

Ahn Y-H (2006) Sustainable nitrogen elimination biotechnologies. *Process Biochemistry* 41: 1709–1721.

Bolster CH and Saiers JE (2002) Development and evaluation of a mathematical model for surface-water flow within Shark River Slough of the Florida Everglade. *Journal of Hydrology* 259: 221–235.

de-Bashan L-E and Bashan Y (2004) Recent advances in removing phosphorus from wastwater and its future use as fertilizers (1997–2003). *Water Research* 38: 4222–4246.

Garnaey KV, van Loosdrecht MCM, Henze M, Lind M, and Jørgensen SB (2004) Activated sludge wastewater treatment plant modelling and simulation: State of the art. *Environmental Modelling and Software* 19: 763–784.

Gilbride KA, Lee D-Y, and Beudette LA (2006) Molecular techniques in wastewater: Understanding microbial communities, detecting pathogens, and real-time processes. *Journal of Microbiological Methods* 66: 1–20.

Hughes J and Heathwaite L (1995) *Hydrology and Geochemistry of British Wetlands*. London: Wiley.

Juretschko S, Loy A, Lehner A, and Wagner M (2002) The microbial community composition of a nitrifying–denitrifying activated sludge from an industrial sewage treatment plant analyzed by the full-cycle rRNA approach. *Systematic and Applied Microbiology* 25: 84–99.

Kadlec RH and Knight RL (1996) *Treatment Wetlands*. New York: CRC Press LLC.

Kelly JJ, Siripong S, McCormack J, et al. (2005) DNA microarray detection of nitrifying bacterial 16S rRNA in wastewater treatment plant samples. *Water Research* 39: 3229–3238.

Kjellin J, Wörman A, Johansson H, and Lindahl A (2007) Controlling factors for water residence time and flow patterns in Ekeby treatment wetland, Sweden. *Advances in Water Research* 30(4): 838–850.

Levenspiel O (1999) *Chemical Reaction Engineering*. New York: Wiley.

Liwarska-Bizukokc E (2005) Application of image techniques in activated sludge wastewater treatment processes. *Biotechnology Letters* 27: 1427–1433.

Rossetti S, Tomei MC, Nielsen PH, and Tandoi V (2005) '*Microthrix parvicella*', a filamentous bacterium causing bulking and foaming in activated sludge systems: A revew of current knowledge. *FEMS Microbiology Reviews* 29: 49–64.

Schmidt I, Sliekers O, Schmidt MS, et al. (2003) New concepts of microbial treatment processes for the nitrogen removal in wastewater. *FEMS Microbiology Reviews* 27: 481–492.

Seviour RJ and Blackall LL (eds.) (1999) *The Microbiology of Activated Sludge*. Dordrecht: Kluwer Academic Publishers.

Van Niftrik LA, Fuerst JA, Sinninghe Damsté JS, et al. (2004) The anammoxosome: An intracytoplasmic compartment in anammox bacteria. *FEMS Microbiology Letters* 233: 7–13.

Wagner M and Loy A (2002) Bacterial community composition and function in sewage treatment systems. *Current Opinion in Biotechnology* 13: 218–227.

Boreal Forest

D L DeAngelis, University of Miami, Coral Gables, FL, USA

Introduction
Climate and Soils
Forest Structure and Species
Animals

Biodiversity
Ecosystem Dynamics
Conservation and Global Issues
Further Reading

Introduction

The boreal forest biome is also referred to as the 'taiga' (Russian for 'swamp forest'). Geographically, the boreal forest is located between latitudes 45° and 70° N, and virtually all of it is in Canada, Alaska, and Siberia, with portions in European Russia and Fennoscandia. The boreal forest is bordered on the north by treeless tundra and on the south by mixed forest. The boreal forest is termed a 'biome' by ecologists, a term that refers to a biogeographic unit that is distinguished from other biomes by the structure of its vegetation and dominant plant species. A biome is the largest scale at which ecologists classify vegetation. All parts of a biome tend to be within the same climatic conditions, but because local conditions differ, a biome may encompass many specific ecosystems (e.g., peatlands, river floodplains, uplands) and plant communities. Despite this diversity within a biome, in referring to the boreal forest we will here use the terms 'biome' and 'ecosystem type' interchangeably.

Climate and Soils

The climate of the boreal forest is continental and, importantly, for the growing season, there tends to be between 30 and 150 days of temperatures above 10°C. Temperature lows can fall below −25°C. Average annual precipitation is 38–50 cm, with the lowest amounts in the northern boreal forest, and greater frequency of precipitation during the summer season. Water is seldom limiting because of the generally flat topography and low rate of evaporation.

Permafrost can occur in the northern parts of this zone, the southern limit coinciding roughly with a mean air temperature of −1°C and snow depth of about 40 cm. The zone of permafrost generally starts at depths ranging from 1.5 to 3 m in the areas of the boreal forest where it occurs. Its occurrence limits soil processes to an upper active layer and impedes water drainage, leading to waterlogged soils. The soil decomposition rate in the taiga is slow, which leads to the accumulation of peat.

Several soil types characterize the boreal forest. The soils of a major part of the boreal forest, under a dense coniferous canopy, are heavily podzolized where the soil is permeable, and so it consists largely of Spodosols. Intense acid leaching forms a light ash-colored eluvial soil horizon leached of most base-forming cations such as calcium. Thus taiga soils tend to be nutrient poor. Gelisols are common in the north, where permafrost occurs. These are young soils with little profile development. Histosols, which are high in organic matter, form in non-permafrost wetlands, where decomposition is slowed by hypoxic conditions. These are often referred to as peatlands.

Forest Structure and Species

Because many hardwood trees are both sensitive to low winter temperatures and require a long and warm summer, the true boreal forest begins where the few remaining hardwoods become a minor part of the forest. Four coniferous genera dominate a major part of the taiga: *Picea* (spruce), *Abies* (fir), *Pinus* (pine), and *Larix* (larch). The hardwoods, which largely occur in dwarf form, include *Alnus* (alders), *Populus* (poplars), *Betula* (birches), and *Salix* (willows). The hardwoods tend to be early successional species following disturbances such as fires or erosion/deposition processes on riverbanks, which are eventually shaded out by slower-growing spruces and firs. Much of the main boreal forest is dominated by a few spruce species. These form a dense canopy in the central and southern taiga, with a ground cover of dwarf shrubs, such as cranberries and bilberries, and mosses and lichens. In northern Siberia, huge areas are covered almost solely by larch, and the canopy is much less dense. Pine species, which can withstand a range of harsh conditions, grow in light, sandy soils and other dry areas. As the boreal forest-tundra boundary is approached, conifers thin out to a woodland, with lichen and moss dominating the ground. Trees become more and more stunted.

The standing stock of biomass of the boreal forest ranges is estimated at 200 (range 60–400) metric tons per hectare (t ha^{-1}). This compares with an estimate of 350 t ha^{-1} for the temperate deciduous forest and 10 t ha^{-1} for the tundra ecosystems. The boreal forest differs from the temperate forest in having a much higher percentage of its total biomass as photosynthetic foliage (7% vs. 1%). It differs from the tundra in having a lower percentage of root biomass (22% vs. 75%).

Animals

Animal life in the boreal forest is far less diverse than in most temperate zone ecosystems. One component of the taiga fauna, conspicuous for its frequent devastating effects on thousands of hectares of forest, is that of phytophagous insects. Populations of these insects, which include pine sawflies, spruce budworms, bark beetles, and many others that attack conifers, are capable of escaping natural enemies and building up to huge population densities. The large monospecific stands of the boreal forest may be especially vulnerable. The high numbers of insects during the warm months is a main explanation for the large numbers of birds that migrate from the south to breed in the taiga, especially large numbers of species of warblers and thrushes. A number of bird species are adapted to being residents of the taiga. Grouses such as the capercaillie of the Old World, are adapted to year-round life in the taiga, as are some owls, woodpeckers, tits, nuthatches, crossbills, and crows. Small mammal herbivores of the boreal forest include the squirrels, chipmunks, voles, and snowshoe hares. These provide food for a small number of predator species, including the red fox (*Vulpes vulpes*) and members of the weasel family. The moose (*Alces alces*) (called elk in the Old World) has a wide geographic distribution in the taiga. They are prey for wolves (*Canis lupus*) and occasionally the brown bear (*Ursus arctos*).

Biodiversity

Tree species richness is far smaller than that in the temperate forests to the south, where more than 100 species are typically observed in 2.5° × 2.5° quadrats in eastern United States. Species richness clearly declines from south to north in the taiga. Whereas 40 or more tree species can be found in the southern taiga in Canada, this declines to 10 or so species near the tundra boundary. Animal species also show strong gradients. Reptile and amphibian species are almost nonexistent above 55°. Mammal species richness declines from close to 40 species to about 20 going northward in the boreal forest biome in North America, while bird species decline from about 130 to less than 100.

Ecosystem Dynamics

In keeping with its position between much warmer climate of the temperate zone and colder climate of the tundra, the boreal forest's indices of production are intermediate between those two ecosystem types. Annual net primary production in the boreal forest has been estimated at 7.5 t ha^{-1} yr^{-1} (range 4–20). This compares with 11.5 t ha^{-1} yr^{-1} for temperate forest and 1.5 t ha^{-1} yr^{-1} for tundra ecosystems. Mean boreal forest litterfall is estimated to be 7.5 t ha^{-1} yr^{-1} compared with 11.5 and 1.5 t ha^{-1} yr^{-1} for the temperature forest and tundra, respectively. Because low temperatures slow decomposition, the rate of litterfall decomposition in the boreal forest, 0.21 t ha^{-1} yr^{-1}, is also intermediate between 0.77 and 0.03 t ha^{-1} yr^{-1} for the temperate forest and tundra. This means that it takes roughly $3 \times (1/0.21) = 14$ years for 95% of a pulse of litter to decompose.

Fire is an inherent factor in the ecosystem dynamics of the boreal forest. Lightning-caused fires occur on a given area at intervals of 20–100 years in drier areas to 200+ years in wetter areas such as floodplains. Because nutrients tend to be tied up in slowly decomposing organic matter, fire may be important for maintaining tree growth by releasing pulses of nutrients periodically. Many taiga plant species have adaptations to fires, such as serotinous cones and early sexual maturity of some conifers, and resprouting capacity of hardwood trees and many herbs and shrubs. Fires also reset the successional cycle, allowing shade-intolerant species like birch and aspen to invade.

Conservation and Global Issues

The boreal forest represents the single largest pool of living biomass on the terrestrial surface (more than 30% of the total terrestrial pool), and is therefore critically important in global carbon dynamics. Much of the carbon is stored in the ground layer. Currently, the taiga is thought to act as a net sink of carbon. However, global climate change, in the form of higher temperatures, may cause significant changes in the carbon dynamics by increasing decomposition rates faster than photosynthetic rates. Fire frequencies may also increase with temperature, as precipitation is not expected to rise, which will further increase the release of carbon stored in the ground layer. According to some studies, the boreal forest will be a net contributor to CO_2 in the atmosphere under the projected climate changes.

Climate-induced changes in the boreal forest will also have an impact on migrant birds that use the region for reproduction. Changes in tree species composition may challenge the capacity of birds to adapt, as has already the increasing fragmentation of the forest due to clear-cutting in many areas within the biome.

See also: Tundra.

Further Reading

Danell K, Lundberg P, and Niemälä P (1996) Species richness in mammalian herbivores: Patterns in the boreal zone. *Ecography* 19: 404–409.

Henry JD (2003) *Canada's Boreal Forest*. Washington, DC: Smithsonian.
Hunter ML, Jr. (1992) Paleoecology, landscape ecology, and conservation of neotropical migrant passerines in boreal forests. In: Hagan JMIII and Johnston DW (eds.) *Ecology and Conservation of neotropical Migrant Landbirds*, pp. 511–523. Washington, DC: Smithsonian Institution Press.
Knystautus A (1987) *The Natural History of the USSR*. New York: McGraw-Hill.
Krebs CJ, Boutin S, and Boonstra R (2001) *Ecosystem Dynamics of the Boreal Forest: The Kluane Project*. New York: Oxford University Press.
Larsen JA (1980) *The Boreal Ecosystem*. New York: Academic Press.
Oechel WC and Lawrence WT (1985) Taiga. In: Chabot BF and Mooney HA (eds.) *Physiological Ecology of North American Plant Communities*, pp. 66–94. New York: Chapman and Hall.

Botanical Gardens

M Soderstrom, Montreal, QC, Canada

Gardens for Systematic Study
The Gardens of the Ancients
Recreating Eden
The Gardens of Discovery

Botanic Gardens in Colonies
The Intrinsic Value of Biodiversity and Nature
Education and the Future
Further Reading

Gardens for Systematic Study

Botanic gardens are gardens where plants are gathered together for systematic study. Often they imitate a number of naturally occurring ecosystems: the San Francisco Botanical Garden has created a cloud forest section while the basement of the Palm House (**Figure 1**) in the Royal Botanical Gardens at Kew (**Figure 2**) features marine and intertidal habitats, for example. But in botanic gardens the term ecology means far more than imitation, and the gardens' ecological impact has changed as philosophies and world views have evolved.

Originally, interest was directed toward collecting and studying plants themselves, with little care taken in recording details of the plants' habitats or in safeguarding the ecosystems. Later, during the period of what might be called the imperial botanic garden, Western countries used botanic gardens to transfer plants from one part of the world to another, with sometimes devastating consequences for the ecosystems receiving the foreign plants. Most recently, botanic gardens have begun to play a major role in conserving endangered plants and preserving threatened habitats. Nearly 2500 botanic gardens are listed with Botanic Gardens Conservation International. To search for gardens by country, refer to http://www.bgci.org.uk/. **Table 1** lists a few selected gardens.

According to Botanic Garden Conservation International, at the beginning of the twenty-first century some 2000 botanic gardens in 148 countries harbored representatives of more than 80 000 plant species, or about one-third of the vascular plant species in the world. The gardens range from large ones like Kew and the New York Botanical Garden, where gorgeous plant displays are coupled with scientific research, to much smaller ones like Nezahat Gokyigit Memorial Park, near Istanbul, Turkey and Bafut Botanic Garden in northwest Cameroon which concentrate on safeguarding and studying local biosystems.

The Gardens of the Ancients

The idea of the modern botanic garden dates from the Renaissance, but it is possible that gardens which resembled them existed long before. Certainly plants valued for their medicinal properties were collected, grown, and studied in gardens in many parts of the world. Chinese tradition says that the emperor Shen Nung experimented to find the medicinal properties of plants as early as the twenty-seventh century BCE. But since no writing existed at the time and his materia medica *Shen Nung Pen T'sao Ching* dates only from the seventh century CE, the possibility of a garden somewhat like a botanic garden in ancient China is only that, a possibility.

Figure 1 The Palm House at Kew is one of its most distinctive features, and inspired many other glasshouses in other botanic gardens. Photograph by M. Soderstrom.

Figure 2 Bluebells growing under trees in the Conservation Area of The Royal Botanic Gardens at Kew. Photograph by M. Soderstrom.

The systematic garden developed by the Greek scholar Theophrastus (372–288 BCE) is much better documented. The author of two major works on plants and botany *Historia de Plantis* (History of Plants or Inquiring into Plants) and *De Causis Plantarums* (The Causes of Plants), he was a trusted associate of Aristotle, who bequeathed to him his library, garden, and the leadership of his school. Among the students was Alexander the Great who appears to have sent back plants from his campaigns through Central Asia, which were then planted in Theophrastus's garden.

Other illustrious gardens featuring plants gathered for study were established in pre-Spanish-conquest Mexico. The Mexican emperor Montezuma's garden brought together plants from tropical regions as well as Mexico's highlands. Hernando Cortez was impressed by them when he and his men overran Mexico in the 1520s. He described the great gardens he found there as unlike anything known in Europe at the time.

Things would soon change, however, in part because of the plants brought back to Europe by explorers like Cortez.

Table 1 Selected botanic gardens

Early botanical gardens
Orto Botanico at Pisa, Italy: founded c. 1545
Orto Botanico at Padua, Italy: founded c. 1545
Hortus Botanicus, Leiden, Netherlands: founded 1590
Le Jardin des plantes de la Université Montpellier, France: founded 1593
Oxford Physic Garden, Oxford University, UK: founded 1621
Le Jardin des plantes, Paris, France: founded 1626

Some other notable European gardens
Botanischer Garten und Botanisches Museum Berlin-Dahlem, Berlin, Germany
Linnaean Garden, Botaniska trädgården, Uppsala, Sweden
Jardín Botánico de Madrid, Spain
Jardim Botanico, University of Coimbra, Portugal
The Royal Botanic Gardens at Kew, London, UK
The Royal Botanic Garden, Edinburgh, Scotland
Eden Project, Cornwall UK
The National Botanic Garden of Wales, Llanarthne, Carmarthenshire, Wales, UK

Some gardens with colonial roots
Amani Nature Reserve, Tanzania
Bogor Botanical Gardens, Bogor, Indonesia
Indian Botanical Gardens, Shibpur, Kolkata, India
Pamplemousse Botanic Gardens, Mauritius
Rimba Ilmu Botanic Gardens, Kuala Lumpur, Malaysia
Royal Botanic Gardens, Trinidad
Singapore Botanic Gardens, Singapore

Some notable New World gardens
USA
  Boyce Thompson Arboretum. Superior, AZ
  Brooklyn Botanic Garden, New York
  Chicago Botanic Garden, Chicago, IL
  Fairchild Tropical Botanic Garden, Fairchild, FL
  Hawaii Tropical Botanical Garden, outside Hilo, Hawaii
  Missouri Botanical Garden, St. Louis, MO
  New York Botanical Gardens, New York
  San Francisco Botanical Garden at Strybing Arboretum, CA

Canada
  Jardin botanique, Montréal, QC
  Royal Botanical Gardens, Hamilton, OM
  UBC Botanical Garden and Centre for Plant Research, Vancouver, BC

Latin America
  Belize Botanic Garden, San Ignacio, Belize
  Jardin Botanico Francisco Javier Clavijero, Xalapa, Veracruz, Mexico
  The UNAM Botanical Garden, Mexico City, Mexico
  Jardim Botânico de São Paulo – São Paulo, Brazil
  Jardim Botânico do Rio de Janeiro – Rio de Janeiro, Brazil

Some Asian gardens
Maharashtra (Mahim) Nature Park in Mumbai, India
Narayana Gurukula Botanical Sanctuary, North Wayanad, Kerala, India
Beijing Botanical Garden, Beijing, China
Lijiang Botanic Garden & Research Station, Yunnan Province, China
Nanjing Botanical Garden, Nanjing, China
Koishikawa Botanical Gardens, Tokyo, Japan

(Continued)

Table 1 (Continued)

Some Southern Hemisphere gardens
Kirstenbosch National Botanical Garden, Cape Town, South Africa
Royal Botanic Gardens – Melbourne, Victoria, Australia
Royal Botanic Gardens – Sydney, New South Wales, Australia
Alice Springs Desert Park and Olive Pink Botanic Garden, Northern Territory, Australia
Bafut Botanic Garden in northwest Cameroon

Recreating Eden

Records from the Middle Ages testify to the interest of Europeans in studying plants for their medicinal properties. By the time of the Renaissance the five volumes of herbal lore prepared by the second-century pharmacist-doctor Dioscorides were used throughout Europe to teach about plants useful for medicine. Many monasteries had little plots of loosestrife and mints, of St. John's wort and chamomile, while untold numbers of midwives and lay healers cultivated medicinal herbs. One record of such a garden is a decree by Pope Nicolas V who in 1447 set aside part of the Vatican grounds as a garden where medicinal plants could be grown and botany taught as a branch of medicine.

A 100 years later Italy saw the establishment of the first botanic gardens in the modern sense at two universities, Padua and Pisa. The two dispute which was first. The Orto Botanico at Padua was established by decree of the Senate of the Venetian Republic in May 1545 and in July the monastery of S. Giustina ceded about 20 000 square meters to the republic and the University of Padua. No such decrees exist for the Orto Botanico of Pisa, but a letter written in early July 1545 by Lucca Ghinni, founder of the garden, suggests that it was already in existence then. What is clear is that these two gardens were places where plants were grown for systematic study, and which were organized to make that study easier. Nor were Pisa and Padua alone: in 1590 the University of Leiden established its botanic garden, while a year later the Jardin des Plantes of the Université Montpellier in southern France was begun.

Today a glimpse of what these gardens were like can be enjoyed at Leiden where a walled garden set apart from the rest of the university's botanic garden, the Hortus Botanicus, is laid out as it was in about 1594 by the pioneer botanist, Clusius. His career also gives a sense of the inquiring spirit which was developing among observers of the natural world. A native of the part of Flanders now in France, he spent his life collecting and describing plants all over Europe. He wrote treatises on the flora of Spain, Austria, and Hungary, corresponded with every botanist of note in Europe, collected and distributed plants and bulbs widely, and wrote the first monographs on both the tulip and the rhododendron. Behind all this lay a belief that the beauty of plants was a reflection of the wonders of God's creation and the harmony of the universe.

The religious impulse was extremely important during this period. Practically no one in Christendom in the sixteenth and seventeenth century doubted that Eden as described in the Bible had once existed. Many hoped that it still did. Part of Portugal's explorations were fired by the desire to find the lost paradise, while Christopher Columbus included a converted Jew in his first crew. The man spoke Hebrew, Arabic, and Aramaic, and so, it was thought, would be able to converse to the inhabitants of Eden, should that splendid garden be discovered on the westering voyages.

Eden, of course, was not found, and many botanists, both religous and secular, began to wonder if Eden might be recreated simply by bringing together all the plants which must have grown in it. Some thought that even if a latter-day Eden were impossible to create, much good would be done by studying as much of God's creation as possible: each plant was a facet of God, so that knowing all plants would mean knowing an important part of God.

There were built-in contradictions in this effort, however, since it coincided with the great age of exploration when plants and animals unimagined by Europeans were brought back from the Americas for study. Questions arose: Were they created at the same time as all the familiar flora and fauna? Or were there perhaps two Creations, or parts of the world which had escaped the Flood? Opinions varied, but one thing was clear: the theological ideas behind the efforts to bring plants together for study would have to be modified.

The Gardens of Discovery

Indeed one of the foremost gardens of the age was located where it was in order to escape the influence of the Roman Catholic church and its educational institutions. The Jardin des Plantes (**Figure 3**) in Paris was chartered in 1626 by

Figure 3 The Jardin des plantes of the Muséum de l'histoire naturelle is now surrounded by Paris, but when it was opened in the seventeenth century it was outside the city's walls. Photograph by M. Soderstrom.

Louis XIII on land a short way outside the wall encircling the city which put it beyond the reach of the Université de Paris and its Faculté de médecine. For the next 150 years during the high tide of French exploration and colonization and throughout the French Enlightenment, Paris's botanic garden was the world's main center for plant collection and study as well as home to sometimes audacious research into other aspects of the natural world.

In England, several medicinal, or physic gardens, were also established in the seventeenth century. The first was the Oxford Physic Garden, set up in 1621 "for the advancement of medicine . . . the promotion of learning and the glorification of the work of God." Spain and Portugal began their royal botanic gardens somewhat later. The Jardín Botánico de Madrid was established in 1755 while the Jardim Botanico of the University of Coimbra dates its roots to 1775. The small botanic garden which was to become the Royal Botanical Gardens at Kew was started a few years before them as the pet project of Frederick, Prince of Wales, and his wife on the royal country estates upstream from London on the Thames. Frederick's son, George III, expanded the garden and saw to it that British explorers under the aegis of Sir Joseph Banks were given mandates to bring back plants for the Royal Gardens.

Botanic Gardens in Colonies

Britain and other colonial powers began not only to increase the size of foreign plant collections in their botanic gardens at home, but also to establish gardens in the countries they were colonizing. The Dutch set up gardens in southern Africa and on Java in what is now Indonesia both to provision their ships and to study and to acclimatize plants which might be useful either at home or in other colonies. The French followed suit on Mauritius in the Indian Ocean and on Martinique in the Caribbean. The British had their own botanic gardens at Calcutta, Singapore, and in what is now Sri Lanka. The Germans, who were late-comers to the colonial game, set up botanic gardens in Africa in what is now Cameroon and Tanzania in the late nineteenth century. In all cases the gardens maintained close ties with the home country, and the home gardens.

There are a number of ways that this network of botanic gardens have had ecological effects. By introducing plants into the home country, they paved the way for exotics to become established in new habitats. Two examples of introductions which appeared initially to have few negative effects are plants brought back to the Jardin des Plantes in Paris. The black locust, a large tree originally found in a relatively limited area of the Appalachians of North America, now grows freely in forests and woodlots all over Europe as well as far beyond its home range in the United States and Canada. Its scientific name *Robinia pseudoacacia* L., honors Jean Robin who was the King's gardener even before

the Jardin des Plantes was opened. A tree Robin planted in the early 1600s was transplanted by his son to the Jardin, and still grows there, the oldest tree in the center of Paris.

Another plant which migrated via the Jardin des Plantes is the butterfly bush, *Buddleia davidii*. This native of China was sent back to France by Abbé Armand David in the nineteenth century. It now thrives in cultivated gardens but also grows wild along railroad lines and in disused land in Europe and North America.

Both of these plants are today considered undesirable alien invaders in some parts of their adopted countries. The black locust can produce thick plantations whose shade does not allow other, native plans to grow, while buddleia frequently forms dense thickets, forcing out native plants along streambeds and in old pastures.

Other transplants produced consequences which took less time to become apparent. Among them is breadfruit, a native of the South Pacific, which Sir Joseph Banks, then director of Kew, thought would be good food for the slaves who worked in the sugarcane plantations in the Caribbean. After a false start in 1791 – the first shipment from Tahiti was on the Bounty when its crew mutinied against Captain William Bligh – breadfruit and the plantain, another import, helped make plantation agriculture profitable by providing cheaply and easily grown food.

Coffee first arrived in the Caribbean directly from a botanic garden. In 1714 Louis XIV obtained a plant from Amsterdam and sent it to the Jardin des Plantes. The intendant of the day had the Jardin's first heated greenhouse constructed for it, where it did very well. By 1721 enough new plants had been propagated from it to risk sending the first offspring to the botanic garden at Martinique in hopes that after acclimatization there, the plants could be established in the French possessions around the Caribbean. It worked: the coffee plantations of the French Antilles as well as of Brazil, Jamaica, Columbia, and Mexico were all initially planted with descendants from that one coffee tree.

Another example is that of rubber. Many plants in tropical Asia, Africa, Central America, and Brazil, produce latex: Columbus may have been the first to mention 'white milk' oozing from the bark of some trees while the French explorer La Condamine brought the first specimens of *caoutchouc* to Europe in the eighteenth century. But it was not until 1839 when Charles Goodyear discovered a process which produced rubber suitable for hoses and other industrial uses that demand increased dramatically.

The only commercial source of rubber for most of the nineteenth century was the wild rubber tree in Amazonia, *Hevea brasiliensis*. So intense was the demand that a direct steamship line ran from Manaus more than 1800 km (1100 miles) upstream on the Amazon to Liverpool, carrying trading goods one way, and latex the other. In 1876 Henry Wickham, a plant collector engaged by Kew's director Joseph Hooker, chartered a ship on the line to rush some 70 000 seeds across the Atlantic. He got permission from

Brazilian authorities for the transfer by convincing them of the need to release "exceedingly delicate botanic specimens specially designated for delivery to Her Britannic Majesty's own Royal Garden at Kew."

Hooker arranged for a night freight train to meet the ship when it docked at Liverpool and cleared space in Kew's glasshouses for the seeds. Within 2 weeks of their arrival in England, some 7000 seedlings had begun to grow, and a year later 1900 plants were sent to the Perdeniya Garden in what is now Sri Lanka. From there, seedlings were distributed to several other tropical botanic gardens. The Singapore Botanic Gardens (**Figure 4**) got 22 seedlings, 11 of which it used for propagation in the garden. By 1917 it is estimated that the Singapore garden and its director Henry 'Rubber' Ridley had distributed seven million seeds and by 1920 the Malaysian peninsula was producing more than half the world's rubber. There is no way of estimating how many native plants disappeared during the rapid transformation of jungle into rubber plantations. Indirectly the cultivation of rubber had other effects on habitats also, since it made the development of trucks and cars – and therefore of the industrialized world's sprawling, petroleum-powered society – much easier.

Those who undertook these transfers of plants felt no guilt at the massive reworkings of ecosystems which ensued. Most people in the nineteenth and early twentieth century believed that God made the world for humans to enjoy so that making plants serve humans was doing God's work.

At the same time, however, many botanic gardens by accident or design preserved part of the native vegetation in the gardens themselves. For example, the New York Botanical Garden (**Figure 5**) includes 16 ha (40 acres) of first growth, mixed hardwood forest. This remnant is a unique reminder of the forest which covered most of what is now the city of New York before Europeans wrested control from the indigenous population.

Other examples of habitat conservation include the Singapore Botanic Gardens' small jungle enclave amid the city's myriad high rises as well as the Conservation Area at Kew. There a part of the garden is being conserved as British farmland, with upkeep and interventions following traditional British agricultural practices.

Figure 4 View of the Palm Valley, the heart of the Singapore Botanic Gardens. Photograph by M. Soderstrom.

Figure 5 The hemlock forest in the New York Botanical Garden preserves a remnant of the forest which once covered much of the New York City region. Photograph by M. Soderstrom.

The Intrinsic Value of Biodiversity and Nature

The philosophical framework in which most botanic gardens operate now places a high value on maintaining what measure of biodiversity exists today. This can be seen as a direct outgrowth of concerns about the natural world which began to develop during the early twentieth century as the damage resulting from industrialization, population growth, and uncontrolled exploitation of natural resources became apparent. Rather than being motivated primarily by a desire to study God in nature, scientists and others began to think that nature was intrinsically valuable, over and above whatever link it might have with a deity or what economic advantage it might bring to human society. Often work begun by botanic gardens directly led to better understanding of the wonderful interplay of organisms in ecological systems, with far reaching philosophical and scientific repercussions.

Take for example the huge water lily *Victoria amazonica*, grown by many botanic gardens today. When it was first described in the early nineteenth century by plant explorers who found it in British Guyana, the accounts caused a sensation: in addition to having lovely flowers, its leaves grew up to 6 ft in diameter, and were strong enough to support an adult man. Seeds were sent back to Europe several times, but it was not until 1849 that Kew was able to raise plants to the stage where they could be set out in ponds. Of these, three flowered, the first being one in Duke of Devonshire's garden at Chatsworth. The one at Kew bloomed the next year after being installed in a special glasshouse, and some 30 000 visitors came to marvel at the flowers and the leaves. The craze was not confined to England: the Hortus Botanicus (**Figure 6**) at Leiden succeeded in getting a plant to flower in 1872, and kept one alive during the coldest days of World War II when there was only enough fuel to keep the water lily greenhouse heated. Early pictures from both the

Figure 6 The Clusius Garden in the Hortus Botanicus at the University of Leiden is arranged much as it was in 1594. Photograph by M. Soderstrom.

Singapore Botanic Gardens and the Missouri Botanic Garden feature the plant too.

But the story of *Victoria amazonica* does not end with special ponds and crowds of visitors. One of the oddities of the flowers is that when they are dissected in the wild, a particular sort of beetle (*Cyclocephala hardyi*) is often found inside. For a long time botanists suspected that the beetles pollinated the plant but were not sure how. The mystery was only unraveled in the 1970s when Ghillean Prance, Kew's director from 1988 to 1999 but then a research biologist for the New York Botanical Garden, spent nights standing hip deep in Brazilian ponds, watching the flowers open and beetles flie in and out. He found that the beetles were attracted to the fragrance of the opening flowers, crawled inside to feed, and were trapped there when the flowers closed as dawn approached. The next evening the beetles, sticky from feeding, crawled back out as the flowers opened again, picking up a load of pollen as they passed. Then they flew away to repeat the process in another flower, and incidentally pollinate it. In so doing they demonstrated the complexity of ecosystems and the intricate way plants and animals are interrelated.

Much of botanic gardens' present-day work in the field, the laboratory, and in the gardens themselves is designed to study these kinds of relationships. Botanic gardens today also actively work for conservation of species by propagating plants and collecting and storing seeds. According to the World Conservation Union, 34 000 taxa around the world are considered threatened with extinction. Of them 10 000 threatened species – or about a third – are growing in one or more botanic gardens. In some cases, the collection of plant specimens comes just in the nick of time. The canyon in Chiapas, Mexico where botanists in the late 1990s found *Deppea splendens*, a shrub with lovely two-inch orange flowers hanging in long clusters, has since been cleared for cornfields: the plant is thought to be now extinct in the wild. But seeds from the shrub flourished in the San Francisco Botanical Garden, and cuttings from the plants are even sold by the Friends of the Garden at their annual sale.

Perhaps the biggest conservation project is Kew's Millennium Seed Bank. The £ 80 million (US$ 160 million) undertaking is housed in new facilities at Kew's Wakehurst auxiliary garden south of London. Its aim is to collect seeds from 24 000 species of plants from all of the world by 2020, and to keep them in secure locations so that they can be used at a later date. When seeds are dried so that they contain only 5% moisture, about 80% of them can successfully be held at −20 °C for periods of up to 200 years. A portion of all seeds will be held in their country of origin in order to avoid repeating 'theft' of plants like that which occurred during the great period of European colonialism. Fortunes were made then by European exploiters of coffee, rubber, and other plants but the countries of origin received no compensation.

As part of the effort to compensate for damage done in the past and to preserve remaining biodiversity, Botanic Gardens Conservation International has set a series of targets to be met by 2010. The overall aims are general – things like protection of plant diversity, conservation of endangered species in botanic gardens and in their native habitat, and public education about the importance of plant diversity. Specific goals in the 20-item to-do list are quite specific, though. For example, at least 10% of each of the world's ecological regions are to be effectively conserved, and the number of trained botanic garden staff working in conservation, research, and education should be doubled. In addition, international databases of such things as which endangered species are cultivated in what botanic garden and what plant introduction has become invasive in what range are under development. Many botanic gardens are already promoting awareness of the problems posed by invasive species through such things as the St. Louis Declaration on Invasive Plant Species, developed at a conference organized by the Missouri Botanical Garden in 2001.

Several new botanic gardens have also been established recently with the principal aim of protecting unique and relatively untouched environments. One of them is the Alice Springs Desert Park in central Australia, opened to the public in 1997, which preserves a section of that continent's desert. Another is the Bafut Botanic Garden in northwest Cameroon, also opened in 1997, a savanna botanic garden and forest reserve.

In addition, two other new botanic gardens point the way to reclaiming landscapes destroyed by human carelessness and greed. The first is the Eden Project in Cornwall UK, where a former clay pit has been transformed into a botanic garden with several distinct ecosystems represented in geodesic buildings sunk into the former mine landscapes. The other is the Maharashtra (Mahim) Nature Park in Mumbai, India, where 15 ha of former garbage dump have been reclaimed. The reconstructed forest is now home to 380 varieties of plants, 84 varieties of birds, and about 34 kinds of butterflies.

Education and the Future

Kings and religious authorities are no longer the patrons behind botanic gardens, so those in charge of them must convince the public, governments, and industry to support the gardens and their work. This is why botanic gardens today devote so much effort to education and public information projects. Some are high tech like the interactive rain forest displays in the Climatron at the Missouri Botanic Garden. Others, like the 12-acre adventure site in the New York Botanical Garden, introduce children to ecological concepts through activities full of action. Still others aim to make botanic gardens places where pleasure goes hand in hand with research and learning. Kew has an ice-skating rink

Figure 7 The Bog and Marsh Garden at the Jardin botanique in Montreal presents wetlands plans – many of them endangered – in a series of basins and ponds. Photograph by M. Soderstrom.

in winter, Montreal's Jardin botanique (**Figure 7**) offers twilights full of Chinese lanterns in the fall, gardens everywhere advertise their spring flowers and their summer splendor to lure people to see their plants, and hear their message. The effectiveness of these educational efforts may mean the difference between governments setting ecologically sound policy or not. Without public recognition that habitat protection and biodiversity are important, governments in democratic countries may drag their feet while in countries where decisions are made from the top down, those in power would not be convinced of the need to do the same.

Further Reading

Brockway L (1979) *Science and Colonial Expansion: The Role of the British Royal Botanic Gardens*. New York and London: Academic Press.
Hyams E (1969) *Great Botanical Gardens of the World* (with photographs by Macquitty W). London: Bloomsbury Books.
Laissus Y (1995) *Le Muséum national d'histoire naturelle*. Paris: Découvertes Gallimard.
McCracken DP (1997) *Gardens of Empire: Botanical Institutions of the Victorian British Empire* London: Leicester University Press.
Prest J (1981) *The Garden of Eden: The Botanic Garden and the Re-Creation of Paradise*. New Haven: Yale University Press.
Soderstrom M (2001) *Recreating Eden: A Natural History of Botanical Gardens*. Montreal: Véhicule Press.

Relevant Websites

http://www.bgci.org – Botanic Gardens Conservation International.
http://www2.ville.montreal.qc.ca – Jardin botanique in Montreal.
http://www.mnhn.fr – Jardin des Plantes of the Muséum de l'histoire naturelle.
http://www.nybg.org – New York Botanical Garden.
http://www.sbg.org.sg – Palm Valley, Singapore Botanic Gardens.
http://www.kew.org – Royal Botanic Gardens, Kew.
http://www.sbg.org.sg – Singapore Botauic Gardeus.
http://www.centerforplantconservation.org – The Saint Louis Declaration on Invasive Plants.
http://www.hortus.leidenuniv.nl – The Hortus Botanicus of Leiden, University of Leiden.

Caves

F G Howarth, Bishop Museum, Honolulu, HI, USA

Caves

Caves are defined as natural subterranean voids that are large enough for humans to enter. They occur in many forms, and cavernous landforms make up a significant portion of the Earth's surface. Limestone caves are the best known. Limestone, calcium carbonate, is mechanically strong yet dissolves in weakly acidic water. Thus over eons great caves can form. Caves form in other soluble rocks, such as dolomite (calcium magnesium carbonate), but they are usually not as extensive as those in limestone. Volcanic eruptions also create caves. The most common are lava tubes that are built by the roofing over and subsequent draining of molten streams of fluid basaltic lava. In addition, cave-like voids form by erosion (e.g., sea caves and talus caves) and by melting water beneath or within glaciers. Depending on their size, shape, and interconnectedness, caves develop unique environments that often support distinct ecosystems.

Cave Environments

The physical environment is rigidly constrained by the geological and environmental settings and can be defined with great precision because it is surrounded and buffered by thick layers of rock. Caves can be water-filled or aerial.

Aquatic Environments

Aquatic systems are best developed in limestone caves since water creates these caves. Debris-laden water in voids in nonsoluble rock will eventually fill caves. A significant exception is found in young basaltic lava that has flowed into the sea. Here, subterranean ecosystems develop in the zone of mixing freshwater and salt water within caves and spaces in the lava. The system is fed by food carried by tides and groundwater flow. Frequent volcanism creates new habitat before the older voids fill or erode away. Aquatic cave environments are dark, three-dimensional (3D) mazes, in which food and mates may be difficult to find. In addition, the water can stagnate, locally becoming hypoxic with high concentrations of toxic gases including carbon dioxide and hydrogen sulfide.

Terrestrial Environments

The terrestrial environment in long caves is buffered from climatic events occurring outside. The temperature stays nearly constant, fluctuating around the mean annual surface temperature (MAST); except passages sloping down from an entrance tend to trap cold air and remain a few degrees cooler than MAST. Passages sloping up are often warmer than MAST. The environment is strongly zonal (**Figure 1**). Three zones are obvious: an entrance zone where the surface and underground habitats overlap; a twilight zone between the limit of photosynthesis and the zone of total darkness. The dark zone can be further subdivided into three distinct zones: a transition zone where climatic events on the surface still affect the atmosphere, especially relative humidity (RH); a deep zone where the RH remains constant at 100%; and an innermost stagnant air zone where air exchange is too slow to flush the buildup of carbon dioxide and other decomposition gasses. The boundary between each zone is often determined by shape or constrictions in the passage. In many caves, the boundaries are dynamic and change with the seasons.

The subterranean aerial environment is stressful for most organisms. It is a perpetually dark, 3D maze with a water-saturated atmosphere and occasional episodes of toxic gas concentrations. Many of the cues used by surface animals are absent or operate abnormally in caves (e.g., light/dark cycles, wind, sound). Passages can flood during rains, and crevices might drop into pools and water-filled traps. If the habitat is so inhospitable, why and how do surface animals forsake the lighted world and adapt to live there? It is the presence of abundant food resources that provides the impetus for colonization and adaptation.

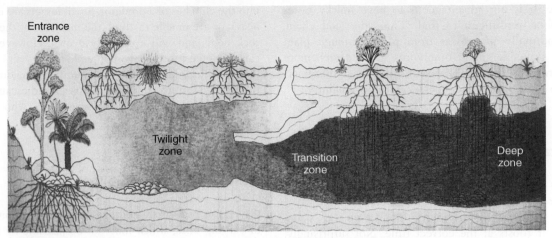

Figure 1 Schematic profile view of the cave habitat showing the location of principal zones.

Food Resources

The main energy source in limestone caves is sinking rivers, which carry-in abundant food not only for aquatic communities but also via flood deposits for terrestrial communities. Rivers are less important in nonsoluble rock, such as lava, but percolating runoff washes surface debris into caves through crevices. Other major energy sources are brought in by animals that habitually visit or roost in caves, plants that send their roots deep underground, chemoautotrophic microorganisms that use minerals in the rock and accidentals that fall or wander into caves and become lost.

Generally in surface habitats, accumulating soil filters water and nutrients and holds these resources near the surface where they are accessible to plant roots and surface-inhabiting organisms. However, in most areas with underlying caves, the soil is thin with areas of exposed bare rock because developing soil is washed or carried into underground voids by water or gravity. Soil formation is limited, and much of the organic matter sinks out of the reach of most surface animals.

Except for guano deposits, flood deposits, scattered root patches, and other point-source food inputs, the defining feature of cave habitats is the appearance of barren wet rock. Visible food resources in the deep cave are often negligible, and what food deposits there are would be difficult for animals to find in the 3D maze. Food resources in the system of smaller spaces is difficult to sample and quantify, but in theory, some foods may be locally concentrated by water transport, plant roots, or micro point source inputs such as through cracks extending to the surface. These deposits would be more easily exploited than would widely scattered deposits.

In each biogeographic region, a few members of the surface and soil fauna have invaded cave habitats and adapted to exploit this deep food resource. The colonists usually were pre-adapted; that is, they already possessed useful characteristics resulting from living in damp, dark habitats on the surface.

Cave Communities

Guano Communities

Many animals live in or use caves. Cave-inhabiting vertebrates are relatively well-known. Cave bats, swiftlets (including the edible-nest swiftlet of Southeast Asia), and the oil bird in South America use echolocation to find their way in darkness. Pack rats in North America, along with cave crickets and other arthropods also roost in caves. Large colonies of these cave-nesting animals carry in huge quantities of organic matter with their guano and dead bodies. This rich food resource forms the basis for specialized communities of microorganisms, scavengers, and predators. Arthropods comprise the dominant group of larger animals in this community, and like their vertebrate associates, most species are able to disperse outside caves to found new colonies.

Deep Cave Communities

In the deeper netherworld, communities of mysterious, obligate cave animals occur. Most are invertebrates, but a few fishes and salamanders have colonized the aquatic realm. Crustaceans (shrimps and their allies) dominate in aquatic ecosystems, and insects and spiders dominate terrestrial systems. Although a few species are specialists on living plant roots or other specific resources, most are generalist predators or scavengers. The relatively high percentage of predators indicates the importance of accidentals as a food resource. However, many presumed predatory species, such as spiders, centipedes, and ground beetles, will also scavenge on dead animals when available. It is not advantageous to have finicky tastes where food is difficult to find. Thus, the food chain, which normally progresses from plants through plant feeders, scavengers, and omnivores to predators, more closely resembles a food web with most species interacting with most of the other species in the community.

Adaptations to Cave Life

Animals roosting or living in caves must adapt to cope with the unusual environment. Paramount for the cave-roosting vertebrates is the ability to find their way to and from their roosts at the correct time. Not surprisingly, the birds and bats display uncanny skill in memorizing the complex maze to and from their cave roosts. Pack rats use trails of their urine to navigate in and out of caves. Species using the twilight and transition zones can use the daily meteorological cycle for cues to wake and leave the cave. Those roosting in the deep zone may rely on accurate internal clocks to know when it is beneficial to leave their roost.

Organisms that adapt to live permanently underground must make changes in behavior, physiology, and structure in order to thrive in the stressful environment. They need to find food and mates and successfully reproduce in total darkness. Their hallmark is the loss or reduction of conspicuous structures such eyes, bodily color, protective armor, and wings. These structures are worthless in total darkness, but they can be lost quickly when selection is relaxed because they are expensive for the body to make and maintain. How such losses could happen quickly is demonstrated by the cave-adapted planthoppers (Cixiidae). The nymphs of surface species feed on plant roots and have reduced eyes and bodily color whereas their adults have big eyes, bold colors, and functional wings. The cave-adapted descendents maintain the nymphal eyes, color, and other structures into adulthood, a phenomenon known as neoteny.

The high relative humidity and occasional episodes of elevated CO_2 concentrations are stressful to cold-blooded organisms. The blood of insects and other invertebrates will absorb water from saturated atmosphere, and the animals literally will drown unless they have adaptations to excrete the excess water. High levels of CO_2 force animals to breathe more, which increases water absorption. Cave-adapted insects often have modified spiracles to prevent or cope with their air passages filling with water.

Most lava tube arthropods have specialized elongated claws to walk on glassy wet-rock surfaces. Many have elongated legs to step across cracks rather than having to descend and climb the other side. Jumping or falling might land a hapless animal in a pool or water-filled pit or into the clutches of a predator. Small insects are often too heavy or are unable to climb the meniscus at the edge of rock pools and will eventually drown. However, many cave-adapted insects have unique knobs or hairs near the base of each elongated claw and modified behavioral traits that allow them to climb the meniscus and escape. Some of the latter are predators or scavengers, who wait on pools for victims.

Other Cave-Like Habitats

Cavernous rock strata contain abundant additional voids of varying sizes, which may not be passable by humans. These voids are interconnected by a vast system of cracks and solution channels. The smaller capillary-sized spaces are less important biologically because their small size limits the amount of food resources they can hold and transport. Voids larger than about 5 cm can transport large volumes of food as well as serve as habitat for animals. In terms of surface area and extent, these intermediate-size voids are the principal habitat for specialized cave animals. Many aspects of their life history may occur only in these spaces. Some cave species (such as the earwig, *Anisolabis howarthi* (**Figure 2**), and sheet web spiders, Linyphiidae, in Hawaiian lava tubes) prefer to live in crevices and are only rarely found in caves. In addition, cave-adapted animals have been found living far from caves in cobble deposits beneath rivers, fractured rock strata, and buried lava clinker in Japan, Hawai'i, Canary Islands, Australia, and Europe. These discoveries corroborate the view that cave adaptation and the development of cave ecosystems can occur wherever there is suitable underground habitat.

Because these smaller voids are isolated from airflow from the surface, the environment resembles the stagnant air zones of caves. Caves serve as entry points and windows in which to observe the fauna living within the voids

Figure 2 The Hawaiian cave earwig, *Anisolabis howarthi* Brindel (family Carcinophoridae). Photo by W. P. Mull.

in the cavernous rock strata. The view is imperfect because the environment is so foreign to human experience.

Case Study: Hawai'i

Food Web

The main energy sources in Hawaiian lava tube ecosystems are tree roots, which penetrate the lava for several decameters; organic matter, which washes in with percolating rainwater; and accidentals, which are surface and soil animals blundering into the cave. Both living and dead roots are utilized, and this source is probably the most important. Furthermore, both rainwater and accidentals often use the same channels as roots to enter caves, so that root patches often provide food for a wide diversity of cave organisms. The importance of roots in the cave ecosystem makes it desirable to identify the major species. This has become possible only recently by using DNA-sequencing technology. The most important source of roots is supplied by the native pioneer tree on young lava flows; *Metrosideros polymorpha*. *Cocculus orbiculatus*, *Dodonaea viscosa*, and *Capparis* are locally important in drier habitats. Several different slimes and oozes occur on wet surfaces and are utilized by scavengers in the cave. They are mostly organic colloids deposited by percolating groundwater, but some may be chemoautotrophic bacteria living on minerals in the lava. Cave-roosting vertebrates do not occur in Hawai'i. Native agrotine moths once roosted in caves in large colonies, but the group has become rare in historic times. The composition of the community their colonies once supported is unknown. Feeding on living roots are cixiid planthoppers (*Oliarus*). Their nymphs suck xylem sap with piercing mouthparts. The blind flightless adults wander through subterranean voids in search of mates and roots. Caterpillars of noctuid moths (*Schrankia*) prefer to feed on succulent flushing root tips, but they also occasionally scavenge on rotting plant and animal matter. Tree crickets (*Thaumatogryllus*), terrestrial amphipods (*Spelaeorchestia*), and isopods (*Hawaiioscia* and *Littorophiloscia*) are omnivores but feed extensively on roots. Cave rock crickets (*Caconemobius*) are also omnivorous as well as being opportunistic predators. Feeding on rotting organic material and associated microorganisms are millipedes (*Nannolene*), springtails (*Neanura*, *Sinella*, and *Hawinella*), and phorid flies (*Megaselia*). Terrestrial water treaders (*Cavaticovelia aaa*) suck juices from long-dead arthropods. Feeding in the organic oozes growing on wet cave walls are larvae of craneflies (*Dicranomyia*) and biting midges (*Forcipomyia pholeter*). The blind predators include spiders (*Lycosa howarthi*, *Adelocosa anops* (**Figure 3**), *Erigone*, *Meioneta*, *Oonops*, and *Theridion*), pseudoscorpions (*Tyrannochthonius*), rock centipedes (*Lithobius*), thread-legged bugs (Nesidiolestes), and beetles (*Nesomedon*, *Tachys*, and *Blackburnia*). Most of the cave predators will also scavenge on dead animal material.

Figure 3 The no-eyed big-eyed hunting spider, *Adelocosa anops* Gertsch (family Lycosidae) from caves on the island of Kaua'i. Photo by the author.

Nonindigenous Species

Several invasive nonindigenous species have invaded cave habitats and are impacting the cave communities. The predatory guild is the most troublesome, with some species being implicated on the reduction of vulnerable native species. Among these, the nemertine worm (*Argonemertes dendyi*) and spiders (*Dysdera*, *Nesticella*, and *Eidmanella*) have successfully invaded the stagnant air zone within the smaller spaces. The colonies of cave-roosting moths disappeared from the depredations of the roof rat (*Rattus rattus*) on their roosts and from parasites purposefully introduced for biological control of their larvae. Many non-native species (such as *Periplaneta* cockroaches, *Loxosceles* spiders, *Porcellio* isopods, and *Oxychilus* snails) survive well in larger accessible cave passages, where they have some impact, but they appear not to be able to survive in the system of smaller crevices. A few alien tree species also send roots into caves, creating a dilemma for reserve managers trying to protect both cave and surface habitats since their roots support some generalist native species but not the host-specific planthoppers.

Succession

Inhabited Hawaiian lava tubes range in age from 1 month on Hawai'i Island to 2.9 million years on O'ahu Island. On Hawai'i Island colonization and succession of cave ecosystems can be observed. Crickets and spiders arrive on new flows within a month of the flow surface cooling. They hide in caves and crevices by day and emerge at night to feed on windborne debris. *Caconemobius* rock crickets are restricted to living only in this aeolian (wind-supported) ecosystem and disappear with the establishment of plants. The obligate cave species begin to arrive within a year after lava stops flowing in the caves. The predatory wolf spider, *Lycosa howarthi*, arrives first and preys on wayward aeolian arthropods. Other predators and scavenging arthropods – including blind, cave-adapted *Caconemobius* crickets – arrive during the next decade. Under rainforest conditions, plants begin to invade the surface after a decade, allowing the root feeding cave animals to colonize the caves. *Oliarus* planthoppers arrive about 15 years after the eruption and only 5 years after its host tree, *Metrosideros polymorpha*. The cave-adapted moth, *Schrankia* species, and the underground tree cricket, *Thaumatogryllus cavicola*, arrive later. The cave species colonize new lava tubes from neighboring older flows via underground cracks and voids in the lava. Caves between 500 and 1000 years old are most diverse in cave species. By this time the surface rainforest community is well-developed and productive, while the lava is still young and maximal amount of energy is sinking underground. As soil formation progresses, less water and energy reaches the caves, and the communities slowly starve. In highest rainfall areas, caves support none or only a few species after 10 000 years. Under desert conditions, succession is prolonged for 100 000 years or more. Mesic regimes are intermediate between these two extremes. New lava flows may rejuvenate some buried habitat as well as create new cave habitat.

Perspective

The fauna of a large percentage of the world's cave habitats remain unknown to science, and new species continue to be discovered in well-studied caves. Additional biological surveys are needed to fill gaps in knowledge and improve our understanding of cave ecosystems. Improved methods for sampling the inaccessible smaller voids are needed. The cave environment is a rigorous, high-stress one, which is difficult for humans to access and envision because it is so foreign to human experience. Working in caves can be physically challenging. However, recent innovations in equipment and exploration techniques allow ecologists to visit the deeper, more rigorous environments.

In spite of the difficulties of working in the stressful environment, several factors make caves ideal natural laboratories for research in evolutionary and physiological ecology.

Since cave habitats are buffered by the surrounding rock, the abiotic factors can be determined with great precision. The number of species in a community is usually manageable and can be studied in total. Questions that are being researched are how organisms adapt to the various environmental stressors; how communities assemble under the influence of resource composition and amount; and how abiotic factors affect ecological processes. For example, a potential overlap between cave and surface ecological studies occurs in some large pit entrances in the tropics. The flora and fauna living in these pits frequently experience CO_2 levels 25–50 times ambient.

See also: Rocky Intertidal Zone.

Further Reading

Camacho AI (ed.) (1992) *The Natural History of Biospeleology*. Madrid: Monografias, Museo Nacional de Ciencias Naturales.

Chapman P (1993) *Caves and Cave Life*. London: Harper Collins Publishers.

Culver DC (1982) *Cave Life*. Cambridge, MA: Harvard University Press.

Culver DC, Master LL, Christman MC, and Hobbs HH, III (2000) Obligate cave fauna of the 48 contiguous United States. *Conservation Biology* 14: 386–401.

Culver DC and White WB (eds.) (2004) *The Encyclopedia of Caves*. Burlington, MA: Academic Press.

Gunn RJ (ed.) (2004) *Encyclopedia of Caves and Karst*. New York: Routledge Press.

Howarth FG (1983) Ecology of cave arthropods. *Annual Review Entomology* 28: 365–389.

Howarth FG (1993) High-stress subterranean habitats and evolutionary change in cave-inhabiting arthropods. *American Naturalist* 142: S65–S77.

Howarth FG, James SA, McDowell W, Preston DJ, and Yamada CT (2007) Identification of roots in lava tube caves using molecular techniques: Implications for conservation of cave faunas. *Journal of Insect Conservation* 11(3): 251–261.

Humphries WF (ed.) (1993) *The Biogeography of Cape Range, Western Australia. Records of the Western Australian Museum, Supplement no.45*. Perth: Western Australian Museum.

Juberthie C and Decu V (eds.) (2001) *Encyclopaedia Biospeologica Vol III*. Moulis, France: Société de Biospéologie.

Moore GW and Sullivan N (1997) *Speleology Caves and the Cave Environment*, 3rd edn. St. Louis, MO: Cave Books.

Wilkins H, Culver DC, and Humphreys WF (eds.) (2000) *Ecosystems of the World, Vol. 30 Subterranean Ecosystems*. Amsterdam: Elsevier Press.

Chaparral

J E Keeley, University of California, Los Angeles, CA, USA

Introduction

Chaparral is the name applied to the evergreen sclero-phyllous (hard-leaved) shrub vegetation of southwestern North America, largely concentrated in the coastal zone of California and adjacent Baja California. It is a dense vegetation often retaining many dead spiny branches making it nearly impenetrable (**Figure 1**). It dominates the foothills of central and southern California but is replaced at higher elevations by forests. On the most arid sites at lower elevations evergreen chaparral is replaced with a lower-stature summer deciduous 'soft chaparral' or sage scrub.

Chaparral owes much of its character to the Mediterranean climate of winter rain and summer drought. The severe summer drought, often lasting 6 months or more, inhibits tree growth and enforces the shrub growth form. Intense winter rains coincide with moderate temperatures that allow for rapid plant growth, producing dense shrublands. These factors combine to make this one of the most fire-prone ecosystems in the world. This Mediterranean climate is the result of a subtropical high-pressure cell that forms over the Pacific Ocean. During the summer, this air mass moves northward and blocks water-laden air masses from reaching land, and in winter this high-pressure cell moves toward the equator and allows winter storms to pass onto land. On the Pacific Coast it is wettest in the north, where the effect of the Pacific High is least, and becomes progressively drier to the south, and consequently chaparral dominates more of the landscape in the southern part. Interestingly, these synoptic weather conditions form globally at this same latitude (30–38° north or south) and on the western sides of continents. As a result similar Mediterranean-climate shrublands occur in the Mediterranean Basin of Europe, central Chile, South Africa, and southern Australia.

The Ecological Community

Chaparral is a shrub-dominated vegetation with other growth forms playing minor or temporary successional roles after fire. More than 100 evergreen shrub species occur in chaparral, although sites may have as few as one or more than 20 species, depending on available moisture, slope aspect, and elevation. The most widely distributed shrub is chamise (*Adenostoma fasciculatum*), ranging from Baja to northern California, occurring in either pure chamise chaparral or in mixed stands. It often dominates at low elevations and on xeric south-facing slopes. The short needle-like leaves produce a sparse foliage, and soil litter layers are poorly developed and result in weak soil horizons. Chamise often forms mixed stands of vegetation with a number of species. These include the bright smooth red-barked manzanita (*Arctostaphylos* spp.), the sometimes spiny ceanothus, also known as buckbrush or California lilac (*Ceanothus* spp.). On more mesic north-facing slopes chaparral is commonly dominated by broader-leaved shrubs, including the acorn-producing scrub oak (*Quercus* spp.), the cathartic coffeeberry (*Rhamnus californica*), redberry (*R. crocea*), the rather bitter chaparral cherry (*Prunus ilicifolia*), and chaparral holly (*Heteromeles arbutifolia*), from whence the film capital Hollywood derives its name.

The most common shrub species and the majority of herbaceous species have fire-dependent regeneration, meaning that seeds remain dormant in the soil until stimulated to germinate after fire (see the section titled 'Fire' below). These include chamise, manzanita, and ceanothus shrubs, which flower and produce seed most years but seldom produce seedlings without fire. Some ceanothus species are relatively short-lived or are easily shaded out by other shrubs and die after several decades. They, however, persist as a living seed pool in the soil. In addition, a large number of annual species live most of their life as dormant seeds in the soil, perhaps as long as a century or more. Also, many perennial herbs with

Figure 1 Chaparral shrubland in California. Photo by J. E. Keeley.

underground bulbs, known as geophytes, may remain dormant for long periods of time between fires.

All of the other shrub species listed above are not fire dependent and produce seeds that germinate soon after dispersal; however, successful reproduction is relatively uncommon. This is because their seedlings are very sensitive to summer drought and because there are a number of herbivores that live in the chaparral understory and prey on seedlings and other herbaceous vegetation. These include deer mice (*Peromyscus maniculatus*), woodrats (*Neotoma fuscipes*), and brush rabbits (*Sylvilagus bachmani*). Both rodents (mice and rats) are nocturnal; however, evidence of woodrats, or packrats as they are sometimes called, is very evident in many older chaparral stands because of the several foot high nests of twigs they make under the shrub canopy. These animals not only affect community structure by consuming most seedlings and herbaceous species, but also are important vectors for disease and other health threats. For example, deer mice are host to the deadly hanta-virus and woodrats are a critical host for kissing bugs (family Reduviidae) that can cause lethal allergic responses in humans. All animals including reptiles act as hosts for Lyme disease-carrying ticks (*Ixodes pacificus*). The browser of mature shrubs is the black-tailed deer (*Odocoileus hemionus*), although many are attacked by specific gall-forming wasps and aphids. Often scrub oak will have large fruit-like structures produced by gall wasp (family Cynipidae). The adult wasp oviposits on a twig, leaf, or flower and the developing larvae hijack the metabolic activities of the plant cells and force it to produce a highly nutritious spongy parenchymous tissue for the developing wasp larva.

These shrubs that reproduce in the absence of fire have successful seedling establishment largely restricted to more mesic plant communities such as adjacent woodlands, or to very old chaparral with deep litter layers that enhance the moisture holding capacity of the soil. When seedlings do establish under the shrub canopy, they typically persist for

decades as stunted saplings in the understory. These saplings are heavily browsed by rodents and rabbits and often will produce a swollen woody basal burl that survives browsing and continually sprouts new shoots. If these saplings survive until fire, they are capable of resprouting from their basal burl after fire and exhibit a growth release that enhances their chances of recruiting into the mature canopy during early succession. Thus, in some sense these shrubs may be indirectly fire dependent for completion of their life cycle.

Chaparral has a number of herbaceous or woody (lianas) vines, including manroot (*Marah macrocarpus*) and chaparral honeysuckle (*Lonicera* spp.). These vines overtop the canopy of the shrubs and flower on an annual or near-annual frequency. The former produce fleshy spiny fruits with very large seeds that are highly vulnerable to predation and the latter dry capsules with light seeds that may be wind borne. Both have weak seed dormancy and often establish seedlings in the understory.

Yucca (*Yucca whipplei*) is a fibrous-leaved species that persists as an aboveground rosette of evergreen leaves. It often survives fire because it prefers open rocky sites with very little vegetation to fuel intense fires. Because they are monocotyledonous species they have a central meristem that is protected by the outside leaves, which can withstand severe scorching. This species flowers prolifically after fire and exhibits a remarkable mutualism with the tiny yucca moth (*Tegiticula maculata*). Moth pupae survive in the soil and emerge in the growing season as adults that fly to yucca flowers where they collect pollen. They then instinctively fly to another yucca plant and pollinate the flower, ensuring cross-pollination, and then oviposit an egg in the base of the ovary. This egg soon hatches and the larva feeds on the developing seeds. Yucca moths only reproduce on yucca flowers and yuccas apparently require the pollinator services of this moth for successful seed production, a classic example of symbiosis.

Community Succession

Chaparral succession following some form of disturbance such as fire is somewhat different than in many other ecological communities. Generally all of the species present before fire in chaparral will be present in the first growing season after fire, and thus chaparral has been described as being 'auto-successional', meaning it replaces itself. In the absence of disturbance chaparral composition appears to remain somewhat static with relatively few changes in species composition or colonization by new species. In part because of the rather static nature of chaparral, old stands have been described with rather pejorative terms such as 'senescent', 'senile', 'decadent', and 'trashy', and considered to be very unproductive with little annual growth. This notion derives largely from wildlife studies done in the mid-twentieth century that concluded, due to the height of shrubs in older stands, there was very little browse production for wildlife. However, if total stand productivity is used as a measure, very old stands of chaparral appear to be very productive and are not justly described as senescent. Also, these older communities appear to retain their resilience to fires and other disturbances, as illustrated by the fact that recovery after fire (see below) in ancient stands (150 years old) recover as well as much younger stands.

Allelopathy

The lack of shrub seedlings and herbaceous plants in the understory of chaparral and related shrublands has led to extensive research on the potential role of allelopathy, which is the chemical suppression by the overstory shrubs of germination (known as enforced dormancy) or growth of understory plants. Often this lack of growth extends to the edge where these shrublands meet grasslands, and forms a distinct bare zone (**Figure 2**). The importance of allelopathy has long been disputed, with some scientists arguing that animals in the shrub understory are the primary mechanism limiting seedlings and herbaceous species from establishing. While research has not completely ruled out the possibility of chemical inhibition, it is known that for a large portion of the flora, allelopathy has no role in seed dormancy but rather dormancy is due to innate characteristics that require signals such as heat and smoke to cue germination to postfire environments rich in nutrients and light.

Fire

The marked seasonal change in climate is conducive to massive wildfires, which are spawned by the very dry shrub

Figure 2 Bare zone between chaparral and grassland. Photo by J. E. Keeley.

foliage in the summer and fall and spread by the dense contiguous nature of these shrublands. Fires have likely been an important ecosystem process since the origin of this vegetation in the late Tertiary Period, more than 10 Ma, if not earlier. Until relatively recently the primary source of ignitions was lightning from summer thunderstorms. Fires would largely have been ignited in high interior mountains and coastal areas would have burned less frequently and only when these interior fires were driven by high winds with an offshore flow. In many parts of California such winds occur every autumn and are called Santa Ana winds in southern California and Diablo winds or Mono winds in northern California. When Native Americans colonized California at the end of the Pleistocene Epoch around 12 000 years ago, they too became a source of fires, and as their populations greatly increased over the past few thousand years humans likely surpassed lightning as a source of fire, at least in coastal California. Today humans account for over 95% of all fires along the coast and foothills of California.

Chaparral fires are described as crown fires because the fires are spread through the shrub canopies and usually kill all aboveground foliage. Normally, following a wet winter, high fuel moisture in chaparral shrubs makes them relatively resistant to fire. The amount of dead branches is important to determining fire spread because they respond rapidly to dry weather and combust more readily than living foliage. As a consequence, fires spread readily in older vegetation with a greater accumulation of dead biomass. However, there is a complex interaction between live and dead fuels, wind, humidity, temperature, and topography. In particular, wind accelerates fire spread primarily by heating living fuels and often can result in rapid fire spread in young vegetation with relatively little dead biomass. Fires burning up steep terrain also spread faster for similar reasons.

Community Recovery from Wildfires

Rate of shrub recovery varies with elevation, slope aspect, inclination, degree of coastal influence, and patterns of precipitation. Recovery of shrub biomass is from basal resprouts (**Figure 3**) and seedling recruitment from a dormant soil-stored seed bank. After a spring or early summer burn, sprouts may arise within a few weeks, whereas after a fall burn, sprout production may be delayed until winter. Regardless of the timing of fire, seed germination is delayed until late winter or early spring and is less common after the first year. Resilience of chaparral to fire disturbance is exemplified by the marked tendency for communities to return rapidly to prefire composition.

Shrub species differ in the extent of postfire regeneration from resprouting versus reproduction from dormant

Figure 3 Postfire resprouts from basal burl of chamise with meter stick. Photo by J. E. Keeley.

seed banks. Most species of manzanita and ceanothus have no ability to resprout from the base of the dead stem and thus are entirely dependent on seed germination. Such shrubs are termed 'obligate-seeders'. A few species of manzanita and ceanothus as well as chamise resprout and reproduce from seeds, and these are referred to as 'facultative-seeders'. The majority of shrubs listed above, however, regenerate after fire entirely from resprouts and are 'obligate-resprouters'.

In the immediate postfire environment the bulk of plant cover is usually made up of herbaceous species present prior to the fire only as a dormant seed bank or as underground bulbs or corms. This postfire community comprises a rich diversity of herbaceous and weakly woody species, the bulk of which form an ephemeral postfire-successional flora. This 'temporary' vegetation is relatively short-lived, and by the fifth year shrubs will have regained dominance of the site and most of the herbaceous species will return to their dormant state. These postfire endemics arise from dormant seed banks that were generated after the previous fire and typically spend most of their life as

dormant seeds. These are termed 'postfire endemics' and they retain viable seed banks for more than a century without fire until germination is triggered by heat or smoke of a fire. Postfire endemics are highly restricted to the immediate postfire conditions and if the second year has sufficient precipitation may persist a second year but usually disappear in subsequent years.

Not all of the postfire annuals are so restricted, rather some are quite opportunistic, taking advantage of the open conditions after fire but persisting in other openings in mature chaparral. Such species often produce polymorphic seed pools with both deeply dormant seeds that remain dormant until fire and nondormant seeds capable of establishing in or around mature chaparral. These species fluctuate in relation to annual precipitation patterns, often not appearing at all in dry years.

Herbaceous perennials that live most of their lives as dormant bulbs in the soil commonly comprise a quarter of the postfire species diversity. Nearly all are obligate resprouters, arising from dormant bulbs, corms, or rhizomes and flowering in unison in the first postfire year. Almost none of them produce fire-dependent seeds; however, reproduction is fire dependent because postfire flowering leads to produce nondormant seeds that readily germinate in the second year.

Diversity in chaparral reaches its highest level in the first year or two after fire. It is made up of a large number of relatively minor species and a few very dominant species and is illustrated by dominance–diversity curves (**Figure 4**). Dominance in chaparral is driven by the fact that a substantial portion of resources are taken by vigorous resprouting shrubs and much less is available for the many annual species regenerating from seed.

Plants are not the only part of the biota that has specialized its life cycle to fire. Smoke beetles

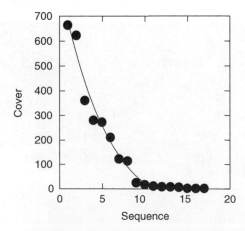

Figure 4 Dominance–diversity curve based on cover of species in sequence from highest to lowest from postfire chaparral.

(*Melanophila* spp.) are widely distributed in the western US and are attracted by the infrared heat given off by fires. Often while stems are still smoldering they will bore into the scorched wood and lay their eggs.

Seed Germination

Many chaparral species have fire-dependent regeneration, meaning that dormant seeds in the soil require a stimulus from fire for germination. A few species have hard seeds that are cracked by the heat of fire and this stimulates germination. Ceanothus seeds are a good example of this germination mode. However, for the majority of species, seeds do not respond to heat but rather to chemicals generated by the burning of plant matter. This can result from exposure to smoke or charred wood. In many of these species seeds will not germinate when placed at room temperature and watered, unless they are first exposed to smoke or charred wood. In natural environments the seeds remain dormant for decades until fire. There is evidence that a variety of chemicals in smoke and charred wood may be responsible for stimulating germination of postfire species, and both inorganic and organic compounds may be involved.

Seeds of many species have a requirement for cold temperatures ($<5\,^{\circ}C$), which is interpreted as a seasonal cue, but in these chaparral species this requirement is not like the cold stratification requirement of many species from colder climates, where the seeds require a certain duration of cold in order to prevent winter germination. In California species just a short burst of cold often will trigger germination; thus, cold is not a cue that winter is over (as with more northern latitude species) but rather that winter has arrived, which is consistent with the winter germination behavior of these Mediterranean-climate species.

Seed Dispersal

Shrubs can be divided into those with temporal dispersal versus those with strong spatial dispersal. The former are the fire-dependent species that accumulate dormant seed banks, which in essence disperse these shrubs in time, from one fire cycle to the next. Within this group there is limited spatial dispersal. Ceanothus have explosive capsules that shoot seeds a short distance of a meter or two from the parent shrub. Manzanitas drop most of their seeds beneath the parent plant because their small dry fruits are not attractive to birds, although a small number of the seeds are distributed further by coyotes (*Canis latrans*) and bears (*Ursus americans*, historically also included *U. horribilis*). Chamise produces small light fruits that may be carried tens of meters or more by the wind

but it appears that most are distributed around the parent shrub.

The postfire endemic annuals also have seeds that are largely dispersed in time rather than in space. Most do not have characteristics suggestive of widespread dispersal. For example, the fire-following whispering bells (*Emmenanthe penduliflora*) derives its name because the flowers and fruits are pendulous and drop seeds directly beneath the parent plant. Postfire endemics in the sunflower family (Asteraceae), a family noted for well-developed dispersal with dandelion-like pappus, commonly have deciduous pappus, which ensures dispersal in time rather than in space.

Shrub species that exhibit fire-free (nonfire-dependent) reproduction have fruits highly attractive to birds and mammals, and the bulk of the seed crop appears to be dispersed by these vectors. Seedling recruitment is sensitive to desiccation and thus it is of some significance that one of the main dispersers of these fruits, the scrub jay (*Aphelocoma californica*), preferentially caches seeds in the shade.

Regional Variation in Fire Regime

California chaparral exhibits regional differences in burning patterns and largely due to regional variation in winds. In much of coastal California autumn winds create severe fire conditions. These occur every year and result in 5–10 days of strong offshore flow with windspeeds of 100 kph or more. These winds result from a high-pressure system in the interior West, and are known as Santa Ana winds in southern California and Diablo or Mono winds in northern California. As these air masses move from the high-pressure cell in the interior to a low-pressure trough off the coast, the air descends and dries adiabatically, resulting in relative humidity below 10%. The fact that these winds occur every year and arrive at the end of an extended drought results in one of the most severe fire conditions in the world. As a consequence only a small portion of southern California landscape has escaped fire during the last century, and much of the lower-elevation chaparral has burned at an unnaturally high frequency.

In contrast, Santa Ana winds are absent from the southern Sierra Nevada and parts of the central coast, in part to mountain barriers that fail to funnel these winds coastward. This, coupled with lower human population density, has resulted in many fewer fires. As a consequence nearly half of the landscape in the southern Sierra Nevada has not had a fire for well over a century. This condition places these landscapes at the upper end of the historical range of variability. Nonetheless, these older stands of chaparral appear to maintain natural ecosystem processes and exhibit no sign of dying out or replacement by other vegetation types. This is particularly evident, following fires in ancient stands of chaparral from the region, that it exhibits vegetative recovery in cover and diversity indistinguishable from postfire recovery in younger stands.

Future Threats and Management

Degradation and type conversion of native shrublands to alien-dominated grasslands has been noted by numerous investigators, some of whom contend that increased frequency of disturbance is the primary factor that favors non-native annuals over woody native species. In the absence of fire, seeds of non-natives have a low residence time in the soil; thus, the presence of these species on burned sites is more often due to colonization after fire. Typically a repeat fire within the first postfire decade is sufficient to provide an initial foothold for aliens. In addition to outcompeting native plant species, non-native grasses alter the fire regime from a crown-fire regime to a mixture of surface- and crown-fires, where highly combustible grass fuels carry fire between shrub patches. This increases the likelihood of fires and ultimately increases fire frequency. As fire frequency increases there is a threshold beyond which the native shrub cover cannot recover.

Fire management practices potentially conflict with natural resource needs. These landscapes currently experience an unnaturally high frequency of fire; thus, much of it is at risk for alien invasion. When fire managers add to this by using prescription burning and other fuel manipulations, they open up these shrublands and expose them to invasion and potential type conversion to non-native grasslands. In managing these landscapes it might be helpful to consider the fact that the vast majority of alien species in California are opportunistic species that capitalize on disturbance. Adding additional disturbance through prescription burning (or grazing) will only exacerbate the alien problem.

Very little chaparral landscape is protected in parks or wilderness areas. Much of it is in private hands or under federal jurisdiction. Historically, it has largely been managed as rangeland by frequent burning to destroy the chaparral cover, or burned to reduce fuels perceived to be hazardous to more desirable forests or urban environments. Today the expansion of urban development has resulted in large portions of urban communities being juxtaposed with watersheds of potentially dangerous chaparral fuels. Historical studies show that large high-

intensity crown fires are a natural part of this ecosystem and there is little reason to believe there will not be more such fires in the future. Fire management has always worked under the philosophy that they can change the vulnerability of communities to wildfires through manipulation of fuels. However, over the past century of such management, every decade has been followed by one of increasing losses to wildfires. Californians need to embrace a different model of how to view fires on these landscapes. Our response needs to be tempered by the realization that these are natural events that cannot be eliminated from the southern California landscape. We can learn much from the science of earthquake or other natural hazard management. No one pretends they can stop earthquakes; rather, they engineer infrastructure to minimize impacts. In the future, living safely with fire is not going to be achieved solely by fire management practices, but will require close integration with urban planning.

See also: Mediterranean.

Further Reading

Arroyo MTK, Zedler PH, and Fox MD (eds.) (1995) *Ecology and Biogeography of Mediterranean Ecosystems in Chile, California and Australia.* New York: Springer.

Christensen NL and Muller CH (1975) Effects of fire on factors controlling plant growth in Adenostoma chaparral. *Ecological Monographs* 45: 29–55.

Halsey RW (2004) *Fire, Chaparral, and Survival in Southern California.* San Diego, CA: Sunbelt Publications.

Halsey RW (2005) In search of allelopathy: An eco-historical view of the investigation of chemical inhibition in California coastal sage scrub and chamise chaparral. *Journal of the Torrey Botanical Society* 131: 343–367.

Keeley JE (2000) Chaparral. In: Barbour MG and Billings WD (eds.) *North American Terrestrial Vegetation*, pp. 203–253. Cambridge: Cambridge University Press.

Keeley JE and Fotheringham CJ (2003) Impact of past, present, and future fire regimes on North American mediterranean shrublands. In: Veblen TT, Baker WL, Montenegro G, and Swetnam TW (eds.) *Fire and Climatic Change in Temperate Ecosystems of the Western Americas*, pp. 218–262. New York: Springer.

Mooney HA (ed.) (1977) *Convergent Evolution of Chile and California Mediterranean Climate Ecosystems.* Stroudsburg, PE: Dowden, Hutchinson and Ross.

Odion DC and Davis FW (2000) Fire, soil heating, and the formation of vegetation patterns in chaparral. *Ecological Monographs* 70: 149–169.

Rundel PW, Montenegro G, and Jaksic FM (eds.) (1998) *Landscape Disturbance and Biodiversity in Mediterranean-Type Ecosystems.* New York: Springer.

Wells PV (1969) The relation between mode of reproduction and extent of speciation in woody genera of the California chaparral. *Evolution* 23: 264–267.

Coral Reefs

D E Burkepile and M E Hay, Georgia Institute of Technology, Atlanta, GA, USA

Introduction

Corals are simple, clonal invertebrates that serve as ecosystem engineers, building living structures (reefs) so large that they can be seen from space. These structures, which rival the greatest feats of human engineering, are powered through symbiosis with single-celled algae that are housed within the coral animal. This coral–algal cooperation facilitates a productive ecosystem that can grow in the nutrient-poor 'desert' of isolated tropical seas. The rich structural complexity provided by the coral's hard bodies gives shelter to many other species of plants and animals making coral reefs among the Earth's most biologically diverse ecosystems, harboring hundreds of thousands to millions of species worldwide (**Figure 1**).

Coral reefs also support human societies by providing critical sources of protein, protecting coasts from damaging waves, attracting tourists, and serving as the backbone of the economies for many tropical islands. In addition, coral reefs are crucial in the fight against human diseases, as many of the plants and animals that live on coral reefs produce chemicals that are useful as pharmaceuticals. Reefs have also fascinated naturalists and scientists for centuries. Before publishing his groundbreaking work on natural selection, Charles Darwin published a treatise on reefs in 1842 hypothesizing that coral atolls (rings of reefs in the deep tropical Pacific) were formed around mountain tops as these mountains sank back into the Earth's crust under their own weight. It was more than 100 years before drilling technologies developed to the point where this hypothesis was tested – like with so many other aspects of Darwin's writings, he proved to be correct. For modern ecologists, reefs serve as a model ecosystem for developing basic hypotheses about the ecology and evolution of population structure, of community organization, and about how species diversity evolves and is maintained. In addition, reefs give us a glimpse of the spectacular record of Earth's history because the hard skeletons of corals fossilize to provide a long record of changes in coral distribution and abundance and also record chemical signals of past climatic events, like temperature and sea-level changes. Thus, reefs not only feed and protect humans and other species, but also provide a valuable window into our past, including how our present activities may be changing our environment, and possibly our future.

In this article, we review the major ecological interactions that shape coral reef ecosystems. We pay particular attention to (1) the dynamic relationship between corals and the symbiotic algae living within their tissues, (2) the role of reef herbivores in protecting corals from being overgrown by seaweeds, (3) the numerous ecological processes such as predation, competition, recruitment of juvenile reef organisms, and disturbance that influence the structure of coral reefs, and (4) the dynamic ecological connections between reefs and nearby ecosystems such as seagrass beds and mangrove forests. Finally, we review the current dangers to coral reefs and how these threats undermine the ecological integrity of these diverse ecosystems.

The Coral–Algal Mutualism

Corals are ecosystem engineers in that the growth of their calcium carbonate skeleton creates the biogenic structure on which the entire ecosystem depends. The calcification and growth of reef corals depends on a mutualism between corals and their intracellular, photosynthetic dinoflagellates, *Symbiodinium* spp., also known as zooxanthellae. Both corals and zooxanthellae benefit from this relationship, as corals can receive up to 95% of their carbon from the zooxanthellae's photosynthesis, while the zooxanthellae acquire the nitrogen and other inorganic nutrients in coral excretion products for their growth. Photosynthesis by zooxanthellae enhances calcification in corals and increases coral growth rates, ultimately leading to reef accretion and the massive reef framework found in many tropical seas. Thus, the physical structure of live and dead corals created by the coral–zooxanthellae mutualism provides heterogeneity

Figure 1 Coral reefs, like this one in the Indo-Pacific, harbor hundreds of thousands to millions of species worldwide. Photo credit M.E. Hay.

and habitat complexity, facilitating the coexistence of diverse plant and animal assemblages.

Although zooxanthellae were initially assumed to represent one species, recent molecular evidence shows that there are at least seven distinct types or clades (referred to as clades A–G). Many corals house multiple clades of zooxanthellae, setting the stage for possible competition among symbionts and for symbiont selectivity by the host. Clades of zooxanthellae differ in their photosynthetic capacity and their tolerance of light, temperature, and other stressors, making them differentially useful to their hosts under changing environmental conditions. When corals are stressed by increasing light levels or temperatures, they often expel their zooxanthellae and become pale in color (called coral bleaching). This process of bleaching may allow corals to take up new clades of zooxanthellae that are better adapted to the new environmental conditions. However, corals that fail to reacquire zooxanthellae or acquire the wrong clades may ultimately die from the stress, suggesting that a failure of corals to acquire appropriate symbionts can be fatal under changing environmental conditions. Such alterations in the coral–zooxanthellae mutualism may allow corals greater flexibility in adapting to global climate change, which is a major threat to the health of coral reefs and the integrity of the coral–zooxanthellae mutualism.

Ecological Interactions on Coral Reefs

Competition

Competition for limiting resources such as nutrients, space, light, or food is often a strong mechanism limiting the distribution and abundance of species in communities. On many coral reefs, the limiting resource for most benthic organisms is space or light, as most of the reef structure is often occupied (**Figure 2**). Consequently, corals have

Figure 2 Competition for space is often an important ecological force structuring coral reefs as corals and other invertebrates cover most of the benthos on healthy coral reefs. Photo credit M.E. Hay.

evolved a variety of competitive mechanisms including sweeper tentacles, digestive filaments, and rapid growth rates that allow them to fight neighbors for new space or protect the space they already occupy. Slow growing, massive corals often have the most potent direct competitive mechanisms (i.e., sweeper tentacles and digestive filaments that can sting and directly harm neighboring corals) while branching corals such as many *Acropora* spp. rely on their high growth rates to overtop and shade competitors. Other reef invertebrates such as sponges exude chemical compounds that are toxic to their neighbors, essentially using chemical warfare (termed allelopathy) to gain new space.

Although most early studies of competition on reefs focused on coral–coral competition, more recent studies have examined coral–seaweed competition because reefs are now more commonly overgrown by seaweeds that periodically seem to be killing corals. The conventional wisdom is that seaweeds are competitively superior and can overgrow and kill most corals. Although not all seaweeds are harmful to corals, most studies of coral–algal competition show that direct competition from seaweeds reduces the growth, survivorship, fecundity, and recruitment of many corals. Contact with the calcareous green seaweed *Halimeda opuntia* has even been shown to induce black band disease in some corals. Small, filamentous seaweeds, which are not as directly harmful to corals as are larger, foliose seaweeds, often trap sediments next to coral tissue, and this can smother and kill corals. Thus, even competition with typically innocuous filamentous seaweeds can be harmful on reefs that receive high sediment loads. However, corals are not uniformly susceptible to competition from seaweeds, and competitive outcomes may vary with coral morphology. Foliose corals such as *Agaricia* spp. are more susceptible to seaweed overgrowth than massive corals such as *Montastrea* spp. In addition, seaweeds have disproportionately high negative effects on smaller coral colonies, particularly newly recruited corals, and large stands of seaweed can prevent juvenile corals from recruiting to reefs at all.

Competition on coral reefs is not limited to sessile invertebrates; mobile animals also compete. Because herbivores are abundant on undisturbed coral reefs and standing biomass of seaweeds is generally low in these conditions, competition between herbivores would be expected. When the herbivorous sea urchin *Diadema antillarum* was removed from Caribbean reefs by either purposeful experimental manipulations or by disease outbreak, feeding rates of herbivorous fishes increased, as did the densities of some species, suggesting that fishes and urchins competed for food. *Diadema* also competed intensely with each other for food. However, competition for limiting algal resources generally did not result in a decrease in the size of *Diadema* populations but an increase in the size of their mouthparts (called the Aristotle's lantern) relative to the size of their body. Basically, *Diadema* bodies would shrink in size when food was limiting as a tradeoff between growth and survival.

Herbivory

Because seaweeds can overgrow and kill corals, herbivores are critical for coral reef function because they keep reefs free of seaweeds, thus facilitating the recruitment, growth, and resilience of corals. Fishes and urchins are typically the dominant herbivores on coral reefs with fishes in some reef areas biting the bottom at rates of >100 000 times per m^2 every day. When in sufficient numbers, either fishes alone or sea urchins alone can remove greater than 90% of the daily primary production on reefs. By feeding on seaweeds that are competitively superior to corals, herbivorous fishes both clear the substrate for settling coral larvae and prevent seaweed overgrowth of established corals. In return, the biogenic structure and topographic complexity of reef corals benefit herbivorous reef fishes and urchins by providing food, habitat, and refuges from predation. When herbivores are removed by experimentation, overfishing, or disease,

seaweeds replace corals and the biogenic structure of the reef degrades. Both reductions in coral structure and increases in seaweeds are associated with losses of herbivorous reef fishes. Interestingly, large-scale manual removals of seaweeds from reefs have resulted in only temporary increases in herbivorous fish abundance with seaweeds becoming the dominant benthic organism once again after several months. Thus, reductions in seaweeds without recovery of corals may inhibit the recovery of many reef fishes, leading to the continued degradation of coral reefs.

The main herbivorous fishes on coral reefs are generally surgeonfishes (Acanthurdiae) and parrotfishes (Scaridae) with rabbitfishes (Siganidae), chubs (Sparidae), and damselfishes (Pomacentridae) also responsible for considerable herbivory in some locations (**Figure 3**). Surgeonfishes typically feed on turf algae and foliose seaweeds with some species feeding primarily on detrital material. Parrotfishes have robust jaws with teeth fused into a beak-like formation (hence the name parrotfish), which allows them to feed on tough, calcified seaweeds in addition to algal turfs and foliose seaweeds. Although the important role of herbivores in influencing reef community is well established, less is known about the importance of individual herbivore species or the role of herbivore diversity in affecting coral reef health. Herbivore diversity should benefit reefs because a more diverse herbivore assemblage should include herbivores with varied attack strategies, which in turn should increase the efficiency of seaweed removal because particular seaweeds are unlikely to be well defended against all types of herbivores. Experimental manipulations of herbivorous fish diversity demonstrate that species-richness is important for reef function because complementary feeding by different herbivorous fishes suppresses upright seaweeds, facilitates crustose corallines and turf algae, reduces coral mortality,

Figure 3 Herbivores, like this mixed-species school of parrotfishes in the Caribbean, are important to coral reef health because they remove seaweeds that would otherwise overgrow and kill corals. Photo credit M.E. Hay.

and promotes coral growth. Hence not only are herbivores critical for coral reefs, but herbivore species-richness is also essential as a range of feeding strategies and physiologies allows efficient removal of seaweeds and promotes coral health.

Predation

Predation is often a strong top-down force in ecosystems mediating coexistence of lower trophic-level species by preventing competitive exclusion among ecologically similar organisms. In fact, predators often maintain species diversity in ecological communities by preventing expansions of certain prey that would otherwise outcompete competitively inferior organisms and come to dominate the community. If important predators are removed from a food web, the absence of their strong effects can ripple throughout the system, fundamentally altering a variety of predator–prey interactions.

The effects of the largest predators on reefs such as sharks, jacks (Carangidae), and large groupers (Serranidae) are virtually unknown due to the logistical problems of studying such large creatures and the fact that the majority of these species were rare before ecologists began studying reef ecology *in situ* (**Figure 4**). Although rigorous study of the roles that these fishes play in communities has been limited, a recent model of a Caribbean reef food web suggests that sharks are often the most strongly interacting species in these webs indicating that their removal may have had strong cascading effects on reefs. Further, surveys of lightly fished reefs in the northwestern Hawaiian Islands showed that large apex predators such as sharks and jacks represented >50% of the total fish biomass as compared to <3% on heavily fished reefs from the main Hawaiian Islands. These abundant apex predators on lightly fished reefs surely exert a strong top-down force on the community structure of these reefs.

However, the human exploitation of medium-sized predatory fishes has given us the best insight into how predation influences reef communities. On many Pacific coral reefs, outbreaks of the crown-of-thorns starfish, *Acanthaster planci*, cause loss of many square kilometers of coral reefs. These starfish are voracious coral predators that forage in large groups of up to hundreds of thousands of individuals that can decimate large stands of coral, and their outbreaks have become more frequent since the 1960s when they were first documented. Research in the Fiji islands has shown that outbreaks of *Acanthaster* are correlated to fishing pressure on reefs. *Acanthaster* are 1000 times more dense around islands that have high fishing pressure and low predatory fish abundance than they are on reefs that have light fishing pressure and high predator abundance. High densities of *Acanthaster*

Figure 4 Apex predators like sharks (top) and groupers (bottom) are now rare on many coral reefs due to overfishing. Photo credit M.E. Hay.

decrease cover of reef-building corals and crustose coralline algae while increasing cover of filamentous algae. Thus, the removal of large predators is associated with explosions of *Acanthaster* populations that then have strong cascading effects on the organization of reef communities.

A similar situation exists on many reefs in eastern Africa where intense fishing of triggerfishes and large wrasses allows population explosions of sea urchins such as *Echinometra mathaei*. Reefs unprotected from fishing have six times more urchins than protected reefs, and feeding by these dense urchin populations physically erodes the reef structure once most of the algal biomass has been consumed. This intense grazing decreases coral cover and diversity as well as increases bioerosion rates up to 20-fold compared to reefs that are protected from fishing and have abundant predators. When formerly fished areas were protected from fishing, urchin-eating fishes became more abundant and predation on urchins increased, suggesting that recovery of predatory fish populations should lead to lower urchin populations and a recovery of reef structure over time.

Disturbance

Although biotic interactions (e.g., competition and herbivory) are emphasized as having important consequences for coral reef structure, abiotic disturbances such as hurricanes, temperature fluctuations, sedimentation stress, and sea-level change also produce long-lasting effects on reefs. Coral reefs are one of the hallmark ecosystems strongly influenced by disturbance as the frequency and intensity of hurricanes or disturbance events determines how many species of corals coexist on reefs. If disturbance is very frequent or very intense, then only species that can recolonize disturbed areas quickly or that can withstand intense disturbances will persist. If disturbance is infrequent and mild, then the most competitive species eliminate the less competitive species and come to dominate. However, if disturbance is of an intermediate frequency and intensity, then species with different life-history characteristics (i.e., good colonizers vs. good competitors) can coexist because the disturbance-intolerant species are not displaced frequently and the poor competitors are not outcompeted.

Reefs often recover from acute disturbances such as storms but infrequently recover from chronic disturbances. The coupling of acute natural disturbances with chronic anthropogenic disturbances often leads to precipitous declines in coral reef health. One of the best examples of compounded disturbances driving coral reef decline is from the reefs of Jamaica. Chronic overfishing of herbivorous fishes compounded with two hurricanes and the mass mortality of the herbivorous sea urchin *Diadema antillarum* acted synergistically to force these once coral-dominated reefs into an alternate state of seaweed dominance (**Figure 5**). In more than two decades, these reefs have shown few signs of recovery. In fact, the episodic effects of natural physical disturbances, coral disease, and coral bleaching along with the constant anthropogenic disturbances of overfishing and pollution have combined to decrease coral cover an average of 80% on reefs throughout the Caribbean over the past few decades. Although disturbance is a natural and integral part of coral reef ecosystems, the compounding of many disturbances over short timescales is often more than reefs can withstand.

Positive Interactions

Ecologists now realize that positive interactions between species can have strong, cascading effects on natural communities and are no less important than negative interactions (i.e., predation or disturbance) in affecting community structure. On reefs, the most obvious positive interaction is the mutualism between corals and their symbiotic algae. Another is the positive feedback between herbivores and corals that maintains a coral-dominated

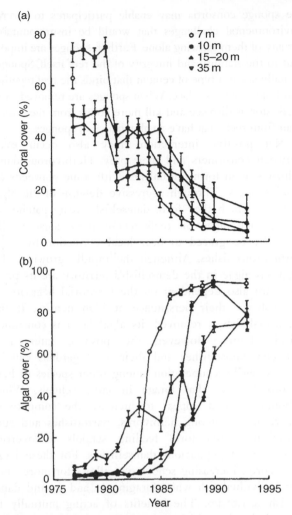

Figure 5 Degradation of Jamaican coral reefs over two decades. Changes in (a) coral cover and (b) seaweed cover at four depths in Discovery Bay, Jamaica. This decline in corals and increase in seaweeds was the result of synergistic interactions of natural and anthropogenic disturbances including overfishing, hurricanes, disease, and eutrophication. A similar decrease in coral cover occurred throughout Jamaica with coral cover nationwide declining from about 60% to about 4%. From Hughes TP (1994) Catastrophes, phase shifts, and large-scale degradation of a Caribbean coral reef. *Science* 265: 1547–1551.

ecosystem. Other crucial positive interactions come from species that are normally thought of as competitors but can mutually benefit each other under the right conditions. Sponges, for example, compete with each other, but can also interact positively. It is more common to find morphologically similar species of sponge growing intermingled in multispecies groups than it is to find a sponge colony growing alone. When growing in these groups, the growth rates of the different species of sponge are often greater than what they would be if these sponges were growing by themselves. This enhancement of growth rates may stem from differences among the species in their susceptibility to predation, pathogens, and physical disturbance. The summed traits of the sponge consortia may enable participants to survive environmental challenges that would be insurmountable for any of them growing alone. Further, sponges are important to the stability and integrity of the reef itself. Sponges actually act as a type of cement that binds the reef together and holds corals in place. When sponges are removed from reefs, storms displace and kill more corals from these reefs than from reefs that have an abundance of sponges.

Net positive interactions may also occur even between consumers and their prey. Herbivorous damselfishes often form mutualisms with some seaweeds on tropical reefs. Through aggressive defense of the algal mats on which they feed, damselfish create patches of species-rich algae on reefs where these algae would normally be grazed to near local extinction by large herbivorous fishes. Although the rapidly growing filamentous algae in the damselfish's territory are its prey, they are also dependent on the territorial behavior of the fish for their persistence at high density. If the territorial fish is removed, its algal lawn is consumed within hours. However, the positive interactions between damselfishes and their algal gardens can be overridden by cooperation among other species of herbivorous fishes that forage in large schools. While schooling would appear to increase the competition for resources among herbivores, parrotfishes and surgeonfishes often form feeding schools to overrun territories of pugnacious damselfishes. For these large herbivores, increasing school size allows for more bites per individual fish when foraging in and around damselfish territories. The benefits of acting mutually to overwhelm damselfish and gain access to resource-rich habitats must outweigh the potential for competitive interactions between the fishes using these schools. Similarly, piscivorous fishes such as grouper and moray eels often hunt in mixed-species or cooperative foraging groups. For these fishes that forage cooperatively, many mouths may be better than one in terms of overall prey yield to each predator when summed over a lifetime of hunts.

Finally, small wrasses (Labridae) and gobies (Gobiidae) act as cleaner fishes on tropical reefs and remove parasites, mucus, and dead or infected tissue from larger fishes. Reef-based cleaner fish are found at specific cleaning stations, usually situated on prominent portions of the reef. These fish can clean up to 2300 individuals of 132 different species in a day, and some client fish visit cleaners over 100 times a day. If these cleaner fish are removed from reefs, diversity of reef fishes declines, especially for large, transient fishes that may visit reefs specifically to be cleaned. Cleaner fishes can, thus, have a strong effect on parasite loads of their client fishes and on fish-usage patterns across patchy reef environments.

Replenishment of Coral Reefs: The Role of Reproduction and Recruitment in the Ecology of Reefs

Most reef organisms are sessile (corals, sponges, seaweeds) or use only a small portion of a much larger reef habitat (most reef fishes). Thus, colonization of new habitats is achieved through recruitment of juvenile organisms that may drift for long distances in the plankton before settling onto, and using a small area of reef. Consequently, the production and recruitment of juvenile organisms is a key factor in the ecology of reefs as the replenishment of plant and animal populations is integral to the resilience and recovery of reefs in the face of natural and anthropogenic disturbances.

Coral species differ considerably in their modes of reproduction and in the ability of their larvae to disperse to new reefs. Many corals reproduce both asexually through fragmentation and sexually by the production of gametes. Important reef-building corals such as acroporids are extremely successful at reproducing asexually and are dispersed when storms break apart parent colonies and spread the fragments to new portions of a reef where they can reattach and grow. Sexual reproduction in corals is also variable in that corals are typically either brooders or spawners. Brooders release fertilized larvae into the water column while spawners release sperm and eggs into the water column, where they fertilize and disperse with the ocean currents. These fertilized larvae will eventually exit the plankton and return to reefs as newly recruited juvenile corals. Research from the Great Barrier Reef, Australia has shown that there is large variation in the abundance of coral recruits across both large and small spatial scales. The best predictor of differences in recruitment rates among reefs was the fecundity, not abundance, of adult corals and explained 72% of the variation in recruitment for acroporid corals. Recruitment rates decreased dramatically as the fecundity of adults decreased, but this decrease was not linear; a small decrease in the fecundity of adults resulted in a dramatic decrease in juvenile recruitment (**Figure 6**). These results suggest that processes such as sedimentation, eutrophication, and competition with seaweeds, all of which reduce the fecundity of adult corals, could dramatically affect the replenishment of coral populations.

Recruitment of juveniles is also important to the replenishment of fish populations and considerable research has gone into determining how recruitment processes affect the assembly of reef fish communities. Most reef fishes, like corals, have planktonic larvae that can disperse wide distances from their point of origin. One of the key questions in the ecology of reef fishes is how the recruitment of juvenile fishes is related to the density of fishes already on the reef (i.e., whether local patterns of recruitment are density dependent or density independent). Despite considerable research on the subject, little

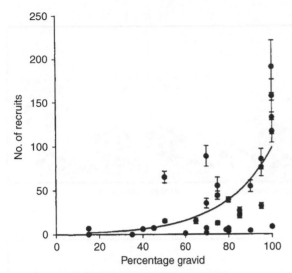

Figure 6 Relationship between the percentage of coral colonies (*Acropora hyacinthus*) that were gravid and the number of coral recruits. Each point is a separate reef on the Great Barrier Reef, Australia. From Hughes TP, Baird AH, Dinsdale EA, *et al*. (2000) Supply-side ecology works both ways: The link between benthic adults, fecundity, and larval recruits. *Ecology* 81: 2241–2249.

consensus has been reached and studies have shown that recruitment rates can be either positively or negatively correlated with adult abundance (positively or negatively density dependent), or show no correlation at all (density independent). These relationships may vary with the species being studied, with location, or with the currents and physical processes prevailing at the time of the test. Continued research is needed to generalize how recruit and adult densities are related and how environmental and biological variables change these relationships.

A key component to the replenishment of populations of coral reef organisms is the extent to which reefs are connected to other reefs (i.e., whether juveniles recruit to reefs from local or distant sources). Coral reefs, and marine ecosystems in general, differ from many terrestrial systems in that juvenile organisms have the potential to ride ocean currents and be dispersed over wide distances potentially connecting geographically distant populations. However, the actual extent that marine populations are connected to each other is still a topic of vigorous debate. This knowledge is crucial to the protection and management of reefs as the connectivity of populations of coral reef organisms will determine whether local populations can be managed with efforts based close to the target population (if the system is fairly closed and recruitment from local populations is frequent) or if management of local populations will necessitate international cooperation (if reefs are fairly open and recruitment is driven by larvae from distant reefs). Thus, solving the question of connectivity among reefs is critical to the preservation of reef health.

Initial models of connectivity for fish populations in the Caribbean suggested that many of the populations were very open and well connected to other populations hundreds of kilometers away. However, these models were based on passive dispersal of fish larvae and simple models of surface currents and did not account for the effect of larval behavior on dispersal or for the effects of fine-scale oceanographic processes. Thus, viewing larvae as passive dispersal agents may overestimate the actual dispersal of larvae and the connections among reefs. Recent models of connectivity in the Caribbean that account for larval behavior suggest that fish populations are less connected than assumed under passive dispersal models and that different regions of the Caribbean are essentially isolated from each other, at least on an ecological timescale. However, there are considerable differences among regions of the Caribbean as to the extent that reefs are connected to distant areas as some regions appear to import a large portion of recruits while other regions are primarily self-seeding. The differences in the relative importance of local and long-distance recruitment of juveniles among regions of the Caribbean underscores the role that careful planning will play in the implementation of marine reserves for protection of coral reefs as understanding how reefs are connected to one another will influence how large reserves should be and where they should be located.

Landscape Ecology of Coral Reefs: Connections of Coral Reefs to Mangrove and Seagrass Systems

Coral reefs are typically found in close proximity to other coastal ecosystems, particularly seagrass beds and mangrove forests. These different ecosystems are often connected to reefs via the movement of animals and nutrients across their boundaries. For example, carnivorous grunts (Haemulidae) forage in seagrass beds at night but school around large coral heads on reefs during the day as a refuge from predation. Coral heads that harbor fish schools receive nutrient supplements from fish excretion, grow up to 23% faster, and have more nitrogen and zooxanthellae per unit area than do corals without resident fishes. Thus, fishes that have no direct trophic link with corals collect nutrients from other ecosystems (seagrass beds) and concentrate these near their host coral. This facilitates coral growth, enhancing the coral's value as a refuge for these fishes and for other reef organisms.

Mangroves and seagrass beds also serve as nursery grounds and provide refuge from predators and an abundance of food for many juvenile fishes that are typically found on coral reefs as adults. Grunts (Haemulidae), snappers (Lutjanidae), barracuda (*Sphyraena barracuda*), and some parrotfishes (Scaridae) are particularly

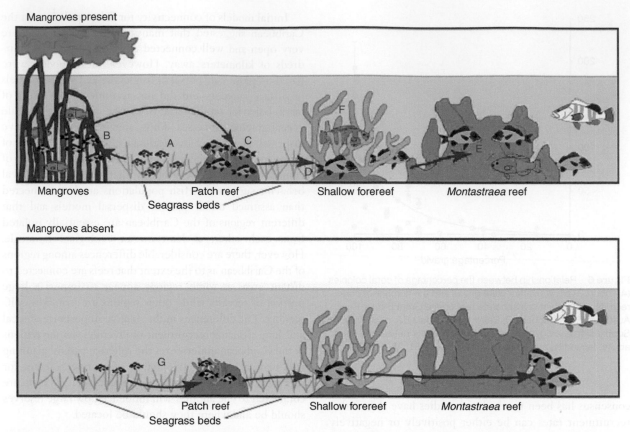

Figure 7 Schematic illustrating the connection between mangroves and coral reefs. Ecosystem connectivity is stylized for *Haemulon sciurus* (gray and black fish) and *Scarus guacamaia* (orange and green fish) although other parrotfish (scarid), grunt (haemulid), and snapper (lutjanid) species also exhibited similar ontogenetic shifts in habitat use. *H. sciurus* show a substantial shift in size frequency from seagrass (A) to mangroves at approximately 6 cm. On reaching a given size in seagrass beds, juvenile fish then move to mangroves (B) which serve as an intermediate nursery habitat before migrating to patch reefs (C). If mangroves are not present, *H. sciurus* move directly from seagrass to patch reefs, appearing on patch reefs (G) at a smaller size and at lower density (260 ha^{-1} compared to 3925 ha^{-1} in mangrove-rich systems). In the presence of mangroves, the biomass of *H. sciurus* is significantly enhanced on patch reefs, shallow forereefs, and *Montastraea* reefs (C, D, E). *S. guacamaia* (F) has a functional dependency on mangroves and is not seen where mangroves are absent. Illustration describes findings from Mumby PJ, Edwards AJ, Arias-Gonzalez JE, *et al.* (2004) Mangroves enhance the biomass of coral reef fish communities in the Caribbean. *Nature* 427: 533–536. Schematic and description courtesy of Peter J. Mumby.

dependent on the presence of nearby mangroves. In Belize, reefs closely associated with mangroves have up to 26 times more biomass of some species of fish than reefs not associated with mangroves. A common species on these reefs, the bluestriped grunt (*Haemulon sciurus*), typically goes through an ontogenetic change in habitat use as it migrates from seagrass beds to mangroves to patch reefs to the forereef as it ages (**Figure 7**). In areas where mangroves are absent, bluestriped grunts move from seagrass beds directly to patch reefs and are typically smaller than grunts that inhabit patch reefs with nearby mangroves. Thus, mangroves provide important habitat where juvenile grunts feed and increase in size before moving to patch reefs which may subsequently decrease the threat of predation once they move to these reefs. Further, the rainbow parrotfish (*Scarus guacamaia*), the largest herbivorous fish in the Caribbean, is functionally dependent on mangroves for shelter; juveniles of

this species live primarily in mangroves, and the species goes locally extinct on reefs when nearby mangroves are removed (**Figure 7**). Interestingly, density of fishes that have no direct link to mangroves at any stage of their life history can still be influenced by the proximity of mangroves, probably via interactions with mangrove-dependent fishes. Thus, the composition of the fish community on reefs is greatly influenced by the proximity of mangroves, and the rapid removal of mangrove forests from coastlines worldwide will certainly have drastic negative impacts on the ecology coral reefs.

Geographic Distribution of Coral Reefs

Coral reefs exist in tropical areas worldwide (**Figure 8**). In general, reefs are abundant in areas with shallow coastlines and clear, warm water where riverine discharge of

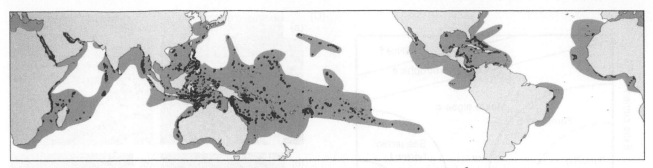

Figure 8 Worldwide distribution of coral reefs. Coral reefs (red dots) cover roughly 250 000 km² of the Earth's surface, but zooxanthellate corals inhabit a far wider range (blue shading). From Veron JEN (2000) *Corals of the World,* vols. 1–3. Townsville, QLD: Australian Institute of Marine Science.

sediments is low. Large coral reefs are rarely found in areas above 29° latitude where ocean temperatures fall below 18 °C for extended periods as this slows coral growth and their capacity to build large reefs; however, zooxanthellate corals can be found in areas with water temperatures as low as 11 °C. In addition, herbivory is often less intense in cooler waters meaning that seaweeds are more abundant in temperate areas and that competition between corals and seaweeds is more intense. The combination of cooler water temperatures and more intense competition with abundant seaweeds likely interact to limit the latitudinal range of large coral reefs. However, when the physical and ecological criteria are met, the results can be phenomenal. For example, the most biologically diverse reefs occur in the tropical Indo-Pacific in the areas around Indonesia and the Philippines and house over 550 species of coral and thousands of species of fish. The Great Barrier Reef off northeastern Australia is the largest reef in the world with more than 2800 individual reefs occupying over 1800 km of the Australian coastline and can be seen from outer space.

Threats to Coral Reefs

Coral reefs are imperiled around the world because of the compounding effects of multiple stressors such as overfishing, pollution, climate change, and change in coastal land use. The decline of reefs is particularly evident in the Caribbean where coral cover has decreased by 80% in recent decades and may drop further as reefs fail to rebound following continued coral bleaching, overfishing, disease outbreaks, and other disturbances. The causes of coral reef decline are many and frequently act synergistically to drive coral reefs to alternate states such as seaweed-dominated reefs or sea urchin barrens (**Figure 9**). We review the major threats to coral reefs and the role that marine reserves can play in stemming the tide of coral reef decline.

Coral Bleaching

Coral bleaching occurs when corals degrade or expel their dinoflagellate symbionts in response to environmental stressors such as elevated sea surface temperature and increased UV radiation. Although corals can reacquire symbionts and recover in weeks to months, recovered corals may grow slower and have reduced fecundity as compared to previously unbleached corals, giving bleaching-resistant corals an ecological advantage after bleaching events. In severe cases, bleaching may occur on the scale of hundreds to thousands of kilometers and radically alter coral cover and composition with coral mortality from bleaching events approaching 100% in extreme cases. Branching corals such as acroporid and pocilloporid corals are often more susceptible to bleaching and mortality than are massive corals, allowing the slower-growing massive corals to be more persistent on reefs after bouts of strong bleaching. Bleaching events not only decrease live coral cover but also provide large areas for seaweed colonization, and these seaweeds can prevent corals from reestablishing if herbivores are not present in sufficient numbers to suppress seaweed colonization and growth. Additionally, large-scale bleaching and mortality of branching corals can suppress fish populations that are dependent on live coral for shelter and food.

Analyses of coral bleaching on Caribbean reefs over the past two decades suggests that small increases in regional sea surface temperature (0.1 °C) result in large increases in the geographic extent and intensity of coral bleaching events. Given that climate change models suggest an increase in sea surface temperatures of 1–3 °C over the next 50–100 years, coral bleaching events may become an intense, annual stress on coral reefs throughout the Caribbean and even the world. Although corals may adapt and their bleaching thresholds may increase over time as sea surface temperatures rise, the threat of repeated, intense bleaching events over the next several decades is a significant concern. If even the conservative predictions of global climate models are realized, these climate changes could result in the fundamental reorganization of the ecology of coral reefs.

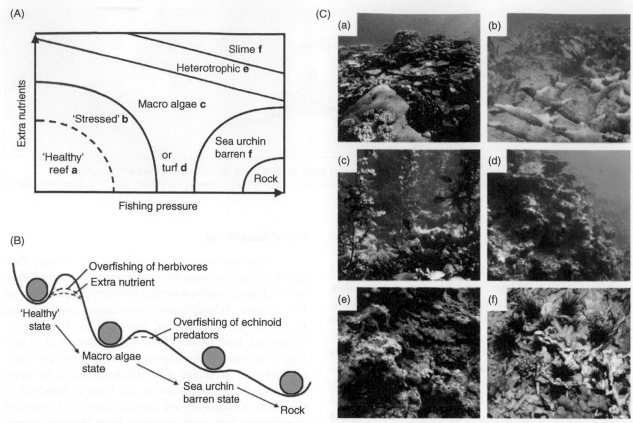

Figure 9 Alternate states in coral reef ecosystems. (A) A conceptual model showing human-induced transitions between alternate ecosystem states based on empirical evidence of the effects from fishing and excess nutrients. The 'stressed' state illustrates loss of resilience and increased vulnerability to phase-shifts. (B) A graphic model depicting transitions between ecosystem states. 'Healthy' resilient coral-dominated reefs become progressively more vulnerable owing to fishing pressure, pollution, disease, and coral bleaching. The dotted lines illustrate the loss of resilience that becomes evident when reefs fail to recover from disturbance and slide into less-desirable states. (C) Pictoral representation of the different reef states shown in (A). From Bellwood DR, Hughes TP, Folke C, and Nystrom M (2004) Confronting the coral reef crisis. *Nature* 429: 827–833.

Disease and the Structure of Coral Reef Communities

The impact of diseases on coral reefs has been realized over only the past two decades. Two of the most extensive disease outbreaks have been on reefs in the Caribbean and have fundamentally changed the ecology of Caribbean reefs. In 1983–84 an unknown pathogen swept through the Caribbean and killed approximately 99% of the then abundant sea urchin *Diadema antillarum*. In many areas of the Caribbean, *D. antillarum* had been the dominant herbivore keeping reefs free of most fleshy seaweeds and facilitating recruitment and growth by corals. After the mass mortality, levels of herbivory plummeted and standing crop of seaweeds dramatically increased on many reefs. *D. antillarum* populations are recovering in some areas of the Caribbean, and in these 'urchin zones', seaweeds cover 0–20% of the reef as opposed to 30–79% of the reef. Juveniles corals are ten times more abundant in some urchin zones. The potential recovery of this critical herbivore gives hope to

Caribbean reefs many of which are still enveloped in a blanket of seaweed.

The other outbreak that altered the structure of Caribbean reefs was the epidemic of white band disease among acroporid corals in the mid to late 1980s. This disease attacked two of the major reef-building corals in the Caribbean *Acorpora palmata* and *A. cervicornis*. These two corals were once so abundant on Caribbean reefs that early coral reef ecologists named characteristic zones on reefs for these dominant corals (i.e., the 'palmata zone' and the 'cervicornis zone'). These corals that had dominated Caribbean reefs for at least a half million years are now rare to absent on most reefs and have declined so dramatically that they are both being listed as threatened species under the Endangered Species Act in the United States. If the prevalence and severity of coral diseases is linked to pollution and climate change as has been demonstrated for some studies, then a continued increase in the effects of diseases on the ecology of reefs can be expected.

Shifting Baselines, Overfishing, and the Altered Food Webs of Coral Reefs

In many regions of the world, coral reefs are mere remnants of what they were only a few decades ago. These changes to reefs are not adequately appreciated due to the problem of the 'shifting baseline syndrome' – reefs that are deemed 'normal' today are not what was 'normal' only a few decades ago, much less a century or more ago. Each new generation of divers or marine ecologists suffers from reduced expectations of what a healthy coral reef should be. For example, as a graduate student and post-doc, one of us (M.E. Hay) dove on Caribbean reefs dominated by luxuriant stands of elkhorn and staghorn coral (*Acropora palmata* and *A. cerviconis*) the size of football fields and saw reefs abundant with grouper, large herbivorous fishes, and *Diadema* urchins that formed 'fields' of gigantic black pincushions on regions of some reefs. In contrast, the younger author here (D.E. Burkepile) has never seen a live stand of elkhorn coral more than a few m^2 and is lucky to see one *Diadema* on most dives. However, both of us dive on reefs that are vastly different from those that the first European colonists would have experienced. Because of this problem of shifting baselines, it is informative for ecologists to explore the history and paleoecology of reefs in order to deduce how reef communities have changed over hundreds, or thousands, of years.

Caribbean reefs were once dominated by sea turtles, crocodiles, manatees, large predatory fishes such as sharks and large groupers, and the now-extinct monk seal. Reefs with such a diversity of charismatic megafauna scarcely exist today anywhere in the world. Centuries of overfishing have made many of these species ecologically extinct and altered the strong trophic interactions that once dominated Caribbean food webs (**Figure 10**). Including humans into the ecological equation began a process of 'fishing down the food web' where large consumers such as sharks and manatees were the primary targets of human harvesting. After the larger animals were depleted, fisheries switched to smaller predators such as groupers and then to herbivores such as parrotfishes. The changes in the connections of these food webs have fundamentally altered the dynamics of these ecosystems and have resulted in cascading effects such as the decline of corals and increase of seaweeds and sponges.

Although the largest megafauna are now largely gone from Caribbean reefs, we have some idea of their historical populations. For example, green turtles were once so abundant that ships' naturalists from the sixteenth to seventeenth centuries remarked that they could navigate to the Cayman Islands via the sounds of turtles swimming and that congregations of turtles seemed so thick as to confound a ship's path. One estimate puts the total pre-Columbian population of green turtles in the Caribbean at greater than 30 million as opposed to the tens of thousands today. Green turtles typically eat seagrasses and seaweeds, but the top-down force that this historical population would have exerted on seagrass production and biomass is unrivaled by any current estimates of herbivory in seagrass beds. Because the biota of coral reefs has changed so dramatically over the past few hundred years, Jeremy Jackson writes that scientists trying to understand the ecological processes that structure coral reefs are "...trying to understand the ecology of the Serengeti by studying the termites and the locusts while ignoring the elephants and the

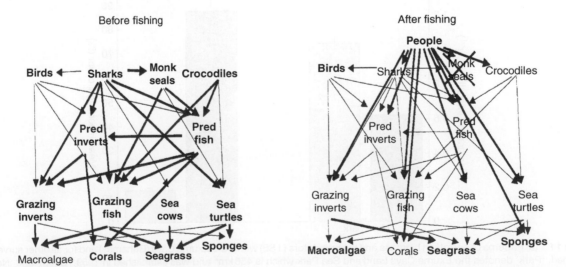

Figure 10 Simplified coastal food web for coral reefs showing changes in some important top-down interactions due to overfishing; before (left side) and after (right side) fishing. Bold font represents abundant; normal font represents rare; 'crossed-out' represents extinct. Thick arrows represent strong interactions; thin arrows represent weak interactions. Modified from Jackson JBC, Kirby MX, Berger WH, *et al.* (2001) Historical overfishing and the recent collapse of coastal ecosystems. *Science* 293: 629–638.

wildebeest." Basically, the biotic forces that impact coral reefs today are mere shadows of what they once were, and humans have radically changed the ecological and evolutionary trajectories that have influenced coral reef ecosystems for millennia.

Protection and Resurrection of Coral Reefs

One of the saving graces of coral reefs over the next few decades may be the creation and enforcement of marine reserves that protect reefs from overfishing. Overfishing is one of the most devastating threats to reefs, as fishers preferentially remove the large-bodied fishes that are the strongest interactors in these ecosystems, resulting in fundamental changes to the food webs of reefs. The establishment of marine reserves limits or prevents the harvesting of fishes and invertebrates from areas of reef and theoretically allows populations of overharvested species to rebound, reestablishing viable populations of fishes and crucial ecosystem processes on reefs. Recent studies of the efficacy of marine reserves show that reducing fishing pressures on reefs allows increases in the

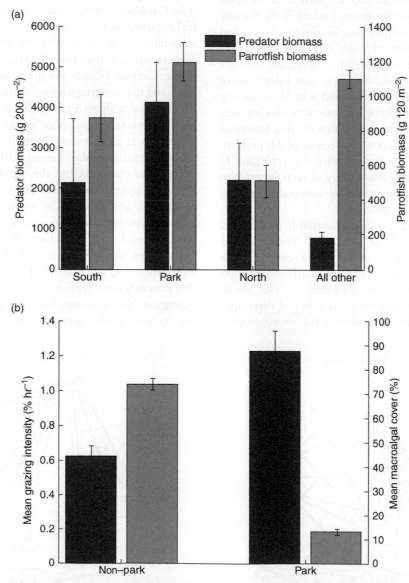

Figure 11   (a) Patterns of parrotfish biomass and their predators (±SE) within the Exuma Cays (Bahamas) and for all other survey areas combined. 'Park' denotes the Exuma Cays Land and Sea Park which is 456 km^2 and was established in 1959. 'South' and 'North' represent reef systems that are near the southern and northern borders of the park. (b) Mean macroalgal cover (gray bars) (±SE) and grazing intensity of parrotfishes (black bars) inside and outside the Exuma Cays Land and Sea Park. Reserve impacts are significant ($P < 0.01$) for each variable. From Mumby PJ, Dahlgren CP, Harborne AR, *et al*. (2006) Fishing, trophic cascades, and the process of grazing on coral reefs. *Science* 311: 98–101.

density, biomass, individual size, and diversity of fishes and invertebrates inside the reserves and that these effects occur rapidly and are longlasting. In addition, these reserves not only allow increases in fish density and biomass within the protected areas but also result in the 'spillover' of fishes as they migrate from the reserves into unprotected areas. Thus, marine reserves may subsidize fish populations on reefs that are not directly protected from fishing, although the extent to which this spillover effect will actually affect unmanaged reefs is equivocal.

Marine reserves can also restore trophic linkages that enhance the recovery of coral reefs. For some reefs in the Bahamas, long-term protection from fishing (i.e., roughly 50 years of enforcement) has led to increases in the abundance of medium-sized predatory fishes such as the Nassau grouper (*Epinephelus striatus*) (**Figure 11**). Increases in grouper abundance resulted in increased predation rates on small herbivorous parrotfishes, which would seemingly decrease the rate of herbivory on reefs. However, the protection from fishing also allowed large parrotfishes to recover and actually increased the overall rate of herbivory in the reserve despite increased predation on smaller herbivores (**Figure 11**). These increased rates of herbivory decreased macroalgal abundance and may increase coral abundance and cover over time if this balance between predation and herbivory can be maintained. Although the benefits of reserves to conservation and fisheries are promising, one of the main challenges to the success of marine reserves is the enforcement of no-harvesting policies once the reserve is established. In many areas, reserves are 'paper parks' or parks in name only as there is insufficient money or political will to achieve the enforcement necessary for the reserves to succeed. However, if marine reserves can be implemented and enforced they will be one of the best tools that conservation science currently has to protect, and hopefully resurrect, many coral reefs.

Summary

Fossil evidence shows that corals have dominated many reefs for over 10 000 years. However, the balance between the ecological forces of predation, competition,

disturbance, and recruitment that allowed thousands of years of uninterrupted reef growth have now been grossly altered by human activities. Consequently, healthy and dynamic reefs have declined dramatically in the last two decades as a result of overfishing, climate change, pollution, and other anthropogenic insults. Although the ecological future of reefs seems bleak, we hope that creative management of these ecosystems has the potential to protect them for future generations.

Further Reading

Bellwood DR, Hughes TP, Folke C, and Nystrom M (2004) Confronting the coral reef crisis. *Nature* 429: 827–833.

Birkeland C (ed.) (1997) *Life and Death of Coral Reefs*. New York: Chapman and Hall.

Burkepile DE and Hay ME (2006) Herbivore versus nutrient control of marine primary producers: Context-dependent effects. *Ecology* 87: 3128–3139.

Cowen RK, Paris CB, and Srinivasan A (2006) Scaling and connectivity in marine populations. *Science* 311: 522–527.

Dulvy NK, Freckleton RP, and Polunin NVC (2004) Coral reef cascade and the indirect effects of predator removal by exploitation. *Ecology Letters* 7: 410–416.

Gardner TA, Cote IM, Gill JA, Grant A, and Watkinson AR (2003) Long-term region-wide declines in Caribbean corals. *Science* 301: 958–960.

Halpern BS (2003) The impact of marine reserves: Do reserves work and does reserve size matter? *Ecological Applications* 13: S117–S137.

Hay ME (1997) The ecology and evolution of seaweed–herbivore interactions on coral reefs. *Coral Reefs* 16: S67–S76.

Hughes TP (1994) Catastrophes, phase shifts, and large-scale degradation of a Caribbean coral reef. *Science* 265: 1547–1551.

Hughes TP, Baird AH, Dinsdale EA, *et al.* (2000) Supply-side ecology works both ways: the link between benthic adults, fecundity, and larval recruits. *Ecology* 81: 2241–2249.

Jackson JBC, Kirby MX, Berger WH, *et al.* (2001) Historical overfishing and the recent collapse of coastal ecosystems. *Science* 293: 629–638.

Knowlton N and Rohwer F (2003) Multispecies microbial mutualisms on coral reefs: The host as a habitat. *American Naturalist* 162: S51–S62.

McClanahan TR and Mangi S (2000) Spillover of exploitable fishes from a marine park and its effect on the adjacent fishery. *Ecological Applications* 10: 1792–1805.

McCook LJ, Jompa J, and Diaz-Pulido G (2001) Competition between corals and algae on coral reefs: A review of evidence and mechanisms. *Coral Reefs* 19: 400–417.

Mumby PJ, Dahlgren CP, Harborne AR, *et al.* (2006) Fishing, trophic cascades, and the process of grazing on coral reefs. *Science* 311: 98–101.

Mumby PJ, Edwards AJ, Arias-Gonzalez JE, *et al.* (2004) Mangroves enhance the biomass of coral reef fish communities in the Caribbean. *Nature* 427: 533–536.

Veron JEN (2000) *Corals of the World*, vols.1–3. Townsville, QLD: Australian Institute of Marine Science.

Desert Streams

T K Harms, R A Sponseller, and N B Grimm, Arizona State University, Tempe, AZ, USA

Distribution and Physical Template	Nutrient Dynamics
Temporal Dynamics	Human Modifications
Biota	Further Reading
Energetics	

Distribution and Physical Template

Desert streams occupy arid and semiarid regions defined by low annual precipitation. Semiarid and arid climate zones are found on all continents and include both hot and cold deserts. Although the range of temperatures varies across desert regions, summer temperatures may exceed 40 °C in hot deserts. Annual precipitation ranges from <100 to 300 mm yr^{-1} and combined with high temperature can result in high rates of evapotranspiration. Higher precipitation in the mountains (up to ~1000 mm yr^{-1}) can feed streamflow in the low deserts, often supporting perennial flows in large basins. Arid and semiarid regions are characterized by distinct seasons defined by precipitation and/or snowmelt and the amount of precipitation that falls during these seasons shows high interannual variability. This results in extreme seasonal and interannual variation in stream discharge. Indeed, streams in some desert regions flow in response to rain events that occur only once in several years or even less.

Arid and semiarid lands account for over one-third of global lands, making desert streams prominent among aquatic ecosystems. The large geographic area covered by deserts results in a wide variation in temperature and precipitation regimes as well as in geomorphology. Thus, the hydrogeomorphic templates and resulting ecological characteristics of desert streams exhibit a great diversity of patterns. Despite this extensive distribution of desert streams, the vast majority of ecological studies of desert streams have occurred in the southwestern United States, Australia, and Antarctica. Our discussion thus draws from results of studies in these ecosystems. Future studies of desert stream ecosystems in other regions are likely to add new dimensions to the state of our understanding presented here.

Desert stream hydrographs are punctuated by events when discharge may exceed baseflow by several orders of magnitude. Precipitation falling on the catchment rapidly reaches the stream and stream discharge rapidly dissipates following floods. Infiltration of desert soils is minimal and, at the scale of whole basins, much of the water that is not returned to the atmosphere by evapotranspiration reaches

streams via overland flow during storms or via infiltration of permeable low-order channel sediments followed by subsurface flow. Resulting flash floods scour the streambed, resulting in downstream export of sediments and aquatic organisms and creating a wide channel. Large floods also deposit alluvial materials in riparian zones and may remove riparian vegetation. These effects vary depending on the scale of the event (see the section titled 'Temporal dynamics').

The boundaries of a stream ecosystem in any climate region extend beyond the wetted channel and comprise a stream–riparian corridor (**Figure 1**). The aquatic ecosystem encompasses surface water as well as the alluvial sediments beneath the streambed where surface and groundwater mix, termed the hyporheic zone. The parafluvial zone is defined by the region of the active channel over which water flows only during floods and in desert streams this region can be much wider than the stream itself. Finally, the riparian zone is the land area surrounding the stream that is significantly influenced by the stream. The availability of water contrasts starkly among these subsystems in desert streams making each subsystem more distinct than in mesic streams. In deserts, the

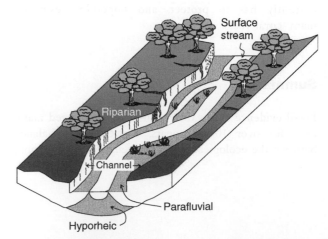

Figure 1 Schematic drawing of a desert stream–riparian corridor.

hyporheic zone often contains water and sustains biological activity in the subsurface even in the absence of surface flow. The parafluvial zone contains surface water only during floods and flow quickly recharges the subsurface through coarse sediments or gravel in this zone, leading to short periods of surface flow but sustained subsurface flow. The riparian zone contrasts starkly with desert uplands owing to the presence of shallow groundwater that is accessible to plants.

Desert streams may contain sections of both gaining and losing hydrologic templates. Gaining sections of streams are those in which the water table is sloped toward the stream channel such that groundwater discharges into the surface stream. Losing reaches are characterized by a water table that slopes away from the stream causing surface flow to recharge groundwater. Along losing reaches, the predominate direction of water flow in desert catchments is from the uplands directly into the stream before recharging riparian groundwater; whereas along gaining reaches, water flowing overland into the riparian zone recharges groundwater there before discharging into the surface stream. These contrasting hydrologic templates can have significant effects on nutrient dynamics, water storage, and biota within stream–riparian corridors. Losing reaches, for example, often have no surface flow during dry seasons, whereas gaining reaches are a more permanent source of surface water. Due to permeable sediments, interactions between surface and subsurface water are dynamic. For example, water moves into the hyporheic zone in regions of downwelling and from the hyporheic zone to the surface in regions of upwelling. Water within the hyporheic zone moves through the interstitial spaces of sediments slowing water velocity compared to the surface stream. Such patterns in hydrologic flows have important implications for nutrient cycling and stream productivity.

Temporal Dynamics

Desert streams are highly variable over time, at a range of temporal scales. In addition to the seasonality that typifies many streams, temporal dynamics of desert streams are strongly influenced by disturbances at two extremes (flash flooding and drying) of a hydrologic spectrum. Researchers have largely focused on temporal dynamics at decadal and lower scales; however, decade- to century-scale channel change establishes a geomorphic template upon which these higher-frequency dynamics play out. Our discussion will consider temporal change from low- to high-frequency events (**Figure 2**).

The concept of disturbance has various meanings, but in stream ecology disturbance is usually associated with hydrologic extremes that change ecosystem structure and processes. Using terms from disturbance ecology, we can characterize a disturbance regime, which has features such

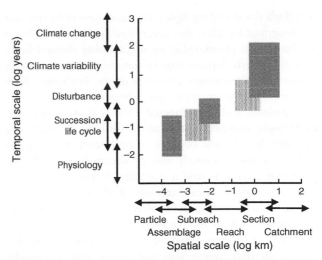

Figure 2 Temporal scales considered in this section (ordinate) are correlated with spatial scales (abscissa), and each phenomenon discussed is associated with a characteristic range of time and space scales.

as interannual (or interdecadal) variability, seasonality, and timing, frequency, and magnitude of individual events. Disturbance is intimately connected to succession, which is most simply defined as the change in ecosystem properties on a site following disturbance. Ecosystem components that are affected by a disturbance are those that undergo succession after the disturbance; for example, a flash flood that removes algae and invertebrates but does not affect streamside vegetation initiates succession in the stream but not in the riparian zone.

Successional patterns depict the temporal changes in stream and riparian communities and processes that are superimposed upon a larger temporal scale of variability. For longer-lived riparian vegetation, successional patterns and time frames may be similar to those of terrestrial communities but for stream biota, succession often plays out against a seasonal backdrop. Thus, successional patterns differ between seasons, with faster increases in biomass during warmer months. Successional patterns also vary depending upon the size and nature of the initiating disturbance as well as antecedent conditions (which are themselves influenced by the disturbance regime – timing and clustering of individual events). Whereas disturbances that occur over short timescales may produce predictable recovery sequences, biota recover from longer-term, infrequent disturbances with less regularity. Effects on biota of pronounced interannual variation that characterizes deserts include shifts in community composition of invertebrates and differences in the relative importance of nitrogen-fixing cyanobacteria versus nonfixer algae in the primary producer assemblages.

Flash flooding and drying are the primary disturbances that characterize desert streams. Flood magnitude is usually described by the peak discharge, but other aspects of a flash flood hydrograph – the steepness of its rise and the length of its tail – also determine flood effects. Floods are important geomorphic agents, shaping channel form, as well as disturbances that initiate biotic succession. In deserts, floods connect elements of the landscape, from ridgetops to large rivers to groundwater, which are otherwise isolated and disconnected. Drying is at the opposite hydrologic extreme but is more difficult to treat as a discrete disturbance because it represents a protracted reduction and ultimately loss of stream flow. As drying progresses, there is first a concentration of mobile biota that precedes a concentration of dissolved materials (through evaporation); there may be isolation of sections of a stream and distinctive patterns of surface-water loss; direction of surface–subsurface water exchange may flip; organisms may move into sediments; and, ultimately, surface flow is lost entirely. Drying ends when surface flow resumes, either during a flood or as a gradual increase in discharge.

At the scale of centuries, events that occur only every 50–100 years or so can shape channel form and initiate riparian succession. For example, in the southwestern USA, a period of erosion occurred forming arroyos or gullies and draining the riverine wetlands that were once characteristic of these desert environments. This period left a geomorphic structure that persists today in many southwestern river–riparian ecosystems. Dramatic changes such as these can affect groundwater–surface water interactions and change species composition of the riparian vegetation. Indeed, such large-scale changes have repercussions for many stream characteristics, underscoring the importance of the hydrogeomorphic template in establishing structure and function of stream–riparian ecosystems.

Decadal variability resulting in relatively wet and dry periods in the southwestern USA is related to quasi-cycles of the Pacific Decadal Oscillation and El Niño Southern Oscillation (ENSO). For the southwestern USA, a strong ENSO signal is seen in decadal patterns of winter runoff from the Puerco and Grande Rivers in New Mexico and Sycamore Creek in Arizona. During wetter periods, frequent high-discharge events remove active-channel vegetation, leaving open gravel bars (parafluvial zone). Although these particular characteristics may be unique to desert streams of the southwestern USA, the important point is that larger-scale forcing from global climatic patterns can result in decadal shifts in near-stream riparian vegetation that have profound consequences for stream ecosystem function.

Given the high degree of interannual variability of desert environments, annual averages often carry little information and long-term trends are masked. Years vary not only in the total amount of runoff but also in the temporal distribution and timing of individual events or clusters of events. During the five wettest years of a 30-year period for Sycamore Creek, frequent floods meant the ecosystem was in an early successional state

most of the year, whereas stream organisms experienced severe drying conditions over much of the year during the five driest years. Furthermore, years with the same total annual runoff may differ substantially in seasonal distribution of that runoff with consequences for the seasonal patterns of drying. A single, large late-summer flood in 1970 in Sycamore Creek carried the same total volume of water as nine total floods distributed more evenly across the spring season in 1988, with the result that much of the stream was dry during the hottest months in 1970 but was undergoing succession during summer 1988 (**Figure 3**).

Although seasonality in desert streams may be strongly dependent upon the distribution of disturbance events, other variables in addition to discharge, such as temperature and day length, can influence the biota of desert streams. Deserts are defined only by low precipitation; thus, there is a broad range across the world's deserts in both flow seasonality and annual temperature distributions. Flow seasonality may vary from highly unpredictable and episodic events to a relatively predictable and sustained increased in discharge associated with a distinct wet season, with consequences for successional patterns. Temperatures of stream water in cool deserts may fluctuate seasonally from near-freezing to 20 °C; in hot deserts, daytime stream temperature can reach >30 °C but may be ameliorated by extensive evaporative cooling.

Temperature variation over the course of a single 24-h period may be nearly as great as seasonal variation. High albedo and low heat capacity of desert land surfaces cause extensive diel fluctuations in air temperatures, leading to wide (though comparatively muted) ranges in stream temperature. Particularly in summer when evapotranspiration rates are very high, streamflow varies measurably over 24 h, causing stranding at stream margins. At points where drying streams sink into the sediments, the end of the stream can migrate up and down the channel by several meters! Desert streams of Antarctica show extreme diel variation in discharge, but by a very different mechanism. Streamflow is generated by solar melting of the vertical walls (ice cliffs) of glaciers. During summer, when the Sun circles around the horizon at a low angle, melting (and streamflow) stops when the cliffs are in shadow.

Biota

Desert stream ecosystems support a diverse assemblage of riparian plants and stream biota. A unifying characteristic of desert stream organisms is the shared evolutionary history in a hydrologically extreme environment. The consequences of this extreme physical template are evident from the variety of adaptations that allow species to thrive in systems prone to flash flooding and prolonged drought. Rather than providing a list of taxonomic names

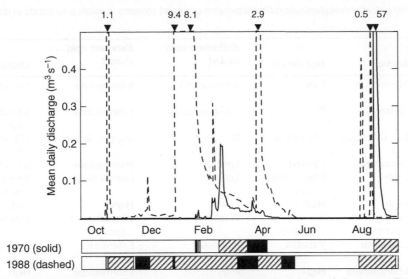

Figure 3 Contrasting seasonal discharge patterns in 2 years with nearly identical total annual discharge. Bars at bottom show time periods likely to be influenced by postflood succession (hatched), drying (open), or neither (solid). Redrawn from Grimm NB (1993) Implications of climate change for stream communities, pp. 293–314. In: Kareiva P, Kingsolver J, and Huey R (eds.) *Biotic Interactions and Global Change*. Sunderland, MA: Sinauer Associates.

for each group, we place emphasis on life history, behavioral, and morphological adaptations for living in hydrologically variable ecosystems.

Desert stream ecosystems house diverse periphyton communities, which include a variety of filamentous green algae, epilithic, epiphytic, and episammic diatoms (attached to rocks, plants, and sediments, respectively), and nitrogen-fixing cyanobacteria. Both flash flooding and prolonged drought decimate algal biomass in desert streams. Rapid drying is particularly lethal, and algae typically die within hours of exposure to the hot, dry desert environment. Algal species often have physiological adaptations that allow for some resistance to drying, however, and can withstand periods of gradual drying. Such adaptations include the production of extracellular mucilage that increases cellular water retention, and intracellular osmoregulatory solutes that also prevent water loss in drying sediments. In addition to these mechanisms, at the onset of drying, algae may also produce spores, cysts, or zygotes that can reactivate upon rewetting. Benthic algae rapidly recolonize stream sediments following floods, whereas recovery following drought is variable and depends on the degree and modes of drought resistance. In Antarctic desert streams, for example, glacial melt is the primary source of streamflow, and primary producers (cyanobacterial and other microbial assemblages) are activated by higher temperature and renewed flows which may occur seasonally or even on a diel basis. However, these organisms are also able to persist for decades in the absence of liquid water.

A productive and diverse invertebrate fauna characterizes many desert streams, consisting of insect and crustacean taxa residing in both benthic and hyporheic habitats. Life-history characteristics of desert stream invertebrates reflect an evolutionary history in a hydrologically variable ecosystem, and are shaped by both flooding and drying disturbances (**Table 1**). Most stream invertebrate larvae have few mechanisms that confer resistance to either type of hydrologic disturbance. Instead, many species have short developmental times (e.g., 1–3 weeks) that increase the probability of offspring surviving to reproductive maturity in ephemeral environments, and ensure that some aerial adults are available for recolonization following floods or upon rewetting previously dry channels. In addition, organisms with longer life cycles exhibit an array of avoidance behaviors to minimize the effects of flooding and drying disturbance. These include timing reproductive activity to periods of low flood probability, as well as ovipositing eggs in sections of stream that are likely to retain water for longer periods of time (e.g., deep pools, riffles). Finally, air breathing insects (e.g., coleopterans, hemipterans) may exhibit more direct avoidance behaviors, including the use of rainfall as a cue for leaving aquatic habitats before floods.

Relative to mesic counterparts, fish assemblages of arid river systems are species poor, and are composed of taxa that also have specific adaptations to life in hydrologically variable systems. These adaptations include large reproductive efforts, multiple clutches per year, and short developmental times. Such life-history features, along with the ability to migrate long distances during periods of sufficient flow, allow native desert fish to rapidly colonize habitats after disturbances, and result in dramatic temporal fluctuations in population size. In addition,

Table 1 Invertebrate colonization/recolonization characteristics of desert streams in relation to floods in different physiographic regions

	Mesic	Hot desert	Endorheic cold desert	Exorheic cold desert	Chapparal	Glacial
Colonization sources[a]	Numerous	Few	Intermediate	Intermediate	Intermediate	Few
Colonization distances[b]	Close	Far	Intermediate	Intermediate	Intermediate/ far	Far
Pathways[c]	DD, um, S, O, H	DD, um, S, O	S	dd, um, S, o, h	DD, um, S, O, h	dd, um, d, O, h
Refugia	Abundant	Limited	Limited	Intermediate	Intermediate	Limited
Species diversity	High	Intermediate/ high	Low	Intermediate	Intermediate/ high	Low
Resilience[d]	High	High	Low	High	High	Unknown
Flood occurrence	Spring/ summer	Winter/ summer	Winter	Winter/spring/ summer	Winter	Spring/ summer
Spatial extent of flood	Extensive	Variable	Extensive	Extensive	Extensive	Extensive
Severity of flood	Intermediate	High	High	Intermediate	High	Intermediate

[a]Refers to sources separate from the perturbed stream.
[b]Refers to distance from other unaffected water bodies.
[c]Status at time of spate: DD/dd, downstream drift; um, upstream migration; S/s, survivors; O/o, oviposition; H/h, hyporheic. Upper and lower case indicates major or lesser importance, respectively.
[d]Refers to number of taxa, not individuals, following recovery.
Reproduced from Cushing CE and Gaines WL (1989) Thoughts on recolonization of endorheic cold desert spring-streams. *Journal of the North American Benthological Society* 8: 277–287.

while intense flash floods can decimate fish populations, many desert fish have morphological adaptations that allow for some resistance to high flows. These include depressed skulls, keeled or humped napes, buttressed fins, narrow caudal peduncles, slim bodies, and reduced scales – all of which act to reduce drag and improve swimming ability in turbulent flow.

As desert stream ecosystems contract during drought, fish become isolated in pools where the physical environment can fluctuate dramatically. Although complete water loss is lethal, as streams contract, individuals of many fish species can survive in small pools, as well as beneath logs, stones, and within beds of algae. As a consequence, native desert fish are able to tolerate a broad range of temperatures (7–37 °C); indeed, desert pupfish of western North America can survive in temperatures that exceed 40 °C. Similarly, most desert fish are able to tolerate high salinity and low dissolved oxygen concentration. Others still, like the African lungfish, can burrow into the stream substrate during dry periods and survive for months by breathing atmospheric air with primitive lungs.

Streamside forests, or riparian zones, stand out as hot spots for aboveground primary productivity in arid landscapes. Arid riparian zones include assemblages of phreatophytic deciduous trees capable of accessing groundwater, as well as shrubs and annual grasses. The overall taxonomic composition of riparian zones is typically in striking contrast to that of the surrounding desert landscape. Deeply rooted riparian trees are well suited to an environment where the water table is temporally variable.

Obligate wetland species appear in desert riparian areas with permanent access to shallow groundwater, whereas those found in areas with strong seasonal fluctuations in water table elevation have structures such as tap roots or root architecture that maximizes water capture during precipitation events. Many riparian tree species in arid landscapes actually require over-bank flooding at specific times of the year to induce germination. Riparian vegetation is thought to play an important role in the overall cycling of nutrients in arid landscapes by taking up nutrients present in shallow groundwater and building organic matter pools in riparian soils. Finally, riparian vegetation serves as critical habitat for invertebrate, vertebrate, and avian taxa within arid landscapes.

Energetics

In contrast with streams of temperate and tropical biomes and because of their flood-shaped channel morphology, desert streams generally are not shaded by adjacent riparian vegetation. As a consequence, incidence of photosynthetically active radiation (PAR) reaching desert streams is high, and rates of instream primary production, the process by which energy is captured and organic matter is produced in ecosystems, are among the highest documented for benthic algae. The accrual of algal biomass in turn represents the energetic basis for stream food webs, and is central to the overall ecosystem dynamics of arid streams. For example, the abundance of high-quality benthic algae, together with warm temperatures and selection for rapid growth, result in among the highest rates of secondary production reported for benthic invertebrates. Moreover, owing to high standing stocks and growth rates, invertebrates play an important role in organic matter dynamics and nutrient cycling in desert streams. Indeed, the quantity of organic matter ingested by stream invertebrates can be 2–6 times greater than primary production. Finally, the emergence of desert stream insects represents an important resource for predators in adjacent terrestrial habitats.

At the ecosystem level, high rates of algal productivity in desert streams set them apart from streams of forested regions with respect to the relative rates of production (P) and respiration (R). Specifically, desert streams are often autotrophic ($P > R$). This is in striking contrast to streams of other biomes that receive the bulk of organic matter from outside the stream ecosystem and are often highly heterotrophic ($P << R$). Productivity of desert streams is also influenced by the disturbance regime. Flash floods scour stream channels, decimate existing organisms, and initiate a suite of algal and macroinvertebrate successional processes that correspond to temporal changes in photosynthesis and respiration (**Figure 4**). Post-flood recovery of heterotrophs is enhanced by availability of organic matter that was stranded or deposited on the stream margins and in the riparian zone during dry periods.

In addition to metabolic changes associated with flash floods, spatial patterns of photosynthesis and respiration and post-flood successional dynamics are further influenced by hydrologic exchange between hyporheic and parafluvial subsystems and the surface stream. Specifically, rates of photosynthesis and the speed of post-flood recovery are greatest where nutrient-rich water from hyporheic and parafluvial sediments enters the surface stream. Conversely, rates of respiration are

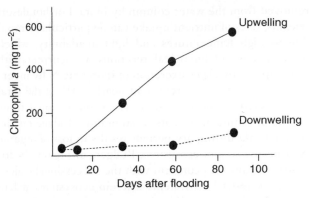

Figure 5 Comparison of algal colonization in zones of upwelling and downwelling following floods in Sycamore Creek, AZ. Valet HM, Fisher SG, Grimm NP, Camill P (1994) Vertical hydrologic exchange and ecological stability of a desert stream ecosystem. *Ecology* 75(2): 548–560.

greatest where oxygen and organic matter from the surface stream enters subsurface and lateral sediments (**Figure 5**). When both surface and hyporheic processes are taken into account, desert streams may more closely approximate a balanced metabolism ($P = R$), which highlights the connection between the two subsystems.

Nutrient Dynamics

Various factors can limit the rate of primary production if demand (requirements of autotrophs) exceeds availability. Limiting factors in streams are typically light and nutrients. Because desert streams often have open canopies and receive abundant light, nutrients are the primary constraint on algal growth. Lack of precipitation in arid and semiarid regions leads to very slow rates of weathering of parent materials, which can lead to phosphorus (P) limitation as rocks are the ultimate source of P in ecosystems. However, in many well-studied arid and semiarid watersheds of the US southwest, volcanic-derived parent materials yield highly dissolved P. Primary production in these desert streams is thus limited by nitrogen (N).

Nutrients enter streams via inputs from upstream, from groundwater and overland flow, in plant materials deposited from the riparian zone, and in the case of N, via fixation of atmospheric N_2 by cyanobacteria. Unidirectional flow of water results in continual input and output of nutrients in dissolved and particulate forms, although inputs of limiting nutrients may be low due to processing that occurred upstream. Nutrient spiraling theory, a set of hypotheses that describe how nutrients move between water column, subsurface, and biotic compartments while being transported downstream, predicts that nutrient uptake should be more efficient under conditions of nutrient limitation. In streams limited by N, for example, inorganic N is rapidly

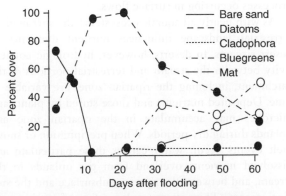

Figure 4 Recolonization of Sycamore Creek, AZ by primary producers. Redrawn from Fisher SG, Gray LJ, Grimm NB, and Busch DE (1982) Temporal succession in a desert stream ecosystem following a flash flood. *Ecological Monographs* 52: 93–110.

removed from the water column by biota. For hot desert streams, rates of nutrient uptake can be particularly rapid due to high temperatures and light availability, which increase rates of biological reactions. Concurrent with rapid uptake by algae, excretion of inorganic N by invertebrate consumers can represent nearly 30% of the total N delivered to the ecosystem. Over successional time, export of rafting algal mats or stranding of algae on the stream banks during dry periods results in loss of organic N from the stream ecosystem. Budgets of organic N for desert streams thus conform with the successional trajectory of terrestrial ecosystems wherein ecosystems at late successional stages tend to lose nutrients. In contrast to terrestrial ecosystems, however, net primary productivity may continue to be positive during late successional stages of desert streams, resulting in continued uptake of inorganic nutrients by primary producers.

In surface water, nutrient cycling is dominated by uptake of nutrients by algae and benthic biofilms. The dominant pathway of nutrients in the surface is therefore from inorganic to organic forms. Regeneration (mineralization) of inorganic nutrients in the subsurface may in turn resupply dissolved inorganic nutrients. Processes occurring in the hyporheic and parafluvial zones thus contribute strongly to patterns of nutrient availability in desert streams. Water flowing through sediments slows in velocity allowing for greater interactions between sediment surfaces and materials delivered in water. Microbes inhabiting the interstitial areas of sediments transform nutrients present in these downwelling zones, influencing the spatial distribution of nutrients in the stream channel.

In coarse sediments where dissolved oxygen remains relatively high as water moves through the hyporheic zone, mineralization often dominates N transformations, resulting in a localized increase in streamwater dissolved inorganic N concentrations at locations of upwelling. Increased nutrient availability in zones of upwelling is often associated with hot spots of algal biomass. These patterns are typical of alluvium-dominated reaches where algae are the predominant primary producers. In patches where macrophytes colonize gravel bars and parafluvial zones or in patches of fine sediment deposition, dissolved oxygen concentration in the subsurface is decreased due to root respiration and decomposition of plant-derived organic matter, and hyporheic flows are slowed, all of which lead to hypoxic or anoxic conditions. Hot spots of denitrification are associated with anoxic conditions in the hyporheic zone and water upwelling downstream of such patches is therefore depleted of inorganic N (**Figure 6**).

Because of these differences in nutrient processing between surface and subsurface flowpaths, streams that undergo drying may exhibit marked spatial variability in nutrient availability. Sections of the stream that dry may continue to harbor subsurface flows and rapid

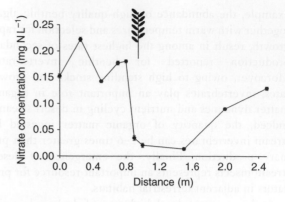

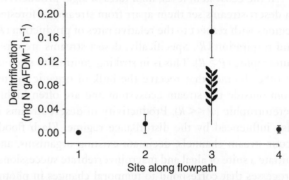

Figure 6 Changes in concentration of nitrate in water flowing through plant-colonized gravel bars (top). Note precipitous drop in concentration when water encounters a plant patch (denoted by branch figure). Denitrification is a likely mechanism accounting for the drop in concentration of nitrate; *in situ* rates of denitrification increase in plant patches (bottom). Redrawn from Schade JD, Fisher SG, Grimm NB, and Seddon JA (2001) The influence of a riparian shrub on nitrogen cycling in a Sonoran desert stream. *Ecology* 82: 3363–3376.

transformation of nutrients for some time after surface flows are depleted. Nutrient inputs and outputs from dry reaches may show strong contrasts in forms or concentration of nutrients. In contrast, reaches characterized by perennial flow tend to show dampened upstream–downstream contrasts due to the homogenizing effects of processes occurring in surface flows.

As with nearly all aquatic ecosystems, the surrounding terrestrial landscape influences nutrient dynamics in desert streams. In deserts, however, hydrologic connectivity between the stream and terrestrial portions of the catchment, including the riparian zone, are variable in time. Deposited nutrients and those stored by plants and microbes may accumulate in the riparian zone and uplands during dry periods. When precipitation or snowmelt events occur, water carries these particulate and dissolved nutrients overland from the uplands to the stream, and between the riparian subsurface and the surface stream. This creates pulses of nutrient transport between desert streams and their watersheds. Pulsed inputs of nutrients result in hot moments of nutrient processing, short time periods with rapid rates of nutrient transformations. Hot moments may account for a significant fraction of annual nutrient processing within riparian zones of desert streams.

Connectivity between terrestrial and aquatic portions of desert stream–riparian corridors may also occur from the stream to the riparian zone. Riparian plants can access water and nutrients from the hyporheic zone as well as from shallow groundwater. Access to these more permanent sources of water and nutrients leads to high productivity in riparian zones relative to desert uplands. Stream biota may transfer nutrients between streams and riparian zones of desert streams. Owing to high rates of primary productivity, insect emergence from desert streams can result in significant exports of nutrients out of the wetted stream. Emerging aquatic insects may thus provide a significant source of nutrients to riparian food webs.

Human Modifications

Desert streams present important challenges to understanding and management of water resources. They exemplify the resource that is most precious to humans inhabiting arid and semiarid regions, yet they are threatened by increasing pressures of human exploitation, agricultural expansion, and urbanization. Direct appropriation of streamflow to support human activities is the most serious threat to desert streams. This takes the form of diversion, interbasin water transfer, and groundwater withdrawal (which reduces baseflow); for example, groundwater withdrawals over the past century have converted the Santa Cruz River in Tucson, Arizona from a perennial to an ephemeral stream. In the Salt River of central Arizona, river diversion into a system of canals that feed agricultural and domestic/industrial demand in Phoenix has left a dry riverbed throughout the metropolitan region. Water extraction, primarily for irrigation, has also resulted in salinization of streams in much of the world and has triggered shifts in the composition of biotic communities.

In their appropriation of water for a variety of uses, people also modify the form and hence the function of desert streams. For example, the creation of canals that are straightened and lined with concrete replaces structurally complex streams with ecosystems that are unlikely to support the ecosystem functions characteristic of unmodified desert streams. Furthermore, impoundment and flow regulation can have profound effects on riparian ecosystems, for example, through the colonization and persistence of exotic plant species that outcompete native

species under conditions of lowered water tables and reduced flow variability.

See also: Ecosystem Ecology; Rivers and Streams: Ecosystem Dynamics and Integrating Paradigms; Rivers and Streams: Physical Setting and Adapted Biota.

Further Reading

Boulton AJ, Peterson CG, Grimm NB, and Fisher SG (1992) Stability of an aquatic macroinvertebrate community in a multi-year hydrologic disturbance regime. *Ecology* 73: 2192–2207.

Cushing CE and Gaines WL (1989) Thoughts on recolonization of endorheic cold desert spring-streams. *Journal of the North American Benthological Society* 8: 277–287.

Fisher SG, Gray LJ, Grimm NB, and Busch DE (1982) Temporal succession in a desert stream ecosystem following a flash flood. *Ecological Monographs* 52: 93–110.

Fisher SG, Grimm NB, Marti E, Holmes RM, and Jones JB (1998) Material spiraling in stream corridors: A telescoping ecosystem model. *Ecosystems* 1: 19–34.

Fountain AG, Lyons WB, Burkins MB, et al. (1999) Physical controls on the Taylor Valley Ecosystem, Antarctica. *Bioscience* 49: 961–971.

Grimm NB (1993) Implications of climate change for stream communities. In: Kareiva P, Kingsolver J, and Huey R (eds.) *Biotic interactions and Global Change*. Sunderland, MA: Sinauer Associates pp. 293–314.

Grimm NB and Fisher SG (1989) Stability of periphyton and macroinvertebrates to disturbance by flash floods in a desert stream. *Journal of the North American Benthological Society* 8: 293–307.

Grimm NB, Arrowsmith RJ, Eisinger C, et al. (2004) Effects of urbanization on nutrient biogeochemistry of aridland streams. In: DeFries R, Asner G, and Houghton R (eds.) *Ecosystem Interactions with Land Use Change*. Geophysical Monograph Series 153, pp. 129–146. Washington, DC: American Geophysical Union.

Hastings JR and Turner RM (1965) *The Changing Mile: An Ecological Study of Vegetation Change with Time in the Lower Mile of an Arid and Semi-Arid Region*. Tucson: University of Arizona Press.

Holmes RM, Jones JB, Fisher SG, and Grimm NB (1996) Denitrification in a nitrogen-limited stream ecosystem. *Biogeochemistry* 33: 125–146.

McKnight DM, Runkel RL, Tate CM, Duff JH, and Moorhead DL (2004) Inorganic N and P dynamics of Antarctic glacial meltwater streams as controlled by hyporheic exchange and benthic autotrophic communities. *Journal of The North American Benthological Society* 23: 171–188.

Minckley WL and Melfe GK (1987) Differential selection by flooding in stream-fish communities of the arid American southwest. In: Matthews WA and Heins DC (eds.) *Ecology and Evolution of North American Stream Fish Communities*, pp. 93–104. Norman, OK: University of Oklahoma Press.

Schade JD, Fisher SG, Grimm NB, and Seddon JA (2001) The influence of a riparian shrub on nitrogen cycling in a Sonoran desert stream. *Ecology* 82: 3363–3376.

Stanley EH, Fisher SG, and Grimm NB (1997) Ecosystem expansion and contraction in streams. *Bioscience* 47: 427–435.

Stromberg J and Tellman B (eds.) (in press) *Ecology and Conservation of Desert Riparian Ecosystems: The San Pedro River Example*. Tucson: University of Arizona Press.

Valet HM, Fisher SG, Grimm NP, and Camill P (1994) Vertical hydrologic exchange and ecological stability of a desert stream ecosystem. *Ecology* 75(2): 548–560.

Deserts

C Holzapfel, Rutgers University, Newark, NJ, USA

Geography
Biogeography and Biodiversity
Ecophysiology and Life Strategies

System Ecology (Ecosystem and Communities)
Human Ecology
Further Reading

What makes the desert beautiful is that somewhere it hides a well. Antoine de Saint Exupéry

Geography

Definition of Deserts

It is common belief that all deserts are hot and sandy places. While this is generally not true, a common factor of deserts is aridity, the temporal and/or spatial scarceness of water. True deserts can be delineated from other biomes based on their aridity. Of the following groups, only the first two are considered as true deserts here.

Aridity can be divided into four groups:

- extreme arid: less than 60–100 mm mean annual precipitation;
- arid: from 60–100 to 150–250 mm;
- semiarid: from 150–250 to 250–500 mm; and
- nonarid (=mesic): above 500 mm.

Since evaporation depends largely on temperature, bioclimatic aridity cannot be defined solely by the amount of precipitation. Therefore, the higher limits given above refer to areas with high evaporativity in the growing season (e.g., in subtropical areas with rainfall in warm seasons). This is taken into consideration in UNESCO's 'World Map of Arid Regions' that defines bioclimatic aridity by P/ET ratios (annual precipitation/mean annual evapotranspiration). P/ET ratios smaller than 0.03 qualify for hyperarid zones (roughly corresponding to the extreme arid zone above) and a ratio of 0.03–0.20 as arid zone (thereby corresponding to the arid zone mentioned above).

Another common way of delineating deserts is based on their vegetation pattern and optional land use. Extreme arid zones typically show contracted vegetation restricted to favorable sites or lack vegetation altogether. Arid zones are characterized by diffuse vegetation. Semiarid zones mostly are characterized by continuous vegetation cover (if edaphic conditions allow for it) and only very locally dry-land farming (without irrigation) is possible. Farming without irrigation becomes a reliable option at larger scale in nonarid zones only.

Based on geographic location and a combination of temperature and geographical causes of aridity, deserts can be separated into five classes:

- *Subtropical deserts.* They are found in the hot dry latitudes between 20° and 30°, both north and south. These deserts lie within the subtropical high pressure belt where the descending part of the Hadley's cell air circulation causes general aridity.
- *Rain shadow deserts.* They are found on the landward side of coastal mountain ranges.
- *Coastal deserts.* Found along coasts bordering very cold ocean currents that typically wring moisture as precipitation from the air before it reaches the land, these deserts are often characterized by fog.
- *Continental interior deserts.* They are found deep within continents and far from major water sources.
- *Polar deserts.* They are found both in the northern and southern cold dry polar regions.

This article focuses on extreme/hyperarid and arid zones and on the first four of geographic desert classes listed above.

Where Are Deserts Found?

True deserts are found on all continents except the European subcontinent (see **Figure 1**). Altogether about 20% of landmass can be classified as desert, making it the largest biome on Earth. **Table 1** gives an overview of the largest deserts. In addition to these major deserts, many smaller, separately named deserts exist; many of these can be classified as local rain shadow deserts. All desert areas of the world border on land semiarid zones. These are either Mediterranean-type climate and vegetation, or dry temperate or tropical grasslands/savannas. The vicinity to these areas is important as many desert organisms are either shared with these transitional biomes or evolved from similar more mesic organisms.

Desert Landforms

According to relief type, two general groups of desert landforms can be distinguished: (1) shield-platform deserts and (2) mountain and basin deserts. The shield-platform

Figure 1 Map of world distribution of deserts. Shown are the arid and hyperarid desert region. Borders are somewhat tentative as a clear separation from semidesert scrublands is often not readily possible. Polar deserts are excluded.

deserts are most common in Africa, the Middle East, India, and Australia and are characterized by tablelands and basin lowlands. Mountain and hill slopes in this type of deserts are restricted to ancient mountains or areas with more recent volcanic activity. The geologically younger mountain and basin deserts (also called mountain and range) are predominant in the Americas and Asia and consist typically of mountain ranges separated by broad alluvially filled valleys. Within the two groups of desert landforms, there are several dominant geomorphological landscape types that are described here briefly.

Desert mountains consist chiefly of sheer rock outcrops and tend to rise abruptly from desert plains. The slopes of these mountains differ according to geological origin of the parent material. Igneous rock mountains tend to be characterized by large debris (boulder fields), while softer sedimentary rocks tend to lack these. Desert mountains (**Figure 2**) dominate the desert of the USA (38% of desert area), the Sahara (43%), and Arabia (45%).

Piedmont bajada formations (**Figure 3**) cover roughly a third of the arid Southwest USA but do less so in other desert areas of the world. These formations are built up from alluvial material that tends to accumulate in fans at the mouth of mountain canyons. Individual alluvial fans often coalesce and form large-scale graded slopes called piedmont bajadas (often only 'bajadas'). Depending on deposition age and location along the bajada, the fill material is very diverse and differs strongly in alluvial particle size and soil structure, thus creating complex gradients and mosaics of distinct geological landforms. These gradients have been studied extensively in the Sonoran and Mojave Deserts and it had been shown that predominantly, the age and consequent erosion of the alluvial material within these mosaics determine the biological communities that can establish on it.

Table 1 List of the major desert areas of the world (larger than 50 000 km^2)[a]

Name	Size (km^2)	Type	Temperature	Countries
Sahara Desert	8 600 000	Subtropical	Hot	Egypt, Libya, Chad, Mauritania, Morocco, Algeria Tunisia.
Kalahari Desert	260 000	Subtropical	Hot	Botswana, Namibia, South Africa
Namib Desert	135 000	Coastal	Hot	Namibia
Arabian Desert	2 330 000	Subtropical	Hot	Saudi Arabia, Jordan, Iraq, Kuwait, Qatar, United Arab Emirates, Oman, Yemen, Israel
Syrian Desert	260 000	Subtropical	Hot	Syria, Jordan, Iraq
Kavir Desert	260 000	Subtropical	Hot	Iran
Thar Desert	200 000	Subtropical	Hot	India, Pakistan
Gobi Desert	1 300 000	Continental	Cold	Mongolia, China
Taklamakan	270 000	Continental	Cold	China
Karakum Desert	350 000	Continental	Cold	Turkmenistan
Kyzyl Kum	300 000	Continental	Cold	Kazakhstan, Uzbekistan
Great Victoria Desert	647 000	Subtropical	Hot	Australia
Great Sandy Desert	400 000	Subtropical	Hot	Australia
Gibson Desert	155 000	Subtropical	Hot	Australia
Simpson Desert	145 000	Subtropical	Hot	Australia
Great Basin Desert	492 000	Continental	Cold	United States
Chihuahuan Desert	450 000	Subtropical	Hot	Mexico, United States
Sonoran Desert	310 000	Subtropical	Hot	United States, Mexico
Mojave Desert	65 000	Subtropical/rain shadow	Hot (cold)	United States
Atacama Desert	140 000	Coastal	Hot	Chile, Peru
Patagonian and Monte Deserts	673 000	Rain shadow	Cold	Argentina
Antarctic Desert	1 400 000	Polar	Very cold	Antarctica

[a]Various sources.

Figure 2 Desert mountains: the Cambrian sandstone formations rise almost vertically from the valley floor filled deeply by sands that locally eroded from the mountain fronts. Wadi Ram, Jordan, October 2003. Photograph by C. Holzapfel.

Figure 4 Desert flats: this large desert basin in the Atacama Desert has very little surface dynamics and fine-textured soil materials are overlain by rocks forming a partial pavement. Among the harshest deserts on Earth, the Atacama receives very little to no rainfall and plant growth is lacking in most years. South of Antofagasta, Chile, October 1994. Photograph by C. Holzapfel.

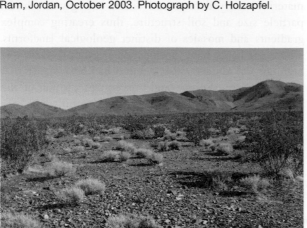

Figure 3 A piedmont bajada in the Mojave Desert: alluvial fan deposits stemming from a nearby mountain range vary in age and structure (here a mixture of Pleistocene and Holocene deposits). The position along the fan and the composition and structure of the deposits determine hydrology and plant growth (here the common desert shrubs *Larrea tridentata* and *Ambrosia dumosa*). Fremont Valley, California, USA, March 2006. Photograph by C. Holzapfel.

Figure 5 Sand dunes: ergs are seas of sands that are constantly on the move. Vast sand deserts are typical for the Sahara (shown here) and Arabian Desert. Douz, Southern Tunisia, March 1986. Photograph by C. Holzapfel.

Desert flats (basins) are another common landscape type (about 20% in the USA and 10–20% in other regions). Often these flats have rather fine-textured soils and with sufficient rainfall vegetation is diffuse and rather evenly spaced across the landscape (**Figure 4**). In more arid regions and when rainfall redistribution is patchy due to minor relief differences, distinct banded vegetation patterns can arise. These bands exist mostly in Africa and Australia but are also present in restricted areas in the Middle East and North America. The open areas produce the runoff of rainfall that accumulates in the

bands, supporting the growth of vegetation. Another type of flat desert region can be differentiated as hammadas (bedrock fields). These bedrock fields develop *in situ* and depending on the size of rock fragments, can build dense pavements (regs) consisting of densely packed surface stones resting on finer-textured subsoil. This desert landscape type is common in the Sahara and the Middle East and accounts for 40% of the area.

Sand dunes (**Figure 5**), known as 'ergs' in Arabic-speaking countries, are dominant desert landscapes only in extreme arid desert areas (25% of Sahara and Arabian Deserts, less than 1% in arid Southwest USA). Characterized by moving sands, they depend on sources of sand, sufficient wind energy, and favorable accumulation areas. Depending on these factors and prevailing wind direction, different dune types arise.

Figure 6 Desert playas are often prehistoric lake beds with fine-textured, alkaline soils. Depending on current rainfall and temperature, playas can be flooded and then resemble the former lakes, as this playa on the altiplano of the Andes at an elevation of 4400 m. Plant life on playas is typically sparse but when microorganisms and invertebrates are active, birds such as these Andean flamingos (*Phoenicopterus andinus*) assemble in large numbers. East of San Pedro de Atacama, Chile, November 1994. Photograph by C. Holzapfel.

Figure 7 Ephemeral stream: in arid areas vegetation concentrates along and in the bed of temporal stream beds. Due to available water in the subsoil that is plant extractable even long after temporal surface flow ceased, most of the primary production and species diversity in extreme deserts is restricted to these habitats. Nahal Zin, Negev Desert, Israel, April 1987. Photograph by C. Holzapfel.

Crescent-shaped dunes (barkhan dunes) form perpendicular to the prevailing wind direction and tend to be highly mobile. Linear dunes form in the direction of the wind and therefore do not move over the desert landscape. This distinction is of biotic importance as the edges of dunes are favorable to plant growth, while the dune crest and upper slopes, due to sand drift and fast erosion, are usually devoid of vegetation. In sandy flats, individual shrubs tend to accumulate sand deposits and eventually form phytogenic hummocks (so-called nebkhas).

Playas (**Figure 6**) are depressions with very fine-textured, often saline soils. Playas are the beds of former lakes that can be flooded in years of abundant rainfalls. These depressions are known under various names (North Africa and Middle East: chotts, sebkas). Even though individual playas can be large, worldwide they cover only 1% of desert.

Badlands form in areas with clay-rich soils and are typically located at the margins of arid lands, although they are found locally in arid regions as well. Depending on the strength of water forced erosion, badlands are areas with extremely high surface relief, typically forming fantastic 'lunar landscapes.'

Dry river beds of ephemeral streams (**Figure 7**) are of little importance with respect to land cover in deserts (only 1–5% worldwide), but are of immense biological importance. In extreme arid regions, these are the only places with vascular plant growth, and almost all animals depend at least at times during their life on the primary production here. This biological importance of ephemeral streams is therefore a foremost feature in deserts, and even though small in size, these landmarks were always distinctly named by human desert dwellers (washes in North America, wadi/*oued* in Arabic-speaking regions, *arroyo seco* in Spanish-speaking regions).

What Is Special about Desert Climates?

Life in desert is limited by the scarceness of water. Secondary limiting factors are correlated to the main factor: the dearth of nutrients for producers and food energy for consumers and for both – at least temporally – high heat stress. Precipitation typically is so low that water becomes the controlling factor for biological processes. Precipitation is also highly variable throughout the year and typically occurs in infrequently defined events (discontinuous input). To make things even worse, precipitation varies randomly between years and is therefore not predictable. The coefficient of variation between years in arid areas is typically larger than 30% of the long-term average (and ranges in some extreme deserts to 70%). For comparison, temperate zones and tropical areas typically have coefficient of variation of less than 20%. Individual precipitation events in deserts can be tremendously large (for instance, 394 mm in a single rainstorm in the Peruvian desert that receives a long-term annual precipitation average of 4 mm) and due to surface runoff, large-scale flash flooding can occur (**Figure 8**). Even though such sudden floods will replenish needed water to desert systems, erosion and direct damage to desert plants are the consequence. Because of the small and temporally highly variable rainfall amounts, deserts have been described by Noy-Meir as "water-controlled ecosystems with infrequent, discrete, and largely unpredictable water inputs."

Adding to and interacting with the pronounced temporal variation is the high spatial variation of rainfall in

Figure 8 A flash flood obstructs traffic on a desert road. Depressions and stream beds quickly flood after strong rainfall events due to the high surface runoff in deserts. As in the case here, the precipitation source of the water can be remote and floodwater travels far distances. Due to high stream velocity and carried erosion material, such sudden floods can be disruptive to biotic communities and dangerous to humans. Sedom, Israel, March 1991. Photograph by C. Holzapfel.

deserts. This variation is caused by: (1) oreographic features (e.g., increase with altitude), (2) differences in degree and direction of slopes, and (3) the typically small size of precipitation fronts (often less than 1 km in diameter).

Depending on the latitudinal geographical location and the origin of rain fronts, deserts receive either precipitation in the cooler season (cyclonic/frontal rainstorms) or in the warmer season (tropical, convective rainstorms). Some transitional desert regions receive both. The seasonality of rainfall is of great bioclimatological importance as evapotranspiration is larger during the warmer season and rainfall therefore tends to have a smaller biological effect. On the other hand, cold deserts receive precipitation during the cold season mostly in the form of snow and biological activity is then limited both by low temperatures and aridity. Snowmelts in spring create deep-reaching wetting fronts that will hold water available for plant uptake during the warmer growing season. Locally important other water inputs are condensations of atmospheric moisture as dew. These are crucial for plant production in the coastal fog deserts that otherwise do not receive direct precipitations. It is less common inland but can be noticeable in high desert areas as well (for instance in the Negev Desert of Israel). Fog water inputs are directly usable for cryptogamic organisms (e.g., lichens) and many arthropods. Foliar uptake of fog by vascular plants has been demonstrated but its relative importance in the water balance remains controversial. Water vapor tends to move along temperature gradients and can be important in dry soils with strong diurnal radiation. An upward movement of water vapor at night causes formation of dew close to the surface. Such water

might sustain germinated plants until they produce roots long enough to reach deeper and wetter soil depths.

Deserts usually experience an extreme diurnal temperature range, with high daylight temperatures (up to 50 °C, the highest temperature recorded in Death Valley was 56.7 °C), and extremely low nighttime temperatures (often dropping below 0 °C). This is caused by very dry air that is transparent to infrared (heat) radiation from both the sun and the ground. Thus during daylight all of the sun's heat reaches the ground. As soon as the sun sets, the desert cools quickly by radiating its heat into space. Clouds reflect ground radiation and desert skies are usually cloudless, thereby increasing the release of heat at night. With intense sun radiation, surface temperatures can be extreme and depending on the color and type of surface can exceed 80 °C.

Desert Soils

The main features of desert soil that affect water and nutrient availability include texture, content of organic matter, pH, and orientation within the landscape. Desert soils show typically little development from parent material and some authors even state that typical developed soils do not exist in deserts. Most desert soils are classified as Aridisols and are differentiated into soils with a clay (argilic) horizon (Argids) and soils without such horizons (Orthids). Other soils, less common in deserts, are mollisols, soils with dark A horizons, and Vertisols, cracking clay soils. Accumulated subsurface horizons with either clays or calcium carbonate (calcic horizons) have clear implication as impediments to water infiltration.

Most desert soils tend to be slightly to highly basic. Such reactivity can negatively affect phosphorous and micronutrient availability as these are generally not in solution at pH > 7.0. Organic matter helps to increase infiltration and via decomposition adds to nutrient availability. It is often distributed unevenly in desert soils (see below).

Soils in deserts have important effects on water inputs as they act as short-term water stores and modify water availability by a number of regulation processes. These regulation processes include direct infiltration and often more importantly runoff and horizontal redistribution of water. Redistribution by runoff tends to be of crucial importance in deserts and contributes to spatially very patchy distribution of water. Relatively impermeable surfaces (e.g., biotic or physical crust in clay-rich soils) create runoff areas that result in catchments that are water rich. Such water redistribution enables patchy plant production even in extreme arid zones, where plant growth would not be possible since evenly distributed sparse rainfalls would not exceed the threshold needed for plant life. Because of sparse plant growth, soil-created redistribution of water is more important than precipitation interception through plant surfaces. However, locally such interception combined with stem flow can create water-rich spots under shrub or tree canopies. In contrast, smaller precipitation events can be locally intercepted and lost by evaporation. This is the reason that soils in the understory of desert shrubs or trees can be either wetter or dryer than the surrounding soil.

Soil texture is of large importance as it affects both infiltration and the movement of wetting fronts. Fine-textured soils that are high in clay and silt fraction tend to impede infiltration, in which wetting fronts move only very slowly, and surface evaporation after rainfalls can be very high. More-coarse-textured soil rich in sand fractions, as for instance sandy loams, is characterized by high infiltration rates and rapid percolation. For this reason, coarse-textured soils are often better for plant growth. As this is in contrast to soils in mesic areas where fine-textured soils are commonly considered to be superior for plant production, this is called the 'inverse texture effect'.

Clearly, the orientation and dynamics of soil surfaces within the landscape plays a large role in arid ecosystems. Exposed southern (or northern, depending on the hemisphere) slopes receive high solar radiation and therefore due to higher evapotranspiration, tend to be drier than opposite slopes (**Figure 9**). These inclination differences are observable on large-scale landscape level or small-scale microtopography level. An example is the sun-exposed sides of shrub hummocks that are often only raised by a few centimeters, but can be bioclimatically and ecologically very different from the less-

exposed side. Slope exposition also plays a role when rainfall directions due to prevailing winds are constant. Rain-exposed slopes can receive up to 80% more water than other slopes.

Biogeography and Biodiversity

General Diversity

The casual observer often assumes that deserts support only low species richness and diversity because of the harsh environmental conditions prevalent in arid areas, but among plants and animals almost all taxa are represented (even aquatic groups like fishes and amphibians) here, and their species richness may be comparable to that of more mesic environments. Even though detailed comparative data are lacking, it has been argued that the diversity in North American deserts is comparable to some grasslands and even temperate forests. In general, however, evidence based on correlations along climate gradients indicates a decrease of species richness in plants and animals with increasing aridity. Regardless of this, specific taxa can be more species rich in deserts than in bordering less arid systems and regionally show negative relationship of richness with increasing precipitation. Examples for these are reptiles and birds in North America, and ants in Australia. Taxonomic groups that are generally species rich in deserts are rodents, reptiles, some insect groups (e.g., ants and termites), solpugids (camel spiders), and scorpions. In the following, an overview of typical desert taxa is given, and some emphasis is given on the ecological role of these groups in deserts. More specific treatment of ecophysiological adaptations follows in the next section.

Ecological Role and Diversity of Microorganisms

Even though obviously not readily observable, microorganisms inhabit all desert areas and in the extreme arid zones are often the only life forms present. Relatively little is known about the diversity within the lower three kingdoms (Fungi, Protista, Monera) in general and even less is known about the species richness of these groups in deserts. A recent survey that uses 'DNA fingerprinting', aiming at resolving bacterial ribosomal DNA, indicated that soils of semiarid sites can harbor higher bacterial richness then mesic sites. Since factors other than water availability are more important (chiefly soil pH) in determining microbial diversity, it can be assumed that true desert can be quite rich as well.

Mycorrhizal fungi seem to be quite important in desert ecosystems, as in more mesic ecosystems. It appears that

Figure 9 Marked phenological and plant composition differences due to slope exposition (southeast facing slope on the left and northwest facing slope on the right). The southeast-facing slope is subject to higher evaporational water losses and receives less direct rainfall compared to the northwest-facing slope. Such abiotic differences result in clear biotic contrast in arid environments, making these systems ecological model cases. Judean Desert, Palestine, December 1989. Photograph by C. Holzapfel.

mycorrhizae of desert plants not only supply the plants with nutrients but also supply moisture during the dry season, at times taking the place of root hairs. Studies conducted in the Chihuahuan Desert indicated that most dominant, perennial species have high arbuscular (AM) fungal infection rates in their coarse roots system, while fine-rooted annual species in comparison show much lower infection rates and are also much less dependent on mycorrhizal associations in general. Worth mentioning are mycorrhizal desert truffles (*Terfezia* and *Tirmania*: *Ascomycetes*), that are host specific to *Helianthemum* species in the arid region of the Middle East and the Mediterranean zones of the Old World. The desert of the American West supports an elusive community of aboveground observable fungi in which the *Gasteromycetes* (puffballs and allies) figure predominately. Another common example is *Podaxis psitillaris* (desert shaggy mane), a species most common in sandy deserts.

Except for their crucial part in mycorrhizal associations, desert microorganisms are noteworthy for their role in three typical desert phenomena: desert crusts, desert varnish, and interstitial communities. Desert crusts are microbiotic communities composed of drought and heat-tolerant algae, cyanobacteria, fungi, lichen, and mosses. These often species-rich communities are held together by sticky polysaccharide secretions and thus form surface crusts. Desiccated crusts are often indiscernible until rainfall or dew moistens the surface and microbial communities become active and green. Under extreme conditions, such crusts can form below the surface. This is possible especially under the protection of semitransparent calcareous or siliceous stones (quartz is a good example) that enables transmission of light up to a depth of 5 cm. The most common life form in crusts (and in some areas also in hot deserts in general) is cyanobacteria. Among their roles in the desert ecosystem are atmospheric fixation of nitrogen and the binding of soil particles. Together with mineral-reducing bacteria, the cyanobacteria are important in soil fertilization and soil formation and thereby have clearly important effects on vascular plants and dependent animal consumers. In hot deserts, cyanobacterial crusts often form smooth surfaces, while in cold deserts, where crust forming interacts with frost heaving, a very rough surface is typical. These different surface types clearly affect vascular plants differently.

Even exposed desert rocks can support life. Clearly the most visible organisms are crustose lichens. However, when conditions become too extreme for growth of lichens, bacteria can still survive on the surface of rocks. Desert varnish, the dark and shiny surface found on sun-exposed, porous stones in hot deserts, is the result of bacterial activity. These bacterial colonies obtain energy from inorganic and organic substances and trap submicroscopic, wind-borne clay particles. These particles accumulate in a thin layer and act as sun protection. Over very long time periods, estimated at thousands of years, these bacterial communities oxidize wind-blown manganese and iron particles and when baked together with clay particles form the dark desert varnish. The color of desert varnish varies depending on the relative proportion of oxidized manganese (dark black) to iron (reddish).

Environmental conditions even more extreme than those that support surface bacterial growth can still allow the formation of interstitial communities. These communities consist mostly of algal species that inhabit the matrix of sedimentary rocks in depth up to 4 mm. These communities can stay dormant for long periods of time and inhabit hot and cold deserts alike (they are known to exist on exposed rocks in Antarctica).

Desert Flora

Even though the geological record indicates that arid conditions existed for a long time (since the Devonian), the current modern desert flora might have originated in the Miocene, expanded in the Pliocene (after restrictions during moist periods in the Cretaceous and Tertiary), and reached its current distribution only during the Pleistocene. Specifically, the deserts in the North American Southwest are relatively young. Overall richness and uniqueness of desert floras reflect size, age, and isolation of desert areas, with larger deserts typically hosting larger numbers of endemic species. Smaller desert regions and edges of larger regions are often characterized by species that evolved in adjacent more mesic areas and partially adapted to arid conditions. A good illustration is the high incidence of Mediterranean plants in desert areas bordering regions with semiarid Mediterranean climates in all parts of the world. In general, desert floras tend to have high affinity to bordering semiarid climate zones, such as Mediterranean climate-type regions and semiarid grassland. Taxonomical studies of many species groups revealed that desert species have evolved (recently) from nondesert species. Biogeographically, strong floristic links exist between old deserts in North Africa, Middle East, and Asia. Floristic similarities among desert regions stretching from North Africa to Central Asia are particularly obvious since no wide barriers of ocean or humid vegetation exist to restrict plant migration; these floristic similarities are present despite strong climatic contrasts ranging from hot environments in North Africa to the much colder, arid Central Asia deserts. Apparent links between the North American Great Basin and Central Asian deserts might be explained by plant migration across the Beringian land bridge. Clear affinities between the deserts in both Americas can be explained by the Panamanian land bridge. In this respect, the distribution of *Larrea* shrubs is remarkable. The two recognized species – *Larrea divaricata* in South America and *L. tridentata* in North America – are taxonomically and phenotypically very close. It appears that the genus *Larrea* evolved in South America and migrated only tens of thousands of years ago (bird assisted?) to North America where it quickly became the dominant shrub in all warmer desert areas. Corresponding to the isolation of the Australian continent, the flora of the Australian desert is very different from all other deserts of the world.

Dominant plant life forms in deserts reflect water stress conditions typical for deserts (for a treatment of drought adaptations see the following section). While trees are relatively rare and restricted to more-mesic microsites, a wide range of plant life forms can be found that include many short-lived and seasonal active plants (e.g., annual or ephemeral plants and bulbous plants/geophytes). The dominant life forms that visually shape the plant formations are perennial woody plants (mostly shrubs) and fleshy succulent plants (cacti and others). Large succulent species can be dominant in some of the hot desert regions (e.g., the saguaro cactus in the Sonoran Desert). A few plant families are predominant in desert areas. The aster family (Asteraceae) is the most diverse plant family in deserts overall; it is especially numerous in Australia, southern Africa, the Middle East, and North America. Some deserts can be dominated by grass species (Poaceae). Some plant families have their global center of diversity in deserts and most likely evolved here. Notable examples are the chenopods (Chenopodiaceae) that are diverse in arid and semiarid regions of Australia, North America, and from the Sahara to Central Asia. The New World cacti (Cactaceae) are another example of a group of species rich in deserts but relatively sparse in other biomes.

Deserts are home to some of physiognomically extremely unusual plant types. Worth mentioning in this respect are plant characters as the Joshua trees of the Mojave Desert (*Yucca brevifolia*), the famous Welwitchia of the Namib (*Welwitschia mirabilis*), and the boojum tree (**Figure 10**) of the Sonoran Desert in Baja California (*Fouquieria columnaris*). Exactly why and how deserts host these exceptional plant types is not clearly understood and such 'Dr. Seussification' of the desert flora deserves systematic study.

Desert Fauna

The faunas of deserts are often biogeographically more distinct between regions than the desert floras are. Despite this, many similarities exist between the different desert regions. Such phylogenetic similarities typical for the African–Asian deserts are explained by the lack of dispersal barriers, and similarities between North American and Asian regions on one hand and North American and South American regions on the other are likely the result

Figure 10 Boojum trees (*Fouquieria columnaris*) with associated shrubs, agave, and cacti on a bajada in the Sonoran Desert of Baja California. Cataviña region, Mexico, October 1997. Photograph by C. Holzapfel.

of existing land bridges. The Australia desert fauna, as its desert flora, is very distinct. As mentioned earlier on, almost all animal taxa are present in deserts, but some groups are more diverse than others, with the major deciding factor for this being the general aridity.

Relative to other insect groups, ants and termites are very diverse in deserts. However, their species richness is lower than it is in the Wet Tropics, where these groups originated. These groups reach high population densities and ecological importance is high. With up to 150 species per hectare, the highest species richness for ants is found in Australian deserts. Most desert arthropods are either detrivores (termites, beetles, etc.) or granivores (mostly ants), or are predators feeding on these (scorpions, spiders, etc.). Due to the lack of constant plant production, herbivores are relatively sparse or show pronounced, often dramatic temporal–spatial fluctuations (e.g., mass flights of desert locusts). Species rich substone communities consisting of protozoa, nematodes, mites, and other microarthropods are typical for deserts, creating a microcosm where grazers and predators feed on bacteria, algae, fungi, and detritus.

Fishes live in almost every aquatic habitat on the globe and small, permanent desert water sources are no exceptions. Obviously richness is extremely low, but species often live in very restricted areas and often under extreme conditions. The desert pupfishes (*Cyprinodon* sp.) in the deserts of North America are among the most species-rich groups in deserts. Some species live at temperatures of 45 °C and salt regimes 4 times that of seawater, while some species are restricted to an area as small as 20 m^2 (e.g., the Devil's Hole pupfish in Nevada). These fishes are opportunistic omnivores.

Likewise, desert amphibian communities are depauperate since at least the juvenile stages depend on water.

Only a small fraction of the world's amphibians, mainly anurans, are able to occupy deserts.

Reptiles are common and widespread in all deserts and, with the exception of crocodilians and amphisbaenians (worm lizards), all orders are represented in deserts. Relatively few tortoises occur in deserts since they are restricted due to their plant diet. Snakes and lizards are well represented (especially in Australia). The extreme high diversity of reptiles in Australian deserts has been explained by low diversity of mammal and birds which resulted in lower competition for food and lesser predation pressure than in other desert regions. It appears that reptiles as endothermic consumers enjoy an advantage over other ectothermic consumers in the deserts of Australia that are characterized by low-quality plant production.

Even though birds have basic adaptation to cope with dry climates, diversity in deserts worldwide is relatively low and a clear positive relationship between rainfall and bird diversity is typical. Despite this, few desert specialist species developed among the avifauna: sand grouse, lark, parrots, etc.

Likewise, mammals are not very diverse in comparison to other biomes, but some taxa evolved to be true desert groups. Among smaller mammals are the heteromyds in North America, the jirds and gerbils in the African–Asian deserts, and the dayurid marsupials in Australia. Some of the desert mammals are rather large and therefore have advantageous low surface-to-volume ratios (see next section). The 'flagships' for this are clearly the camel species (Camelidae) that originated in the Americas in the Miocene and are now naturally found in desert regions of the Old and New Worlds; they are clearly the largest animals in all desert regions. It is of significance that most large herbivorous mammals, including camels, donkeys, goats, sheep, and horses, have been domesticated historically in deserts and semiarid regions and are common as domesticated livestock today. Other large, nondomesticated ungulates such as gazelles, ibexes, and oryxes are generally extinct or at least rare and endangered.

Convergence of Desert Life Forms

Most desert plants and animals initially evolved from ancestors in moister habitats, an evolution that occurred mostly independently on each continent. Despite this phylogenetic divergence, a high degree of similarity of body shape and life form exists among the floras and faunas of different desert regions. Since desert environments are defined by their water limitation and have similar landscapes worldwide, it is not surprising that many organisms show convergent evolution and are morphologically and functionally alike. Similar pressures of natural selection have resulted in similar life forms.

In fact, many of these analogous species groups became textbook examples of evolutionary convergence:

- Stem and leaf succulence is found in nonrelated plant taxa: cacti in New World, milkweeds and *Euphorbia* species in the Old World (however, this form is lacking in Australia).
- Bipedal locomotion is found in unrelated small rodent groups: jerboa (family Dipodidae) in the Old World, kangaroo rats (family Heteromyidae) in the New World.
- Bipedal locomotion is shared in a few larger mammals: African springhare (genus *Pedetes*), desert-living kangaroos (Macropodidae) in Australia.
- North American horned lizards (genus *Phrynosoma*) and the Australian thorny devil, the unrelated agamid lizard *Moloch horridus*, share similar grotesque spiny body armors. This has been explained as an adaptive suit that facilitates their need of having a large body due to their specialization on ants. Ants as eusocial insects present a clumped however low digestible source of food (formic acid, chitin). Both lizard groups are in need of a larger digestive system and therefore large bodies that in turn makes them slow moving and in need of protection.

Many adaptations that are discussed in the following sections are typical for all desert regions of the world. A combination of these traits creates the 'typical' desert life form that to some extent is similar worldwide.

Ecophysiology and Life Strategies

Strategies for Coping with Drought

All life originated in the sea and all organisms that have left their ancestral home depend on an 'inner sea', high internal water content. This phylogenetic inheritance restricts life in many habitats, and obviously deserts are among the harshest in this respect. Even though deserts are not only water limited (they are also low in nutrients and energy resources), adaptations to cope with the spatiotemporal scarcity of water are predominant of most (if not all) true desert organisms.

All desert life forms, animals, plants, and microorganisms alike, employ one or more of three basic strategies to cope with the dearth of water: (1) drought evasion, a strategy of avoiding water stress temporarily in inactive states; (2) drought endurance, a suit of adaptations that reduce actual stress and enable being active during drought; and (3) drought resistance, a suit of adaptations evolved to avoid water stress altogether. Note that water and heat stresses are coupled, thus many of the adaptations mentioned below can be understood as strategies to cope with both.

1. Drought-evading organisms 'choose' to pass exceedingly dry periods in dormant stages. Predominant examples are short-lived (ephemeral) plants that survive the dry season or longer periods of drought in the dormant seed stage. Such annual plants are indeed very common in many deserts of the world and compose a large portion of the plant diversity in many areas (up to 80% of species richness). An equivalent for animals can be found in cryptobiosis of invertebrate eggs and larvae. Such aridopassivity can be found in fully developed organisms as well; examples are bulbous geophytes and desert animals that pass dry season belowground inactive (estivation). Choosing of less arid microsites is another way of avoiding drought. In animals, these are typically behavioral space choices (e.g., permanent habitation or temporary use of stress-protected microsites: below shrubs or stones, rock fissures, litter, below tree and shrub canopies, or even soaring in high air). Likewise, many plants are restricted to favorable microsites (e.g., under tree and shrub canopies, runon microsites, algae growing under stones). Some organisms, mostly plants, are able to lose water almost completely and 'resurrect' once water becomes available again (poikilohydry: *Selaginella* species, algae, lichens, and moss species).

2. Drought endurance is a main strategy common among the dominant desert organisms worldwide. A suit of ecophysiological, morphological, and behavioral adaptations work together to reduce the most detrimental impacts of water stress.

Reducing water expenditure. Evergreen desert shrubs are capable of fine-tuned regulation of stomatal movement. Specialized photosynthetic pathways evolved in desert plants that minimize water loss and maximize carboxylation. C_4 and crassulacean acid metabolism (CAM) pathways are adaptations to hot temperatures, compared to the C_3 pathway adapted to colder conditions. Animals of arid regions are able to regulate and restrict water loss by concentrating urine. Birds and reptiles excrete urinary waste as uric acid that can be concentrated and allow reabsorption of water in the urinary tract, a trait not available to mammals. Desert mammals and most other taxa excrete dry feces and reduce the urine flow rate. Water loss through surfaces is reduced in plants through an increase in thick lipid cuticulae, epidermal hair cover, sunken stomata, small surface/volume ratio (leafless plants with photosynthesizing stems – xenomorphic). Animals employ a variety of adaptations that reduce water loss: impermeable integuments (e.g., in arthropods), changes of lipid structure in the epidermis that create diffusion barriers to water vapor (some desert birds), denser hair or feather cover, and small surface-to-volume ratios (common in large mammals).

Prevention of overheating. High-temperature stress is closely connected to water stress as many of the ways of coping with higher temperatures involve expenditure of water, thereby exacerbating water stress. Examples are transpiration cooling in plants and evaporative cooling in animals (including humans; see below). Desert organisms typically have high heat tolerance and capability to function at high temperatures. The comparatively high-temperature optima and temperature compensation points of photosynthesis in plants and high body temperatures and high lethal temperatures in animals attest that. Among the most thermotolerant species are desert-dwelling ants that forage on extremely hot surfaces. A Saharan desert ant species (*Cataglyphis bicolor*) is noted to hold the record with a critical thermal maximum of $55 \pm 1\,°C$.

Apart from tolerating high temperatures, an array of mechanisms evolved to decrease or dissipate heat loads both in plant and animals. The formation of sheltering boundary layers, employment of insulating structures, and increase of reflection (white color, glossiness) are among these mechanisms. Behavioral space and temporal choices are a contribution to the prevention of overheating. Seeking of sheltered microhabitats and nocturnal activity of many (if not most) desert animals are obvious examples. The nocturnal CO_2 uptake in CAM plants is an interesting analog to this.

3. Drought-resisting organisms employ adaptations that allow them to pass dry periods in an active state without experiencing physiological water stress. The succulence of many typical desert plants worldwide is a form of water storage that enables these plants to use water during dry periods. Examples for taxa that are rich in succulent species are the cacti (Cactaceae) and yuccas (Agavaeae) in the New World and some members of Euphorbiacea and Crassulaceae in the Old World. Succulent plants typically cannot become dormant and therefore require at least periodically predictable precipitation, a requirement that explains the general lack of succulent plants in extreme arid environments where prolonged droughts are common. Most succulent plants have fairly shallow root systems that react very quickly following larger rainfall events. An analog to plant succulence in animals can be found in desert snails that can store large amount of water. The ability of desert mammals (notably the camel) to store large amount of water in the blood is another analogous trait. The accumulation of fat tissue that can be metabolically transformed into water (see below) as a water storage mechanism is somewhat controversial and is more universally understood as being merely an energy source (e.g., fat reserves in desert reptile tails, body of rodents, and the famous camel's hump).

Water Uptake in Deserts

Animals

Vertebrates are able to obtain water from three sources: (1) free water, (2) moisture contained in food, and

(3) metabolic water formed during the process of cellular respiration. Some are able to receive water from all three sources, while others are able to exploit only one or two methods. Highly mobile animals tend to be restricted to the use of open water sources that are often sparse and far between. Typical examples are desert birds that fly in regular intervals to the few bodies of water available. To mention are the desert-adapted orders of sand grouse (Pteroclidiformes) and some doves (Columbiformes) that tend to visit standing water in large flocks at dawn and/or dusk. The former are even known to transport water soaked in their specialized belly feathers to their flightless chicks. Many desert animals are able to use available water opportunistically by drinking large quantities in short time. This ability is proverbial in the camel that can take up to 30% of its body weight in a few minutes. Camels and other desert mammals have resistant blood cells that can withstand osmotic imbalance. Animals living in more mesic environments (including humans) would destroy their red blood cell at such high water content in their blood. Much of the free available water has high salinity, and so it is not a surprise that many desert animals show high salt tolerance, for instance by employing salt-excreting glands. Other animals, mostly the ones that are restricted in their mobility (e.g., mammals, reptiles, and insects), rely on water obtained from their food. Carnivorous and insectivorous animals typically receive enough water from their prey. Herbivores do so as well, as long as the moisture content of the consumed plant material is relatively high (>15% of fresh weight: fresh shoots and leaves, fruits, and berries). The ultimate desert-adapted method however is the extraction of metabolic water. Especially seed-eating (granivorous) animals are able to metabolically oxidize fat, carbohydrate, or protein. Rodents and some groups of desert birds (e.g., larks, Old World and New World sparrows) are able to convert these energy sources into water: 1 g of fat produces 1.1 g of water, 1 g of protein produces 0.4 g of water, and 1 g of carbohydrates produces 0.6 g of water. Schmidt-Nielson has shown that kangaroo rats (genus *Dipodomys*) are able to obtain 90% of their water balance from metabolic water derived from consumed seeds. The remaining 10% is obtained from moisture stored in seeds. The use of already stored body fat as source of water is controversial. It has been argued that metabolizing fat and other storage sources into water requires increased ventilation and therefore increases water loss by transpiration from lung tissue. At the most, no net gain of water will be the result. According to this, the camel's hump might function simply as a fat energy storage facility, one that is situated in one place in order to reduce isolation and allow dissipation of heat.

In areas with high humidity, animals are able to receive water from dew. Such direct uptake as the main

Figure 11 Desert sand rat (*Psammomys obesus*). As the scientific name implies, this day-active desert rodent can store large amounts of body fat as reserves during unproductive seasons. Like other desert rodents, it obtains all of its needed water through its plant diet. Negev Desert, Mitzpe Ramon, Israel, May 2003. Photograph by C. Holzapfel.

source of water is probably restricted to arthropods and some mollusks (snails). There is some evidence that rodents can utilize condensation by water enrichment of stored food (**Figure 11**).

Plants (and microorganisms)

Plants, with few exceptions, depend on water uptake by their roots from the soil. Due to low soil matrix water potentials and high salinity in arid regions, such soil water is often not readily available. One way for desert plants to overcome this restriction physiologically is to osmoregulate the plant cell water potentials to overcome the low potentials of desert soils, a mechanism that also aids them in extracting water from saline solutions. Indeed, some of the lowest water potentials have been measured in desert shrubs (−8 to −16 MPa (mesic plants rarely go below −2 to −3 MPa)) and salt-tolerant (halophytes) desert perennials (as low as −9 MPa). In general, many desert plants tend be deep rooted and are therefore able to exploit water reserves that tend to be available in the deeper soil layers. Due to the need of desert plants to forage extensively for water, root-to-shoot ratio of desert plants is typically high and rooting depths are larger than in other ecosystems. In extreme cases, as in phreatopytes, rooting depth can exceed 50 m. This was found for mesquite trees (genus *Prosopis*) that are practically independent from local precipitation and are able to maintain very high transpiration rates for prolonged periods. In contrast and as mentioned before, many succulent plants that store water in their tissues tend to be shallowly rooted and are able to intercept even light summer rains that do not cause a deeper recharge of soils and would otherwise be lost to evaporation. Annual plants and most grasses also benefit from being shallowly rooted. In general, many desert plants can react quickly to available water by deploying fast-growing 'water roots' from special dormant root meristems. Shallow rooting plants show temporally intensive water exploitation patterns while plants with deeper root systems are characterized by spatially extensive water exploitation patterns.

Some deep-rooted perennial plants exhibit hydraulic redistribution from deeper soils to shallow soils. Water is absorbed from the soil at greater depth during the day and moves via the transpiration stream upward into shallower roots and the aboveground parts of the plant. At night when the air is more humid and plant stomata are closed, plants become often fully hydrated and water may be exuded from the root into the dry shallow soil. This pattern, described as hydraulic lift, may have nutritional benefits for the perennial plant itself, as it enables it to utilize the nutrients from what would have otherwise been dry soil. Released water – on the other hand – might become available for competing plants. Hydraulic lift has been described in almost all of the dominant shrubs of the arid Western US (e.g., *Artemisia tridentata*, *Larrea tridentata*, *Ambrosia dumosa*) and might be prevalent all over the world's arid zones.

Plants of saline habitats, halophytes, must be able to acquire water with high salt concentrations. They need to overcome the high osmotic pressure of saline solutions and need to avoid the potential toxicity of some ions (Na^+, Cl^-). In order to achieve such a high salt tolerance, halophytes employ strategies as osmoregulation, dilution of inner cell salt concentration by succulence, and use specialized salt-excreting glands.

Special water-rich habitats within deserts, for instance, permanent stream sides and springs, attract extrazonal plants that often possess only few aridity adaptations. Found in these oases are wetland plants and some salt-tolerant tree species that can be characterized as 'water spenders'. Good examples are palm trees (the date palm *Phoenix dactylifera* and the Californian palm *Washingtonia filifera*) and salt cedars (*Tamarix* species).

Direct uptake of condensed atmospheric water (dew and fog) and water vapor is generally possible only for some specialized poikilohydrous vascular plants, but is of much greater importance for microbiotic organisms such as lichens and cyanobacteria.

Strategies to Cope with Unpredictable Water Resources

A wealth of adaptations arose in desert organisms that allows them to utilize the pronounced spatiotemporal stochasticity of water availability typical to deserts. As detailed before, mobile organisms are able to use spatially patchy water sources that are not available to less-mobile organisms. These sessile organisms often have dormant dispersal units that can reach good microsites where they

can establish, reproduce, and eventually send their own diaspores onto other favorable microsites. Such life cycles are typical for short-lived plants (annuals, ephemerals) and some invertebrates. These diaspores typically remain viable for long periods and can 'sit and wait' for years with sufficient precipitation. Most annual desert plants form such extensive seed banks. Seeds within such seed banks tend not to germinate equally and even after strong precipitation events, a fraction of the seed will remain dormant. Such fractional dormancy might serve as avoidance of sibling competition as it will reduce densities, but more importantly has been explained as a bet-hedging adaptation in order to cope with rainfall stochasticity. When no supplemental rainfall follows an initial germination triggered by a rainfall event, at least a fraction of the seeds will be available in the following years, thereby ensuring the long-term survival of the population. In addition to dormancy, many desert plants develop some water-sensing adaptation (so called 'water clocks') that controls both dispersal and germination. Dry inflorescences of the famous rose of Jericho (*Anastatica hierochuntica*) and other annual plants (e.g., the New World *Chorizanthe rigida*) open up only after abundant rainfall and release only some of their seeds (**Figure 12**). Many desert plants have morphologically different seeds that differ in dispersal ability and have different germination requirements (amphicarpic plants). In general, a high proportion of desert plants suppress seed dispersal altogether (atelechory). This has been interpreted as an adaptation to remain on the mother site, as it has already been proved to be a favorable location.

Most perennial plants suppress flowering (aridopassive shrubs) or sprouting altogether (e.g., geophytes) in drought

Figure 12 Dry dead plant of the rose of Jericho (*Anastatica hierochuntica* – mustard family Brassicaceae). Seed pods of this annual plant are contained within curled branches forming a ball that opens when moistened and seeds are released only after rainfall events. Dead Sea region, Israel, March 1987. Photograph by C. Holzapfel.

years. This is analogous to many desert animals that shift sexual maturity and mating to synchronize with favorable conditions. Similar to plants, sterility is typical for extreme drought years and dispersal and migration (nomadism) are triggered by precipitation regimes. There is some indication that insects and desert shrubs can shift their sex expression with changing rainfall regimes. Especially, monoecious shrubs, plants that have male and female reproductive units on the same individual, can shift their sex ratio with water availability. The male function requires fewer resources from the plant ('cheaper sex'), and is typically the predominant sex in dry years.

Many desert shrubs tend to break apart into separate shoot sections over time (axial disintegration). This so-called 'clonal splitting' is very common for desert shrubs worldwide and has been explained as a risk-spreading adaptation. In time of severe drought, instead of the death of the whole original individual, some segments of the original shrub may survive. The consequence of this growth strategy is often the formation of shrub rings that grow outward and have a dieback zone in the center. Age estimations have been made based on this growth form. Large creosote bush (*Larrea tridentata*) rings in the Mojave Desert, for instance, have been determined to be of an age exceeding 11 000 years (e.g., the famous 'King Clone' located by Vasek in 1980).

System Ecology (Ecosystem and Communities)

The leading question in desert ecology is whether aridity alone can explain all aspects of biological systems. If so, desert environments could be understood simply by characterizing the harsh, abiotic environmental factors that prevail in desert systems. Thus, desert systems do not follow the typical ecosystem view and can be described as simplified systems that react to discrete rain events (triggers) by short-term growth production (pulse), interspersed by long-term storage of organic material (seeds, roots, and stems – reserves). This pulse–reserve conceptual desert model is clearly too simplistic; however, it provides an important framework for the description of major ecological components of deserts.

In contrast to this basic view of deserts, two major alternative hypotheses have been developed in regard to the driving factors defining communities and populations in deserts. One hypothesis states that only the primary producers are water limited and all other trophic levels (consumers) are determined by the magnitude of this water-dependent primary production. Another hypothesis postulates that water shortage affects organisms only individually and has no direct effect on higher-order species interactions. According to this view, aridity effects on ecosystems and communities are rather the indirect outcomes of direct physiological and behavioral responses of individual organisms (and their populations) to scarcity of water. Despite the fact that the temporal and spatial lack of water is clearly the driving force behind the individual ecologies of desert species, current research makes it clear that species interactions, including both negative and positive ones, can be strong in deserts. The following sections strive to provide a brief summary of the types of interactions typical to deserts.

Production

Net annual primary production (NPP) is lower in deserts than in most major biomes. However, when taking into account that deserts typically are also characterized by low amounts of permanent plant mass (standing phytomass), relative primary production (the ratio of NPP/standing phytomass) is among the highest worldwide (see **Table 2**). As rainfall fluctuates strongly within and between years, it is no wonder that there is a tremendous spatiotemporal variation in the amount of primary production. However, due to the lack of responsive vegetation structure and typically low levels of soil fertility, deserts are somewhat limited in their biological

Table 2 Phytomass and primary production of deserts in comparison to some other major biomes of the world[a]

Plant formation	Phytomass of mature stands ($t\,ha^{-1}$)	Net annual primary production ($t\,ha^{-1}\,yr^{-1}$)	Relative primary production
Tropical forests	60–800	10–50	0.004–0.05
Deciduous forest	370–450	12–20	0.03–0.06
Boreal forest	60–400	2–20	0.03–0.05
Savanna	20–150	2–20	0.1–0.14
Temperate grassland	20–50	1.5–15	0.08–0.3
Tundra	1–30	0.7–4	0.09–0.1
Deserts	1–4.5	0.5–1.5	0.33–0.5

[a]Modified from Evenari M, Schulze E-D, Lange O, Kappen L, and Buschbom U (1976) Plant production in arid and semiarid areas. In: Lange OL, Kappen L, and Schulze E-D (eds.) *Water and Plant Life – Problems and Modern Approaches*, pp. 439–451: Berlin: Springer; and other sources.

potential to react to extremely wet years. Semiarid grass-lands, rich in very plastic perennial plant structures and therefore exhibiting high potential growth rates, show much larger fluctuations in response to changing water availability (**Figure 13**). Also water-use efficiency (NPP divided by annual water loss) in deserts is lower than it is in dry grasslands (0.1–0.3 g per 1000 g water in deserts compared to up to 0.7 g in dry grasslands and 1.8 g in forests).

During brief periods when water is available in excess, the typically short supply of nitrogen (and other plant macronutrients) is limiting. Even though nitrogen is limit-ing in almost all terrestrial ecosystems, deserts are typically more limited due to four reasons: (1) plant growth is triggered by available water faster than nutri-ents can be replenished by decomposition; (2) desert soils typically have little nutrient-holding capacities; (3) the nutrient-rich organic matter is located in the upper layers of soils, a layer that is typically too dry for root growth to occur, rendering the nutrients inaccessible; and (4) detritus and other organic material is deposited and accu-mulated unevenly across the desert surface. Plant debris

Figure 13 Extreme, 30-fold differences of plant growth on a rocky desert slope during (a) a dry (precipitation 40 mm, NPP 0.03 t ha^{-1} yr^{-1}) and (b) an extremely wet year (193 mm, 0.87 t ha^{-1} yr^{-1}). Northern Dead Sea, Palestine, March 1991 and March 1992. Photographs by C. Holzapfel.

typically accumulates passively under the canopy of shrubs or is concentrated in nests of animals such as harvester ants and termites. Thus the desert is an 'infertile sea' with interspersed islands of fertility.

Resource–Consumer Relationships (Trophic Interactions)

In contrast to some ecosystems, food chains in deserts can be characterized by the importance of the link between producers and consumers via decomposition. Less than in most mesic environments, plant material is typically not directly consumed alive; some estimation puts the amount of energy that moves via decomposition into the food web as above 90% of total primary production. Since food resources are unpredictable, many animals can opportunistically switch from one mode of consumption to another (e.g., many arthropods are either herbivores or decomposers).

Decomposition

Microbial decomposition is often limited by low water availability, resulting in the accumulation of dry plant material and seeds. For that reason, animal detrivores are more important in deserts than in more mesic envir-onments. Examples are darkling beetles, termites, and isopods. Termites are abundant in most of the warmer deserts and are often the dominant decomposers of dead plant material (above- and belowground) that play an extraordinarily important role in nutrient cycling. Since most termites live belowground, they are also important in the formation of soils. A similar phenomenon is dis-played by scavenging animals, which are comparatively abundant among the desert fauna. Examples are large mammals (hyenas, coyotes, and jackals) and many birds (Old World and New World vultures, ravens, etc.). Like smaller detrivores in the desert fauna, many of these scavengers can switch to a predatory diet when needed.

Herbivory

Similar to other ecosystems, deserts host a large variety of herbivorous animals that potentially utilize every part of the plants. Some of the drought adaptations of plants, discussed before, also function to deter herbivores. Tough outer layers, spines, and elevated leaf chemicals, all typical for desert plants, can therefore also be under-stood as mechanisms to protect low and therefore costly primary production. Some plants appear to employ growth forms that make them less conspicuous for herbi-vores. Remarkable examples are the living stones (*Lithops* species, Aizoaceae) of South Africa that blend with the surrounding rocky desert pavement.

Predation

Abundant detrivorous arthropods are the most important prey source in the desert and provide the base for a relatively large assembly of smaller (e.g., spiders, scorpions) and larger predatory animals (e.g., reptiles, birds). The abundance of long-term stored seeds and fruits in desert systems supports an assembly of a diverse guild of granivores (seed predators). These granivores are recruited from taxonomically much differentiated groups (e.g., ants, birds, rodents), all of them potentially competing for similar food sources. Carnivorous predators can be abundant as well. These predators are mammals, birds, and reptiles (mostly snakes). Because of the relative openness of the desert terrain, prey organisms rely on a number of predator avoidance strategies. Examples are general crypsis (camouflage), 'freezing behaviors', and nocturnal activity pattern. Active deterrents are spines (desert hedgehogs, horned lizards), hard shells (desert tortoise), and poisons that can be employed in active predation as well. Strong predator pressure combined with the need for efficient predation in a desert environment poor in prey might be the reason that some of the most poisonous animals we know (e.g., snakes, scorpions, Gila monster) are true desert animals.

Parasitism

Parasitic interactions are often very conspicuous in desert environments. Many desert shrubs show abundant signs of an attack by gall-forming insects. For instance, *Larrea tridentata*, the dominant shrub in all the hot deserts of North America, is attacked by 16 specialized species of gall-forming insects. Parasitic plants, stem and root parasites alike, are common in deserts worldwide. Though detailed studies are lacking, these parasites seem to have the potential of reducing host plant production and performance (**Figure 14**).

Nontrophic Species Interactions

Competition among and within species has been recognized as an important force that shaped the communities in all mesic environments and the question whether this is also true for deserts is a natural one, however one that has not been answered univocally. Some researchers conclude that biomass production and densities in desert are typically below a threshold that would necessitate competition for resources. Observing the same density pattern, other researchers state that because such low densities indicate strong resource limitation in desert, strong competition should ensue.

Based on studies of spatial plant community structure, it appears that current competition in deserts is rare; most studies show clumped or neutral patterns – itself a sign of the lack of competition – while only few studies show a clear regular pattern (a sign of past competition). Experimental removal of individual plants in the Mojave Desert, on the

Figure 14 Heavy infestation of a desert shrub (*Ambrosia dumosa*) by an epiphytic parasite (*Cuscuta* sp.). Parasitic plants can be common in deserts and their effects can add to the abiotic stresses of aridity. Panamint Valley, California, USA, April 1995. Photograph by C. Holzapfel.

other hand, demonstrated interspecific competition among dominant desert shrubs. Spatial studies that assess the size distributions in dependence of distance between desert shrubs typically detect signs of negative association; larger shrubs tend to be spaced farther from each other than smaller ones. Removal experiments with granivorous rodents commonly result in density increase of the remaining species, thereby indicating current competition. The fact that character displacement, the evolution of divergent body features in coexisting species, has been demonstrated for desert rodents is another sign that competition has been of importance at least at one time.

Ecological theory predicts that negative interactions (such as resource competition) decrease in importance with increasing abiotic stress, and positive interactions (such as facilitation) increase. Following this, it should be possible to observe along a mesic to arid gradient a waning of competitive interaction and an increase of facilitative interactions. Indeed, a clear indication of this has been observed in a survey of positive effects among plants that resulted in a proportionally large number of cases from arid regions. In many deserts of the world, one can easily observe the positive association of either young perennials with adult perennials or herbaceous plants with larger perennial plants. Experimentally, it had been shown that the perennials had net positive effect on the smaller sheltered plants. Examples for these so-called 'nurse plant effects' are the associations of young succulent plants (often cacti), trees, and shrubs and the prevalent, close association of annual plants with desert shrubs. Typically the larger nurse plant provides canopy shading and increased soil fertility (see above discussion on islands of fertility), and sometimes protection from herbivorous animals to the sheltered plants. In accordance with this prediction, shrub–annual associations tend to be

Figure 15 Clear associations of annual plants with shrubs (here *Ambrosia dumosa*) are common in deserts. Annual plants benefit from nutrient enrichment and shade provided by the shrub canopy and since they usually only provide little benefit to the shrub (e.g., thatch-induced increase in water infiltration and lower soil surface evaporation), they can compete with the shrub for resources. Owens Valley, California, USA, March 1997. Photograph by C. Holzapfel.

strongly positive in arid sites and less so (or even negative) in less arid sites (**Figure 15**). As nothing ever in nature is one-sided, these unidirectional facilitative effects are countered by negative effects as the nursed plants can have negative, competitive effects on their benefactor. Competition for water has been shown between annuals and sheltering shrubs and such negative effects are typical once sheltered young succulents outgrow the nurse plant.

Tradeoffs in competitive/facilitative interactions are also found between taxonomically very distant groups. One example is the complex nature of interaction between microbial crusts and vascular plants. For one, these crusts can have very contrasting effects on seed placement. Cold deserts tend to have very rough crust surfaces that facilitate seed deposition and establishment, while the smooth crusts typical to hot deserts decrease such seed entrapment. Because of these differences, no general effect of desert crusts on the performance of vascular plants has been recognized. Nitrogen fixation by cyanobacteria increases nitrogen availability, thereby favoring plant growth; however, the creation of crusts can result in runoff and water redistribution that in turn locally reduces plant performance.

Human Ecology

Origin and History

Humans have lived at the edge of desert and in the desert proper for ever and there are some indications that modern *Homo sapiens* evolved when the world climate turned to be more arid at the end of the Pleistocene. Though lacking

many of the physical adaptations of true desert dwellers, we humans might be a desert species after all. One of the adaptations humans bring to live in the desert is a rather high heat tolerance. The combination of upright position that minimizes direct sun exposure during hottest times of the day, the profusion of sweat glands all over the body, and the lack of body hair, together with an energetically conservative way of movements, contributes to our ability to cope with hot deserts. As long as water and salt balances are maintained, humans can perform relatively well under heat stress. This is evidenced by the success of persistence-hunting practices in desert and semideserts, which involves tracking large ungulate prey on foot during midday heat. Such persistence hunting, today only employed by hunter-gatherers in the Kalahari Desert, has been the most successful mode of hunting prior to the domestication of dogs, and uses the relative heat balance advance that a well-hydrated and trained human can have over animal quadrupeds. Recent data show that contemporary hunters run for 2–5 h over distances of 15–35 km at temperatures of 39–42 °C until prey items (mostly antelopes) overheat and can be overcome.

Deserts have been important throughout human history and the first civilizations arose in or close to deserts (Mesopotamia and Egypt). Agriculture practices, often involving irrigation, are sometimes interpreted as cultural ways to deal with the stochasticity of the desert climate. It is interesting to note that the first written law, the codex written by the Babylonian King Hammurabi dating back to 1750 BC, was designed to manage such crucial irrigation systems. It is basically the same set of laws that gave rise to our modern laws. Since ancient history, deserts have been the cradle of great civilizations on one hand and the theater of fierce armed conflict on the other (**Figure 16**). One wonders whether the nuclear weapons

Figure 16 Many ancient sites thrived near or in deserts. The former Nabatean capital Petra is located in a desert valley surrounded by steep mountains. From here the Nabateans, an Arabic tribe, controlled the trade through the deserts of the Middle East. Petra, Jordan, October 2003. Photograph by C. Holzapfel.

tests that have been conducted in the deserts of New Mexico and Nevada (among other desert sites worldwide) symbolize that deserts can foster both the beginning and the end of civilization.

Desert Economy

For humans, there are traditionally only three basic ways to sustain themselves in deserts: hunting-gathering, pastoralism, and to some extent agriculture.

Ever since the rise of agriculture in the Neolithic era, foraging as the exclusive mode of production (hunter-gatherers) became limited to areas that were marginal to agriculture or animal husbandry. Naturally deserts are among these zones. Examples of peoples who foraged as hunter-gatherers are the aborigines in Australian deserts (this practice receded since the European discovery of the continent), and the !Kung (bushmen) of the Kalahari, who remain foragers in our times. Recent research on the !Kung people showed that hunting-gathering is a suitable lifestyle that can sustain healthy populations that are even able to spend sufficient leisure time, all this as long as population densities are low. Some Amerindian people employed hunting-gathering in deserts as well. There are some evidences that a later immigration wave of people, the Nadene, linguistically distinct from the first Clovis people, were culturally better adapted to harsh environments and settled first in semiarid grasslands and eventually in deserts (the Navajo and Apache might be the descendents of the Nadene).

Pastoralism is a true desert activity that is also typical for semiarid grasslands. It is obvious that many of the livestock animals that were and are herded by pastoralists originated from arid and semiarid areas and therefore are well adapted to such environments. The ancestors of horses, sheep, and goats evolved in semiarid environments and donkeys and camels in arid environments. People who live as pastoralists in deserts often combine animal husbandry with some scale of horticulture; this combination is called transhumance. In order to use the stochastic desert environment optimally, many pastoralists have to follow rainfall events and are partly or truly nomadic, as is exemplified by the traditional lifestyle of the Bedouin of the Arabian Peninsula (**Figure 17**).

The use of agriculture most likely did not evolve in the desert proper, but it has to be mentioned that the first cultured plants, annual grasses, and legumes were domesticated near the edge of the desert in the Middle East (10 000–8000 BC Natufian culture). Independently, in likewise semiarid areas in Mexico (Tehuacan Valley, before 7200 BC), the domestication of Teosinte into corn (*Zea mays*) took place. Deserts harbored in historical times small-scale horticulture near springs and elaborately designed irrigation systems that utilized the effects of runoff and water redistribution. Water-harvesting systems in

Figure 17 The nomadic lifestyle is a cultural adaptation of desert-dwelling people to the unpredictability of the desert environment. As still seen here in the Sahara Desert, traditionally camels were essential for transport between grazing areas and arable oases. Douz, Southern Tunisia, March 1986. Photograph by C. Holzapfel.

runoff farms have been found and partly recreated in the Negev Desert (e.g., the Nabatean system in Avdat and Shifta) and in the arid southwest of North America. Large-scale agricultural enterprises depend on permanent water courses. As along the Nile in Egypt and along the Tigris and Euphrates in Mesopotamia, these water sources originated from areas far beyond the desert region. Modern, often large-scale irrigation projects are mostly independent from surface water and use deeper aquifers.

In history, many large cities were established in desert areas (Egypt, Middle East, South America) and there are many cites in deserts in our times (Phoenix, Tucson, Las Vegas). Incidentally, the climate and ecology of urban areas even in the temperate, nonarid zones has many similarities to true deserts (e.g., water limitation due to surface sealing and runoff, high temperatures, etc.).

Human Impact on Deserts

As all ecosystems with low productivity, deserts are fragile to disturbance. Some ecologists go as far as to state that no direct succession occurs at all after disturbance but it is at least obvious that regeneration times after perturbations can be very long. The few long-term studies following disturbance, as for instance the vegetation recovery of ghost towns in the American West, demonstrate these long recovery times that often exceed many decades. It can be generalized that any human impact that changes the soil structure will last very long. Unlike in mesic environments, abandoned agricultural fields in deserts will recover only very slowly (if at all) to natural desert vegetation. Additionally, formerly irrigated fields will have elevated salt concentrations for long periods of time. Soil surface disturbance caused by off-road vehicles inflict severe changes in hydrological characteristics of soils, which might remain permanently. The increase in off-road vehicles in the North American deserts, and increasingly also in the Middle East, is a serious threat to deserts and desert biotas.

Desertification, largely the human-caused extension of the desert, is one the most serious problems facing the globe. Causes of the growth of the desert regions are multifaceted and are a combination of natural long-term variation in the weather, climate destabilization, and human mismanagement due to overpopulation and land-use change. Under the UN Convention to Combat Desertification, desertification is defined as land degradation in arid, semiarid, and dry subhumid areas resulting from various factors, including climatic variations and human activities. The effects of desertification promote poverty among rural people, and by placing stronger pressure on natural resources, such poverty tends to reinforce existing trends toward desertification. It is now clear that in several regions, desert environments are expanding. This process includes general land degradation in arid, semiarid, and also in dry and subhumid areas. Clearly in areas where the vegetation is already under stress from natural or anthropogenic factors, periods of drier-than-average weather may cause degradation of the vegetation. If such pressures are maintained, soil loss and irreversible change in the ecosystem may ensue, so that areas that were formerly savanna or scrubland vegetation are reduced to human-made desert. To counter this process that will increasingly endanger lives and livelihood of millions of people (not to speak of drastic effects on the biodiversity of the planet), synoptic management approaches are needed that combine understanding of the process and investigation into the regional causes of the process, in order to comprehend the effects on the Earth's overall system. It is important to emphasize that desert border areas that undergo desertification will not simply convert into natural deserts. Disturbed and overused semiarid zones are characterized by lower biodiversity than original, natural deserts. Therefore desertification will not simply increase the global area of deserts; it will create large tracts of devastated lands.

Human activity and human-caused climate change will facilitate the migration of ruderal (disturbance-adapted) plant species into locally favorable microsites within the desert. This has been shown for the vegetation along roadsides in the Middle East and in the North American southwest. Even though this might enhance local, small-scale species richness, an overall reduction in regional diversity and a loss of desert-adapted species might follow. Such a strong mixing of former distinct biotic zones has been observed along the edges of deserts in the context of human-caused disturbances and climate change. A wide variety of 'extrazonal' plants are crossing zonal borderlines, a process that will potentially lead to a marked decrease in large-scale species diversity. This migration by species that are native to the general geographic area but are now spreading into new climatic or biogeographic zones is an overlooked aspect of species invasion.

Due to typically strong abiotic stress, desert areas have been in the past remarkably resistant to invasions by non-native organisms. Notable exceptions have been biological invasions by deliberately introduced organisms in Australian deserts (e.g., rabbits and *Opuntia* species). However, invasion seems to increase rapidly worldwide and many desert areas today show a dramatic increase in the arrival and spread of non-native species. At present, the deserts of the American South West seem to be affected most. Plants originated from the Old World, mostly grasses (e.g., annual *Bromus* species, some perennial grasses), but increasingly members of other plant families also have invaded many desert communities and can have strong impacts on native desert communities. Among the detrimental effects are dramatic changes in fire regimes and direct competition with recruiting shrub seedlings and native annual plants, and even negative effects on adult desert perennials have been demonstrated. The main reason for these trends is due to general land-use changes in desert and desert margins. In the Southwestern US, disturbances due to increasing suburbanization of deserts, besides increases in nutrient depositions, seem to be central agents of these changes.

Endangered Species

Many of the larger vertebrate desert species are threatened and a number of species have been lost to global extinction. The openness of the desert habitat and naturally small population size makes large mammals and birds conspicuous and thus very vulnerable to overhunting. Threatened species include the central Asian wild Bactrian camel (*Camelus bactrianus*), the onager (*Equus hemionus*, a wild ass of southwestern and central Asia), and large antelope species as the addax (*Addax nasomaculatus*) of North Africa and the Arabian oryx (*Oryx leucoryx*, **Figure 18**). Hunting is also the main reason that larger birds are endangered. Among birds many bustard species are threatened (e.g., the houbara, *Chlamydotis* sp.) or are already extinct (e.g., the Arabian subspecies of the ostrich: *Struthio camelus syriacus*). Large predators have been and continue to be extensively hunted since they are perceived to be a threat to livestock (e.g., desert subspecies of the Old World leopards, *Panthera pardus jarvisi*). International efforts to save many of the larger endangered animals are currently ongoing; many of these efforts involve reintroductions.

Figure 18 Many larger desert animals became extinct in the wild due to hunting pressure. These captive Arabian oryx (*Oryx leucoryx*) are part of a breeding effort that led to release operations into formerly occupied desert ranges (Oman, Bahrain, Jordan, Saudi Arabia). Wadi Araba, Israel, May 2003. Photograph by C. Holzapfel.

Invasive species can have detrimental effect on threatened species as well. An example is the increased fire frequency caused by annual, non-native grasses, which is threatening populations of the desert tortoise (*Gopherus agassizii*) in the deserts of North America.

Desert Research

One of the major attractions of desert ecosystems for scientists lies in their simplicity. Spatial patterns of life are often visible and clear cut and ecologists tend to feel empowered by the sense of ecological understanding. As any desert scholar will have to attest though, this simplicity is only relative. In comparison to more complex systems, deserts seem to invite ecological questions with greater ease than for instance tropical rainforests would. Therefore much of basic ecological knowledge has been founded in desert research and these dry places more often than not were used as simplified models for the green and (forbiddingly) complex world. Thus it is no wonder that the desert has spawned many research efforts, notably among them large, coordinated ecological research enterprises. The permanent research sites established worldwide during the International Biological Program (IBP) are good examples; in the US, many of these continue to be monitored under the Long Term Ecological Research (LTER) program.

See also: Mediterranean; Steppes and Prairies; Temperate Forest.

Further Reading

Belnap J, Prasse R, and Harper KT (2001) Influence of biological soil crusts on soil environments and vascular plants. In: Belnap J and Lange OL (eds.) *Biological Soil Crusts: Structure, Function, and Management*, pp. 281–300. Berlin: Springer.

Evenari M, Schulze E-D, Lange O, Kappen L, and Buschbom U (1976) Plant production in arid and semi-arid areas. In: Lange OL, Kappen L, and Schulze E-D (eds.) *Water and Plant Life – Problems and Modern Approaches*, pp. 439–451. Berlin: Springer.

Evenari M, Shanan L, and Tadmor N (1971) *The Negev. The Challenge of a Desert*. Cambridge, MA: Harvard University Press.

Fonteyn J and Mahall BE (1981) An experimental analysis of structure in a desert plant community. *Journal of Ecology* 69: 883–896.

Fowler N (1986) The role of competition in plant communities in arid and semiarid regions. *Annual Review of Ecology and Systematics* 17: 89–110.

McAuliffe JR (1994) Landscape evolution, soil formation, and ecological patterns and processes in Sonoran Desert bajadas. *Ecological Monographs* 64: 111–148.

Noy-Meir I (1973) Desert ecosystems: Environment and producers. *Annual Review of Ecology and Systematics* 4: 25–41.

Petrov MP (1976) *Deserts of the World*. New York: Wiley.

Polis GA (ed.) (1991) *The Ecology of Desert Communities*. Tucson, AZ: University of Arizona Press.

Rundel PW and Gibson AC (1996) *Ecological Communities and Processes in a Mojave Desert Ecosystem: Rock Valley, Nevada*. Cambridge: Cambridge University Press.

Schmidt-Nielsen K (1964) *Desert Animals: Physiological Problems of Heat and Water*. London: Oxford University Press.

Shmida A (1985) Biogeography of the desert flora. In: Evenari M, Noy-Meir I, and Goodall DW (eds.) *Hot Deserts and Arid Shrublands*, pp. 23–88. Amsterdam: Elsevier.

Smith SD, Monson RK, and Anderson JE (1997) *Physiological Ecology of North American Desert Plants*. Berlin: Springer.

Sowell J (2001) *Desert Ecology: An Introduction to Life in the Arid Southwest*. Salt Lake City, UT: University of Utah Press.

Whitford WG (2002) *Ecology of Desert Systems*. San Diego: Academic Press.

Dunes

P Moreno-Casasola, Institute of Ecology AC, Xalapa, Mexico

Published by Elsevier B.V.

Introduction

Coastal beaches and dunes have a worldwide distribution. They are common in both temperate and humid tropical areas, in arid climates, and in regions covered by snow during the winter. Beaches and dunes are considered two of the most dynamic systems. They are not permanent structures, but rather huge sand deposits that move and have an episodic supply of sand.

They can be found in deserts as well as on dissipative coasts with a plentiful supply of sediments and where there are strong onshore winds or winds that are parallel to the coastline. Sand dunes are eolian bedforms and beaches are marine geomorphic structures. Dunes form from marine sand delivered to the beach from the nearshore by waves. The exposed sediment is dried by the sun and the wind then transports sand inland to form incipient dunes and foredunes. Tidal range is important in this process since a high range exposes a large intertidal area that often dries out between the tides. These sediments constitute a major source of wind-blown sand given that sand-sized sediments are more easily transported by wind.

Dune size varies considerably. Some of the biggest dunes are found in deserts such as Badain Jaran Desert in the Gobi Desert in China (approximately 500 m), the Sossuvlei Dunes, Namib Desert (380 m), and the Great Sand Dunes National Park Preserve in Colorado, USA (230 m). Along the coast, on the Bassin d'Arcachon, France, is Europe's largest sand dune, the Dune du Pyla, nearly 3 km long, reaching 107 m in height, and moving inland at a rate of $5 \, \text{m yr}^{-1}$.

Dune Origin and Formation

Wind is the main agent forming sand dunes. There is an exchange of sediments between beaches and dunes and this is part of a natural process that maintains both morphological stability and ecological diversity. Once exposed, sand is vulnerable to aerodynamic processes.

Wide beaches are formed in the summer and narrower beaches during the winter. Storms erode beaches and transport sand out of the system or to other beaches. Bonding, both by moisture and chemical precipitates, may cause surface adhesions, raising thresholds and reducing erosion. Sometimes salt forms a whitish crust on the sand surface, also bonding sand grains.

Sand grains come in a wide variety of shapes, colors, and densities, depending on their origin and on how long they have been rolling in water currents and wind. Silicate sand and calcium carbonate sand (formed by fractured shells and skeletons) are the more common components of coastal dunes. Sand texture, as well as shape and density, affect transport. Smaller particles are easier to move than larger ones. Sediment size is measured on the Wentworth scale. It is harder for angular grains to become airborne but they may move further once they have. Denser grains are harder to move and often accumulate as lag deposits on the upper beach.

Almost all wind-blown sand travels quite close to the ground, through a mechanism called saltation. Individual grains move in a series of continuous leaps. Once airborne, a grain describes a curve path, and lands hitting the ground at a low angle, but with sufficient force to rebound into the air again. It hits other sand grains that become airborne and do the same thing. In a short time, there is a considerable amount of sand in the air. Under most circumstances, deposition takes place within a short distance although sometimes sand may be transported long distances alongshore where the wind blows parallel to the coast. Deposition is favored by obstacles such as driftwood, clumps of vegetation, boulders, and plastic objects which perturb air flow and create a shelter zone. Small dunes are formed with their tails – called trailing ridges – stretching downwind.

Changes in wind strength and direction cause rapid resedimentation. Often a dune's surface changes by the hour, creating complex stochastic patterns. Over time, these processes create recognizable dune bedforms such as ripples, sand waves, and barchans.

Most coastal dunes form in the presence of vegetation. An important determinant of dune form is the drag imposed by the vegetation on the air flow. Dunes can be classified according to the percentage vegetation cover. At one extreme are dunes that have been stabilized by their vegetation cover (fixed, shore-parallel ridges and parabolic dunes) and at the other are the free wind forms (barchan or sand wave dunes, transverse dunes). Transitional forms are typified by a fragmented topography (hummock dunes).

There is a strong interaction between vegetation and dune form, and there are several patterns of incipient dunes. Plant form modifies sand deposition, forming a leading edge (as in the case of *Ammophila arenaria*), a trailing edge (*Spinifex hirsutus*), or intermittent deposition in clumped vegetation. Perennial grasses such as *Agropyron junceiforme* and *Elymus arenarius* as well as tropical long-branched creepers (*Canavalia rosea* and *Ipomoea pes-caprae*) grow laterally and vertically and are able to raise a dune a meter or two high.

Sand dunes act as a buffer to extreme winds and waves and they also shelter landward communities. They replenish the depleted beach and near-shore during and after storms, and are important in the retention of fresh-water tables against salt intrusion. They filter rain water and are also important habitats for plants and animals. People have always appreciated their beauty and recreational value.

R. W. Carter wrote that "Of all the coastal systems, sand dunes have suffered the greatest degree of human pressure." Many have been irreversibly altered by human activities such as tourist developments, golf courses, and urban growth.

Abiotic Factors

Dune ecosystems may be viewed as a series of gradients related to various environmental factors, which operate on different spatial and temporal scales. If we view a profile from the sea landward, we first have the beach (near-shore and back-shore), the embryo or incipient dunes, and the foredune. The first dune ridge (the next inland from the foredune) is normally the highest and forms a continuous sand structure. The second is an older dune ridge, frequently lower because of the reduction in sand supply and the gradual loss of sand. This formation occurs when we have a series of parallel ridges, formed by onshore winds, each ridge lower than the previous. Sand is trapped by vegetation and saltation cannot be initiated beneath the vegetation, unless a blow-out forms. Older dune ridges become fragmented when blowouts and parabolic dunes develop. Parabolic dunes are formed when prevailing winds blow at right angles to the dune ridges. Poorly stabilized regions are rapidly eroded but the more vegetated areas on either side remain covered by plants for a longer time. As the bare sand of the central region moves inland, the two horns or tips of the parabola remain attached to the relatively stabilized sand of the trailing ridges. A slack (a dune depression where sand has been blown away until the water table is exposed) may be formed in the middle, between the parabola arms. Parabolic dunes can also be formed in transverse dunes.

Throughout the dune field, there are gradients in salinity, sedimentation, nutrients, flooding, and shelter. Dune vegetation forms a complex spatial mosaic, mainly because of variations in physical gradients which depend on the distance to the sea and topography. Disturbances also result in temporal successions that add another dimension of complexity to the spatial mosaic described.

Sand Movement

Dune movement has only been measured in a few dune systems and most of the published records are based on estimates from maps, the height of sand on fence posts, houses, and trees, etc. The results show that the rate of movement varies considerably among systems, varying from a few centimeters per year to 70 m per month, the latter in New Zealand (personal observation of Patrick Hesp). Dune formation depends on an adequate supply of sand and the wind to transport it. The interaction of wind and vegetation is of primary importance for dune growth. Colonization by plants accelerates dune growth, because surface roughness created by vegetation decreases wind flow and increases sand deposition. Several plants show an inherent capacity to bind sand and are able to develop extensive horizontal and vertical rhizome systems. The growth form and the ecological dynamics of dune plants are important contributors to foredune growth. Rhizomatous growth (as in the grass *Ammophila*) or stoloniferous growth (as in *Ipomoea* or *Spinifex*) can extend the foredune depositional area by 5–15 m in a few months. *Elymus arenaria* (Europe) develops vertical rhizomes 150 cm long and *Ipomoea pes-caprae* (pantropical) can have 25-m-long branches that are buried two or three times along their length. **Figure 1** shows species that are able to survive and reproduce successfully under high rates of sand mobility in different parts of the world. In each region, sand-tolerating species have evolved, and they play a very important role in dune formation. Sand deposition produces vigorous growth in some of these species; both plant height and plant cover increase, making these species excellent dune fixers. Many hypotheses have been suggested to explain this response of sand dune plants, but there are few studies in which the explanations are based on experimental evidence. Changes in soil temperature, increased space for root development, higher nutrient and moisture availability, a response to darkness, meristem stimulation, and interactions with endomycorrhizae and nematodes are probably factors that play an important role in this response.

Nutrients

There are great differences in the soil properties of young dunes (formed by recently blown sand), and those of more mature dunes in which vegetation has dominated. Newly

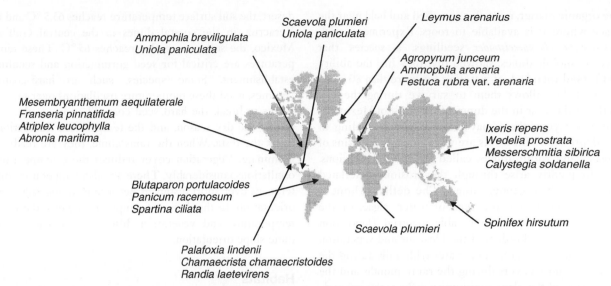

Figure 1 Species that are able to survive and reproduce successfully under high rates of sand mobility in different parts of the world. Many regions have their own set of species that play important roles in stabilizing sand dunes locally.

blown sand from the beach is low in mineral nutrients. Dune soils show marked changes as they age. Pioneer species that initiate dune stabilization are able to live in very poor soils. On fully vegetated dunes, organic matter and nutrients accumulate, and the leaching effects of rainfall decrease. Leaching dissolves carbonate and moves it downward to the water table. With time, the organic matter of nutritionally poor soils of younger dunes increases, and pH decreases. The increase in organic matter content varies among dune systems, depending on the climate and colonizing species. In high-rainfall climates such as Southport (Lancashire, Great Britain), organic matter increases slowly at first but much faster after about 200 years. In Studland, Dorset, the early invasion of *Calluna* is largely responsible for a very rapid increase in organic matter. Primary productivity and the competitive abilities of coastal plants are frequently limited by nutrient availability, with nitrogen deficiency the most severe. As succession advances, plants increase their cover, communities change from grasslands to thickets, and then to tropical or temperate forests, adding nutrients and organic matter to soils. In dune depressions, where water is not a limiting factor for plant establishment, the accumulation of organic matter is faster. Experiments with dune plants have shown that many species are slow-growing and generally show growth responses characteristic of plants from infertile habitats.

Salinity

Soil salinity comes from salt spray and foam blown inland, and the amount of salt usually correlates well with the distance inland or degree of protection from the wind. In some regions with a Mediterranean climate, such as California, soil salinity follows a seasonal progression. Late summer additions by fog and salt spray result in high values at this time of the year. Winter rains leach salt away, salinity decreases, and in early spring reaches its lowest level. Salinity gradients affect species distribution, especially for those plants sensitive to salinity. Germination and growth might be difficult when soil salinity is high. Salts in the soil affect plants by making water less available, and high salinity is considered a physiological drought. Frequently, there are no shared species between the beach and the more sheltered or inland areas of the dunes. Experiments on sea rockets (*Cakile maritima*) and lupines (*Lupinus* spp.) which were sprayed with seawater showed that lupine seedlings were not tolerant of salt spray. The level of salt spray in a Californian beach may be $1 \, mg \, cm^{-2} \, d^{-1}$ on a calm day, but is much higher on a windy day. On other beaches and dunes, where onshore winds are not as strong, airborne spray is very low and plants are not subjected to these conditions.

Water

The primary source of water for dune plants is rainfall. Radiation causes considerable diurnal and nocturnal temperature variations. These fluctuations in soil temperature are sufficient to cause the periodical condensation of water vapor in the soil. This produces an increase in water availability from dew that is sufficient to maintain plants in rainless periods. Fog can be another source of water, but in some areas it contains salt. Studies in open dune communities have shown that soil moisture increases to depths of about 60 cm below the dune surface and then tends to fall off. In closed dune communities, where the soil has some organic matter, rainfall is absorbed and held near the surface where it is available to roots. Experiments with *Chamaecrista chamaecristoides* seedlings, a species that thrives in mobile dunes, showed that they had the ability to withstand total lack of watering for more than 80 days. This probably allows them to survive during the dry months of the year in the dunes of the Gulf of Mexico.

In a wet year, there may be widespread flooding in dune depressions. Blowouts are wind hollows or basins of exposed sand within dunes, called slacks or depressions. They frequently arise through the erosion of deflated areas in poorly vegetated dunes. The deflation limit is reached owing to the presence of water, algae, or the accumulation of coarse immovable material. Deposition occurs around the borders of the blowout and vegetation may recolonize the area. The water table falls during the dry season and recovers during the rainy months and the composition of the plant community reflects this ground-water regime. When the soil is completely flooded, the prevailing anaerobic conditions can influence its chemical composition and the concentration of nutrients, affecting plant survival and growth. Flooding can cause the local extinction of some non-wetland species and facilitate the establishment of others.

The frequency and duration of slack inundation are factors that can alter the distribution of vegetation and plant community composition. When flooding takes place occasionally, on very wet years, many of the plants die, and when the water recedes, colonization takes place again. In wet slacks that flood every year, water-loving plants establish and a completely different set of species is found in these areas. Thus community composition will depend on the differential tolerance of plants to the environmental conditions associated with inundation, particularly anoxia. Species are good indicators of the water table depth. In temperate areas, *Erica tetralix*, *Glyceria maxima*, *Carex nigra*, and *Juncus effusus* are some of the more common species. In Europe, slacks are very important for endemic and rare species. In tropical regions, *Cyperus articulatus*, *Lippia nodiflora*, *Hydrocotyle bonariensis* (Mexico) and *Paspalum maritimum*, *Fimbrisitylis bahiensis*, *Marcetia taxifolia* (Brazil) are frequently found in these depressions. Thickets are also common and in Mexico they are formed by *Pluchea odorata*, *Chrysobalanus icaco*, and *Randia laetevirens*. In Brazil, there is Ericaceae scrub dominated by *Humiria balsmifera*, *Protium icicariba*, and *Leucothoe revoluta*.

Temperature

On open sand dunes, there are considerable diurnal and nocturnal temperature variations. In California, on an August day, when the air temperature was above 15.5 °C 1 m above the ground, the soil surface was at 38 °C and soil 15 cm below the surface was at 19 °C. In a Nevada desert, the soil surface temperature reaches 65.5 °C and in Veracruz, in the coastal dunes in the central Gulf of Mexico, the soil surface also reaches 65 °C. These temperatures are critical for seed germination and seedling establishment. Some species, such as hard-coated legumes, need these temperature oscillations over several weeks to break the hard seed coat. They lie on the soil during the dry season, and the temperature fluctuations break the testa. When the rains come, they are ready to germinate. Vegetation cover reduces these temperature oscillations considerably. There are also temperature differences over short distances because of topography and orientation. In the dunes of temperate regions, there are temperature and vegetation differences depending on dune slope orientation.

Habitats

Coastal dunes are very dynamic systems offering a wide variety of habitats with different physical and biotic conditions, and this allows for the existence of species with very diverse life-history traits. They can be visualized as a permanently changing environment with distinct degrees of stabilization that is closely correlated with topography, the disturbance produced by sand movement, and distance to the sea. Dune habitats can be classified into three types: (1) those where sand movement dominates, sea spray is sometimes important, and nutritionally poor soils prevail (they are formed by the sandy beach, embryo or incipient dunes, foredunes, blowouts, and active dunes); (2) humid and wet slacks or depressions, that is, those habitats which become inundated during the rainy season when the water table rises and they sometimes may even form dune lakes with wetland vegetation; (3) stabilized habitats, which show no sand movement, conditions are less stressful, and there is more organic matter in the soil. Vegetation cover is more continuous – grasslands, thickets, woodlands, and tropical forests.

Figure 2 shows a beach and dune topographic profile as well as the intensity of some of the abiotic factors mentioned and the areas where they affect the dune system.

Biological Factors

Dune plants are found all over the world, from the frosty regions of Canada and Patagonia, to the tropical areas of the Caribbean, Africa, and the South East, and the dry regions of Australia, Peru, and California. They are subjected to very different climatic conditions, they share few species, and life forms vary. Raunkaier developed an ecologically valuable system of plant classification, based on the position of the vegetative perennating buds or the persistent stem apices in relation to the ground level

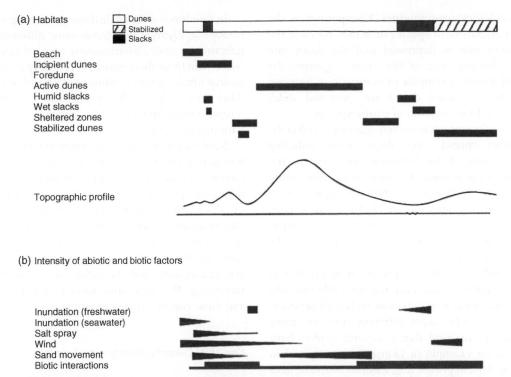

Figure 2 (a) Beach and dune topographic profile showing each of the habitats. (b) Intensity, indicated by the width of the line, of some of the abiotic factors mentioned along with the areas where they affect the dune system. Reproduced with permission from Moreno-Casasola P and Vázquez G (2006) Las comunidades de las dunas. In: Moreno-Casasola P (ed.) *Entornos veracruzanos*: *La costa de La Mancha*. Xalapa, Mexico: Instituto de Ecología AC.

during the unfavorable season of the year, which can be either the cold winter or the dry summer. There is a strong correlation between the climate of an area and the life forms of the plants present. This system allows comparisons to be made between particular areas or regions. The biological spectrum found in a dune system is an expression of the number of species in each life-form class as a percentage of all the species present. A comparison of the biological spectrum of a dune system in Braunton Burrows (North Devon, Great Britain) and one in La Mancha (Veracruz, Gulf of Mexico) was made. Braunton Burrows is dominated by hemicryptophytic plants (perennating buds are at the surface of the sand) and therophytes (annuals that survive the unfavorable season as seeds); La Mancha is dominated by phanerophytic types (these grow continually, forming stems that often have naked buds projecting into the air, such as in *Hippophae rhamnoides* or *Chamaecrista chamaecristoides*).

Facilitation and Succession

Ecological succession refers to a more-or-less predictable and orderly change in the composition or structure of an ecological community. Facilitation is one of the mechanisms by which succession takes place. It occurs when plant establishment is favored or facilitated by previously established plant communities that ameliorate

environmental extremes. Physical factors in dune environments produce a very harsh environment where few plants can survive. Several studies show that facilitation takes place in the early stages of colonization and succession in coastal dunes. As succession proceeds, pioneer species will tend to be replaced by more competitive species, the abiotic environment will become less harsh, and biotic interactions such as competition and predation will be more common.

Dune succession is comprised of a pioneer (also called yellow dunes, associated with the most seaward dunes that are still receiving a significant input of wind-blown sand), intermediate, and mature stages (gray dunes or inland dunes with little or no sand, a high humus content, and where soil development has occurred). The rate of succession varies with the harshness of the environment. This is related to the abiotic factors mentioned and the vegetation stock. Detailed studies have been undertaken in the Lake Michigan dunes, a salt-free system, in the coastal dunes of Newborough Warren, several sites in Holland, and La Mancha, among others.

Competition, Predation, Disease

Biotic interactions among plants are an important determinant of structure and dynamics. Competition is recognized as one of the most important forces structuring ecological communities. Competition is the interaction of organisms or species such that, for each, the birth or growth rate is depressed and the death rate increased by the presence of the other organisms (or species). Well-known examples of competition between plants growing on coastal dunes are grass and shrub encroachment and the invasion of exotic species.

Grass encroachment occurs when aggressive and competitive grasses spread over dune areas, reducing biodiversity because of the dominance of a few species. Grass encroachment is found in many dune areas, where grasslands become the dominant community type. Among the more aggressive species are *Calamagrotis epigejos*, *Ammophila arenaria*, and *Schizachyrium scoparium*. Shrub encroachment is also common, for example, in the Caribbean (*Coccoloba uvifera*).

Species introduction has been a common practice in dunes, both for dune stabilization and for cattle ranching activities. European marram grass was widely dispersed in other regions that were quite different from its native Europe, mainly to fix sand dunes. Several conifers have also been used for example in Doñana's dune system in southern Spain. African grasses (e.g., *Panicum maximum*) have been brought to America and used to replace local grass species because they have been considered better fodder.

Neither the effects of fauna nor those of grazing animals (especially rabbits) on dunes have received the attention they deserve. The importance of herbivory by rabbits was seen in Great Britain during the outbreak of myxomatosis, a viral disease which infects rabbits. The disappearance of rabbits led to profound changes in the structure of the vegetation, mainly the development of scrub in several dune areas. Rabbits also produce nitrogen and phosphorus enrichment beneath the scrub species under which they find shelter, causing N-fixing root nodules to invade.

Lethal yellowing is a specialized bacterium, an obligate parasite that attacks many species of palms, including the coconut palm (which has become the symbol of tropical beaches). Extensive coconut plantations in the Tropics have been abandoned because of coconut dieback, and shrub encroachment has taken place.

Symbiotic Relations

Symbiotic associations involving nitrogen fixation by microorganisms are frequent in dunes. There are nitrogen-fixing bacteria such as *Rhizobium*, which form a symbiosis with numerous forbs and shrubs in temperate and tropical dune systems. Some of the plants showing nodules are *Ulex europaeous*, *Trifolium* spp., *Lupinus arboreus*, and *Hippophae rahmnoides* in Europe, *Acacia* shrubs in South Africa, and *Chamaecrista chamaecristoides* in Mexico.

In foredunes and mobile dunes, pioneer grasses such as *Ammophila*, *Elytrigia*, and *Uniola* show different degrees of infection by vesicular-arbuscular mycorrhizae (VA). The major benefit to these grasses is probably enhanced phosphorus uptake under conditions of phosphorus limitation. They also help in the aggregation of sand particles. Tropical sand dune plants also frequently show symbiosis with mycorrhizae.

Sand dunes are harsh environments where abiotic factors act as filters that determine species survival. The interactions between abiotic and biotic factors in sand dunes change as dunes mature. They are delicate systems in which plant cover, formed by different vegetation structures and species assemblages, maintains the system in a stabilized condition. Higher diversity is found when there are different habitats. Today, these fragile systems are endangered and the urbanization of the coast is increasing. We must find ways to make our activities and dune conservation compatible.

See also: Deserts; Floodplains; Landfills.

Further Reading

Barbour MG, Craig RB, Drysdale FR, and Ghiselin MT (1973) *Coastal Ecology: Bodega Head*. Los Angeles, CA: University of California Press.

Carter RWG (1988) *Coastal Environments: An Introduction to the Physical, Ecological and Cultural Systems of the Coastlines*. New York: Academic Press.

Hesp PA (2000) Coastal sand dunes: Form and function. Massey University Coastal Dune Vegetation Network, New Zealand, Technical Bulletin No. 4, 28pp.

Lortie CJ and Cushman JH (2007) Effects of a directional abiotic gradient on plant community dynamics and invasion in a coastal dune system. *Journal of Ecology* 95(3): 468–481.

Martínez ML and Psuty NP (eds.) (2004) *Coastal Dunes: Ecology and Conservation*. Berlin: Springer.

Moreno-Casasola P and Vázquez G (2006) Las comunidades de las dunas. In: Moreno-Casasola P (ed.) *Entornos veracruzanos: La costa de La Mancha*. Xalapa, Mexico: Instituto de Ecología AC.

Olson JS (1956) Rates of succession and soil changes on southern Lake Michigan sand dunes. *Botanical Gazette* 199: 125–170.

Packham JR and Willis AJ (1997) *Ecology of Dunes, Salt Marshes and Shingle*. London: Chapman and Hall.

Pilkey OH, Neal WJ, Riggs SR, *et al.* (1998) *The North Carolina Shore and Its Barrier Islands: Restless Ribbons of Sand*. London: Duke University Press.

Ranwell DS (1972) *Ecology of Salt Marshes and Sand Dunes*. London: Chapman and Hall.

Rico-Gray V (2001) *Encyclopedia of Life Sciences: Interspecific Interaction*. New York: Macmillan Publishers.

Seeliger U (ed.) (1992) *Coastal Plant Communities of Latin America*. New York: Academic Press.

Van der Maarel E (1993) (ed.) *Dry Coastal Ecosystems*, vol. 2A. Amsterdam: Elsevier.

Van der Maarel E (1994) (ed.) *Dry Coastal Ecosystems*, vol. 2B. Amsterdam: Elsevier.

Van der Maarel E (1997) (ed.) *Dry Coastal Ecosystems*, vol. 2C. Amsterdam: Elsevier.

Estuaries

R F Dame, Charleston, SC, USA

Introduction

Estuarine ecosystems are among the most complex and complicated systems in the biosphere. Because they are at the interface of terrestrial, freshwater, and marine systems, estuaries are subject to massive fluxes of materials and energy. Further, as a large percentage of the human population lives in close proximity to estuarine and coastal environments, anthropogenic impacts and stress are major driving factors in determining the health and functional status of estuarine ecosystems. In this section, the structure and function of estuarine ecosystems are examined.

Definitions of Estuarine Ecosystems

To set the stage for any discussion of estuarine ecosystems, a clear working definition is needed. One of the simplest and most utilized definitions of an estuarine ecosystem is:

the zone where freshwater from land runoff mixes with seawater.

Another common definition is:

An estuary is a semi-enclosed coastal body of water that has free connection with the sea where seawater is diluted by freshwater derived from land drainage.

The preceding definitions focus on the geomorphological and hydrological aspects of estuaries with no mention of the abiotic or physical driving sources of energy, that is, tides and solar insolation. Nor are any biotic components or processes utilized. Thus, the following definition is proposed:

An estuarine ecosystem is a system composed of relatively heterogeneous biologically diverse subsystems, i.e., water column, mud and sand flats, bivalve reefs and beds, and seagrass meadows as well as salt marshes. These subsystems are connected by mobile animals and tidal water

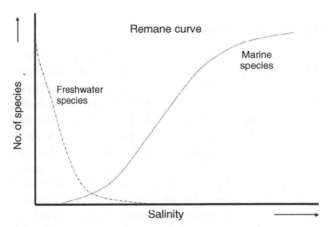

Figure 1 A generalized Remane curve with number of species plotted versus estuarine salinity gradient.

flows that are embedded in the geomorphological structure of creeks as well as channels, and together form one of the most productive natural systems in the biosphere.

Recent quantitative studies indicate that estuaries are ecoclines that are composed of gradients containing relatively heterogeneous subsystems that are environmentally more stable than ecotones (**Figure 1**). Ecoclines are boundaries with more gradual, progressive change between freshwater and the sea. In this view, the organisms in the estuary are either from freshwater or from marine environments; there are no brackish water species. Each estuarine system will respond to at least a freshwater and a marine gradient as well as have its own particular combination of biological and physical components and processes. Thus, every estuarine ecosystem is unique.

Geomorphological Types of Estuaries

Bar-Built and Lagoonal

Bar-built or lagoonal estuaries form in the areas behind sandy barrier islands and usually drain relatively small watersheds. The exchange of water between the estuary and the sea occurs through tidal inlets. Astronomical tides and winds are the major forces controlling water circulation and water height. The areas behind barrier islands are generally subject to less wave action and this promotes the development of extensive wetlands. Bar-built estuaries are generally smaller than other estuarine types, suggesting that they have a higher surface area to volume ratio and, therefore, play a greater role in ecological processes than was previously thought. Well-studied examples of bar-built estuaries are found along the temperate and subtropical coasts of eastern North America, Europe, Asia, and the southern and eastern shores of Australia.

Riverine Estuaries

There are two fundamentally different riverine systems (**Figure 2**). First, are those that arise in the piedmont, have extensive watersheds, receive substantial freshwater discharge, but only a small portion of their watershed is covered by wetlands. Chesapeake Bay and San Francisco Bay in North America as well as the Eastern Scheldt in Northern Europe are well-studied examples of this type of riverine system. A second type of riverine system known as coastal plain estuaries are characterized by a much gentler slope with proportionally more wetlands than piedmont estuaries. Generally, these systems are less studied, smaller and have a lower, more sluggish flows.

Estuarine Ecosystems and Maturity

In an attempt to place a more ecosystems oriented emphasis on estuaries, the 'geohydrologic continuum theory of Marsh-Estuarine ecosystem development' was proposed as a scheme for categorizing estuarine ecosystems. In this theory, the tidal channels of the estuarine ecosystem represent a physical or geohydrologic model of how the ecosystem adapts until there is a change of state. Mature portions of the system are at the ocean–estuary interface, mid-aged components are intermediate

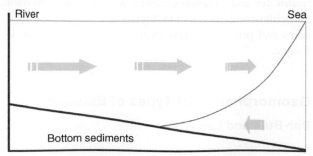

Figure 2 Generalized material flux patterns in a Riverine estuary.

within the longitudinal distribution of the system, and young or immature areas are at the land–estuary interface. Mature systems export particulate and dissolved materials, mid-aged areas import particulate and export dissolved materials, and immature systems import both particulate and dissolved materials. Some estuarine ecosystems may have all three types while others may have only one or two.

Estuaries as Complex Systems

While it is generally acknowledged that ecosystems are complex systems, it is appropriate to describe estuarine ecosystems in the context of the complex systems approach. Complexity as used here can be defined by (1) the nonlinear relationships between components; (2) the internal structure created by the connectivity between the subcomponents; (3) the persistence of the internal structure as a form of system memory; (4) the emergence or the capacity of a complex system to be greater than the sum of its parts; (5) the reality that complex systems constantly change and evolve in response to self-organization and dissipation; and (6) behaviors that often lead to multiple alternative states. Thus, estuarine ecosystems are open nonequilibrium systems that exchange matter and energy as well as information with terrestrial and marine ecosystems as well as internal subsystems. These exchanges not only connect various components, but are the essential elements of feedback loops that generate nonlinear behavior and the emergence of structures and behaviors whose sum is greater than the whole. These systems do exhibit alternate states, for example, Chesapeake Bay appears to have a benthic state dominated by oysters and a water column state dominated by plankton.

Major Estuarine Subsystems or Habitats

The landscape approach to estuarine ecosystems focuses on subsystems or habitats as major components within estuaries. Because organisms respond to the amount of change in the physical (abiotic) environment, their reaction to their environment results in subsystems or habitats composed of specific groups of species that are adapted to that particular set of abiotic factors. In estuarine ecosystems, the major abiotic factors are salinity, water velocity, intertidal exposure, and depth.

Water Column

Water is the primary medium for the transport of matter and information in estuarine ecosystems. Freshwater enters the estuary either as precipitation or as an accumulation driven by gravity down-slope through streams and rivers to the estuary. Salt water enters the estuary from the sea via tidal forcing. The gradient of increasing salt concentration from freshwater to marine divides the estuary into zones of salt stress and subsequently into different pelagic subsystems (**Figure 2**).

Phytoplankton are small chlorophytic eukaryotes that drift as single cells or chains of cells in estuarine currents. Diatoms and dinoflagellates are the dominant groups while species composition of a specific system is usually determined by salinity, nutrients, and light. They are a major component of the estuarine water column and provide food for many suspension-feeding animals. Planktonic primary production is seasonal and varies from distinct peaks in the arctic to spring and autumn blooms in temperate systems and almost no peaks in tropical estuaries. Average annual planktonic primary production in estuaries is about $200-300\,\mathrm{gC\,m^{-2}\,yr^{-1}}$ and is mainly a function of light, nutrient availability, and herbivore grazing.

There are two major categories of zooplankton: holoplankton that in most estuaries are dominated by calanoid copepods which spend their entire life in the planktonic state and the diverse meroplankton that only spend their larval state in the plankton. Most estuarine zooplankton are believed to be herbivores and play a major role in connecting carnivores to phytoplankton. They are also thought to be major sources of inorganic nutrients that are available to phytoplankton.

The microbial loop in estuaries is composed of micro- and nano-planktonic bacteria, protozoans, and flagellates. Initially, the microbial loop was thought to play a major role in recycling nutrients with dissolved organic matter (DOM) a major product. However, the recent finding that a sizable proportion of DOM is made up of viruses has forced a major change in the microbial loop model (**Figure 3**). The current paradigm of the microbial–viral

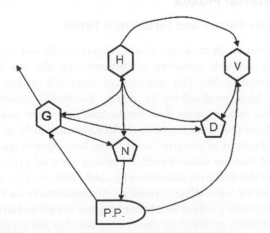

Figure 3 A simple microbial–viral loop food web for an estuarine system. D, dissolved organic matter; G, grazers; H, heterotrophs; N, nutrients; P.P., primary producers; V, viruses.

loop envisions the viruses $(10^{10}\,\mathrm{l^{-1}})$ as 10 times more abundant than bacteria $(10^9\,\mathrm{l^{-1}})$ and controllers of bacterial diversity and abundance. The viruses are small (20–200 nm), ubiquitous particles that use the process of cell lysis to attack and kill bacteria. As a result, more bacterial biomass is shunted into DOM and away from the macroplankton and suspension-feeding macrobenthos. The much more rapid viral recycling of nutrients also has the potential to generate more stability in the system.

Large mobile animals, birds, terrestrial and aquatic mammals, and fish, shrimps and crabs, are common residents as well as transients in estuarine systems. These animals transform and translocate materials both within the estuary and between the estuary and other systems. The nekton organisms, in particular, use the tidally forced water column as a pathway between deeper channels and intertidal habitats where they seek refuge, feed, and develop.

Marshes and Mangroves

Emergent vascular plant-dominated intertidal wetlands are major subsystems in most estuaries. The two most common habitats are geographically zoned latitudinally with marshes dominating the temperate zone and mangroves the frost free subtropical and tropical zones. Both are found in low-energy wave-protected, sedimentary, high-salinity, and intertidal environments near the mouth of the estuary. While wetlands in the high-salinity portion of estuaries are low in species diversity (almost monocultures) of vascular plants, diversity is much higher in the freshwater reaches.

Salt marshes reach their greatest extent and productivity along the Gulf and southeast Atlantic coast of North America where the cord grass *Spartina alterniflora* dominates. This high production is the result of near ideal conditions of temperature, salinity, light, sediment texture, nutrients, and tidal range. Marsh grasses produce large quantities of both above- and belowground biomass that accumulates in the surrounding sediments (**Table 1**). The stems and leaves of the grasses also provide a structural base for an epiphytic community that further increases production. Decomposition processes in the organically rich sediments generate a strongly anaerobic reducing environment making the salt marsh a major

Table 1 Primary production in estuaries

Primary producer	Annual primary production $(\mathrm{gC\,m^{-2}})$
Macrophytes	
Spartina	400–2480
Rhizophora	696–2100
Microphytobenthos	50–200
Epiphytes	12–260
Phytoplankton	25–150

center for nutrient cycling. The nutrient uptake mechanisms of vascular plants are poisoned by the reducing environment; however, air passages in the roots, rhizomes, and stems of these grasses aerate the surrounding sediments so that nutrient uptake can be maintained. The vertical stems and leaves of *Spartina* also serve as a passive filter that slows water flow, can remove via deposition suspended sediments from the water column, and allows many marshes to maintain their elevation with respect to rising sea level. This same environment provides food and refuge for many economically important nekton.

Mangroves are intertidal, tropical, and subtropical woody vascular plants that fill a niche similar to that of *Spartina*. In the high-salinity portions of the estuary, the red mangrove, *Rhizophora*, dominates. Red mangroves have prop roots that lift the plants above the reducing environment of the surrounding sediments. There is a gradient from high production in riverine swamps to low production in high-salinity scrub areas. On a global scale of increasing light with decreasing latitude, the closer a system to the equator, the higher the mangrove productivity. Nutrients have also been implicated as a major limiting factor on mangrove productivity. There is evidence that mangrove production is enhanced by flushing action of storms. In addition to being a nursery for many fish, shrimps, and crabs, the structural mass of mangroves may form a protective buffer to the impacts of storm surges and tsunamis on coastal and estuarine systems.

Seagrasses

Seagrasses are submerged vascular plants that are found in aerobic, clear-water, high-salinity systems with moderate water flow. Cold water systems are dominated by eel grass, *Zostera*, and in the tropics turtle grass, *Thalassia*, is the major group. These grasses are not found in estuaries with high suspended sediment loads, that is, Georgia and South Carolina where there is insufficient light penetration to support their growth. They are also limited to the upper 20 m of water because water pressure compresses their vascular tissues. Maximum seagrass production can approach $15–20\,gC\,m^{-2}\,d^{-1}$. The high productivity of the seagrass is almost equaled by the productivity of the epiphytes on their leaves; however, the sediment trapping abilities of seagrasses give them an advantage over phytoplankton and epiphytes in nutrient limiting conditions. The structure of the seagrasses provides feeding habitat for many mobile animals as well as deposit feeding and suspension-feeding benthos.

Invertebrate Reefs and Beds

Suspension-feeding benthic animals are common in most estuaries because of the high availability of suspended phytoplankton. A number of bivalves and a few worms can aggregate in very dense, high biomass beds or reefs. These structures are found both intertidally and subtidally in high to moderate salinities. The eastern oyster, *Crassostrea virginica*, in its intertidal form builds some of the most extensive aclonal reefs known. Intertidal beds of *Crassostrea* and *Mytilus* can have biomass densities exceeding $1000\,gdb\,m^{-2}$. Depending on the estuary, suspension feeders such as oysters and mussels have been shown to control phytoplankton populations in some systems and influence nutrient cycling by short-circuiting planktonic food webs and reducing the recycle time for essential nutrients. There is evidence that the presence of a significant bivalve suspension-feeder component in estuarine ecosystems enhances system stability.

Mud and Sand Flats

Mud and sand flats are common to the intertidal zone of most estuaries. The major biotic components of tidal flats are bacteria, microbenthic algae, small crustaceans, and burrowing deposit feeders. As in the water column, the microbial–viral loop is thought to play a major role in the decomposition of organic matter in tidal flat sediments. In some estuaries, the microbial–viral loop utilizing a variety of electron acceptors may represent a significant sink for matter and energy. Thus, the prevailing processes on these flats can potentially redirect the fluxes of matter and energy away from macrofaunal food webs to those dominated by microbial processes. The occurrence of tidal flats was originally attributed to the hydrodynamics and sediment sources in tidal creeks; however, with the application of complexity theory to ecological systems, these flats are also being described as alternative states of salt marshes and bivalve beds.

Material Fluxes

Water Fluxes and Residence Times

Interest in the exchange of nonliving materials and organisms between estuarine ecosystems and the sea was initiated by the first quantitative metabolic studies on the high productivity of marsh dominated estuaries. These studies were first synthesized in simple energy budgets that were found to explain less than 50% of the productivity of estuarine ecosystems. Investigators speculated that the unaccounted-for energy must be exported from the estuarine ecosystem by tidal currents. This idea led to the 'outwelling hypothesis' that states that estuarine ecosystems produce much more organic material than can be utilized or stored by the system and that the excess is exported to the coastal ocean where it supports near coastal ocean productivity. While the energy budget or mass balance approach is a cheaper and quicker method of determining the direction of material fluxes, in recent years the direct measurement of material fluxes is favored because this approach provides statistically meaningful results.

Another aspect to the fluxes of materials in estuarine systems is the time the water mass remains in the system or residence time (also known as flushing time or turn-over time). Residence time can provide essential information to resource managers on the retention and dispersal of toxins, the incubation of invasive species, and the carrying capacity of a system for benthic suspension feeders (**Figure 4**). Recent studies on the physics and geomorphology of water in estuarine tidal channels suggest that the residence time of water may vary greatly from place to place within some estuaries. Such variations have been used to explain growth variations in bivalves in different locations within the same estuary. Traditional estimates of an estuarine system's residence time can be computed from measurements of system volume, tidal prism, and water input to the system. The advent of fast computers and numerical models, however, now allows for much more modeling of these systems with the potential for more sophisticated spatial and temporal management strategies.

In riverine systems, river flow is the main physical cause of material and organismic transport from estuaries to the sea. Each of these systems are a unique and changing feature on the present landscape because rising sea level is drowning their basins and sediments are gradually filling their channels. For example in Chesapeake Bay, 35% of the particulate nitrogen and most of the phosphorus is buried in the sediments of the bay. Of the nitrogen in the bay water column, 31% was exported to the sea and 8.9% was removed from the system as commercial fish harvest. In general, the nutrients transported and exported by riverine estuaries are thought to be a significant source for generating

new organic production in the coastal ocean. As many of these systems have dams or have them proposed, managers must take into account the direct and indirect effects of these structures on recreational and coastal fisheries.

In bar-built estuaries, tides are usually the major source of energy for the transport of materials into and out of the estuary. If the fastest currents are on the flooding tides, then the system tends to import suspended particulate material. In contrast, if ebbing tides have the fastest currents, then the system usually exports suspended particulate materials. The Wadden Sea of Northern Europe is a flood-dominated system and North Inlet in South Carolina is an ebb-dominated system.

In shallow, high-insolation, low-precipitation, warm systems, evaporation can dictate the direction of transport. This is the case in some small tropical systems where water loss due to evaporation is replaced by the influx of water and nutrients from the adjacent sea.

Organismic Transport

In addition to inanimate materials, the larval and adult stages of many organisms are exchanged between the estuary and the sea. Some organisms may be passively carried by estuarine currents while others may actively swim or take advantage of the direction of tidal flows to move across the estuary–ocean interface.

Primary producers, including phytoplankton and resuspended benthic microalge, depend on passive transport between estuaries and the sea. Most flux studies show that these organisms have a net seasonal or annual transport into the estuary from the coastal ocean. This import has been explained by passive filtration by estuarine wetlands and by active filtration by suspension-feeding animals within the estuary. Protozoans, bacteria, and viruses are also found in the estuarine water column and while they most certainly are passively transported by estuarine currents, the direction of their net flux is yet to be determined.

The exchange of invertebrate larvae between estuaries and the coastal ocean has been explained by two competing schools of thought, the passive and active hypotheses. In the passive hypothesis, the horizontal movements of larvae are mainly a function of current direction and velocity. The active transport school contends that invertebrate larvae swim both vertically and horizontally to take advantage of tidal currents. In one group that includes oysters, the early stage larvae stay high in the water column with later stages sinking to lower depths. This strategy allows downstream movement of early larvae with some exiting the estuary to the sea, while older larvae are entrained in inflowing bottom currents and effectively retained in the estuary. A second

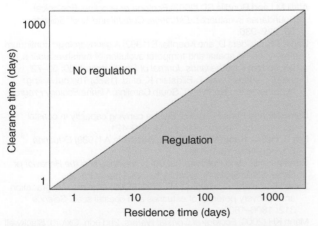

Figure 4 A plot of water volume residence time versus bivalve clearance time showing areas of potential control by suspension-feeders.

approach is used by larvae that migrate vertically in the water column in synchrony with tidal cycles. This strategy allows larvae to maximize upstream transport and retention. A third group has larvae that are immediately transported to the coastal ocean where they stay for weeks before returning into the estuary using wind and tidal currents. A final group uses the coastal ocean during their adult and larval life. In this case, the postlarvae enter the estuary maintaining their position by swimming against tidal currents.

Nekton organisms (fish, crabs, and shrimps) are mobile links between the various subsystems of estuarine ecosystems as well as links between the estuary and the sea. These animals feed and accumulate biomass while in the estuary and then move back to the coastal ocean, thus exporting biomass and inorganic wastes.

Global Climate

While seasonal and latitudinal climatic effects on coastal and estuarine systems have long been documented, the impacts of global climate change (warming or cooling) on estuarine systems have only recently been quantified. Major storms, El Nino-Southern oscillation (ENSO) events, seismic sea waves, or tsunamis and sea level rise (SLR) are global effects that can significantly influence water and material fluxes in estuaries.

Hurricanes and major storms generally influence estuaries through storm surges and short-term increases in precipitation. These enormous pulsed fluxes of water can change the geomorphology of estuaries and their watersheds, massively resuspend sediments, and flush materials off the landscape and into the estuary. Tsunamis can be even larger than storm surges and can have similar impacts to even greater areas of the coastal ocean and estuaries. However, extensive marsh and mangrove wetlands common to estuaries can buffer these pulses of water and reduce the damage they can cause to the coastal landscape.

ENSO events only affect some estuaries. The effect is usually a drought or higher than average precipitation. For example in some South Carolina estuaries, ENSO-induced precipitation and upland runoff can depress salinity up to 75% for as much as 3 months.

SLR is an example of global change on both seasonal and annual time scales that directly influences estuarine systems. Seasonal changes in sea level are the result of air pressure changes at the water's surface and the expansion or contraction of water mass due to heating and cooling. In estuarine systems, these changes are reflected in the depth of the system, but more importantly in the area and time of exposure or submergence in the intertidal zone. SLR will gradually force the transgression of estuaries upslope along the coastal plain. Eventually, SLR will compete with human development for the coastal landscape.

Estuarine Ecosystem Resilience and Restoration

Estuarine ecosystems and subsystems can and do exhibit alternate or multiple states of existence. The ability of an ecosystem to absorb disturbance and resist a change in state is termed ecological resilience, as opposed to engineering resilience, which is the time it takes a system to return to its original state. In the last decades of the twentieth century, ecologists observed that ecosystems were not static entities, but appeared to change in response to external and internal forces. In the Chesapeake Bay estuary, for example, some of the factors causing a state change were over-fishing, increased suspended sediment load, eutrophication, species invasion, and disease. The bay's responses to these forces were slow at first, but with the steady increase in the human population in the bay watershed and with its adherent development, the signs of a state change were dramatically evident. The oyster reefs, a major benthic subsystem or habitat that had dominated the bay for centuries, began to decline rapidly or crash. The benthic-dominated food web was replaced by a planktonic food web. Management efforts to restore the initial oyster-dominated system did not work, probably because they had a single species focus and because ecosystems are strongly nonlinear which means the path to restoration is different from that leading to the initial change of state and many more components of the ecosystem are involved in addition to the oysters.

See also: Mangrove Wetlands; Salt Marshes.

Further Reading

Alongi DL (1998) *Coastal Ecosystem Processes*. Boca Raton, FL: CRC Press.

Attrill MJ and Rundlle SD (2002) Ecotone or ecocline: Ecological boundaries in estuaries. *Estuarine, Coastal and Shelf Science* 55: 929–936.

Dame RF, Childers D, and Koepfler E (1992) A geohydrologic continuum theory for the spatial and temporal evolution of marsh-estuarine ecosystems. *Netherlands Journal of Sea Research* 30: 63–72.

Dame RF, Chrzanowski T, Bildstein K, et al. (1986) The outwelling hypothesis and North Inlet, South Carolina. *Marine Ecology Progress Series* 33: 217–229.

Dame RF and Prins TC (1998) Bivalve carrying capacity in coastal ecosystems. *Aquatic Ecology* 31: 409–421.

Day J, Hall C, Kemp W, and Yanez-Arabcibia A (1989) *Estuarine Ecology*. New York: Wiley.

Gunderson LH and Pritchard L (2002) *Resilience and the Behavior of Large-Scale Systems*. Washington, DC: Island Press.

Lotze HK, Lenihan H, Bourque B, et al. (2006) Depletion, degradation, and recovery potential of estuaries and coastal seas. *Science* 312: 1806–1809.

Mann KH (2000) *Ecology of Coastal Waters*, 2nd edn. Oxford: Blackwell Science.

Floodplains

B G Lockaby, Auburn University, Auburn, AL, USA

W H Conner, Baruch Institute of Coastal Ecology and Forest Science, Georgetown, SC, USA

J Mitchell, Auburn University, Auburn, AL, USA

Introduction

Globally, floodplains may be of greater value to society than any other ecosystem type. This is because of the critical role that interactions between floodplains and associated streams play in maintaining supplies of clean water. While that role is conceptually simple, the processes which define interactions (i.e., floodplain functions) in aquatic–terrestrial ecotones are exceedingly complex. Consequently, it is necessary to develop some understanding of the ecological mechanisms behind those interactions in order to fully appreciate the importance of floodplain ecosystems. To that end, the goal of this article is to provide a first-iteration overview of floodplain form and function (**Figure 1**).

A key concept is that floodplains and associated streams are both causes and reflections of the other's characteristics and functioning. As an example, the

Figure 1 Panoramic view of the Timpisque River and the Palo Verde Marsh in Costa Rica.

climate and geomorphology of a landscape will define the hydrology and initial chemistry of streams. Stream characteristics determine the hydrology, soil characteristics, flora and fauna, and biogeochemistry of the floodplain. In turn, biogeochemical feedback from floodplains to streams helps define the environment seen by aquatic flora and fauna. Thus, a strong interdependency exists between aquatic and terrestrial components of riparian ecotones.

It is critical to understand that land clearing and development, construction of dams and impoundments, pollutant export, and other human activities constitute major influences on streams and floodplains. In some cases, these will override the original hydrology, biogeochemistry, and ecology. As an example, the original hydrology of a riparian system could be dramatically altered by the construction of bermed roadways that cross streams without adequate provision for through flow. Since hydrology is the primary driver of all floodplain functions, corresponding changes in net primary productivity (NPP), species composition of animal and plant communities, and biogeochemistry could be expected to follow.

Geomorphic Origins

Streams in steep topography tend to undergo continual downcutting and, consequently, act as sources of fine and coarse material with little to no opportunity for deposition. Sediment loads are easily carried downstream because the high gradient of the channel imparts sufficient energy for water to retain particles. In many cases, as streams emerge from steeper terrain and move into flatter areas such as coastal plains, the gradient of the channel

(a) Nonincised stream

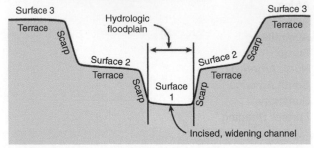

(b) Incised stream (early widening phase)

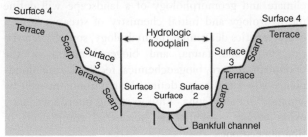

(c) Incised stream (widening phase complete)

Figure 2 Terraces in (a) nonincised and (b and c) incised streams. In Stream Corridor Restoration: Princples, Policies, and Practices (10/98). Interagency Stream Restoration Working Group (15 federal agencies) (FISRWG).

Figure 3 Cross-sectional view of a floodplain topographic positions. Adapted from Hodges 1998.

decreases and flows may spread and lose energy. This promotes the occurrence of overbank flow and creation of deposition surfaces or sediment sinks. However, a floodplain may shift between being a sediment sink or source depending on hydrologic changes induced by climate, anthropogenic activities, or other influences. Downcutting also occurs as older floodplains are abandoned by streams and become terraces which resemble stair steps in cross-section (**Figure 2**).

Sedimentation occurs as particles settle during sheetflow and is highly variable both temporally and spatially. Sediment deposition or alluviation makes possible the high soil fertility that is generally associated with many floodplains although there are notable exceptions. Rates of sediment accumulation vary markedly among floodplains and, in the southeastern United States for example, range from 1 to 6 mm yr^{-1}.

Deposition and scouring may often occur simultaneously on different portions of floodplains and at different times in individual locations. Consequently, the scale at which sedimentation is assessed is very important

in gaining an accurate assessment of net changes. The result of the spatial irregularities is a pattern of swales and berms that generally runs parallel to the stream course. The convex and concave microrelief may represent elevation differentials of only a few centimeters. Nonetheless, those minor differences have major importance in defining soil environments for vegetation and in influencing the extent of contact between floodwaters and the floodplain surface. In many cases, the microtopography of major floodplains is somewhat predictable (**Figure 3**) and, similarly, drives spatial patterns of species composition and NPP of vegetation communities. However, changes in microrelief may be much less apparent on some floodplains due to either prolonged or infrequent flooding.

Hydrology

Hydrology is the foremost determinant of vegetation species occurrence, NPP, biogeochemistry, floral and faunal habitat, and all floodplain functions and traits. Consequently, any insights into the nature of floodplain ecosystems and the basis of their societal value are predicated upon an understanding of hydrology. The 'flood-pulse' concept provides one framework within which to develop this understanding.

In this concept, the river and floodplain are considered as a single system and the 'rhythm of the pulse' (i.e., the hydroperiod) is the controlling mechanism which regulates exchange of energy and material between the river and floodplain. An influx of sediment and nutrients and export of organic carbon from the floodplain will occur at intervals dependent on the pulse rhythm. Examples of common rhythms include single, long duration and multiple, short duration which might be stereotypic of high-order river floodplains and low-order headwater streams, respectively.

In general, flood frequency and duration may decrease and increase, respectively, as stream order rises. When headwaters originate in mountainous terrain, narrow V-shaped valleys form and hydroperiods may be characterized as flashy (i.e., frequent flooding, with sharp rises and drops associated with stage levels). Hydroperiods reflect the integration of rainfall patterns, water storage capacities, and many other factors across the associated catchments. Consequently, stage level rises and falls are slower due to the 'buffering' that is provided by high storage capacities and the greater variability of other factors. Conversely, small catchments have much less storage capacity and, consequently, streams respond rapidly to precipitation events. As a result, floodplains of large rivers can stay flooded for significant portions of a year while low-order floodplains may be inundated frequently but for much shorter periods.

Interchange of water between floodplains and rivers is very complex and involves mutualistic influences. The nuances of those interactions form the basis of the role of floodplains as ecotones and regulators of energy and nutrient exchange. At low stage levels, water within swales and depressions may have originated with the river, precipitation, an upwelling of groundwater, or some combination. From a biogeochemical standpoint, the origin is significant in terms of the degree of spatial and temporal contact with the floodplain. At low stage levels, there is less opportunity for river water to contact the floodplain and, consequently, biogeochemical and dissolved organic carbon exchanges are minimal. As stage levels rise, the potential for the floodplain to influence the biogeochemistry of sheetflow increases as well. However, at some point, increasing floodwater volumes and higher velocities reduce contact with the floodplain. This is because a decreasing proportion of the sheetflow volume is in contact with the floodplain as volumes increase. Similarly, temporal contact is reduced as sheetflow velocities rise.

There is also significant interaction between the river and floodplain in terms of groundwater. Channel waters often generate a head pressure which declines with distance from the stream bank. Groundwater transmittance will decline as hydraulic conductivity of alluvium decreases (e.g., clays have reduced conductance compared to sands). In humid regions, groundwater near the channel moves under pressure and will contact and mix with water that has seeped into the alluvium from adjacent uplands. As a result, groundwater mixing can be quite active during periods of low evapotranspiration.

Biogeochemistry

Once considered purely as nutrient sinks, floodplains are now known to play multiple roles from a geochemical perspective. Based on the type of floodplain, associated vegetation, and the degree and nature of disturbance, floodplains may also serve as sources or transformation zones for nutrients. The widely held perception of floodplains as fertility hot spots belies the complexity associated with input–output budgets as well as the biogeochemical processes within the floodplain ecosystem. In particular, the impact of hydroperiod on

biogeochemical processes sets floodplain biogeochemistry apart from that of non-wetland ecosystems. Periodic flooding makes possible nutrient exchange across the aquatic–terrestrial ecotone and controls the nature of decomposition, nutrient uptake and release by vegetation, and many other processes. As an example, the process of denitrification or the anaerobic conversion of nitrate to gaseous forms of nitrogen is very important on floodplains. In addition, the interaction of hydrology and biogeochemistry necessitates the development of unique approaches to the study of nutrient cycling in these ecosystems.

As previously mentioned, floodplains may serve as sinks, sources, or transformation zones for geochemical inputs of nutrients derived from inflow, precipitation, nitrogen fixation, and soil weathering. Multiple roles may proceed simultaneously on the same floodplain if spatial heterogeneity in hydrology, vegetation, disturbance, and nutrient influx so dictate. The use of a geochemical budget allows net inputs to be compared to net outputs and is based on the perspective of the ecosystem as an integrated system.

In general, the factors that promote nutrient sink activity on floodplains include (1) presence of aggrading vegetation; (2) wide carbon: nutrient ratios in living vegetation and detritus; (3) topographic positions conducive to somewhat frequent, short duration, and low-energy flooding; (4) basin geomorphology that promotes significant sediment loads in streams (e.g., redwater, brownwater, or whitewater based on the color of suspended clay); (5) high occurrence of nitrogen-fixers; and (6) until nutrient saturation is approached, association with a river subjected to high anthropogenic nutrient loadings.

Alternatively, rivers draining low gradient basins with sandy soils are often referred to as blackwater systems because their waters are stained with organic substances (**Figure 4**). These tend to carry low sediment loads and, consequently, alluviation (i.e., sink activity) is less pronounced. Also, floodplains occupied by mature vegetation communities may act as transformers of nutrients (e.g., inorganic inputs of nitrogen converted to organic outputs) rather than a sink or source. The latter is a key facet of the 'kidney function' of these systems and has great significance for maintenance of water quality. Sink activity, such as the filtration and accumulation of sediments (and associated nutrients) from sheetflow also plays a major role in cleansing water (**Figure 5**). Finally, floodplains that have been altered in some way by disturbance may function as nutrient sources. The longevity of the source activity could be short-term (e.g., a well-planned forest harvest followed by rapid forest regeneration) or long-term (e.g., conversion to agricultural or urban uses, impoundments, or climate change).

Similarly, all biogeochemical processes within floodplain ecosystems reflect the overriding influence of

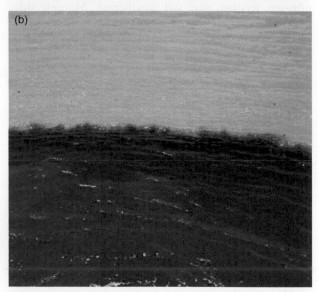

Figure 4   Amazon River: (a) a broad and (b) a close-up view. The formation of the Amazon River at the 'o encontros das aquas' or mixing of the Rio Negro and Rio Solimoes near Manaus, Brazil. The blackwater Rio Negro is contrasted with the sediment-laden Rio Solimoes.

Figure 5   Flint River – sediment accumulation on the Flint River floodplain near Ft. Valley, GA during floodwater drawdown.

hydrology. As an example, the timing of litterfall is heavily affected by hydroperiod because different vegetation communities occur under different hydrologic regimes. In the southeastern United States, forest species associated with *Nyssa* may grow under wetter conditions than communities dominated by some species of *Quercus*. On wetter sites, *Nyssa* foliage tends to senesce earlier in the autumn than other floodplain tree species and, consequently, the senesced foliage is exposed to a different microenvironment than litter that falls later in the year. As a result, nutrient release and immobilization sequences are likely to differ among sites.

Mass loss and nutrient dynamics during decomposition are a function of both litter quality and the decomposition microenvironment. Litter quality (the biochemical composition of detritus) is defined by the conditions under which a plant is growing as well as genetics and has been shown to be closely linked to variation in hydroperiod. Also, the frequency and duration of flooding play a dominate role in determining biomass and composition of microbial populations. Key determinants of shifts between nutrient mineralization and immobilization include hydroperiod and nutrient inflow. In the southeastern United States, mass loss rates of foliar litter (with litter quality held constant) are maximized by moderate durations of flooding followed by several months of noninundation.

In general, rates of litter mass loss in forested floodplains exceed those of uplands. Globally, decay constants for temperate floodplain forests average approximately 1.00 while the mean for all temperate deciduous forests is less than 0.80. This differential is partly due to the greater availability of soil moisture (better habitat for microbial populations) during parts of the year. However, mass loss, as measured by disappearance of confined litter, includes both mechanical disintegration as well as metabolic conversion of organic carbon and, consequently, periodic inundation offers greater opportunities for disintegration and export.

The general perception that floodplains are very fertile has led to misconceptions regarding the degree to which insufficient nutrient availability may constrain floodplain NPP. In many cases, it is true that floodplain soils are more fertile than upland counterparts. However, vegetation species found in many floodplains often have higher annual nutrient requirements compared to species adapted to uplands. Consequently, forest vegetation on many floodplains is likely to be nitrogen deficient and, in some cases such as blackwater systems, deficient in phosphorus and base cations as well. An example would be the nutrient-demanding *Populus deltoides* Batr. plantations that grow in extraordinarily fertile soils of the Southern Mississippi Alluvial Valley, USA. In spite of fertile soils and high aboveground NPP ($20–25 \, t \, ha^{-1} \, yr^{-1}$), those systems would increase in NPP if supplied with additional nitrogen.

The degree to which a floodplain ecosystem is deficient or nondeficient for particular nutrients is critical in regard to that system's potential to act as a nutrient sink. As previously mentioned, the kidney function is enhanced if floodplain vegetation can assimilate incoming nutrients from sources such as polluted water or atmospheric inputs. Once a deficiency is eliminated, it is still possible for floodplain vegetation to assimilate particular nutrients such as nitrogen through luxury consumption. However, a level may be reached after which the vegetation's capacity to retain nutrients is saturated. The latter condition reflects a high degree of biotic stress and is a serious threat to floodplain vegetation associated with eutrophic streams.

Vegetation Community Structure and Composition

Vegetation communities in floodplain systems have developed over hundreds of years as a function of soil type, topography, and hydrology. The type of vegetation growing on a particular floodplain will be dominated by trees or shrubs adapted to the environmental conditions of that floodplain. Hydroperiod is the most important local environmental condition determining composition, and the species found respond to elevation differences relative to the river's flooding regime. Typical floodplain forests begin at the natural levee where coarse-grained deposits result in quickly draining soils and continue as surface elevations decrease away from the river and become more poorly drained.

Structural characteristics of floodplain forests vary depending upon location (**Table 1**). Stem density and basal area are generally greater in the southeastern

United States and the humid tropics than in arid areas, but in arid areas basal area can still exceed $50\,m^2\,ha^{-1}$. Basal areas in floodplain forests tend to be as high as or higher than that of upland forests. Almost without exception, the number of tree species increases as flooding decreases. The greatest number of tree species occurs in wet, tropical floodplains such as the Amazon. The understory of floodplain forests is generally lower in density and species numbers, probably due to reduced light levels and the extended flooding conditions.

Adaptations of Floodplain Vegetation

Due to the alternating wet–dry environment experienced by trees growing on floodplains, they have developed a variety of physiological and morphological adaptations that allow survival during flooding. Initially, stimulation of alcohol dehydrogenase (ADH), enzyme activity may provide a temporary means to support essential metabolic functions. The anaerobic pathway is less efficient than the aerobic pathway (39 moles ATP per mole hexose vs. 3 moles ATP per mole hexose), but provides an energy resource while anatomical changes are occurring.

The seeds of floodplain tree species require oxygen for germination, and even those species that can grow in permanently to nearly permanent flooded conditions (e.g., *Taxodium* and *Nyssa*) require moist, but not flooded, soil for germination and establishment. Occasional drawdowns are necessary for the survival of tree species. Rapid stem elongation, such as been observed with *Nyssa aquatica*, allows the seedling to get its crown above the water surface of subsequent floods. The dispersal and survival of many wetland tree seeds is dependent upon hydrologic conditions. *Taxodium* and *Nyssa* seeds are produced in the fall and winter between the periods of lowest and highest streamflows, giving the seeds the widest possible range of

Table 1 Mean structural and aboveground productivity characteristics of floodplain forests

| Area | No. of species | Density (no. ha^{-1}) | Basal area (m^{2}ha^{-1}) | Biomass (t ha^{-1}) | Aboveground NPP | | |
					Leaf	Wood (t ha yr^{-1})	Total[a]
Southeastern USA	13	1242	45.0	302	5.36	7.78	13.26
Northeastern USA	10	970	26.1	150			
North Central USA	5	546	29.5				
Western USA	5	310	27.5				
Central USA	12	405	33.5	290	4.20	2.50	8.70
Europe		1237	26.5	314	3.48	17.88	
Central America	10	726	49.9	118	11.61		
Caribbean	27	3359	42.4	224	15.55		
South America	89	687	33.0	413			
Africa	26						
Southeast Asia					9.15		
Australia	12	493		260			

[a]Total NPP does not always equal leaf plus wood as some sources only report total.

hydrologic conditions. Overall, seed production of many wetland species seems to be linked to the timing and magnitude of hydrologic events.

Stem hypertrophy, commonly called butt swell or buttressing, is characterized by an increase in diameter of the basal portion of the stem and is common in *Taxodium*, *Fraxinus*, *Nyssa*, and *Pinus* species. Basal swelling can extend from just above the ground level to several meters depending upon the depth and duration of flooding. Swelling generally occurs along that portion of the trunk that is flooded seasonally. Increased air space in the swollen portion of the stem allows increased movement of gases within the plant. Ethylene production has been documented to play a regulatory role in altering growth and stem anatomy of woody plants, and has been found to be higher in flooded *Fraxinus* stems with well-defined hypertrophy than those without stem hypertrophy. Lenticel hypertrophy has long been associated with flooding and acts to increase internal gas transport from the stem to the roots. Duration of flooding does not appear to affect the number of lenticels formed but does affect the size. The formation of hypertrophied lenticels under anoxic conditions also appears to be induced by ethylene. Other commonly observed features in flooded environments include buttress roots and knees. Buttress roots appear as fluted projections at the base of mature trees and extend for several feet from the trunk outward and down into the soil. Because of the shallow nature of root systems in saturated or flooded soils, these buttress roots are thought to provide additional support to the tree. Knees are common in *Taxodium* spp. in the southeastern United States. Their function has not been confirmed, although there is some speculation that they also serve in stability of the tree. In Australia, *Melaleuca* trees on floodplain sites have modified bark structures such as papery bark with internal longitudinal air passages that allow them to tolerate flooded conditions.

Productivity

Riverine floodplains are typically characterized by high productivity. Productivity is enhanced in many floodplain areas by the continued import and retention of nutrient-rich sediments from headwater regions and lateral sources, increased water supply (especially in arid regions), and more oxygenated root zones as a result of flowing waters. The flood pulse advantage has long been recognized, with ancient Egyptians setting taxes based on the extent of the annual flood.

Primary productivity of unaltered, seasonally flooded ecosystems is generally higher than that of floodplain forests that are permanently flooded or those with stagnant waters. Despite the theoretical basis for increased floodplain productivity due to pulsing, it has been difficult to confirm. More recent studies tend to point toward the idea that

seasonal flooding can be both a subsidy and a stress. In the southeastern United States, aboveground NPP was similar for upland hardwood, bottomland hardwood, and *Taxodium–Nyssa* forests. The reason for this may be that for some sites, subsidies and stresses occur simultaneously and cancel one another. As a result, flood intensity and duration affect soil moisture, available nutrients, anaerobiosis, and even length of growing season in a complex and nonlinear 'push–pull' arrangement. When hydrology is altered rapidly, aboveground productivity is less than in natural forest communities with nearly continuous flooding (**Figure 6**).

Aboveground biomass in floodplain forests ranges between 100 and 300 t ha^{-1}, although there is one report of a forest in Florida where biomass exceeds 600 t ha^{-1}. Leaves account for only 1–10% of the total aboveground biomass. Belowground biomass has been sampled rarely and varies greatly, but reported values tend to be somewhat lower than the 20% of total biomass often cited for upland species. Total aboveground biomass production (leaves plus stem wood) ranges from 668 to 2136 g m^{-2} yr^{-1}, with leaves accounting for approximately 47% of the production. Although it has been reported that there are no latitudinal patterns in NPP, litterfall production of *Taxodium* forests in the United States shows a curvilinear relationship with latitude with a maximum occurring at about 31.9° N. In northern Australia, litterfall in *Melaleuca* forests has been reported to be 2–3 times greater than that in forests in the southern part of the continent. Changes to natural hydrologic regimes decrease litter production by half. As a result of the high productivity, generally associated with floodplain forests, carbon sequestration is particularly important there.

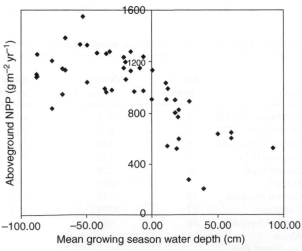

Figure 6 Relationship between aboveground net primary productivity (NPP) of floodplain forests of the southeastern United States and mean water depth during the growing season.

Anthropogenic Impacts

Rivers and associated floodplains have been vitally linked to civilization throughout history for food production. In order to make farming easier and more productive, rivers have been diverted and floodplains have been deforested and drained or leveed to provide fertile land. A major consequence of the widespread use of floodplains and adjacent uplands for agriculture has been the generation of large sediment loads in associated streams and rivers. As a result, much sediment has been deposited on streambeds and floodplains with negative consequences for aquatic habitat and floodplain vegetation. More recently, impoundments have become commonplace for energy production and water storage and levees continue to be built to provide space for development as well as for farming. Globally, it is estimated that, at a minimum, 75% of total floodplain area has been lost.

Floodplain function is dependent on connectivity between the river and its riparian area. Unfortunately, many anthropogenic impacts eliminate or reduce that connectivity so that key functions such as water filtration are much reduced at the landscape level. Similarly, alterations in hydroperiod caused by human activity often drive changes in composition and productivity of vegetation communities as those species adapted to the former conditions decline and are replaced by others.

Additional impacts include fragmentation of riparian vegetation communities and stimulation of invasive non-native plant invasion. Fragmentation often results in reduced habitat quality while successful invasion by non-native species may cause major alterations in community composition, structure, and function. While ecological restoration of floodplains has attracted widespread interest, economic constraints have primarily limited restoration applications to localized areas. However, notable exceptions include restoration of the portions of the Pantanal River Basin in South America and the Kissimmee River Corridor in Florida, USA.

More recently, urbanization has led to significant and growing impacts on floodplains in many parts of the world. As catchments become developed, the concomitant rise in impervious surface drives major increases in runoff volume and velocity. As a result, rising limbs during flood events become much steeper, a condition that is often associated with higher in-channel velocities. Higher flow velocity increases the rate of channel incision resulting in a lowered groundwater table and reduced connectivity between the stream and floodplain. In addition, urbanization stimulates loadings of nutrients (particularly, nitrogen) and causes a considerable degree of water pollution in general.

Further anthropogenic impacts include channelization of river systems. Channelization has benefited farming and waterborne transportation by reducing flooding and removing obstacles to barge and other water traffic. However, water quality has suffered in many instances since there is again less opportunity for river waters to contact floodplain surfaces and undergo pollutant reduction.

Africa

The African continent has approximately 99 large wetlands, of which 43 are floodplain systems. Some of the larger floodplain systems include the Zaire Swamps ($200\,000\,km^2$), the Inner Niger Delta of Mali ($320\,000\,km^2$ when flooded), the Sudd of the Upper Nile ($16\,500\,km^2$ of permanent swamp and $15\,000\,km^2$ of seasonal floodplain), and the Okavango ($14\,000\,km^2$ of permanent swamp and $14\,000\,km^2$ of seasonal floodplain). These floodplain systems are in dynamic equilibrium with the constant flux of pulsing events occurring within them at different spatial and temporal scales. Goods and services resulting from pulsing events include floodplain recession agriculture, fish production, wildlife habitat, livestock grazing, ecotourism, and biodiversity, as well as natural products and medicine.

In semiarid and arid regions of Africa, floodplains are often the only source of year-round water. As in other floodplains around the world, vegetation distribution is strongly related to flooding frequency and duration and microtopography. Dense evergreen tree growth occurs on higher well-drained areas like levees and termite mounds, while grasslands tend to dominate lower, more frequently flooded areas. Typical grasses found growing in these frequently flooded areas (called swamp) include *Phragmites*, *Typha*, and *Polygonum*. Tree and bush genera in less frequently flooded areas include *Hyphanene*, *Borassus*, *Acacia*, *Ficus*, and *Kigelia*.

Floodplain areas are centers of high diversity of animal and plant life. These floodplain areas are of profound importance for fish production and probably serve as spawning and recruitment areas. Interannual fluctuations in fish production have been correlated with the flooding regime. Numerous bird species (over 400 in some floodplains) can be found in these areas, including bee-eaters, jacanas, malachite kingfishers, grey herons, egrets, African fish eagles, and Zaire peacocks. The birds share the floodplains with antelope (sitatunga, waterbuck, puku, and lechwe), hippopotamus, zebra, and buffalo; vegetation ranges from water lilies and papyrus to floodplain forests with minor topographical variations playing an important role in distribution of forest and grassland. Climatic variations are also important, with forest only occurring near rivers in drier areas while in wetter areas forests can extend for a considerable distance away from the river.

River meanders tend to be cut off during flooding periods, adding diversity to the floodplain topography.

Unfortunately, very few studies of the ecology of many of the African floodplain systems have been carried out. The most studied floodplain system in Africa is the Okavango Delta. Annual floods travel uninterrupted down the Okavango River and inundate the Okavango Delta from April to September. River water is characterized by moderate levels of nutrients, but when it enters the floodplain it becomes strongly enriched by nutrients via leaching from soil, detritus, and feces. Organic carbon enrichment comes from leaching of floodplain leaf litter and soil, although dissolved organic carbon release from leaf litter is over 2 orders of magnitude greater than for leached soils. This nutrient enrichment has a major impact on aquatic productivity in the delta and illustrates the strong links between terrestrial and aquatic ecosystems.

African floodplains face a different set of challenges as opposed to those in developed countries. In Africa, floodplains generally occur in semiarid areas to arid regions, and flooding is the driving force behind the high productivity of these areas. From as early as the 900s, people have inhabited these areas, and pastoral and agricultural economies are dependent upon the continued presence of the floodplains. Continued pressures from agricultural practices within the floodplains themselves and population growth that demands the transfer of water to alleviate shortages outside of the floodplain need to be addressed to ensure survival of these important ecosystems.

Asia

In northern Asia, there are extensive productive wetlands along the floodplains of rivers. In western Siberia, the river Ob extends over $50\,000\,km^2$ and supports what is called the largest waterfowl breeding and moulting area in Euroasia. The Ob Valley is a labyrinth of intricately arranged channels and floodplain lakes. As in other seasonal floodplains, the region is a land of fluctuating water levels, with seasonal and annual fluctuations in river discharge and flooding patterns. This area avoided any serious human impacts for centuries, but oil and gas exploration has resulted in significant pollution and transformation of the landscape.

The Indus River has long been the lifeblood of arid Pakistan. In earlier times, people used the river's water to cultivate the floodplain, but during the last 100 years, the river has been dammed and diverted into one of the largest and most complex irrigation systems in the world. In the absence of a drainage system to remove irrigation water, evaporation leaves salt in the soil. As a result of this salinization of the soil, combined with waterlogging, over $400\,km^2$ of irrigated land is lost each year.

Many of the large river systems in South Asia display considerable annual variation in discharge and, during the rainy season, may flood very large expanses of land (e.g., approximately 200 km on each side of the Ganges). In some cases, entire deltaic areas may be inundated. The prolonged, monomodal flooding promotes extensive spatial and temporal contact with floodplains and, consequently, dominates the socioeconomics of the large human populations near those systems. Agricultural activities often cause significant sediment export from upper reaches of many rivers and, as a result, delta tributaries may become clogged. Due to the subsequent reduced flow, salinity can increase in soils and alter species composition in the delta forests.

About 80% of Bangladesh ($115\,000\,km^2$) is formed by floodplains of the Ganges, Brahmaputra, and Meghna rivers. In major floods, 57% of the country can be flooded. Availability of water during the dry season makes it possible to grow three crops a year in some areas. Deposition of waterborne sediments keeps the soils fertile and algal growth enriches the soil by fixing nitrogen. As in many parts of the world, forest vegetation of South Asian floodplains strongly reflects variations in hydroperiod and soil. In the Ganges and Brahmaputra River Valleys, within areas with heavy clay soils where flooding occurs for most of the year or permanently, forest vegetation may be only 5–10 m in height and occur in conjunction with numerous vines. However, combinations of similar flooding regimes and lighter, fertile soils may increase canopy heights by 10 m or more. Many of these riverine forests exhibit a prevalence of evergreen or semievergreen species, although at higher altitudes alders may dominate. Some lowlands, in particular many river deltas such as those of the Ganges–Brahmaputra and Irrawady, are occupied by mangrove forests.

In China, 95% of the population is concentrated in the eastern half of the country, mainly in the vast alluvial plains of the major rivers, the Yellow and Yangtze Rivers primarily. High population densities coupled with high growth rates, rapid urbanization, and industrialization play a major role in most Asian countries. Water resources in this region are under increasing pressure as the demand for domestic supplies, agricultural use, and hydroelectric power increases. Past water resource and agricultural management practices have resulted in rapid loss and degradation of natural wetlands throughout the region. The regulation of rivers and streams through embankments and dams has eliminated floodplains and reduced groundwater recharge. Changing hydrological regimes have increased flooding during the rainy season and reduced availability during dry periods. Water resource management has often resulted in numerous man-made wetlands such as reservoirs and paddy fields that have very different functions and values than natural wetlands, and are in no way a substitute for natural wetlands, particularly floodplain wetlands. In short naturally occurring floodplains in these regions are threatened by numerous human activities, including mining, aquaculture, unsustainable forestry or fisheries practices, and conversion of forests to urban or agricultural land.

Australia

Australia is distinctive in that there are few permanent wetlands due to high evaporation rates and low rainfall. Most wetlands on the continent are intermittent and seasonal. Common features of floodplains are waterholes and lagoons called billabongs that retain water seasonally or permanently, providing important habitat for many animals at different times of the year. Floodplain wetlands tend to be sites of extraordinary biological diversity of waterbirds, native fish, invertebrate species, aquatic plants, and microbes. Key drivers of this biodiversity are the lateral connectivity to the river of the floodplain wetland and the unpredictable flows that create wide ranges of temporally and spatially different aquatic ecosystems.

Humid coastal areas are drained by short, perennial streams, while much of the streamflow in the rest of the country is intermittent or nonexistent because of low and unreliable rainfall, high evaporation, and flat topography. Even under these conditions, forested wetlands can be found throughout Australia, but they can only be classed as true forests in the wettest localities. The largest area of floodplain forested wetland (over 60 000 ha) occurs on the Murray River. Floodplain forests are generally composed of *Melaleuca* or *Eucalyptus* species, but they cannot survive very long periods (>5 months) of flooding. If flooding exceeds several weeks during the growing season, forest canopy cover declines to between 10% and 70%, creating open woodlands.

Tropical floodplain wetlands are found across northern Australia, covering an estimated 98 700 km^2. Vegetation of these wetlands has been mapped at various scales, but there are few specific or long-term analyses of the distribution or successional changes of the plants. The Ord River floodplain in northern Australia encompasses approximately 102 000 ha and is a large system of river, tidal mudflat and floodplain wetlands that supports extensive stands of mangroves, large numbers of waterbirds, and significant numbers of saltwater crocodiles. In southeastern Australia, the Murray–Darling river system drains 14% of the continent and contains the greatest amount of floodplain wetlands on the continent.

In recent years, floodplain areas have undergone considerable change because of animal (buffalo, pigs, cane toads) and plant (mimosa, salvinia, paragrass) invasions, changes in fire regimes, water resource management, and saline intrusion. Dams and the cumulative impact of diversions and upstream river management have turned many floodplains into terrestrial ecosystems. The effect of this change in flooding has not been well studied and data exist only for a fraction of the area affected. Floodplain loss will continue until there is a better understanding of the long-term ecological effects of dams and diversions.

Europe

Floodplains in Europe have been influenced by humans for thousands of years. Civilizations often were established near rivers and frequently utilized floodplain resources for food (agriculture or hunting), power (wood or water mills), and shelter. As communities grew there was an increased need to control the flooding that naturally occurred in the floodplains with the use of dams, dikes, and ditching. These structures altered the hydrology which in turn has altered the forest composition in these areas. Furthermore, channel straightening has caused major hydrologic changes resulting from faster flow and increased groundwater depth. In some of the Danube watersheds, there has been an 80% decrease in first-order streams from 1780 to 1980.

In many areas, depth to groundwater has increased due to the 'drying' of the floodplains and this has driven shifts in the composition of vegetation communities. In particular, species such as *Quercus robur*, *Fraxinus* spp., and *Ulmus* spp. are becoming rarer due to the altered hydrology. Forestry practices have induced a further shift from natural systems to faster growing *Populus* clones in many of the floodplains across central Europe. However, reestablishment of the more traditional forest composition of uneven-aged oak, ash, and maple mixes has been achieved in some areas as recently as the past 50 years. Large portions of the forests remain monocultures of even-aged *Acer* or *Fraxinus*.

The Danube Delta represents one of the largest wetlands in Europe and is undergoing eutrophication as a result of increasing nutrient inputs, decreased riparian vegetation, and loss of the filtration function. One major difference between European floodplains and others worldwide is that increased flow and flooding often occurs in the spring as a result of snowmelt in high altitudes.

North America

In the dry climate of the Western United States, water is a limited resource not only for the wildlife but also for the human inhabitants. Although wetland areas comprise a very small portion of total land area (i.e., less than 2%), over 80% of wildlife is dependent on their presence. Rainfall in this region varies from less than 15 cm yr^{-1} in the desert regions to greater than 140 cm in the mountains. In the mountainous regions, rainfall and snowmelt are greater than losses and, therefore, wetlands rarely dry out. However, evapotranspiration in the basin areas is 3–4 times greater than precipitation and, consequently, soil salinization is a stress to which vegetation must adapt. In the driest regions soil salinization prevents vegetative establishment. Ephemeral drains are prevalent in the intermountain west with snow melt and high rain contributing to their flow.

At higher elevations in the United States, where soils are semipermanently inundated or saturated, associations of *Populus*, *Salix*, and *Acer* are found. Floodplain areas flooded or saturated 1–2 months during the growing season are comprised of a wide array of hardwood trees. Common species in the United States include *Fraxinus* spp., *Tilia* spp., *Ulmus* spp., *Liquidambar* spp., *Celtis* spp., *Acer* spp., *Plantanus* spp., and some *Quercus* spp. At the highest elevations, flooding occurs for less than a week to about a month during the growing season. Typical tree species include a variety of *Quercus* spp. and *Carya* spp., with some *Pinus* spp.

Floodplains of the southeastern United States occur within three physiographic regions: (1) coastal plains, (2) piedmont, or (3) Appalachian Mountains. Rainfall is sufficiently prevalent during all seasons except for brief periods of drought. Successional patterns of southern forested floodplains are often dictated by hurricanes, tornados, catastrophic ice storms, and extended drought. Soils are typically acidic, with the exception of near neutral pH soils across much of the Southern Mississippi Alluvial Floodplain and the Selma Chalk geologic region of Alabama and Mississippi. In many floodplains, as one moves in a direction perpendicular to the river, soil textures range from coarse sands near stream channels, fine sands in natural levees, to loams and clays in backwater areas. This separation pattern is a result of particle size and sheet flow velocity.

The lowest elevation, nearly always flooded sites on floodplains in the southeastern United States are occupied by *Taxodium–Nyssa* swamps. In other parts of the world, it appears there are no similar tree species that can survive permanent or long periods of inundation. As long as the floodplain channel remains stable and flooding frequency remains constant, these species should dominate the stands indefinitely.

South America

Much of our current knowledge about forested floodplains has been derived from extensive studies performed in the sub-basins of the Amazon River. In particular, our understanding of floodplain biogeochemistry, NPP, vegetation dynamics, geomorphology, and faunal relationships has been greatly influenced by Amazonian research.

In comparison to river basins in other parts of the world, the water balance of Amazonia lowlands is roughly evenly divided between evapotranspiration and runoff. This contrasts with systems in Asia where runoff dominates due to generally steeper terrain and many African systems where broad floodplains and high potential evapotranspiration result in low runoff. Floodplain forests in South America are typically composed of a small number of fast-growing, early-successional species capable of surviving periodic floods and large amounts of sediment deposition (e.g., *Salix* and *Inga* spp.).

The 'flood pulse' concept was originally conceptualized in relation to the Amazon and similar floodplains and can be applied worldwide. The major river floodplains of South America such as the Amazon, Orinoco, and the Parana display singular, river-borne flood pulses of large amplitude and duration. In contrast, inundation on floodplains situated within large depressions such as the Pantanal is generally rainfed (as opposed to overbank flow from rivers), and also displays a singular periodicity but with lower amplitude. Finally, multiple flood pulses that are less predictable in terms of occurrence and amplitude are characteristic of floodplains associated with smaller order streams.

Some of the classic research that defined global variation among floodplains took place in Amazonia and was associated with contrasts between blackwater versus brownwater or whitewater rivers. Similar types of floodplain systems occur in many parts of the world. The color of the river waters is reflective of the geomorphology of particular systems and is a strong indicator of floodplain biogeochemistry, vegetation dynamics, and NPP. Whitewater rivers in the Amazon Basin derive their color from white clay sediments that originate in the Andes. The suspended clays contain higher levels of nutrients (particularly base cations) which, when deposited, often create fertile floodplains labeled varzea.

In contrast, blackwaters are stained by fulvic acids and other organic compounds and are more acidic than whitewater counterparts (pH <5.0 vs. >6.0 for blackwater and whitewater, respectively). Due to the low sediment loads, floodplains associated with blackwater streams are often nutrient poor and are referred to as igapo. Consequently, forest litterfall production on varzea floodplains is often considerably higher than that of the igapo (approximately 10 vs. 5 t ha^{-1} yr^{-1} for the respective system types). Also, the standing crop of fine roots is much higher in igapo soils compared to varzea, a reflection of greater belowground allocation of biomass as would be expected in resource-poor soils. Such adaptations increase the likelihood of nutrient capture from decomposing igapo litter. The distinctions in hydrology and biogeochemistry between the igapo and varzea also drive major differences in vegetation species occurrence, root, shoot, and reproductive phenology, and community structure.

Distinctions between floodplains types are also important in regard to animal populations. This is particularly true for fish which depend on interactions with inundated floodplains for resource acquisition, reproductive habitats, and other factors. As an example, the lower NPP on igapo floodplains may translate to lower food resources for fish. While the amount of plant detritus exported from varzea floodplains is higher, phytoplankton production also depends on settling of the clay sediments so that sufficient light can penetrate the waters. Although more difficult to document in riverine systems, fish catches are generally much lower in igapo lakes compared to varzea counterparts.

As is the case in much of the world, South American floodplain ecosystems are under pressure from an array of human activities. As an example, the lower reaches of the Parana' River have undergone changes in hydrology due to construction of dams and upstream expansion of agriculture. The altered hydrology, along with increased concentrations of sediment and other contaminants have resulted in heavy impacts to fish populations and concomitant economic declines in local fishing communities. Although there is a growing voice for conservation and protection of natural resources, it is unclear to what extent anthropogenic impacts may be curtailed.

Summary

As pathways between aquatic and terrestrial ecosystems, floodplains perform a myriad of functions that are critical to humanity and all other components of the biosphere. Because of the vital need of all organisms for clean water, the kidney or filtration function is the most important attribute of healthy floodplain systems.

The filtration function entails sediment and nutrient deposition and, consequently, has long made floodplains very attractive for exploitation as agricultural sites. It is ironic that the very function that makes floodplains so important attracts major disturbances which, in turn, result in destruction of the kidney function in those systems. Globally, that destruction is reflected in the magnitude of floodplain loss (i.e., 75%).

While the primary cause of floodplain destruction is shifting from agriculture to urban development, it would be unrealistic to expect that the general magnitude of anthropogenic pressures on these systems will abate. Consequently, an answer to the critical question of whether adequate supplies of clean water exist will become increasingly uncertain. In order to provide a positive answer and, subsequently, protect human health and well-being, it is vital that we more clearly understand how these ecotones operate so that functional floodplains can be maintained and integrated into evolving landscapes.

See also: Ecosystem Ecology; Ecosystems; Riparian Wetlands; Rivers and Streams: Ecosystem Dynamics and Integrating Paradigms; Rivers and Streams: Physical Setting and Adapted Biota; Swamps.

Further Reading

Brinson MM (1990) Riverine forests. In: Lugo AE, Brinson MM, and Brown SL (eds.) *Forested Wetlands, Vol. 15: Ecosystems of the World*, pp. 87–141. Amsterdam: Elsevier Science Publishers.

Cavalcanti GG and Lockaby BG (2005) Effects of sediment deposition on fine root dynamics in riparian forests. *Soil Science Society of America Journal* 69: 729–737.

Groffman PM, Bain DJ, Band LE, *et al.* (2003) Down by the riverside: Urban riparian ecology. *Frontiers in Ecology and the Environment* 6: 315–321.

Hupp CR (2000) Hydrology, geomorphology and vegetation of coastal plain rivers in the south-eastern USA. *Hydrological Processes* 14: 2991–3010.

Junk WJ (1997) *Ecological Studies 126: The Central Amazon Floodplain: Ecology of a Pulsing System*. Berlin: Springer.

Lewis WM, Jr., Hamilton SK, Lasi MA, Rodriguez M, and Saunders JF, III (2000) Ecological determinism on the Orinoco floodplain. *Bioscience* 50: 681–692.

McClain ME, Victoria RL, and Richey JE (2001) *The Biogeochemistry of the Amazon Basin*. New York, NY: Oxford University Press.

Megonigal JP, Conner WH, Kroeger S, and Sharitz RR (1997) Aboveground production in southeastern floodplain forests: A test of the subsidy-stress hypothesis. *Ecology* 78: 370–384.

Messina MG and Conner WH (1998) *Southern Forested Wetlands: Ecology and Management*. Boca Raton, FL: CRC Press.

Mitsch WJ and Gosselink JG (2000) *Wetlands*, 3rd edn. New York, NY: Wiley.

Naiman RJ and Decamps H (1997) The ecology of interfaces: Riparian zones. *Annual Review of Ecology Systematics* 28: 621–658.

National Academy of Science (2002) *Riparian Areas. Functions and Strategies for Management*. Washington, DC: National Academy Press.

Paul MJ and Meyer JL (2001) Streams in the urban landscape. *Annual Review of Ecology and Systematics* 32: 333–365.

van Splunder I, Coops H, Voesenek LACJ, and Blom CWPM (1995) Establishment of alluvial forest species in floodplains: The role of dispersal timing, germination characteristics and water level fluctuations. *Acta Botanica Neerlandica* 44: 269–278.

Forest Plantations

D Zhang, Auburn University, Auburn, AL, USA

J Stanturf, Center for Forest Disturbance Science, Athens, GA, USA

An Overview and Economic Explanation of Global
   Forest Plantation Development
Factors Influencing Forest Plantation Development

Forest Plantations and Conservation of Natural Forests
Direct Ecological Effects of Forest Plantation
Further Reading

Between the extremes of afforestation and unaided natural regeneration of natural forests, there is a range of forest conditions in which human intervention occurs. Previously, forest plantations were defined as those forest stands established by planting and/or seeding in the process of afforestation or reforestation. Within plantations, there is a gradient in conditions. At one extreme is the traditional forest plantation concept of a single introduced or indigenous species, planted at uniform density and managed as a single age class (the so-called monoculture). At the other extreme is the planted or seeded mixture of native species, managed for nonconsumptive uses such as biodiversity enhancement. To further complicate matters, many forests established as plantations come to be regarded as secondary or seminatural forests and no longer are classed as plantations. For example, European forests have long traditions of human intervention in site preparation, tree establishment, silviculture, and protection; yet they are not always defined as forest plantations.

Further refinement of the plantation concept is necessary in order to encompass the full range of actual conditions. A useful typology is based on purpose, stand structure, and composition of plantations. Thus, an industrial plantation is established to provide marketable products, which can include timber, biomass feedstock, food, or other products such as rubber. Industrial plantations usually are regularly spaced with even age classes. Home and farm plantations are managed forests but at a smaller scale than industrial plantations, producing fuelwood, fodder, orchard, and garden products but still with regular spacing and even age classes. A wide range of agroforestry systems exist, distinguishable as a complex of treed areas within a dominantly agricultural matrix. Environmental plantations are established to stabilize or improve degraded areas (commonly due to soil erosion, salinization, or dune movement) or to capture amenity values. Environmental plantations differ from industrial plantations by virtue of their purpose; they may still be characterized as regularly spaced with even age classes. Efforts to restore forest ecosystems are increasing and often utilize the technology of plantation establishment, at least initially.

Recently, FAO defined 'planted forests' as forests in which trees have been established through planting or seeding by human intervention. This definition is broader than plantations and includes some seminatural forests that are established through assisted natural regeneration, planting or seeding (as many planted forests in Europe that resembled natural forests of the same species mix) and all forest plantations which are established through planting or seeding. Planted forests of native species are classified as forest plantations if characterized by few species, straight, regularly spaced rows, and/or even-aged stands. Forest plantations may be established for different purposes and were divided by FAO into two classes: protective forest plantations which are typically unavailable for wood supply (or at least having wood production as a secondary objective only) and often consist of a mix of species managed on long rotations or under continuous cover; and productive plantation forests which are primarily for timber production purposes.

Figure 1 shows that, in 2005, some 36% of global forests (about 4 billion ha, covering 30% of total global land area) are natural forests, 53% are modified natural forests, 7% are seminatural forests, and the remaining 4% are forest plantations. Of these forest plantations, productive forest plantations account for 78% and protective forest plantations account for 22%. While natural forests and modified natural forests declined between 1990 and 2005, seminatural forests and forest plantations increased (**Figure 2**).

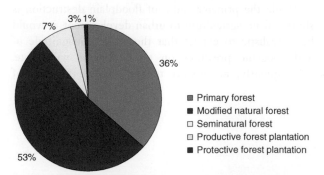

Figure 1 Global forest characteristics 2005. Modified from FAO (2005) Global forest resources assessment 2005. *FAO Forestry Paper 147.* Rome, Italy.

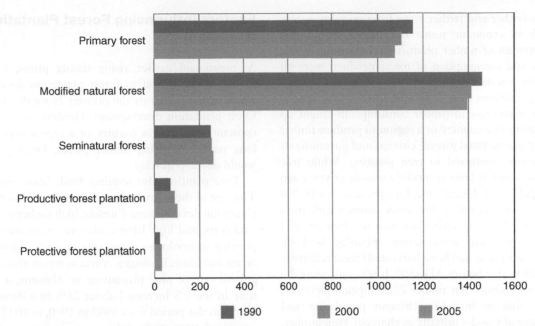

Figure 2 Global trends in forest characteristics 1990–2005 (million ha). Modified from FAO (2005) global forest resources assessment 2005. *FAO Forestry Paper 147*. Rome, Italy.

This article provides an overview and economic explanation of global forest plantation development. It also presents factors influencing global forest plantation development and lists the usefulness of forest plantations, including their roles in the conservation of natural forests. Finally, it summarizes the impact of forest plantations on biodiversity and other ecological functions.

An Overview and Economic Explanation of Global Forest Plantation Development

Currently, there are about 109 million ha of productive forest plantations in the world. Productive forest plantations represented 1.9% of global forest area in 1990, 2.4% in 2000, and 2.8% in 2005. The Asia region accounted for 41%; Europe 20%; North and Central America 16%; South America and Africa 10% each; and Oceania 3%.

Forest plantations have been increasing at an increased rate. The area of forest plantations increased about 14 million ha between 2000 and 2005 or about 2.8 million ha per year, 87% of which are in the productive class. The area of productive forest plantations increased by 2.0 million ha per year during 1990–2000 and by 2.5 million ha per year during 2000–05, an increase of 23% compared with the 1990–2000 period. All regions in the world showed an increase in plantation area, with the highest plantation rates found in Asia, particularly in China. The ten countries with the greatest area of productive forest plantations accounted for 79.5 million ha or 73% of the

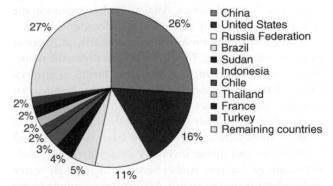

Figure 3 Ten countries with largest area of productive forest plantations in 2005. Modified from FAO (2005) Global forest resources assessment 2005. *FAO Forestry Paper 147*. Rome, Italy.

total global area of productive forest plantations (**Figure 3**). China, the United States, and the Russian Federation together accounted for more than half of the world's productive plantations.

Forest plantations, productive or protective, develop in response to a relative scarcity of timber and other goods and services associated with forests. In the early part of modern human history, population was sparse, forests were abundant, and survival, economic development, and territorial control were the primary concerns of governments and society. As forest resources declined, assuring an adequate timber supply gradually caught the attention of rulers and planners and became state policy. Often, the very first policy implemented would be to regulate timber harvesting schedule and intensity. Society also responded by moving to frontiers farther and farther away from population centers, which in economic terms is called a shift in the extensive margin of timber production. In a nutshell, the production and consumption of forest products were all from natural forests in the early part of human history, and forest plantations were not needed.

When the increase in timber consumption caught up with the ability of a country or a region to produce timber in naturally regenerated forests, citizens and governments would become interested in tree planting. While tree planting occurred at least several thousands of years ago in the Middle East, China, and Europe, and nearly 200 years ago in the Americas, the areas planted with trees through afforestation (planting land that was formerly in a nonforest cover) and reforestation (planting land on which a former forest had been harvested) were relatively insignificant in size before AD 1800. It was only after the industrial revolution that timber consumption increased drastically, due to increasing human population and industrial use of wood – initially as charcoal, then lumber, other solid wood products including mine props and railroad ties, and pulp and paper, and finally for conservation uses – that large-scale forest plantations started to emerge in Europe, North America, Asia, and other regions in the last century, especially in the last few decades.

Thus, forest plantations develop primarily in response to economic necessity. Timber depletion drives the transition of human consumption of natural forests to artificial forests. Early in the development of North America, for example, timber prices were low, and forest lands were more valuable for other uses, especially the production of food. So trees were removed, forest lands were converted to other use, and timber inventory declined. As the standing inventory declines, timber becomes increasingly scarce and timber prices start to rise. As the prices continue to rise for timber in natural forests, the purposeful husbandry of planted forests becomes economically attractive, and productive forest plantations begin to emerge.

Further, timber depletion affects the supply and demand balance for environmental services from natural forests, whether or not these services go through formal markets. Related to this balance is the fact that the demand for most environmental services such as clean water, clean air, and esthetics, which are often produced from or protected by forests, is highly correlated with personal income. As personal income increases, society demands more environmental services from forests, as well as more wood commodities. When natural forests are depleted to the extent that they cannot adequately provide these services, protective forest plantations emerge. In some developing countries, subsistence farming requires forests to protect farming and grass land from potential flooding, dust storms, soil erosion, and desertification, and trees are thus planted for protective purposes whether or not their personal incomes actually grow over time.

Factors Influencing Forest Plantation Development

As mentioned earlier, rising timber prices, caused by timber scarcity, lead to forest plantation development. Thus, timber prices are the primary factor that influences forest plantation development. Holding everything else constant, whenever a country or a region experiences a long period of rising timber prices, forest plantations would develop quickly.

Tree planting also requires land, labor, and capital. The cost of these production factors thus influence forest plantation development. Further, high timber prices, high land costs, and high labor costs force innovation in tree-growing technologies in conventional silvicultural treatments and biotechnologies. A recent report shows that the growth rate of pine plantations in Alabama, a southern state in the US increased about 25% in a decade (from 8.20% in the period from 1982 to 1990, to 10.17% in the period of 1990–2000). This increase in growth rate is attributed to advancement in tree-growing technologies as well as an increase in management intensity.

Government policies influence forest plantation developments as well. Taxes on land and forest-related income, cash subsidies to plant trees, regulations on land use and labor, and free education and extension services to forest farmers all have an impact, positively or negatively, on tree planting. In general, the primary motivation for the private sector to plant trees is to generate financial (or other) benefits from their investment. In some cases, government policies (positive or pervasive incentives in taxes, subsidies) provide or take away a significant proportion of the financial benefits from forest plantation development. Where governments own land, they could conduct afforestation and reforestation activities directly, for purely financial reasons or for social and environmental benefits or both.

The US South is perhaps an important region in timber supply as it produces some 18% of the world's industrial round wood with just 2% of the world's forestlands and 2% of the world's forest inventory. Some 90% of forest lands in the southern US are owned by nonindustrial private and industrial owners, and timber markets are competitive. A study of tree planting showed that tree planting by both forest industry and nonindustrial private landowners was positively related to the availability (measured as previous-year harvest) and the price of land. Planting by forest industry and nonindustrial private landowners was responsive to market signals, positively to softwood pulpwood prices and negatively to planting costs and interest rates. Finally, government subsidy programs, which increase the total plantation area, might have substitution effects on nonindustrial private tree planting. The federal income tax break for reforestation expenses promoted reforestation in the southern US.

Since forests often have a long production cycle, perhaps the most important government policy in promoting forest plantation development is to provide long-term and secure property rights (private property or land tenure) to private landowners or forest farmers. Many theoretical and empirical studies substantiate that long-term and secure property rights promote tree planting activities in both developed and developing countries. For example, in British Columbia, Canada tree planting was done more often and more promptly following harvest when forest property rights were secure. In Ghana, reforestation was significantly influenced by the form of forest tenure, and more intensive resource management was fostered by more secure forms of tenure.

Forest Plantations and Conservation of Natural Forests

Plantation forests can provide most goods and services that are provided by natural forests. These include timber, nontimber forest products, protection of clean water and clean air, soil erosion control, biodiversity, esthetics, carbon sequestration, and climate control. Nonetheless, as the value of environmental services from natural forests is higher than that from forest plantations, the demand for conservation of natural forests is stronger. It is possible that a division of land, with some land specialized in timber production and other land in providing environmental services, would produce more forest-related goods and services to society. Because forest plantations grow much faster than natural forests, forest plantations are seen as an increasingly important source of timber supply. Should more forest plantations be developed, more natural forests might be saved.

In 1995, natural forests contributed some 78% of global industrial timber supply, and the remaining was from forest plantations. With growing concerns about the status and loss of natural forests, the rapid expansion of protected areas, and large areas of forest unavailable for wood supply, plantations are increasingly expected to serve as a source of timber. The general trend of the sector is for timber supply to shift from natural forests to plantations.

A simple simulation of global timber supply and demand, allowing forest plantations and their productivity to extend at the current rate, has shown that logging on natural forests could fall by half, from about 1.3 billion m^3 in 2000 to about 600 million m^3 in 2025. Thus, forest plantations will have an increasingly significant role in substituting products from natural forests, even if they cannot replace harvests from natural forests for a long period of time.

One side impact of forest plantation development is that the supply of large quantities of low-cost timber could perhaps undermine the value of natural forest stands, leading to more rapid destruction, especially where legal frameworks and law enforcement are inadequate. Therefore from a global perspective, the transition from natural forests as the primary source of timber supply to forest plantations will take a long time. Nonetheless, the transition has been completed in some countries such as New Zealand and Chile.

Direct Ecological Effects of Forest Plantation

Forest plantations have direct ecological effects in addition to the positive impact of reducing pressure on natural forests. Generalizations are difficult, however, in part because plantation management regimes are diverse and the appropriate comparison is not always to unmanaged natural forests. In worst-case scenarios, natural forests or savannas on fragile soils are converted to plantations of exotic species that lower groundwater tables, decrease biodiversity, and develop extreme nutrient deficiencies in successive rotations. While this scenario overstates the impact of plantations, their generally monoculture nature and intensive management raises concerns about the effect of plantations on biodiversity, water, long-term productivity and nutrient cycling, and susceptibility to insects and diseases.

Biodiversity illustrates the complicated ecological impact of forest plantations; although biodiversity encompasses genetic, species, structural, and functional diversity, much of the focus in discussions about diversity has been at the genetic, species, and local ecosystem levels. As has occurred in agriculture, the introduction of genetically improved exotic or native species in forestry increases productivity and carbon-fixation efficiency. In some regions, this introduction has also increased interspecies diversity at landscape and regional scales. In France, compared with 70 natural forest tree species, 30 introduced species that are commonly used in forest plantations have helped increase the interspecies genetic diversity of forests at the local level. In Europe, at least, there is no doubt that the introduction of new tree species has increased the species richness of forests. Nevertheless, exotic species, even those long naturalized species such as Douglas-fir (*Pseudotsuga menziesii*) are unacceptable in nature conservation schemes.

Exotic species can have negative impacts on native species and communities. For example, fast-growing species can replace native forest species because of their natural invasive potential, as have been observed with *Eucalyptus* in northwestern Spain and Portugal. As the introduction of exotic species has potential risks, confirmation of long-term adaptation to local environmental conditions and pest resistance is necessarily the first step for the use of exotic species in extensive plantation programs.

Plantations tend to be even-aged and managed on relatively short rotations; thus, simple stand structures are common. When repeated across a landscape, large areas of similar species and low structural complexity result in a loss of habitat for taxa that require the kind of conditions provided by naturally regenerated stands or old forests. It has been reported that the bird fauna of single-species plantation forests is less diverse than that of natural and seminatural forests. In other cases, however, bird species diversity in plantation forests is comparable with that in naturally generated stands. For example, cottonwood (*Populus deltoides*) plantations in the Mississippi River Valley in the southern United States are intensively managed (rotation lengths of 10–15 years), reaching crown closure in 2 years. In comparison to natural stands, bird species diversity and abundances are similar for all guilds except cavity nesters.

Where avian diversity is decreased in managed forests generally, loss of structure following harvest is usually the cause. In plantations, simplified structure may be exacerbated further by use of exotic species or by monoculture. Because plantations are harvested at or near economic optima, rather than at biological maturity, plantations seldom develop much beyond the stem exclusion stage of stand development and do not re-establish characteristics of old forests or complex stand structures such as snags and coarse woody debris. Strategies to compensate for the simplifying tendencies of plantations and integrate biodiversity considerations include complex plantations composed of multiple species, varying planting spacing, thinning to variable densities, and retaining uncut patches and snags after harvest. Such biological legacies should benefit invertebrates such as saproxylic beetles as well as fungi, small mammals, and birds.

Silvicultural and site management practices of site preparation, competing vegetation control, and fertilization may reduce understory and groundcover vegetation diversity, although the effects of previous land use such as agriculture may play a larger role. For example, in southern United States industrial pine plantations, understory diversity was correlated with previous land use; lower diversity of native forest species occurred in plantations established on former farmland and higher diversity in plantations on cutover forest land.

Some species can benefit from forest plantations. For example, clear-cutting and short rotations favor the occurrence of ruderal plant species over some long-lived climax species. Forest plantations accommodate edge-specialist bird species and generalist forest species such as deer. Some rare and threatened species have been found to occupy forest plantations, especially when they lost most of their habitat to agricultural and urbanized land uses. For example in the UK, the native red squirrel is out-competed in native woodlands by the gray squirrel introduced from North America but the red squirrel thrives in conifer plantations, which are poor habitat for the gray squirrel.

Spatial considerations play a role in maintaining biodiversity at the landscape scale. Landscape diversity can meet the habitat needs of wildlife and be achieved by varying the size and shape of plantations and incorporating adjacency constraints into harvest scheduling models (i.e., a plantation adjacent to a recently harvested or young stand cannot be harvested until the adjacent stand reaches a certain age or crown height). Retaining areas of naturally regenerated forest, riparian buffers, or open habitat creates a landscape mosaic that combined with prescribed burning in fire-affected ecosystems, adds to landscape diversity. Landscape connectivity that provides dispersal corridors for mobile species is fostered by careful placement of forest roads and firebreaks.

Concerns about plantations and water are as varied as the issues surrounding biodiversity but generally relate to water use, water quality, or alteration of natural drainage. Species of *Eucalyptus* planted outside their native Australia have attracted the most negative attention for their putative excessive water use, especially in Africa and India but *Populus* species have similarly been accused in China of lowering local water tables and adding to drought. Species such as *Eucalyptus camaldulensis*, *E. tereticornis*, and *E. robusta* (and hybrids of these and other eucalypts) are drought tolerant and able to transpire even under considerable moisture stress. On balance they probably do not use more water than adjacent natural forests but certainly use more of the available water than grasslands or agricultural crops. There is little evidence that they can abstract groundwater; however, there is no recharge below the root zone. In the Wheatbelt of Western Australia, removal of the deep-rooted native vegetation including eucalypts and conversion to cereal crops has caused water tables to rise with subsequent salinization of soils and surface water bodies. Plantations of oil mallee crops (*E. polybractea*, *E. kochii* subsp. *plenissima*, and *E. horistes*) are planted to restore natural hydrology and counteract salinization.

Negative effects of plantations on water quality and aquatic resources are more due to intensive management than to use of exotic species. Intensive mechanical site preparation, especially on sloping sites, can result in sediment movement into streams. Chemical herbicides are used to control competing vegetation at various stages in the plantation growth cycle, but usually for site preparation in place of mechanical treatments or early in the life of the stand to release crop species from competitors. Less intense site preparation, formulations of herbicides that are not toxic to insects or other aquatic organisms and break down in soil, careful placement of chemicals to avoid direct application to water bodies, and designation of riparian buffers all have contributed to protection of water quality.

Harvesting practices, especially placement and construction of harvest roads and layout of skidding trails, potentially can degrade water quality. In developed nations, forest practices such as site preparation, harvesting, use of herbicides, and even choice of species may be regulated to some extent. In the United States, best management practices (BMPs) to address non-point source pollution and protect water quality have been codified by state agencies and landowners follow them voluntarily. Research shows generally high rates of compliance. Certification schemes substitute the coercive power of the marketplace for that of government; the various certification bodies differ in how they regard plantations, especially with regard to the use of herbicides, exotic species, or genetically modified trees.

Use of inorganic fertilizers to overcome fertility deficiencies, promote rapid growth, and sustain biomass accumulation generally has been found to have little impact on aquatic systems unless fertilizers are applied directly to streams, lakes, rivers, or adjacent riparian zones. Greater attention has focused on nutrient removals in harvests and the potential for intensive management to reduce site fertility and cause a fall-off in productivity of subsequent rotations. Claims of later-rotation productivity declines have been hard to substantiate, however, as general improvements in seed and seedling quality, genetic makeup, site preparation and competition control, and more careful harvesting that conserves site fertility have raised, rather than lowered, yields. Nevertheless, there exist documented cases of lowered fertility caused by export of nutrients in the harvested wood. These localized cases have been caused by low initial fertility, often of phosphorus, potassium, or micronutrient deficiencies inherent in the soil parent material that are easily overcome by application of inorganic fertilizers.

In the most intensive management of pine plantations for pulpwood in the southern United States, some companies routinely apply complete nutrient mixes containing all macro- and micronutrients as a precaution, despite lack of demonstrated deficiency of most nutrients except phosphorus and a responsiveness to added nitrogen. A stand may be fertilized with nitrogen up to five times in a 25-year rotation, sometimes in combination with phosphorus. These stands occur mostly on relatively infertile Ultisols and Spodosols developed on old marine sediments. On better soils (Alfisols, Entisols, and Vertisols), cottonwood plantations managed on 10-year rotations receive only an initial application of nitrogen at planting to promote rapid height growth to better compete with herbaceous competitors. Management of site nutrients in intensive plantations is critical to high yields as well as to protect long-term productivity and may require attention to

retaining soil organic matter, especially on sandy soils. Factors to consider include inherent soil fertility (nutrient stocks as well as transformations and fluxes), plant demand and utilization efficiency, and nutrients export in products removed as well as leakages.

It is common wisdom that monoculture plantations are more susceptible than natural forests to insect and disease attacks, yet there is little evidence this is generally true. On the one hand, single-species stands occur naturally and some of these natural vegetation types are the product of periodic, catastrophic disturbances such as pine bark beetles or spruce budworm. On the other hand, one explanation for the often greater productivity of exotic tree species than attained in their native habitat is the lack of yield-reducing insects and diseases. But diversity in the abstract is not a guarantor of lessened risk; diverse, multiple-species stands themselves are not immune to devastating attack by introduced pests, a situation likely to increase in frequency as a result of globalization of trade in timber products.

Often the practices associated with intensive management are the causes of insect and disease problems. For example, the desire to maximize wood production may set the level of tolerable damage from native pests lower than the stable equilibrium levels for the pest; attempts to control the pest at lower levels may cause unstable population growth cycles. The potential risks of plantations stem from their uniformity: the same or a few species, planted closely together, on the same site, over large areas. Pests and pathogens adapted to the dominant species may build up quickly due to food supply and abundant sites for breeding or infection. Proximity of the branches and stems in closely spaced stands may favor buildup of species with low dispersal rates or small effective spread distances. Conversely, the same uniformity of plantations that contributed to the risks of insects and diseases also confers some advantages. Species can be chosen that have resistance to diseases, for example, the greater resistance of loblolly pine (*Pinus taeda*) compared to slash pine (*P. elliottii*) to *Cronartium* rust was one reason loblolly was favored by forest industry in the US South. The shorter rotation length of plantations relative to naturally regenerated stands means trees are fallen before they become overmature and become infected. The compact shape and uniform conditions in plantations facilitate detection and treatment of economically important pests and pathogens.

Plantations may negatively impact adjacent communities – because of invasive natural regeneration of planted trees in adjacent habitat or alteration of local and regional hydrologic cycles and poor management practices may damage aquatic systems. Plantations are certainly simpler and more uniform than naturally regenerated stands or native grasslands, and may support a less diverse flora and fauna. Nevertheless, plantations can contribute to biodiversity conservation at the landscape level by adding structural complexity to otherwise simple grasslands or agricultural landscapes and by fostering the dispersal of forest-dwelling species across these areas.

Further, comparisons of plantations to unmanaged native forests or even naturally regenerated secondary forests are not necessarily the most appropriate comparisons to make. Although the conversion of old-growth forests, native grasslands, or some other natural ecosystem to forest plantations rarely will be desirable from a biodiversity point of view, in that forest plantations often replace other land uses including degraded lands and abandoned agricultural areas. Objective assessments of the potential or actual impacts of forest plantations on biological diversity at different temporal and spatial scales require appropriate reference points. Forest plantations can have either positive or negative impacts on biodiversity at the tree, stand, or landscape level depending on the ecological context in which they found. Impacts on water quantity and quality can be minimized if sustainable practices are followed; similarly with soil resources and long-term site productivity. Both complex plantations for wood production and environmental plantations can beneficially impact local and regional environments.

Lastly, managing forest plantations to produce goods such as timber while at the same time enhancing ecological services such as biodiversity involves tradeoffs; this can be made only with a clear understanding of the ecological context of plantations in the broader landscape. Tradeoffs also require agreement among stakeholders on the desired balance of goods and ecological services from plantations. Thus, there is no single or simple answer to the question of whether forest plantations are 'good' or 'bad' for the environment.

See also: Boreal Forest; Temperate Forest; Tropical Rainforest.

Further Reading

Binkley CS (2003) Forestry in the long sweep of history. In: Teeter LD, Cashore BW, and Zhang D (eds.) *Forest Policy for Private Forestry: Global and Regional Challenges*, pp. 1–8. Wallingford: CABI Publishing.

Brown C (2000) The global outlook for future wood supply from forest plantations. *FAO Working Paper GFPOS/WP/03*. Rome, Italy.

Carnus J-M, Parrotta J, Brockerhoff E, *et al.* (2006) Planted Forests and Biodiversity. *Journal of Forestry* 104(2): 65–77.

Clawson M (1979) Forests in the long sweep of history. *Science* 204: 1168–1174.

Evans J and Turnbull JW (2004) *Plantation Forestry in the Tropics: The Role, Silviculture and Use of Planted Forests for Industrial, Social, Environmental and Agroforestry Purposes*, 3rd edn. Oxford: Oxford University Press.

FAO (2001) Global forest resources assessment 2000. *FAO Forestry Paper 140*. Rome, Italy.

FAO (2005) Global forest resources assessment 2005. *FAO Forestry Paper 147*. Rome, Italy.

Harris TG, Baldwin S, and Hopkins AJ (2004) The south's position in a global forest economy. *Forest Landowner* 63(4): 9–11.

Hartsell AJ and Brown MJ (2002) Forest statistics for Alabama, 2000. *Resource Bulletin SRS-67*, 76pp. Ashville, NC: USDA Forest Service Southern Research Station.

Li Y and Zhang D (2007) Tree planting in the US South: A panel data analysis. *Southern Journal of Applied Forestry* 31(4): 192–198.

Royer JP and Moulton RJ (1987) Reforestation incentives: Tax incentives and cost sharing in the South. *Journal of Forestry* 85(8): 45–47.

Stanturf JA (2005) What is forest restoration? In: Stanturf JA and Madsen P (eds.) *Restoration of Boreal and Temperate Forests*, pp. 3–11. Boca Raton, FL: CRC Press.

Stanturf JA, Kellison RC, Broerman FS, and Jones SB (2003) Pine productivity: Where are we and how did we get here? *Journal of Forestry* 101(3): 26–31.

Zhang D (2001) Why so much forestland in China would not grow trees? *Management World* (in Chinese) 3: 120–125.

Zhang D and Flick W (2001) Sticks, carrots, and reforestation investment. *Land Economics* 77(3): 443–56.

Zhang D and Oweridu E (2007) Land tenure, market and the establishment of forest plantations in Ghana. *Forest Policy and Economics* 9: 602–610.

Zhang D and Pearse PH (1997) The Influence of the form of tenure on reforestation in British Columbia. *Forest Ecology and Management* 98: 239–250.

Freshwater Lakes

S E Jørgensen, Copenhagen University, Copenhagen, Denmark

Introduction

Freshwater lakes and reservoirs are basins filled with freshwater. Only 2.53% of the global water is freshwater; 1.76% of the global water is stored in ice caps, glaciers, and permafrost, while all fresh groundwater makes up 0.76% of the global water. It leaves 0.01% only for the surface freshwater, of which 70% or 0.007% of the global water is stored in the freshwater lakes. As surface water is easily accessible water, the storage of water in lakes and reservoirs becomes very important for the water supply and represents a large proportion of the world's readily accessible water (see **Figures 1** and **2**). Lake water is not only used for human consumption. Other water uses include industrial applications and processes and transportation and generation of hydropower.

Figure 1 Lake Baikal, the deepest lake in the world. The volume of Lake Baikal corresponds to almost 20% of all global surface freshwater.

Figure 2 Crater Lake, Oregon State, the lake famous throughout the world for its clarity. The Secchi disk transparency is 42 m.

The World's Freshwater Lakes

Table 1 gives an overview of 12 important freshwater lakes, including the deepest lake, the lake with the largest surface area, and the lake with the biggest volume. The lakes are not equally distributed in the world. About 10% of the total land is occupied by lakes in Scandinavia, while lakes occupy less than 1% of the land area in Argentina and China.

Importance of Lakes

The lake and reservoir water uses are becoming more intensive and multipurpose, particularly for lakes in heavily populated areas and intensively utilized regions. We can distinguish nine functions of lakes and reservoirs:

1. drinking water supply,
2. irrigation,
3. flood control,
4. aquatic production and fishery,
5. fire and ice ponds,
6. transportation,
7. hydropower,
8. conservation of biodiversity, and
9. recreation.

The multipurpose and extensive use of lakes and reservoirs can often lead to abuse and conflicts. There are numerous examples of such conflicts which are often rooted in inappropriate and insufficient water management.

Water Quality Problems of Lakes and Reservoirs

Nine problems associated with the extensive use of lakes and reservoirs can be identified.

Table 1 Major freshwater lakes

Lake	Volume (km³)	Area (km²)	Max. depth (m)
Lake Baikal	22 995	31 500	1 741
Lake Tanganyika	18 140	32 000	1 471
Lake Superior	12 100	82 100	170
Lake Malawi	6140	22 490	706
Lake Michigan	4920	57 750	110
Lake Huron	3540	59 500	92
Lake Victoria	2700	62 940	80
Lake Titicaca	903	8 559	283
Lake Erie	484	25 700	64
Lake Constance	48.5	571	254
Lake Biwa	27.5	674	104
Lake Maggiore	37.5	213	370

Eutrophication

This is the most pervasive water quality problem on a global scale, being a primary cause of lake deterioration. Eutrophication (nutrient enrichment) represents the natural aging process of many lakes in which they gradually become filled with sediments and organic materials over a typically geologic timescale. Human activities in a drainage basin can, however, dramatically accelerate this process. Its primary cause is the excessive inflow of nutrients (mainly phosphorus, sometimes nitrogen, sometimes both) to a water body from municipal wastewater treatment plants and industries, as well as drainage or runoff from urban areas and agricultural fields. Most lakes in densely inhabited regions of the world suffer from eutrophication, both in industrialized and developing countries. The impacts of the eutrophication process include heavy blooms of phytoplankton in a water body. These blooms will inevitably result in (1) reduced water transparency; (2) decreased oxygen concentration in the water column, particularly in the bottom layer (hypolimnion), which can cause fish kills and the remobilization or resuspension of heavy metals and nutrients into the water column; and (3) significant declines in the biodiversity of the lakes, including the disappearance of sensitive aquatic species. In shallow lakes, eutrophication can also cause an enormous increase in the growth of submerged and emergent rooted aquatic plants, as well as floating plants. This can lead to dramatic changes in the ecosystem structure.

If the sources of nutrients are removed or reduced significantly, the eutrophication problems can be fully controlled (see **Figures 3** and **4**). Lake Constance, also known as Bodensee, gives very illustrative examples.

Figure 4 Lake Bled, where restoration by siphoning hypolimnic water has been applied.

After the Second World War, the phosphorus concentration in the lake was about $0.01 \, mg \, l^{-1}$ and the lake was oligotrophic. In the year 1980, the lake was mesotrophic to eutrophic and the phosphorus concentration was about $0.08 \, mg \, l^{-1}$. Due to a massive reduction in the discharge of phosphorus from all sources, wastewater, agricultural drainage water, and septic tanks, it has been possible to reduce the phosphorus concentration to about $0.013 \, mg \, l^{-1}$ today. Lake Biwa, Japan, is illustrative of a partial solution of the problem (**Figure 5**). The discharge of phosphorus from wastewater has been significantly reduced since the 1970s, but due to almost no reduction in the phosphorus coming from agricultural drainage water, it has only been possible to stabilize the eutrophication level at a phosphorus concentration about $0.035 \, mg \, l^{-1}$. If on the other hand, the phosphorus in wastewater would not have been reduced, the eutrophication level would have increased.

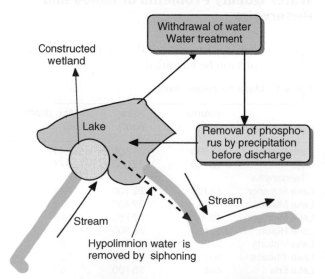

Figure 3 Abatement of eutrophication requires often the use of several methods at the same time, as shown here: removal of phosphorus from wastewater, construction of wetland to remove phosphorus from the inflowing tributary, and removal of hypolimnic (bottom) water by siphoning.

Figure 5 Lake Biwa in Japan is a very important recreational area for the population. A museum has been erected to present for the population all aspects of the lake: the culture, the limnology, the geology, and the history.

Acidification

This process of lake deterioration is caused mainly by acid precipitation and deposition. The nitrogen and sulfur compounds that cause this problem are emitted by industrial activities and by the consumption of fossil fuels, and fall to the land surface. The water in a lake can become acidic over time if its drainage basin does not contain the appropriate soil and geologic characteristics to neutralize the acidic water prior to its inflow into the lake. The primary consequence of acidification of lake water is the significant reduction of species diversity, the extinction of fish populations, and the disruption of lake ecosystem equilibrium. Other causes of lake acidification also exist, including water discharges from mining activities and the direct discharge of industrial waste effluents containing acidic components. Natural sources of acidifying substances include volcanic activities and natural emissions of gases. This problem, because of the geological characteristics, has been a major problem in Scandinavia (except the most southern Scandinavia) and the northeastern United States.

Toxic Contamination

This problem can have direct and dramatic impacts on both human and ecosystem health. Toxic substances originate not only from industrial activities and mining, but also as a result of intensive agriculture practices. Identification of the number of lakes and reservoirs exhibiting toxic contamination will doubtlessly increase in future years as we obtain more information on their concentrations in the environment, particularly in developing countries. Major impacts of toxic substances include the disappearance of sensitive species, as well as their accumulation in lake sediments and biota. The latter can directly and indirectly impact human health. Because the number of risk assessments applied to already-existing chemicals is currently extremely low (~500), a complete solution of this problem will take many years.

Water-Level Changes

Significant changes in water levels, particularly dropping levels, can be caused by

1. excessive withdrawal of water from lakes and/or their inflowing or out flowing rivers, and
2. the diversion of the inflowing water.

The consequences of water-level changes include: decrease in lake volume and/or surface area; unstable shoreline area communities; changes in lake ecosystem structure; reduced fish spawning areas; increased water retention time (decreased flushing rate), which can accelerate other negative lake processes (e.g., eutrophication,

retention of toxic substances); and increased salt concentration, leading to reduced water quality for human uses.

Lake Aral is probably the most illustrative example of this problem. Due to uncontrolled use of the inflowing river water for irrigation, the water level in the lake was reduced by almost 20 m. The lake was divided in two lakes, Large and Small Aral, with together less than half of the original lake area and with a salinity 10 times what it was 40 years ago.

Salinization

This process is an increase in the concentration of salts (all ions, not just sodium and chloride) in lake water, caused by such factors as (1) decreased in-lake water levels; (2) overuse of water in the drainage basin (e.g., cooling water, irrigation); and (3) global climate change. The effects of water salinization include (1) dramatic changes in lake biological structure; (2) lower fish production; and (3) reduced biodiversity. Human utilization of lake water with a high salt concentration also can become very problematic. This problem, however, can at least be partly addressed with the implementation of appropriate environmental management and agricultural practices in a lake's drainage basin.

Siltation

Accelerated soil erosion, resulting from such activities as the overuse or misuse of arable land, mining and/or deforestation in a drainage basin, can lead to the excessive loading of suspended solids (sediment) to lakes. The consequences of these increased loads include the rapid accumulation of sediment within the lake basin, and the increased turbidity (decreased transparency) of the water in the lake. The immediate impacts can be a significant reduction in the number of living organisms in a lake, decreased biodiversity, and reduced fisheries.

Introduction of Exotic Species

The intentional introduction of exotic (nonresident) species has become an almost common practice in some fisheries to increase the production of commercially important species. The introduction of Nile perch into Lake Victoria is a primary example. However, the intentional or unintentional (or sometimes illegal) introduction of exotic species can cause very serious problems in a given lake. The accidental introduction of zebra mussels in Lake Erie and water hyacinths in several lakes of China provides a dramatic example of this phenomenon. The introduction of exotic species can provoke very dramatic changes in the ecosystem structure not only at the biological community level, but also in a lake's chemical–physical environment. The major negative consequences of exotic species include the

(1) disappearance of native species; (2) alteration of trophic equilibrium; (3) significant reduction in species diversity; and (4) reduction of water transparency and changes in algae bloom patterns, via chemical–physical feedback processes in a lake.

Overfishing

Unsustainable fishing practices, sometimes combined with other problems, can lead to the collapse of fisheries. It seems to be an increasing problem for many African lakes, particularly Lake Victoria.

Pathogenic Contamination

This problem is caused by discharge of untreated sewage or runoff from livestock farms, a problem in both developing and developed countries. A *Cryptosporidium* outbreak in Lake Michigan (Milwaukee) sickened 400 000 people in the early 1990s. It is likely that many deaths in developing countries are due to consumption of dirty lake water.

Further Reading

ILEC (2005) Managing Lakes and Their Basins for Sustainable Use: A Report for Lake Basin Managers and Stakeholders, 146pp. Kusatsu, Japan: International Lake Environment Committee Foundation. http://www.ilec.or.jp/eg/lbmi/reports/LBMI_Main_Report_22February2006.pdf (accessed October 2007).

ILEC and UNEP (2003) World Lake Vision: A Call to Action, 37pp. Kusatsu, Japan: World Lake Vision Committee. http://www.ilec.or.jp/eg/wlv/complete/wlv_c_english.PDF (accessed October 2007).

Jørgensen SE, de Bernard R, Ballatore TJ, and Muhandiki VS (2003) *Lake Watch 2003. The Changing State of the World's Lakes*, 73pp. Kusatsu, Japan: ILEC.

Jørgensen SE, Löffler H, Rast, and Straškraba M (2005) *Lake and Reservoir Mangement*, 502pp. Amsterdam: Elsevier.

O'Sullivan PE and Reynolds CS (2004, 2005) *The Lakes Handbook*, vols. 1 and 2, 700pp. and 560pp. Blackwell Publishing.

Freshwater Marshes

P Keddy, Southeastern Louisiana University, Hammond, LA, USA

Six Types of Wetlands	Human Impacts
The Distribution of Marshes	Wetland Restoration
Water as the Critical Factor	Summary
Other Environmental Factors Affecting Marshes	Further Reading
Plant and Animal Diversity in Wetlands	

Wetlands are produced by flooding, and as a consequence, have distinctive soils, microorganisms, plants, and animals. The soils are usually anoxic or hypoxic, as water contains less oxygen than air, and any oxygen that is dissolved in the water is rapidly consumed by soil microorganisms. Vast numbers of microorganisms, particularly bacteria, thrive under the wet and hypoxic conditions found in marsh soils. These microbes transform elements including nitrogen, phosphorus, and sulfur among different chemical states. Therefore, wetlands are closely connected to major biogeochemical cycles. The plants in wetlands often have hollow stems to permit movement of atmospheric oxygen downward into their rhizomes and roots. Many species of animals are adapted to living in shallow water, and in habitats that frequently flood. Some of these are small invertebrates (e.g., plankton, shrimp, and clams), while others are larger and more conspicuous (fish, salamanders, frogs, turtles, snakes, alligators, birds, and mammals).

Six Types of Wetlands

There are six major types of wetlands: swamp, marsh, fen, bog, wet meadow, and shallow water (aquatic). These six types are produced by different combinations of flooding, soil nutrients, and climate. A seventh group, saline wetlands, which includes salt marshes and mangroves, is often treated as a distinct wetland type. Saline wetlands occur mostly along coastlines (see Mangrove Wetlands), but also occasionally in noncoastal areas where evaporation exceeds rainfall, such as in arid western North America, northern Africa, or central Eurasia.

Swamps and marshes have mineral soils with sand, silt, or clay. Swamps are dominated by trees or shrubs (see Swamps), whereas marshes are dominated by herbaceous plants such as cattails and reeds (**Figure 1**). Such wetlands tend to occur along the margins of rivers (**Figure 2**) or lakes, and often receive fresh layers of sediment during annual spring flooding. Marshes are among the world's most biologically productive ecosystems. As a consequence, they are very important for producing wildlife, and for producing human food in the form of shrimp, fish, and waterfowl.

Fens and bogs have organic soils (peat), formed from the accumulation of partially decayed plants. Most

Figure 1 Marshes occur in flooded areas, such as this depression flooded by beavers in Ontario, Canada. As the photo illustrates, marshes form at the interface of land and water. Courtesy of Paul Keddy.

Figure 2 Extensive bulrush (*Schoenoplectus* spp.) marshes along the Ottawa River in central Canada. The stalks of purple flowers indicate the invasion of this marsh by purple loosestrife (*Lythrum salicaria*), a native of Eurasia. Courtesy of Paul Keddy.

peatlands occur at high latitudes in landscapes that were glaciated during the last ice ages. In fens, the layer of peat is relatively thin, allowing the longer roots of the plants to reach the mineral soil beneath. In bogs, plants are entirely rooted in the peat. As peat becomes deeper (the natural trend as fens become bogs), plants become increasingly dependent upon nutrients dissolved in rainwater, eventually producing an 'ombrotrophic' bog. The large amounts of organic carbon stored in peatlands help reduce global warming.

Wet meadows occur where land is flooded in some seasons and moist in others, such as along the shores of rivers or lakes. Wet meadows often have high plant diversity, including carnivorous plants and orchids. Examples of wet meadows include wet prairies, slacks between sand dunes, and wet pine savannas. Pine savannas may have up to 40 species of plants in a single square meter, and hundreds of species in a hundred hectares.

Aquatic wetlands are covered in water, usually with plants rooted in the sediment but possessing leaves that extend into the atmosphere. Grasses, sedges, and reeds emerge from shallow water, whereas water lilies and pondweeds with floating leaves occur in deeper water. Aquatic wetlands provide important habitat for breeding fish and migratory waterfowl. Animals can create aquatic wetlands: beavers build dams to flood stream valleys, and alligators dig small ponds in marshes or wet meadows.

The Distribution of Marshes

Wetlands can occur wherever water affects the soil. Not only are there therefore many kinds of wetlands, but their size and shape is very variable. Wetlands can include small pools in deserts and seepage areas on mountainsides, or they can be long but narrow strips on shorelines of large lakes (**Figure 3**), or they may cover vast river floodplains (**Figure 4**) and expanses of northern plains. The two largest wetlands in the world (both $>750\,000\,\mathrm{km^2}$) are the West Siberian lowland and the Amazon River basin. The West Siberian Lowland consists largely of fens and bogs, but marshes occur along rivers, particularly in the more southern regions (**Figure 5**). The Amazon is a tropical lowland with freshwater swamps and marshes containing more kinds of trees and fish than any other region of the world.

Water as the Critical Factor

Water is a critical factor in all marshes. The duration of flooding is the most important factor determining the kind of wetland that occurs. Water can arrive as short pulses of flooding by rivers, as rainfall, or as slow and steady seepage. Each mode of arrival produces different kinds of

Figure 3 Sedges, grasses, and forbs compose this marsh on the leeward side of a narrow peninsula projecting into one of the Great Lakes (Lake Michigan), Michigan, USA. Courtesy of Cathy Keddy.

Figure 5 The largest wetland in the world occurs in the Western Siberian Lowland. Although much of this is peatland, marshes occur along the watercourses, particularly in the southern areas. Courtesy of M. Teliatnikov.

Figure 4 Extensive marshes of bulltongue (*Sagittaria lancifolia*) and American bulrush (*Schoenoplectus americanus*) now occur in coastal Louisiana, USA, where logging destroyed baldcypress forest. Courtesy of Paul Keddy.

Figure 6 Southern marshes on the coastal plain of North America may be dominated by a single grass, maidencane (*Panicum hemitomon*). This marsh occupies an opening within a baldcypress swamp, Louisiana, USA. Courtesy of Cathy Keddy.

wetlands. In order to better understand marshes, let us consider four examples of wetlands with very different flooding regimes.

Floodplains. Wetlands along rivers are often flooded by annual pulses of water (see Riparian Wetlands). These pulses may deposit thick layers of sediment or dissolved nutrients that stimulate plant growth. In floodplains (see Floodplains), animal life cycles are often precisely determined by the timing of the flood. Fish may depend upon feeding and breeding in the shallow warm pools left by retreating floodwaters. Birds may time their nesting to be able to feed their young on the fish and amphibians left behind by receding water. Marshes are often intermixed

with swamps, depending upon the duration of flooding (**Figure 6**). Early human civilizations developed in this type of habitat, along the Nile, Indus, Euphrates and Hwang Ho, where the annual flooding provided fertilized soil and free irrigation.

Peat bogs. Some peat bogs receive water only as rainfall. As a consequence, the water moves slowly, if at all, and contains very few nutrients. Hence, these types of wetlands often are dominated by slow growing mosses and evergreen plants (see Peatlands). Most such wetlands occur in the far north in glaciated landscapes. Humans have developed a number of uses for the peat – in Ireland, the peat is cut into blocks and used for fuel. In Canada, the peat is harvested and bagged for sale to gardeners. In Russia, peat is used to fuel electrical plants. Marshes may form on the edges of bogs where nutrients accumulate from runoff, or along river courses where nutrients are more available.

Seepage wetlands. In gently sloping landscapes water can seep slowly through the soil. In northern glaciated landscapes, such seepage can produce fens, which have distinctive species of mosses and plants, and may develop in distinctive parallel ridges. In more southern landscapes, seepage can produce pitcher plant savannas or wet prairies. Often these seepage areas are rather small (only a few hectares in extent) but are locally important because of the rare plants and animals they support. Seepage areas can be larger, and when the water flow is sufficiently abundant, shallow water can move across a landscape in a phenomenon known as sheet flow. The vast Everglades, with its distinctive animals, is a product of sheet flow of water from Lake Okeechobee in south central Florida southward to the ocean.

Temporary wetlands. In many parts of the world, small temporary (or ephemeral) pools form after heavy rain or when snow melts. These pools can go by a variety of local names including vernal pools, woodland ponds, playas or potholes (see Temporary Waters). The aquatic life in these pools is forced to adopt a life cycle that is closely tied to the water levels. Many species of frogs and salamanders breed in such pools, and the young must metamorphose before the pond dries up. Wetland plants may produce large numbers of seeds that remain dormant until rain refills the pond.

Since water has such a critical effect on wetlands, where water levels change, plant and animal communities will change as well. A typical shoreline marsh will often show distinct bands of vegetation ('zonation'), with each kind of plant occupying a narrow range of water depths (**Figure 7**). Most kinds of animals, including frogs and birds, also have their own set of preferred water depths. Wading birds (egrets, ibis, herons) may feed in different depths of water depending upon the length of their legs. Ducks, geese, and swans can feed at different water depths depending upon the length of their necks. Some water birds (Northern Shoveler, flamingos) strain microorganism from shallow water, while others (cormorants, loons) dive to feed further below the surface. Some ducks prefer wetlands that are densely vegetated, while others prefer more open water. Hence, even small changes in the duration of flooding or depth of water can produce very different plant and animal communities.

Many marsh plants adapt to flooding by producing hollow shoots, which allow oxygen to be transmitted to the rooting zone. The tissue that allows the flow of oxygen is known as aerenchyma. Not only can oxygen move by diffusion, but there are a number of methods in which oxygen moves more rapidly through large clones of plants, entering at one shoot and leaving at another. Consequently, plants can play an important role in oxidizing the soil around their rhizomes, allowing distinctive microbial communities to form. Some marsh plants also have floating leaves (e.g., water lilies) or even float entirely on the surface (e.g., duckweeds). The largest floating leaves in the world (**Figure 8**) are those of the

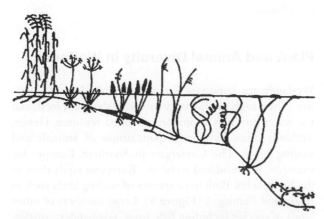

Figure 7 Different marsh plants tolerate different water levels. Hence, as the water level changes from shallow water (left; seasonally flooded) to deeper water (right; permanently flooded), the plants appear to occur in different zones. Courtesy of Rochelle Lawson.

Figure 8 The Amazon water lily has the largest floating leaves of any wetland plant. Note the prominent ribs on the underside of the leaf. Courtesy of Corbis.

Amazon water lily (*Victoria amazonica*). The gargantuan leaves can be 2 m in diameter with an elevated lip around the circumference. There are two gaps in the lip to allow water to drain, and large spines to protect the underwater sections of the foliage.

Other Environmental Factors Affecting Marshes

Nutrients

The main nutrients that affect the growth of marsh plants, and plants in general, are nitrogen and phosphorus. As described above, flood pulses that carry sediment down river courses can produce particularly fertile and productive marshes. Floodplains can therefore be thought of as one natural extreme along a gradient of nutrient supply. At the other end of the gradient lie peat bogs, which depend partly or entirely upon rainfall, and which therefore receive few nutrients. Sphagnum moss is well adapted to peatlands, and often comprises a large portion of the peat. In between the natural extremes of river floodplains (high nutrients) and peat bogs (low nutrients), one can arrange most other types of wetlands. The type of plants, and their rate of growth, will depend where along this gradient they occur, but most marshes generally occur in more fertile conditions.

While nutrients enhance productivity, paradoxically they can often reduce the diversity of plants and animals. Often, the high productivity is channeled into a few dominant species. One finds large numbers of common species, while the rarer species disappear. Humans often increase nutrient levels in watersheds and wetlands, thereby changing the species present and reducing their diversity. Carnivorous plants are known for tolerating low nutrient levels, because they can obtain added nutrients from their prey. Common examples include pitcher plants (*Sarracenia* spp.), bladderworts (*Utricularia* spp.), and butterworts (*Pinguicula* spp.). Cattails (*Typha* spp.) and certain grasses (*Phalaris arundinacea*) are particularly well known for rapid growth and an ability to dominate marshes at higher nutrient levels.

Disturbance

A disturbance can be narrowly defined as any factor that removes biomass from a plant. In marshes, sources of disturbance may include waves in lakes, fire, grazing, or (in the north) scouring by winter ice. One of the principal effects of disturbances is the creation of gaps in the vegetation, allowing new kinds of plants to establish from buried seeds. Most marshes have large densities of buried seeds, often more than 1000 seeds m^{-2}. After disturbance, marsh plants can also re-emerge from buried rhizomes.

Hence, cycles of disturbance play an important role in creating marshes.

Although the presence of fire in wetlands may seem paradoxical, fire can often occur during periods of drought. Northern peatlands, cattail marshes on lakeshores, wet prairies, and seepage areas in savannas can burn under the appropriate conditions. In northern peatlands, a fire can remove thousands of years of peat accumulation in a few days, even uncovering boulders and rock ridges that were buried beneath the peat. In marshes, fire can selectively remove shrubs and small trees, preventing the marsh from turning into a swamp. In the Everglades, burning can create depressions that then cause marshes to revert to aquatic conditions.

Animals that feed upon plants often cause only small and local effects. Think of a moose grazing on water lilies, a muskrat feeding on grasses, or a hippopotamus feeding on water hyacinth. Often the small patch of removed foliage is quickly replaced by new growth. But when herbivores become overly abundant, they can destroy the marsh vegetation entirely. In northern North America along Hudson Bay, Canada geese (*Branta canadensis*) are now so abundant that they remove all vegetation from expanses of coastal marsh. In southern North America, along the Gulf of Mexico, an introduced mammal, nutria (*Myocastor coypus*), similarly can strip marsh vegetation to coastal mudflats. To some extent, disturbance by herbivores is a natural phenomenon, one that has occurred cyclically throughout history. However, in the above two examples, one suspects humans may be the ultimate cause of the large-scale overgrazing (see the next section).

Periodic droughts may at times function like a natural disturbance by killing adult plants, and allowing new species to re-establish from buried seeds. Vernal pools and prairie potholes both have plant and animal species that are adapted to this kind of cyclical disturbance.

Plant and Animal Diversity in Wetlands

Wetlands are important for protecting biological diversity. Their high productivity provides abundant food, and the water provides an important added resource. Hence, wetlands often have large populations of animals and wading birds. The Camargue in Southern Europe, for example, is considered to be the European equivalent of the Everglades. Both have species of wading birds such as storks and flamingos (**Figure 9**). Large numbers of other kinds of species including fish, frogs, salamanders, turtles, alligators (**Figure 10**), crocodiles, and mammals require flooded conditions for all or at least part of the year. If the wetlands are drained, all of the species dependent upon them will disappear.

Figure 9 Marshes provide essential habitat for many kinds of wading birds including flamingos, Jurong Bird Park, Singapore. Courtesy of Corbis.

Figure 10 Alligators are one of the many species that benefit from protected marshes such as the Everglades, Loxahatchee National Wildlife Refuge, Florida, USA. Courtesy of Paul Keddy.

All wetlands, however, do not support the same species. Often, as already noted in the section entitled 'Disturbance', small differences in water level or nutrient supplies will produce distinctive types of wetlands. Hence, wetlands that are variable in water levels and fertility will frequently support more kinds of species than wetlands that are uniform. Along the Amazon River floodplain, for example, different kinds of swamp, marsh, grassland, and aquatic communities form in response to different flooding regimes, and each has its own complement of animal species. In the Great Lakes, different flood durations similarly produce different types of wetlands, from aquatic situations in deeper water, to marshes and wet meadows in shallower water. Some types of frogs, such as bullfrogs, require deeper water, while others, such as gray tree frogs, require shrubs.

Human Impacts

Humans have had, and continue to have, serious impacts upon wetlands in general, and marshes in particular. Some human impacts include draining, damming, eutrophication, and alteration of food webs. Let us consider these in turn.

One of the most obvious ways in which humans affect wetlands is by draining them. When the wetlands are drained, the soil becomes oxidized, and terrestrial plants and animals replace the wetland plants and animals. Often, drainage is followed by conversion to agriculture or human settlement, entirely removing the marshes that once existed. Vast areas of farmland in Europe, Asia, and North America were once marshes and have now been converted to crops for human consumption. Many countries now have laws to protect wetlands from further development, although the degree of protection provided, and the degree of enforcement, varies from one region of the world to another. Wetlands are also often included in protected areas such as national parks and ecological reserves.

Construction of dams can also have severe negative effects upon wetlands. The dams may be built for flood control, irrigation, or generating electricity. The wetland behind the dam may be destroyed by the prolonged flooding, whereas the wetlands downstream are disrupted by the lack of normal flood pulses. A single dam can therefore affect a vast area of wetlands. The degree of damage depends upon the pattern of water level fluctuations in the reservoir behind the dam, but in general large areas of marsh are lost both upstream and downstream from the dam. Sediment that would have expanded and fertilized wetlands during periodic floods becomes trapped behind the dam. Most of the world's large rivers have now been significantly affected by dams. To protect wetlands, it is necessary to identify rivers that are still relatively natural and to prevent further dams from being constructed. In other cases, it is possible to remove dams and allow natural processes to resume. An artificial levee can be considered a special type of dam that is built parallel to a river to prevent it from flooding into adjoining lands. Levees harm marshes by preventing the annual flooding, and by allowing cropland and cities to move into floodplains.

Humans can also affect wetlands by changing the nutrients in the water. Sewage from cities provides a specific 'point source' of nutrients, particularly nitrogen and phosphorus, that enter water courses then spread into wetlands. Activities such as agriculture and forestry provide 'diffuse sources' of nutrients, where runoff from large areas carries dissolved nutrients, and nutrients attached to clay particles, into the water and into adjoining marshes. The added nutrients can stimulate plant growth, which may seem to be beneficial – but it often leads to significant changes in the biota. Rarer plants and animals that are adapted to low fertility are replaced by more common plants and animals that exploit fertile conditions. Rapid growth of algae, followed by decay, can eliminate oxygen from lakes, causing fish kills. Protecting the quality of marshes therefore requires two sets of actions. First, it is necessary to control the obvious point sources of pollution by building sewage treatment plants. Second, it is necessary to use entire landscapes with care, with the broad objective of reducing nutrients in runoff. This can involve carefully timing the fertilization of crops, maintaining areas of natural vegetation along watercourses, fencing cattle away from stream valleys, minimizing construction of new logging roads, and avoiding construction on steep hill sides.

Herbivores are common in wetlands, and a natural part of energy flow from plants to carnivores. Common examples of large herbivores include moose, geese, muskrats, and hippopotamuses. Humans can disrupt wetlands by disrupting the natural balance between herbivores and plants. Herbivores can increase to destructive levels in several ways. When humans introduce new species of herbivores, rates of damage to plants may increase greatly – for example, nutria introduced from South America are causing significant damage to coastal wetlands in Louisiana. When humans reduce predation on herbivores, they may also increase to higher than natural levels. Killing alligators may damage wetlands by allowing herbivores such as nutria to reach high population densities; similarly, the loss of natural predators may be one of the reasons that Canada geese have multiplied to levels where they can destroy wetlands around Hudson Bay. There is also evidence that when humans harvest blue crabs, snails that the crabs normally eat begin to multiply and damage coastal marshes. These types of effects are difficult to study, since the effects may be indirect and take place over the long term.

Road networks are a final cause of damage to wetlands. The obvious effects of roads include the filling of wetlands, and the blocking of lateral flow of water into or out of wetlands. But there are many other effects. When amphibians migrate across roads to breeding sites, vast numbers can be killed by cars. In northern climates, the road salt put on roads as a de-icer can flow into adjoining wetlands. Snakes may be attracted to the warm asphalt and killed by passing cars. Invasive plant species can arrive along newly constructed ditches. Overall, roads change a landscape by accelerating logging, agriculture, hunting, and urban development. As a consequence, the quality of the marshes in a landscape is linked to two factors: the abundance of roads (a negative effect) and the abundance of forest (a positive effect). Although it may not be obvious, halting road construction (or removing unwanted roads) and protecting forests (or replanting new areas of forest) may have important consequences for all the marshes in a landscape.

Wetland Restoration

Humans have caused much damage to wetlands over the past thousand years, and the effects have increased as human populations and technological power have grown. We have seen some examples of damage in the preceding section. In response to such past abuses, humans have also begun consciously re-creating wetlands. There are a growing number of efforts to create new wetlands and enhance existing wetlands. Along both the Rhine River and the Mississippi River, some levees have been breached, allowing floodwater to return and marshes to recover. Depressions left by mining, or deliberately constructed for wetlands, can be flooded to recreate small marshes in highly developed landscapes. Construction of dams and roads has been more carefully regulated.

The future of marshes will likely depend upon two human activities: our success at protecting existing marshes from damage and our success at restoring marshes that have already been damaged. The list of the world's largest wetlands in **Table 1** provides an important set of targets for global conservation.

Summary

Marshes are produced by flooding, and, as a consequence, have distinctive soils, microorganisms, plants, and animals. The soils are usually anoxic or hypoxic, allowing vast numbers of microorganisms, particularly bacteria, to transform elements including nitrogen, phosphorus, and sulfur among different chemical states. Marsh plants often have hollow stems to permit movement of atmospheric oxygen downward into their rhizomes and roots. Marshes are some of the most biologically productive habitats in the world, and therefore support large numbers of animals, from shrimps and fish through to birds and mammals. Marshes are one of six types of wetlands, the others being swamp, fen, bog, wet meadow, and shallow water. Humans can affect marshes by changing water levels with drainage ditches, canals, dams, or levees. Other human impacts can arise from pollution by added nutrients, overharvesting of selected species, or building road networks in landscapes.

Table 1 The world's largest wetlands (areas rounded to the nearest 1000 km^2)

Rank	Continent	Wetland	Description	Area (km^2)	Source
1	Eurasia	West Siberian Lowland	Bogs, mires, fens	2 745 000	Solomeshch, chapter 2
2	South America	Amazon River basin	Savanna and forested floodplain	1 738 000	Junk and Piedade, chapter 3
3	North America	Hudson Bay Lowland	Bogs, fens, swamps, marshes	374 000	Abraham and Keddy, chapter 4
4	Africa	Congo River basin	Swamps, riverine forest, wet prairie	189 000	Campbell, chapter 5
5	North America	Mackenzie River basin	Bogs, fens, swamps, marshes	166 000	Vitt et al., chapter 6
6	South America	Pantanal	Savannas, grasslands, riverine forest	138 000	Alho, chapter 7
7	North America	Mississippi River basin	Bottomland hardwood forest, swamps, marshes	108 000	Shaffer et al., chapter 8
8	Africa	Lake Chad basin	Grass and shrub savanna, shrub steppe, marshes	106 000	Lemoalle, chapter 9
9	Africa	River Nile basin	Swamps, marshes	92 000	Springuel and Ali, chapter 10
10	North America	Prairie potholes	Marshes, meadows	63 000	van der Valk, chapter 11
11	South America	Magellanic moorland	Peatlands	44 000	Arroyo et al., chapter 12

Modified from Fraser LH and Keddy PA (eds.) (2005) *The World's Largest Wetlands: Ecology and Conservation*. Cambridge: Cambridge University Press.

See also: Floodplains; Mangrove Wetlands; Peatlands; Riparian Wetlands; Swamps; Temporary Waters.

Further Reading

Fraser LH and Keddy PA (eds.) (2005) *The World's Largest Wetlands: Ecology and Conservation*. Cambridge: Cambridge University Press.
Keddy PA (2000) *Wetland Ecology*. Cambridge: Cambridge University Press.

Middleton BA (ed.) (2002) *Flood Pulsing in Wetlands: Restoring the Natural Hydrological Balance*. New York: Wiley.
Mitsch WJ and Gosselink JG (2000) *Wetlands*, 3rd edn. New York: Wiley.
Patten BC (ed.) (1990) *Wetlands and Shallow Continental Water Bodies, Vol. 1: Natural and Human Relationships*. The Hague: SPB Academic Publishing.
Whigham DF, Dykyjova D, and Hejnyt S (eds.) (1992) *Wetlands of the World 1*. Dordrecht: Kluwer Academic Publishers.

Greenhouses, Microcosms, and Mesocosms

W H Adey, Smithsonian Institution, Washington, DC, USA

P C Kangas, University of Maryland, College Park, MD, USA

Introduction
Physical/Chemical Control Parameters
Biotic Parameters
The Operational Imperative

Case Study: Coral Reef Microcosm
Case Study: Florida Everglades Mesocosm
Case Study: Biosphere 2
Further Reading

Introduction

An ecosystem is an assemblage of organisms living together and interacting with each other and their environment. An element of biodiversity and biogeochemical

time stability is implied, although dimensions are optional, ranging from the biosphere subset, the biome, to perhaps a field or pond. Ecosystems with their complex food webs and biotic physical/chemical relationships are self-organizing due to the genetic information existing in the genome of each species. Even when spatially well bounded, ecosystems are not closed. At the very least, they are subject to energy input and energy and materials exchange with adjacent ecosystems. Often, ecosystems demonstrate biotic exchange with adjacent ecosystems that can be complex and include reproductive and seasonal phases.

The development of an ecosystem in a greenhouse implies that the ecosystem is solar driven, and thus no deeper than the photic zone of the ocean, although the basic principles discussed could generally apply to deep ocean ecosystems. Greenhouse placement generally requires spatial limitation, and scaling of model to analog is necessary for many physical and biotic factors. In some cases, it is intuitive, and in others must be empirically demonstrated by trial and error in comparison with analog function. All of the following case studies demonstrate aspects of this necessary scaling exercise. Normal, biogeochemical, and biotic exchange with adjacent ecosystems must also be simulated.

The reasons for placing or developing ecosystems within greenhouses for research or educational purposes have varied enormously and have ranged from the strongly funded, multiscientist research endeavors, to the classroom aquarium or terrarium. Even the naming of the research field of endeavor has varied widely: to name a few, synthetic ecology, ecological engineering, controlled ecology, closed systems ecology, ecosystem modeling, etc. The systems themselves are called living systems models, microcosms, mesocosms, macrocosms, ecotaria, living machines, closed ecological life support systems (CELSS), etc.

In this article, we limit our discussions to those serious research efforts in which significant effort is expended to match biodiversity, food web and symbiotic relationships, as well as biogeochemical function to analog wild ecosystems. By definition, such systems cannot be closed; however, the known interchanges, biotic and biogeochemical, with adjacent ecosystems must be known, studied in the wild, and be simulated so that the essential functional characteristic of the analog ecosystem can be maintained.

Hundreds, perhaps thousands, of microcosm studies of liter or few liter dimensions of a very limited biodiversity have been undertaken to elucidate component ecosystem function, often related to toxic compound effect. Rarely could these studies be regarded as the modeling of an ecosystem. At the other extreme, perhaps the most complex ecosystem modeling effort ever undertaken was the Biosphere II project in Arizona during the 1980s and 1990s. Biosphere II was an ecologically well-conceived collection of interacting terrestrial marine and freshwater ecosystems. However, it was intentionally operated as a closed system because of its planned space station future. Several decades ago, it was widely regarded among ecologists that even though greenhouse enclosure provided a critical element of control over variables, the difficulties inherent in enclosure and operation were too great to allow ecosystem modeling. As the examples we provide below show, this judgment was only minimally correct and perhaps no more severe than the breakup of wild ecosystems by development or farming expansion. Human expansion and perturbation has severely altered many of the ecosystems on Earth, and has altered all ecosystems in at least minor ways. The entire biosphere has been in effect placed in a poorly operated greenhouse, with the atmosphere serving as its upper 'glass' roof. There can be no valid argument against greenhouse enclosure of ecosystems for research and education. It is simply one end of a complex spectrum of interacting biota and biogeochemistry that we seek to understand. Indeed, in many ways, such model ecosystems may be 'purer' than their wild counterparts.

Physical/Chemical Control Parameters

The Enclosure

The shape of an ecosystem relative to its controlling physical and energy parameters can be crucial. In the case of aquatic systems, the relative thickness of the water mass and its relationship to the bottom establish the basic character of an ecosystem. A large body of water would be dominated by true plankters, normally living most of their lives suspended in mid and surface waters, with little benthic (or bottom) influence, whereas the shallow stream or narrow lagoon of a few meters in depth is benthic dominated. Light enters only through the air–water interface of a water ecosystem, and the shape of the containing body of water relative to depth, as well as water turbidity, determines the photosynthetic versus heterotrophic character of the ecosystem. The direction of current flow and wave action through an aquatic system relative to the position and orientation of its communities is critical to simulate in any model. The direction, frequency, and strength of wind relative to forest or field size can also be critical to systems function, as can be the physical dimension and density of such ecosystems.

The all-glass or acrylic aquarium box, ranging from about 40 l (10 gallons) to 1000 l (250 gallons), is a standard piece of equipment in terrestrial and aquatic modeling, and by drilling holes to attach pipes and linking all-glass tanks in complex arrays, many aspects of wild ecosystems can be modeled with reasonable accuracy.

The construction of molded fiberglass tanks or poured concrete or concrete block tanks, sealed with a wide variety of newer sealants, has considerable advantages for larger systems.

Ideally, the ecosystem envelope would be like that of the boundary of the mathematical modeler, a theoretical boundary allowing the controlling of exchange but not having any inherent characteristics. Walls, whatever their nature, unless rather esoteric measures are used to prevent organisms and organic molecules from using their surfaces, or blocking wind or current, are intrusions into the model ecosystem that may or may not be acceptable. For a small model of a planktonic system, the presence of uncleaned walls may prevent the system from being plankton dominated. To some degree, walls also interact with the water and atmosphere of the ecosystem they contain. For most purposes, glass and many plastics are ideal in this respect.

Greenhouse walls and roofs can block ultraviolet light and, for most ecosystem models, a component of artificial light is probably essential to achieve both the intensity and spectral veracity of natural light. Reinforced cement block or concrete can be valuable construction materials for large systems; however, concrete interacts with both water and atmosphere, being one of the limiters of veracity in of Biosphere II (as we describe below), and must be sealed with epoxy or other, carefully considered resins (**Figure 1**).

Many chemical elements and compounds used in construction are toxic. Some of these are only mildly poisonous and are often required by organisms as elements in small quantities and only become toxic in excess. Others are always toxic and only concentration determines effect. Glass, acrylics, epoxies, polyesters,

Figure 1 Florida Everglades mesocosm during construction. The butyl rubber-lined concrete block walls were used to constrain the entire system as well as to physically separate the salinity subcomponents thereby creating a salinity gradient. The plastic box at the lower right contains the tide controller, which determines the tide level in the estuary (center tank and the four smaller units behind).

polypropylenes, polyethylenes, nylons, Teflon, and silicones, among others, are structural materials commonly used in model/greenhouse construction. When properly cured these materials are generally inert, nonbiodegradable, and nontoxic. Many metals and organic additives easily find their way into construction processes and must be avoided or sealed off.

Physical/Chemical Environment

Many of the physical/chemical parameters of ecosystem, such as temperature, salinity, pH, hardness, and oxygen, are more or less obvious and generally accepted as crucial. Others such as light, wind, tides, currents, and wave action, have often been neglected or at least minimally considered in their effects on ecosystem models.

Light

Whole communities or parts of ecosystems, where plants are major components, typically capture a maximum of 6% of the incident light energy in photosynthesis. Nevertheless, full light is often required to achieve that peak transfer of energy. Also much higher capture efficiencies are possible when forcing energy such as wind and wave are present. In many cases, if greenhouse roofs cannot be opened, artificial lighting will have to be introduced to achieve the correct spectrum and intensity to drive the primary production characteristics of an analog ecosystem.

Water Supply/Water Environment

Whether a terrestrial or aquatic ecosystem is planned, the supply and internal transfer of water is critical. Air- and water-handling systems need to be carefully designed to prevent water contamination. Since water sequestration and loss is more or less inevitable, the water quality of both initial water and later top-ups must be carefully controlled. Rarely would tap water be acceptable. Water is the universal solvent, whether in liquid or gaseous form, and often 'sequesters' gases. Most ecosystems in greenhouses require the dedicated monitoring and control of atmospheric and water quality. Managed aquatic plant systems, such as algal turf scrubbers (ATSs), have been successfully used to manage water quality of adjacent ecosystems interaction, as we describe in some of the examples. Such systems can also control atmospheric quality (**Figure 2**).

Water and Air Movement

In virtually all water ecosystems, the water flows, and in most shallow water systems it oscillates (surges) as well. In models, this flow and surge are developed, at least

Figure 2 Red mangrove community in the Florida Everglades mesocosm from the engineering control pad. The large fan in the upper center provides wind for the mangrove communities. The box in the lower right is one of a bank of five algal turf scrubbers (ATSs) that control water quality (nutrients, pH, O$_2$) in the coastal system.

Figure 3 Engineering/control pad in the Florida Everglades mesocosm. The green diagonal tube in the center is an Archimedes screw that lifts water from the coastal tank (far right) for distribution to the estuary (back right), the ATS (left-foreground), and the wave generator (out of view to lower right).

initially by pumps. However, standard impellor pumps destroy or damage many plankters, particularly larger zooplankton and swimming invertebrate larvae. Several approaches are available to solve this problem, including using slow-moving piston pumps, membrane pumps, and Archimedes screws (**Figure 3**). All of these devices can work well, though relative performance is not fully quantified.

In terrestrial environments, fans for wind and air handlers for heat and air conditioning, as well as the cooling or heating surfaces employed have the same effect on flying insects and birds. On the other hand, in the wild, ultraviolet light, wind, and rain have critical controlling effects on many plant predators. These factors cannot be omitted.

Biotic Parameters

Ecosystem Structuring Elements

Some communities of organisms are structured by physical elements – a sandy beach, or rock, for example. However, in most terrestrial environments and in many shallow aquatic environments, plants and algae are the structuring elements. They not only provide the food and water and atmospheric chemistry but also greatly increase surface for attachment and cover. In general, plants also provide a spatial heterogeneity (spatial surface) that does not exist in the physical world. Particularly in the marine environment, where calcification is enhanced, many animals join plants to provide a community structure that consists of reef or shell framework. This framework is calcium carbonate (or other organic solid such as chiton) instead of (or along with) plant cellulose. In constructing any living ecosystem, it is essential that these structuring elements be first developed as 'colonial' stages soon after the physical environment is formed.

Ecosystem Subunits

In the construction of greenhouse ecosystems, subunit installation can be utilized. However, it would be impossible to individually extract and emplace the tens to hundreds of species amounting to hundreds to millions of individuals that occur in these subunits. Installation of sub-blocks of wild ecosystem includes the microspecies and keeps their relationships intact. For example, soil blocks, or in the marine or aquatic habitat, mud or rock blocks, can be introduced into the preexisting physical/chemical elements of the model ecosystem.

Repeated efforts must be taken to install rock, soil, mud, or 'planktonic blocks'. These injections should be periodically carried out during system stocking; at completion of development, they should be followed by several final injections. The process of cutting out, or otherwise extracting, an ecological block or ecosystem subunit and transporting it to the waiting model can be stressful to the community of organisms within the block. Even in the model, the block meets conditions that at least initially consist of the raw physical/chemical environment unameliorated by the effects of a functioning community of organisms. The first block injections are likely to lose species. However, with each addition, the diversity of reproductively successful species increases.

All ecological communities are patchy. An island, coral reef, a large salt marsh, a field, even a forest, all differ from place to place. Chance factors of organism settlement, negative and positive interaction between species, the local effect of environment, and real differences of environment (wave exposure, current, etc.) all lead to patchiness within a community. The model itself, no

Figure 4 Salt marsh community 1 year after establishment of the Florida Everglades mesocosm. Young white mangrove tree to left. After 5 years of self-organization the system followed a succession to a white mangrove/buttonwood swamp.

matter how accurate, is a patch, or several patches, that the modeler hopes represents a 'mean' of most wild patches.

After the structuring elements are established, and the entire pool of available species from the type community given a chance at immigration into the model, the model will self-organize. In the form of the genetic codes of its constituent species, the ecosystem carries a tremendous quantity of information with regard to its structure and function. Particularly since we know and understand only a small part of this information, we should be loath to subvert ecosystem self-organization (**Figure 4**).

Care in adhering to wild density levels will help prevent overstocking, overgrazing, and overpredation until the model is better understood. Single members of species guilds can be selected to perform a function, and thus reproductive density is achieved without exceeding ecosystem density requirements. In general, the larger the population of any single species, the more likely it is that breeding success will be achieved.

Ecosystem Interchange

However, one arbitrarily draws the boundary, no ecosystem occurs in total isolation. In many cases when such boundaries are arbitrary, major survival effects must be provided by adjacent ecosystems. For example, coral reefs and most shallow benthic communities are greatly dependent on the effects of adjacent open bodies of water for food, oxygen, and wave and current 'drive'. Typically, filters, the core element of aquarium science, have been devised to fill the need for the larger, less animal-dense body of adjacent open water that has been 'filtered' by the settling or loss of organic particles to deep water. However, such filters to a large extent usurp the role

normally provided by plants in most of the communities that are modeled. Unfortunately, in so doing they do not add oxygen as the plants do, and they raise nutrient levels. Both bacterial and foam fractionation methods remove organic particulates and swimming plankters, including reproductive stages that should be part of ecosystem function. Managed aquatic plant systems, such as ATSs, have been successfully used to manage water quality of adjacent ecosystems interaction, as we describe in some of the examples.

Although terrestrial systems, in general, may be less difficult in this regard, simulation of biotic interchange may be crucial. For example, birds and mammals often change ecosystems seasonally and even diurnally and the effects may be critical. Many insects are seasonal, some for very short periods, and often cross ecosystem boundaries. In some cases, it may be possible to provide these interactions through a human manager; however, a refugium, or alternate ecosystem may be necessary to achieve veracity.

The Operational Imperative

Successful enclosed ecosystem operation requires the monitoring of a large number of physical and chemical factors. To a large extent, this can be automated with electronic sensors, and the data can be logged and the system computer controlled. Some chemical parameters require wet chemistry, though a once-a-week analysis is usually sufficient in a well-run system. Like any piece of complex laboratory equipment (a scanning electron microscope, for example) a dedicated and highly trained technician is needed to manage the monitoring equipment, though in a well-tuned system, considerable time can be available for other duties.

An operational feature that is rarely discussed, and in practice is mostly anecdotal is that of population instability. A mesocosm, in effect, is a few-square-meter patch of a larger ecosystem. In the wild, ecosystem patches of a few square meters can be subject to considerable short-term variability, though stability is achieved to some extent by the smoothing effect of the larger ecosystem that may be measured in square kilometers.

Microcosms and mesocosms require an ecologist, fully acquainted with 'normal' community structure of the 'wild-type' system. Effectively, that ecologist/operator performs as the highest, and most omnivorous, predator. In the cases of algal or insect 'explosions' the operator's function is obvious, a once-a-week cropping or 'grazing' (i.e., hand harvest) until the explosion tendency subsides. In other cases, the short-term introduction of predator to carry out the limited cropping or grazing role can be quite successful.

These 'managed predators' can be kept in a refugium unit where they are readily available for such service.

Case Study: Coral Reef Microcosm

The Caribbean coral reef ecosystem model shown in **Figure 5** received natural sunlight from one side, south-facing at 37.5° N latitude; the metabolic unit had six 160 W VHF flourescent lamps (to match tropical intensity), step-cycled to bring mid-day peak intensity to approximately $800 \, uE \, m^{-2} \, s^{-1}$ and total incoming light to 220 Langleys/day (**Figure 6**). The ATS, lighted at night, had three 100 W metal halide lamps. The discussion presented represents data accumulated throughout the 9th year of 10 years of operation.

The physical and chemical components of the microcosm were measured in the metabolic unit and closely match those of the St. Croix analog (**Table 1**). The pH of the microcosm ranges from 7.96 ± 0.01 ($n = 62$) in the morning to 8.29 ± 0.10 ($n = 39$) in the late afternoon. Because of linked interacting photosynthesis and calcification in the ecosystem, calcium concentrations and alkalinity continually fall during the day and are stable or rise slightly at night. Calcium was added each morning as a solution of aragonite dissolved in HCl at approximately $24\,000 \, mg \, l^{-1}$. To keep microcosm concentrations above $420 \, mg \, l^{-1}$, after a full day of calcification, the mean concentration of calcium in the system was maintained at $491 \pm 6 \, mg \, l^{-1}$. Bicarbonate, was added as either $NaHCO_3$ or $KHCO_3$ dissolved in distilled water. The mean alkalinity was $2.88 \, meq \, l^{-1}$ ($n = 59$), in order to maintain levels above $2.40 \, meq \, l^{-1}$. Water quality

(nutrients, oxygen, and pH) of the system was controlled by algal turf scrubbing.

The mean oxygen concentration of the microcosm as shown in **Figure 6** is very close to that of the analog St. Croix reef. Net primary productivity (NPP) and respiration (R) were calculated based on the rate of oxygen increase and decrease, respectively, across the point of saturation ($6.5 \, mg \, l^{-1} \, O_2$), to avoid atmospheric fluxes. This gave a mean gross primary productivity (GPP) of

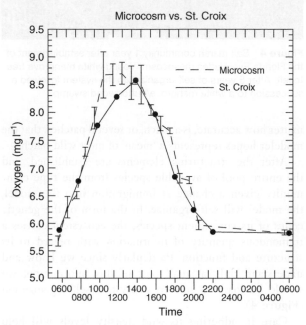

Figure 6 Comparison of mean daily oxygen concentration in the coral reef microcosm in comparison with that over the wild analog reef (1-year means).

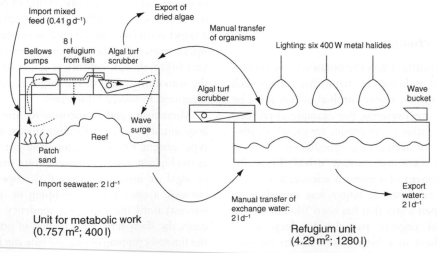

Figure 5 Diagram of coral reef microcosm with its refugium.

Table 1 Comparison of physical/chemical parameters between coral reef microcosm and the wild analog reef

	Microcosm	St. Croix Reefs (fore-reef)[a]
Temperature (°C) (am–pm)	26.5 ± 0.03 ($n = 365$)–27.4 ± 0.02 ($n = 362$)	24.0–28.5
Salinities (ppt)	35.8 ± 0.02 ($n = 365$)	35.5[b]
pH (am–pm)	7.96 ± 0.01 ($n = 62$)–8.29 ± 0.02 ($n = 39$)	8.05–8.35[c]
Oxygen concentration (mg l^{-1}) (am–pm)	5.7 ± 0.1 ($n = 14$)–8.7 ± 0.2 ($n = 11$)	5.8–8.5
GPP (g O_2 m^{-2} d^{-1}); (mmol O_2 m^{-2} d^{-1})	14.2 ± 1.0 ($n = 4$); 444 ± 3 ($n = 4$)	15.7; 491
Daytime NPP (g O_2 m^{-2} day^{-1}); (mmol O_2 m^{-2} d^{-1})	7.3 ± 0.3 ($n = 4$); 228 ± 9 ($n = 4$)	8.9; 278
Respiration (g O_2 m^{-2} h^{-1}); (mmol O_2 m^{-2} h^{-1})	0.49 ± 0.04 ($n = 4$); 15.3 ± 1.3 ($n = 4$)	0.67; 20.9
N–NO_2^- + NO_3^2 (µmol)	0.56 ± 0.07 ($n = 6$)	0.28
Calcium (mg l^{-1}); (mmol l^{-1})	491 ± 6 ($n = 33$); 12.3 ± 0.2 ($n = 33$)	417.2[d]; 10.4
Alkalinity (meq l^{-1})	2.88 ± 0.04 ($n = 59$)	2.47[b]
Light[e] (Langleys d^{-1})	220	430 (surface); 220 (5 m deep in fore-reef)

[a]The St. Croix data is from Adey and Steneck (1985).
[b]Tropical Atlantic means from Millero and Sohn (1992); no data available for St. Croix.
[c]Values from Enewetak and Moorea (Odum and Odum, 1955; Gattuso et al., 1997).
[d]Tropical Atlantic means from Sverdrup et al. (1942); no data available for St. Croix.
[e]The light levels of the system were measured with a pyranograph. All of the physical and chemical components of the microcosm are compared to the fore-reef of St. Croix since light levels are equivalent (Kirk, 1983; Adey and Steneck, 1985).
For references, see Small A and Adey W (2001) Reef corals, zooxanthellae and free-living algae: A microcosm study that demonstrates synergy between calcification and primary production. *Ecological Engineering* 16: 443–457.

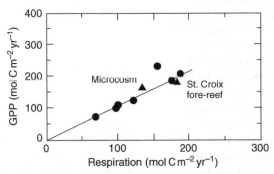

Figure 7 GPP as a function of respiration in the coral reef microcosm and its wild analog reef in comparison with selected worldwide reefs.

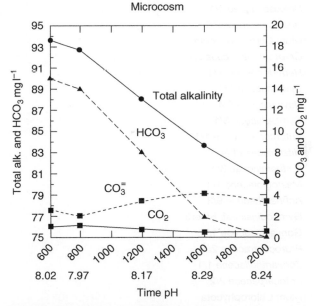

Figure 8 Mean daytime carbonate cycle in coral reef microcosm calculated by namograph from pH and total alkalinity data.

14.2 ± 1.0 gO_2 m^{-2} d^{-1}, as compared to the mean GPP for the analog fore-reef at 15.7 gO_2 m^{-2} d^{-1}. The difference between the microcosm and reefs *in situ* can be accounted for by the difference in spatial heterogeneity; topographic relief on the St. Croix fore-reef typically ranges from 1 to 2 m, while in the microcosm only 10–30 cm is possible.

In **Figure 7**, GPP versus *R* for the microcosm and its analog are plotted, showing that both are well within the range of typical wild reefs. Even though primary productivity of the microcosm is very close to the wild analog, the fact that respiration is somewhat lower probably relates to the proportionally lower spatial heterogeneity in the microcosm.

Whole ecosystem calcification in the coral reef model, at 4.0 ± 0.2 kg CaCO$_3$ m^{-2} yr^{-1}, is related to its primary components (stony coral 17.6%, *Halimeda* 7.4%, *Tridacna* 9.0%, algal turf, coralline and foraminifera 29.4%, and miscellaneous invertebrates 36%). Through analysis of

the microcosm's daily carbonate system, it is demonstrated that bicarbonate ion (**Figure 8**), not carbonate ion, is the principal component of total alkalinity reduction in the water column.

This coral reef microcosm contained 534 identified species within 27 phyla (**Table 2**), with an estimated 30% unaccounted for due to lack of taxonomic specialists. Because of the length of time that this model system was closed to biotic interchange, virtually all of the biotic composition of the system (over 95%) had to be maintained by

Table 2   Families of organisms, with numbers of species and genera found in the Coral reef microcosm after 10 years of operation, 7 years in closure

Plants, algae, and cyanobacteria
Division Cyanophota
  Chroococcaceae 6/5
  Pleurocapsaceae 4/2
  UID Family 4/4
  Oscillatoriaceae 8/6
  Rivulariaceae 4/1
  Scytonemataceae 1/1
Phylum Rhodophyta
  Goniotrichaceae 2/2
  Acrochaetiaceae 2/2
  Gelidiaceae 1/1
  Wurdemanniaceae 1/1
  Peysonneliaceae 3/1
  Corallinaceae 11/8
  Hypneaceae 1/1
  Rhodymeniaceae 3/2
  Champiaceae 1/1
  Ceramiaceae 3/3
  Delesseriaceae 1/1
  Rhodomelaceae 7/6
Phylum Chromophycota
  Cryptomonadaceae 2/2
  Hemidiscaceae 1/1
  Diatomaceae 6/4
  Naviculaceae 9/4
  Cymbellaceae 3/1
  Entomoneidaceae 1/1
  Nitzchiaceae 6/4
  Epithemiaceae 3/1
  Mastogloiaceae 1/1
  Achnanthaceae 9/3
  Gymnodiniaceae 6/4or5
  Gonyaulacaceae 1/1
  Prorocentraceae 2/1
  Zooxanthellaceae 1/1
  Ectocarpaceae 2/2
Phylum Chlorophycota
  Ulvaceae 1/1
  Cladophoraceae 4/2
  Valoniaceae 2/2
  Derbesiaceae 3/1
  Caulerpaceae 3/1
  Codiaceae 6/2
  Colochaetaceae 1/1
Phylum Magnoliophyta
  Hydrocharitaceae 1/1
Kingdom Protista
Phylum Percolozoa
  Vahlkampfiidae 2/1
  UID Family 2/2

  Stephanopogonidae 2/1
Phylum Euglenozoa
  UID Family 4/3
  Bondonidae 7/1
Phylum Choanozoa
  Codosigidae 2/2
  Salpingoecidae 1/1
Phylum Rhizopoda
  Acanthamoebidae 1/1
  Hartmannellidae 1/1
  Hyalodiscidae 1/1
  Mayorellidae 2/2
  Reticulosidae 2/2
  Saccamoebidae 1/1
  Thecamoebidae 1/1
  Trichosphaeridae 1/1
  Vampyrellidae 1/1
  Allogromiidae 1/1
  Ammodiscidae 1/1
  Astrorhizidae 1/1
  Ataxophragmiidae 1/1
  Bolivinitidae 3/1
  Cibicidiidae 1/1
  Cymbaloporidae 1/1
  Discorbidae 5/2
  Homotremidae 1/1
  Peneroplidae 1/1
  Miliolidae 10/2
  Planorbulinidae 2/2
  Siphonidae 1/1
  Soritidae 4/4
  Textulariidae 1/1
Phylum Ciliophora
  Kentrophoridae 1/1
  Blepharismidae 2/2
  Condylostomatidae 1/1
  Folliculinidae 4/3
  Peritromidae 2/1
  Protocruziidae 2/1
  Aspidiscidae 7/1
  Chaetospiridae 1/1
  Discocephalidae 1/1
  Euplotidae 11/3
  Keronidae 7/2
  Oxytrichidae 1/1
  Psilotrichidae 1/1
  Ptycocyclidae 2/1
  Spirofilidae 1/1
  Strombidiidae 1/1
  Uronychiidae 2/1
  Urostylidae 4/2
  Cinetochilidae 1/1
  Cyclidiidae 3/1
  Pleuronematidae 3/1

(Continued)

Table 2 (Continued)

Uronematidae 1/1
Vaginicolidae 1/1
Vorticellidae 2/1
Parameciidae 1/1
Colepidae 2/1
Metacystidae 3/2
Prorodontidae 1/1
Amphileptidae 3/3
Enchelyidae 1/1
Lacrymariidae 4/1
Phylum Heliozoa
Actinophyridae 2/1
Phylum Placozoa
Family UID 5
Phylum Porifera
Plakinidae 2/1
Geodiidae 5/2
Pachastrellidae 1/1
Tetillidae 1/1
Suberitidae 1/1
Spirastrellidae 2/2
Clionidae 4/2
Tethyidae 2/1
Chonrdrosiidae 1/1
Axinellidae 1/1
Agelasidae 1/1
Haliclonidae 4/1
Oceanapiidae 1/1
Mycalidae 1/1
Dexmoxyidae 1/1
Halichondridae 2/1
Clathrinidae 1/1
Leucettidae 1/1
UIDFamily 2/?
Eumetazoa
Phylum Cnidaria
UID Family 3/?
Eudendridae 1/1
Olindiiae 1/1
Plexauridae 1/1
Anthothelidae 1/1
Briareidae 1/1
Alcyoniidae 2/2
Actiniidae 3/2
Aiptasiidae 1/1
Stichodactylidae 1/1
Actinodiscidae 4/3
Corallimorphidae 3/2
Acroporidae 2/2
Caryophylliidae 1/1
Faviidae 3/2
Mussidae 1/1
Poritidae 3/1

Zoanthidae 3/2
Cerianthidae 1/1
Phylum Platyhelminthes
UID Family 1/1
Anaperidae 3/2
Nemertodermatidae 1/1
Kalyptorychidae 1/1
Phylum Nemertea
UID Family 2/2
Micruridae 1/1
Lineidae 1/1
Phylum Gastrotricha
Chaetonotidae 3/1
Phylum Rotifera
UID Family 2/?
Phylum Tardigrada
Batillipedidae 1/1
Phylum Nemata
Draconematidae 3/1
Phylum Mollusca
Acanthochitonidae 1/1
Fissurellidae 2/2
Acmaeidae 1/1
Trochidae 1/1
Turbinidae 1/1
Phasianellidae 1/1
Neritidae 1/1
Rissoidae 1/1
Rissoellidae 1/1
Vitrinellidae 1/1
Vermetidae 1/1
Phyramidellidae 1/1
Fasciolariidae 2/2
Olividae 1/1
Marginellidae 1/1
Mitridae 1/1
Bullidae 1/1
UID Family 4/?
Mytilidae 2/1
Arcidae 2/1
Glycymeridae 1/1
Isognomonidae 1/1
Limidae 1/1
Pectinidae 1/1
Chamidae 1/1
Lucinidae 2/2
Carditidae 1/1
Tridacnidae 2/1
Tellinidae 1/1
Phylum Annelida
Syllidae 3/2
Amphinomidae 1/1
Eunicidae 3/1

(*Continued*)

Table 2   (Continued)

Lumbrineridae 1/1	*Diogenidae* 1/1
Dorvilleidae 1/1	*Xanthidae* 2/?
Orbiniidae 1/1	Phylum Echinodermata
Spionidae 1/1	*Ophiocomidae* 1/1
Chaetopteridae 1/1	*Ophiactidae* 1/1
Paraonidae 1/1	*Cidaroidae* 1/1
Cirratulidae 4/3	*Toxopneustidae* 1/1
Ctenodrilidae 4/3	*Holothuriidae* 1/1
Capitellidae 3/3	*Chirotidae* 1/1
Muldanidae 1/1	Phylum Chordata
Oweniidae 1/1	*Ascidiacea* UID Fam..1/1
Terebellidae 2/1	*Grammidae* 1/1
Sabellidae 14/4	*Chaetodontidae* 1/1
Serpulidae 6/6	*Pomacentridae* 5/4
Spirorbidae 2/2	*Acanthuridae* 1/1
Dinophilidae 1/1	

Phylum Sipuncula
  Golfingiidae 1/1
  Phascolosomatidae 3/2
  Phascolionidae 1/1
  Aspidosiphonidae 3/2
Phylum Arthropoda
  Halacaridae 1/1
  UID Family 2/?
  Cyprididae 2/2
  Bairdiiaae 1/1
  Paradoxostomatidae 1/1
  Pseudocyclopidae 1/1
  Ridgewayiidae 2/1
  Ambunguipedidae 1/1
  Argestidae 1/1
  Diosaccidae 1/1
  Harpacticidae 1/1
  Louriniidae 1/1
  Thalestridae 1/1
  Tisbidae 1/1
  Mysidae 1/1
  Apseudidae 2/1
  Paratanaidae 1/1
  Tanaidae 1/1
  Paranthuridae 1/1
  Sphaeromatidae 1/1
  Stenetriidae 1/1
  Juniridae 1/1
  Lysianassidae 1/1
  Gammaridae 4/4
  Leucothoidae 1/1
  Anamixidae 1/1
  Corophiidae 1/1
  Amphithoidae 2/2
  Alpheidae 2/2
  Hippolytidae 2/1
  Nephropidae 1/1

reproduction. Based on standard species/area relationships ($S = kA^z$, where S = species richness and A = area), the predicted pan tropic coral reef biodiversity calculated from the model biodiversity (at three million species) exceeds that of recent estimates for wild coral reefs.

Case Study: Florida Everglades Mesocosm

This greenhouse-scale mesocosm is a 98 500 l butyl-lined, concrete block tank divided into seven connected sections of varying salinity (**Figure 9**). Each section contains water, algae, animals, sediments, and wetland–coastal plants representative of habitats along a transect from the full salinity Gulf of Mexico through the estuarine Ten-Thousand Islands and into the freshwater Florida Everglades (**Figure 10**).

As in the wild analog, the Gulf Shore and estuary are part of the same dynamic water mass. Here, the estuarine salinity gradient is created by pump-driven tidal inflow interacting through open weir constrictions and against downstream freshwater flow. Tank #1, the Gulf Shore, acts as a tidal reservoir for the estuary, thereby saving the need for a blank reservoir (**Figure 11**). The primary pump is an 800 lpm, Discflo^TM unit that utilizes a rotating disk, rather than plankton-destructive impellers. The freshwater is derived from rain and from reverse osmosis extraction from the Gulf Shore (the equivalent of Gulf evaporation and resulting rainfall in the wild). All aquatic organisms, including adult invertebrates, can move from the estuary to the Gulf Shore. All organisms that can survive Discflo^TM pumping (including small fish) can return to the estuary via tidal inflow. The freshwater system, at times, flows directly into the

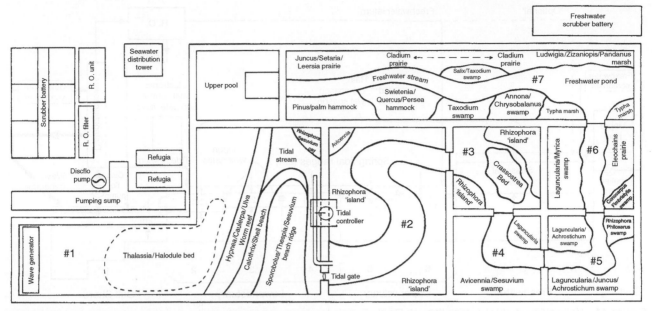

Figure 9 Plan view of the Florida Everglades mesocosm and its critical engineering components.

Figure 10 Florida Everglades mesocosm approximately 4 years after construction showing salt marsh, black mangrove, and red mangrove communities (from front to background at left) and lower freshwater stream at right. At this point the greenhouse roof is providing a significant constraint to community succession by limiting vertical growth of mangrove and hammock trees.

uppermost estuary and technically all organisms can enter the estuary from freshwater.

The initial stocking of the mesocosm was completed in mid-1988, and small collections continued to be injected through 1990. During this period, partial censuses for key organisms were undertaken, and, where required, additional stocking was carried out. From late 1990 to late 1994, the system was operated as a

biotically closed system, with minor human interaction, functioning as an omnivorous predator.

Major physical/chemical parameters are shown in **Table 3**. Dissolved nitrogen was monitored as nitrite plus nitrate in each of the community units (tanks); these were typically at levels of 5–8 μM $(NO_2 + NO_3)$ through the middle of the estuary, and at 3–5 μM $(NO_2 + NO_3)$, in the Gulf (#1) system. Levels average a few μM higher in winter than in summer. Nutrient flow-through is achieved by algal export, in the ATS banks. When levels drop below 1–2 μM in the Gulf (#1) system (typically in summer), the dried scrubber algae are redistributed to the system.

After 4 years of biotically closed operation, the Florida Everglades mesocosm was censused for organisms. The abundance of the principal higher plants, algae, invertebrates, and fish are shown in **Figures 12–14**. A total of 369 species (not including bacteria, fungi and the minor 'worm' phyla) was tallied. Excluding algae, protists, and small invertebrates, which could not be censused during introduction, it can be estimated that approximately 20–40% of the introduced species survived through the 4 years of biotic closure. In most cases these were the dominating species in the analog ecosystems. At the time of termination of the system as a carefully monitored mesocosm, only 15–30% of the originally introduced species were reproductively maintaining populations. However, in most cases, as **Figures 12–14** show, these

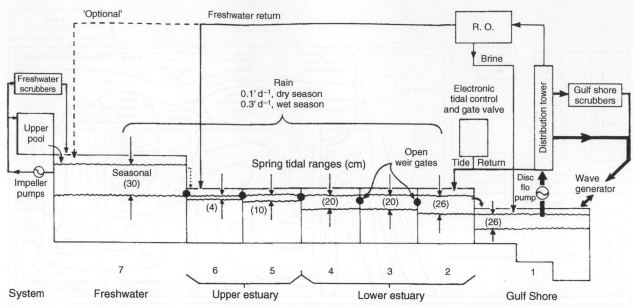

Figure 11 Vertical/longitudinal section through Florida Everglades mesocosm showing water management system and tide levels.

Table 3 Physical/chemical parameters of the Florida Everglades mesocosm

Parameter	Tank #1	Tank #2	Tank #3	Tank #4	Tank #5	Tank #6	Tank #7
Temperature °C							
Spring	23.4	23.2	22.5	22.6	22.0	21.3	23.2
Summer	25.7	25.5	25.4	25.7	25.6	25.1	25.1
Fall	22.2	22.3	21.8	22.0	21.7	21.3	22.1
Winter	21.0	20.9	19.9	19.6	19.0	18.4	21.9
Salinity, ppt	31.6	31.2	30.5	28.7	19.7	0.7	0.1
$[NO_2 + NO_3]$ μM							
Tap H_2O as top up[a]	7.2	7.6	8.2	6.3	5.4	6.6	6.7
Milli RO as top up[b]	1.4	1.7	2.3	1.8	0.9	1.7	1.4
Tidal range cm/0.5 day	13–26	13–26	13–20	11–20	6–10	0–4	0
Hydroperiod cm yr^{-1}	0	0	0	0	0	0	30.5

[a]Nutrient levels in system as $(NO_2 + NO_3)$ μM when 'Tap H_2O as top up'.
[b]'Milli RO as top up' refers to mean system values when reverse osmosis water from Milli RO™ is used as evaporative replacement.

were the species that provided primary structure and metabolism in the analog ecosystem.

Case Study: Biosphere 2

Biosphere 2, located near Tucson, AZ, USA, is the largest greenhouse system ever built with nearly three acres (1.2 ha) of enclosed space. It is unique in surpassing any other greenhouse ecosystem in size, complexity, and duration of operation. The system was originally intended as a model of the Earth's biosphere (e.g., biosphere 1) with several tropical and

subtropical ecosystems, an agricultural area, wastewater treatment wetlands, and a human habitat, along with a factory-sized machinery area for maintaining physical–chemical conditions. It was built to develop bioregenerative technology for future space travel, to educate the public about biosphere-scale issues and for basic ecological research. Atmospheric closure of gas cycles was part of the system design, which was tested with a prototype module of 11 000 ft^3 (312 m^3) from 1988 to 1990.

A number of ecologists were consulted for the creation of the greenhouse's ecosystems which included plots of rainforest, desert, savanna, mangrove estuary,

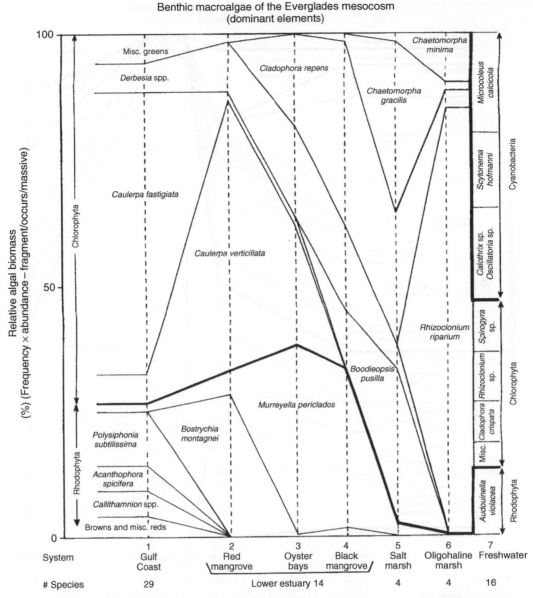

Figure 12 Relative biomass of dominant benthic algae in Florida Everglades mesocosm.

and ocean with coral reef. Thousands of species were added to the greenhouse intentionally and unintentionally (i.e., in ecosystem sub-blocks, as described above), from existing tropical systems as distant as Venezuela and from the local Arizona desert. After construction the ecosystems self-organized and many of the added species went extinct within the system as expected. Success, in terms of replication of the analog ecosystems in nature, has varied among the different model ecosystems, but most have developed and sustained a significant degree of ecological integrity.

Two experiments were conducted in Biosphere II during which humans were enclosed inside the system:

the first for 2 years (1991–93) and the second for 6 months (1994). These experiments tested concepts of sustainability at a very basic level since the humans had to rely on the overall greenhouse system for life support function. However, changes in the gas cycles within the greenhouse caused the human experiments to be modified and ultimately terminated. During the first human experiment oxygen concentration in the atmosphere decreased dramatically because high rates of soil respiration released more carbon dioxide than was taken up in photosynthesis; some of the carbon dioxide was absorbed as carbonates in the concrete of the greenhouse foundation. Oxygen had to be pumped into the system to maintain the humans so that the 2-year test could be completed. During the second

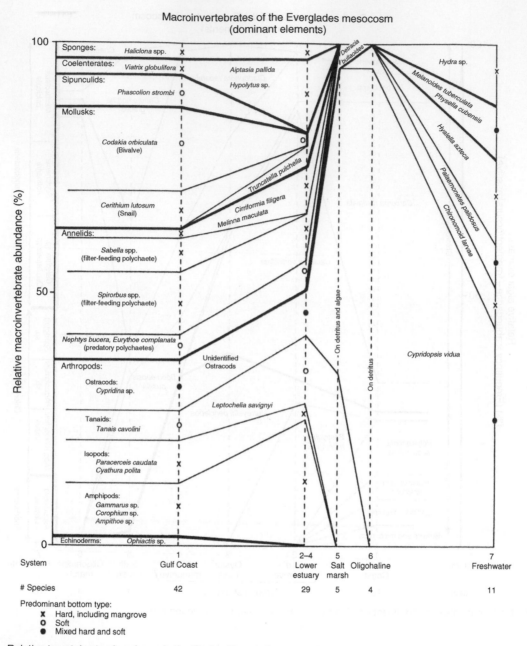

Figure 13 Relative invertebrate abundance in the Florida Everglades mesocosm.

human experiment, buildup of noxious concentrations of nitrous oxide in the atmosphere from microbial metabolism caused the experiment to be shut down ahead of the planned schedule.

At least two of the basic principles of ecosystem modeling discussed in the introduction were violated in this system. Incoming light was greatly reduced, due to the glass and significant support structure, resulting in insufficient photosynthesis and primary productive to balance respiration. This could have been offset by introducing a subset of highly efficient photosynthesis (such as provided by an ATS),

using artificial lighting; indeed, some ATS systems were used, but only as a minor element of control on the ocean system. Monitored exchange with the external environment could also have been employed. Also, the concrete as an atmospheric reactant should have been sealed with a non-reactive material, such as glass or plastic.

Much controversy developed during these human experiments. Colombia University took over management of the system from 1996 to 2003. During this time period the research program changed from human enclosure experiments to work on global climate change.

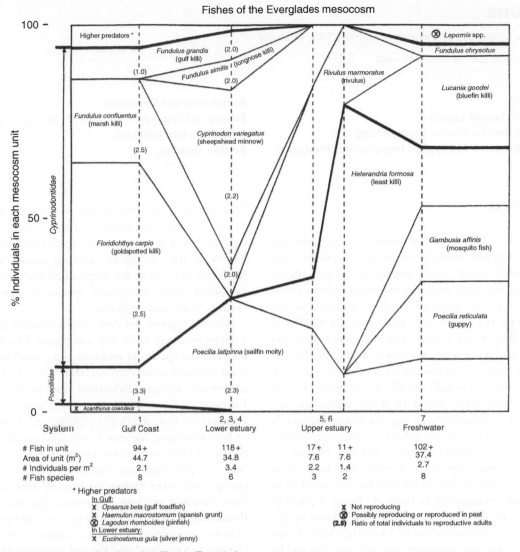

Fishes of the Everglades mesocosm

System	Gulf Coast	Lower estuary	Upper estuary		Freshwater
	1	2, 3, 4	5, 6	7	
# Fish in unit	94+	118+	17+	11+	102+
Area of unit (m²)	44.7	34.8	7.6	7.6	37.4
# Individuals per m²	2.1	3.4	2.2	1.4	2.7
# Fish species	8	6	3	2	8

* Higher predators
In Gulf:
X *Opsanus beta* (gulf toadfish)
X *Haemulon macrostomum* (spanish grunt)
⊗ *Lagodon rhomboides* (pinfish)
In Lower estuary:
X *Eucinostomus gula* (silver jenny)

X Not reproducing
⊗ Possibly reproducing or reproduced in past
(2.5) Ratio of total individuals to reproductive adults

Figure 14 Distribution of fish (% of total) in Florida Everglades mesocosm.

See also: Freshwater Lakes; Freshwater Marshes.

Further Reading

Adey W, Finn M, Kangas P, *et al.* (1996) A Florida Everglades mesocosm – model veracity after four years of self-organization. *Ecological Engineering* 6: 171–224.

Adey W and Loveland K (2007) *Dynamic Aquaria: Building and Restoring Living Ecosytems,* 3rd edn., 505pp. San Diego: Elsevier/ Academic Press.

Kangas P (2004) *Ecological Engineering: Principles and Practice,* 452pp. Boca Raton: Lewis Publishers, CRC Press.

Körner C and Arnone J, III (1992) Responses to elevated carbon dioxide in artificial tropical ecosystems. *Science* 257: 1672–1675.

Marino BDV and Odum HT (1999) *Biosphere 2: Research Past and Present,* 358pp. Amsterdam: Elsevier (Also Special Issue of *Ecological Engineering* 13: 3–14).

Osmund B, Aranyev G, Berry J, *et al.* (2004) Changing the way we think about global change research: Scaling up in experimental ecosystem science. *Global Change Biology* 10: 393–407.

Petersen J, Kemp WM, Bartleson R, *et al.* (2003) Multi-scale experiments in coastal ecology: Improving realism and advancing theory. *Bioscience* 53: 1181–1197.

Small A and Adey W (2001) Reef corals, zooxanthellae and free-living algae: A microcosm study that demonstrates synergy between calcification and primary production. *Ecological Engineering* 16: 443–457.

Walter A and Carmen Lambrecht S (2004) Biosphere 2, center as a unique tool for environmental studies. *Journal of Environmental Monitoring* 6: 267–277.

Lagoons

G Harris, University of Tasmania, Hobart, TAS, Australia

Background

Coastal lagoons are estuarine basins where freshwater inflows are trapped behind coastal dune systems, sand spits, or barrier islands which impede exchange with the ocean. They are most frequent in regions where freshwater inflows to the coast are small or seasonal, so that exchange with the ocean may not occur for months or years at a time. Many occupy shallow drowned valleys formed when the sea level was lower during the last ice age and subsequently flooded by postglacial sea level rise. The tidal range is usually small. Accordingly, coastal lagoons are frequently found in warm temperate, dry subtropical, or Mediterranean regions along moderately sheltered coasts. Lagoons are infrequent in wetter temperate and tropical regions where freshwater inflows are sufficient to scour out river mouths and keep them open. Here estuaries are dominated by salt marshes in temperate and mangroves in tropical climes. A particularly good example is the series of coastal habitats on the southern and eastern coastline of Australia which change from open temperate estuaries and salt marshes in the wetter southern regions of Tasmania, through a series of coastal lagoons of varying sizes and ecologies along the south and east coasts, to open subtropical and tropical estuaries, reefs, and mangroves in the warmer and wetter north. A similar, although inverted, sequence can be seen running south along the east coasts of Canada, and the northeastern, central, and southeastern coasts of the USA. The resulting lagoons have varying water residence times, depending on volume, climate, freshwater inflow volumes, and the tidal prism.

Some lagoons are predominantly freshwater or brackish, while others are predominantly marine; so the dominant organisms in coastal lagoons reflect the balance of freshwater and marine influences. All are influenced by the local biogeography. Thus, the dominant species in Northern Hemisphere lagoons are quite different from those in their Southern Hemisphere equivalents. Different coastal regions of the globe differ in their biodiversity; for example, the endemic biodiversity of seagrasses is very high in Australian waters.

Nevertheless, two points are worthy of note. First, there is great functional similarity between systems despite differing in the actual species involved. Second, human activity is quickly moving species around the world so that there are large numbers of what might be called 'feral' introduced species in coastal waters close to ports and large cities.

Coastal lagoons are ecologically diverse and provide habitats for many birds, fish, and plants. The interactions between the species in estuaries and coastal lagoons produce valuable ecosystem services. Indeed, the value of ecosystem services calculated for such systems by Costanza *et al.* was the highest of any ecosystem studied. Lagoons are also esthetically pleasing and desirable places to live, providing harbors, fertile catchments, and ocean access for cities and towns; thus, they have long been the sites of rapid urban and industrial development. Habitat change and other threats to lagoons now compromise these valuable services. All around the world they are threatened by land-use change in their catchments, urbanization, agriculture, fisheries, transport, tourism, climate change, and sea level rise. Coastal waters and lagoons are therefore definitive examples of the problems of multiple use management. Rapid population growth in coastal areas is common in many western countries (particularly the common 'sea-change' phenomenon, in which there is a trend toward rapid population growth along coasts), so the threats and challenges are increasing rapidly. Climate change and sea level rise are also becoming issues to be dealt with. In tropical and subtropical regions there is both evidence of rapid coastal habitat loss and population growth as well as an increased frequency of severe hurricanes. Modified systems impacted by severe hurricanes and tsunamis appear to be more fragile in the face of extreme events and certainly do not degrade gracefully.

Research and the management of coastal systems require a synthesis of social, economic, and ecological disciplines. Around the world there are a number of major research and management programs which aim to apply ecosystem knowledge to the effective management of coastal resources. Current examples include work in Chesapeake Bay and the Comprehensive Everglades

Restoration Plan in the USA. In Italy the lagoon of Venice is a classic example. In Australia major programs have been undertaken in coastal embayments and lagoons in Adelaide (Gulf of St. Vincent), Brisbane (Moreton Bay), and Melbourne (Port Phillip Bay). (For details on these programs and useful links, see www.chesapeakebay.net, www.evergladesplan.org, and www.healthywaterways.org.) Land-use change (both urbanization and agriculture) in catchments, together with the use of coastal lagoons for transport and tourism, has led to a combination of changes in physical structures (both dredging and construction of seawalls and other barriers), altered hydrology and tidal exchanges, increased nutrient loads, and inputs of toxicants. The resulting symptoms of environmental degradation include algal blooms (which may be toxic), loss of biodiversity, and ecological integrity (including the loss of seagrasses and other important functional groups), anoxia in bottom waters, loss of important biogeochemical functions (denitrification efficiency), and the disturbances caused by introduced, 'feral' species from ships and ballast water.

Inputs – Catchment Loads

Land-use change in catchments changes the hydrology of rivers and streams and increases nutrient loads to lagoons. Rivers draining clear catchments, or those with extensive urbanization, show 'flashier' flow patterns with water levels rising and falling quickly after rainfall. The hydrological balance and water residence times of the lagoons are altered as a result. While nutrient loads are generally proportional to catchment area (**Figure 1**), loads from cleared agricultural or urban catchments are higher than those from forested catchments, the nutrient loads being proportional to the amount of cleared land or the human population in the catchment. Carbon, nitrogen, and phosphorus loads all increase; C loads from wastewaters may lead to biochemical oxygen demands (BODs) and anoxia, while increased N and P loads stimulate algal blooms and the growth of epiphytes in seagrasses. A further problem is the fact that forested catchments tend to export organic forms of N and P (which are less biologically active in receiving waters), whereas cleared and developed catchments tend to export biologically available inorganic forms of N and P. Thus, both nutrient loads and the availability of those loads increase when catchments are cleared and developed.

N is in many cases (particularly in warmer coastal waters) the key limiting element in lagoons because of high denitrification efficiencies in sediments and long water residence times in summer. In temperate waters N and P may be co-limiting or the limitation may vary seasonally and on an event basis. Overall the climate regime, geomorphology, and biogeochemistry of coastal lagoons seem to lead to extensive N limitation and

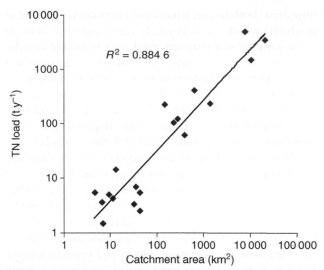

Figure 1 The empirical relationship between catchment area (km^2) and the total nitrogen load (tonnes per year) to their associated coastal lagoons. Data from the catchments of 19 coastal lagoons on the east coast of Australia. Details of data sources are given in Harris GP (1999) Comparison of the biogeochemistry of lakes and estuaries: Ecosystem processes, functional groups, hysteresis effects and interactions between macro- and microbiology. *Marine and Freshwater Research* 50: 791–811.

denitrification is an important process which determines many ecological outcomes. The effect of land-use change on N loads is therefore a key area of concern. A considerable amount of work has been done on the export of N from catchments around the world. Catchments tend to retain on average about 25% of the N applied to them and export about 75%. There are both latitudinal and seasonal factors which affect this figure. Catchment exports on the eastern coast of North America show an effect of latitude, with warmer, southern catchments with perennial vegetation exporting about 10% of applied N and more northerly catchments with seasonal vegetation growth exporting as much as 40% of applied N, particularly in winter. P exports tend to come primarily from sewage and other wastewater discharges, and also from erosion and agricultural runoff. Catchment loads show evidence of self-organized pattern and process in catchments – nutrient loads and stoichiometries change over time at all scales and the distribution of inflowing nutrients may be fractal.

Fates and Effects – Physics and Mixing

Water movement and mixing are driven by the effects of wind and tide on coastal lagoons. The basic hydrodynamics of coastal systems are well represented by physics-based simulation models of various kinds. A number of two- and three-dimensional (2D and 3D) models now exist (both research tools and commercially available products) which can adequately represent wind-induced wave patterns and currents, tidal exchanges and circulation, and changes in surface elevations due to tides and winds. (For an introduction to a variety of models, see www.estuary-guide.net/toolbox or www.smig.usgs.gov; models by Delft Hydraulics at www.wldelft.nl and DHI www.dhigroup.com/) Input data required are basic meteorological data: wind speed and direction, plus solar insolation, and a detailed knowledge of the morphometry and bathymetry of the lagoon in question. Based on the conservation of mass and momentum and various turbulence closure schemes, it is possible to adequately model and predict both velocity fields and turbulent diffusion in the water column. Calibration and validation data are obtained from *in situ* current meters and pressure sensors. Bottom stress, sediment resuspension, and wave-induced erosion can also be represented. It is thus possible to model the effects of various climate and engineering scenarios, everything from sea level rise to construction projects of various kinds. These models are widely used to develop environmental impact statements (EIS) for major projects and to manage major dredging projects around the world. Only some of these models are capable of long-term predictions of water balance and of water residence times. Such predictions require careful analysis of long-term meteorological records and good predictive models of inflows and evaporation. Nevertheless such models also exist.

Fates and Effects – Ecological Impacts and Prediction

Given the nature of the threats, the value of ecosystem services delivered, and the importance of ecosystem management, there have been many studies of ecosystems in coastal lagoons. As noted above some of the ecosystem studies have been in the form of major multidisciplinary programs. The knowledge obtained has then frequently been encapsulated in various kinds of predictive ecological models which attempt to provide answers to 'what if' questions from environmental managers and engineers seeking to implement catchment works or reductions in wastewater discharges. The ecological models are driven by the hydrodynamic models described above – the physical setting provides the basic context for the ecological response. In many cases the knowledge has also been built into a variety of EIS and risk assessments which attempt to judge the possible detrimental effects of land-use change, port construction, harbor dredging, and other engineering developments in urban and industrial areas.

Empirical Knowledge and Models

Despite the pandemonium of interactions between species in coastal marine systems (or perhaps because of it), there are some high-level empirical relationships which can be used for diagnosis and management. Much as Vollenwieder discovered in lakes there are some predictable high-level properties of coastal marine systems. For example, the total algal biomass (as chlorophyll *a*) responds to N loads just as lakes respond to P loads. This is further evidence of the importance of N as a limiting element in marine systems, and for the key of P as the limiting element in freshwater systems. The differing biogeochemistry of marine and freshwater ecosystems is explicable on the basis of the evolutionary history and geochemistry of the two systems. The existence of a relationship between N and algal biomass is evidence for a kind of 'envelope dynamics' of these diverse systems. N does not limit growth rates of the plankton so much as the overall biomass. As a result of high growth rates, grazing, and rapid nutrient regeneration in surface waters, the total community biomass reaches an upper limit set by the overall rate of supply of N. This is a form of 'extremal principle' of these pelagic ecosystems which indicates that with sufficient biodiversity then an upper limit to maximum nutrient use efficiencies can be reached. A similar model of high-level ecosystem properties has been developed in which some fundamental physiological properties of phytoplankton (the slope of the P vs. I curve at low light and the maximum photosynthetic rate) are used to develop a production model based on biomass, photosynthetic properties, and incident light. This amounts to saying that even in shallow coastal systems it is possible to get some reasonable empirical predictions of the physiological (photosynthetic parameters and nutrient uptake efficiencies) and ecosystem responses to some driving forces (nutrient loads and incident light).

A second form of empirical determinant of system function is set by the stoichiometry and biogeochemistry of these systems. The characteristic elemental ratios in the key organisms (algae, grazers, bacteria, macrophytes) and the ratios of elemental turnover set limits on the overall system performance. The predominant element ration in pelagic marine organisms is the Redfield ratio (106C:15N:1P). This aspect of the biogeochemistry of coastal lagoons has been used in a global comparison of the biogeochemistry of these systems by the IGBP LOICZ program. Knowing the loading rates of major nutrients, the concentrations of nutrients in the water column, and the rates of tidal exchange allows simple mass balance models of C, N, and P to be constructed. The salt and water budget of these systems can be used to obtain bulk hydrological fluxes. Making stoichiometric assumptions via the Redfield ratio about fluxes of C, N, and P (as well as oxygen) in the plankton and across the sediment interface allows estimates to be made of the overall autotrophic–heterotrophic balance of the system as well as nitrogen fixation and denitrification rates (essentially by estimating the 'missing N' based on the C, N, and P stoichiometry). These techniques have made it possible to do global comparisons of the biogeochemistry of lagoons around the world and to examine the effects of inflows, tidal exchanges, and latitude or climate. This has been a major contribution to the knowledge of the ways in which major elements are processed and transported from the land to the ocean through the coastal zone.

The overall impression is that pristine lagoons (loaded by largely organic forms of C, N, and P) are mostly net heterotrophic and strong sinks for N through denitrification. More eutrophic systems with higher N and P loads (and more of those in inorganic forms) tend to be net autotrophic and, if dominated by cyanobacterial blooms, net N fixing systems. Decomposition of these blooms may be sufficiently rapid to cause anoxia in bottom waters and lead to the cessation of denitrification and the export of N (as ammonia) on the falling tide. Warm temperate and subtropical lagoons – with low hydrological and nutrient loads – seem to have higher denitrification efficiencies than temperate systems. They are often heterotrophic and strongly N limited systems. An extreme is Port Phillip Bay in Melbourne which has low freshwater inflows, high evaporation, a long water residence time (*c.* 1 year), high denitrification efficiency (60–80%) and is so N limited that it imports N from the coastal ocean on the rising tide. Temperate lagoons and estuaries have higher freshwater and nutrient inflows, are more eutrophic (autotrophic), and are exporters of N. Temperate systems are therefore more likely to show occasional P limitation. Overall, the cycling of the major elements is driven by the stoichiometry of the major functional groups of organisms.

Thus in biodiverse ecosystems it is possible to obtain some high-level state predictors from a knowledge of key drivers and the basic physiology and stoichiometry of the dominant organisms. The predictions so produced are not perfect but they do capture a large fraction of the behavior of these systems. At this level these models can be used for the management of nutrient loads to coastal lagoons.

Detailed Simulation Models of Ecosystems, Functional Groups, and Major Species

Many of the questions that are asked of ecologists studying coastal systems are of a more detailed nature and relate to loss or recovery of major species, functions, or functional groups – ecosystem services and assets if you like. Examples would be dominant algal groups, seagrasses, macroalgae, denitrification rates, benthic biodiversity, fish recruitment, etc. At this level a large

number of dynamical ecological simulation models of shallow marine systems have been constructed. There is much more uncertainty in the ecological models than there is in the physical models. Much of the required ecological detail is unknown, key parameters can be ill-defined, the data are usually sparse in space and time, and the computational resources are not adequate to the task of a complete simulation of the entire system. Ecological models are therefore abstractions which attempt to represent the major ecological features and functions of the greatest relevance to the task at hand. Nevertheless, 30 years of research in lagoons and coastal systems around the world have uncovered a number of major functional groups and ecosystem services which, when coupled together in models, give some guide as to the overall ecological responses.

The generic models of coastal systems use two basic functional components. A nutrient, phytoplankton, zooplankton (NPZ) model for the water column, and a benthic model incorporating the necessary functional groups – macroalgae, zoo- and phytobenthos, seagrasses – with the groups chosen to represent the particular system of interest. All functional groups are represented by their basic physiologies and stoichiometries and the interconnections (grazing, trophic closure, decomposition, and denitrification rates) are represented by established relationships. The NPZ models adequately predict the average chlorophyll of lagoons and, when coupled with 3D physical models, can give predictions of the spatial distribution of algal biomass in response to climate and catchment drivers. For reasons which will become clear below, these models only predict average biomass levels and cannot predict all the dynamics of the various trophic levels. The coupling between the plankton and the benthos in lagoons is nonlinear and results in some strongly nonlinear responses of the overall system to changes in nutrient loads. Basically, there is competition between the plankton and the benthos for light and nutrients which can drive switches in system state. Thus, lagoons, much like shallow lakes, may show state switches between clear, seagrass-dominated states and turbid, plankton-dominated states.

The major driver of the state switches is the high denitrification efficiencies exhibited by the diverse phyto- and zoobenthos in lagoons with strong marine influences. As long as there is sufficient oxygen in bottom waters, diverse zoobenthos burrow and churn over the sediments causing extensive bioturbation and 3D structure in the sediments. Clams, prawns, polychaete worms, crabs, and other invertebrates set up a complex system of burrows and ventilate the sediments through feeding currents and respiratory activity. Given sufficient light at the sediment surface the phytobenthos (particularly diatoms, the microphytobenthos, MPB) photosynthesize rapidly and set up strong gradient of oxygen in the top few millimeters of the sediment. These gradients, together with the strong 3D microstructure of the sediments set up by the zoobenthos, favor the co-occurrence of adjacent oxic and anoxic microzones which are required for efficient denitrification. N taken up by the plankton sinks is actively denitrified by the sediment system. In marine systems the abundance of sulfate in seawater ensures that P is not strongly sequestered by the sediments. Thus, the basis of the LOICZ models lies in the efficiency of denitrification of N in sediments and the more or less conservative behavior of P in these systems. These ecosystem services are supported by the high biodiversity of the coastal marine benthos.

In lagoons with higher nutrient loads, the entire ecosystem may switch to an alternative state. Increased N loads stimulate the growth of plankton in the water column and shade off the MPB. The increased planktonic production sinks to the bottom depleting oxygen and reducing the diversity of zoobenthos, restricting the community structure to those species resistant to low oxygen concentrations. Active decomposition in anaerobic sediments together with reduced bioturbation leads to the cessation of denitrification and the release of ammonia from the sediments. So instead of actively denitrifying and eliminating the N load, the system becomes internally fertilized and algal production rises further. This is analogous to the internal fertilization of eutrophic lakes through the release of P from anoxic sediments. In both cases the switch is caused by a change in redox conditions and the change in performance of suites of microbial populations. Once switched to a more eutrophic state (algal bloom dominated), these lagoons do not easily revert to their clear and macrophyte dominated state. Loads must be strongly reduced to get them to switch back – something which may not be possible if the catchment has been modified by urban or agricultural development. There is thus evidence for strong hysteresis in the response of these ecosystems to various impacts.

The overall biodiversity and nutrient cycling performance of coastal lagoons therefore depends on the relative influences of marine and freshwaters, the differing biodiversity of marine and freshwater ecosystems, the relative C, N, and P loads to the plankton and the benthos, and on seasonality, latitude, and climate drivers. Nevertheless, at least the broad features of their behavior can be explained and predicted on the basis of sediment geochemistry, and the stoichiometry and physiology of the major functional groups in these ecosystems. Empirical work on a number of lagoons up the east coast of Australia allowed Scanes et al. to effectively determine the response of 'titrating' these systems with nutrients. As the N load to the lagoons was increased, seagrasses were lost and algal blooms were stimulated. Even at a crude level of visual assessments it was possible to rank these systems in order of loading and to show that

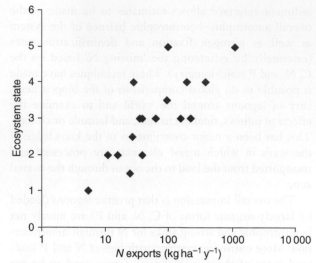

Figure 2 The empirical 'ecosystem titration' relationship between catchment N exports and the resulting ecosystem state in 17 coastal lagoons on the east coast of Australia. Ecosystem state is defined as 1, pristine; 3, showing marked seagrass loss and the growth of macrophytic algae; 5–6, dominated by nuisance algal blooms (some of which may be toxic). Data from personal observations and reworked from Scanes P, Coade G, Large D, and Roach T (1998) Developing criteria for acceptable loads of nutrients from catchments. In: *Proceedings of the Coastal Nutrients Workshop, Sydney (October 1997)*, pp. 89–99. Artarmon, Sydney: Australian Water and Wastewater Association.

the pattern of response was entirely similar to that predicted by the models (**Figure 2**). Thus, despite difference in biogeochemistry and biodiversity, shallow lakes and coastal lagoons have broadly similar response to increased nutrient loads and other forms of human impact. Even broad indicators of system state reveal consistent patterns of change.

So oligotrophic lagoons with a Mediterranean climate (warm temperatures in summer and long water residence times) and strong marine influences can be strong sinks for N, whereas cooler, temperate lagoons and estuaries with larger freshwater inflows and higher productivity may export N and be frequently P limited. As the LOICZ program intended, we have managed a broad understanding of the ways in which the coastal zone influences the transport of major elements from land to ocean.

Nonequilibrium Dynamics

If more detailed descriptions and predictions are required (e.g., the diversity and abundance of individual species and other specific ecosystem services and assets), then the predictive ability is less. One of the reasons for this is the fact, alluded to above, that these are nonequilibrium systems which respond to individual events (storms and engineering works) over long time periods. The elimination and invasion of species may take decades and the responses of freshwater lagoons, for example, to salt incursions may also take decades. A particularly good example is Lake Wellington in the Gippsland Lakes system in Victoria, Australia. The entire system is slowly responding to the ingress of salt made possible by the opening of the lagoon system mouth (Lakes Entrance) in 1883. Lake Wellington, the lake farthest inland, remained fresh until after the 1967 drought when a combination of high N and P loads from agriculture, the extraction of water from the inflowing La Trobe River for power station cooling and irrigation, and the incursion of salt killed all the freshwater macrophytes in the Lake. In a few years the lake switched from its previous clear and macrophyte dominated state to being turbid and dominated by toxic algal blooms. It does not appear to be possible to switch it back.

The response of these lagoon systems to climate and other perturbations is nonlinear and complex because of the interactions between the major functional groups and because the timescales of response of the major groups differ strongly. Phytoplankton may respond to changes in loads and water residence times in a matter of days, whereas seagrasses take decades or longer to recover. By perturbing a simple coupled plankton-benthos model with storm events and 'spiked' N loads, Webster and Harris showed that the threshold load for the elimination of seagrasses could be altered considerably depending on the characteristics of the input loads. So the response of the system was a function of the overall load and the frequency and magnitude of events. Climate change and catchment development both alter the overall C, N, and P load to lagoons as well as the characteristics of that load, so that ecological responses by lagoons are highly complex and change over time depending on a variety of modifications and management actions. Consequently, lagoons are always responding to the last storm or intervention and the abundance of key species drifts to and fro over time as the entire plankton–sediment system responds.

The picture is made more complex by the evidence for strong trophic cascades in marine as well as freshwater systems. Coastal ecosystems are frequently over-fished; larger predators and grazers are removed by human hand. Removal of the 'charismatic megafauna' of coastal systems, together with beds of shellfish and other edible species, has changed the ecology of many lagoons and estuaries. Coastal ecosystems around the world have also been strongly modified by the removal of natural physical structures (mangroves and reefs) which confer resilience in the face of extreme events. We have removed both larger fish and benthic filter feeders from many systems compromising function and the ability to respond to changes in catchment loads. Overall there has been a consistent simplification of both physical and ecosystem structures (removal of reefs and macrobiota, simplification of food chains, etc.) and a trend toward more eutrophic (nutrient rich) and simplified systems dominated by microbiota, especially algae and bacteria. We know less about the response of ecosystems to changes in the 'top down' trophic structure than we do about the responses to 'bottom up' catchment drivers; nevertheless, there is good evidence for similar nonlinearities and state switches in response. A nonequilibrium view of coastal lagoons changes the way we look at them. Overall there is a need to pay attention to the 'precariousness' of these systems and manage them adaptively for resilience and response to natural and anthropogenic impacts. Despite being over-fished and highly modified, there is still a need for the ecosystem services they produce.

Emerging Concepts – Multifractal Distributions of Species and Biomass

The underlying complexity of interactions and species distributions is displayed when detailed (high-frequency) observations are made of the spatial and temporal distributions of biomass and species. There is now much evidence to show that the underlying distribution of the plankton and the MPB are fractal or multifractal. Similarly, high-frequency observations in catchments show similar multifractal and even paradoxical properties of hydrological and nutrient loads. So underlying all the generalizations discussed above lies a pattern of behavior which gives strong evidence of self-generated complexity which arises from the pandemonium of interactions between species and functional groups. Indeed, we can probably argue that the kinds of general, system level, responses described above would not occur if it were not for the underlying complexity. While making high-level statements about ecosystem behavior possible, these small-scale, multifractal properties (and the possibilities created by emergence) cause problems when we wish to make predictions at the meso-scale level of dominant species and functional groups. Because of the work that has been done across the levels of organization, coastal lagoons are very good examples of a new kind of ecology – an ecology of resilience and change, rather than an ecology and equilibrium and stasis.

One fundamental problem that these new insights reveal is that most of the data we presently use for the analysis of coastal lagoons are collected too infrequently to be useful for anything other than the analysis of broad trends. Data collected weekly or less frequently are strongly aliased and cannot reveal the true scales of pattern and process. It is just possible to analyze daily data for new insights and processes but high-frequency data – collected at scales of hours and minutes – reveal a wealth of new information. Aliased data combined with frequentist statistical techniques that 'control error' actually remove information from multifractally distributed data and raise the possibility of serious type I and II errors in ecological interpretations. Most importantly, there is information contained in the time series of multivariate data that can be collected from coastal systems. Most analyses of ecological data from ecological systems use univariate data and because of the infrequent data collection schedules – including gaps and irregular time intervals – time series analyses are not possible.

We are just beginning to find new technologies and techniques to study the high-frequency multivariate behavior of these systems using moorings and other *in situ* instruments. New electrode technologies make on-line access to data possible and throw up new possibilities for new kinds of observations of system state. We are beginning to realize that in addition to the 'top down' causation of climate and trophic interactions, there is also a 'bottom up' driver of complexity and the strong possibility of the emergence of high-level properties from the interactions between individuals. New forms of statistical analyses display information in time series of complex and emergent systems. This emerging understanding of complexity and emergent properties changes the ways in which we should approach EIS and risk assessments. We now know that interactions and self-generated complexity, together with hysteresis effects at the system level, can cause surprising things to happen as a result of anthropogenic change. Coastal lagoons are now classic examples of this. That means that risk assessments and EIS cannot look at impacts and changes in isolation; somehow we must develop integrated risk assessment tools that examine the interactive and synergistic effects of human impacts on coastal ecosystems. A further level of complexity is contained in the similar complex and emergent properties of the interactions between agents in the coupled environmental and socioeconomic (ESE) system in which all coastal lagoons are set. Multiple use management decisions are set in a complex web of ESE interactions across scales. Decisions made about industrial and engineering developments for financial capital reasons influence both social capital and ecological (natural capital) outcomes. Feedbacks ensure that this is also a highly nonlinear set of interactions. What we do know is that the prevalent practices of coastal management and exploitation are not resilient in the face of extreme events and that they do not degrade 'gracefully' when impacted by hurricanes and tsunamis. New management practices will be required.

See also: Mangrove Wetlands.

Further Reading

Adger WN, Hughes TP, Folke C, Carpenter SR, and Rockström J (2005) Socio-ecological resilience to coastal disasters. *Science* 309: 1036–1039.

Aksnes DL (1995) Ecological modelling in coastal waters: Towards predictive physical–chemical–biological simulation models. *Ophelia* 41: 5–35.

Berelson WM, Townsend T, Heggie D, *et al.* (1999) Modelling bio-irrigation rates in the sediments of Port Phillip Bay. *Marine and Freshwater Research* 50: 573–579.

Brawley JW, Brush MJ, Kremer JN, and Nixon SW (2003) Potential applications of an empirical phytoplankton production model to shallow water ecosystems. *Ecological Modelling* 160: 55–61.

Costanza R, d'Arge R, de Groot R, *et al.* (1998) The value of ecosystem services: Putting the issues in perspective. *Ecological Economics* 25: 67–72.

Fasham MJR, Ducklow HW, and Mckelvie SM (1990) A nitrogen-based model of plankton dynamics in the oceanic mixed layer. *Journal of Marine Research* 48: 591–639.

Flynn KJ (2001) A mechanistic model for describing dynamic multi-nutrient, light, temperature interactions in phytoplankton. *Journal of Plankton Research* 23: 977–997.

Gordon DC, Boudreau PR, Mann KH, *et al.* (1996) LOICZ biogeochemical modelling guidelines. *LOICZ Reports and Studies, No. 5.* Texel: LOICZ.

Griffiths SP (2001) Factors influencing fish composition in an Australian intermittently open estuary. Is stability salinity-dependent? *Estuarine, Coastal and Shelf Science* 52: 739–751.

Harris GP (1999) Comparison of the biogeochemistry of lakes and estuaries: Ecosystem processes, functional groups, hysteresis effects and interactions between macro- and microbiology. *Marine and Freshwater Research* 50: 791–811.

Harris GP (2001) The biogeochemistry of nitrogen and phosphorus in Australian catchments, rivers and estuaries: Effects of land use and flow regulation and comparisons with global patterns. *Marine and Freshwater Research* 52: 139–149.

Harris GP (2006) *Seeking Sustainability in a World of Complexity.* Cambridge: Cambridge University Press.

Harris GP and Heathwaite AL (2005) Inadmissible evidence: Knowledge and prediction in land and waterscapes. *Journal of Hydrology* 304: 3–19.

Hinga KR, Jeon H, and Lewis NF (1995) Marine eutrophication review. Part 1: Quantifying the effects of nitrogen enrichment on phytoplankton in coastal ecosystems. Part 2: Bibliography with abstracts. *NOAA Coastal Ocean program, Decision Analysis Series, No 4.* Silver Spring, MD: US Dept of Commerce, NOAA Coastal Ocean Office.

Howarth RW (1998) An assessment of human influences on fluxes of nitrogen from the terrestrial landscape to the estuaries and continental shelves of the North Atlantic Ocean. *Nutrient Cycling in Agroecosystems* 52: 213–223.

Howarth RW, Billen G, Swaney D, *et al.* (1996) Regional nitrogen budgets and the riverine N and P fluxes for the drainages to the North Atlantic Ocean – Natural and human influences. *Biogeochemistry* 35: 75–139.

Lotze HK, Lenihan HS, Bourque BJ, *et al.* (2006) Depletion, degradation and recovery potential of estuaries and coastal seas. *Science* 312: 1806–1809.

McComb AJ (1995) *Eutrophic Shallow Estuaries and Lagoons.* Boca Raton: CRC Press.

Mitra A (2006) A multi-nutrient model for the description of stoichiometric modulation of predation in micro- and mesozooplankton. *Journal of Plankton Research* 28: 597–611.

Moll A and Radach G (2003) Review of three-dimensional ecological modelling related to the North Sea shelf system. Part 1: Models and their results. *Progress in Oceanography* 57: 175–217.

Murray AG and Parslow JS (1999) Modelling of nutrient impacts in Port Phillip Bay – A semi-enclosed marine Australian ecosystem. *Marine and Freshwater Research* 50: 597–611.

Nicholson GJ and Longmore AR (1999) Causes of observed temporal variability of nutrient fluxes from a southern Australian marine embayment. *Marine and Freshwater Research* 50: 581–588.

Occhipinti-Ambrogi A and Savini D (2003) Biological invasions as a component of global change in stressed marine ecosystems. *Marine Pollution Bulletin* 46: 542–551.

Pollard DA (1994) A comparison of fish assemblages and fisheries in intermittently open and permanently open coastal lagoons on the south coast of New South Wales, south-eastern Australia. *Estuaries* 17: 631–646.

Roy PS, Williams RJ, Jones AR, *et al.* (2001) Structure and function of south-east Australian estuaries. *Estuarine, Coastal and Shelf Science* 53: 351–384.

Scanes P, Coade G, Large D, and Roach T (1998) Developing criteria for acceptable loads of nutrients from catchments. In: *Proceedings of the Coastal Nutrients Workshop, Sydney (October 1997)*, pp. 89–99. Artarmon, Sydney: Australian Water and Wastewater Association.

Scheffer M (1998) *Shallow Lakes*. London: Chapman and Hall.

Scheffer M, Carpenter S, and de Young B (2005) Cascading effects of overfishing marine systems. *Trends in Ecology and Evolution* 20: 579–581.

Seitzinger SP (1987) Nitrogen biogeochemistry in an unpolluted estuary: The importance of benthic denitrification. *Marine Ecology – Progress Series* 41: 177–186.

Seitzinger SP (1988) Denitrification in freshwater and coastal marine systems: Ecological and geochemical significance. *Limnology and Oceanography* 33: 702–724.

Seuront L, Gentilhomme V, and Lagadeuc Y (2002) Small-scale nutrient patches in tidally mixed coastal waters. *Marine Ecology-Progress Series* 232: 29–44.

Seuront L and Spilmont N (2002) Self-organized criticality in intertidal microphytobenthos patterns. *Physica A* 313: 513–539.

Smith SV and Crossland CJ (1999) Australasian estuarine systems: Carbon, nitrogen and phosphorus fluxes. *LOICZ Reports and Studies, No. 12.* Texel: LOICZ.

Sterner RW and Elser JJ (2002) *Ecological Stoichiometry: The Biology of Elements from Molecules to the Biosphere*. Princeton, NJ: Princeton University Press.

Vollenweider RA (1968) Scientific fundamentals of the eutrophication of lakes and flowing waters, with particular reference to nitrogen and phosphorus as factors in eutrophication. *Technical Report DAS/SCI/68.27*, 182pp. Paris: OECD.

Walker DI and Prince RIT (1987) Distribution and biogeography of seagrass species on the northwest coast of Australia. *Aquatic Botany* 29: 19–32.

Walker SJ (1999) Coupled hydrodynamic and transport models of Port Phillip Bay, a semi-enclosed bay in south-eastern Australia. *Marine and Freshwater Research* 50: 469–481.

Webster I and Harris GP (2004) Anthropogenic impacts on the ecosystems of coastal lagoons: Modelling fundamental biogeochemical processes and management implications. *Marine and Freshwater Research* 55: 67–78.

Relevant Websites

http://www.chesapeakebay.net – Chesapeake Bay Programme.
http://www.dhigroup.com – DHI.
http://www.evergladesplan.org – Everglades.
http://www.healthywaterways.org – Healthy Waterways.
http://www.estuary-guide.net – Toolbox, The Estuary Guide.
http://www.wldelft.nl – wl delft hydraulics.

Landfills

L M Chu, The Chinese University of Hong Kong, Hong Kong SAR, People's Republic of China

Introduction
Postclosure End Uses
Soil Cover
Vegetation

Fauna
Ecological Approach
Further Reading

Introduction

Landfills are seminatural terrestrial ecosystems reconstructed on lands degraded by waste disposal. They are unique in terms of site formation, nature of stratum, and biological activities, but vary according to their age, waste composition, engineering design, and ecological practice. From an environmental perspective, landfills are depositories for municipal solid wastes (sanitary landfills) and less frequently hazardous wastes (secure landfills). Landfills are ubiquitous, as sanitary landfilling is the most common method of municipal solid waste management worldwide. Landfill leachate is formed when rainwater infiltrates and percolates through the degrading waste, while landfill gas is a microbial degradation byproduct under anaerobic conditions. Modern landfills are designed and engineered to restrict the formation and movement of landfill leachate and gas, and to minimize environmental nuisance caused by wind-blown litter, pests, and odor during operation. These landfills, either the containment or entombment type, have buried waste that is isolated from the environment. Older landfills are of the dilution and attenuation type that makes use of the substratum for pollution mitigation; they are unconfined with no facilities for leachate treatment and gas extraction. With dilution and attenuation landfills, problems associated with leachate and gas are common. In terms of environmental biotechnology, landfills can be regarded as large-scale bioreactors in which the organic matter in the buried waste is anaerobically degraded to produce landfill gas which is methane-rich and can be used for electricity generation.

Postclosure End Uses

Once fill capacity is reached, landfills are closed for rehabilitation. With the exception of older landfills that are left abandoned with minimal human interference, most postclosure landfills are rehabilitated using engineering and ecological approaches. Landfilled wastes are isolated physically from the biosphere by bottom barrier layers and surface cap technologies. Barrier systems can be sophisticated with multiple layers of geotextiles and impermeable synthetic membranes. The surface is usually covered with soil of 1–2 m thickness. Subsequent site development entails the establishment of a vegetation cover on the landfill soil, with the primary aims of minimizing environmental impact and making good value of its designated afteruse. Technically, closed landfills can be rehabilitated by either spontaneous ecological development in the absence of human intervention, manipulated succession followed by natural development, or habitat creation which involves intensive and prolonged management. Sole natural development is unreliable and slow, and lacks control of the ecological outcome. The aftercare period for a landfill can be as long as 30 years, but public safety and engineering concerns are usually of higher priority than the ecological function of the reclaimed site.

The criteria for selecting afteruses for former landfills include landuse planning policies, site characteristics, soil resource availability, social needs, and cost consideration. As construction on postclosure landfills is generally prohibited due to severe subsidence as a result of organic matter decomposition, and fire hazards associated with landfill gas, it is a usual practice to reclaim urban sites for soft end uses in order to provide amenity facilities such as parks, botanical gardens, golf courses, and playing fields that are safe for use by the public. Alternatives end uses for agriculture, nature conservation, and forestry are also common. Grassland has been one of the most popular end uses for rural sites, but agricultural conversion is not always appropriate because of the lack of quality topsoil. Nature conservation is sometimes a more suitable afteruse as it requires less intensive aftercare and is more flexible on the postclosure ecological

design, though the transformation to wildlife habitat is not imperative. End use after closure can be mixed landscapes as in the Fresh Kills Landfill in New York, USA, which is converted to an amenity parkland with a range of landuses which include forests, dry lowlands, tidal wetlands, freshwater wetlands, waterways, and wildlife habitats.

Soil Cover

As it is the final soil cover which supports vegetation establishment and ecosystem development, the quality of the soil material and the thickness of the soil cover are of fundamental importance in affecting rehabilitation success. However, as good soil is usually not available or expensive, the soil used is usually derived from *ex situ* substandard soil or subsoil that is nutrient deficient and poor structured. There are inevitable problems of soil compaction and waterlogging for clay soil, drought when coarse soil is used, as well as infertility, which can be amended by conventional measures such as plowing, organic matter amendment, nurse species planting, and fertilizer application.

Many old landfills have experienced revegetation failure to various extents as a result of leachate seepage, landfill gas evolution, poor soil management, and minimal aftercare. Landfill gas is the major cause, among other constraints such as low fertility, high soil temperature, drought, and toxicity from leachate contamination. Unless it is vented to the atmosphere or extracted for energy production, it will displace oxygen and suffocate plant roots, which usually results in the death of vegetation, and gas production can last for 75 years after the deposition of wastes. Even for an engineered site, gas problems may still exist if an impermeable layer is not formed for the entire site or the soil cap is cracked by uneven subsidence of the site. Landfill gas creates a reducing soil condition which severely impairs microbial processes such as decomposition and symbiotic nitrogen fixation; this together with elevated soil temperature of over 40 °C is detrimental to plants, and plant growth is impeded under the adverse impact of these landfill-associated factors. Localized pollution hot spots reduce plant coverage and result in patchy greenness. A thin soil cover will exacerbate the problem of gas and leachate contamination.

The revegetation success of closed landfills depends heavily upon the quality of the soil cover material, adaptation of the planted vegetation to the landfill environment, and aftercare management strategy. In containment and entombment landfills, contamination by landfill gas and leachate is usually greatly alleviated, though not necessarily eliminated. However, the final soil cover may remain stressful for plant growth, and there is also concern that the containment design may elevate nutrient and water stresses on these landfills. Thin soil cover, poor soil quality, and unfavorable landfill conditions will result in poor vegetation growth, especially in the initial phase of ecosystem development.

It is important for rehabilitated landfills to develop a functional soil–plant system, as shortage of nutrients, in particular nitrogen, is common in most imported soils for use as the final top layer on completed landfills. This can be achieved by the addition of chemical fertilizers at the onset of postclosure revegetation works. However, as repeated application is costly, revegetated sites are usually left to nature for the accumulation of nutrients needed for the establishment of self-perpetuating nutrient cycle. This has to be achieved to allow good vegetation growth, the establishment of a fully functional soil–plant system, and ecosystem development. Plant growth during the early phase of ecosystem rehabilitation is usually limited by the rate of nutrient turnover, and the use of poor soil material as the final cover will inevitably result in rehabilitated sites that are neither productive nor sustainable. There is a paucity of information on the nutrient fluxes and compartmentation in landfill cover soils, and there is only partial idea of nutrient mobilization and immobilization as a function of soil status, and stage of soil development and vegetation succession. Shortage of mineral nutrients could be either due to a lack of sufficient nutrient capital or a failure in mineralization processes. Therefore, litterfall, litter quality, mineralization rate, and the level of biological activity are important determinants of landfill soil quality. Slow decomposition rate implies that nutrients are trapped in organic matter and are not available to nutrient transformation.

Nutrients such as nitrogen and phosphorus accumulate in landfill soil as the ecosystem develops, and their levels have a positive correlation with vegetation establishment. In abandoned landfills, without much aftercare, litter from invaded vegetation is the primary source of organic matter and nutrients in the absence of biological fixation. However, there is a lack of information regarding the nitrogen capital of landfill soils. Nitrogen is supplied from fertilizer application, decomposition, biological fixation, and rainfall. It is susceptible to immobilization on the youngest sites, and the primary production of newly established grassy vegetation cannot rely on decomposition, even though the rate is comparatively high for a sustainable nitrogen turnover. Total amount of nitrogen mineralized on more mature woodlands is high, but it is unclear as to how much nitrogen accumulated in the soil is sufficient to create a self-perpetuating ecosystem on closed landfills.

Within the soil, the microflora, fauna, and the abiotic components are all important and interrelated compartments of the landfill ecosystem. Former landfills support diverse soil and litter fauna which have an active role in the detritus food web. They comprise of high diversity and populations of saprohagous arthropods and macroinvertebrates such as isopods, millipedes, and centipedes that are tolerant of the landfill environment. Springtails and mites are abundant in landfills with gas problems. Earthworms are also adaptive to landfill conditions and have been inoculated to landfills for soil amelioration, but natural colonization and soil improvement appear to be slow, and it takes 3–14 years for earthworm species to invade landfills. Low accumulation of organic matter and patchy coverage of vegetation can hinder the recruitment as well as the mobility of earthworms in landfills.

The best soil cover on landfills should support diverse communities of soil microflora and invertebrates which play crucial roles in organic matter decomposition and nutrient cycling. Active populations of microorganisms and invertebrates will improve the physicochemical status of the soil, which in turn encourage the colonization of plants to support more diverse animal species, thus forming a community of a greater structural complexity and functional stability. This is important not only for the success of revegetation but also the successional development afterwards. In the long run, this will facilitate autogenic change which is the result of the recruitment of late-successional species and the development of ecosystem processes on these man-made habitats.

Vegetation

Plant cover on landfills contributes to its landscape and assists in the reduction of leachate discharge through evapotranspiration. The latter function is particularly important if the landfill is not capped with an impermeable layer to control infiltration. Other benefits provided by the vegetation cover include visual improvement of the site, creation of wildlife habitat, and the sequestration of greenhouse gases. The species chosen for revegetation purpose depends on the afteruse of the site, climatic conditions, nursery stock availability, and hardiness of the species.

Despite the tremendous efforts and investment devoted to site engineering, the inclusion of a soil cover does not guarantee the successful establishment of vegetation. The depth and quality of the soil layer affect revegetation as a thicker soil cover is required for woody species which have deeper root systems.

Poor vegetation performance is a common feature of many old landfills. In the US, a nationwide survey conducted in the early 1980s showed that the major cause for plant failure was the high concentration of landfill gas in the root zone. Negative correlation was found between landfill gas concentration and plant coverage or tree growth in municipal landfills because tree growth was hampered by high landfill gas content, and to a certain extent by high soil temperature and drought. In addition, root development, and hence plant growth at landfills was also adversely affected by pedoclimatic conditions such as high underground temperature, drought, soil acidity, and contamination by leachate. To counteract these problems, the planting of species tolerant to the above adverse conditions is recommended. This is why earlier studies on landfill revegetation focused on the adaptability of plant species to landfill gas. Leguminous trees are better than nonlegumes in their tolerance to high landfill gas that prevailed in old landfills.

Rehabilitation is traditionally initiated by hydroseeding with grass species and/or planting of tree species for erosion control and esthetic improvement. Landscaping and artificial revegetation are the initial rehabilitation works, irrespective of the afteruse of the site, as this accelerates ecological development. The site is revegetated preferentially using grasses which grow fast and provide good immediate ground cover to control erosion and reduce visual impact. Grass swards also survive better than trees on landfills with gas influence, a feature which is attributed to their shallow rooting depth. Tree planting is less popular especially on the top platform of a landfill, because of the negative effects of tree growth on landfills. Following initial revegetation, the rehabilitated site is left for secondary succession to take place.

Grassland can be a versatile habitat option for closed landfills as it can be established on a wide range of soil types. While pasture or arable grassland is more demanding on soil quality and requires greater fertilizer input, low-maintenance grassland can be established on infertile soils. A seed mix of more species and the inclusion of wildflowers can increase the species richness of the vegetated sites. Open grasslands are good habitats for many animal species (e.g., butterflies), but others prefer scattered scrubs and trees for shelter. Planting trees had not usually been recommended as it was believed that tree roots would perforate and crack by drying out the landfill cap. In addition, tree growth on landfill soil may be difficult because of poor soil quality. However, as woodlands have the greatest conservation value, it seems desirable to plant trees to form woodlands which have the benefits of increasing forest resources, habitat connectivity, wildlife biodiversity, and landscape integration.

Vegetation is an integral part of the landfill ecosystem, and flora composition of vegetated sites differs with respect to landfill technology (i.e., gas and leachate control), hydrometeorological conditions, as well as the quality and depth of the soil cover. Vegetation composition is also directly affected by the species planted, survival of the planted species, replanting/enrichment planting, natural invasion of other species, and the seed bank in the soil cover material. A suitable species will very much adapt to and survive in the landfill conditions, at least for a certain period of time, and facilitate the growth of late-successional species. With differential site availability, species availability, and species performance, rehabilitation can be directed by using different soil and planting strategies to achieve successional intervention. A good choice of species for revegetation could enhance the sustainability of ecosystem development. Nitrogen-fixers and those pioneer species usually outcompete other species in the first 10–20 years of ecosystem development after rehabilitation. Nitrogen-fixing trees such as the tropical species of *Acacia confusa*, *A. auriculiformis*, *A. mangium*, *Albizia lebbeck*, *Casuarina equisetifolia*, *Leucaena leucocephala* and temperate species such as *Alnus glutinosa* are widely used for planting on closed landfills. These species assist in nitrogen accumulation in the landfill soil and are very important in the successional development of the soil cover. Therefore, enrichment planting of late-successional species is sometimes necessary at a later stage of development to enhance plant density and maintain species-rich vegetation in secondary forests on closed landfills.

The establishment of woodland communities is the result of gradual ecological development, which cannot be achieved simply by tree planting. The success and speed of succession rely on the availability of appropriate seeds with the proper dispersal mechanism and the presence of effective animal dispersers, and species with the appropriate ecological characteristics. The seed bank, in the cover soil, supplies the species for the early vegetative colonization, which resembles the floristic composition of the areas where the soil is obtained. Soil seed densities decline with landfill age, a trend similar to the course of old-field succession. Young landfills have more r-selected species, which tend to produce more seeds, whereas older sites have more K-selected species, which produce fewer seeds but a higher population of perennials. Some woody plants that are more adaptive can invade gaps and establish slowly. Trees that are either early-successional species or leguminous species should be planted in greater proportion to accelerate succession in landfills, preserve the biodiversity of local flora, and provide more favorable habitats for wildlife conservation. Planting more native or exotic species has been debated; natives, though not necessarily fast growing, are adaptive to local environmental conditions, and provide indigenous characters that are not found in artificial revegetation, but whether natives or exotics are better choices depends on their adaptation to landfills conditions and the quality of soil for revegetation.

Postclosure landfills can be a good refuge for rare species including wild orchids, and are important to the conservation of native flora. Older sites are better developed in terms of soil quality and vegetation coverage. Ecosystem development on closed landfills can be rapid and is accelerated by artificial planting and good management practices.

Fauna

The landfill cover supports vegetation which serves as a habitat for native fauna, but ecologically, it is useless if rehabilitated landfills fail to provide suitable grounds for faunal colonization. Not much has been done on the faunal assemblages on closed landfills, but rehabilitated landfills are potential sites for faunal colonization because they attract insects and herpetofauna and have an important role to play in wildlife conservation.

Open grasslands developed on abandoned landfills are an important insect habitat, and some closed landfills which have been converted into woodlands or grasslands provide valuable habitat for butterflies, especially those species which are declining in population and distribution. However, butterfly community composition and structure have stronger links with vegetation that are either a source of nectar or host plants for larvae, and do not necessarily reflect the successional development of closed landfills.

Closed landfills could also be colonized by amphibians and reptiles within a few years after revegetation, and herpetofaunal diversity and abundance increase with time after closure. Constructed wetlands, though not a conventional option for habitat creation on landfills, provide refuges for amphibians and reptiles. An example of this are ponds that have been designed and constructed on a landfill in Cheshire, England, specifically for great crested newts that were originally present on the site before landfilling.

Birds play a very vital role in the secondary succession on landfills as seed dispersers. It has been reported that birds introduced 20 new plant species to a landfill annually via endozoochory. This increases the floral diversity and contributes to vegetation development. However, only species that produce fleshy fruits will be spread by frugivorous species. It is generally advocated that more fleshy-fruited natives should be planted to attract birds, and even small mammals such as bats for full restoration of the ecological function of landfill as a wildlife habitat.

The reestablishment of faunal communities is closely related to that of vegetation. Closed landfills are potential refuges for uncommon and rare species, and it is suggested that planting of more natives can aid in the creation of a more favorable habitat for ecological diversity. Rehabilitated landfills may not be as ecologically diverse as natural areas, but their conservation values should not be overlooked, as they can be good wildlife habitat and connecting links to enhance remnant fragmented areas. Sites with relatively high biodiversity and rare-species records should be designated conservation areas, especially for those which are not suitable for other alternative development.

Ecological Approach

The basic ecological principles of successional development are totally applicable to rehabilitated landfills, and rehabilitation success depends on the reestablishment of biological activities of surface horizons in the long term. The natural succession of grassland to woodland ecosystem is slow and may take up to 50 years. It is generally accepted that intervention of ecosystem reconstruction followed by natural succession is the best practicable option for landfills. If closed landfills were reclaimed properly, they could provide an attractive source of land for nature conservation, forestry, and recreation. However, the success of reclamation depends much upon the growth of plants and the efficient cycling of nutrients in the cover material. An integrated approach which includes gas control, soil management, and directed succession can accelerate the development of a sustainable ecosystem in terms of structure and function on closed landfills.

See also: Biological Wastewater Treatment Systems.

Further Reading

Chan YSG, Wong MH, and Whitton BA (1996) Effects of landfill factors on tree cover: A field survey at 13 landfill sites in Hong Kong. *Land Contamination and Reclamation* 2: 115–128.

Chan YSG, Chu LM, and Wong MH (1997) Influence of landfill factors on plants and soil fauna: An ecological perspective. *Environmental Pollution* 97: 39–44.

Dobson MC and Moffat AJ (1993) *The Potential for Woodland Establishment on Landfill Sites*, 88pp. London: Department of the Environment, HMSO.

Dobson MC and Moffat AJ (1995) A re-evaluation of objections to tree planting on containment landfills. *Waste Management and Research* 13: 579–600.

Ecoscope (2000) *Wildlife Management and Habitat Creation on Landfill Sites: A Manual of Best Practice.* Muker, UK: Ecoscope Applied Ecologists.

Ettala MO, Yrjonen KM, and Rossi EJ (1988) Vegetation coverage at sanitary landfills in Finland. *Waste Management and Research* 6: 281–289.

Flower FB, Leone IA, Gilman EF, and Arthur JJ (1978) *A Study of Vegetation Problems Associated with Refuse Landfills*, EPA-600/2-78-094, 130pp. Cincinnati: USEPA.

Handel SN, Robinson GR, Parsons WFJ, and Mattei JH (1997) Restoration of woody plants to capped landfills: Root dynamics in an engineered soil. *Restoration Ecology* 5: 178–186.

Moffat AJ and Houston TJ (1991) Tree establishment and growth at Pitsea landfill site, Essex, U.K. *Waste Management and Research* 9: 35–46.

Neumann U and Christensen TH (1996) Effects of landfill gas on vegetation. In: Christensen TH, Cossu R, and Stegmann R (eds.) *Landfilling of Waste: Biogas*, pp. 155–162. London: E & FN Spon.

Robinson GR and Handel SN (1993) Forest restoration on a closed landfill: Rapid addition of new species by bird dispersal. *Conservation Biology* 7: 271–278.

Simmons E (1999) Restoration of landfill sites for ecological diversity. *Waste Management and Research* 17: 511–519.

Wong MH (1988) Soil and plant characteristics of landfill sites near Merseyside, England. *Environmental Management* 12: 491–499.

Wong MH (1995) Growing trees on landfills. In: Moo-Young M, Anderson WA, and Chakrabarty AM (eds.) *Environmental Biotechnology: Principles and Applications*, pp. 63–77. Amsterdam: Kluwer Academic.

Mangrove Wetlands

R R Twilley, Louisiana State University, Baton Rouge, LA, USA

Published by Elsevier B.V.

Introduction
Ecogeomorphology of Mangroves
Biodiversity
Ecosystem Processes

Impacts of Environmental Change
Management and Restoration
Further Reading

Introduction

Mangroves refer to a unique group of forested wetlands that dominate the intertidal zone of tropical and subtropical coastal landscapes generally between 25° N and 25° S latitude. These tropical forests grow along continental margins between land and the sea across the entire salinity spectrum from nearly freshwater (oligohaline) to marine (euhaline) conditions. The coastal forests also inhabit nearly every type of coastal geomorphic formation from riverine deltas to oceanic reefs – another example of the tremendous 'biodiversity' of mangrove ecosystems. Mangroves are trees considered as a group of halophytes with species from 12 genera in eight different families. A total of 36 species has been described from the Indo-West-Pacific area, but fewer than ten species are found in the new world tropics. The term mangroves may best define a specific type of tree, whereas mangrove wetlands refers to whole-plant associations with other community assemblages in the intertidal zone, similar to the term 'mangal' introduced by Macnae to refer to swamp ecosystems. In addition, the habitats of tropical estuaries consist of a variety of primary producers and secondary consumers distributed in bays and lagoons that have the intertidal zone dominated by mangrove wetlands. These may be referred to as mangrove-dominated estuaries.

There are numerous reviews and books that describe the ecology and management of mangroves around the world, including references describing techniques to study the ecology of mangrove wetlands.

Ecogeomorphology of Mangroves

The environmental settings of mangroves are a complex behavior of regional climate, tides, river discharge, wind, and oceanographic currents (**Figure 1**). There are about $240 \times 10^3 \, \text{km}^2$ of mangroves that dominate tropical continental margins from river deltas, lagoons, and estuarine settings to islands in oceanic formations (noncontinental). The landform characteristics of a coastal region together with geophysical processes control the basic patterns in forest structure and growth. These coastal geomorphic settings can be found in a variety of life zones that depend on regional climate and oceanographic processes. Hydroperiod of mangroves resulting from gradients in microtopography and tidal hydrology (**Figure 1**) can influence the zonation of mangroves from shoreline to more inland locations forming ecological types of mangrove wetlands. Lugo and Snedaker identified ecological types of mangroves based on topographic location and patterns of inundation at local scales (riverine, fringe, basin, and dwarf; **Figure 1**) that Woodroffe summarized into basically three geomorphic types (riverine, fringe, and inland). A combination of ecological types of mangroves can occur within any one of the geomorphic settings occurring at a hierarchy of spatial scales that can be used to classify mangrove wetlands.

Various combinations of geophysical processes and geomorphologic landscapes produce gradients of regulators, resources, and hydroperiod that control mangrove growth (**Figure 2**). Regulator gradients include salinity, sulfide, pH, and redox that are nonresource variables that influence mangrove growth. Resource gradients include nutrients, light, space, and other variables that are consumed and contribute to mangrove productivity. The third gradient, hydroperiod, is one of the critical characteristics of wetland landscapes that controls wetland productivity. The interactions of these three gradients have been proposed as a constraint envelope for defining the structure and productivity of mangrove wetlands based on the relative degree of stress conditions (**Figure 2**). At low levels of stress for all three environmental gradients (such as low salinity, high nutrients, and intermediate flooding), mangrove wetlands reach their maximum levels of biomass and net ecosystem productivity.

Soil nutrients are not uniformly distributed within mangrove ecosystems, resulting in multiple patterns of nutrient limitation. Along a microtidal gradient in carbonate reef islands, trees were generally N-limited in the fringe zone and P-limited in the interior or scrub zone. Fertilization studies demonstrated that not all ecological processes respond similarly to or are limited by the

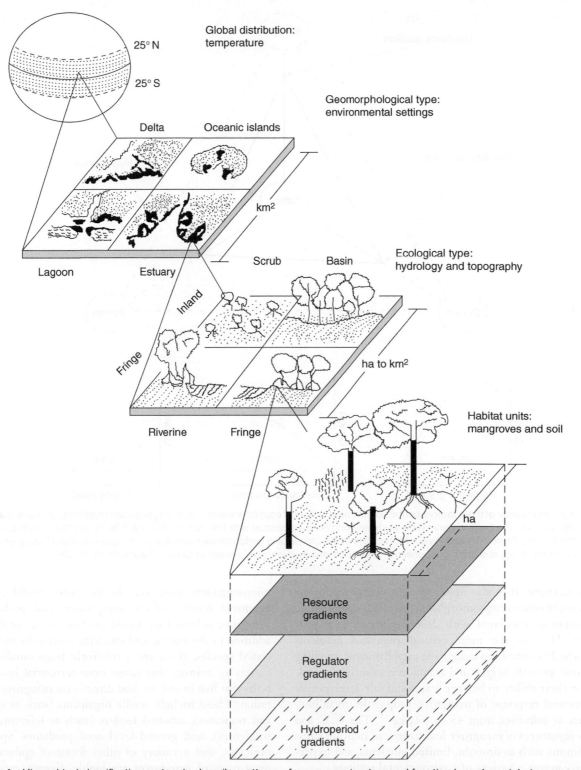

Figure 1 Hierarchical classification system to describe patterns of mangrove structure and function based on global, geomorphic (regional), and ecological (local) factors that control the concentration of nutrient resources and regulators in soil along gradients from fringe to more interior locations from shore. Modified from Twilley RR, Gottfried RR, Rivera-Monroy VH, Armijos MM, and Bodero A (1998) An approach and preliminary model of integrating ecological and economic constraints of environmental quality in the Guayas River estuary, Ecuador. *Environmental Science and Policy* 1: 271–288 and Twilley RR and Rivera-Monroy VH (2005) Developing performance measures of mangrove wetlands using simulation models of hydrology, nutrient biogeochemistry, and community dynamics. *Journal of Coastal Research* 40: 79–93.

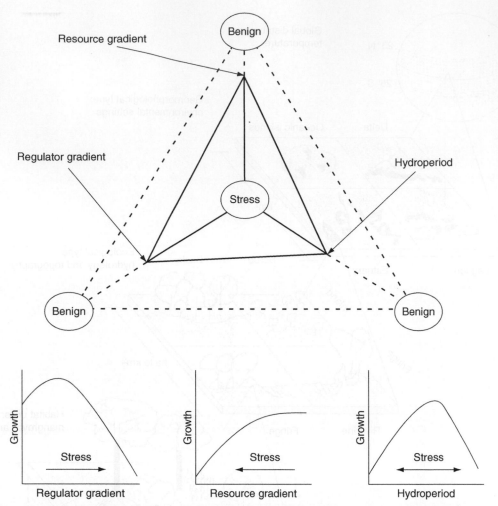

Figure 2 Interaction of three factors controlling the productivity of coastal wetlands, including regulator gradients, resource gradients, and hydroperiod. The bottom panel defines stress conditions associated with how gradients in each factor control growth of wetland vegetation. From Twilley RR and Rivera-Monroy VH (2005) Developing performance measures of mangrove wetlands using simulation models of hydrology, nutrient biogeochemistry, and community dynamics. *Journal of Coastal Research* 40: 79–93.

same nutrient. It is also apparent that mangrove forests growing in other ecogeomorphic settings are also prone to P-limitation associated with different geophysical processes. One of the most critical regulator gradients (**Figure 2**) controlling mangrove establishment, seedling survival, growth, height, and zonation is salinity, depending on their ability to balance water and salt. Interspecific differential response of mangrove propagules to salinity occurs at salinities from 45 to 60 g kg^{-1}. The δ^{13}C and δ^{15}N signatures of mangrove leaf tissue can indicate stress conditions such as drought, limited nutrients, and hypersalinity across a variety of environmental settings.

Biodiversity

Mangrove ecosystems support a variety of marine and estuarine food webs involving an extraordinarily large number of animal species and complex heterotrophic

microorganism food web. In the New World tropics, extensive surveys of the composition and ecology of mangrove nekton have found 26–114 species of fish. In addition to the marine and estuarine food webs and associated species, there are a relatively large number and variety of animals that range from terrestrial insects to birds that live in and/or feed directly on mangrove vegetation. These include sessile organisms (such as oysters and tunicates), arboreal feeders (such as foliovores and frugivores), and ground-level seed predators. Sponges, tunicates, and a variety of other forms of epibionts on prop roots of mangroves are highly diverse, especially along mangrove shorelines with little terrigenous input. Over 200 species of insects have been documented in mangroves in the Florida Keys, similar to the richness of insects and faunal biota observed in other parts of the Caribbean. One of the most published links between mangrove biodiversity and ecosystem function may be the presence of crabs in mangrove wetlands. Crabs can influence forest structure, litter dynamics, and nutrient cycling of mangrove wetlands, suggesting that they are a keystone guild in these forested ecosystems.

Ecosystem Processes

Succession

Succession in mangroves has often been equated with zonation, wherein 'pioneer species' would be found in the fringe zones, and zones of vegetation more landward would 'recapitulate' the successional sequence toward terrestrial communities. Zonation in mangrove communities has variously been accounted for by a number of biological factors, including salinity tolerance of individual species, seedling dispersal patterns resulting from different sizes of mangrove propagules, differential consumption by grapsid crabs and other consumers, and interspecific competition. Snedaker proposed the establishment of stable monospecific zones wherein each species is best adapted to flourish due to the interaction of physiological tolerances of species with environmental conditions. Geological surveys of the intertidal zone of Tabasco, Mexico, demonstrated that the zonation and structure of mangrove wetlands are responsive to eustatic changes in sea level, and that mangrove zones can be viewed as steady-state zones migrating toward or away from the sea, depending on its level. Thus, both monospecific and mixed vegetation zones of mangrove wetlands represent steady-state adjustments rather than successional stages. Many models of mangrove succession are based on how gap dynamics influence spatial patches of community dynamics across the landscape.

Productivity and Litter Dynamics

Tree height and aboveground biomass of mangrove wetlands throughout the tropics decrease at higher latitudes, indicating the constraint of climate on forest development in the subtropical climates. In addition, mangrove biomass can vary dramatically within any given latitude, an indication that local effects of regulators, resources, or hydroperiod may significantly limit the potential for forest development at all latitudes. The primary productivity of mangroves is most often evaluated by measuring the rate of litter fall, as recorded for other forested wetlands. Regional rates in litter production in mangroves are a function of water turnover within the forest, and rank among the ecological types is as follows: riverine > fringe > basin > scrub.

The dynamics of mangrove litter, including productivity, decomposition, and export, can determine the coupling of mangroves to the secondary productivity and biogeochemistry of coastal ecosystems. Patterns of leaf-litter turnover have been proposed to vary among ecological types of mangroves with greater litter export in sites with increasing tidal inundation (riverine > fringe > basin). However, several studies in the Old World tropics in higher-energy coastal environments of Australia and Malaysia have emphasized the influence of crabs on the fate of mangrove leaf litter, rather than geophysical processes. In these coastal environments, crabs consume 28–79% of the annual leaf fall. A similar biological factor was observed in the neotropics where the crab *Ucides occidentalis* in the Guayas River estuary (Ecuador) processed leaf litter at similar rates observed in Old World tropics. Differences in litter turnover rates among mangrove wetlands are a combination of species-specific degradation rates, hydrology (tidal frequency), soil fertility, and biological factors such as crabs.

Nutrient Biogeochemistry

The nutrient biogeochemistry of mangrove wetlands as either a nutrient source or sink depends on the process of material exchange at the interface between mangrove wetlands and the estuary, which is largely controlled by tides (tidal exchange, TE, in **Figure 3**). Nutrient exchanges may occur either with coastal waters (TE) or with the atmosphere (atmosphere exchange, AE), depending on whether the nutrient has a gas phase or not (**Figure 3**). Substantial amounts of carbon and nitrogen can exchange with the atmosphere, resulting in very complex mechanisms both at the interface with coastal waters and with the atmosphere that influence the mass balance of these nutrients. In addition, there are internal processes, including root uptake (UT), retranslocation (RT) in the canopy, litter fall (LF), regeneration (RG), immobilization (IM), and sedimentation (SD) (**Figure 3**). The balance of these nutrient flows will determine the exchanges across the wetland boundary.

There are very few comprehensive budgets of carbon, nitrogen, or phosphorus for mangrove ecosystems. Mangrove sediments have a high potential in the removal of N from surface waters, yet estimates of denitrification have a large range from a low of $0.53\,\mu mol\,m^{-2}\,h^{-1}$, to $9.7–261\,\mu mol\,N\,m^{-2}\,h^{-1}$ in mangrove forests receiving effluents from sewage treatment plants. Small amendments of $^{15}NO_3$ followed by direct measures of $^{15}N_2$ production have shown that denitrification accounts for <10% of the applied isotope suggesting that NO_3 is accumulated in the litter via immobilization on the forest floor rather than a sink to the atmosphere. The other nutrient sink in mangrove wetlands is the burial of nitrogen and phosphorus associated with sedimentation. A survey of sedimentation and nutrient accumulation among five sites in south Florida and Mexico indicates patterns associated with the ecological types of mangroves, with rates of about $5.5\,g\,m^{-2}\,yr^{-1}$. This rate is higher than nitrogen loss via denitrification, indicating the significance of burial as nitrogen sink in mangrove ecosystems. Intrasystem nutrient recycling mechanisms in the

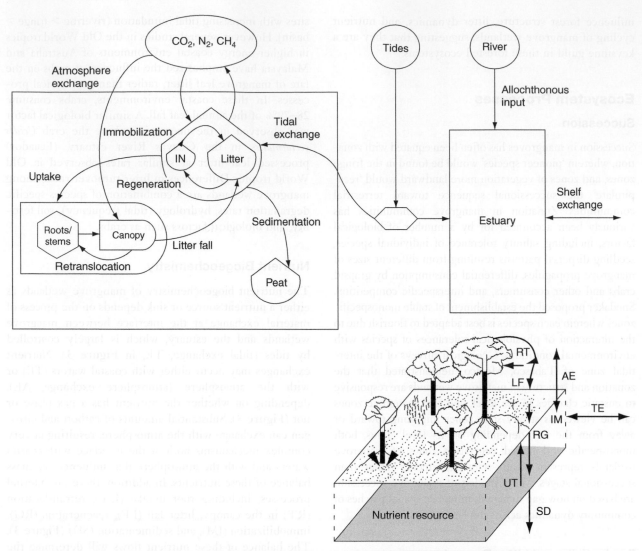

Figure 3 Upper panel: Schematic of the various fluxes of organic matter and nutrients in a mangrove ecosystem, including exchange with the estuary (IN = inorganic nutrients). Lower panel: A diagram of a mangrove wetland with soil nutrient resources describing the various processes associated with intrasystem cycling and exchange. From Twilley RR (1997) Mangrove wetlands. In: Messina M and Connor W b0665 (eds.) *Southern Forested Wetlands: Ecology and Management*, pp. 445–473. Boca Raton, FL: CRC Press.

canopy may be a site of nitrogen conservation in mangroves and, together with leaf longevity, could influence the nitrogen demand of these ecosystems. The significance of this ecological process to the nutrient budget of different mangrove wetlands has not been determined.

Surveys of nitrogen exchange demonstrate some of the principles of determining the function of mangrove wetlands as a nutrient sink. The largest nitrogen flux of nitrogen from sites in Mexico and Australia is export of particulate nitrogen, consistent with organic carbon representing the largest flux from most mangroves (**Figure 3**). Compared to other flux studies of mangroves, there seems to be a pattern of net inorganic fluxes into the wetlands and corresponding flux of organic nutrients out. The best summary may be that mangrove wetlands transform the tidal

import of inorganic nutrients into organic nutrients that are then exported to coastal waters. Carbon export from mangrove ecosystems ranges from 1.86 to $401\,g\,Cm^{-2}\,yr^{-1}$, with an average rate of about $210\,g\,Cm^{-2}\,yr^{-1}$. Carbon export from mangrove wetlands is nearly double the rate of average carbon export from salt marshes, which may be associated with the more buoyant mangrove leaf litter, higher precipitation in tropical wetlands, and greater tidal amplitude in mangrove systems studied.

Mangrove Food Webs

The function of mangrove wetlands as a source of habitat and food to estuarine-dependent fisheries is one of the most celebrated values of forested wetlands. There are several excellent reviews that describe the secondary productivity of tropical mangrove ecosystems. The original 'outwelling hypothesis' of mangroves has been revised from the original paradigms based on comparisons among different mangrove estuaries using natural isotope abundance to trace mangrove organic matter through estuarine food chains. There are seasonal and spatial differences in the amount of mangrove detritus that can be measured in shrimp and fish that inhabit mangrove estuaries. If the distance from the source of mangrove detritus increases, the proportion of carbon in the tissue of shrimp from mangrove detritus decreases as the signal of carbon phytoplankton increases. The seasonal timing of mangrove export of detritus relative to the migration of estuarine-dependent fisheries may also dilute the contribution of mangrove detritus from the food webs among diverse sites. The migratory nature of many of the nekton communities and the seasonal pulsing of both organic detritus input and *in situ* productivity result in very complex linkages of mangroves with estuarine-dependent fisheries. In addition, mangrove detritus low in nitrogen relative to carbon may be modified by the microbial community and then utilized by higher trophic levels, masking the direct utilization of this organic matter as an energy source.

Impacts of Environmental Change

Mangroves are arguably an excellent indicator of how ecosystems will respond to the manifold impacts of global environmental change and land-use disturbance. Given present patterns, the combined effects of climate and land-use change will be noticeably evident in reduced goods and services of mangroves to human systems throughout the tropics in the twenty-first century. For example, accelerated rates in sea-level rise have been speculated as the most critical environmental change affecting the continued existence of mangrove ecosystems. Numerous processes contribute to vertical accretion of mangroves at a rate that balances the increase in regional sea-level rise. Critical rates in sea-level rise have been estimated above which there is a projected collapse of mangrove ecosystems. While some speculation suggests that mangroves cannot sustain existence at sea-level rise >1.2–2.3 mm yr^{-1}, there is evidence that mangroves located in particular environmental settings existed through periods of accelerated sea level rise. Mangroves in Australia can keep pace with changes in sea-level rise with rates ranging from 0.2 to 6 mm yr^{-1} in the south Alligator tidal river. Also, mangrove forests in many estuaries in northern Australia tolerated sea-level rise of 8–10 mm yr^{-1} in the early Holocene. Many of these mangroves receive terrigenous sediments and exist in macrotidal environments, with critical rates that are

much different than for mangroves in microtidal and carbonate environments. In addition, mangrove areas can be sustained along the coastline by migrating inland under conditions of increased sea-level rise. But this inland migration will depend on whether suitable inshore landscapes are available. The most significant recent restriction to mangrove colonization is human land use of available landscapes.

Mangroves in many coastal regions such as Gulf of Mexico and Caribbean are distributed in latitudes where the frequency of hurricanes and cyclones is high, resulting in strong effect on mangrove forest structure and community dynamics. Several patterns have been observed in Florida, Puerto Rico, Mauritius, and British Honduras. Species attributes and availability of propagules are important factors along with the severity of storm and sediment disturbance in projecting recovery patterns. Frequent storm disturbance tends to favor species capable of constant or timely flowering, abundant seedling or sprouting, fast growth in open conditions, and early reproductive maturity. Woody debris resulting from these disturbances have an important role in biogeochemical properties of disturbed mangrove forests. Although mangrove trees show these 'traits', it is important to consider the cumulative impact of human activities on these ecosystems in conjunction with the complex natural cycle of regeneration and growth of mangrove forests. Cyclonic disturbance in areas with higher rates of sea-level rise has been demonstrated to cause sediment collapse (drop in surface elevation) that reduces the ability of mangroves to recolonize disturbed areas. Yet this potential impact may vary across ecogeomorphic types of mangroves.

River (and surface runoff) diversions that deprive tropical coastal deltas of freshwater and silt result in losses of mangrove species diversity and organic production, and alter the terrestrial and aquatic food webs that mangrove ecosystems support. Freshwater diversion of the Indus River to agriculture in Sind Province over the last several hundred years has reduced the once species-rich Indus River delta to a sparse community dominated by *Avicennia marina*. It is also responsible for causing significant erosion of the seafront due to sediment starvation and the silting-in of the abandoned spill rivers. A similar phenomenon has been observed in southwestern Bangladesh following natural changes in river channels of the Ganges and the construction of the Farakka barrage that reduced the dry season flow of freshwater into the mangrove-dominated western Sundarbans. Freshwater starvation, both natural and human-induced, has had negative impacts on the biodiversity of mangroves in the Ganges River delta as well along the dry coastal life zone of Colombia (the Ciénaga Grande de Santa Marta lagoon).

Deforestation of mangrove wetlands is associated with many uses of coastal environments, including urban, agriculture, and aquaculture reclamation, as well as the use of forest timber for furniture, energy, chip wood, and construction materials. Two reclamation activities that have contributed to examples of massive mangrove deforestation are agriculture and aquaculture enterprises. Agriculture impacts on mangroves are most noted in West Africa and parts of Indonesia. Many of the large agricultural uses are found in humid coastal areas or deltas where freshwater is abundant and intertidal lands are seasonally available for crop production. Mariculture use of the tropical intertidal zones, in the construction and operation of shrimp ponds, has become one of the most significant environmental changes of mangrove wetlands and water quality of tropical estuaries in the last several decades.

Oil spills represent contaminants to mangroves that can alter the succession, productivity, and nutrient cycling of these coastal forested wetlands. These impacts have been well documented in ecological studies in Puerto Rico, Panama, and Gulf of Mexico. An oil slick in a mangrove wetland will cause a certain mortality of trees depending on the concentration of hydrocarbons and species of trees, as well as the edaphic stress levels already existing at the site. Thus, those mangroves in dry coastal environments may be more vulnerable to oil spills than those in more humid environments.

Management and Restoration

Mangroves produce a variety of forest products, support the productivity of economically important estuarine-dependent fisheries, and modify the water quality in warm-temperate and tropical estuarine ecosystems. These goods and services lead to increased human utilization of mangrove resources that vary throughout the tropics depending on economic and cultural constraints (**Figure 4**). Economic constraints are usually in the form of available capital to fund land-use changes in coastal regions, as well as river-basin development. Cultural constraints are complex and determine the degree of environmental management and natural resource utilization. However, the sustainable utilization of coastal resources, to a large degree, is controlled by these two social conditions of a region. Human use and value of mangrove wetlands are therefore a combination of both the ecological properties of these coastal ecosystems together with patterns of social exploitation. Therefore, any best management plan designed to provide for the sustainable utilization of mangrove wetlands has to consider both the ecological and social constraints of the region. Humans are part of all ecosystems, and management of natural resources is a combination of policies that seek to regulate the actions of societies within limitations

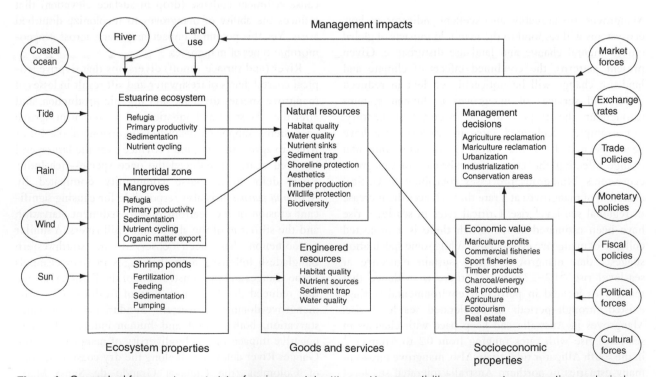

Figure 4 Conceptual framework constraints of environmental setting and human activities on ecosystem properties, ecological functions, and uses of mangrove ecosystems that determine management decisions in coastal environments. From Twilley RR, Gottfried RR, Rivera-Monroy VH, Armijos MM, and Bodero A (1998) An approach and preliminary model of integrating ecological and economic constraints of environmental quality in the Guayas River estuary, Ecuador. *Environmental Science and Policy* 1: 271–288.

that are imposed by the environment. Recent emphasis has been placed on comprehensive ecosystem restoration programs that represent changes in management of landscapes to reduce impacts on natural processes that enhance system recovery.

There have been several reviews of mangrove restoration, which collectively have alluded to the concept that since these forested wetlands are adapted to stressed environments, they are relatively amenable to restoration efforts. The success of mangrove restoration is the establishment of the proper environmental settings that control the characteristic structure and function of mangrove wetlands. The goal of ecological restoration is to return a degraded mangrove site back to either the natural condition (restoration) or to some other new condition (rehabilitation). The rates of change in the ecological characteristics of mangrove wetlands between natural, degraded, and some rehabilitated condition will depend on the type of environmental impact, the magnitude of the impact, and the ecogeomorphic type of mangrove wetland that is impacted. The success of any mangrove restoration project depends on the establishment of proper site conditions (geophysical processes and geomorphic features) along with ecological processes of the site such as the availability of propagules and the recruitment of these individuals to sapling stage of development. Some of the key parameters of a restoration project include the elevation of the landscape to provide the proper hydrology of the site, recognizing the significance of natural processes to sustaining the restored condition, and proper planting techniques to enhance recruitment. Several models of different properties of mangroves have been developed during the last decade to help facilitate planning and design of mangrove restoration projects and improve our management of these critical features of coastal landscape.

See also: Lagoons; Mediterranean.

Further Reading

Alexander TR (1967) Effect of Hurricane Betsy on the southeastern Everglades. *Quarterly Journal of the Florida Academy of Sciences* 30: 10–24.

Allen JA, Ewel KC, Keeland BD, Tara T, and Smith TJ (2000) Downed wood in Micronesian mangrove forests. *Wetlands* 20: 169–176.

Alongi DM, Christoffersen P, Tirendi F, and Robertson AI (1992) The influence of freshwater and material export on sedimentary facies and benthic processes within the Fly Delta and adjacent Gulf of Papua (Papua New Guinea). *Continental Shelf Research* 12: 287–326.

Bacon PR (1990) The ecology and management of swamp forests in the Guianas and Caribbean region. In: Lugo AE, Brinson M, and Brown S (eds.) *Ecosystems of the World 15: Forested Wetlands*, pp. 213–250. Amsterdam: Elsevier Press.

Bacon PR (1994) Template for evaluation of impacts of sea level rise on Caribbean coastal wetlands. *Ecological Engineering* 3: 171–186.

Baldwin A, Egnotovich M, Ford M, and Platt W (2001) Regeneration in fringe mangrove forests damaged by Hurricane Andrew. *Plant Ecology* 157: 151–164.

Ball MC (1980) Patterns of secondary succession in a mangrove forest of southern Florida. *Oecologia* 44: 226–235.

Ball MC (1988) Ecophysiology of mangroves. *Trees* 2: 129–142.

Berger U and Hildenbrandt H (2000) A new approach to spatially explicit modelling of forest dynamics: Spacing, ageing and neighborhood competition of mangrove trees. *Ecological Modelling* 132: 287–302.

Blasco F (1984) Climatic factors and the biology of mangrove plants. In: Snedaker SC and Snedaker JG (eds.) *The Mangrove Ecosystem: Research Methods*, pp. 18–35. Paris: UNESCO.

Botero L (1990) Massive mangrove mortality on the Caribbean coast of Colombia. *Vida Silvestre Neotropical* 2: 77–78.

Boto KG, Saffingna P, and Clough B (1985) Role of nitrate in nitrogen nutrition of the mangrove *Avicennia Marina*. *Marine Ecology Progress Series* 21: 259–265.

Boto KG and Wellington JT (1988) Seasonal variations in concentrations and fluxes of dissolved organic and inorganic materials in a tropical, tidally-dominated, mangrove waterway. *Marine Ecology Progress Series* 50: 151–160.

Brown S and Lugo AE (1994) Rehabilitation of tropical lands: A key to sustaining development. *Restoration Ecology* 2: 97–111.

Camilleri JC (1992) Leaf-litter processing by invertebrates in a mangrove forest in Queensland. *Marine Biology* 114: 139–145.

Carlton JM (1974) Land-building and stabilization by mangroves. *Environmental Conservation* 1: 285.

Chapman VJ (1976) *Mangrove Vegetation*. Vaduz, Germany: J. Cramer.

Chen R and Twilley RR (1998) A gap dynamic model of mangrove forest development along gradients of soil salinity and nutrient resources. *Journal of Ecology* 86: 1–12.

Chen R and Twilley RR (1998) A simulation model of organic matter and nutrient accumulation in mangrove wetland soils. *Biogeochemistry* 44: 93–118.

Chen R and Twilley RR (1999) Patterns of mangrove forest structure and soil nutrient dynamics along the Shark River Estuary, Florida. *Estuaries* 22: 955–970.

Cintrón G (1990) Restoration of mangrove systems. *Symposium on Habitat Restoration*. Washington, DC: National Oceanic and Atmospheric Administration.

Cintrón G, Lugo AE, Martinez R, Cintrón BB, and Encarnacion L (1981) Impact of oil in the tropical marine environment, pp. 18–27. Technical Publication. Division of Marine Resources, Department of Natural Resources of Puerto Rico.

Corredor JE and Morell MJ (1994) Nitrate depuration of secondary sewage effluents in mangrove sediments. *Estuaries* 17: 295–300.

Craighead FC and Gilbert VC (1962) The effects of Hurricane Donna on the vegetation of southern Florida. *Quarterly Journal of the Florida Academy of Sciences* 25: 1–28.

Davis JH (1940) The ecology and geologic role of mangroves in Florida. In: *Carnegie Institution, Publication No. 517*, pp. 303–412. Washington, DC: Carneigie Institution.

Davis S, Childers DL, Day JWJ, Rudnick D, and Sklar F (2001) Wetland-water column exchanges of carbon, nitrogen, and phosphorus in a southern everglades dwarf mangrove. *Estuaries* 24: 610–622.

Davis S, Childers DL, Day JW, Rudnick D, and Sklar F (2003) Factors affecting the concentration and flux of materials in two southern everglades mangrove wetlands. *Marine Ecology Progress Series* 253: 85–96.

Duke NC (2001) Gap creation and regenerative process driving diversity and structure of mangrove ecosystems. *Wetlands Ecology and Management* 9: 257–269.

Duke NC and Pinzon Z (1993) Mangrove forests. In: Keller BD and Jackson JBC (eds.) *Long-Term Assessment of the Oil Spill at Bahia las Minas, Panama, Synthesis Report, Volumn II, Technical Report*, pp. 447–553. New Orleans, LA: US Dept of the Interior, Minerals Management Service, Gulf of Mexico OCS Regional Office.

Ellison JC (1993) Mangrove retreat with rising sea-level Bermuda. *Estuarine, Coastal and Shelf Science* 37: 75–87.

Ellison AM (2000) Mangrove restoration: Do we know enough? *Restoration Ecology* 8: 219–229.

Ellison AM and Farnsworth EJ (1992) The ecology of Belizean mangrove-root fouling communities: Patterns of epibiont distribution and abundance, and effects on root growth. *Hydrobiologia* 20: 1–12.

Ellison JC and Stoddart DR (1991) Mangrove ecosystem collapse during predicted sea-level rise: Holocene analogues and implications. *Journal of Coastal Research* 7: 151–165.

Ewe SML, Gaiser EE, Childers DL, *et al.* (2006) Spatial and temporal patterns of aboveground net primary productivity (ANPP) in the Florida Coastal Everglades. *Hydrobiologia* 569: 459–474.

Ewel KC, Ong JE, and Twilley R (1998) Different kinds of mangrove swamps provide different goods and services. *Global Ecology and Biogeography Letters* 7: 83–94.

Ewel KC, Zheng SF, Pinzon ZS, and Bourgeois JA (1998) Environmental effects of canopy gap formation in high-rainfall mangrove forests. *Biotropica* 30: 510–518.

Farnsworth EJ and Ellison AM (1991) Patterns of herbivory in Belizean mangrove swamps. *Biotropica* 23: 555–567.

Farnsworth EJ and Ellison AM (1993) Dynamics of herbivory in Belizean mangal. *Journal of Tropical Ecology* 9: 435–453.

Farquhar GD, Ball MC, von Caemmerer S, and Roksandic Z (1982) Effect of salinity and humidity on $\delta^{13}C$ values of halophytes – evidence for diffusional isotope fractionation determined by the ratios of intercellular/atmospheric CO_2 under different environmental conditions. *Oecologia (Berlin)* 52: 121–137.

Fell JW and Master IM (1973) Fungi associated with the degradation of mangrove (*Rhizophora mangle* L.) leaves in south Florida. In: Stevenson LH and Colwell RR (eds.) *Estuarine Microbial Ecology*, pp. 455–465. Columbia, SC: University of South Carolina Press.

Feller IC (1993) *Effects of Nutrient Enrichment on Growth and Herbivory of Dwarf Red mangrove*. PhD Dissertation, Georgetown University.

Feller IC (1995) Effects of nutrient enrichment on growth and herbivory of dwarf red mangrove (*Rhizophora mangle*). *Ecological Monographs* 65: 477–505.

Feller IC and McKee KL (1999) Small gap creation in Belizean mangrove forests by a wood-boring insect. *Biotropica* 31: 607–617.

Feller IC, Whigham DF, McKee KL, and Lovelock CE (2003) Nitrogen limitation of growth and nutrient dynamics in a disturbed mangrove forest, Indian River Lagoon, Florida. *Oecologia* 134: 405–414.

Feller IC, Whigham DF, O'Neill JP, and McKee KL (1999) Effects of nutrient enrichment on within-stand cycling in a mangrove forest. *Ecology* 80: 2193–2205.

Field CD (1996) Restoration of mangrove ecosystems. In: *International Society for Mangrove Ecosystems*. Hong Kong: South China Printing.

Fry B, Bern AL, Ross MS, and Meeder JF (2000) $\delta^{15}N$ studies of nitrogen use by the red mangrove, *Rhizophora mangle* L., in south Florida. *Estuarine, Coastal and Shelf Science* 50: 723–735.

Fry B and Smith TJ, III (2002) Stable isotope studies of red mangroves and filter feeders from the Shark River estuary, Florida. *Bulletin of Marine Sciences* 70: 871–890.

Garrity SD, Levings SC, and Burns KA (1994) The Galeta oil spill. I. Long-term effects on the physical structure of the mangrove fringe. *Estuarine, Coastal and Shelf Science* 38: 327–348.

Getter CD, Scott GI, and Michel J (1981) The effects of oil spills on mangrove forests: A comparison of five oil spill sites in the Gulf of Mexico and the Caribbean Sea. *Proceedings of the 1981 Oil Spill Conference*, pp. 65–11. Washington, DC: API/EPA/USCG.

Gilmore RG, Jr. and Snedaker SC (1993) Mangrove forests. In: Martin WH, Boyce SG, and Echternacht AC (eds.) *Biodiversity of the Southeastern United States/Lowland Terrestrial Communities*, pp. 165–198. New York: Wiley.

Glynn PW, Almodovar LR, and Gonzalez JG (1964) Effects of hurricane Edith on marine life in La Parguera, Pueto Rico. *Caribbean Journal of Science* 4: 335–345.

Gosselink JG and Turner RE (1978) The role of hydrology in freshwater wetland ecosystems. In: Good DFWRE and Simpson RL (eds.) *Freshwater Wetlands: Ecological Processes and Management Potential*, pp. 633–678. New York: Academic Press.

Hedgpeth JW (1957) Classification of marine environments. *Geological Society of America, Memoir* 67(1): 17–28.

Huston MA (1994) *Biological Diversity*. Cambridge: Cambridge University Press.

Iizumi H (1986) Soil nutrient dynamics. In: Cragg S and Polunin N (eds.) *Workshop on Mangrove Ecosystem Dynamics*, p.171. New Delhi: UNDP/UNESCO Regional Project (RAS/79/002).

Jones DA (1984) Crabs of the mangal ecosystem. In: Por FD and Dor I (eds.) *Hydrobiology of the Mangal*, pp. 89–109. The Hague: Dr. W. Junk Publishers.

Koch MS and Snedaker SC (1997) Factors influencing *Rhizophora mangle* L. seedlings development into the sapling stage across resource and stress gradients in subtropical Florida. *Biotropica* 29: 427–439.

Krauss KW, Allen JA, and Cahoon DR (2003) Differential rates of vertical accretion and elevation change among aerial root types in Micronesian mangrove forests. *Estuarine Coastal and Shelf Science* 56: 251–259.

Krauss KW, Doyle TW, Twilley RR, Smith TJ, Whelan KRT, and Sullivan JK (2005) Woody debris in the mangrove forests of south Florida. *Biotropica* 37: 9–15.

Kristensen E, Andersen FØ, and Kofoed LH (1988) Preliminary assessment of benthic community metabolism in a Southeast Asian mangrove swamp. *Marine Ecology Progress Series* 48: 137–145.

Lee SY (1989) Litter production and turnover of the mangrove *Kandelia candel* (L.) Druce in a Hong Kong tidal shrimp pond. *Estuarine, Coastal and Shelf Science* 29: 75–87.

Leh CMU and Sasekumar A (1985) The food of sesarmid crabs in Malaysian mangrove forests. *Malay Naturalist Journal* 39: 135–145.

Lewis RR (1982) Mangrove forests. In: Lewis RR (ed.) *Creation and Restoration of Coastal Plant Communities*, pp. 153–171. Boca Raton, FL: CRC Press.

Lewis RR (1990) Creation and restoration of coastal plain wetlands in Florida. In: Kusler JA and Kentula ME (eds.) *Wetland Creation and Restoration*, pp. 73–101. Washington, DC: Island Press.

Lewis RR (1990) Creation and restoration of coastal wetlands in Puerto Rico and the US Virgin Islands. In: Kusler JA and Kentula ME (eds.) *Wetland Creation and Restoration*, pp. 103–123. Washington, DC: Island Press.

Lin G and Sternberg LSL (1992) Differences in morphology, carbon isotope ratios, and photosynthesis between scrub and fringe mangroves in Florida, USA. *Aquatic Botany* 42: 303–313.

Lin G and Sternberg LSL (1992) Effect of growth form, salinity, nutrient and sulfide on photosynthesis, carbon isotope discrimination and growth of red mangrove (*Rhizophora mangle* L.). *Australian Journal of Plant Physiology* 19: 509–517.

Lovelock CE, Feller IC, Mckee KL, Engelbrecht BMJ, and Ball MC (2004) The effect of nutrient enrichment on growth, photosynthesis and hydraulic conductance of dwarf mangroves in Panama. *Functional Ecology* 18: 25–33.

Lugo AE (1980) Mangrove ecosystems: Successional or steady state? *Biotropica* 12: 65–72.

Lugo AE (1998) Mangrove forests: A tough system to invade but an easy one to rehabilitate. *Marine Pollution Bulletin* 37: 427–430.

Lugo AE and Snedaker SC (1974) The ecology of mangroves. *Annual Review of Ecology and Systematics* 5: 39–64.

Lynch JC, Meriwether JR, McKee BA, Vera-Herrera F, and Twilley RR (1989) Recent accretion in mangrove ecosystems based on ^{137}Cs and [210]Pb. *Estuaries* 12: 284–299.

Macnae W (1968) A general account of the fauna and flora of mangrove swamps and forests in the Indo-West-Pacific region. *Advances in Marine Biology* 6: 73–270.

Malley DF (1978) Degradation of mangrove leaf litter by the tropical sesarmid crab *Chiromanthes onychophorum*. *Marine Biology* 49: 377–386.

McKee KL (1993) Soil physicochemical patterns and mangrove species distribution – Reciprocal effects? *Journal of Ecology* 81: 477–487.

McKee KL, Feller IC, Popp M, and Wanek W (2002) Mangrove isotopic ($\delta^{15}N$ and $\delta^{13}C$) fractionation across a nitrogen vs. phosphorus limitation gradient. *Ecology* 83: 1065–1075.

Medina E and Francisco M (1997) Osmolality and $\delta^{13}C$ of leaf tissues of mangrove species from environments of contrasting rainfall and salinity. *Estuarine, Coastal and Shelf Science* 45: 337–344.

Naidoo G (1985) Effects of waterlogging and salinity on plant–water relations and on the accumulation of solutes in three mangrove species. *Aquatic Botany* 22: 133–143.

Nedwell DB (1975) Inorganic nitrogen metabolism in a eutrophicated tropical mangrove estuary. *Water Research* 9: 221–231.

Nixon SW (1980) Between coastal marshes and coastal waters – A review of twenty years of speculation and research on the role of salt marshes in estuarine productivity and water chemistry. In: Hamilton P and MacDonald KB (eds.) *Estuarine and Wetland Processes with Emphasis on Modeling*, pp. 437–525. New York: Plenum Press.

Odum WE and Heald EJ (1972) Trophic analysis of an estuarine mangrove community. *Bulletin Marine Science* 22: 671–738.

Odum WE and McIvor CC (1990) Mangroves. In: Myers RL and Ewel JJ (eds.) *Ecosystems of Florida*, pp. 517–548. Orlando, FL: University of Central Florida Press.

Odum WE, McIvor CC, and Smith TJ (1982) *The Ecology of the Mangroves of South Florida: A Community Profile. FWS/OBS-81/24.* Washington, DC: US Fish and Wildlife Service, Office of Biological Resources.

Parkinson RW, DeLaune RD, and White JR (1994) Holocene sea-level rise and the fate of mangrove forests within the wider Caribbean region. *Journal of Coastal Research* 10: 1077–1086.

Pinzon ZS, Ewel KC, and Putz FE (2003) Gap formation and forest regeneration in a Micronesian mangrove forest. *Journal of Tropical Ecology* 19: 143–153.

Ponnamperuma FN (1984) Mangrove swamps in south and Southeast Asia as potential rice lands. In: Soepadmo E, Rao AN, and McIntosh DJ (eds.) *Proceedings Asian Mangrove Symposium*, pp. 672–683. Kuala Lumpur: University of Malaya.

Pool DJ, Lugo AE, and Snedaker SC (1975) Litter production in mangrove forests of southern Florida and Puerto Rico. In: Walsh G, Snedaker S, and Teas H (eds.) *Proceedings of the International Symposium on the Biology and Management of Mangroves*, pp. 213–237. Gainesville, FL: Institute of Food and Agricultural Sciences. University of Florida.

Rabinowitz D (1978) Early growth of mangrove seedlings in Panama, and an hypothesis concerning the relationship of dispersal and zonation. *Journal of Biogeography* 5: 113–133.

Rivera-Monroy VH, Day JW, Twilley RR, Vera-Herrera F, and Coronado-Molina C (1995) Flux of nitrogen and sediment in a fringe mangrove forest in Terminos Lagoon, Mexico. *Estuarine, Coastal and Shelf Science* 40: 139–160.

Rivera-Monroy VH and Twilley RR (1996) The relative role of denitrification and immobilization in the fate of inorganic nitrogen in mangrove sediments. *Limnology and Oceanography* 41: 284–296.

Rivera-Monroy VH, Twilley RR, Boustany RG, Day JW, Vera-Herrera F, and Ramirez MdC (1995) Direct denitrification in mangrove sediments in Terminos Lagoon, Mexico. *Marine Ecology Progress Series* 97: 97–109.

Rivera-Monroy VH, Twilley RR, Bone D, *et al.* (2004) A conceptual framework to develop long-term ecological research and management objectives in the wider Caribbean region. *Bioscience* 54: 843–856.

Robertson AI (1986) Leaf-burying crabs: Their influence on energy flow and export from mixed mangrove forests (*Rhizophora* spp.) in northeastern Australia. *Journal of Experimental Marine Biology and Ecology* 102: 237–248.

Robertson AI and Alongi DM (1992) *Tropical Mangrove Ecosystems*, vol. 41. Washington, DC: American Geophysical Union.

Robertson AI, Alongi DM, and Boto KG (1992) Food chains and carbon fluxes. In: Robertson AI and Alongi DM (eds.) *Tropical Mangrove Ecosystems*, pp. 293–326. Washington, DC: American Geophysical Union.

Robertson AI and Blaber SJM (1992) Plankton, epibenthos and fish communities. In: Robertson AI and Alongi DM (eds.) *Tropical Mangrove Ecosystems*, pp. 173–224. Washington, DC: American Geophysical Union.

Robertson AI and Daniel PA (1989) The influence of crabs on litter processing in high intertidal mangrove forests in tropical Australia. *Oecologia* 78: 191–198.

Robertson AI and Duke NC (1990) Mangrove fish-communities in tropical Queensland, Australia: Spatial and temporal patterns in densities, biomass and community structure. *Marine Biology* 104: 369–379.

Rodelli MR, Gearing JN, Gearing PJ, Marshall N, and Sasekumar A (1984) Stable isotope ratio as a tracer of mangrove carbon in Malaysian ecosystems. *Oecologia* 61: 326–333.

Rojas-Galaviz JL, Yáñez-Arancibia A, Day JW, Jr., and Vera-Herrera FR (1992) Estuarine primary producers: Laguna de Terminos-a study case. In: Seeliger U (ed.) *Coastal Plant Communities of Latin America*, pp. 141–154. San Diego, CA: Academic Press.

Romero LM, Smith TJ, and Fourqurean JW (2005) Changes in mass and nutrient content of wood during decomposition in a south Florida mangrove forest. *Journal of Ecology* 93: 618–631.

Ross MS, Meeder JF, Sah JP, Ruiz LP, and Telesnicki GJ (2000) The Southeast saline Everglades revisited: 50 Years of coastal vegetation change. *Journal of Vegetation Science* 11: 101–112.

Roth LC (1992) Hurricanes and mangrove regeneration: Effects of Hurricane Juan, October 1988, on the vegetation of Isla del Venado, Bluefields, Nicaragua. *Biotropica* 24: 375–384.

Rützler K and Feller C (1988) Mangrove swamp communities. *Oceanus* 30: 16–24.

Rützler K and Feller C (1996) Caribbean mangrove swamps. *Scientific American* 274: 94–99.

Saenger P, Hegerl EJ, and Davie JDS (1983) *Global Status of Mangrove Ecosystems. Commission on Ecology Paper No. 3*, pp. 83. International Union for the Conservation of Nature (IUCN).

Saenger P and Snedaker SC (1993) Pantropical trends in mangrove above-ground biomass and annual litterfall. *Oecologia* 96: 293–299.

Sauer JD (1962) Effects of recent tropical cyclones on the coastal vegetation of Mauritius. *Journal of Ecology* 50: 275–290.

Scholander PF, Hammel HT, Hemmingsen E, and Garay W (1962) Salt balance in mangroves. *Plant Physiology* 37: 722–729.

Sherman RE, Fahey TJ, and Battles JJ (2000) Small-scale disturbance and regeneration dynamics in a neotropical mangrove forest. *Journal of Ecology* 88: 165 178.

Simberloff DS and Wilson EO (1969) Experimental zoogeography of islands: The colonization of empty islands. *Ecology* 50: 278–289.

Smith TJ, III (1987) Seed predation in relation to tree dominance and distribution in mangrove forests. *Ecology* 68: 266–273.

Smith TJ, III (1992) Forest structure. In: Robertson AI and Alongi DM (eds.) *Tropical Mangrove Ecosystems*, pp. 101–136. Washington, DC: American Geophysical Union.

Smith TJ, Boto KG, Frusher SD, and Giddins RL (1991) Keystone species and mangrove forest dynamics: The influence of burrowing by crabs on soil nutrient status and forest productivity. *Estuarine Coastal and Shelf Science* 33: 419–432.

Smith TJ, III, Robblee MB, Wanless HR, and Doyle TW (1994) Mangroves, hurricanes, and lightning strikes. *BioScience* 44: 256–262.

Snedaker S (1982) Mangrove species zonation: Why? In: Sen DN and Rajpurohit KS (eds.) *Tasks for Vegetation Science*, vol. 2, pp. 111–125. The Hague: Junk.

Snedaker SC (1986) Traditional uses of South American mangrove resources and the socio-economic effect of ecosystem changes. In: Kunstadter P, Bird ECF, and Sabhasri S (eds.) *Proceedings, Workshop on Man in the Mangroves*, pp. 104–112. Tokyo: United Nations University.

Snedaker SC (1989) Overview of ecology of mangroves and information needs for Florida Bay. *Bulletin of Marine Science* 44: 341–347.

Snedaker SC, Meeder JF, Ross MS, and Ford RG (1994) Discussion of Ellison, JC and Stoddart, DR 1991. Mangrove ecosystem collapse during predicted sea-level rise: Holocene analogues and implications. *Journal of Coastal Research* 7: 151–165, *Journal of Coastal Research* 10: 497–498.

Snedaker SC and Snedaker JG (1984) *The Mangrove Ecosystem: Research Methods*. London: UNESCO.

Sousa WP, Quek SP, and Mitchell BJ (2003) Regeneration of Rhizophora mangle in a Caribbean mangrove forest: Interacting effects of canopy disturbance and a stem-boring beetle. *Oecologia* 137: 436–445.

Sutherland JP (1980) Dynamics of the epibenthic community on roots of the mangrove Rhizophora mangle, at Bahia de Buche, Venezuela. *Marine Biology* 58: 75–84.

Teas HJ (1981) Restoration of mangrove ecosystems. In: Carey RC, Markovits PS, and Kirkwood JB (eds.) *Proceedings of Workshop on Coastal Ecosystems of the Southeastern United States*, pp. 95–103. Reno, NV: US Fish and Wildlife Service, Office of Biological Services, FWS/OBS-80/59.

Thayer GW, Colby DR, and Hettler WF, Jr. (1987) Utilization of the red mangrove prop root habitat by fishes in south Florida. *Marine Ecology Progress Series* 35: 25–38.

Thom B (1967) Mangrove ecology and deltaic morphology: Tabasco, Mexico. *Journal of Ecology* 55: 301–343.

Thom BG (1982) Mangrove ecology – A geomorphological perspective. In: Clough BF (ed.) *Mangrove Ecosystems in Australia*, pp. 3–17. Canberra: Australian National University Press.

Thom BG (1984) Coastal landforms and geomorphic processes. In: Sneadeker SC and Sneadeker JG (eds.) *The Mangrove Ecosystem: Research Methods*, pp. 3–17. Paris: UNESCO.

Tilman D (1982) *Resource Competition*. Princeton, NJ: Princeton University Press.

Tomlinson PB (1995) *The Botany of Mangroves*. New York: Cambridge University Press.

Twilley RR (1988) Coupling of mangroves to the productivity of estuarine and coastal waters. In: Jansson BO (ed.) *Coastal-Offshore Ecosystems: Interactions*, pp. 155–180. Berlin: Springer.

Twilley RR (1995) Properties of mangroves ecosystems and their relation to the energy signature of coastal environments. In: Hall CAS (ed.) *Maximum Power*, pp. 43–62. Denver, CO: Colorado Press.

Twilley RR (1997) Mangrove wetlands. In: Messina M and Connor W (eds.) *Southern Forested Wetlands: Ecology and Management*, pp. 445–473. Boca Raton, FL: CRC Press.

Twilley RR, Cárdenas W, Rivera-Monroy VH, *et al.* (2000) Ecology of the Gulf of Guayaquil and the Guayas River Estuary. In: Seeliger U and Kjerve BJ (eds.) *Coastal Marine Ecosystems of Latin America*, pp. 245–263. New York: Springer.

Twilley RR and Chen RH (1998) A water budget and hydrology model of a basin mangrove forest in Rookery Bay, Florida. *Marine and Freshwater Research* 49: 309–323.

Twilley RR, Chen RH, and Hargis T (1992) Carbon sinks in mangroves and their implications to carbon budget of tropical coastal ecosystems. *Water, Air and Soil Pollution* 64: 265–288.

Twilley RR, Gottfried RR, Rivera-Monroy VH, Armijos MM, and Bodero A (1998) An approach and preliminary model of integrating ecological and economic constraints of environmental quality in the Guayas River estuary, Ecuador. *Environmental Science and Policy* 1: 271–288.

Twilley RR, Lugo AE, and Patterson-Zucca C (1986) Production, standing crop, and decomposition of litter in basin mangrove forests in southwest Florida. *Ecology* 67: 670–683.

Twilley RR, Pozo M, Garcia VH, Rivera-Monroy VH, Zambrano R, and Bodero A (1997) Litter dynamics in riverine mangrove forests in the Guayas River estuary, Ecuador. *Oecologia* 111: 109–122.

Twilley RR and Rivera-Monroy VH (2005) Developing performance measures of mangrove wetlands using simulation models of hydrology, nutrient biogeochemistry, and community dynamics. *Journal of Coastal Research* 40: 79–93.

Twilley RR, Rivera-Monroy VH, Chen R, and Botero L (1998) Adapting and ecological mangrove model to simulate trajectories in restoration ecology. *Marine Pollution Bulletin* 37: 404–419.

Twilley RR, Snedaker SC, Yañez-Arancibia A, and Medina E (1996) Biodiversity and ecosystem processes in tropical estuaries: Perspectives from mangrove ecosystems. In: Mooney H, Cushman H, and Medina E (eds.) *Biodiversity and Ecosystem Functions: A Global Perspective*, pp. 327–370. New York: Wiley.

Vermeer DE (1963) Effects of Hurricane Hattie, 1961, on the cays of British Honduras. *Zeitschrift fur Geomorphologie* 7: 332–354.

Wadsworth FH (1959) Growth and regeneration of white mangrove in Puerto Rico. *Caribbean Forester* 20: 59–69.

Waisel Y (1972) *Biology of Halophytes*, 395pp. New York: Academic Press.

Walsh GE (1974) Mangroves: A review. In: Reimold R and Queen W (eds.) *Ecology of Halophytes*, pp. 51–174. New York: Academic Press.

Wanless HR, Parkinson RW, and Tedesco LP (1994) Sea level control on stability of Everglades wetlands. In: Davis S and Ogden J (eds.) *Everglades: The Ecosystem and Its Restoration*, pp. 199–223. Delray Beach, FL: St. Lucie Press.

Watson J (1928) *Mangrove Forests of the Malay Peninsula*. Singapore: Fraser & Neave.

Woodroffe CD (1990) The impact of sea-level rise on mangrove shoreline. *Progress in Physical Geography* 14: 483–520.

Woodroffe C (1992) Mangrove sediments and geomorphology. In: Robertson AI and Alongi DM (eds.) *Tropical Mangrove Ecosystems*, pp. 7–42. Washington, DC: American Geophysical Union.

Woodroffe CD, Chappell J, Thom BG, and Wallensky E (1986) Geomorphological dynamics and evolution of the South Alligator tidal river and plains. In: *ANU, North Australia Research Unit Monograph 3*. Darwin: North Australian Research Unit.

Yañez-Arancibia A (1985) *Fish Community Ecology in Estuaries and Coastal Lagoons: Towards an Ecosystem Integration*. Mexico City, UNAM Press.

Yañez-Arancibia A and Day JW, Jr. (1982) Ecological characterization of Terminos Lagoon, a tropical lagoon-estuarine system in the Southern Gulf of Mexico. *Oceanologica Acta* SP: 431–440.

Yañez-Arancibia A and Day JW, Jr. (1988) *Ecology of Coastal Ecosystems in the Southern Gulf of Mexico: The Terminos Lagoon Region*. Mexico City: Universidad Nacional Autonoma de Mexico, Ciudad Universitaria, Mexico.

Yáñez-Arancibia A, Lara-Domínguez AL, and Day JW (1993) Interactions between mangrove and seagrass habitats mediated by estuarine nekton assemblages: Coupling of primary and secondary production. *Hydrobiologia* 264: 1–12.

Yáñez-Arancibia A, Lara-Domínguez AL, Rojas-Galaviz JL, *et al.* (1988) Seasonal biomass and diversity of estuarine fishes coupled with tropical habitat heterogeneity (southern Gulf of Mexico). *Journal of Fish Biology* 33(supplement A): 191–200.

Mediterranean

F Médail, IMEP Aix-Marseille University, Aix-en-Provence, France

Introduction
Main Environmental Characteristics of
    the Mediterranean Ecoregions
Patterns and Determinants of Mediterranean
    Biodiversity
Historical Biogeography and Evolution of
    Mediterranean Biodiversity

Convergence versus Nonconvergence of
    Mediterranean Ecosystems
Ecosystem Characteristics and Processes
Disturbances and Ecosystem Dynamics
Conclusion: Current Evolution of Mediterranean
    Ecosystems under Global Changes
Further Reading

Introduction

Mediterranean-type ecosystems occur in areas characterized by winter rainfall and summer drought. Five ecoregions of the world possess a Mediterranean climate and form the Mediterranean biome: the Mediterranean Basin, California (see Chaparral), central Chile, the southern and southwestern Cape Province of South Africa (SW Cape), the southwestern and parts of southern Australia (SW Australia) (**Table 1**). These Mediterranean ecoregions are all centered between 30° and 40° north or south of the equator, and are exposed to similar atmospheric and oceanic circulation patterns with cool ocean currents. Mediterranean ecoregions occur only along the western sides of continents, and occupy limited areas between deserts and temperate regions.

The most typical characteristics of Mediterranean ecosystems, compared to temperate or boreal biomes, are their spatial and temporal complexity inducing strong heterogeneities, in terms of physical factors (geography, geology, geomorphology, pedology, bioclimate) and of their biological components and species life-history traits. Paleogeographical and historical episodes, current geographical and climatic contrasts have molded both an unusually high biodiversity and ecological complexity, and favored the emergence of a functional uniqueness for several ecosystems. High species-richness and endemism due to contrasted biogeographical origins, and original functional dynamics at local and landscape levels linked to stress effects, represent indeed key components of these ecosystems.

Main Environmental Characteristics of the Mediterranean Ecoregions

Climate

The Mediterranean ecoregions are usually defined by their particular climates, which are transitional between temperate and dry tropical climates. The main characteristic is the existence of a combined dry and hot summer period of variable length, which imprints a strong water stress on species and ecosystems during summer. A high unpredictability characterizes these Mediterranean climates, with high yearly variation in timing as well as amount of rainfall or occurrence of extreme temperatures.

Rainfall is extremely variable, with mean annual values ranging from 100 to 2000 mm. The lowest values are found at desert margins, especially in North Africa and the Near East. The isohyet of $100 \, \mathrm{mm} \, \mathrm{yr}^{-1}$ represents the borderline between the Mediterranean and the Saharan climates. Rainfalls higher than 1500 mm are mostly found at medium altitudes of some coastal mountain ranges. But Mediterranean-type climates differ markedly between and within the Mediterranean regions, in terms of total rainfall and seasonality. For example, annual rainfall of parts of southern California and central Chile are comprised between 250 and 350 mm, whereas some Mediterranean montane sites of SW Africa receive as much as 3000 mm by year, and the summer rainfall is similar here to annual totals for California or Chile.

Mean minimum temperatures of the coldest month (m) are often used to define climatic subdivisions in the Mediterranean Basin (**Table 2**). These values are correlated to elevation and to a lesser extent to increased latitude and continentality. In most places, m is between 0 and +3 °C although extremes can reach +8 to +9 °C in desert margins and −8 to −10 °C on the highest mountains.

Aridity and temperature play an essential role in the structure and composition of Mediterranean ecosystems. The Emberger pluviothermic quotient (Q_2) constitutes the most utilized index for classifying Mediterranean climates:

$$Q_2 = \frac{2000P}{M^2 - m^2}$$

where P is the annual rainfall (in mm), M is the mean maximum temperatures of the warmest month of the year, m is the mean minimum temperatures of the coldest month of the year.

Table 1 Main environmental characteristics and major ecosystem-types of the Mediterranean ecoregions

	North Hemisphere		South Hemisphere		
	Mediterranean Basin	California	Central Chile	SW Australia	Cape Region
Surface (km^2)	2 300 000	324 000	140 000	310 000	90 000
Topographic heterogeneity	High	High	Very high	Low	Moderate
Climatic heterogeneity	Very high	Very high	High	Moderate	High
Rainfall reliability	Moderate	Low	High	Very high	High
Main lithological substrates	Calcareous rocks, occasional siliceous rocks	Argillaceous and mafic igneous rocks, occasional ultramafic rocks	Argillaceous and mafic igneous rocks	Siliceous rocks (sandstones, quartzites)	Siliceous, argillaceous and mafic igneous rocks
Soil fertility	High–moderate	Moderate	High	Very low–low	Very low–moderate
Natural fire frequency (year)	25–50	40–60	Fire free	10–15	10–20
Forests and woodlands	Very diverse and heterogeneous; with many sclerophyllous broad-leaved oaks (Quercus ilex, Q. suber) and broad-leaved oaks (Quercus pubescens, Q. faginea, Q. ithaburensis), and conifers (Pinus halepensis, P. brutia, Cedrus atlantica, C. libani, Abies, Juniperus)	Diverse forests with thermophilous conifers (Pinus attenuata, P. sabiniana, Cupressus macrocarpa) and oaks (Quercus douglasii, Q. agrifolia, Q. lobata), and mesophilous conifers (Abies, Pinus . . .) at higher altitudes; coast redwood (Sequoia sempervirens)	Very diverse, with semiarid Acacia caven and Prosopis chilensis forests in the north; subtropical broad-leaved and sclerophyllous forests with Peumus boldus and Cryptocarya alba in the central region; deciduous Nothofagus forests farther south, with Araucaria araucana	Patchy and open woodlands dominated by Eucalyptus (E. diversicolor, E. marginata); low woodlands with Banksia; thickets with Acacia, Melaleuca, and Allocasuarina	Very patchy and scarce forests; composed of cool and humid Afromontane plants, with warm subtropical elements; sclerophyllous trees and conifers (Afrocarpus, Podocarpus)
Shrublands	Maquis with Erica, Arbutus on siliceous soils; garrigues with Quercus coccifera, Cistus, Ulex, on calcareous soils; phryganas with spiny shrubs (Sarcopoterium, Astragalus, Genista)	Chaparrals with Adenostoma (chamisal), Arctostaphylos (manzanita chaparral), Ceanothus, scrub oaks (Quercus dumosa); coastal scrubs with Artemisia, Baccharis, Salvia	Open shrubland with Acacia caven (espinal); matorrals with Lithraea caustica, Quillaja saponaria; coastal matorrals with cacti (Trichocereus) and bromeliads (Puya)	Kwongan and scrub-heaths with Proteaceae (Banksia, Grevillea, Hakea) and ericoids plants (Epacridaceae); mallee dominated by shrubby Eucalyptus (E. incrassata, E. oleosa, E. socialis)	Fynbos with major plant types: restioids (Restionaceae), ericoids, proteoids (Proteaceae) and geophytes; renosterveld dominated by ericoids (renosterbos: Elytropappus); succulent karoo with Aizoaceae
Grasslands	Very diverse grasslands with numerous annuals and perennials herbs (Poaceae, Fabaceae, Asteraceae); steppes with Stipa tenacissima and Lygeum spartum in North Africa	Native perennial bunchgrasses with Stipa, Poa and Koeleria, replaced by annual grasslands with forbs (Avena, Bromus, Lolium, Erodium)	Anthropogenic prairies with numerous European herbs and grasses; wet grasslands with native Juncus procerus	Very scarce and patchy; grasslands on granite outcrops with annual everlastings (Helichrysum; Helipterum) or perennial Lechenaultia	Very scarce, fire-prone grasslands and grassy shrublands dominated by geophytes

Data from Davis GW and Richardson DM (1995) *Ecological Studies, Vol. 109: Biodiversity and Ecosystem Function in Mediterranean-Type Ecosystems*. Berlin and Heidelberg: Springer; Cowling RM, Rundel PW, Lamont BB, Arroyo MK, and Arianoutsou M (1996) Plant diversity in Mediterranean-climate region. *Trends in Ecology and Evolution* 11: 362–366; Cowling RM, Ojeda F, Lamont BB, Rundel PW, and Lechmere-Oertel R (2005) Rainfall reliability: A neglected factor in explaining convergence and divergence of plant traits in fire-prone Mediterranean-climate ecosystems. *Global Ecology and Biogeography* 14: 509–519; Dalmann PR (1998) *Plant Life in the World's Mediterranean Climates*. Oxford: Oxford University Press. Medail, ined.

Table 2 Vegetation levels showing the correspondence between thermal variants and dominant woody types of the Mediterranean Basin

Vegetation level	Thermal variant	m (°C)	T (°C)	Dominant woody species
Infra-Mediterranean	Very hot	> +7 °C	> +17 °C	Argania, Acacia gummifera
Thermo-Mediterranean	Hot	+3 to +7 °C	> +17 °C	Olea, Ceratonia, Pinus halepensis and P. brutia, Tetraclinis, (Quercus)
Meso-Mediterranean	Temperate	0 to +3 °C	+13 to +17 °C	Sclerophyllous Quercus, Pinus halepensis and P. brutia
Supra-Mediterranean	Cool	−3 to 0 °C	+8 to +13 °C	Deciduous Quercus, Ostrya, Carpinus orientalis (Pinus brutia)
Mountain-Mediterranean	Cold	−7 to −3 °C	+4 to +8 °C	Pinus nigra, Cedrus, Abies, Fagus Juniperus,
Oro-Mediterranean	Very cold	< −7 °C	< +4 °C	prostrate spiny xerophytes

m, mean minimum temperatures of the coldest month; T, mean annual temperature.
Modified from Quézel P and Médail F (2003) *Ecologie et biogéographie des forêts du bassin méditerranéen*. Paris: Elsevier.

Table 3 Main types of bioclimates and their theoretical correspondence with the dominant vegetation types of the Mediterranean Basin

Bioclimate	Mean annual rainfall (for m = 0 °C)	Number of months without rainfall	Main vegetation type
Per-Arid	< 100 mm	11–12	Saharan
Arid	100–400 mm	7–10	Steppe and pre-steppe (Juniperus turbinata, Pinus halepensis, Pistacia atlantica)
Semi-Arid	400–600 mm	5–7	Pre-forest (Pinus halepensis, P. brutia, Juniperus spp., Quercus)
Sub-Humid	600–800 mm	3–5	Forest (Mostly sclerophyllous Quercus, Pinus halepensis, P. brutia, P. pinaster, P. pinea, P. nigra, Cedrus)
Humid	800–1000 mm	1–3	Forest (Mostly deciduous Quercus, Pinus brutia, P. pinaster, P. nigra, Cedrus, Abies, Fagus)
Per-Humid	>1000 mm	<1	Forest (Deciduous Quercus, Cedrus, Abies, Fagus)

m: mean minimum temperatures of the coldest month.
Modified from Quézel P and Médail F (2003) *Ecologie et biogéographie des forêts du bassin méditerranéen*. Paris: Elsevier.

According to the levels of humidity and the winter severity, several bioclimatic zones and thermal variants are respectively defined, and they can be included in the climagram of Emberger; their combination permits to define six main bioclimatic types (**Table 3**).

Soils and Nutrients Availability

The five Mediterranean ecoregions are characterized by different geologies and soils characteristics due to their contrasted physiographic histories (**Table 1**). Landscapes of SW Australia and SW Africa consist of inland mass of geologically older origins than the three other ecoregions where mountain-building events occurred as recently as the Tertiary and the Quaternary.

In the two South Hemisphere ecoregions, soils on older substrata of uplands are generally highly leached lithosols in South Africa and by laterites and the process of podzolization in southern Australia; they have been exposed to weathering since the Paleozoic or even the Precambrian. Coastal deposits are younger and determine calcareous sands, decalcified humus podzols, and bleached sands.

Calcareous soils of limestone origin are scarce and occur only in few places of the South African coasts and in the central–southern region of Mediterranean Australia.

In California and Chile, the violent tectonic activity down the west coasts during the Late Tertiary and Early Quaternary has given rise to rugged landscapes of the Cordilleran mountain chain. Upland areas possess generally coarse-textured lithosols, whereas the major inland valleys (Great Valley of California and Central Valley of Chile) have more fertile soils linked to alluvial deposits.

In the Mediterranean Basin, the diverse tectonic and orogenic activities, and also the consequences of Pleistocene glaciations, induced a complex patchwork of landscapes and a mosaic of soil types. The predominantly limestone rocks have given rise to the terra rossa soil, a clay-rich and relatively fertile soil of the lowland areas. Soils of the humid uplands are often leached podzols or brown forest soils occurring within forested landscapes.

In spite of these sometimes large differences in substratum geology, there are several similarities between soils of the Mediterranean ecoregions, due to similar pedogenic processes linked to water-driven erosion and leaching. These seasonally droughted and moderately to strongly leached soils are indeed characterized by a low availability of several nutrients, especially phosphorous, and nitrogen which is greatly affected by fire.

Patterns and Determinants of Mediterranean Biodiversity

To compare the biodiversity of Mediterranean-type ecosystems, it is useful to partition species-richness and diversity into three main types of spatial scales, and to consider successively regional diversity, differentiation diversity, and local diversity (**Table 4**). Regional diversity is the product of local richness and turnover along habitat and geographic gradients.

The five Mediterranean-climate regions harbor a remarkable and huge regional biodiversity, among the highest in the world. With only 2% of the world's terrestrial surface, the Mediterranean biome contains nearly 20% of the Earth's total plant diversity, making very significant biodiversity hot spots, second only after tropical ones. The Mediterranean Basin exhibits the greatest diversity of plant species, both general and endemic, but with the much greater surface area (84% of the total of Mediterranean ecoregions). This ecoregion possess a higher tree richness (290 indigenous trees with 201 endemics) than the California Floristic Province (173 trees with 77 endemics), although its surface is seven times larger. In fact, the latter ecoregion has more or less the same surface area and plant biodiversity as Morocco, but California is four times richer in strictly local endemics. The case of the Cape Floristic Region is even more remarkable, since the endemic plant richness reaches close to 70% and the total plant species

reaches 9090 taxa, which makes it one of the world's richest areas. The interplay between diverse processes of historical biogeography and heterogeneous environmental conditions has promoted these considerable species-richness and endemism levels in the different Mediterranean ecoregions. Regional diversity peaks generally in areas with high topographical and climatic heterogeneity. However, the two highest plant species-rich regions of the SW Cape and SW Australia are characterized by global topographically and climatically uniform lowlands. Edaphic complexity, and more recently rainfall reliability (measured as interannual variation in seasonal and monthly rainfall and as the frequency of different-sized rainfall events), have been invoked as main determinants of this exceptionally high biodiversity. The regional-scale plant richness is indeed twofold higher in the western Cape Region, which receives reliable winter rainfall than the less reliable and nonseasonal zone in the eastern Cape. Reliable rainfall regimes are argued to promote higher and rapid speciation and lower extinction rates, and this pattern could partly explain the overall highest plant diversity of the SW Cape and SW Australia which have significantly more reliable regime than the other three Mediterranean ecoregions. If we consider diverse groups of vertebrates (**Table 4**), biodiversity patterns are more contrasted, and species-richness and endemism are often attenuated compared to plants, notably for birds. Nevertheless, for reptiles, amphibians, and freshwater fishes, the uniqueness of Mediterranean biotas is again noteworthy since the endemism rate is generally comprised between 30% and 60% (**Table 4**).

The differentiation diversity refers to changes of species composition along habitat gradients (beta diversity) or geographical gradients (gamma diversity). Highest levels of differentiation diversity are recorded for plant species in

Table 4 Main biodiversity components of the five Mediterranean ecoregions

| Biodiversity components | North Hemisphere | | South Hemisphere | | |
	Mediterranean Basin	California	Central Chile	SW Australia	Cape Region
Local diversity	Low–very high	Low–moderate	Low–?high	Low–high	Moderate–high
Differentiation diversity	Moderate	Moderate	Low–moderate?	High	High
Regional diversity	Moderate	Moderate	Low	High	High
Plant richness/endemism	c. 25 000/12 500 (50%)	3488/2128 (61%)	3539/1769 (50%)	5710/3000 (52.5%)	9086/6226 (68.5%)
Mammal richness/ endemism	224/25 (11%)	151/18 (12%)	65/14 (22%)	57/12 (21%)	90/4 (4%)
Bird richness/endemism	497/32 (6%)	341/8 (2%)	226/12 (5%)	285/10 (4%)	324/6 (2%)
Reptile richness/endemism	228/77 (34%)	69/4 (6%)	41/27 (66%)	177/27 (15%)	100/22 (22%)
Amphibian richness/ endemism	86/27 (31%)	54/25 (46%)	43/29 (67%)	33/19 (58%)	51/16 (31%)
Freshwater fish richness/ endemism	216/63 (29%)	73/15 (21%)	43/24 (56%)	20/10 (50%)	34/14 (41%)

Data from Cowling RM, Rundel PW, Lamont BB, Arroyo MK, and Arianoutsou M (1996) Plant diversity in Mediterranean-climate region. *Trends in Ecology and Evolution* 11: 362–366; and Mittermeier RA, Robles Gil, Hoffmann M, *et al.* (2004) *Hotspots Revisited: Earth's Biologically Richest and Most Endangered Terrestrial Ecoregions*. Monterrey: CEMEX, Washington: Conservation International, Mexico: Agrupación Sierra Madre.

the winter rainfall zones of SW Cape and SW Australia, with a high turnover for fire-killed shrub lineages. Disproportionate radiation of several shrub genera (e.g., 667 species of *Erica* in the Cape Floristic Region with 96.5% of endemic heaths) explains the strong dissimilarities in species composition between morphologically close communities. A similar, but attenuated, pattern is found in California and in the Mediterranean Basin for relatively recent shrub lineages and annual herbs, which are respectively the keystone species of matorrals and grasslands.

At the local scale, that is, less than 0.1 ha (alpha diversity), Mediterranean biodiversity is two times lower than that of tropical regions. However, a great variation exists within each ecoregion and between different habitats. Open and frequently burned shrublands and heaths on nutrient-poor soils, in particular, fynbos in SW Cape and kwongan in SW Australia, xerophytic rocky grasslands, and temporary pools encompass the highest plant diversity. Postfires communities in dense shrublands, notably chaparral in California (see Chaparral), and maquis in the Mediterranean Basin, are also characterized by a rich fire-ephemeral flora with numerous annual plants. Mean local plant richness of Mediterranean forests is comprised between 10 species per m^2 and 25–110 species per $1000\,m^2$. At this spatial scale, woody plant communities of the Mediterranean Basin are both very heterogeneous, but also among the richest types, ahead of the alpha diversity found in the SW Cape. Several nonexclusive determinants have been invoked to explain this high local diversity and species coexistence: an important regional species pool linked to complex historical biogeography, differentiation, and character displacement along structural niche axes, spatiotemporal variations in resource availability, recurrent disturbances (fire, grazing), neighborhood effects, and lottery processes. Finally, few generalizations are available from the numerous studies on local plant diversity in the Mediterranean vegetation, and the strongest evidence is that diversity represents an unimodal function of productivity or nutrient supply of soils. Species–area relationships fit a power function model for the majority of Mediterranean plant communities, but communities with a preponderance of perennials and paucity of annuals (e.g., Australian heathlands, mature Californian matorrals) are fitted by the exponential species–area model.

Historical Biogeography and Evolution of Mediterranean Biodiversity

Mediterranean floras and faunas constitute highly complex assemblages of species of different biogeographical origins. These huge species diversities and levels of endemicity are in great part related to the contrasted historical biogeography of each Mediterranean ecoregion, with different evolutionary patterns and processes before and after the Pleistocene period.

Pre-Pleistocene History: The Diversification and Mixing of Different Lineages

A general picture suggests that humid and mesothermic climates predominated throughout most of the Late Cretaceous and Tertiary in Mediterranean ecoregions where subtropical forests were dominating. The onset of Mediterranean climates is relatively young, and paleoclimatic reconstructions demonstrate that cooling and drying, with the combination of summer drought and mild winter temperatures rose quite rapidly in Late Miocene or Early Pliocene (c. 5–3 Ma). In the Mediterranean Basin, a gradual but deep climatic change occurs during the Pliocene (3.5–2.4 Ma), with a significant drop in temperature and a marked seasonality in thermal and rainfall regimes; the stabilization of the summer drought arises here at c. 2.6 Ma. Concomitant with the appearance of Mediterranean-climate conditions, a massive ecological radiation was initiated in the Late Miocene–Pliocene. However, there are also phylogenetic evidences of earlier radiation events during the Oligocene involving geographic speciation; this is the case of the shrub genus *Protea* (Proteaceae) in South African fynbos and the geophytic genus *Androcymbium* (Colchicaceae) in xerophytic grasslands of South Africa and the Mediterranean Basin. The deterioration of Tertiary warm climates resulted in the extinction of several subtropical and warm-temperate species during Plio-Pleistocene, but historical discrepancies exist between the Mediterranean ecoregions, from the point of view of the impact of environmental changes and the macroscale process of species recolonization after drastic climatic events.

Mediterranean ecoregions closely situated to subtropical regions and without major geographical barriers between these two biomes have significantly experienced moderate cases of species extinctions, because of possible latitudinal shifts in species range or a higher climatic stability. This is the case of the eastern lowlands of the Cape Floristic Region, where patches of subtropical thickets and warm-temperate forests are still dominated by trees and shrubs of paleotropical and Gondwanian affinities. These highly species- and genus-rich formations are extremely ancient, with many elements phylogenetically basal to the western Cape species. The precise timing of this lineage diversification is still unknown but geographic speciation was probably the key process.

In south–central Chile, the actual coastal forests at mid-latitudes remained fairly stable despite major climatic and tectonic changes in southern South America during the Pleistocene. These forests are indeed characterized by a notable evolutionary stability, favoring the conservation of ancient species assemblages. The nearest ancestor of the Mediterranean sclerophyllous vegetation of Chile corresponds probably to the Neogene subtropical paleoflora that occured in the Proto-Andean foothills of this area during the lower to mid-Miocene (*c.* 20–15 Ma), and the pre-Pleistocene paleofloras were developed under a more humid and warmer paleoclimate induced by an Andean rain shadow effect.

In California, shrublands (chaparrals) are composed of temperate, subtropical, and desert elements. Many of these Californian chaparral genera appeared in the Early Tertiary, following the trends since the Eocene toward cooler and drier climates. This sclerophyllous flora seems to have arisen as an understory component of evergreen woodlands and form only the chaparral in response to the latter appearance of the Mediterranean climate and related increased disturbance by fire. During unfavorable periods, southward migration of species in Baja California or central Mexico, or their location in southern coastal refugia and subsequent recolonization explain the actual persistence in California of several outstanding trees (*Sequoia, Sequoiadendron, Tsuga, Umbellularia*, etc.), whereas these genera have disappeared around the Mediterranean sea because of the existence of several east–west orographic and maritime barriers which prevent putative latitudinal migrations of species.

Moreover, in the Mediterranean ecoregion, climatic coolings during the Neogene have provoked severe extinctions, for example, of 45 genera of megathermous and warm-temperate ligneous species distributed in the northwestern part of the Mediterranean sea. The main diversification of the Mediterranean Basin species took place during the Miocene, notably in conjunction with the collision of the African and Eurasian platforms through the Arabian Plate. Recurrent episodes of species dispersal and vicariance have occurred between the eastern and western Mediterranean regions due to several marine regressions/transgressions and the ongoing rise of Alpine and Atlas orogenesis. Another major biogeographic event of the end of the Tertiary is the closure of the Mediterranean–Atlantic gateways (now the Gibraltar Strait). This episode known as the Messinian salinity crisis (5.77–5.33 Ma) induced a considerable evaporation of the Mediterranean Sea and the formation of several land bridges suitable for dispersal–vicariance events and species radiations. Thus, the Mediterranean species pool arises from diverse biogeographic origins and if it includes still some subtropical species, mostly originated from African and Asian lineages, the extratropical species of autochthonous and northern lineages predominate.

Influence of Pleistocene Glacial Refugia to Current Patterns of Mediterranean Biodiversity

Pleistocene climatic cycles have profoundly affected the biogeographical footprint of several Mediterranean species, notably in the Mediterranean Basin, California, and Chile. Recent researches combining genetics and biogeography (i.e., phylogeography) and paleoecology underline that glacial refugia represent crucial areas for the long-term persistence and dynamics of modern biodiversity in temperate regions. Glacial refugia constitute territories sheltered from the strong climatic deteriorations during Ice Ages, and where species survived the drastic consequences of severe cold and aridity. A major and noteworthy glacial event is the Last Glacial Maximum (LGM) that occurred *c.* 20 000 years BP. First, there exists a clear influence of Pleistocene climatic cycles on patterns of species-richness and endemism. Second, full-glacial refugia have also had a powerful influence in shaping current patterns of genetic diversity in several temperate and Mediterranean ecoregions. Finally, these refugia played an important role on vegetation dynamics during previous interglacial periods of the Pleistocene; these areas contributed to the forest recolonization process that started approximately 13 000 years ago in the Mediterranean Basin and lasted throughout the Holocene. Once climatic conditions became truly favorable, the expansion of a highly diversified deciduous forest could happen rapidly over large territories, such as in the Pindos mountains in northwestern Greece where over 20 deciduous woody plants could already be found in around 10 000 yr BP.

There is growing evidence, both at a global-scale and for Mediterranean-type ecosystems, indicating that biodiversity hot spots coincide generally with areas that were buffered against climatic extremes. Reduced impacts of Milankovitch climate oscillations and smaller-amplitude climatic changes during the LGM constitute the best descriptors to explain both the survival of paleoendemics and the speciation of neoendemics. Therefore, the climate stability–diversity pattern accounts for the location of most of the endemic-rich Mediterranean areas, and this pattern appears clearly when we compare two ecoregions with distinct historical biogeography, the Mediterranean Basin and SW Africa.

Around the Mediterranean, the alternation of humid and hyperarid phases in North Africa or interglacial and glacial episodes in Europe have induced profound shifts in the evolution and geographical distribution of species lineages, resulting in, respectively, expansion from Mediterranean refugia or extinction–reduction of populations during unfavorable periods. The major Mediterranean areas where temperate and thermophilous species survived are the three Iberian, Italian, and Balkan peninsulas, but also the largest Mediterranean islands and the submontane and mountain margins of North Africa, Turkey, and Catalonia-Provence. The glacial events induced the extinction of several subtropical lineages and paleoendemics, and speciation by radiation was quite reduced.

In South Africa, glacial climates of Pleistocene were moister in western Cape and drier in eastern Cape than present. It seems that the western Cape elements have extended northward into the present succulent karoo and Namib desert during glacial episodes, whereas drier conditions in the east have induced the restriction of Cape vegetation and species to some mesic refugia. Due to the peninsula configuration surrounded by ocean, southwestern Africa was one of the mildest continental landmasses of the world, and Pleistocene climates there were exceptionally stable favoring higher rates of speciation and lower rates of extinction compared to eastern Cape or to northern Mediterranean ecoregions.

Convergence versus Nonconvergence of Mediterranean Ecosystems

Mediterranean-climate ecosystems have been often cited as classic examples of convergence in ecosystem structure and function owing to their similar environments. But comparative studies during the 1980s have demonstrated that this convergence pattern is too simplifying, and several divergences exist also.

Most discussions of the Mediterranean convergence hypothesis have focused on similarities among communities and on ecological similarity of distantly related taxa. Several cases of ecological convergence deserve attention. SW Australia and SW Cape are characterized by larger amounts of summer rain, lower soil fertility, and more frequent fires, compared to the three other Mediterranean regions. These two Southern Hemisphere Mediterranean ecosystems show remarkable convergence of plant traits and community structure on climatically and edaphically matched sites. Patterns of plant biodiversity are also similar on nutrient-poor soils in South African fynbos and Mediterranean heathlands of Spain. Striking convergences exist in the two Northern Hemisphere Mediterranean ecoregions, and several forest-types of the Mediterranean Basin are relatively similar from a biogeographic and ecological point of view, to those of the California Floristic Province. Ancient landmass connections which lasted until the Eocene (Madro-Tethysian flora) explain the existence of some common tree genera (*Pinus, Quercus, Arbutus, Cupressus, Juniperus, Platanus*) between these regions. At low and medium altitudes, physiognomical similarities exist for sclerophyll oak forests. The existence of thermophilous coniferous forests at the thermo- and meso-Mediterranean (*Pinus halepensis, P. brutia, P. pinaster*) and the thermo- and meso-Californian (*Pinus attenuata, P. sabiniana, Cupressus macrocarpa*) levels, but also the presence of several mesophilous coniferous (*Abies, Pinus,* etc.) at higher altitude constitutes another major resemblance.

Life-history traits shared by many sclerophyllous woody plants of Mediterranean ecosystems represent an outstanding example of convergence, with evergreen, small, or even needle-like (ericoid) leaves, low specific leaf area (SLA: ratio of fresh leaf area to dry mass), strongly seasonal photosynthesis and growth, vigorous resprouting following fire or cutting. Sclerophylly (evergreen and thick leathery leaves) is the most common and widespread strategy of plants in these Mediterranean ecosystems.

Convergence can be explained by similar abiotic factors (climate and notably the intensity of summer drought, soil nutrient status fertility, fire regimes), and phylogenetic composition of lineages induced by common historical biogeography. Rainfall reliability was also recently suggested as another important factor; the two more ecologically similar regions (SW Australia and the SW Cape) have indeed significantly more reliable regimes than California and the Mediterranean Basin.

If we consider plant traits, and notably leaf evolution and resprouting capacity of evergreen woody plants, two hypotheses were formulated to explain the observed convergences: (1) evolutionary adaptation, that is, the production of new phenotypes by the action of natural selection; (2) niche conservatism, with a relative stasis in trait evolution, induced by the ecological match between organisms and their environment caused by the spatial and temporal sorting of existing lineages. Methods of phylogenetic comparative biology have recently demonstrated that similarity between traits of Mediterranean woody plants could be due more to phylogenetical inertia than to common adaptive strategies under Mediterranean climate. The absence of deep morphological changes in leaf size and SLA suggests that most of the ancestors of shrubland taxa had already acquired plant life-history traits that contributed to their success under Mediterranean climates. Thus, these subtropical 'phantoms' predate the onset of the mediterraneity during the mid-Pliocene.

More precisely, there exists a clear co-variation of life-history traits with regards to the lineage age, and two groups with distinctive characters associations can be defined for Mediterranean angiosperms:

1. A pre-Pliocene group, consisting of mostly sclerophyllous, vertebrate-dispersed, fleshy-fruited, and large-seeded plants which resprout from stump after disturbance (fire, clearing) and are often late colonizers in successional stages of ecosystem dynamics (e.g., *Arbutus unedo, Olea* spp., *Quercus* spp.). This resprouting 'strategy' often considered as typically 'Mediterranean' represents in fact an ancient trait that emerged under a subtropical climatic regime, well before the advent of the Mediterranean climates.

2. A post-Pliocene group, including nonsclerophyllous plants which are obligate seeders after disturbance (e.g., *Cistus* spp., *Lavandula* spp.). These taxa successfully diversified and competed with taxa of the pre-Pliocene group due to their short life cycle, high seed production with small and dry-fruited anemochorous seeds. Seeders grow significantly faster and allocate more leaf biomass than resprouters. These set traits account for the important

ecological plasticity of these plants which are associated with earlier successional stages.

Therefore, historical biogeography could largely explain the physiognomic similarities of some ecosystems or species found also in non-Mediterranean-type climate regions. This is the case of several modern matorral-type communities present in the American Southwest, Mexico, or eastern Australia, whose sclerophyllous vegetation is roughly similar from a structural and functional point of view to the true Mediterranean shrublands. Even in southwestern China, some sclerophyllous species possess striking morphological convergence with Mediterranean trees; for example, *Quercus phillyroides* in the Danxiashan escarpments (Guangdong Province) is astonishingly similar to the evergreen Mediterranean oaks (*Q. ilex* and *Q. coccifera*), whereas the matorrals with *Olea* and *Pistacia* recently discovered in some gorges of the Yunnan Province are globally alike to the Mediterranean ones. These similarities support the view that sclerophylly represents more a general response of semiarid conditions with a severe dry season, few intense frost episodes, and low-nutrient soils, rather than a specific adaptation to the summer-drought pattern *per se*.

Evolutionary convergence among bird communities was also tested between California, Chile, and the Mediterranean region, along matched habitat gradients of increasingly complex vegetation structure (from shrublands to forests), and compared with non-Mediterranean communities. Results of these ecomorphological comparisons indicate that the Mediterranean bird communities do not resemble each other any more than the non-Mediterranean ones. To summarize, convergence depends on the ecological compartments considered; Mediterranean convergence may exist for some community attributes (e.g., species-richness, community structure), and is more likely to occur among organisms such as plants, invertebrates, or reptiles that depend on seasonal patterns of climate and nutrient cycling, whereas homeotherm vertebrates are more deeply influenced by the structure of ecosystems.

Ecosystem Characteristics and Processes

Main Ecosystem Types

The vegetation types usually considered as 'typically Mediterranean' are the evergreen and sclerophyllous shrublands or heathlands, named maquis and garrigue in the Mediterranean Basin, chaparral in California, matorral in Chile, fynbos in SW Africa, kwongan and mallee in SW Australia. But there exists considerable differences in the composition and structure of these shrublands (**Table 1**). Depending on the type of shrubland chosen, it is possible to find strong similarities or dissimilarities,

between and within the diverse Mediterranean ecoregions. The similarity most frequently cited is that between kwongan and fynbos which show a striking ecological convergence with an open shrub cover and a high shrub diversity with the quasi-absence of annuals, a dominance of postfire seeders, and serotinous shrubs, and the frequent occurrence of seed dispersal by ants (myrmecochory). Shrublands occur generally in a mosaic with xerophytic grasslands, steppes, woodlands, or forests.

True Mediterranean forests are rare and they represent 1.8% of world's forest area. Northern Hemisphere Mediterranean forests show a higher structural and species diversity than those of the Southern Hemisphere, because the latter cover less extensive areas and, as in South Africa, may be outside the range of Mediterranean bioclimate. The Southern Cape forests are very patchy with mainly subtropical sclerophyllous trees and conifers (*Afrocarpus*, *Podocarpus*). Forests of Mediterranean Chile are more diverse due to the strong latitudinal gradient and the increase in rainfall from north to south; semiarid *Acacia caven* and *Prosopis chilensis* forests in the north are succeeded by subtropical broad-leaved and sclerophyllous forests in central Chile, and by deciduous *Nothofagus* forests farther south. Together with species-rich sclerophyllous shrublands (kwongan and mallee), the forests and woodlands of SW Australia are dominated by *Eucalyptus*, *Acacia*, and *Casuarina* on poor sandy soils, where mean annual rainfall exceeds 400 mm; several types of Australian woody vegetation are distinguished according to the foliage cover of tallest stratum and the high of the trees.

In the Mediterranean Basin, the current diversity of forest structures can be organized into three major structural types based on bioclimatic and/or human impact criteria (**Tables 2** and **3**).

True forest vegetation types are related to metastable equilibrium of vegetation structures. They represent the potential structures at the end of a dynamic ecological cycle, which can be achieved where soil and climate conditions are favorable and where the impact of man is not too strong. Dominant species are sclerophyllous oaks in semiarid bioclimates and deciduous oaks in more humid conditions.

Preforest types can be divided into two categories. Under perhumid, humid, and subhumid bioclimates, they consist of vegetation structures that have undergone severe human impact, although their soil is still relatively well preserved. They are transitory structures from true forests to more open systems. Under semiarid bioclimatic conditions, or under particularly stressful conditions (e.g., ultramafic substrates) in any bioclimate, preforests are comprised of shrub-dominated vegetation structures with scattered trees (matorrals). Conifer species (*Pinus*, *Tetraclinis*) play an important role in these structures.

Presteppic forest types, very frequent in southern and eastern Mediterranean, consist of open-vegetation structures dominated by nonforest plant species under scattered trees. Nonforest species are steppe-type perennial species that can eventually be replaced by ruderal annual species when grazing occurs. Presteppes are most frequent under warm and hot temperature variants of arid (and sometimes semiarid) bioclimates. They gradually merge into steppes under hotter and drier conditions. On mountains, presteppes are a transitional vegetation structure from forests (or preforests) to high-elevation steppes dominated by low and scattered cushion-like spiny xerophytes.

Annual grasslands represent also key ecosystems in the two northern Mediterranean ecoregions. The composition and structure of grassland communities are strongly controlled by disturbances, which create a complex pattern of microsites and canopy gaps. Therefore, the heavily grazed grasslands of the Mediterranean Basin have probably the greatest alpha diversity of any temperate plant community, and annuals represent half of the total species found in this region. Nearly one-fifth of California is covered by grasslands, but most of them are dominated by non-native annuals (*Bromus, Avena, Erodium*, etc.) originated from the Mediterranean Basin.

Strategies of Resistance to Climatic Stress

Mediterranean climates induce severe and contrasted stresses to habitats and species. These stresses are compounded by the unpredictable nature of weather patterns, and organisms have to cope with this temporal variation in climate and resource availability. Ecological and ecophysiological studies indicate that Mediterranean species demonstrate similar strategies to resist climatic and edaphic stress. Drought stress proves to be the essential climatic factor responsible for the restriction of productivity, growth, and survival of several groups of Mediterranean plant species, notably the evergreen woody plants. Sclerophyllous leaves also exhibit other water conservation features such as sunken stomata and low cuticular conductance. Other strategies to cope with water stress are related to complex root systems, cellular tolerance to low water potentials or high secondary compound production (e.g., terpenes, tannins).

The drought summer season also induces original physiological strategies for the two main groups of the Mediterranean soil microfauna; oribatid mites are more resistant to dryness and only migrate into the deep soil layers when the soil water content becomes too low; collembolas cross the summer in an egg stage, and several species can surmount the dryness by a deshydratation process similar to anhydrobiosis. There exists a balance of the collembola composition between, on one hand, the winter populations which are composed of common

species but highly diversified qualitatively and quantitatively, and the other hand, the 'reserve populations' present in a latent state in the soil which are expressed only when exceptional summer rainfalls occur or during the onset of the wet season.

On the community level, a striking example is represented by the biotic assemblages of vernal pools, precipitation-filled seasonal wetlands found mainly in the Mediterranean climate regions. Inundation during the growing season largely eliminates colonization by upland species in the pools whereas the terrestrial phase is sufficiently desiccating to prevent establishment of typical wetland species. Several cosmopolitan aquatic plant genera are shared between the five Mediterranean ecoregions, such as ferns (*Pilularia, Marsilea, Isoetes*) or dicots (*Callitriche, Elatine, Ranunculus*), whereas vernal pool specialists are often derived from genera of terrestrial origin. The essential ecological characteristic of these temporary ponds is the stochastic alternation of flooded and dry ecophases within and between years. This induces strong year-to-year differences not only in pool hydrology, but also in the composition and dynamics of vernal pool communities, with a strong temporal and spatial segregation which limits competition as it is the case of the larvae of anuran amphibians. The succession of contrasted phases favors the emergence of varied and highly specialized plant and animal communities, particularly adapted to this high habitat instability. These environmental factors also played a significant selective force in shaping life strategies of temporary pool species. Dormancy provides a determinant means of enduring prolonged unfavorable dry periods, and several typical vernal pool species possess mechanisms that keep them from emerging under unsuitable conditions: drought-resistant reproductive organs such as seeds or oospores for the well-represented annual plants (*c.* 80% of the whole vernal pool specialists in California and the Mediterranean Basin), and eggs or cysts for crustaceans (e.g., cladoceran and anostracan branchiopods). A great adaptability of the life cycle often exists, and when water levels are shortened and water temperature increases, invertebrates and amphibians can present an advanced metamorphosis and several annual plants known as ephemerophytes can complete their entire life cycle within only a few weeks.

If we consider the main ecological processes linked to climatic stress, several studies have demonstrated that competition increases with aridity. Summer drought and poor soil nutrients, coupled with frequent disturbances, explain the reduced rates of competitive displacement observed in most of the Mediterranean communities. Changes in species interactions along water gradients were also demonstrated in arid Mediterranean environments. The importance of positive interactions through facilitation in arid plant communities represents a complex and unrecognized process. Nevertheless, some recent studies suggest that facilitation decreases with elevation in dry Mediterranean mountains (Sierra Nevada in Spain, central Andes of Chile) because of the prevalence of water stress over temperature stress compared to mesic alpine communities.

In addition to summer drought, low winter temperatures and episodic frost exert a strong influence in limiting distribution range and causing alteration in species composition and productivity, notably for ecosystems situated in the northern limits of the Californian and Mediterranean Basin ecoregions. From freezing tests, three groups of plants can be distinguished with regard to frost susceptibility in the Mediterranean Basin: (1) the most sensitive species with 50% frost injury to the leaves at -6 to $-8\,^{\circ}$C and to the shoots at -9 to $-15\,^{\circ}$C (*Ceratonia siliqua*, *Myrtus communis*); (2) the medium sensitive species with serious damage to the foliage at -12 to $-14\,^{\circ}$C, and to the stems at -15 to $-20\,^{\circ}$C (*Olea europaea*, *Quercus coccifera*, *Pinus halepensis*); (3) the resistant species, not seriously damaged until -15 to $-25\,^{\circ}$C (*Quercus ilex*, *Cupressus sempervirens*). The distribution of Mediterranean species of the Northern Hemisphere is indeed shaped by frost events such as late spring below freezing temperatures and absolute minimum temperatures. For example, the extremely cold winters of 1956 and 1985 in the northern Mediterranean contributed to the determination of the distribution of some keystone trees such as olive tree, Aleppo pine, and holm oak.

Links between Biodiversity and Ecosystem Function

The relationship between diversity and productivity of ecosystems remains a controversial issue, and there are still few experimental evidences for Mediterranean ecosystems. Nevertheless, some results related to grasslands and shrublands of the Mediterranean Basin suggest that plant diversity is positively correlated with ecosystem function, notably with primary production. However, in mixed grassland communities, where several growth forms coexist, one of a few dominant species (keystone species) may hide or reverse the observed diversity–biomass production relationship: the inclusion of a dominant grass in experimental low-diversity plots can produce a constant level of productivity along the gradient of species diversity. Thus, the species composition appears to be the main determinant of the productivity performance at each diversity level. Shrubs, grasses, and geophytes have different attributes in these communities and they can significantly affect the value of the ecosystem functioning. Each growth form or functional type contributes significantly to the overall productivity performance of the studied Mediterranean ecosystems.

Several mechanisms were proposed to explain these patterns: (1) the 'sampling effect', related to the higher probability to include highly productive species in species-rich communities; (2) niche complementarity between species, leading to more complete utilization of resource in intact ecosystems relative to depauperate ones: space filling, for example, by herbaceous plants occupying the gaps in shrub canopies, favors the exploitation of different soil layers and resource supply by different root systems and this phenomenon increases productivity; (3) the increase of positive interactions between species in complex assemblages.

Disturbances and Ecosystem Dynamics

Natural disturbances represent a determinant component of the Mediterranean ecosystem dynamics and for the maintenance of their high biodiversity. Indeed, the long-established mature forests which constitute the typical final stage of ecological successions in the Mediterranean Basin, harbor few typical Mediterranean species and the more architecturally complex vegetation exhibit a high degree of convergence in species-richness and composition, with a domination of boreal and eurasiatic forest species. On the contrary, nonmature ecosystems with a more reduced architectural complexity and a higher structural heterogeneity comprise a dominance of Mediterranean species. Thus, natural disturbances such as fire or herbivory can play a major role in the expression of the mediterraneity by rejuvenating mature ecosystems.

Ecosystem Responses to Fire

Fire determines major vegetation structure and composition of several world biomes, as well as temperature, precipitation, and water balance. The fire-prone vegetation cover constitutes at present 40% of the world's land surface, including especially the Mediterranean regions, except central Chile where fires are less frequent (**Table 1**). In Mediterranean-type ecosystems, most of the areas occupied by shrublands have the climate potential to be forests and these shrublands are generally fire-maintained. But in contrast with current opinion that often attributes the existence of these ecosystems only to anthropogenic burning, Mediterranean shrublands seem to have naturally expanded in the Late Tertiary, with flammable C_4 grassy biomes. In the Mediterranean Basin, natural development of pine forests has also played a determinant role in increasing fire frequency since the Miocene (*c.* 9 Ma).

Forest fires are generally considered as catastrophic events in the Mediterranean Basin and California where big fires occur, but the evolutionary and ecological influences of fire also represent key parameters for driving landscape diversity, ecosystem heterogeneity, vegetation dynamics, and species differentiation. Fire determines indirect environmental changes with greater fluctuations in temperature and increasing oxygen concentration in soils, increases light and water availability, reduces aboveground competition and determines a proper regeneration niche for fire-adapted plants. In burned ecosystems, a higher abundance and species-richness is quickly observed several months, or even weeks, after fire and important changes in species occurrence and diversity occur. These immediate and profound ecological modifications are considerably attenuated during the second postfire season because Mediterranean ecosystems are particularly resilient and fire-adapted. Nevertheless, these brief structural changes are of particular importance for the regeneration window of fire-adapted species.

Fire represents a major selective force that shapes the evolution of plant reproductive traits in fire-prone Mediterranean environments, and thus ecosystem dynamics. Strong serotiny, fire-stimulated flowering or germination, smoke- or charred wood-induced germination, resprouting ability through lignotubers, stumps or burls, and seed dormancy favor plant persistence and these traits are more common in Mediterranean ecoregions with a longer fire history or a higher fire frequency. Several lines of evidence corroborated by recent phylogenetic analysis indicate that there has been stronger selective pressure for fire persistence traits in California than in the Mediterranean Basin. Because of a more drastic fire occurrence in California, plant species have evolved here more frequently toward the association of resprouting capacity and propagule-persistence than Mediterranean Basin plants. Furthermore, another peculiarity is the specialization of Californian annual plants which encompass numerous fire endemics that can persist as dormant seed banks for many decades between fires and that occur only 1 or 2 years after a fire.

Influence of Herbivory and Grazing

Herbivore pressures also constitute a crucial feature in some Mediterranean ecosystems, and grazing affects often seriously the structure and diversity of Mediterranean communities, notably grasslands. In the Mediterranean Basin, grasslands and shrublands have experienced frequent man-induced grazing for c. 9000 years and ungulate browsing. These ancient selective pressures explain the widespread existence of distinct plant species life attributes, which can contribute efficiently to reduce the cumulative damage induced by herbivores and linked to the long leaf life span of evergreen plants. Morphological changes following herbivory include the reduction of leaf size, low SLA, physical toughness, modification of branch density, developments of thorns and hairs, and increase of leaf chemical defenses by phenolic or tannin compounds. Variations of phenolic levels in relation to leaf age and season suggest an evolutionary adaptation of Mediterranean evergreen plants to high densities of large herbivores.

Without any disturbance for a few years, grasslands become dominated by a few species of perennial grasses, forbs, or tall large-seeded annuals, which form closed swards with high biomass, cover, and height. Thus, species regeneration will be only successful in gaps created by light grazing, and the observed peaks of species-richness with moderated grazing are consistent with the classic intermediate disturbance hypothesis. The barren areas between mature shrublands or grasslands are also possibly driven by rodent and rabbit activities, which control not only plant community boundaries and structure but also dynamics by influencing nutrient linkages between communities. The complex mosaic of microsites and the constant grazing pressure shift advantage to smaller plants that can occupy the microheterogeneity finely, and the diversities observed can be considerable with more than 50 plant species on $1 m^2$ plots in moderately grazed grasslands or steppes of the Mediterranean Basin.

Fire and grazing must be regarded as two disturbances with not always additive consequences, but often with distinct and interactive effects on structure and dynamics of communities. Moreover, the ability of Mediterranean woody species to resprout after fire seems not to originate from an adaptation to recurrent fires, but rather from an older adaptation to losses of aboveground biomass mainly induced by herbivory.

Conclusion: Current Evolution of Mediterranean Ecosystems under Global Changes

The originality of Mediterranean-climate ecosystems can be explained by complex interaction between historical biogeography patterns and unique ecological processes. Insights of paleoecology and phylogeography indicate the importance of paleogeographical and paleoclimatic events in shaping this massive and unique Mediterranean biodiversity and ecosystem types. Recent phylogenetic methods of comparative biology have demonstrated that the evolution of evergreen sclerophylls, formerly designated as 'typically Mediterranean', may have occurred long before the origin of Mediterranean climates. The predominance of sclerophylly and related conserved life-history traits in numerous keystone ligneous plants of Mediterranean ecosystems reflect mainly the ecological success of these taxa under climatic stress and fires, following by some evolutionary refinements in physiology, rather than a real origin of these traits under Mediterranean climates. But if these species and ecosystems were able, to a great extent, to surmount past environmental changes, their immediate future seems now quite worrying.

At present, these Mediterranean ecosystems are faced with rapid and previously unknown global environmental changes with important repercussions in structure and function. Several models predict that the greatest changes on a world scale are expected in Mediterranean regions, with the highest vulnerability in mountain areas. Climate projections for Mediterranean-type ecosystems suggest drier and warmer conditions, which have probably already triggered species distribution shifts, ecophysiology, phenology, and species interactions. The elevation of atmospheric CO_2 linked to anthropogenic causes could induce an increase of the productivity of many Mediterranean trees, as has already been measured for *Quercus ilex* and *Quercus pubescens* in the northern Mediterranean. On the other side, experimental studies have demonstrated that a decrease in water availability and an increase in temperature might affect the growth pattern and annual productivity of dominant shrubs, with alteration of competitive abilities. These modifications could change, in turn, species composition and structure of several Mediterranean habitats. Furthermore, the generalized and deep magnitude of human impact in the Mediterranean regions accentuates the effects of climatic change, and could obliterate the efficient capacity of ecological resilience of Mediterranean-type ecosystems, even if they have been submitted in the past to other drastic and rapid changes.

See also: Chaparral.

Further Reading

Ackerly DD (2004) Adaptation, niche conservatism, and convergence: Comparative studies of leaf evolution in the California chaparral. *American Naturalist* 163: 654–671.
Arianoutsou M and Groves RH (1994) *Plant–Animal Interactions in Mediterranean-Type Ecosystems*. Dordrecht: Kluwer Academic Publishers.
Arroyo MTK, Zedler PH, and Fox MD (1995) *Ecological Studies, Vol. 108: Ecology and Biogeography of Mediterranean Ecosystems in Chile, California and Australia*. New York: Springer.
Blondel J and Aronson J (1999) *Biology and Wildlife of the Mediterranean Region*. Oxford: Oxford University Press.
Cowling RM (1992) *Fynbos, Nutrients, Fire and Diversity*. Cape Town, South Africa: Oxford University Press.
Cowling RM, Rundel PW, Lamont BB, Arroyo MK, and Arianoutsou M (1996) Plant diversity in Mediterranean-climate region. *Trends in Ecology and Evolution* 11: 362–366.
Cowling RM, Ojeda F, Lamont BB, Rundel PW, and Lechmere-Oertel R (2005) Rainfall reliability: A neglected factor in explaining convergence and divergence of plant traits in fire-prone Mediterranean-climate ecosystems. *Global Ecology and Biogeography* 14: 509–519.
Dalmann PR (1998) *Plant Life in the World's Mediterranean Climates*. Oxford: Oxford University Press.
Davis GW and Richardson DM (1995) *Ecological Studies, Vol. 109: Biodiversity and Ecosystem Function in Mediterranean-Type Ecosystems*. Berlin and Heidelberg: Springer.
Di Castri F and Mooney HA (1973) *Ecological Studies, Vol. 7: Mediterranean-Type Ecosystems. Origin and Structure*. Berlin, Heidelberg, and New York: Springer.
Di Castri F, Goodall DW, and Specht RL (1981) *Ecosystems of the World, Vol. 11: Mediterranean-Type Shrublands*. Amsterdam: Elsevier.
Hinojosa LF, Armesto JJ, and Villagrán C (2006) Are Chilean coastal forests pre-Pleistocene relicts? Evidence from foliar physiognomy, palaeoclimate, and phytogeography. *Journal of Biogeography* 33: 331–341.
Keeley JE and Fotheringham CJ (2003) Species–area relationships in Mediterranean-climate plant communities. *Journal of Biogeography* 30: 1629–1657.
Mazzoleni S, Di Pascale G, Di Martino P, Rego F, and Mulligan M (2004) *Recent Dynamics of Mediterranean Vegetation and Landscape*. London: Wiley.
Mittermeier RA, Robles Gil P, Hoffmann M, *et al.* (2004) *Hotspots Revisited: Earth's Biologically Richest and Most Endangered Terrestrial Ecoregions*. Monterrey: CEMEX, Washington: Conservation International, Mexico: Agrupación Sierra Madre.
Moreno JM and Oechel WC (1994) *Ecological Studies, Vol. 107: The Role of Fire in Mediterranean-Type Ecosystems*. New York: Springer.
Ornduff R, Faber PM, and Keeler-Wolf T (2003) *California Natural History Guides, Vol. 69: Introduction to California Plant Life*. Berkeley and Los Angeles: University of California Press.
Quézel P and Médail F (2003) *Ecologie et biogéographie des forêts du bassin méditerranéen*. Paris: Elsevier.
Smith-Ramírez C, Armesto JJ, and Valdovinos C (2005) *Historia, biodiversidad y ecología de los bosques costeros de Chile*. Santiago de Chile: Editorial Universitaria.

Peatlands

D H Vitt, Southern Illinois University, Carbondale, IL, USA

Introduction
Occurrence
Environmental Limiting Factors
Peatland Types
Important Processes in Peatlands
Initiation and Development of Peatlands
Peatlands as Carbon Sinks
Further Reading

Introduction

Peatlands, or mires as they are sometimes called, are characterized by often deep accumulations of incompletely decomposed organic material, or peat. Peat accumulates when carbon that is sequestered in plant biomass through the process of photosynthesis exceeds the long-term loss of this carbon to the atmosphere via decomposition plus losses of carbon dissolved in water removed from the peatland through hydrological flow. Globally, peatlands contain about 30% of the world's terrestrial soil carbon, while covering only about 3–4% of the Earth's surface, and as such their carbon storage is considerably greater than their land surface area might indicate. Peatlands, in general, are relatively species poor when compared to upland communities in the same geographic region. However, due to the specialized environmental conditions often associated with peatlands, plants, and animals found only in these ecosystems are sometimes present. Peatlands are especially known for the presence of carnivorous plants such as *Sarracenia* and *Drosera* and for the occurrence of a large number of species of peat mosses (the genus *Sphagnum*).

Occurrence

Globally, peatlands occupy about 4 million km^2, with the boreal and subarctic peatland area estimated to be approximately 3 460 000 km^2, or about 87% of the world's peatlands. Six countries have greater than 50 000 km^2 of peatland and these account for 93% of the world's peatlands – five of these countries are predominantly boreal. Russia contains 1.42 million km^2, Canada 1.235 million km^2, the US 625 000 km^2, Finland 96 000 km^2, and Sweden 70 000 km^2; Indonesia has an estimated 270 000 km^2 as well. Although peat-forming plant communities occur in most of the world's nine zonobiomes, they are most prevalent in zonobiome VIII (cold temperate), or more commonly termed the boreal forest or taiga (**Figure 1**). The world's largest peatland complex is located in western Siberia (especially noteworthy is the Great Vasyugan

Mire located between the Ob and Irtysh Rivers at about 58° N and 75° W). Two other large peatland complexes are the Hudson Bay Lowland in eastern Canada and the Mackenzie River Basin in northwestern Canada. Although peatlands have long been associated with cool, oceanic climatic regimes such as those in Britain and Ireland and indeed peatlands are common in these areas (in fact peatlands are most abundant in areas where the regional climate is continental with short cool summers and long cold winters), the vegetation is coniferous and evergreen, and the upland soils are podzolic.

Environmental Limiting Factors

The initiation, development, and succession of peatland ecosystems are influenced by a number of regional, external factors. Especially important are hydrological and landscape position, climate, and substrate chemistry. These regional allogenic factors determine a number of site-specific factors that influence individual peatland sites. These local factors include rate of water flow, quantity of nutrient inputs, the overall chemistry of the water in contact with the peatland, and the amount of water level fluctuation. Additionally, there are a number of internal, or autogenic, processes that help regulate peatland form and function (**Figure 2**). These allogenic and autogenic factors operate in an everchanging world of disturbance that includes natural disturbances, especially wildfire, as well as anthropogenic disturbances such as mining, forestry, and agriculture.

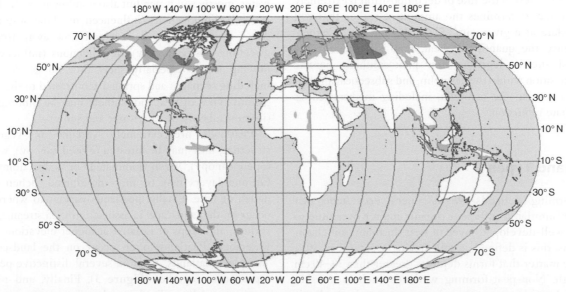

Figure 1 Estimated global distribution of peatlands. Areas colored in light green are those having >10% peat cover. The orange areas in North America and Siberia are the world's largest peatland complexes. The dot in western Siberia is the location for the Vasyugan peatland.

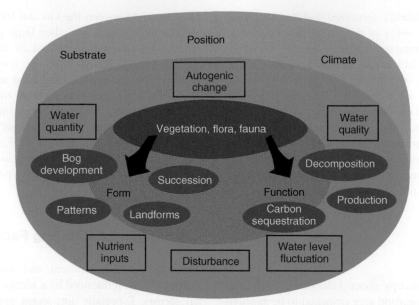

Figure 2 Substrate, position, and climate are regional factors that influence six local factors that are shown in boxes in the diagram. These local drivers direct both the form and function of bogs and fens. Adapted from Vitt DH (2006) Peatlands: Canada's past and future carbon legacy. In: Bhatti J, Lal R, Price M, and Apps MJ (eds.) *Climate Change and Carbon in Managed Forests*, pp. 201–216. Boca Raton, FL: CRC Press.

Peatland form and function are dependent on the process of peat accumulation and the pattern of loss or gain of carbon from habitats. Peat accumulation is dependent on the input of organic matter produced by photosynthesis. This organic matter is first accumulated in the upper, aerobic (or acrotelm) peat column wherein relatively rapid rates of decomposition occur. The rate at which this partially decomposed organic matter is deposited into the water-saturated, anaerobic peat column (the catotelm), wherein the rate of decomposition is extremely slow, largely determines the amount of carbon that will accumulate at a given site. Thus, the amount of carbon, and hence the quantity of peat, that is deposited at a peatland site is dependent on photosynthesis, aerobic decomposition within the acrotelm, and subsequent anaerobic processes in the catotelm, including methenogenesis and sulfate reduction.

Peatland Types

Peat-forming wetlands are in general ecosystems that have accumulated sufficient organic matter over time to have a well-developed layer of peat. In many soil classifications, this is defined as soils having greater than 30% organic matter that forms deposits greater than 30–40 cm in depth. Non-peat-forming wetlands such as marshes (wetlands without trees) and swamps (wetlands dominated by a tree layer) mostly have less than 30–40 cm of accumulated organic material and over time have not

been able to sustain continued accumulation of a carbon-rich peat deposit. Numerous classifications have been proposed that distinguish between various peatland types. For example, peatlands have been classified based on the source of water that has the primary influence on the peatland. Thus, peatlands that are influenced by water that has been in contact with soil or lake waters are termed geogenous and are divided into three types. Peatlands may be topogenous (influenced by stagnant water, mostly soil water, but also nonflowing water bodies as well), limnogenous (influenced by flood water from water courses resulting in lateral flow away from the direction of stream flow), or soligenous (influenced by flowing water, especially sheet flow on gentle slopes, including seepages and springs). Contrasted to these geogenous types of peatlands, others may be ombrogenous (influenced only by rain water and snow).

Peatlands are extremely variable in vegetation structure; they may be forested (closed canopy), wooded (open canopy), shrub dominated, or sedge dominated. Ground layers may be moss dominated, lichen dominated, or bare. Finally, peatlands vary as to where they occur on the landscape: in association with streams, lakes, springs, and seeps or isolated at higher elevations in the watershed. Peatlands often occur on the landscape as 'complex peatlands', wherein several distinctive peatland types occur together (**Figure 3**). Finally, and perhaps most universally utilized, is a classification that combines aspects of hydrology, vegetation, and chemistry into a functional classification of peat-forming wetlands. In

Figure 3 Peatland complex in northern Alberta, Canada. Patterned fen in left foreground, bog island with localized permafrost (large trees) and melted internal lawns to left, and curved treed bog island to right background. Small treed, oval island in center is upland.

general, this view of peatlands would consider hydrology as fundamental to peatland function and recognize two peatland types – fens and bogs.

Fens are peatlands that develop under the influence of geogenous waters (or waters that influence the peatland after being in contact with surrounding mineral, or upland, substrates). Waters contacting individual peatlands have variable amounts of dissolved minerals (especially base cations (Na^+, K^+, Ca^{2+}, Mg^{2+}) and associated anions (HCO_3^-, SO_4^{2-}, Cl^-)), and may also vary in the amount of nutrients (N and P) as well as the number of hydrogen ions. Further complicating this minerotrophy is variation in the flow of water, including amount of flow and as well as source of the water (surface, ground, lake, or stream). Peatlands receiving water only from the atmosphere via precipitation are hydrologically isolated from the surrounding landscape. These ombrogenous peatlands, or bogs, are ombrotrophic ecosystems receiving nutrients and minerals only from atmospherically deposited sources.

In summary and from a hydrological perspective, in fens water flows into and through the peatland after it has been in contact with surrounding materials, whereas in bogs water is deposited directly on the peatland surface and then flows through and out of the bog directly onto the surrounding landscape. Thus, fens are always lower in elevation than the surrounding landscape, while bogs are slightly raised about the connecting upland areas.

The recognition that hydrology is the prime factor for dividing peatlands into fens and bogs dates back to the 1800s. However, in the 1940s, Einar DuReitz recognized that vegetation composition and floristic indicators could be used to further characterize bogs and fens. Somewhat later, Hugo Sjörs associated these floristic indicators with variation in pH and electrical conductivity (as a surrogate for total ionic content of the water). The results of these

early field studies in Sweden provided an overarching view of how hydrology, water chemistry, and flora are associated, and more recent studies delineate how these combined attributes together form a functional classification of northern peatlands that provides an ecosystem perspective.

Bogs

Bogs are functionally ombrotrophic. At least in the Northern Hemisphere, they have ground layers dominated by the bryophyte genus *Sphagnum* (**Figure 4**). Sedges (*Carex* spp.) are absent or nearly so. The shrub layer is well developed and trees may or may not be present. Nearly, all of the vascular plants have associations with mycorrhizal fungi. Microrelief of raised mounds (hummocks) and depressions (hollows) is generally well developed. The peat column consists of a deep anaerobic layer (the catotelm), wherein decompositional processes are extremely slow and a surficial layer of 1–10 dm of the peat column that occupies an aerobic zone (the acrotelm). The acrotelm extends upward from the anaerobic catotelm and is mostly made up of living and dead components of *Sphagnum* plants, wherein vascular plant roots and fallen vascular plant aboveground litter occur. Well-developed acrotelms are unique to ombrotrophic bogs and provide opportunities to study atmospheric deposition and ecosystem response to such deposition.

Bogs are acidic ecosystems that have pH's of around 3.5–4.5. Base cations are limited owing to the ombrogenous source of water and to the cation exchange abilities of *Sphagnum* (see below). Bicarbonate is lacking in bogs and carbon is dissolved in the water column only as CO_2. The lack of geogenous waters limits nutrient inputs to those derived only from atmospheric deposition, and thus nitrogen and phosphorus are in short supply.

Figure 4 Mixed lawn of the peatmosses: *Sphagnum angustifolium*, mostly to the left, and *S. magellanicum* (red), mostly to right.

Bogs appear to be limited in distribution to areas where precipitation exceeds potential evapotranspiration. In many oceanic regions of the Northern Hemisphere (especially Britain, Ireland, Fennoscandia, and coastal eastern Canada), bogs form large treeless expanses. In Europe, the Ericaceous shrub, *Calluna vulgaris*, forms a characteristic component of these treeless landscapes. Many of these oceanic bogs are patterned, with a series of pools of waters separated by raised linear ridges. This sometimes spectacular pool/ridge topography forms either concentric or eccentric patterns (**Figure 5**), with water flowing from the highest raised center of the bog to the lower surrounding edges. Runoff from the surrounding upland (and from the raised bog itself) is concentrated at the margins of these raised bogs and due to increased nutrients, decomposition processes are greater and peat accumulation somewhat less. Thus, the central, open, raised 'mire expanse' part of a bog is surrounded by a wetter, often shaded lagg, or moat, and this 'mire margin' zone may be dominated by plants indicative of fens. Some oceanic bogs have a rather flat mire expanse, with occasional pools of water. Whereas the mire expanse surface of these raised bogs is flat, the dome of water contained within the bog peat is convex and thus the driest part of the bog is at the edges just before contact with the fen lagg. This marginal, relatively dry upslope to the mire expanse is usually treed and is termed the 'rand'.

In continental areas, bogs have a very different appearance (**Figure 6**). These continental bogs have a conspicuous tree layer and abundant shrubs (mostly *Ledum* spp. or *Chamaedaphne calyculata*) while pools of water are not present. In North America, the endemic tree species, *Picea mariana*, dominates these continental bogs, while in Russia bogs have scattered individuals of *Pinus sylvestris*. Farther north in the subarctic and northern boreal zones, peat soils contain permafrost. When entire bog landforms are frozen, the bog becomes drier and dominated by lichens (especially species of the reindeer lichen, *Cladina*). Unfrozen or melted areas contained within these peat plateaus are easily

Figure 6 A continental ombrotrophic bog from western Canada. Tree species is *Picea mariana* (black spruce).

recognized features termed collapse scars (**Figure 7**). Peat plateaus form extensive landscapes across the subarctic zone of both North America and Siberia. Farther south in the boreal zone, bog landforms may contain only scattered pockets of permafrost (frost mounds), that over the past several decades have been actively melting. Recent melting of the raised frost mounds results in collapse of the mound and active revegetation by fen vegetation to form wet, internal lawns with associated dead and leaning trees (**Figure 8**).

Fens

Fens are peatlands that are minerotrophic that when compared to bogs have higher amounts of base cations and associated anions. All fens have an abundance of *Carex* and *Eriophorum* spp. and water levels at or near the surface of the peat (thus acrotelms are poorly developed). Unlike bogs that are characterized by high microrelief of hummocks and hollows, fens feature a more level topography

Figure 5 An oceanic eccentric bog. Maine, USA. Highest elevation of bog is to center left, with elongate axis sloping to distant right. Photo is courtesy of Ronald B. Davis.

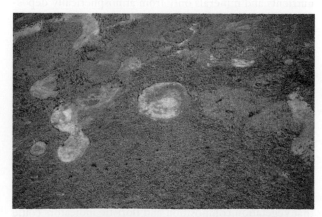

Figure 7 Extensive peat plateaus with permafrost (whitish areas dominated by the reindeer lichens in the genus *Cladina*), with isolated collapse scars (without permafrost – greenish circular to oblong areas), and with lush growth of *Sphagnum* species and sedges.

Figure 8 Bog dominated by *Picea mariana* in background, with dead snags in foreground, indicating recent permafrost collapse and the formation of an internal lawn and dominated by carpet and lawn species of *Sphagnum*.

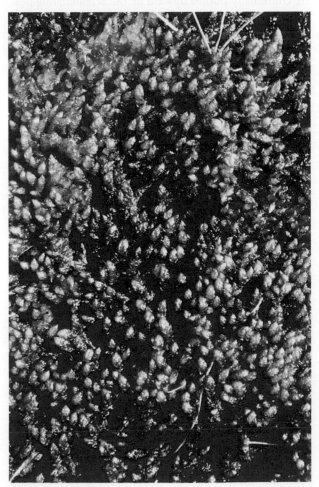

Figure 9 A carpet of the brown moss, *Scorpidium scorpioides*, a characteristic species of rich fens.

of extensive carpets and lawns dominated by species of mosses (**Figure 9**). Depending on the characteristics of the surrounding water, fens can by divided into three types.

Poor fens

These *Sphagnum*-dominated peatlands are associated with acidic waters (pH 3.5–5.5) that contain the least amount of base cations and little or no bicarbonate alkalinity.

Rich fens

True mosses dominate the ground layer of rich fens, especially a series of species that are red-brown in color and often termed 'brown mosses'. Examples of important species would be *Drepanocladus*, *Hamatocaulis*, *Warnstorfia*, *Meesia*, *Campylium*, *Calliergon*, and *Scorpidium*. Waters have pH varying from 5.5 to more than 8.0 and base cations are relatively abundant, especially calcium. Alkalinity varies from very little to extremely high amounts of bicarbonate. Rich fens occur as two types centered on the chemistry of the pore waters. 'Moderate-rich fens' have pH values between 5.5 and 7.0, with little alkalinity. Both brown mosses and some mesotrophic species of *Sphagnum* (e.g., *S. subsecundum*, *S. teres*, and *S. warnstorfii*) dominate the ground layer. 'Extreme-rich fens' are bicarbonate-rich peatlands, often with deposits of marl (precipitated $CaCO_3$) and pH ranging from around neutral to over 8.0. Species of *Scorpidium*, *Campylium*, and *Hamatocaulis* dominate the ground layer.

Whereas water quality (= chemistry) is the main factor controlling fen type and flora, water quantity (= flow) controls vegetation structure and surface topography. Fens, whether poor or rich, are vegetationally extremely variable, ranging from sites having abundant trees (dominated by *Larix laricina* in North America), to sites dominated by shrubs (mostly *Betula*, *Alnus*, and *Salix*), to sites having only sedges and mosses. Topographically, fens may be homogeneous and dominated by lawns and carpets. However, as water flowing through the fen increases, the surface vegetation develops a reticulation of wet pools and carpets separated by slightly raised ridges. Further increase in flow of water directs the patterns into linear pools (some filled with floating vegetation = carpets), sometimes termed flarks, alternating with linear ridges (termed strings; **Figure 10**). These pool/string complexes are oriented perpendicular to water flow, with smaller pools always upstream from the larger ones. Especially prevalent in Scandinavia and Russia, these patterned fens and associated bog islands form extensive peatlands termed aapamires.

Important Processes in Peatlands

Acidification

Sphagnum species have cell walls rich in uronic acids that in aqueous solution readily exchange a hydrogen ion for a base cation. The base cations that are in solution in bogs and poor fens are received by the peatland from atmospheric deposition or inflowing water and are always associated with an inorganic anion (HCO_3^-, SO_4^{2-}, Cl^-).

Figure 10 A patterned fen in western Canada characterized by elongate pools (flarks) separated by raised ridges (strings), oriented perpendicular to water flow.

When the base cation is exchanged for the organically produced H^+, acidity of the peatland waters is produced. This acidity thus originated through the exchange of an inorganic base cation for an H^+ produced by *Sphagnum* growth – hence this is termed inorganic acidity. Inorganic acidity relies on the presence of base cations and can only produce acidity when base cations are present in the pore water to exchange. Inorganic acidity is an extremely powerful process when abundant base cations are present such as in rich fens transitional to poor fens and in poor fens. In bogs, with limited supplies of base cations due to their ombrogenous water supply, inorganic acidity is less important.

Organic material produced by plants is decomposed and carbon mineralized through bacterial and fungal respiration. Under aerobic conditions, bacteria break down long cellulose chains and in doing so eventually produce short-chained molecules that are small enough to be dissolved in the pore waters. This dissolved organic carbon (DOC) may be lost to the peatlands via runoff or may remain suspended in the pore waters for some length of time. These decompositional processes produce acidity through dissociation of humic acids, acidity that is completely produced via organic processes; hence, peatland acidity produced via decompositional processes, and extremely important in ombrotrophic bogs, is termed organic acidity.

Rich fens, with pH above 7.0, also accumulate deep deposits of peat and are well buffered by large inputs of bicarbonate alkalinity. With continued inputs of bicarbonate, rich fens may remain stable for millennia, dominated by brown mosses that have little capacity for inorganic acidification, but strong tolerance for the alkaline peatland waters. However, as rich fens accumulate peat to depths of several meters, there is the possibility that the active surface layer will become more isolated from the bicarbonate inputs and alkalinity may decrease to the point that some tolerant species of *Sphagnum* may invade.

If *Sphagnum* species establish, then cation exchange proceeds, acidity increases while alkalinity decreases, and rich fen plant species are replaced by poor fen species tolerating acidic conditions. This acidification of rich fens has been documented in the paleorecord wherein the change from rich fen to poor fen vegetation takes place extremely rapidly, perhaps in the order of 100–300 years. As a result, these transitional rich fen–poor fen communities are short-lived on the landscape and among the most rare of peatland types.

Water Retention

The surface of a peatland lies on a column of water contained within the peat column. The peatland surface consists of a nearly complete cover of mosses (either peat mosses (*Sphagnum*) or true mosses (brown mosses)) that are continually pushed upward by the accumulating peat. This upward growth is limited only by the abilities of the peat and living moss layer to maintain a continuous water column that allows the living moss layer to grow. The vascular plants that grow in this water-soaked peat column produce roots that are largely contained in the small upper aerobic part of the peat. The mosses, however, alive and growing only from their uppermost stem apices, must maintain contact with the water column; thus, wicking and retaining of water above the saturated water column is paramount for maintenance of the moss layer. Peatland mosses have special modifications that help in this regard. Although some brown mosses have adaptations for water retention, such as the development of a tomentum of rhizoids along the stems, numerous branches along the stem that provide small spaces for capillarity, and leaves that have enlarged bases that retain water, it is in species of *Sphagnum* where water retention (up to 20 times dry plant weight) is greatly enhanced through a number of morphological modifications. *Sphagnum* has unistratose (one-celled thick) leaves consisting of alternating, large, dead, hyaline cells and small, partially enclosed, living, green cells. The walls of the hyaline cells are perforated with pores and are strengthened by the presence of cross-fibrils. Stems and branches are often encased in an outer layer of one or more rows of dead, enlarged cells. All of these hyaline cells have lost their living cell contents very early on in development and as a result the ratio of carbon to nitrogen is high. In addition to the features that allow the plants to hold water internally, the entire *Sphagnum* plant is a series of tiny spaces that serve as reservoirs for capillarity. The branches are surrounded by numerous, overlapping, very concave branch leaves (one-cell thick). The branches are attached to the stem in fascicles of three to five branches, half of which hang along the stem and half extend outward at more or less 90°. The fascicles of branches originate at the stem apex, and slowly develop while still close together at the apex of the stem. This

Figure 11 A longitudinal view of the canopy of *Sphagnum*; each stem is terminated by a capitulum of young branches. The branches along the stem are covered with numerous overlapping leaves and organized into fascicles that have branches that hang down along the stem as well as branches that spread outward from the stem allowing the individual stems to be evenly spaced from one another.

group of maturing branches, the capitulum, along with the top 1–5 cm of mature stem and associated branches form a dense canopy. In total, this canopy (**Figure 11**) consists of numerous small spaces of different sizes and, along with the dead hyaline cells of the leaves and branches, provides the mechanism for wicking and retention of capillary water far above the actual water table, which in turn provides the framework for the aerobic peat column that is so characteristic of bogs.

Nutrient Sequestration (Oligotrophification)

Peat forms due to slow decompositional processes that allow organic materials to be deposited as peat. As organic material is deposited, it contains within its carbon matrix nutrients, especially nitrogen and phosphorus, which were originally incorporated in the cell structure of the living plants, especially those of *Sphagnum* and brown mosses. Relatively rapid decomposition in the acrotelm mineralizes only a portion of the total nutrients tied up in the plant material, making these available for further plant growth as well as fungal and bacterial processing. However, upon entry to the catotelm, almost all decompositional activity stops and the nutrients become tied to organic materials in unavailable forms. Thus, rather than being recycled and remaining available for new plant growth, nitrogen and phosphorus become part of long-term unavailable nutrient pools. The lack of ability to utilize this unavailable pool of nutrients causes peatlands over time to become more oligotrophic at their surface yet also having large amounts of stored nitrogen and phosphorus. For example, *Sphagnum* peat is generally about 1% nitrogen; however, almost all of this catotelmic

nitrogen is unavailable for plant and microorganism use while in place in the peat deposit. When exposed to the atmosphere (e.g., as a garden amendment), the carbon is oxidized to CO_2 and the nitrogen is mineralized to NO_3^- and NH_4^+ and available for plant uptake. Although the actual percent of nitrogen, and other nutrients, may not be as high as that in inorganic soils, the total amount in the soil within any one square meter surface area of the peatland is greater in peat soils due to the depth of the peat present. This oligotrophification, and consequently nutrient storage, is autogenetically enhanced through the buildup of the peat column, placing the peat surface farther from the source of the nutrient inputs. The long-term result of oligotrophication is the regional storage of large pools of both carbon as well as important nutrients, especially nitrogen and phosphorus.

Methane Production

Methane is a highly potent greenhouse gas that originates from both natural and anthropogenic origins. On a weight basis, methane is 21 times more efficient at trapping heat and warming the planet than carbon dioxide. Methane emissions from wetlands account for more than 75% of the global emissions from all natural sources. Methane is a highly reduced compound produced as the end product of anaerobic decomposition by a group of microorganisms called methanogens, which phylogenetically belong to *Archaea*. These strict anaerobes can utilize only a limited variety of substrates with H_2–CO_2 and acetate being the most important too. The H_2–CO_2 dependent methanogenesis is considered the dominant pathway of methane production in boreal peatlands. However, acetate-dependent methanogenesis sometimes dominates in fens. In rich fens, higher nutrient availability promotes the growth of vascular plants (primarily sedges). Roots of these vascular plants penetrate deep into the peat column and therefore transport potential carbon-rich substrates, such as acetate, into the anaerobic layer. Rapid decomposition of organic matter also provides abundant substrates for methanogens. Poor fens, with lower vascular plant cover than that of rich fens, generally have lower potentials for CH_4, and a higher portion of the produced CH_4 comes from H_2–CO_2. Similar to poor fens, *Sphagnum*-dominated bogs also have a higher proportion of CH_4 produced from H_2–CO_2, and it may be that the dominance of mosses (without roots) and mycorrhizal vascular plants (without deep carbon-rich roots), along with the reduced abundance of sedges with well-developed deep roots, prohibit movement of labile carbon substrates to the anaerobic peat layer. Low decomposition rates in acidic bogs also limit the amount of acetate that can be produced during peat decomposition, which in turn limits the acetoclastic pathway. Methanogen diversity in bogs is very low and the composition of the methanogen community in bogs also differs greatly from that characteristic of fens. In general, higher CH_4 production is found in peatlands with higher vascular plant cover, and higher water tables are found in rich fens.

Sulfate Reduction

In peatlands, sulfur occurs in several different redox states (S valences ranging from +6 in SO_4^{2-} to −2 in hydrogen sulfide (H_2S), S-containing amino acids, and other compounds), and conversions between these states are the direct result of microbially mediated transformations. In bogs, the sole sulfur input is via atmospheric deposition, while in fens atmospheric deposition can be augmented by surface and/or groundwater inputs, which may contain sulfur derived from weathering of minerals in rock and soil. Regardless of the sulfur source, when sulfur enters a peatland, there are a variety of pathways through which it can cycle. In the aerobic zone, sulfate can be adsorbed onto soil particles, or assimilated by both plants and microbes. In the anaerobic zone, sulfate can also be adsorbed onto soil particles, assimilated by plants or microbes, or reduced by sulfate-reducing bacteria through the process of dissimilatory sulfate reduction. Dissimilatory sulfate reduction is a chemoheterotrophic process whereby bacteria in at least 19 different genera oxidize organic matter to meet their energy requirements using sulfate as the terminal electron acceptor. Thus, this process is one way in which carbon is lost from the catotelm. If the sulfate is reduced by sulfate-reducing bacteria, the end product (S^{2-}) can have several different fates. In the catotelm, where S^{2-} is formed, it can react with hydrogen, to produce H_2S gas, which can diffuse upwardly into or through the acrotelm where it can be either oxidized to sulfate, or lost to the atmosphere. Alternatively, H_2S can react by nucleophilic attack with organic matter to form organic or C-bonded sulfur (CBS). If Fe is present, S^{2-} can react with Fe to form FeS and FeS_2 (pyrite), which is referred to as reduced inorganic sulfur (RIS). The RIS pool tends to be unstable in peat and can be reoxidized aerobically with oxygen if the water table falls, or anaerobically probably using Fe_3^+ as an anaerobic electron acceptor. If Hg is present, and combines with S^{2-} to form neutrally charged HgS, then Hg sulfide is capable of passive diffusion across cell membranes of bacteria that methylate Hg. Alternatively, bacteria can transfer the methoxy groups of naturally occurring compounds, such as syringic acid, to S^{2-}, and form methyl sulfide (MeSH) or dimethyl sulfide (DMS), although the exact mechanisms by which this occurs are still unknown.

Initiation and Development of Peatlands

Peatlands initiate in one of four ways. The first, the most common, appears to through paludification (or swamping), wherein peat forms on previously drier, vegetated habitats on inorganic soils and in the absence of a body of water, generally due to regional water table rise and associated climatic moderation. Additionally, local site factors also have strong influences on paludification. Second, peat may form directly on fresh, moist, nonvegetated mineral soils. This primary peat formation occurs directly after glacial retreat or on former inundated land that has risen due to isostatic rebound. Third, shallow bodies of water may gradually be filled in by vegetation that develops floating and grounded mats – thus terrestrializing the former aquatic habitat. Both lake chemistry and morphometry as well as species of plants in the local area influence the rates and vegetative succession. Fourth, peat may form and be deposited on shallow basins once occupied by extinct Early Holocene lakes. These former lake basins, lined with vegetated impervious lake clays, provide hydrologically suitable sites for subsequent peat development.

Across the boreal zone, peatland initiation appears to be extremely sensitive to climatic controls. For example, in oceanic areas, peatlands often initiated soon after glacial retreat some 10 000–12 000 years ago. Many of these oceanic peatlands began as bogs and have maintained bog vegetation throughout their entire development. In more continental conditions, most peatlands were largely initiated through paludification. In areas where the bedrock is acidic, most of these early peatlands were poor fens, whereas in areas where soils are base rich and alkaline, rich fens dominated the early stages. Like oceanic peatlands, subcontinental peatlands initiated soon after glacial retreat; however, throughout most of the large expanses of boreal Canada and Siberia, peatland initiation was delayed until after the Early Holocene dry period, initiating 6000–7000 years ago. Many of these peatlands initiated as rich or poor fens and have remained as fens for their entire existence, whereas others have undergone succession and today are truly ombrotrophic bogs. A recent study in western Canada correlated peak times of peatland initiation to Holocene climatic events that are evident in US Midwest lakes, North Atlantic cold cycles, and differing rates of peat accumulation in the one rich fen studied in western Canada.

Peatlands as Carbon Sinks

Peat is about 51% carbon and peatlands hold about 270–370 Pg (petagram) of carbon or about one-third of the world's soil carbon. For example in Alberta (Canada), where peatlands cover about 21% of the provincial landscape, the carbon in peatlands amounts to 13.5 Pg compared to 0.8 Pg in agricultural soils, 2.3 Pg in lake sediments, and 2.7 Pg in the province's forests. Estimates for apparent long-term carbon accumulation in oceanic, boreal, and subarctic peatlands range from around 19 to $25\,g\,C\,m^{-2}\,yr^{-2}$. However, disturbances can have a dramatic effect on carbon accumulation. Wildfire, peat extraction, dams and associated flooding, mining, oil and gas extraction, and other disturbances all reduce the potential for peatlands to sequester carbon, while only permafrost melting of frost mounds in boreal peatlands has been documented to have a positive effect on carbon sequestration. One recent study has suggested that effects from disturbance in Canada's western boreal region have reduced the regional carbon flux (amount of carbon sequestered in the regional peatlands) from about 8940 Gg (gigagram) $C\,yr^{-1}$ under undisturbed conditions to 1319 Gg carbon sequestered per year under the present disturbance regime, yet only 13% of the peatlands have been affected by recent disturbance. These data suggest that although for the long-term peatlands in the boreal forest region have been a carbon sink and have been removing carbon from the atmosphere, at the present time, due to disturbance, this capacity is greatly diminished. Furthermore, when disturbance is examined in more detail, it is wildfire that is the single greatest contributor to loss of carbon sequestration, both from a direct loss as a result of the fire itself as well as from a loss of carbon accumulation due to post-fire recovery losses. If wildfire greatly increases as is predicted by climate change models, then the effectiveness of peatlands to sequester carbon may be greatly reduced and it has been proposed that an increase of only 17% in the area burned annually could convert these peatlands to a regional net source of carbon to the atmosphere. If boreal peatlands become a source for atmospheric carbon, then the carbon contained within the current boreal peatland pool, in total, is approximately two-thirds of all the carbon in the atmosphere.

See also: Boreal Forest; Botanical Gardens; Chaparral.

Further Reading

Bauerochse A and Haßmann H (eds.) (2003) Peatlands: archaeological sites–archives of nature–nature conservation-wise use. *Proceedings of the Peatland Conference 2002 in Hanover, Germany,* Hanover: Verlag Marie Leidorf GmbH (Rahden/Westf.).

Davis RB and Anderson DS (1991) *The Eccentric Bogs of Maine: A Rare Wetland Type in the United States,* Technical Bulletin 146. Orono: Maine Agricultural Experiment Station.

Feehan J (1996) *The Bogs of Ireland: An Introduction to the Natural, Cultural and Industrial Heritage of Irish Peatlands.* Dublin: Dublin Environmental Institute.

Fraser LH and Kelly PA (eds.) (2005) *The World's Largest Wetlands: Their Ecology and Conservation.* Cambridge: Cambridge University Press.

Gore AJP (1983) *Ecosystems of the World. Mires – Swamp, Bog, Fen and Moor,* 2 vols. Amsterdam: Elsevier Scientific.

Joosten H and Clarke D (2002) *Wise Use of Mires and Peatlands – Background and Principles Including a Framework for Decision-Making.* Jyväskylä, Finland: International Mire Conservation Group andInternational Peat Society (http://www.mirewiseuse.com).

Larsen JA (1982) *The Ecology of the Northern Lowland Bogs and Conifer Forests.* New York: Academic Press.

Moore PD (ed.) (1984) *European Mires.* New York: Academic Press.

Moore PD and Bellamy DJ (1974) *Peatlands.* London: Elek Scientific.

National Wetlands Working Group (1988) *Wetlands of Canada. Ecological Land Classification Series, No. 24.* Ottawa: Sustainable Development Branch, Environment Canada, and Montreal: Polyscience Publications.

Parkyn L, Stoneman RE, and Ingram HAP (1997) *Conserving Peatlands.* NowYork: CAB International.

Vitt DH (2000) Peatlands: Ecosystems dominated by bryophytes. In: Shaw AJ and Goffinet B (eds.) *Bryophyte Biology,* pp. 312–343. Cambridge: Cambridge University Press.

Vitt DH (2006) Peatlands: Canada's past and future carbon legacy. In: Bhatti J, Lal R, Price M, and Apps MJ (eds.) *Climate Change and Carbon in Managed Forests,* pp. 201–216. Boca Raton, FL: CRC Press.

Wieder RK and Vitt DH (eds.) (2006) *Boreal Peatland Ecosystems.* Berlin, Heidelburg, New York: Springer.

Wright HE, Jr., Coffin BA, and Aaseng NE (1992) *The Patterned Peatlands of Minnesota.* Minneapolis: University of Minnesota Press.

Polar Terrestrial Ecology

T V Callaghan, Royal Swedish Academy of Sciences Abisko Scientific Research Station, Abisko, Sweden

The Future: Polar Regions and Climate Change	Further Reading

The polar regions are situated at latitudes beyond which the Earth's angle to the Sun is shallow and the input of thermal radiation is low. During the winter period, the Sun is below the horizon and there are prolonged periods of darkness. The resulting low-temperature regimes dominate ecological processes, either directly by affecting plant growth, microbial activity, animal behavior, organism reproduction and survival, or indirectly by controlling the length of the snow- and ice-free periods in which most primary production and dependent biological activity occurs, the availability of water in liquid form, and the expansion and contraction, and other active layer properties in generally primitive soils underlain by permafrost. Feedback mechanisms from polar regions and their ecological systems to the climate system affect local, regional, and global climate. The balance between greenhouse gas emissions from decomposition, particularly soil microbial respiration, and photosynthesis has resulted in a large net accumulation of carbon in arctic soils while ice and snow that cover low, tundra vegetation reflect incoming radiation. Both mechanisms lead to cooling. In contrast, global ocean circulation leads to the redistribution of the Earth heat by cooling the tropics and warming the high latitudes.

Both the Arctic and the Antarctic are characterized by vast wilderness areas that are generally young, as most land areas, with some extensive exceptions in the Arctic, were glaciated in the Pleistocene. Polar regions host some of the Earth's most extreme environments and organisms such as snow algae, lichens that inhabit the crevices within crystalline rocks, and the simple communities of soil fauna in the dry valleys of Antarctica.

Polar environments vary between the Arctic and the Antarctic and also within each region (**Figures 1** and **2**).

The Arctic is dominated by a polar ocean surrounded by continental land masses and islands, whereas the Antarctic is dominated by a polar, largely ice-covered, land mass surrounded by oceans. Terrestrial ecosystems are extensive (7.5 million km^2) and varied and stretch from the closed canopy northern boreal forests in the south, through the latitudinal treeline ecotone and tundra wetlands to the polar desert in the north. Along this latitudinal gradient, mean July temperature varies from about 12 °C in the south to 2 °C in the north, total annual precipitation varies from about 250 to 75 mm (mainly as snow), and net primary production varies from about 1000 to 1 g m^{-2} yr^{-1}. Approximately 6000 animal and 5800 plant species inhabit arctic lands (3 and 5% of global biodiversity, respectively). Biodiversity decreases geometrically along this gradient. Although plant biodiversity is low in comparison with many biomes, it is surprisingly high per square meter because of the small scale of plants, and over 6000 species of animals and plants have been cataloged in and around Svalbard at about 79° N. There is also large environmental variation associated with the climatic effects of northern ocean currents: in arctic Norway, Sweden, and Finland, forests grow north of the Arctic Circle (66.7° N) because of the warming effect of the northward-flowing Gulf Stream, whereas polar bears and tundra vegetation are found at about 51° N in eastern Canada because of the cooling influence of southward-flowing cold ocean currents. In the Arctic, indigenous and other arctic peoples have been part of the ecosystem for millennia.

The large land masses of the Arctic have great connectivity with land masses further south: great rivers flow from low latitudes to the Arctic Ocean, and mammals and hundreds of millions of birds migrate between the summer breeding grounds in the Arctic and overwintering areas in boreal or temperate regions. Food chains in the Arctic are more complex than those in the Antarctic and at the top of the chain are mammalian carnivores such as the polar bears, wolves, and arctic foxes. Population cycles characteristic of arctic animals together with relatively

Figure 1 Ecosystems of the Arctic.

Figure 2 Coastal ecosystem, sub-Antarctic South Georgia.

few species in each trophic level can result in ecological instability and ecological cascades: increasing numbers of snow geese in arctic Canada have denuded vegetation resulting in habitat hypersalinity.

The Antarctic land mass covers some 12.4 million km^2 but less than 1% is seasonally ice free. In the Antarctic, the major environmental variation is associated with the relatively moist and 'warm' maritime climate of the west coast of the Antarctic Peninsula (temperatures are between 0 and 2 °C for 2–4 months in summer) contrasted with the cold, dry polar desert climate of the continental land mass. Consequently, most biological activity and most species are found on the west coast of the Antarctic Peninsula. Vegetation is dominated by relatively simple plant communities of lichens, mosses, and liverworts that support simple soil invertebrate communities. Only two species of higher plant and higher insects occur. Terrestrial mammals are absent and this short trophic structure, together with the isolation of the land mass, has enabled the establishment of a highly specialized, and commonly endemic fauna of ground-nesting birds (e.g., penguins) and seals that depend on the coastal land areas for breeding and moulting, and the sea, for food. Nutrients for plant growth in these areas are mainly derived from the sea and are deposited on land by wind or birds. In contrast, over much of the tundra, low nutrient availability to plants limits primary production. There are no indigenous peoples in the Antarctic and human activities there have been restricted to the past 200 years.

Human activity has, until recently, influenced both Arctic and Antarctic ecosystems less than most biomes on Earth. However, the polar amplification of global climate change together with the inherent sensitivity of polar ecological systems to invasion by species from warmer latitudes has resulted in the vulnerability of polar ecosystems which are now under threat of rapid change.

The Future: Polar Regions and Climate Change

The polar regions are undergoing rapid climate change. There is a general amplification of global warming in the Arctic: surface air temperatures have warmed at approximately twice the global rate, although there are local variations. The average warming north of 60° N has been 1–2 °C since a temperature minimum in the 1960s and 1970s with the largest increase (*c.* 1 °C per decade) in winter and spring. Continental arctic land masses together with the Antarctic Peninsula are the most rapidly warming areas of the globe. Precipitation in the Arctic shows trends of a small increase over the past century (about 1% per decade), but the trends vary greatly from place to place and measurements are very uncertain. There are reductions in Arctic sea ice, river and lake ice in much of the sub-Arctic, and Arctic glaciers. Reduction in Arctic sea ice has occurred at a rate of 8.9% per decade for September relative to the 1979 values and there was an un-predicted extreme reduction in 2007. Permafrost has warmed. Although changes in the active layer depth have no general trend, in some sub-Arctic locations, discontinuous permafrost is rapidly disappearing and changes in permafrost are driving changes in hydrology and ecosystems. In Arctic Russia, ponds are drying in the continuous permafrost zone and waterlogging is occurring where there is discontinuous permafrost.

In Antarctica, temperature trends show considerable spatial variability: the Antarctic Peninsula shows significant warming over the last 50 years, whereas cooling has occurred around the Amundsen-Scott Station at the South Pole and in the Dry Valleys. Consequently, there is no continent-wide polar amplification of global change in Antarctica.

Current polar warming is leading to changes in species' ranges and abundance and a northward and upward extension of the sub-Arctic treelines. Forest is projected to displace considerable areas of tundra in some places. Species tend to relocate, as they have in the past, rather than adapt to new climate regimes. However, this process is likely to lead to the loss of some species: polar bears and other ice-dependent organisms are particularly at threat. In other areas, where rates of species relocation are slower than climate change, the incidence of pests, disease, and fire is likely to increase. Changes in vegetation, particularly a transition from grasses to shrubs, have been reported in the North American Arctic, and satellite imagery has indicated an increase in the 'normalized difference vegetation index' (a measure of photosynthetically active biomass) over much of the Arctic. This index has increased by an average of about 10% for all tundra regions of North America, probably because of a longer growing season. However, such increases in productivity and changes in plant functional types have been shown experimentally to displace mosses and lichens that are now major components of Arctic vegetation.

In Antarctica, warming has caused major regional changes in terrestrial and marine ecosystems. The abundances of krill, Adelie, and Emperor penguins and Weddell seals have declined but the abundances of the only two native higher plants has increased. On continental Antarctica, climate change is affecting the vegetation composed of algae, lichens, and mosses. Introductions of alien species, facilitated by increased warming and increased human activity, are particular threats to southern ecosystems. Recent studies on sub-Antarctic islands have shown increases in the abundance of alien species and negative impacts on the local biota. In contrast, cooling has caused clear local impacts in the Dry Valleys where a 6–9% reduction in lake primary production and a 10% per year decline in soil invertebrates has occurred.

The responses of polar environments to climatic warming include feedbacks to the global climate system and other global impacts. Increased runoff from arctic rivers could affect the thermohaline circulation that redistributes the Earth's heat, thereby causing cooling in the North Atlantic and further warming in the tropics. Reductions in sea ice extent and snow cover together with a shift in vegetation from tundra to shrubs or forests are likely to reduce albedo (reflectivity of the surface) and lead to further warming despite the increased uptake of carbon dioxide by a more productive vegetation. Thawing permafrost is likely to release methane, a particularly powerful greenhouse gas, and evidence of this is already available from various arctic areas.

Not all impacts of climate warming in polar regions are disadvantageous to society: the reduction of sea ice in the Arctic is likely to lead to increased marine access to resources and new fisheries and reduced length of sea routes, while warming on land will probably lead to increased productivity and increased potential for forestry and agriculture.

See also: Alpine Ecosystems and the High-Elevation Treeline; Biological Wastewater Treatment Systems.

Further Reading

Anisimor OA, Vaughan DG, Callaghan TV, *et al.* (2007) Polar regions (Arctic and Antarctic). In: Parry ML, Canziani OF, Palutikof JP, Hanson CE, and Van der Linder PJ (eds.) *Climate Change 2007: Impacts, Adaptation and Vulnerability. Contribution of Working Group II to the Fourth Assessment Report of the Intergovernmental Panel on Climate Change,* pp. 655–685. Cambridge: Cambridge University Press.

Callaghan TV, Björn LO, Chapin FS, III, *et al.* (2005) Tundra and polar desert ecosystems. In: *ACIA. Arctic Climate Impacts Assessment,* pp. 243–352. Cambridge: Cambridge University Press.

Chapin FS, III, Berman M, Callaghan TV, *et al.* (2005) Polar ecosystems. In: Hassan R, Scholes R, and Ash N (eds.) *Ecosystems and Human Well-Being: Current State and Trends,* vol. 1, pp. 719–743. Washington, DC: Island Press.

Convey P (2001) Antarctic ecosystems. In: Levin SA (ed.) *Encyclopaedia of Biodiversity,* vol. 1, pp. 171–184. San Diego: Academic Press.

Nutall M and Callaghan TV (2000) *The Arctic: Environment, People, Policy,* 647pp. Reading: Harwood Academic Publishers.

Richter-Menge J, Overland J, Hanna E, *et al.* (2007) State of the Arctic Report.

Walther GR, Post E, Convey P, *et al.* (2002) Ecological responses to recent climate change. *Nature* 416(6879): 389–395.

Riparian Wetlands

K M Wantzen, University of Konstanz, Konstanz, Germany

W J Junk, Max Planck Institute for Limnology, Plön, Germany

Introduction	Typical Biota and Biodiversity in Riparian Wetlands
Definitions and Concepts	Ecological Services of Riparian Wetlands
Environmental Conditions Determining Riparian Wetlands	Conservation
Types of Riparian Wetlands	Further Reading

Introduction

The riparian zone of running water systems is a site of intensive ecological interactions between the aquatic and the terrestrial parts of the stream valley. Wetlands that occur in this zone exchange water with the aquifer and with the main channel during flood events (**Figure 1**). Riparian wetlands are buffer zones for the water budget of the landscape: they take up excess water from flood events and release it gradually afterwards.

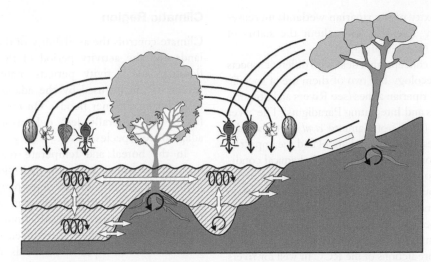

Figure 1 Inputs, turnover, and exchange of organic matter in the stream channel (left) and a riparian wetland water body (center) at low and high water levels. Black arrows indicate organic matter inputs, white arrows indicate water exchange pathways, spirals indicate nutrient spiralling or downriver transport, and circular arrows indicate sites of organic matter turnover *in situ*. Curly brace indicates water-level fluctuations during flood events. Modified from Wantzen KM, Yule C, Tockner K, and Junk WJ (2006) Riparian wetlands. In: Dudgeon D (ed.) *Tropical Stream Ecology*, pp. 199–217. Amsterdam: Elsevier.

Modern ecological theory recognizes the important role riparian wetlands play for biodiversity and for the energy and matter budgets along the whole range of river courses. The carbon and nutrient budgets are influenced by dissolved and particulate substances from the bordering terrestrial ecosystems, by the autochthonous production from the wetland plants, and by allochthonous organic matter delivered by the floodwater. The proportions between these sources are defined by the hydrological patterns, landscape morphology, and climatic conditions. (see Rivers and Streams: Physical Setting and Adapted Biota and Rivers and Streams: Ecosystem Dynamics and Integrating Paradigms).

The crossover between humid and dry conditions creates habitats for organisms coming from either aquatic or terrestrial ecosystems, and for those biota that are specialized on wetland conditions. As the transversal dimension of streamside wetlands is generally small, their overall importance for landscape ecology, biogeochemistry, and biodiversity is often overlooked. However, the total size of these wetlands can be considerable in areas with dense stream networks. Moreover, the corridor-shaped extension of riparian wetlands makes them perfect pathways for the gene flow between remote populations of aquatic and terrestrial biota. Many ecological services are uniquely provided by riparian wetlands, including erosion control, filtering of nutrients and pesticides from adjacent cropland, mitigation of floods, and recreation, which increases their conservation value in a socioeconomic context.

There is a large array of environmental conditions that vary between the different types of riparian wetlands, especially climatic region and prevailing vegetation type, and landscape morphology and hydrologic patterns. This article deals with the different types of riparian wetlands, their deterministic environmental conditions, prevailing ecological processes, typical biota, and aspects of conservation.

Definitions and Concepts

There are many definitions of riparian wetlands. A hydrological definition defines riparian wetlands as

lowland terrestrial ecotones which derive their high water tables and alluvial soils from drainage and erosion of adjacent uplands on the one side or from periodic flooding from aquatic ecosystems on the other (McCormick, 1979)

A functional definition states that

riparian areas are three-dimensional ecotones of interaction that include terrestrial and aquatic ecosystems, that extend down to the groundwater, up above the canopy, outward across the floodplain, up the near-slopes that drain to the water, laterally into the terrestrial ecosystem, and along the water course at a variable with (Ilhard *et al.*, 2000).

Both definitions point to the ecotonal character of riparian wetlands between water bodies on one side and the upland on the other. Riparian wetlands can be, at the smallest scale, the immediate water's edge where some aquatic plants and animals form a distinct community, and pass to periodically flooded areas of a few tens of meters width. At medium scale they form bands of vegetation, and at the largest scale they form extended floodplains of tens of kilometers width along large rivers.

In this case, complexity of the riparian wetlands increases so much that many scientists give them the status of specific ecosystems (see Floodplains).

There are several concepts that deal with different aspects of stream and river ecology but two of them are of specific interest to rivers and riparian zones (see Rivers and Streams: Ecosystem Dynamics and Integrating Paradigms). The 'river continuum concept' (RCC) of Vannote *et al.*, describes the longitudinal processes in the river channel and the impact of the riparian vegetation on the physical and chemical conditions and as carbon source to the aquatic communities in the channel. The 'flood pulse concept' (FPC) of Junk *et al.* stresses the lateral interaction between the floodplain and the river channel and describes the specific physical, chemical, and biological processes and plant and animal communities inside the floodplain. The predictions of the RCC fit well for rivers with narrow riparian zones but with increasing lateral extent and complexity of the riparian zone the FPC becomes more important. Here, we restrict our discussion to riparian wetlands along streams and low-order rivers. Since lateral extent of the riparian zone along low-order rivers can vary considerably in different parts of the same river or between different rivers of the same river order, the applicability of the concepts may also vary.

Environmental Conditions Determining Riparian Wetlands

Riparian habitats are integral parts of a larger landscape and therefore influenced by factors operating at various special and temporal scales. The physical setting that determines rivers and streams basically defines the riparian wetlands (see Rivers and Streams: Ecosystem Dynamics and Integrating Paradigms); however, some environmental features have specific importance on the wetlands that will be dealt with in the following.

Spatial and Temporal Scales

At the regional scale, geomorphology, climate, and vegetation affect channel morphology, sediment input, stream hydrology, and nutrient inputs. At the local scale, land use and related alteration to stream habitats, but also the activity of bioengineers such as beavers, can be of significant influence. At short timescales, individual heavy rainfall events affect the riparian systems; at an annual basis climate-induced changes in light, temperature, and precipitation trigger important cyclic biological events, such as autochthonous primary and secondary production, litterfall, decomposition, and spawning and hatching of animals. On multiannual timescales, extreme flood and drought events, debris-torrents, landslides, heavy storms or fire can have dramatic effects on the riparian zone and its biota.

Climatic Region

Climate controls the availability of the water in the wetlands and the activity period of the organisms. If the flooding and activity periods match, the floodborne resources can be used by the adapted floodplain biota (e.g., during summer floods). On the other hand, winter floods are generally less deleterious for little-flood-adapted tree species.

In the boreal and temperate regions, freezing and drought in winter and snowmelt floods in spring are predictable drivers of the interplay between surface water and groundwater in riparian wetland hydrology. Ice jams may cause stochastic flood events in winter. Normally, stream runoff is reduced during winter, and groundwater-fed riparian wetlands discharge into the stream channel as long as possible. In wetlands with organic sediments, this water is often loaded with large amounts of dissolved organic carbon. In shallow streams that freeze completely during winter, riparian wetlands may serve as refuges for the aquatic fauna, for example, for amphibians and turtles. Spring snowmelt events generally provoke prolonged flood events that exceed the duration of rain-driven floods. These long floods can connect the riparian wetland water bodies to the stream, so that organic matter and biota become exchanged. At the same time, there is often an infiltration (downwelling) of surface water into the riparian groundwater body.

In seasonal wet-and-dry climates (both Mediterranean and tropical savanna climates) water supply by rainfall is limited to a period of several months during which very strong rainstorms may occur. These events, albeit short, are of great importance for the release of dissolved substances and for the exchange of organic substances and biota between wetland and main water course. Moreover, energy-rich organic matter (e.g., fruits) may become flushed from the terrestrial parts of the catchment into riparian wetlands. On the other hand, flash-floods can cause scouring and erosion of fine sediments (including organic matter). During the dry season, groundwater levels are lower and may cause a seasonal drought in the riparian wetlands. In these periods, the aquatic biota either estivate or migrate into the permanent water bodies, and large parts of the stocked organic matter become mineralized. However, even in strongly seasonal zones, like the Brazilian Cerrado, groundwater supply may be large enough to support permanent deposition of undecomposed organic matter.

The distribution of water-conductive (coarse) and impermeable substrate (bedrock and loam) of the valley bottom influences the thickness of the stagnant water body in the riparian zone and thus the extension of organic matter layers. Permanently humid conditions are found in many riparian wetlands of the boreal zone and in the humid tropics. These permanent riparian wetlands can accumulate large amounts of organic carbon. In tropical Southeast Asia (Malayan Peninsula and parts of Borneo), a special case of riparian wetland occurs, the peat swamps. These swamps develop when mangrove forests proceed seawards, and the hinterland soils lose their salt content. Here, large amounts of organic matter from the trees become deposited and the streams flow within these accumulations (see Peatlands).

Valley Size, Morphology, and Connectivity

The common textbook pattern of steep valleys in the upper sections of the streams and open, shallow floodplains in the lower river sections holds true only for very few cases in nature. Rather, we find these two valley types interspersed in an alternating pattern like 'beads on a string'. Shallow areas are more likely to bear extended riparian wetlands; however, if groundwater levels are high enough, even steep valleys may be covered with wetlands. The morphology of riparian wetlands can be described by the entrenchment ratio (i.e., the ratio of valley width at 50-years flood level to stream width at bankfull level) or by the belt width ratio, that is the distance between opposing meander bends over a stream section to stream width at bankfull level. Fifty-years flood often intersect the terrace slope.

Riparian wetlands of different catchments may be linked with each other through swamp areas (e.g., in old eroded landscapes of the Brazilian and Guyana Shields in South America) so that biogeographical barriers can be overcome by aquatic biota even without a permanent connection between the water courses. The term connectivity describes the degree by which a floodplain water body is linked to the main channel. Riparian wetlands may also be connected to the stream, either in a direct connection by a short channel, or indirectly by a longer channel which may be intercepted by a pond. In some cases, these channels can be cryptic/hidden when they are formed by macropores in the organic soils. Alluvial riparian wetlands may be connected to the stream via the hyporheic interstitial zone provided that the sediments are coarse enough to conduct water. Wetlands without any of these pathways exchange water, biota, and organic matter with the main channel during overbank flow of the stream. Purely aquatic organisms depend on the existence of connection channels to migrate between wetland and main water body. For example, amphibia are especially sensitive to fish predation, so that the highest biodiversity of amphibia is found at riparian wetland habitats with the lowest accessibility for fish.

Hydrology and Substrate Type

The slope of the landscape and the rock characteristics of the catchment define the physical habitat characteristics of the stream–wetland system. Riparian wetlands provide habitats with different hydraulic and substrate conditions than the stream channel. Although flooding in streams is generally shorter, less predictable and 'spikier' than in large rivers, there is a large number of exchange processes between the main channel and the riparian zone during these flood events. Major flood events, albeit rare, act as 'reset mechanism' in the floodplain that rejuvenates the sediment structure and the successional stage of the vegetation. Between these rare events, riparian wetlands act as sinks for fine particles and organic sediments that were washed out of the stream channel, the terrestrial zone of the catchment, or derive from an autochthonous biomass production.

Vegetation

Vegetation bordering to and growing within riparian wetlands fulfils many functions: it delivers both substrate for colonization and food resources for aquatic animals, it strips nutrients from the incoming water, and it provides raw material for the organic soils. It retards nutrient loss, filters nutrient input from the upland, reduces runoff by evapotranspiration, and buffers water-level fluctuations. Shading by tree canopies reduces light conditions for algal and macrophyte primary production and it equilibrates soil temperatures. Therefore, riparian wetlands differ completely according to their vegetation cover.

Unvegetated riparian wetlands occur at sites where establishment of higher plants is hampered by strong sediment movement (e.g., high-gradient and braided rivers), low temperatures (high elevation and polar zones), rocky surfaces, or periodical drought (desert rivers). The lack of shading and nutrient competition by higher plants favors growth of algae on the inorganic sediments, and productivity may be high, at least periodically.

High altitudes and/or elevated groundwater levels may preclude tree growth but allow the development of grass or herbal vegetation on riparian wetlands. Hillside swamp springs (helokrenes) can coalesce and form extensive marshes far above the flood level of the stream channel, so that the distinction between 'riparian' and 'common' wetland is difficult.

The tree species of forested riparian wetland are adapted to periodical or permanent waterlogging of the soils. They contribute an important input of organic carbon to the stream system. Large tree logs shape habitat structure by controlling flow and routing of water and sediment between stream channel and wetland. Tree roots increase sediment stability, sequester nutrients, and form habitats.

Types of Riparian Wetlands

Riparian wetlands are very variable in size and environmental characteristics. In the following, we list the most common types according to their hydrological and substrate characteristics (**Figures 2** and **3**).

Hygropetric Zone

At sites where groundwater outflows run over rocky surfaces, hygropetric zones develop. In the thin water film, there is a vivid algal production and a diverse, however, less-studied fauna of invertebrates (mostly aquatic moths, chironomids, and other dipterans). Biota of the hygropetric zone need to be adapted to harsh environmental conditions such as periodical freezing and drying of the surfaces.

Rockpools

Many streams run through bedrock or large boulders which have slots that fill with flood or rain water. Biota colonizing these pools have to be adapted to relatively short filling periods, high water temperatures, and solar radiation. High algal production and low predator pressure (at least at the beginning of the filling period) attract many invertebrate grazers.

Parafluvial and Orthofluvial Ponds

In alluvial stream floodplains, permanent or temporary ponds develop from riverine dynamics either within the active channel (parafluvial pond) or in the riparian zone (orthofluvial pond). They are fed by both surface water and groundwater. In coarse-grained sediments,

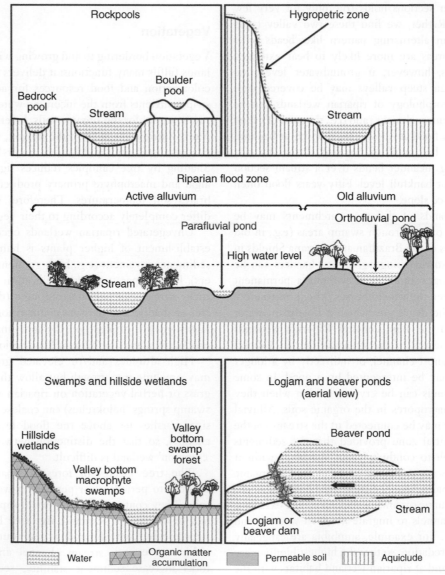

Figure 2 Types of riparian wetlands.

Figure 3 Photographs of riparian wetlands (Tenente Amaral Stream, Mato Grosso, Brazil): (a) Stream channel with hygropetric zone (foreground) and floodplain forest (background), (b) Rockpool carved into the sandstone bedrock, (c) moist organic soil colonized by many aquatic invertebrate taxa. Leaf litter was removed. All photographs by K. M. Wantzen.

these ponds are connected to the main channel by the hyporheic interstitial zone, that is, an ecotone between groundwater and surface water that extends below and at either side of the stream channel. In fine-grained sediments (including organic soils), the contribution of groundwater is much more important, and these ponds are often brownish from dissolved organic matter (humic acids and yellow substances). Para- and orthofluvial ponds contribute disproportionately to total species richness along riparian corridors.

Riparian Flood Zones

Even if no basin-like structures are present, flooding events create wetted zones on either side of the stream, independent of sediment type. Extension and permanence of the wetted zone depends on the valley shape, the porosity of the sediments, and eventual backflooding from tributary streams. In temporarily flooded forests with thick organic layers and in stranded debris dams, the moisture conditions may be long enough to bridge the gap between two flood events, so that many aquatic biota such as chironomids and other midges can complete their larval development in these semiaquatic habitats.

Riparian Valley Swamps

Swamps occur on soils that are waterlogged for most of the year. The lack of oxygen in the sediments allows accumulation of organic matter and selects for tree or herb species that have specific adaptations to these conditions, for example, pressure ventilation in the roots. The vegetation consists of either macrophytes or trees. Due to the shading and oxygen consumption during decomposition of organic matter, some of these riparian wetlands are hostile environments for aquatic metazoa that depend on dissolved oxygen. Some trees such as the Australian gum (*Melaleuca* sp.) shed bark which release secondary compounds that influence biota.

Hillside Wetlands

In areas where the aquiclude extends laterally from the stream, the riparian swamps can merge into hillside wetlands far above the flood level. Given that waterloggedness is permanently provided, these ecosystems tend to develop black organic soil layers from undecomposed plant material. The anoxic conditions in these soils favor denitrification and nitrogen may become a limiting factor for plant growth. Carnivorous plants (Droseraceae, Lentibulariacea, Sarraceniaceae) that replenish their nitrogen budget with animal protein are commonly found in these habitats. At sites where drainage is better, woody plants invade these natural meadows. The soft texture of the soils and their position in hill slope gradients makes these ecosystems highly vulnerable to gully erosion.

Logjam Ponds and Beaver Ponds

Falling riparian trees are stochastic events which may have dramatic consequences for the hydraulics of a stream system. Many tree species are soft-wooded, and tree dynamics are generally high in riparian wetlands. A fallen log blocks the current and creates a dam that accumulates fine particles. These natural reservoirs often extend far into the riparian zone.

Dams built by beaver (*Castor* sp.) can significantly alter the hydrological and biogeochemical characteristics of entire headwater drainage networks in Northern America and Eurasia. Fur trade led to the regional extinction of beavers. Few decades after reintroduction of beavers on a peninsula in Minnesota, they converted a large part of the area into wetlands, which led to a manifold increase in the soil nutrient concentrations. The activity of beavers considerably enhances the biodiversity of wetland-depending species. Beavers increase regional habitat heterogeneity because they regularly abandon impounded areas when the food supply is exhausted and colonize new ones, thereby creating a shifting mosaic of patches in variable stages of plant succession.

Typical Biota and Biodiversity in Riparian Wetlands

The importance of riparian wetland habitats for the conservation of biodiversity is well documented for several watersheds. Riparian areas generally have more water available to plants and animals than adjacent uplands. This is of specific importance in regions with a pronounced dry season, where lack of water affects plant growth. Abundance and richness of plant and animal species tend to be greater than in adjacent uplands because they share characteristics with the adjacent upland and aquatic ecosystems and harbor a set of specific riparian species. Because of their richness and their spatial distribution, the relative contribution of riparian ecosystems to total compositional diversity far exceeds the proportion of the landscape they occupy.

Apart from beavers, several other biota act as 'ecological engineers' that create and modify riparian wetlands. African hippopotamus deepen pools and form trails that increase the ponding of the water. Several crocodilians maintain open water channels. Digging mammals, freshwater crabs, and insects like mole crickets increase the pore space in riparian soils and enhance the water exchange between wetland and stream channel. Similar macropores develop from fouling tree roots. Plants also strongly modify the habitat characteristics in riparian wetlands, either actively, by influencing soil, moisture, and light conditions or, passively, by changing the hydraulic conditions through tree fall or organic debris dams.

Typical wetland species are adapted to the amphibious characteristics of the habitats. They are either permanent wetland dwellers that cope with aquatic and dry conditions or they temporarily colonize the wetlands during either the dry or the wet phase. There are many animal species that permanently colonize riparian wetlands, especially anurans, snakes, turtles, racoons, otters, and many smaller mammals, like muskrats, voles, and shrews. Aquatic insects have developed special adaptations to survive periodical droughts, for example, by having short larval periods or drought resistance. Many birds profit by the rich food offered from the aquatic habitats like dippers, kingfishers, jacamars, warblers, and rails. Periodical colonizers from terrestrial ecosystems are bats, elks, moose, and several carnivorous mammals and birds. Many aquatic species like fish and aquatic invertebrates periodically colonize riparian wetlands. Riparian wetland biota belong to the most threatened species as they suffer from both the impacts on the terrestrial and aquatic systems, and many riparian species are threatened with extinction. The effects of extinction of a species are especially high if it is an ecological engineer or a keystone species, for example, a top predator. Extinction of wolves in the Yellowstone National Park in the US led to overbrowsing of broad-leaved riparian trees by increased elk populations.

Ecological Services of Riparian Wetlands

Riparian wetlands are intrinsically linked to both the stream and the surrounding terrestrial ecosystems of the catchment. In many places of the world, however, riparian zones have remained the only remnants of both wetland and woody habitats available for wildlife. They are surrounded by intensively used areas for either agriculture or urban colonization. The performance of riparian wetlands to provide ecological services becomes reduced by the same degree as these bordering ecosystems become degraded. However, even in degraded landscapes, the beneficial effects of the riparian wetland ecosystems are astonishingly high. For humans, healthy riparian wetlands are vital as filters and nutrient attenuators to protect water quality for drinking, fisheries, and recreation.

Nutrient Buffering

Riparian wetlands are natural traps for fine sediments and for organic matter, but they may vary from a nutrient sink to a nutrient source at different times of a year depending on high or low water levels. Particle-bound nutrients, such as orthophosphate ions, become deposited in the riparian wetlands during spates and may accumulate there. This may increase the amount of phosphate that becomes released during the following flood event. Therefore, technical plans for phosphorus retention in artificial wetlands in agricultural landscapes include a hydraulic design which hampers the release of particles from the wetland, for example, by providing continuous, and sufficiently broad wetland buffer strips along the streams.

For the removal of nitrogen inputs from floodwater and from lateral groundwater inputs, riparian wetlands are very efficient. Generally it can be taken for granted that the slower the water flow (both ground and surface water) the higher is the nitrate uptake rate; however, the precise flow pathways in the sediments have to be considered. In anoxic soils, reduction and denitrification processes transform inorganic nitrogen forms into nitrogen gas which is then released into the atmosphere. Once the nitrate has been completely reduced, sulphate is also reduced in the anoxic sediments. Nitrogen also becomes immobilized by bacterial growth and/or condensation of cleaved phenolics during the aerobic decay of organic matter. Aquatic macrophytes and trees growing in the riparian wetlands are very efficient in nitrogen stripping by incorporating mineral nitrogen forms into their biomass. They can represent the most important nitrogen sinks in riparian systems. Some riparian wetland plants (e.g., alder, *Alnus* sp., and several leguminous trees) have symbiotic bacteria associated to their roots that can fix atmospheric nitrogen when this nutrient is scarce in the soils. Thus, not all riparian wetlands exclusively remove nitrogen.

Carbon Cycle

Like other wetlands, riparian wetlands are important players in the carbon cycle of the watershed. They accumulate large amounts of coarse particulate organic matter (CPOM) and they release dissolved organic matter into the stream and gaseous carbon compounds into the atmosphere (**Figure 1**).

In the boreal zone, the spring snowmelt runoff contributes to more than half of the annual total organic carbon (TOC) export. The larger the riparian wetland zone, the bigger the amount of exported TOC. On the other hand, riparian wetlands receive large amounts of dissolved carbon from litter leachates from the surrounding forests, especially during the leaf-fall period. These leachates can be an important source for phosphorous and other nutrients, as well as for labile carbon compounds. These substances enhance heterotrophic microbial (bacterial and fungal) activity.

Spring snowmelt also carries large amounts of fine particulate organic matter (POM). Riparian wetlands often provide surface structures that act like a comb to accumulate these particles (e.g., macrophytes),

and enhance the production of detritivores. Additional POM is produced by riparian trees. The general trend for litter production to increase with decreasing latitude (valid in forests) is overlain by species-specific productivity and physiological constraints due to the waterloggedness in riparian wetlands. Here, the litter production is generally higher in periodically flooded, than in permanently flooded, wetlands. Depending on the oxygen content of the soils, the chemical composition of the leaves, and the activity of detritivores, more or less dense layers of 'leaf peat' can accumulate in the sediments. This organic matter stock can be increased by undecomposed tree logs and bark. A reduction of the water level in the riparian wetlands leads to an increased mineralization of the carbon stocks and enhances the release of carbon dioxide.

Hydrological Buffering and Local Climate

Riparian wetlands have an equilibrating effect on hydrological budgets. Riparian vegetation dissipates the kinetic energy of surface flows during spates. Riparian wetlands store stormwater and release it gradually to the stream channel or to the aquifer between rainstorm events. Moreover, they are important recharge areas for aquifers. Several current restoration programmes try to increase this recharge function of riparian wetlands in order to stabilize the groundwater stocks for drinking water purposes.

Riparian wetland trees and macrophytes contribute considerably to evapotranspiration and to local and regional climate conditions. The rate of vapor release depends on the plant functional group which needs to be considered for basin-scale water budgets.

Corridor Function for Migrating Species

Riverine wetlands represent a web of ecological corridors and stepstones. In intense agricultural areas they can be considered as 'green veins' that maintain contact and gene flow between isolated forested patches. Providing shadow, balanced air temperatures and moisture, shelter, resting places, food and water supply, they cover the requirements of a great deal of amphibian, reptile, bird, and mammal species. These not only use the longitudinal connection but also migrate laterally and thus reach the next corridor aside. Moreover, long-range migrating birds use the green corridors of riparian zones in general as landmarks for migration. Networks of riparian corridors also facilitate the movement of non-native species. In some US riparian zones, their richness was about one-third greater in riparian zones than on uplands and the mean number and the cover of non-native plant species were more than 50% greater than in uplands.

Refugia and Feeding Ground for Riverine Biota

During flood, drought, and freezing events, but also during pollution accidents in the stream channel, connected riparian wetland habitats represent refugia for riverine animals. In extreme cases, residual populations from the wetlands may contribute to the recolonization of defaunated stream reaches. Riparian wetlands also act as traps and storage sites for seeds both from the upstream and from the uphill areas. The seed banks contain propagules from plants that represent a large range of moisture tolerances, life spans, and growth forms. These seeds may also become mobilized and transported during spate events.

Riparian wetlands offer a large variety of food sources. Connected wetland water bodies 'comb out' fine organic particles including drifting algae from the stream water, they receive aerial and lateral inputs of the vegetation, and they have a proper primary productivity which profits by the increased nutrient input and storage from the surroundings. Many riverine fish and invertebrate species are known to migrate actively into the riparian wetlands in order to profit by the terrestrial resources that are available during flood periods. In analogy to the 'floodpulse advantage' of fish in large river floodplains, stream biota that temporarily colonize riparian wetlands have better growth conditions than those that remain permanently in the stream channel. For example, the macroinvertebrate community of riparian sedge-meadows in Maine (USA) is dominated by detritivorous mayfly larvae (over 80% of the invertebrate biomass) during a 2-month period in spring. The larvae use the stream channel as a refuge and use the riparian wetland as feeding ground where they perform over 80% of their growth.

Reciprocal Subsidies between Aquatic and Terrestrial Ecosystems

Many aquatic species profit by the terrestrial production and vice versa. Apart from leaf litter, large quantities of fruits, flowers, seeds, as well as insects and feces fall from the tree canopies into the streams where they represent important energy and nutrient sources for the biota. In Amazonian low-order rainforest streams, terrestrial invertebrates make up a major portion of the gut content of most fish species. Fruits and seeds are preferred food items for larger fish species that colonize medium- and high-order rivers. Riparian wetlands increase the area of this active exchange zone, and they retain these energy-rich resources for a longer period than a stream bank alone would do.

Aquatic organisms also contribute to the terrestrial food webs. For example, bats are known to forage on the secondary production of emerging insects in riparian wetlands, and the shoreline harbors a large number of terrestrial predators, such as spiders, tiger beetles, and riparian lizards. Experimental interruption of these linkages (e.g., by covering whole streams with greenhouses) has shown that the alteration of riparian habitats may reduce the energy transfer between the channel and the riparian zone.

Recreation

The sound of the nearby stream, the equilibrated climate, and the occurrence of attractive animal and plant species render riparian wetlands highly attractive for recreation purposes such as hiking, bird-watching, or meditation. These can be combined with 'in-channel' recreation activities such as canoeing, rafting, or fishing, and represent an economically valuable ecosystem service, that should be considered in management and conservation plans.

Conservation

Water is becoming scarce in many areas worldwide. Water mining reduces water levels, but high and stable groundwater tables are a prerequisite for the existence of riparian wetlands. In addition to direct water withdrawal, predictions about climatic changes include other threats. Increased stochasticity of the runoff patterns and reduced snowmelt floods are severe threats to the existence of riparian wetlands. The riparian zones of streams and rivers have been sought after by humans since early days. High productivity, reliable water supply, and climatic stability make these ecosystems suitable for a range of human-use types, such as wood extraction, hunting, aquaculture, and agriculture. In areas of intensive agriculture, riparian zones including their wetlands have shrunk to narrow strips or have completely vanished. On the other hand, the ecosystem services are good socioeconomical arguments to restore and enlarge riparian wetlands.

For conservation planning, it is very important to bear in mind that riparian wetlands are very diverse and have typical regional characteristics. Secondly, the whole riparian zone is very dynamic. Many tree species are relatively short-lived and well adapted to changes in the floodplain morphology or in the hydrology of the wetland. The existence of variable hydrological patterns is a prerequisite for the coexistence of annually varying plant and animal communities. Often, large-scale projects restore riparian zones including wetlands according to a single pattern that does not consider these dynamic changes in habitat and species diversity. If large flood events are precluded by dam constructions in the upstream region, the natural habitat dynamics are blocked and the vegetation will develop towards a late-successional stage without pioneer vegetation, and with a reduced range of moisture tolerance. Several studies could prove that once the hydrological fluctuations become reduced by water-level regulation, exotic species can invade river valleys more efficiently.

While many animal species depend exclusively on the specific habitat conditions of wetlands, most riparian amphibians and reptiles migrate into the drier zones of the aquatic–terrestrial ecotones for a part of their life cycle. This makes them vulnerable to increased mortality in the neighboring ecosystems, especially if these have been converted into agricultural or urban use. Therefore, a buffer zone considering the home range of these species is needed to fully protect these species.

See also: Floodplains; Rivers and Streams: Ecosystem Dynamics and Integrating Paradigms; Rivers and Streams: Physical Setting and Adapted Biota.

Further Reading

Ilhardt BL, Verry ES, and Palik BJ (2000) Defining riparian areas. In: Verry ES, Hornbeck JW, and Dolloff CA (eds.) *Riparian Management in Forests of the Continental Eastern United States*, pp. 23–42. Boca Raton, London, New York, Washington, DC: Lewis Publishers.Junk WJ and Wantzen KM (2004) The flood pulse concept: New aspects, approaches, and applications – An update. In: Welcomme RL and Petr T (eds.) *Proceedings of the Second International Symposium on the Management of Large Rivers for Fisheries*, vol. 2, pp. 117–149. Bangkok: FAO Regional Office for Asia and the Pacific.

Lachavanne J-B and Juge R (eds.) (1997) *Man and the Biosphere Series, Vol. 18: Biodiversity in Land–Inland Water Ecotones*. Paris: UNESCO and The Parthenon Publishing Group.

McCormick JF (1979) A summary of the national riparian symposium. In: U.S. Department of Agriculture, Forest Service (ed.) *General Technical Report WO-12 Strategies for Protection and Management of Floodplain Wetlands and Other Riparian Ecosystems*, pp. 362–363pp. Washington, DC: US Department of Agriculture, Forest Service.

Mitsch WJ and Gosselink JG (2000) *Wetlands*, 3rd edn. New York: Chichester, Weinheim, Brisbane, Singapore Toronto: Wiley.

Naiman RJ, Décamps H, and McClain ME (2005) *Riparia – Ecology, Conservation, and Management of Streamside Communities*. Amsterdam: Elsevier.

Peterjohn WT and Correll DL (1984) Nutrient dynamics in an agricultural watershed: Observations on the role of a riparian watershed. *Ecology* 65: 1466–1475.

Verry ES, Hornbeck JW, and Dolloff CA (eds.) (2000) *Riparian Management in Forests of the Continental Eastern United States*. Boca Raton, London, New York, Washington, DC: Lewis Publishers.

Wantzen KM, Yule C, Tockner K, and Junk WJ (2006) Riparian wetlands. In: Dudgeon D (ed.) *Tropical Stream Ecology*, pp. 199–217. Amsterdam: Elsevier.

Rivers and Streams: Ecosystem Dynamics and Integrating Paradigms

K W Cummins and M A Wilzbach, Humboldt State University, Arcata, CA, USA

Introduction

Research scientists, watershed managers, and conservationists alike agree that following an ecosystem perspective is the most productive way to examine streams and rivers. The integration of physical–chemical with biological processes, which is the study of ecosystems, has largely replaced single physical factor or single-species approaches to management and rehabilitation of running waters. In the discussion that follows, fluxes of energy and matter into, through, and out of lotic ecosystems are used as basic processes embraced by the integrating paradigms (conceptual models) that presently underlie inquiry into the structure and function of streams and rivers.

Energy Flux

Energy Sources

Streams and rivers are driven almost entirely by two alternate energy sources: (1) sunlight that fuels the in-stream growth of aquatic plants (primary production), and (2) plant litter from stream-side (riparian) vegetation. The relationship between these two energy drivers is essentially inverse. The heavier the riparian cover over the stream/river channel, the greater the plant litter inputs and the greater the limitation of light reaching the water and therefore in-stream algae and vascular plant growth. In contrast to nonfilamentous algae, very few stream/river consumers utilize macrophytes, filamentous algae, and rooted vascular plants. Rather, the macrophytes enter the energy transfer to consumers after they die as detritus. Stream and river systems in which the majority of the energy transfer is from in-stream plant growth to consumers are termed autotrophic. Those systems dominated by the detrital pathways of energy transfer are heterotrophic. As discussed in the 'river continuum concept' (RCC), the relative importance of these two energy sources changes with stream size. Smaller streams in forested catchments are usually dominated by litter energy sources and wider, mid-sized stream segments are dominated by plant growth in the water. Larger rivers are dependent upon organic matter (OM) delivered from the upstream tributary network.

A model of energy flux, that is, the transfer of energy between trophic levels of plants and animals, was produced by Lindeman in the early 1940s and various forms of this model have more or less been the basis for the investigation of energy flux in running waters ever since. These studies most frequently take the form of energy budgets; an accounting of the energy in and the energy out of a given ecosystem (**Figure 1**) or biological population (**Figure 2**) or community within the system. OM budgets are useful in identifying the sources, magnitude, and fates of energy and provide insight into internal dynamics of a river system. At the system level, inputs include autotrophic production plus energy originating from the surrounding terrestrial environment (allochthonous) that is brought in by various physical vectors.

Outputs include community respiration and losses by downstream transport. Energy retained within a stream reach over a given time interval is referred to as storage. Comparisons of energy flux among and between trophic levels commonly express biomass as caloric equivalents. Animal ingestion, egestion, and growth (increase in mass) are all measured as biomass. Respiration (metabolism) is readily converted to calories consumed using an oxycalorific equivalent. Tables are available that provide conversions of mass to calories for freshwater organisms.

Feeding Roles and Food Webs

Feeding studies of benthic macroinvertebrates have shown that, based on food ingested, most taxa are omnivorous. For example, invertebrates that chew riparian-derived leaf litter in streams, termed 'shredders', ingest not only the leaf tissue and associated microbiota, (e.g., fungi, bacteria, protozoans, and microarthropods), but also diatoms and other algae that may be attached to the leaf surface, as well as very small macroinvertebrates (e.g., first-instar midge larvae). For this reason, trophic-level analysis does not lend itself well to simple trophic categorization of stream macroinvertebrates.

An alternate classification technique, originally described by Cummins in the early 1970s, involves the functional analysis of stream/river invertebrate feeding. The method is based on the combined morphological and

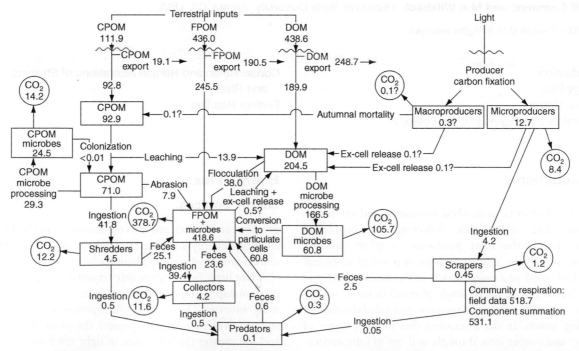

Figure 1 Example of an energy budget for a small woodland stream ecosystem (Augusta Creek watershed, Michigan, USA). All values are in grams ash-free dry mass $m^{-2} yr^{-1}$. Squares represent pools of organic matter in various states; arrows represent transfers and circles represent respiratory consumption of organic matter. From Saunders GW *et al.* (1980) In: LeCren ED and McConnell RH (eds.) *The Functioning of Freshwater Ecosystems.* Great Britain: Cambridge University Press.

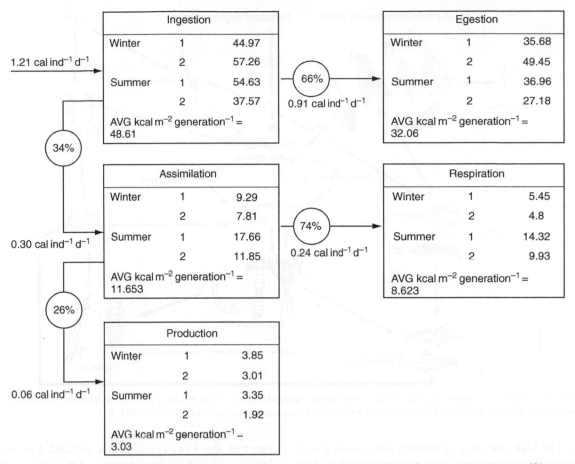

Figure 2   Example of an energy budget constructed over 2 years of study for a population of a stream invertebrate (*Glossosoma nigrior*, Trichoptera) from Augusta Creek, Michigan, USA The budget is based on independent measurements of ingestion, production, and respiration. Modified from Cummins KW (1975) Macroinvertebrates. In: Whitton BA (ed.) *River Ecology*. Berkeley: University of California Press.

behavioral mechanisms of food acquisition used by the invertebrates and four fundamental categories of their food found in running waters (**Figure 3**). There is a direct correspondence between the availability of categories of nutritional resources and the relative abundance of invertebrate populations that are adapted to efficiently harvest a given food resource. Five invertebrate functional feeding groups (FFG) have been designated. These include shredders, filtering collectors, gathering collectors, scrapers, and predators. These partition four food resource categories in running waters that are defined on the basis of particle size and type: (1) coarse particulate organic matter (CPOM), which is primarily riparian litter that has been conditioned, that is, microbially colonized, within the stream; (2) fine particulate organic matter (FPOM), which are particles generally smaller than 1 mm in diameter that are largely derived from the biological and physical breakdown of CPOM and whose surfaces are colonized by bacteria; (3) periphyton, that is, tightly accreted algae and associated organic material; and (4) prey, that is, invertebrate species or larval/

nymphal stages small enough to be captured and consumed by invertebrate predators. As the relative availability of the basic food resources changes, there is a concomitant change in the corresponding ratios of the FFGs of freshwater invertebrates adapted to specific resource categories.

Obligate and facultative members occur within each FFG. These can be different species or different stages in the growth period of the life cycle of a given species. For example, it is likely that most aquatic insects, including predators, are facultative gathering collectors as first-instars newly hatched from the egg. It is with obligate forms that linkages between invertebrates with their food resource categories are most reliable. The distinction between obligate and facultative status is best described by the efficiency with which a given invertebrate converts the resource acquired to growth; that is, obligate forms are more efficient consumers of a given resource, such as conditioned leaf litter, than are facultative forms. For example, shredders feeding on litter consume the fungal-rich leaf matrix, whereas scrapers only abrade the much less nutritious leaf

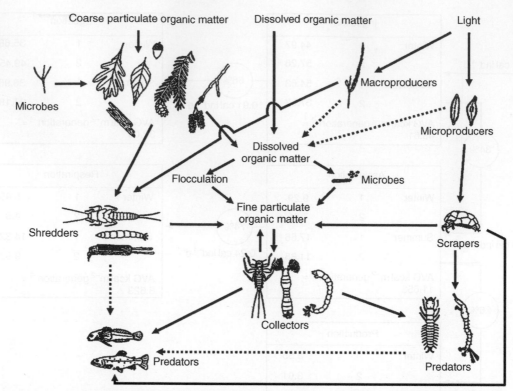

Figure 3 Conceptual model of invertebrate functional feeding groups and their food resources in a small, forested stream ecosystem. Modified from Cummins KW (1974) Structure and function of stream ecosystems. *Bioscience* 24: 631–641.

cuticle. The high efficiency of obligate forms feeding on a particular resource category is in contrast with the wider array of food types consumed by facultative forms, but with lower efficiency. The same morpho-behavioral mechanisms can result in the ingestion of a wide range of food items, the intake of which constitutes herbivory (consumption of living plants), detritivory (consumption of dead OM), or carnivory (consumption of live animal prey).

Although intake of food types changes from season to season, habitat to habitat, and with growth stage, limitations in food acquisition mechanisms have been shaped over evolutionary time and these are relatively more fixed than the food items ingested. Morphological structures that enable aquatic insects to harvest a given food resource category exhibit significant similarities across diverse taxa. This convergent or parallel evolution lies at the

heart of the FFG classification method. For example, larvae of the 26 North American caddisfly (Trichoptera) families are spread among the four major nonpredaceous FFGs. The less highly evolved mayflies (Ephemeroptera) and stoneflies (Plecoptera) are adapted to acquire fewer food resource categories (**Table 1**).

An advantage of the FFG procedure is that it does not require detailed taxonomic separations of the invertebrates. Broad, easily distinguished characteristics allow FFG classification, preferably in the field with live specimens. Separations usually involve systematic distinctions at the level of family or higher, and cut across taxonomic lines. As an example, two groups of case-bearing larval caddisflies (Trichoptera) are sufficient to separate FFG categories at better than 90% efficiency. All families, or genera within

Table 1 Numbers of families and functional group assignment of some representative orders of benthic macroinvertebrates in running-water ecosystems

Order	Total number of families	Number of families by dominant functional feeding groups					
		Shredders	Scrapers	Filtering collectors	Gathering collectors	Predators	Filamentous algal piercers
Ephemeroptera	21		2	5	10	4	
Plecoptera	9	6				3	
Trichoptera	26	5	8	6	4	2	1

families, of Trichoptera that construct mineral cases are scrapers. Those that construct organic cases are shredders.

Given the coupling of FFGs and food resource categories, ratios of the different groups can serve as surrogates for ecosystem parameters. For example, the ratio of the functional groups linked to in-stream primary production (scrapers plus those shredders that may harvest live plant tissue) to those groups dependent upon the CPOM and FPOM heterotrophic food resources (shredders of detrital material plus gathering and filtering collectors) provides an index of the ratio of autotrophy to heterotrophy at the lotic ecosystem level. When measured directly, an ecosystem ratio of autotrophy/heterotrophy ≥ 1 indicates an autotrophic system. A surrogate FFG ratio of 0.75 has been measured in such autotrophic stream/river systems.

Flux of Matter

Nutrient Cycles and Spiraling

The limnological study of standing waters has always been dominated by a conceptual model of closed ecosystems, in which nutrients recycle seasonally, totally within the system. The unidirectional flow of running waters necessitated

modifying this view of closed cycles in lakes to an open-cycle model; that is, the open nutrient cycles in streams and rivers follow a spiraling pattern in which nutrients generated (or delivered) at one point along a stream or river complete the recycling to their initial state at a displaced location downstream (**Figure 4**). Total spiral length represents the sum of the distance traveled by an element as an inorganic solute until its uptake by the biota, plus the distance traveled within the biota until its release back into the water column. If nutrients such as nitrogen or phosphorous are cycled rapidly, the spirals are 'tight', that is, the downstream completion of the cycle is short. If cycling is slow, the closing of the loop is displaced a longer distance downstream and the spirals are more open. The tighter the spiraling cycling loops, the more retentive (conservative) is the stream or river reach.

Transport and Storage of OM

The transport and storage of OM in running-water ecosystems involves complex interactions between (1) the state of the OM, (2) the source of the OM, and (3) the physical, chemical, and biological retention potential for any given reach of stream or river.

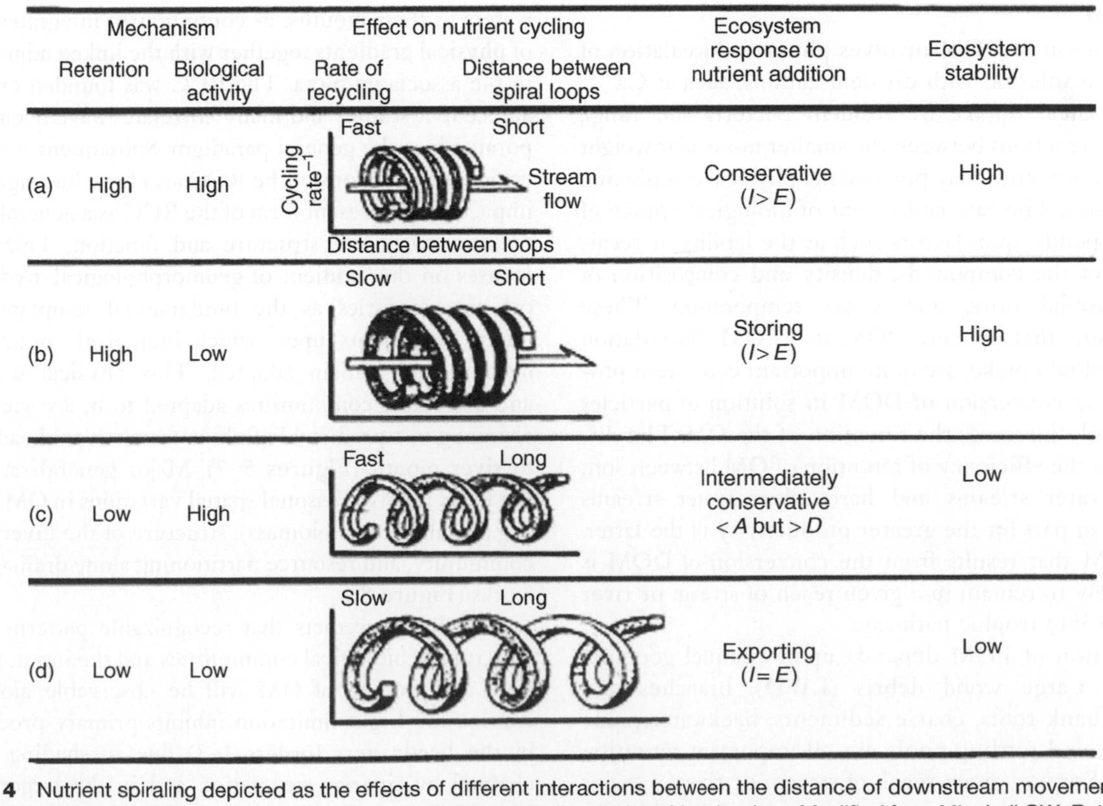

	Mechanism		Effect on nutrient cycling		Ecosystem response to nutrient addition	Ecosystem stability
	Retention	Biological activity	Rate of recycling	Distance between spiral loops		
(a)	High	High	Fast	Short	Conservative $(I > E)$	High
(b)	High	Low	Slow	Short	Storing $(I > E)$	High
(c)	Low	High	Fast	Long	Intermediately conservative $< A$ but $> D$	Low
(d)	Low	Low	Slow	Long	Exporting $(I = E)$	Low

Figure 4 Nutrient spiraling depicted as the effects of different interactions between the distance of downstream movement (velocity × time) and measures of biological activity such as metabolism by benthic microbes. Modified from Minshall GW, Petersen RC, Cummins KW, *et al.* (1983) Interbiome comparison of stream ecosystem dynamics. *Ecological Monographs* 53: 1–25.

State of the OM

Three broad categories of OM are dissolved (DOM, size range <0.45 µm), fine particles (FPOM, size range >0.45 µm to 1 mm), and coarse particles (CPOM, size range >1 mm). While FPOM particles are colonized primarily on the surface by bacteria, CPOM is colonized by fungus, bacteria, and microzoans that penetrate the matrix of the material. Aquatic hyphomycete fungi usually penetrate the CPOM leaf and needle litter first. Bacteria and microzoans follow the fungal hyphal tracks into the matrix of the CPOM. The OM in solution (DOM) includes a full range of molecules from simple very labile ones such as sugars and amino acids to complex recalcitrant compounds such as phenolic compounds.

Sources of the OM

A major source of OM in streams (orders 0–5) is the riparian zone. This border of stream-side vegetation produces litter (e.g., leaves, needles, bud and flower scales, seeds and fruits, small wood and bark) that enters on a seasonal schedule depending upon the relative proportions of deciduous and evergreen species. Other sources of OM are solutions and particles from bank erosion, DOM leachates from litter, exudates, and leachates from periphytic algae and vascular aquatic plants together with their physical fragmentation and mortality.

Physical, chemical, and biological retention potential

The retention of DOM involves physical flocculation of the OM in solution with divalent cations, such at Ca^{++}, and biological uptake by resident bacteria and fungi. Chemical reactions between the smaller molecular weight organic compounds may precede the physical complexing with cations. The rate and extent of biological uptake of DOM depends upon factors such as the lability or recalcitrance of the compounds, density and composition of the microbial flora, and water temperature. These mechanisms that convert DOM to FPOM, flocculation and microbial uptake, are quite important ecosystem processes. The conversion of DOM in solution to particles significantly increases the retention of the OM. The difference in the efficiency of retention of OM between soft, stained water streams and hard, clear water streams accounts in part for the greater productivity of the latter. The POM that results from the conversion of DOM is more likely to remain in a given reach of stream or river and enter into trophic pathways.

Retention of POM depends upon channel geomorphology. Large wood debris (LWD), branches and exposed bank roots, coarse sediments, backwaters, side channels, and settling pools are all important retention features. For any given reach of stream or river, a major source of OM is transport from upstream. In addition, OM is retained when bankfull-flow is exceeded and material is deposited on the upper banks or on the floodplain. OM is returned to the channel when water levels recede. Whether these off-channel areas serve as sources or sinks for OM over an annual cycle depends upon the configuration of the upper banks and floodplains and the patterns of the flood flows. The general fertility of floodplains suggests that they are largely sinks.

Integrative Paradigms in Lotic Ecology

Paradigms, or conceptual models, have continued to be developed, modified, and integrated since the 1980s. The RCC, arguably the most encompassing of these, has guided a large portion of the research on lotic ecosystems in the interim. However, a number of other models have served to elucidate specific components of running-water structure and function or have proposed alternative broad integrating principles.

The RCC

The major goal of the architects of the RCC was to examine the patterns of biological adaptation that overlay the physical setting (template) of stream/river channels in a watershed.

The RCC views entire fluvial systems, from headwaters to their mouths, as continuously integrated series of physical gradients together with the linked adjustments in the associated biota. The RCC was founded on many antecedent studies and many correlates have been incorporated into the general paradigm. Subsequent views and critiques of portions of the RCC also have had significant impact on the present form of the RCC as a general model of lotic ecosystem structure and function. This model focuses on the gradient of geomorphological–hydrological characteristics as the fundamental template along intact catchments upon which biological communities become and remain adapted. This physical template, and biological communities adapted to it, are viewed as changing in a predictable fashion from stream headwaters to river mouth (**Figures 5–7**). Major generalizations of the RCC involve seasonal spatial variations in OM supply (e.g., algal/detrital biomass), structure of the invertebrate community, and resource partitioning along drainage networks (**Figure 8**).

The RCC predicts that recognizable patterns in the structure of biological communities and the input, utilization, and storage of OM will be observable along the continuum. Light limitation inhibits primary production in the headwaters (orders 1–3) due to shading of the channel by riparian vegetation and in the larger rivers (orders greater than 7 or 8) because of light attenuation through the turbid water column that is typical of the

Figure 5   A headwater stream in the Cascade Mountains of Oregon, USA at winter base flow. The coniferous riparian zone provides partial canopy closure and supplies large woody debris to the channel.

Figure 6   The Firehole River, a mid-sized woodland-meadow stream in Yellowstone National Park, USA. Reduced riparian shading enables abundant growth of in-stream algae.

Figure 7   The Smith River in coastal northern California USA. This high order river with a canyon-controlled channel is dependent upon organic matter delivered from the upstream tributary network.

waters (orders 4–6). High biotic diversity is supported in rivers in this size range both because of the variety of habitats and food resources for consumer organisms, and because of overlapping ranges of organisms with evolutionary terrestrial origins (such as the insects) that are dominant in the headwaters with those of marine origins (e.g., annelids and mollusks) that are more prevalent downstream (**Figure 9**).

The RCC has been widely utilized as an organizing principle and has been the subject of many studies resulting in various tests of the concept. As would be expected, a degree of unpredictability in the physical template leads to correspondingly less predictability in the overlay of biological communities. This lack of predictability is often a function of the spatial and temporal scales of reference employed and it can also be induced by human interference. For example, systems may appear more or less variable over time spans of less than a decade, the time period of observation, but long-term variability, at the scale of centuries or greater, is usually obscured by short-term variations.

Other Paradigms

At least eight paradigms, other than the RCC, continue to guide the development of running-water ecosystem theory (**Table 2**). These include serial discontinuity, hierarchical scales, riparian zone influences, flood pulse, hyporheic dynamics, hydraulic stream ecology, patch dynamics, and network dynamics.

Serial discontinuity

Interruptions in the longitudinal continuum, as proposed by the RCC, are caused by engineered impoundments which serve to reset the general patterns of biotic organization. Above the dam, the system exhibits characteristics of a higher order than the impounded stream. Below

lower portions of stream/river networks. The cumulative effect of drainage networks tends to increase nutrient levels in the down stream/river direction. Overall periphyton and rooted vascular plant biomass, and insect and fish diversity are all maximized in the mid-sized running

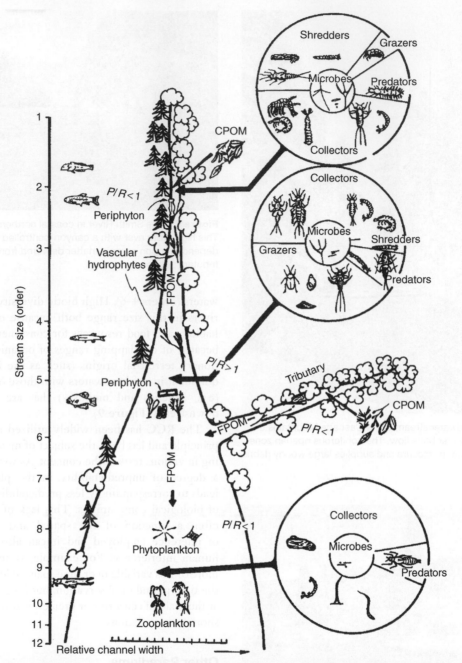

Figure 8 The 'river continuum concept' (RCC). A proposed relationship between stream size (order) and the progressive shift in structural and functional attributes of lotic biotic communities. The heterotrophic headwaters and the large rivers are both characterized by an autotrophic index, or P/R (ratio of gross primary production to total community respiration) of less than 1 ($P/R = <1$). The largely unshaded mid-sized rivers are generally classified as autotrophic with a $P/R = >1$. The invertebrate communities of the headwaters are dominated by shredders and collectors, the mid-sized rivers by grazers (= scrapers) and collectors. The large rivers are dominated by FPOM-feeding collectors. Fish community structure grades from invertivores in the headwaters to invertivores and piscivores in the mid-sized rivers to planktivores and bottom-feeding detritivores and invertivores in the largest rivers. From Vannote RL, Minshall GW, Cummins KW, Sedell JR, and Cushing CE (1980) The river continuum concept. *Canadian Journal of Fisheries and Aquatic Sciences* 37: 130–137.

the dam, the regulated flows often completely alter the seasonal hydrological patterns of the receiving channel. For example, the normal pattern imposed by the discontinuity resulting from a dam is to decrease the flows during natural high-water periods (reservoir storage phase) and release the water during natural low-flow periods (reservoir release phase). The storage phase retains water to prevent flooding and to provide a later water supply during dry periods. During the release phase, water is delivered for irrigation, drinking,

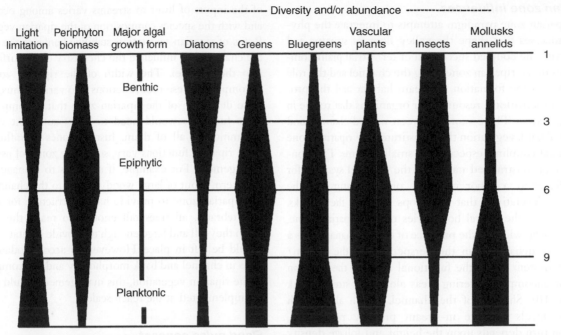

Figure 9 Patterns in categories of biotic diversity, from small streams to large rivers, compared on a relative scale for each parameter, as predicted by the 'river continuum concept'. Numbers at the right are general stream/river order ranges. Modified from Cummins KW (1997) Stream ecosystem paradigms. In: *CNR – Instituto di Ricerca Sulle Acque. Prospettive di recerca in ecologia delle acque. Roma, Italia.*

Table 2 Comparison of most appropriate scales of application for eight commonly used paradigms (conceptual models) for running-water ecosystem analysis

Basin or reach scale		Stream orders or reach length	RCC	HS	RZI	FPC	HD	HSE	PD	ND
Basin	Macro	0/1 Order to estuary	a	a	b	a	c	c	c	b
	Meso	0/1 Order to order 6	a	a	b	a	c	c	c	b
	Micro	0/1 Order to order 2–5	a	a	a,e	d	c	c	c,a	b
Reach	Macro	>1000 m	d	a	b	d	a	c	a	c
	Meso	100–1000 m	d	a	a	c	a	a	a	a
	Micro	<10 m	d	a	a	c	a	a	a	d

RCC, River Continuum Concept; HS, hierarchical scales; RZI, riparian zone influences; FPC, flood pulse concept; HD, hyporheic dynamics; HSE, hydraulic stream ecology; PD, patch dynamics; ND, network dynamics.
a, most direct influence on stream biota and ecosystem processes; b, if channels are braided, ranking moves down in c\scale to a lower order; c, beyond the scale to detect specific (local) differences; d, influence too local to detect general, large-scale patterns; e, may be of less direct importance in naturally deranged (lake interrupted) or beaver-influenced stream systems.

recreation, and, in some cases, to improve fish habitat. In some basins, interruptions in the longitudinal profile of river networks occur in the form of natural lakes or impoundments. Although these may change the sequencing of stream orders, they do not change the annual hydrograph.

Hierarchical scales

The hierarchical scales paradigm addresses a weakness of the RCC. The relative significance of the factors driving the physical, chemical, and biological components of running waters changes with scale. The data on which the RCC is based were all collected at the reach scale and during only several seasons. The hierarchical approach recognizes that ecosystem processes operating at the

reach scale and over short time periods do not adequately represent the patterns viewed over greater spatial and temporal scales. Therefore, descriptions of stream/river ecosystem structure and function must be placed in the appropriate context of space and time.

Ecotones

Several paradigms focus on the ecotones that bridge between the stream/river channel and its surroundings and underpinnings. These include the riparian border primarily along small streams, the aquatic terrestrial interface of large rivers with floodplains, and the subsurface region of the sediments beneath running waters (the hyporheic zone).

Riparian zone influences

The riparian zone paradigm attempts to integrate the physical processes that shape the valley floor of streams and rivers with the coupled succession of terrestrial plant communities in the riparian zone along the channel and the role they play in the formation of stream habitat and the production of nutritional resources for organisms that reside in running waters. The ecotone between wetted channel and terrestrial bank vegetation that constitutes the riparian zone is a critical coupling, especially in small streams. The confined or unconstrainted nature of the channel system, for example, exert a major influence on the nature of the riparian vegetation that develops along the banks. Definition of the lateral boundaries of the riparian zone, or buffer strip width in the parlance of timber managers, is a continuing debate. From the perspective of the stream/river ecosystem *per se*, the functional roles of the riparian corridor encompass differing areas along the stream bank (**Figure 10**). Shading of the channel, which along with nutrient levels regulate in-stream primary production, which in turn depends upon the height and foliage density of the vegetation, steepness of the side slopes, and aspect (compass direction) of the channel. The width of the riparian zone that yields litter inputs and large woody debris to the channel can also vary with height and species composition of the stream-side vegetation. Seasonal timing of litter drop and its introduction into the stream produces patterns around which the life cycles of many steam invertebrates have become adapted. This coupling between riparian litter inputs and stream invertebrates is most direct for invertebrate shredders that feed on conditioned litter. The timing of the inputs of litter to streams varies among ecoregions and with the species composition of the riparian vegetation. Roots of riparian vegetation stabilize banks at the edge of the channel and influence the chemistry of subsurface flow into the channel. The width of the riparian zone that encompasses these root functions also varies. Thus, a complete definition of the riparian zone that encompasses all these functions would need to be of sufficient width to accommodate all of them. Just as zones of influence of these riparian functions vary, so do the zones of associated management. For example, if a goal is to manage for the long-term input of large woody debris to the channel from the riparian zone to provide habitat structure for fish and invertebrates, all trees tall enough to reach the channel when they fall and large enough to provide habitat structure should be left in place. However, to accommodate variations in channel and bank morphology and the composition of the riparian vegetation, this management would need to be implemented at the reach scale.

Flood pulse concept

The flood pulse concept addresses the ecotones between rivers and their floodplains. Unlike the lateral riparian influence on stream ecosystem processes where the impact is largely from the landscape to the stream, the flood pulse concept emphasizes the reciprocal exchange between the major river channel and its floodplain. A consequence of this distinction is that the overwhelming bulk of riverine animal biomass derives directly or indirectly from production on the floodplain and not from downstream transport of OM produced higher in the watershed. Although the importance

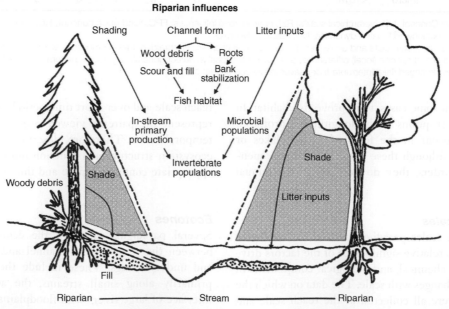

Riparian influences

Figure 10 Influences of the riparian zone on streams. Redrawn from Cummins KW (1988) The study of stream ecosystems: A functional view. In: Pomeroy LR and Alberts JJ (eds.). New York: Springer.

of this aquatic/terrestrial transition zone, the floodplain ecotones, is widely acknowledged, there are few hard data to indicate whether over annual cycles, or longer periods, the primary movement of nutrients and biomass is onto or off the floodplain, or in balance. The general perception of 'fertile floodplains' suggests that the periodically inundated floodplains are sinks relative to the river channel. However, the high productivity of adult fish in many floodplain rivers and the concentration of reproductive activity on the floodplain supports the notion that floodplains are sources and the river is a sink, gradually exporting to the sea. At any rate, the seasonal pulsing of river discharge, the flood pulse, is the major force controlling existence, productivity, and interactions of biota in river–floodplain ecosystems.

For any given storm or series of storms, the movement of material and organisms on to the floodplain follows the rising limbs of the hydrographs, and the return to the river channel follows the falling limbs. Unfortunately, application of the flood pulse concept is restricted because of the wide-scale engineering modifications that have isolated rivers from their floodplains. Natural exceptions to the flood pulse concept are rivers flowing through deeply incised canyons.

Hyporheic dynamics

The hyporheos is the subsurface region beneath and adjacent to stream/river channels that exchanges water with the surface water. This surface water–ground water ecotone is spatially and temporally very dynamic. The conceptual framework of the hyporheic dynamics paradigm has resulted in the incorporation of channel-aquifer dynamics into the general model of the RCC. The hyporheic zone occurs, at least to some extent, beneath and lateral to the active channel from the headwaters to river mouths, except in bedrock channels. Water, solutes, inorganic and organic complexes, and uniquely adapted biota move through interstitial pathways into and out of the sediments. These flowpaths are determined by the bedform of the channel. Where the bedform is convex, there is groundwater recharge from the channel into the sediments. These locations where oxygenated water is driven into the sediments are often spawning sites of salmonid fishes. Concave bedforms are sites of groundwater upwelling into the channel. Invertebrates found in the hyporheos include the small early instars of a wide range of taxa and larger forms at times of extremes inflow, either high or low. In addition, there are some microbenthic forms that are specifically adapted to a groundwater existence. The presence of hyporheic invertebrates is determined largely by siltation and the availability of oxygen. If interstitial sediment spaces are filled with fine sediments and/or conditions are anaerobic, the fauna will be excluded.

Hydraulic stream ecology

The hydraulic stream ecology model emphasizes that the local responses of stream organisms to flow conditions can serve as an organizing principle for running waters. Lotic animals are not well adapted to hydraulic stress and can sustain exposure to such stress for only short periods of time. Common patterns of diurnal and seasonal drift of stream invertebrates along vertical gradients of sediment and current velocity are a manifestation of this stress response. The model identifies mean water velocity and depth as more critical than characteristics of substrates in determining the distributions of stream animals. The model has yet to adequately incorporate the extensive data that implicates the diurnal light cycle as the major control parameter of invertebrate stream drift.

Patch dynamics

The patch dynamics concept emphasizes the patchy distribution of riverine habitats in space and time, and argues that an ever-shifting mosaic of patches enables a greater number of species to co-occur than would be the case under greater environmental constancy. Environmental conditions are predictable in aggregate, but not within a particular patch, and these aggregate conditions confer some regularity in species composition. In this model, particular patch types can be found at any point along the general longitudinal gradient proposed by the RCC. However, there are clear examples of invertebrate 'patches' that at least change in abundance from headwaters along the continuum to large rivers. For example, small headwater streams (orders 1–3) are generally better shaded than higher orders (>3) and sustain less suitable algal periphyton to support scrapers. Further, the dominance of the CPOM–detrital shredder linkage correlates with stream width and the close availability of riparian tree and/or shrub litter, and this generally matches with stream orders 1–3. The extension of the shading of periphyton growth and the riparian CPOM–shredder linkage to larger rivers can occur along braided channels, but these 'patches' will always be more abundant in the headwaters than in mid-sized or larger rivers.

Network dynamics

The network dynamics hypothesis, which combines the hierarchical scales and patch dynamics models, is based on the observation that there are abrupt changes that occur at the confluences of tributaries with the receiving channel. Changes in water and sediment flux at these locations result in changes in the morphology of the receiving channel and its floodplain. In this view, the branching nature of river channel network, together with infrequent natural disturbances, such as fire, storms, and floods, are the formative elements of the spatial and temporal organization of the nonuniform distribution of riverine habitats. Further, the tributary junctions are proposed hot spots of biological activity. Some data show increased fish diversity and abundance at these junctions, but the influence on other components of the biota has yet to be investigated. The 'network dynamics hypothesis' does not address 'patches' represented by braided channels.

Whether hydraulic characteristics, tributary junctions, or other patch phenomena, represent local conditions that need to be integrated along river continua to account for whole-profile trends that are clearly apparent, or whether such phenomena are localized specific modifiers that differentially affect stream orders along profiles has yet to be demonstrated clearly.

Conservation and Human Alterations of Streams and Rivers

A great challenge for stream and river ecology in the twenty-first century will be the restoration of degraded running-water ecosystems while preserving those systems that still remain in good condition. Restoration will dominate in more developed regions where modifications of running waters and their watersheds have been more extensive. In less-developed regions, preservation of many running waters may still be possible, but the distinction between pristine and degraded systems is disappearing rapidly. The historical scientific databases for running waters are generally poor, with largely anecdotal or very incomplete information available. The lotic ecosystem paradigms described above can serve as tools for evaluating present conditions of running waters, surmising their likely antecedent condition, and developing targets and strategies for restoration. Because the majority of degraded streams and rivers have changed beyond our ability to return them to their historical state, it is more logical to use the term rehabilitation. Often the actions will take the form of returning certain organisms or processes to a condition that addresses societal objectives.

In the context of preserving and rehabilitating streams and rivers, it will be important to enlist the best scientific understanding of the structure and function of running-water ecosystems. For example, regulations governing the protection and width of riparian buffer strips, designed to protect stream organisms (usually fish) vary from one area to another, wider in some areas, narrower in others. However, managers and environmentalists should not limit their view of riparian buffers as only a matter of vegetative composition and buffer width with the sole aim of providing shading to reduce water tem-

peratures, a source of large woody debris, or stream bank stabilization. This view of riparian buffers ignores the often completely different in-stream trophic role played by the coupled riparian ecosystem. The buffer width required to produce shade, litter, large wood, nutrients, and bank stabilization are often quite different. Thus, the management and rehabilitation of a given reach of running water requires an integrated approach that acknowledges all the riparian functions and places the actions within the context of the larger watershed.

See also: Desert Streams; Estuaries; Floodplains; Freshwater Lakes; Riparian Wetlands.

Further Reading

Benda L, Poff NL, Miller D, et al. (2004) The network dynamics hypothesis: How channel networks structure riverine habitats. *Bioscience* 54: 413–427.

Cummins KW (1974) Structure and function of stream ecosystems. *Bioscience* 24: 631–641.

Cummins KW (1975) Macroinvertebrates. In: Whitton BA (ed.) *River Ecology.* Berkeley: University of California Press.

Cummins KW (1988) The study of stream ecosystem: A functional view. In: Pomeroy LR and Alberts JJ (eds.). New York: Springer.

Cummins KW (1997) Stream ecosystem paradigms. In: *CNR – Instituto di Ricerca Sulle Acque. Prospettive di recerca in ecologia delle acque.* Rome, Italy.

Frissell CA, Liss WJ, Warren CE, and Hurley MD (1986) A hierarchical framework for stream classification: Viewing streams in a watershed context. *Environmental Management* 10: 199–214.

Gregory SV, Swanson FJ, McKee WA, and Cummins KW (1991) An ecosystem perspective of riparian zones. *Bioscience* 41(8): 540–551.

Junk WJ, Bayley PB, and Sparks RE (1989) The flood pulse concept in river- floodplain systems. *Canadian Journal of Fisheries and Aquatic Sciences, Special Publication* 106: 110–127.

Minshall GW, Petersen RC, Cummins KW, et al. (1983) Interbiome comparison of stream ecosystem dynamics. *Ecological Monographs* 53: 1–25.

Stanford JA and Ward JV (1993) An ecosystem perspective of alluvial rivers: connectivity and the hyporheic corridor. *Journal of The North American Benthological Society* 12: 48–60.

Statzner B and Higler B (1986) Stream hydraulics as a major determinant of benthic invertebrate zonation patterns. *Freshwater Biology* 16: 127–139.

Saunders GW, et al. (1980) In: LeCren ED and McConnell RH (eds.) *The Functioning of Freshwater Ecosystems.* Great Britain: Cambridge University Press

Townsend CR (1989) The patch dynamics concept of stream community ecology. *Journal of the North American Benthological Society* 8: 36–50.

Vannote RL, Minshall GW, Cummins KW, Sedell JR, and Cushing CE (1980) The river continuum concept. *Canadian Journal of Fisheries and Aquatic Sciences* 37: 130–137.

Ward JV and Stanford JA (1983) The serial discontinuity concept of river ecosystems. In: Fontaine TD and Bartell SM (eds.) *Dynamics of Lotic Ecosystems,* pp. 29–42. Ann Arbor, MI, USA: Ann Arbor Science Publications.

Rivers and Streams: Physical Setting and Adapted Biota

M A Wilzbach and K W Cummins, Humboldt State University, Arcata, CA, USA

Introduction
History of the Discipline of Stream and River Ecology
The Physical and Chemical Setting

The Adapted Biota
Further Reading

Introduction

Streams and rivers are enormously important ecologically, economically, recreationally, and esthetically. This importance far outweighs their proportional significance on the landscape. Running waters constitute less than 1/1000th of the land surface and of freshwater resources of Earth and contribute only 2/10 000th of annual global freshwater budgets. Streams and rivers are significant agents of erosion and serve a range of human needs, including transportation, waste disposal, recreation, and water for drinking, irrigation, hydropower, cooling, and cleaning. At the same time, flooding of streams and rivers pose potential natural hazards to human populations. Irrespective of their impact on man, streams and rivers are rich, complex ecosystems that are diagnostic of the integrity of the watersheds through which they course.

There has always been a general anecdotal notion of what constitutes a stream and what constitutes a river; that is, streams are small, narrow, and shallow while rivers are large, wide, and deep. However, the difference between them is without clear distinction in the literature of the last 100 years. For the purposes of this article, streams refer to channels in drainage networks of orders 0–5 and rivers as orders 6–12 and above (see definition of stream order under the section titled 'Channel morphology'). In this article, the history of stream ecology is discussed followed by a treatment of the physical and chemical setting and biological features of major groups of lotic organisms. In a companion article, ecosystem dynamics and integrating paradigms in stream and river ecology are covered.

History of the Discipline of Stream and River Ecology

The formal published beginning of the study of flowing waters (lotic ecology) dates to the early twentieth century in Europe, where initial work focused on the distribution, abundance, and taxonomic composition of lotic organisms. In North America, the ecological stream studies began shortly after. In the 1930s, North American stream ecology was dominated by fishery biology. Stream and River studies worldwide remained descriptive

through the 1950s and this period also marked the beginning of a focus by lotic ecologists on human impacts. Descriptive studies detailed the taxonomic composition and density of the benthic invertebrate fauna found in reaches of streams and rivers variously affected by human impacts. Beginning in the 1960s and 1970s there was a shift to more holistic views of flowing-water ecosystems, with research concentrated on a synthetic view of lotic ecosystems, on energy flow, and on organic matter budgets for first-order catchments. In 1970, Noel Hynes, father of modern stream ecology, published his landmark book *The Ecology of Running Waters* which summarized concepts and literature to that point. With the 1980s, came the realization that running-water dynamics could be fully understood only through an integrated spatial and temporal perspective, and that whole catchments were the basic units of stream/river ecology. For example, holistic organic budget analyses of running-water ecosystems cannot be constructed unless both spatial and temporal scales are applied.

The hallmark of lotic research during the 1980s and 1990s was its interdisciplinary nature. Interactions involved stream ecologists, fishery biologists, aquatic entomologists, algologists, hydrologists, geomorphologists, microbiologists, and terrestrial plant ecologists. It was these interactions between the disciplines that focused the attention of stream biologists on physical processes and greater spatial and temporal scales. This perspective of stream ecosystems continues to direct the science in the twenty-first century, aided immensely by the incorporation of geographic information systems (GIS) analysis.

Although there is general acceptance that the logical basic unit for the study of streams and rivers is the watershed or catchment, most measurements of lotic ecosystem structure and function are still made at the reach or microscale level. Recently, there has been strong impetus to extend the scope of understanding to the watershed mesoscale and beyond because ecosystem processes exhibit effects of differing importance at different spatial and temporal scales and these processes interact across scales. The concern for issues of global climate change in regard to streams has provided additional motivation to analyze entire basins or all the basins in continental regions. Thus, a challenge for lotic ecologists in the twenty-first century remains the integration of data-rich studies at the reach level to entire watersheds and finally the coarse resolution of regional basin analysis relying on satellite imagery. The 'river continuum concept' and other stream/river conceptual models described below should continue to aid in the integration of knowledge about lotic ecosystems along whole catchments, from micro- to macroscale levels.

The Physical and Chemical Setting

Stream and river biota evolved in response to, and in concert with, the physical and chemical setting. Although traditionally the domain of hydrologists, geomorphologists, and chemists, study of processes driving the physical and chemical templates have been embraced by stream ecologists for interpreting patterns in organismic distributions and lotic ecosystem structure and function. From a purely physical perspective, the primary function of rivers is to transfer runoff and move weathering products away from terrestrial portions of the Earth for delivery to the oceans. Despite tremendous variability in the morphology and behavior of rivers, each results from the interaction between geomorphic and hydrologic processes. These processes and their effect on river morphology are summarized, followed by a discussion of major physical (current, substrate, and temperature) and chemical factors that affect the functioning of river ecosystems and the adaptations of stream organisms.

Hydrologic Processes

The total amount of the Earth's water does not change, and is continuously recycled among various storage compartments within the biosphere in a process referred to as the hydrologic cycle (**Figure 1**). The cycle involves evaporation from land and evapotranspiration from terrestrial vegetation driven by solar energy, cloud formation, and precipitation.

Annual global precipitation averages about 100 cm, but the majority evaporates and little falls directly into streams. The remainder either infiltrates into the soil or becomes surface runoff. The relative contributions of different pathways by which water enters streams and rivers varies with climate, geology, watershed physiography, soils, vegetation, and land use.

Water that infiltrates becomes groundwater, which makes up the largest supply of unfrozen freshwater. Groundwater discharges gradually to stream channels through springs or direct seepage when a channel intersects the groundwater table. Baseflow describes the proportion of total stream flow contributed from groundwater, and sustains streams during periods of little or no precipitation. Running waters may be categorized by the balance and timing of stormflow versus baseflow. Ephemeral streams carry water only in the wettest years and never intersect the water table. Intermittent streams flow predictably every year only when they receive surface runoff (**Figure 2**). Perennial streams flow continuously during wet and dry periods, receiving both stormflow and baseflow. The duration, timing, and predictability of flow greatly affect the composition and life-history attributes of stream communities.

Stream and river discharge, the most fundamental of hydrological measurements, describes the volume of water passing a channel cross-section per unit time.

Any increase in discharge must result in an increase in channel width, depth, velocity, or some combination of these. Discharge increases in a downstream direction through tributary inputs and groundwater addition and is accompanied by increases in channel width, depth, and velocity. An estimated 35 000 km^3 of water is discharged annually by rivers to the world's oceans, with the Amazon River alone accounting for nearly 15% of the total.

Hydrographs depict changes in discharge over time. Individual storm events display a steep rising limb from direct runoff, a peak, and a gradually falling recession limb as the stream returns to baseflow conditions (**Figure 3**). Variability in the shapes of hydrographs among streams reflects differences in the climatic, geomorphic, and geologic attributes of their watersheds and differences in the distribution of runoff sources.

Discharge records of sufficient duration allow prediction of the magnitude and frequency of flood events for a given river and year. Recurrence interval (T, in years) for an individual flood may be estimated as

$$T = (n + 1)/m$$

where n is the number of years of record, and m is ranked magnitude of the flood over the period of record, with the largest event scored as $m = 1$.

The reciprocal of T is the exceedance probability, which describes that statistical likelihood that a certain discharge will be equaled or exceeded in any given year. Thus a 1-in-100-year flood has a probability of 1% of occurring in any given year. The probability that a 100-year flood will occur in a river is the same every year, regardless of how long it has been since the last 100-year flood. Recurrence interval information provides an extremely important context for studies of lotic organisms.

Geomorphic Processes

Discharge and sediment supply represent the physical energy and matter that move through river systems,

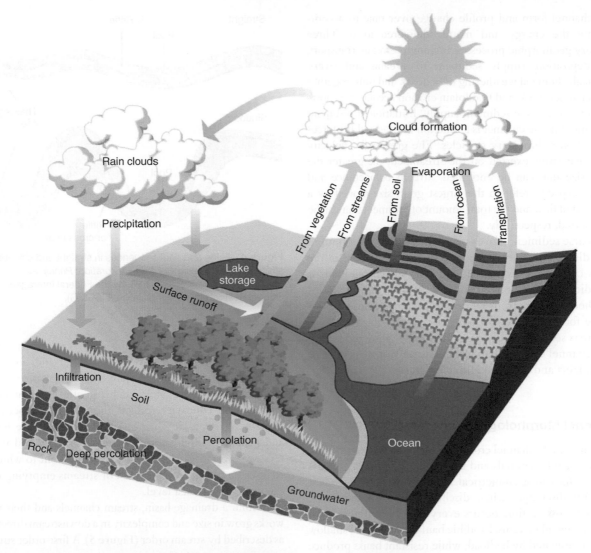

Figure 1 The hydrologic cycle. From *Stream Corridor Restoration: Principles, Processes, and Practices*, 10/98, by the Federal Interagency Stream Restoration Working Group (FISRWG).

Figure 2 An intermittent stream at 3.4 km elevation in the Andes Mountains in Chile, bordered by riparian vegetation of herbs and grasses. Intermittent streams are often important in exporting invertebrates and organic detritus to downstream fish-bearing reaches.

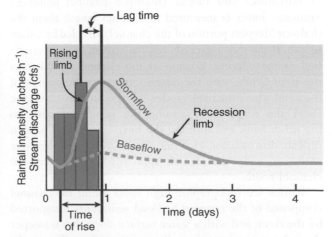

Figure 3 Stream hydrograph from a rainstorm event. From *Stream Corridor Restoration: Principles, Processes, and Practices*, 10/98, by the Federal Interagency Stream Restoration Working Group (FISRWG).

and channel form and profile change over time to accommodate the energy and matter delivered to it. Three primary geomorphic processes, including erosion, transport, and deposition, supply sediment to streams and rivers. Physical/chemical weathering of bedrock and soils, together with channel, bank, and floodplain erosion account for short- and long-term lotic sediment supply. Initiation of sediment movement in the channel is a function of drag and lift forces exerted on sedimentary particles. The greater the velocity and shear stress exerted on the streambed, the greater the grain size that can be entrained. Stream competence and stream capacity refer to the largest grain size moved by a given set of flow and the total amount of sediment that can be transported, respectively.

Coarse sediment moves along the stream/river bottom as bedload, and fine sediment moves downstream in the water column as suspended load. The suspended load, or turbidity, screens out light and scours off organisms attached to the bottom while the organic fraction serves as the food resource for invertebrate filtering collectors. Whereas sediments may be temporarily deposited within mid-channel or point bars, longer-term storage occurs on floodplains and elevated alluvial terraces.

Channel Morphology

Within a reach, channel cross sections reflect the interaction between bank materials and flow and vary from symmetrical in riffles to asymmetrical in pools as flow meanders. Bankfull discharge, when discharge just fills the entire channel cross-section, occurs every 1.5–2 years on average in unregulated systems. Erodible banks lead to wide shallow rivers dominated by bedload, while resistant banks produce narrow, deep channels transporting high suspended loads.

Channel pattern is described by its sinuosity (amount of curvature) and thread (multiple channel braiding). Sinuosity index is measured as channel length along the thalweg (deepest portion of the channel), divided by valley length. If the index exceeds 1.5, the stream/river is classified as meandering. Erosion of the channel bank carves the river bends, with the fastest current at the outside of the bend where the bank erodes. The greater the curve, the faster the water flows around the bend, deflecting to the other bank and forming the next curve. This pattern repeats downstream, creating regular swings in the river with a meander wavelength approximately 11 times the channel width.

Riffles are topographic high spots along the channel composed of the coarsest bedload sediments transported by the river, and with a water surface slope that is steeper than the mean stream gradient at low flow (**Figure 4**). They are typically spaced every five to seven channel widths. Pools are topographic depressions with fine sediments and reduced velocity.

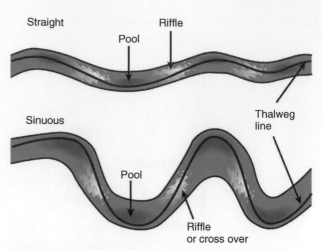

Figure 4 Riffle and pool sequences in straight and sinuous streams. From *Stream Corridor Restoration: Principles, Processes, and Practices*, 10/98, by the Federal Interagency Stream Restoration Working Group (FISRWG).

The longitudinal profile of a river is relatively stable over time, adjusting slowly to discharge and sediment supply. The profile is generally concave, with a steep gradient in its headwaters, and a gentle gradient at its mouth. The concavity reflects the adjustment between climate and tectonic setting (land relief and base level) and geology, which controls sediment supply and resistance to erosion. Base level describes the limit to which a river cannot erode its channel. For streams emptying into the ocean, this is sea level.

Within a drainage basin, stream channels and their networks grow in size and complexity in a downstream direction as described by stream order (**Figure 5**). A first-order stream

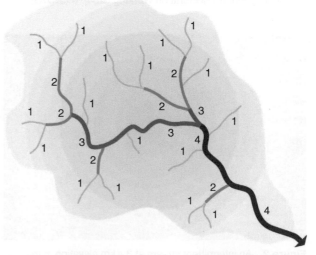

Figure 5 Ordering of stream segments within a drainage network. From *Stream Corridor Restoration: Principles, Processes, and Practices*, 10/98, by the Federal Interagency Stream Restoration Working Group (FISRWG).

lacks permanently flowing upstream tributaries and order number increases only where two stream of equal order join together. Employing this system, the Mississippi and the Nile Rivers at their mouths are order 10. There are usually 3–4 times as many streams of order $n-1$ as of order n, each of which is roughly half as long, and drains a little more than one-fifth of the land area. In the United States, nearly half of the approximately 5 200 000 km total river length are first order. As discussed later, many features of stream ecosystem structure and function are correlated with stream order.

Drainage basins, or watersheds, are the total area of land draining water, sediment, and dissolved materials to a common outlet. Watersheds occur at multiple scales, ranging from the largest river basins to first-order watersheds measuring only a few hectares in size. Larger watersheds are comprised of smaller watersheds and stream segments in a nested hierarchy of ecosystem units. The size and shape of the watershed, and the pattern of the drainage network within the watershed, exerts a strong influence on the flux of energy, matter, and organisms in river systems. Because some movement of energy, matter, and organisms move across and through landscapes independently of drainage basins, a more complete perspective of stream ecology requires consideration of landscape ecology.

Physical Factors

Current

Current ($m\,s^{-1}$ of flow) is the central defining physical variable in running-water systems. Velocity and associated flow forces exert major effects on stream organisms. Current shapes the nature of the substrate, delivers dissolved oxygen, nutrients, and food, removes waste materials, and exerts direct physical forces on organisms on streambed and in the water column, resulting, for example, in the dislodgement and displacement of organisms downstream. Current velocity, which rarely exceeds $3\,m\,s^{-1}$ in running waters, is influenced by the river slope, average flow depth, and resistance of bed and bank materials.

Flow in running waters is complex and highly variable in space and time. At a given velocity, flow may be laminar, moving in parallel layers which slide past each other at differing speeds with little mixing, or turbulent, where flow is chaotic and vertically mixed. The dimensionless Reynolds number, the ratio of inertial to viscous forces, predicts the occurrence of laminar versus turbulent flow. High inertia promotes turbulence. Viscosity is the resistance of water to deformation, due to coherence of molecules. At Reynolds numbers <500, flow is laminar; at >2000 flow is turbulent with intermediate values transitional. Although laminar flow is rare in running waters, microenvironments may contain laminar flow environments, even within turbulent, high-flow settings.

In cross section, a vertical velocity gradient decreases exponentially with depth. Highest velocities are at the surface where friction is least, and zero at the deepest point of the bottom where friction is the greatest. Mean current velocity is at about 60% of the depth from the surface to bottom. A boundary layer extends from the streambed to a depth where velocity is no longer reduced by friction and a thin viscous sublayer of laminar flow exists at its base.

Microorganisms and small benthic macroinvertebrates experience shelter from fluid forces within the sublayer. However, most stream organisms must contend with complex, turbulent flow where they exhibit a variety of morphological and behavioral adaptations for reducing drag and lift. Adaptations of macroinvertebrates and fishes may include small size, dorsoventral flattening to reduce exposure to current, streamlining to reduce current drag, the development of silk, claws, hooks, suckers, and friction pads as holdfasts, and behavioral movement away from high-velocity areas.

Substrate

In running waters, substrate provides food or a surface where food accumulates, a refuge from flow and predators, a location for carrying out activities such as resting, reproduction, and movement, and material for construction of cases and tubes. Algal growth, invertebrate growth and development, and fish egg incubation largely occur on or within the substrate. Substrate includes both inorganic and organic materials, often in a heterogeneous mixture. Mineral composition of the substrate is determined by parent geology, modified by the current. Organic materials include aquatic plants and terrestrial inputs from the surrounding catchment ranging from minute fragments and leaves to fallen trees (**Figure 6**).

Figure 6 Small headwater stream in old-growth Douglas-fir forest in Oregon, showing large woody debris spanning the channel. This spanner log forms a retention structure for organic detritus and sediment as well as refugia and habitat when the channel is inundated by high flows.

Table 1   Size categories of inorganic substrates in streams and rivers

Size category	Particle diameter (range in mm)
Boulder	>256
Cobble	
Large	128–256
Small	64–128
Pebble	
Large	32–64
Small	16–32
Gravel	
Coarse	8–16
Medium	4–8
Fine	2–4
Sand	
Very coarse	1–2
Coarse	0.5–1
Medium	0.25–0.5
Fine	0.125–0.25
Very Fine	0.063–0.125
Silt	<0.063

Modified from Cummins KW (1962) An evaluation of some techniques for the collection and analysis of benthic samples with special emphasis on lotic waters. *American Midland Naturalist* 67: 477–504.

Inorganic and organic materials are often classified by size according the Wentworth scale (**Table 1**). A broad classification of organic materials is discussed in Rivers and Streams: Ecosystem Dynamics and Integrating Paradigms. Organic particles <1 mm in diameter and >0.45 μm (fine particulate organic matter or FPOM) often function as food rather than substrate, and larger organic materials (CPOM) serve as substrate or food, for example, for litter-feeding invertebrates (**Figure 7**). Other substrate attributes, including shape, surface texture, sorting, and

Figure 7   Accumulation of leaf litter in a second-order stream in Oregon (USA) flowing through a second-growth forest with a red alder riparian zone. The litter that is retained at the leading edge of the cobbles provides the major food resource for stream invertebrate shredders and habitat for other invertebrates.

stability, are also determinants of benthic community structure, but these are less easily quantified. In general, larger, more stable rocks support greater diversity and numbers of individual organisms than smaller rocks, but smaller rocks with a higher ratio of surface area to volume support higher densities.

Evaluation of the ecological role of substrate is difficult because of its heterogeneity and covariance with velocity and oxygen supply. Heterogeneity is expressed along the length of a river as decreasing particle size in a downstream direction and at a reach scale as pool and riffle sequences, meandering, and point bar development. Substrate embeddedness describes the degree to which larger sediments, such as cobbles, are surrounded or covered by fine sand and silt. Significant embeddedness reduces streambed surface area and organic matter storage, the flow of oxygen and nutrients to incubating fish eggs and aquatic invertebrates, and entrance to and movement within the streambed by invertebrates.

Temperature

Temperature affects all life processes, including those in running waters. For example, decomposition, primary production and community respiration, and nutrient cycling are all temperature dependent. Most stream organisms are ectothermic, and their metabolism, growth rates, life cycles, and overall productivities are all temperature dependent. Annual temperature changes often serve as environmental cues for and/or regulate life-history events of invertebrates and fishes, especially emergence or spawning. The temperature regime sets limits to where species can live, and many species are adapted to certain thermal regimes. Increasing water temperature decreases dissolved oxygen solubility at the same time that it increases metabolic demand. Thus preferences of such organisms as salmonid fishes for cold water may have as much to do with temperature effects on oxygen availability as with effects of temperature *per se.*

Stream temperature is the net result of heat exchange via (1) net solar radiation, which reflects direct beam solar radiation, modified by cloud cover, day length, sun angle, vegetation, and topographic shading; (2) evaporation and convection, which are affected by vapor pressure and air temperature differentials as well as wind speed; (3) conduction, or heat exchange with streambed; and (4) advection from upstreamwater inputs, including groundwater and tributaries. On a diel basis, stream temperature varies less than air temperature because of the high specific heat of water. The greatest daily fluxes occur in summer in temperate regions, with a minimum before dawn. These fluxes are greatly affected by canopy cover and contributions of groundwater, which usually enters the channel at a temperature within 1 °C of mean annual air temperature. At the catchment level, daily temperature flux increases with distance from the headwaters, with a maximum in mid-order segments. Thermal stratification is rare except in large rivers and at tributary junctions. Seasonal variations in temperature mirror trends in mean monthly air temperature. The timing of the summer maximum often lags the timing of maximum solar radiation. Year-to-year variation in monthly temperatures is low, typically less than $2\,°C$. The annual temperature range of temperate streams is generally $0–25\,°C$, and $0–40\,°C$ in intermittent desert streams. The lower Amazon River is always within one or two degrees of $29\,°C$. Extremes in temperature occur in hot water springs, which can exceed $80\,°C$, and in subarctic and arctic streams that may completely freeze in winter. Surface freezing is usually prevented by snow and ice bridging, but underwater ice may form on streambeds as anchor ice or in the water column as slush or frazil.

Stream ecologists often evaluate temperature effects on stream organisms and ecosystem processes on the basis of degree-day accumulation rather than temperature maxima or minima. Degree days, which are calculated by summing daily mean temperatures above $0\,°C$, can differ among streams with similar maximum or minimum average daily temperatures. Such differences can affect voltinism (number of annual generations) of some species of aquatic insects.

Water Chemistry

Constituents of river water can be divided into five categories, which include dissolved gases, dissolved inorganic ions and compounds, particulate inorganic material, particulate organic material, and dissolved organic ions and compounds. Dissolved gases include oxygen, carbon dioxide, and nitrogen. Dissolved inorganic ions and compounds include major and minor ion groups and trace elements, such as copper, zinc, iron, and aluminum which occur in minute quantities. Nitrogen and phosphorus are minor ions, which function as nutrients essential to plant and animal growth. Major ion groups include cations of calcium, magnesium, sodium, and potassium and the anions bicarbonate, sulfate, and chloride.

The pH, which measures hydrogen ion activity, is affected by concentrations of dissolved gases and major ions and determines the solubility and biological availability of nutrients and heavy metals. Hardness is a measure of calcium and magnesium concentrations normally used to assess the quality of water supplies. Hardness is associated with, but not identical to alkalinity, which measures the ability of streamwater to absorb hydrogen ions, thus buffering changes in pH. Alkalinity is primarily due to bicarbonate and carbonate ions. Total dissolved solids, the sum of the concentrations of major cations and anions, are often estimated as specific conductance. Hardness, alkalinity, and ionic concentrations are frequently positively correlated with stream productivity and taxonomic

richness. Particulate inorganic and organic materials together make up the suspended load in lotic systems, and contribute to turbidity.

Carbon dioxide and oxygen are the most biologically important dissolved gases. Diffusion from the atmosphere maintains concentrations of both oxygen (O_2) and carbon dioxide (CO_2) in streams at close to equilibrium. However, CO_2 is more soluble in water than is O_2, which is 30 times less available in water than air. Groundwater and sites of organic matter decomposition are low in O_2 and enriched in CO_2. Photosynthesis and respiration can alter diel concentrations of oxygen and carbon dioxide in productive systems, with O_2 elevated and CO_2 reduced during day, and the reverse occurring at night. If production is high relative to diffusion, diel changes in O_2 are used to estimate photosynthesis and respiration. Because current and turbulence continually renew O_2 supply, its concentrations are problematic for stream organisms only in sites severely contaminated with organic pollutants or through a combination of high temperatures, drought, and dense populations of aquatic plants. Low O_2 concentrations are better tolerated by stream animals at faster current speeds.

Typical rivers have been described as essentially dilute calcium bicarbonate solutions dominated by a few cations and anions. The considerable natural spatial variability in lotic chemistry largely reflects the type of rocks available for weathering and the amount, chemical composition, and distribution of precipitation. For example, total dissolved solids are approximately twice as great in rivers draining sedimentary terrain compared with igneous and metamorphic rock. Most rivers contain 0.01–0.02% dissolved minerals, about $1/20–1/40$th the salt concentration of the oceans, with an average concentration of $100\ \mathrm{mg\ l^{-1}}$. Generally $\geq50\%$ of this is bicarbonate and 10–30% is chloride and sulfate. River water contains more dissolved solids than does rainwater, because of evaporation, weathering, and anthropogenic inputs. Rainwater, although nearly pure, contains dissolved minerals from dust particles and droplets of ocean spray.

Rainwater is also naturally acidic due to atmospheric carbon dioxide dissolving in the water droplets, forming a weak carbonic acid (H_2CO_3). In catchments with hard rocks resistant to weathering, little buffering capacity, or where decaying plant matter is abundant, streamwater can be acidic even in absence of pollution. Water percolating through the soil enters the stream and is enriched with CO_2 from plant and microbial respiration and forms carbonic acid. The carbonic acid dissolves the calcium carbonate in rocks, producing calcium bicarbonate, which is soluble in water and the source of carbon atoms for aquatic photosynthesis. The dissolution of calcium carbonate increases the amount of stream calcium and bicarbonate ions and the latter dissociates to carbonate ions. At equilibrium, bicarbonate and carbonate ions dissociate, forming hydroxyl ions and resulting in weak alkaline waters, with a pH > 7. At equilibrium, water resists changes in pH because the addition of hydrogen ions is neutralized by the hydroxyl ions formed by dissociation of bicarbonate and carbonate, and added hydroxyl ions react with bicarbonate to form carbonate and water. Thus the buffering capacity of a stream is largely determined by its calcium bicarbonate content. The pH of most natural running waters ranges between 6.5 and 8.5, with values below 5 or above 9 being harmful to most stream organisms. Industrially derived sulfuric and nitric acids have seriously lowered pH in surface waters of large areas of Europe and North America, resulting in reduced species diversity and density.

The Adapted Biota

Many taxonomic groupings inhabit running waters. Key biological attributes, life histories, and distribution patterns of organisms that play a central role in energy flux within lotic ecosystems or that are of significant human interest – namely algae, macrophytes, benthic macroinvertebrates, and fishes – are summarized below.

Algae

Algae are the most important primary producers in running-water ecosystems and because of their sessile nature and short life cycles, their assemblages are used to evaluate stream ecosystem health. Algae are thalloid organisms, bearing chlorophyll *a* and lacking multicellular gametangia. Algal evolution radiated from a common ancestry to several diverse kingdoms. For example, blue-green algae are classified as bacteria, and dinoflagellate algae as protozoans. Algal taxonomy is based on pigmentation, the chemistry and structure of internal storage products and cell walls, and number and type of flagellae. Five major divisions of algae are common in streams, including the Bacillariophyta (diatoms), Chlorophyta (green algae), Cyanobacteria (blue-green algae), Chrysophyta (yellow-green algae), and Rhodophyta (red algae). Of these, the diatoms, green algae, and cyanobacteria are most prevalent. Assemblages of algae attached to the substrate are referred to as periphyton or aufwuchs. Periphyton attached to submerged substrates is a complex assemblage of algae, bacteria, fungi, and meiofauna bound together with a polysaccharide matrix referred to as biofilm. Algae of the water column are phytoplankton, occurring chiefly in slowly moving lowland rivers as sloughed benthic cells or exports from connected standing waters within the watershed.

Diatoms are extremely abundant in freshwater as well as in saltwater, and typically comprise of majority of species within the periphyton. Generally microscopic, diatoms are brownish-colored single-celled algae constructed of two overlapping siliceous cell walls, or valves, fit together like the halves of a petri dish. Valves are connected to each other by one or more 'girdle' bands. The two valves form the frustule, which is uniquely decorated with pores (punctae), lines (striae), or ribs (costae). The symmetry of these decorations defines two groups: radially symmetrical centric diatoms and bilaterally symmetrical pennate diatoms. Diatoms may occur individually, in chains, or in colonies, and those with a divided cell wall (raphe) are able to move. In temperate streams, diatoms exhibit two growth blooms: in spring prior to shading by deciduous canopies as water temperatures rise and nutrients are plentiful; and in fall following leaf abscission, when nutrients released from decaying green algae and deciduous litter are available. Diatoms constitute a high-quality, rapid-turnover food resource for macroinvertebrate scrapers and collectors. Representative diatoms common in stream periphyton are shown in **Figure 8**.

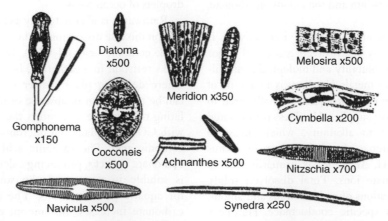

Figure 8 Representative diatoms common in stream periphyton. From Hynes HBN (1970) *The Ecology of Running Waters*. Liverpool: Liverpool University Press.

Green algae occur in a variety of habitats, and are distinguished by the number and arrangement of flagella, their method of cell division, and their habitat. In streams, distinctions are made between micro- and macroforms. Macroalgae occurs as a thallus or as filaments. Filamentous forms may be branched or unbranched. Green algae provide attachment sites for diatoms, and are a source of FPOM and photosynthetic oxygen, but are fed upon by few invertebrates.

Blue-green algae, or cyanobacteria, are prokaryotic organisms of ancient lineage which contain the photosynthetic pigment phycocyanin, used to capture light for photosynthesis. They occur in a variety of habitats and are one of very few groups of organisms that can convert inert atmospheric nitrogen into an organic form. Bluegreen algae may be filamentous or nonfilamentous, and only filamentous forms with heterocysts are capable of nitrogen fixation in aerobic settings. Several of the heterocyst-containing filamentous taxa, (e.g., *Anabaena*, *Aphanizomenon*, and *Microcystis*) can form dense blooms and produce toxins in warm, nutrient-rich waters. Nitrogen-fixing *Nostoc*, common in small streams, forms a unique commensal association with the chironomid midge *Cricotopus*.

Macrophytes

Macrophytes include vascular flowering plants, mosses and liverworts, some encrusting lichens, and a few large algal forms such as the Charales and the filamentous green alga *Cladophora*. Light and current are among the most important factors limiting the occurrence of macrophytes in running waters. Major plant nutrients, particularly phosphorus, can be limiting in nutrient-poor waters but are likely to be present in excess in eutrophic lowland rivers. Three ecological categories include those that are attached to the substrate, those that are rooted into the substrate, and free-floating plants. Attached plants include the mosses and liverworts, certain lichens, and some flowering plants of the tropics. These are all largely found in cool, headwater streams. The mosses are unusual in their requirement for free CO_2, rather than bicarbonate, as their carbon source. In shaded, turbulent streams, their contribution to primary production may override that of the periphyton. Mosses also support very high densities of macroinvertebrates. Rooted plants include submerged (e.g., Hydrocharitaceae, Ceratophyllaceae, and Halorgidaceae) and emergent (e.g., Potamogetonaceae, Ranunculaceae, and Cruciferae) forms and require slow currents, moderate depth, low turbidity, and fine sediments for rooting. They are most common in mid-sized rivers and along the margins of larger rivers where they reduce current velocity, increase sedimentation, and provide substrate for epiphytic microflora. Tough, flexible stems and leaves, attachment by adventitious roots, rhizomes or stolons, and vegetative reproduction are important adaptations. Free-floating plants (e.g., Lemnaceae and Pontederiaceae) are of minor importance in running waters at temperate latitudes as they depend largely on lacustrine conditions. They may accumulate significant biomass in subtropical and tropical settings. Macrophytes in lotic ecosystems contribute to energy flow predominantly through decomposer food chains, as few macroinvertebrates feed on the living plants.

Benthic Macroinvertebrates

The major groups of invertebrates in running waters include three phyla: Annelida (worms) and Mollusca (snails, clams, and mussels) of marine evolutionary origin that are most abundant and diverse in larger rivers, and Arthropoda (crustaceans and insects) that dominate the headwaters, but are abundant all along drainage networks. Representative taxa are illustrated in **Figure 9**.

The Oligochaeta is the most abundant and diverse group of annelids, and are notable for their ability to inhabit low-oxygen environments. Oligochaetes inhabit the sediments, some in tubes, and are almost all gathering-collector detritivores. The worms are segmented with two pairs of stout, lateral chetae on each segment. Annelid leeches (Hirudinea), a minor group occurring in small streams to mid-sized rivers, are gathering collectors or predators.

Gastropod (limpets and snails) and bivalve (clams and mussels mollusks) are restricted in their occurrence in streams and rivers by their calcium requirement for shell formation. Limpets, such as *Ferrissia* (Ancylidae), frequent small, fast-flowing streams where their hydrodynamic shape and sucker formed by the mantle allow them to move over rocks in the current and scrape loose attached algal food with a rasping radula. Snails, such as *Physa*, are abundant scrapers in river macrophyte beds where they employ their radulas to rasp vascular plant surfaces, removing periphyton and epidermal plant tissue. Clams and mussels (Bivalvia = Pelecypoda) are filtering collectors that burrow in the sediments with their incurrent and excurrent siphons exposed. They pump water in to extract dissolved oxygen and FPOM, and out to eliminate wastes. Because bivalve mollusks are sensitive to water quality, they have been used worldwide as indicators of lotic ecosystem health. However the small, ubiquitous fingernail clams (Sphaeridae) are more tolerant, inhabiting a wide range of stream and rivers.

The common Crustacea of running waters include Amphipoda (scuds), Isopoda (aquatic pill bugs), benthic Copepoda (Harpactacoida), and Decapoda (crayfish and freshwater shrimps). Most isopods and amphipods (except *Hyallela*) are detrital shredders feeding on stream-conditioned riparian litter in headwater streams. Although decapod shrimps and crayfish have species found in all sizes of running waters, the former tend to be more abundant in streams, the latter in mid-sized

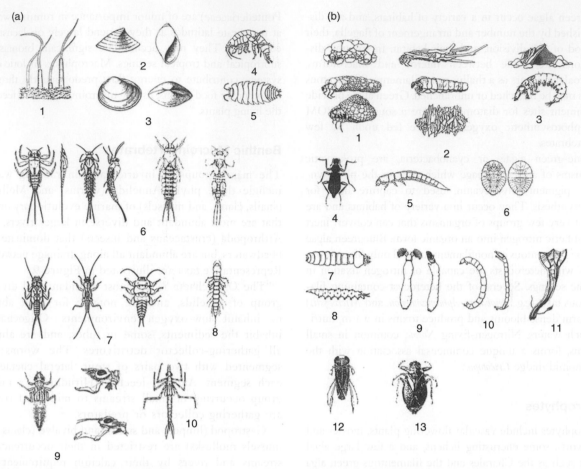

Figure 9 (a) Examples of lotic benthic invertebrates. 1, Annelida, Oligochaeta (Tubificidae); 2, Mollusca, Gastropoda (*left*, Ancylidae; *right*, Physidae); 3, Mollusca, Bivalvia (Spaeridae: *left*, lateral view; *right*, dorsal view); 4, Crustacea, Amphipoda; 5, Crustacea, Isopoda; 6, Insecta, Ephemeroptera; 7, Insecta, Plecoptera; 8, Insecta, Megaloptera (Sialidae); 9, Insecta, Odonata, Anisoptera (*left*, nymph; *right upper*, lateral view of head with extended labium; 10, Insecta, Odonata, Zygoptera (right, nymph; left lower, lateral view of head with extended labium). (b) Examples of lotic benthic invertebrates. 1, Insecta, Trichoptera (mineral case bearers); 2, Insecta, Trichoptera (organic case bearers); 3, Insecta, Trichoptera (net spinner, fixed retreat above); 4, Insecta, Coleoptera (Elmidae adult); 5, Insecta, Coleoptera (Elmidae larvae); 6, Insecta, Coleoptera, Psphenidae larvae (*left*, ventral; right, dorsal); 7, Insecta, Diptera, Tipulidae; 8, Insecta, Diptera, Athericidae; 9, Insecta, Diptera, Simuliidae (*left*, dorsal; *right*, lateral view); 10, Insecta, Diptera, Chironomidae (*left*, Chironominae; *right*, Tanypodinae); 11, Insecta, Diptera, Chironomidae (filtering tube of *Rheotanytarsus*); 12, Insecta, Hemiptera, Corixidae; 13, Insecta, Hemiptera, Belastomatidae.

rivers. Decapods are scavengers, but are usually classified as facultative shredders of plant litter. These crustaceans have always been of interest because of their large size, commercial food and bait value, and importance as food for large game fish. The minute harpactacoid copepods are poorly known, but are often in small streams to large rivers where they are gathering collectors inhabiting accumulations of benthic FPOM.

Aquatic insects (Arthropoda) are the most conspicuous and best-studied invertebrates of running waters. They can be subdivided into the more primitive hemimetabolous orders, in which immature nymphs gradually metamorphose into mature winged adults, and the more evolved holometabolus orders that have a larval and pupal stage. Insect growth is accomplished by the nymphs

or larvae and lasts for weeks to years, while the adults feed little and are short lived (a day to weeks). Terrestrial insects are much more abundant and diverse than lotic forms but there are 13 orders of aquatic or semiaquatic (occurring at lotic margins) taxa. The orders in which all larvae are aquatic are as follows: the hemimetabolous mayflies (Ephemeroptera), stoneflies (Plecoptera), dragon- and damselflies (Odonata), and the holometabolous caddisflies (Trichoptera), and dobson- and alderflies (Megaloptera). These are signature taxa represented in almost all unpolluted lotic ecosystems. Mayflies, which are the only insects that molt as winged subadults (subimagos) to sexually mature adults (imagos), are of immense importance to sport flyfishing. All the odonate and about half of the plecopteran nymphs are predaceous.

The dragonflies and damselflies occur in small streams to large rivers, with many species associated with aquatic vascular plants. The nonpredaceous stonefly nymphs are shredders feeding upon conditioned riparian litter.

Caddisflies are a large aquatic order in which a majority of species construct portable cases made of plant pieces (the shredders) or mineral particles (the scrapers) held together with silk extruded from glands in the head. All the cases are lined with silk into which hooks on the hind prolegs are hooked to maintain the larvae in the case. Larvae circulate water through the case by undulating the abdomen to irrigate the gills and integument and facilitate respiration. Five families of Trichoptera larvae, and all families in the pupal stage, construct nonportable, fixed retreats of organic and mineral material. Most larvae of the five families spin silk nets with which they filter out FPOM food from the flowing water. Species of the family Rhyacophilidae are free ranging without cases and almost exclusively predaceous. Some of the predaceous Megaloptera are among the largest of the lotic aquatic insects, and they are typical of slow-flowing areas and often associated with submerged woody debris.

The holometabolous Coleoptera (beetles), Diptera (true flies), Lepidoptera (aquatic moths), and Hymenoptera (aquatic wasps) constitute the largest insect orders and have some aquatic or semiaquatic representatives, as do the spongeflies of the Neuroptera. The beetles are the only aquatic insects with representatives in which both the larvae and adults live in the water. One family of Diptera, the midges (Chironomidae), is usually more abundant and diverse in running waters than all other aquatic insects combined. Chironomid species are represented in all lotic habitats and all functional feeding groups. Their use in ecological studies has been hampered by the difficulty of identifying the larvae. Very few aquatic moths are found in running waters. A few are scrapers inhabiting fast-flowing streams, but the majority live and feed on the leaves of aquatic macrophytes. Hymenoptera, a large terrestrial order containing many social species, has some parasitic forms in which the females enter the water to oviposit in the immatures of aquatic and semiaquatic orders. The larvae of spongeflies inhabit freshwater sponges where they are either predators or feed directly on sponge tissue.

The hemimetabolous Hemiptera (true bugs), Orthoptera (grasshoppers, etc.), and Collembola (springtails) have aquatic or semiaquatic species. All the widely distributed hemipterans are active predators, occupying the full range of slow water and marginal habitats where they capture prey and imbibe their body fluids using piercing mouth parts. All the Orthoptera and Collembola of running waters are semiaquatic and function as detrital gathering collectors. Functional feeding roles are explained in greater detail in Rivers and Streams: Ecosystem Dynamics and Integrating Paradigms.

Fishes

Fishes, the principal group of vertebrates found in running waters, are of great human interest because of their commercial and recreational value. Approximately 41% (about 8500 species) of the world's fishes live in freshwater. Of these, almost all have representatives that occur in running waters, although with varying degrees of river dependency and saltwater tolerance. Groups with little or no tolerance for saltwater (e.g., Cyprinidae, Centrarchidae, and Characidae) are considered to be primary freshwater fishes, and have dispersed through freshwater routes or evolved in place from distant marine ancestors. Secondary freshwater fishes (e.g., Cichlidae and Poeciliidae) are usually restricted to freshwater but have some tolerance to saltwater. Diadromous fishes migrate between freshwater and saltwater. Anadromous fishes, including many salmonids, lampreys, shad, and sturgeon, spend most of their lives in the sea and migrate to freshwater to reproduce. American and European eels are catadromous fishes, which spend most of their lives in freshwater and migrate to the sea to reproduce. Catadromy appears to be more prevalent in the tropics, and anadromy more common at higher latitudes.

Longitudinal gradients of fish assemblages are common within river systems, and have resulted in several attempts to classify stream zones by the dominant fish species or assemblage found. Because fish faunas vary considerably among geographic and climatic regions, zonation schemes can usually be applied only locally except in Europe. Longitudinal gradients arise as the result of species addition and/or replacement, and reflect adaptations to the type and volume of habitat and available food along the river continuum. Upstream fishes, typified by salmonids and sculpins, have high metabolic rates and consequent high demands for oxygen. Salmonids are active, streamlined fishes with strong powers of locomotion that can maintain position in swift water to feed upon drifting invertebrates. Sculpins, with depressed heads and large pectoral fins, hold close to the streambed and forage for invertebrates among stones on the bottom. Upstream fishes are usually solitary in habit and may exhibit territoriality associated with both breeding and spatial resources. They may extend downstream where oxygen and temperatures are suitable, to join deeper-bodied fishes more tolerant of warmer temperatures and reduced oxygen. Species richness is usually greatest in the mid-order segments, in association with increased pool development and overall habitat heterogeneity. The Cyprinidae, one of the largest and most widespread of primary fish families, is characteristic of moderate gradient streams. Shoaling behavior is common within this group. In high-order reaches, fish assemblages include larger, deep-bodied fishes such as suckers and catfishes that feed on bottom deposits, invertivorous sun-fishes, and predatory pike.

See also: Desert Streams; Rivers and Streams: Ecosystem Dynamics and Integrating Paradigms.

Further Reading

Allan JD (1995) *Stream Ecology: Structure and Function of Running Waters*. London: Chapman and Hall.

Cummins KW (1962) An evaluation of some techniques for the collection and analysis of benthic samples with special emphasis on lotic waters. *American Midland Naturalist* 67: 477–504.
Giller PS and Malmqvist B (1998) *The Biology of Streams and Rivers*. Oxford: Oxford University Press.
Hauer FR and Lamberti GA (1996) *Methods in Stream Ecology*. San Diego: Academic Press.
Hynes HBN (1970) *The Ecology of Running Waters*. Liverpool: Liverpool University Press.
Knighton D (1998) *Fluvial Forms and Processes: A New Perspective*. London: Arnold Publishers.
Leopold LB (1994) *A View of the River*. Cambridge, MA: Harvard University Press.

Rocky Intertidal Zone

P S Petraitis and J A D Fisher, University of Pennsylvania, Philadelphia, PA, USA

S Dudgeon, California State University, Northridge, CA, USA

Introduction
Physical Aspects of the Shore
Attached Organisms
Mobile Organisms
Zonation

Rocky Intertidal Shores as an Important System in Development of Ecology
Unresolved Problems and Future Directions
Further Reading

Introduction

The British ecologist A. J. Southward described the intertidal zone as "the region of the shore between the highest level washed by the waves and the lowest level uncovered by the tide," and thus communities on rocky intertidal shores are primarily defined by the tides and the presence of hard surfaces. The types of organisms, the number of species, and the distribution and abundance of individual species found in a particular rocky intertidal community also depend on the physical aspects of the shore, the supply of resources, food and larvae from overlying water, the biological interactions among the species present, and the regional pool of species. Although rocky intertidal shores cover only a small fraction of the Earth's surface, they contain a large diversity of organisms – ranging from highly productive microalgae to transient vertebrate predators (**Figure 1**).

Physical Aspects of the Shore

Tides

Tides are caused by the gravitational effects of the Moon and Sun, which ideally produce a cycle of two high tides and two low tides per day. However, the amplitude and frequency of the tides are altered by the phases of the Moon, the Earth's orbit and declination, latitude, and the configurations of the shoreline and the seafloor. The tidal range tends to be smaller toward the equator and can vary from several meters in high latitudes to less than tens of centimeters near the equator. Configuration of the coast and the ocean basin can cause harmonic resonances and create tides that vary dramatically in amplitude and frequency. In extreme cases, the reinforcing and canceling effects can produce a single high and low tide per day or almost no change over the course of a day.

The timing of low tides can have a profound effect by exposing organisms to extreme conditions. For example, the lowest tides in the Gulf of Maine, USA tend to occur near dusk or dawn, and so organisms are rarely exposed to mid-day sun in the summer but are often exposed to below freezing temperatures on winter mornings. In contrast, the lowest summer tides in southeastern Australia occur mid-day and expose organisms to extraordinarily high temperatures.

Characteristics of the Shore

Any firm stable surface in the intertidal zone has the potential to support the organisms that commonly occur in rocky intertidal communities, and at low tide, intertidal habitats can range from dry rock to filled tide pools. Rock

Figure 1 Closeup of predatory snails, mussels, barnacles, and brown algae in Maine, USA. Photo by P. S. Petraitis.

surfaces can vary from very hard to relatively soft rock such as from granite to sandstone and can range from smooth platforms to irregular fields of stone cobbles and boulders. Topography, inclination, color, and texture of the rock affect rate of drying and surface temperature, which can limit the distribution and abundance of species. Man-made surfaces such as rock jetties and wooden pier pilings and biogenic surfaces such as mangrove roots can also support communities that are indistinguishable from the communities found on nearby rocky shores.

Tide pools can be very different than the surrounding shore because of thermal variability, changes in salinity from evaporation and runoff, and changes in pH, nutrients, and oxygen levels caused by algae. Pools often support residents such as sea urchins, snails, and fish that would otherwise be restricted to subtidal areas.

The amount of wave surge affects the types of organisms found on the shore and their distribution. Wave surge and breaking waves tend to expand the extent of the intertidal zone and distribution of species by continually wetting the shore and allowing species to extend farther up the shore. Wave surge can also cause mobile animals to seek refuge and can limit the distribution of slow moving species, and the force of breaking waves can damage and sweep away organisms. Sand and debris such as logs swept up by the waves can scour organisms off the surface. In areas of low wave surge, sedimentation of sand and silt may bury organisms or clog gills and other filter-feeding structures.

Attached Organisms

Unlike terrestrial habitats, which depend largely on local plant material to support resident animal populations, rocky intertidal assemblages are supported not only by algal primary production but also by secondary production from suspension feeders, such as barnacles and mussels, which link the ocean's productivity to the shore.

Algae

The term 'algae' refers to an extraordinarily diverse and heterogeneous group comprising about seven major lineages, or roughly 41% of the kingdom-level branches in the Eukarya domain. Most lineages consist of unicellular microalgae, but the multicellular macroalgae that dominate many rocky shores worldwide occur in only three groups (Rhodophyta, Chlorophyta, and Phaeophyta) (**Figure 2**).

Microalgae are ubiquitous and although inconspicuous, they are important members of rocky intertidal communities. For example, diatoms are the primary food source of many grazing gastropods and form biofilms, which facilitate settlement of invertebrate larvae and stabilize meiofaunal assemblages.

Benthic macroalgae (i.e., seaweeds) dominate many rocky shores, especially the low- and mid-intertidal zones of temperate regions, and many exhibit morphologies adaptive for life on wave swept shores. The idealized body plan of a seaweed consists of a holdfast, a stipe, and one or more blades. The holdfast usually attaches the alga either by thin encrusting layers of cells tightly appressed to the rock surface or by a massive, thick proliferation of tissue that often produce mucilaginous 'glues' to adhere the tissue to the rock. The stipes are analogous to plant stems and display remarkable material properties that enable seaweeds to withstand the tremendous hydrodynamic forces imposed by breaking waves. The blade is the principal structure for the exchange of gases and nutrients, and the capture of light for photosynthesis. Blades also contain reproductive tissue, either within a vegetative blade, or in sporophylls (i.e., special blades for reproduction). Some larger brown seaweeds, such as

Figure 2 Extensive brown algal beds in Maine, USA. Photo by P. S. Petraitis.

fucoids and kelps, have gas-filled floats called pneumatocysts that buoy the blade so that it remains closer to the surface where light intensity is greater.

The diversity and complexity of the life cycles of most seaweeds contributes to their great abundance on rocky shores. The life cycle of most seaweeds consists of an alternation of separate gametophyte and sporophyte generations. The two generations can either look the same (i.e., isomorphic) or different (heteromorphic). In some species, the heteromorphic generations are so different that they were originally described as different species. Heteromorphic life histories are hypothesized to represent an adaptation to grazing pressure, and heteromorphic generations clearly show tradeoffs with respect to competitive ability, resistance to disturbance and longevity associated with upright foliose and flat encrusting morphologies.

Sessile Invertebrates

Adults of many invertebrate species are attached permanently to the rock or other organisms (epibiota). These include members of the phyla Porifera (sponges), Cnidaria (hydroids and sea anemones), Annelida (tube-building polychates), Arthropoda (barnacles), Mollusca (mussels and clams), Bryozoa (moss animals), and Chordata (tunicates). Suspension feeding – either by pumping water through a sieve structure or trapping particles carried on induced or external currents – is a common feature of sessile animals and serves to transfer inputs of energy and nutrients produced in the water column into the intertidal zone via the ingestion of plankton. Additionally, by feeding on locally derived detritus, suspension feeders capture some of the nutrients that are produced by neighboring inhabitants.

Sessile intertidal animals are often physically or chemically defended against predation and display plastic phenotypes in response to changing environmental conditions because they are fixed in place and cannot move to avoid predators. For example, the presence of the predatory gastropod *Acanthina angelica* induces change in the shell shape of its barnacle prey *Chthamalus anisopoma*, and the barnacle forms a curved shell making it more difficult for the predator to attack.

Mobile Organisms

Mobile invertebrates and vertebrates that are found on rocky intertidal shores are typically divided into two categories based on the amount of time spent between tidemarks. Resident species remain in the intertidal zone throughout most of their life and face a large range of local physical conditions that they mitigate by a variety of behavioral and physiological adaptations. Many residents find shelter during low tides, either between rocks, under

algae, or in tide pools, while other species attach to exposed rock surfaces just ahead of the incoming tide. Transient species are those that spend only a small part of their life cycles in the intertidal zone (e.g., as juveniles) or are those that enter and leave the intertidal zone during low or high tide.

Invertebrates

Large, mobile invertebrate consumers are ecologically the most intensively studied guild on rocky shores and include species from Turbellaria (flatworms), Crustacea (e.g., crabs, shrimp, amphipods, and isopods), Annelida (e.g., polychaetes), Gastropoda (e.g., snails, nudibranchs, and chitons), and Echinodermata (sea urchins, brittle stars, and sea stars). Herbivores range from grazers of diatom films to browsers of macroalgae, and predators exploit a variety of methods (crushing, stinging, drilling, and partial consumption) to overcome the defenses of their prey.

Small mobile metazoans (roughly 0.1–1 mm and collectively termed meiofauna) thrive on and among the algae, animals, and the trapped sediments on rocky shores. Meiofauna include consumers from many invertebrate phyla, that – due to their small sizes, extremely high abundances, and high turnover rates – are an important guild of consumers whose effects have largely been neglected in comparison to studies of larger invertebrates.

Vertebrates

Vertebrates tend to be transient species that use the intertidal zone to feed or hide and include fish and marine mammals that enter at high tide and birds and terrestrial mammals that enter at low tide (**Figure 3**). For instance, marine iguanas (*Amblyrhynchus cristatus*) of the Galápagos Islands, Ecuador forage extensively on intertidal algae on

Figure 3 Rocky shore in Central California, USA with elephant seals on the beach. Photo by S. Dudgeon.

lava reefs during low tides. The major exceptions are resident intertidal fishes, which are often cryptic and less than 10 cm in length. Resident and transient fishes include hundreds of species from dozens of families, though members of the families Blenniidae, Gobiidae, and Labridae are the most common.

Birds and mammals, characterized by high endothermic metabolic rates and large body sizes, have significant impacts on intertidal communities even at low densities. Birds include locally nesting and migratory species and can remove millions of invertebrates during a season. In addition, birds in some communities provide major inputs of nutrients via guano and prey remains. More than two dozen terrestrial mammals, mostly carnivores, rodents, and artiodactyls, have been reported as consumers or scavengers of rocky intertidal organisms on every continent except Antarctica. Most recorded prey species are mollusks, crabs, or fish. Probably one of the most unusual cases is a population of feral rabbits on a small island off the coast of South Africa that forage on seaweeds in the intertidal zone. Given the mobility of vertebrates, their impact on rocky intertidal shores has been difficult to assess and intertidal activity is often discovered by finding exclusively intertidal animals or algae in the gut contents of otherwise pelagic or terrestrial species.

Little is known about the effects of harvesting by humans in the rocky intertidal zone. Results from a few large-scale studies in Australia, Chile, and South Africa, however, have demonstrated that harvesting has had significant effects on intertidal assemblages.

Zonation

Patterns

Rocky intertidal shores often display a vertical zonation of fauna and flora associated with the strong environmental gradient produced by the rise and fall of the tides. For example, most moderately exposed rocky shores of the northern hemisphere have kelps at the littoral sublittoral interface, followed by rhodophyte algae dominating the low intertidal zone, by fucoid algae, mussels, and barnacles dominating the mid-intertidal zone, and by cyanobacteria, lichens, and a variety of small tufted, encrusting, or filamentous ephemeral seaweeds occurring in the high intertidal zone. While species from many phyla may be found together, often a single species or group is so common; vertical zones are named according to the dominant group (e.g., the intertidal balanoid zone named after barnacles in the family Balanidae).

Combinations of various physical factors acting upon different inhabitants in intertidal zones that vary in their exposure to waves can lead to complex patterns of distribution and abundance along shorelines in a particular region. Nevertheless, some general patterns are evident at a regional scale. Geographically, vertical zonation patterns are most pronounced on temperate rocky shores where species diversity is high and tidal amplitudes tend to be greatest. On rocky shores in the tropics, biotic zones are compressed into narrow vertical bands because of small tidal amplitudes. In polar regions, annual ice scour and low species diversity tend to obscure any conspicuous vertical zonation.

Causes

It is often stated that the upper limits of organisms are set by physical factors, whereas the lower limits are set by biological interactions but there are many exceptions to this rule. The specific causes of the zonation seen on most rocky shorelines vary with geographic location, but zonation results primarily from behavior of larvae and adults, tolerance to physiological stress, the effects of consumers, and the interplay between production and the presence of neighbors.

Adult movements and larval behavior during settlement from the plankton onto rocky shores have major effects on the distribution of animals. For example, studies of barnacles have shown that vertical zonation of larvae in the water column contributes to corresponding vertical zonations of both larval settlement and adults on the shore, a pattern previously ascribed solely to interspecific competition. For seaweeds, behavior is a relatively unimportant cause of their zonation since adult seaweeds are sessile and settling spores are mostly passively transported.

Marine organisms living higher on the shore are faced with more frequent and extreme physiological challenges than their lower shore counterparts, and the upper limits of intertidal distributions for most species are set by cellular dehydration. Dehydration can occur either from freezing during winter or simply desiccation associated with long emersion times. High temperatures and wind, which accelerate the rate of water loss from tissues, exacerbate the effects of desiccation.

Primary and secondary production by sessile organisms can be limited at higher tidal elevations because nutrients and other resources can be acquired only when immersed. Respiration rates of seaweeds and invertebrates are temperature dependent and thus can be greater when an organism is exposed at low tide. For seaweeds, prolonged exposure to dehydration also reduces photosynthesis.

The reduced productivity associated with increased exposure at higher tidal elevations modifies intra- and interspecific interactions. For instance, competition between seaweeds, which may be intense lower on the shore, is reduced at higher tidal elevations and enables coexistence. Competition among intertidal seaweeds is hierarchical with lower shore species dominating those of the higher shore. Thus, fucoid species of the mid intertidal zone are outcompeted for space in the low zone by foliose red seaweeds that pre-empt space with an encrusting perennial holdfast. There is also a competitive hierarchy among mid intertidal zone fucoids with those typically occurring lower on the shore competitively dominant to those higher up. This is most apparent on European rocky shores where the diversity of intertidal fucoids is greatest.

Grazing rates tend to be greater lower on the shore, although there are cases of herbivory by insects setting the upper limits of ephemeral green algae. Grazing by sea urchins at the interface with the sub-littoral zone can limit the lower distributions of macroalgae, but there is little evidence for grazing on perennial seaweeds setting the lower limits of those taxa within the intertidal zone. Grazing of perennial seaweeds is most intense at the sporeling stage soon after settlement. Grazing by gastropods and small crustaceans certainly contributes to losses of biomass of established individuals, but does not affect distributions within the intertidal zone. In contrast, the grazing of established ephemeral species both on emergent rock and tidepools is intense during spring and summer in many regions eventually eliminating those algae from their respective habitats. There are also many examples of consumers using seaweeds as habitat as well as food.

Rocky Intertidal Shores as an Important System in Development of Ecology

The rocky intertidal zone has been a stronghold for ecological research, and the success of intertidal experiments stems in part from the fact that intertidal assemblages are often comprised of the few species that are able to survive the environmental variation associated with the cycling of tides. In addition, many resident intertidal species are small, common, and slow moving or fixed in one place. Thus rocky intertidal shores historically appeared as simple, well-defined habitats in which easily observed and manipulated local interactions control the dynamics of the assemblages. Such initial appearances, however, have been deceiving, and variation in recruitment of offspring from the plankton, a characteristic of many marine species, has stimulated an increased appreciation of the role of oceanographic conditions.

Descriptive Studies: Research Prior to 1960

Descriptions of rocky shores and speculation about the causes of vertical zonation go back more than 195 years. Before the 1960s, ecologists had published descriptions of intertidal areas from more than a dozen large geographical regions that spanned much of the globe and included both sides of the North Pacific and North Atlantic; Greenland;

the West Indies; South and Central America; the coasts of Africa; the Mediterranean; the Black Sea; Indian Ocean Islands; Singapore; Pacific Islands, Australia, and Tazmania. These early accounts of the rocky intertidal remain a potentially valuable source for comparison to contemporary patterns of species distributions due to local species extinctions and introductions.

The Rise of Experimental Studies: 1960–80

Direct experimental manipulation of intertidal organisms accelerated in the 1960s with the groundbreaking work of J. H. Connell and R. T. Paine. Connell manipulated the presence of two species of barnacles in Scotland by selectively removing individuals from small tiles fashioned from the sandstone rock from the shore. He showed that the lower limit of the high intertidal species *Chthamalus stellatus* was set by competition with the mid zone species *Balanus* (now *Semibalanus*) *balanoides* and that the upper limit of *S. balanoides* was set by physical factors. Paine removed the predatory seastar *Pisaster ochraceus* from an area of the intertidal shore in Washington and showed that *Pisaster* was responsible for controlling mussels, which are successful competitors for space and dominate the intertidal shore in the absence of *Pisaster*. These early investigations provided a framework for the rapid growth of experimental studies that characterized the field in recent decades (**Figure 4**).

In general, the observation and experimental manipulations of mobile consumers and their prey has often revealed predation by mobile consumers as an important factor that contributes to the structure of rocky intertidal assemblages. Consumers have been repeatedly shown to be prey species- and prey size-selective, while algal grazing consumers can inadvertently

Figure 4 Grindstone Neck in Maine, USA with Mount Desert Island in the background. This site was used by Menge and Lubchenco in their groundbreaking work in the 1970s. Photo by P. S. Petraitis.

remove newly settled animals and algae as well as their intended prey.

Supply-Side Ecology and External Drivers: 1980–2005

Marine ecologists have known for a long time that success of many intertidal species depend on the supply of propagules (larvae, zygotes, and spores) from the plankton, but it was not until the 1980s that experiments were executed to assess how the supply of propagules influenced the patterns of distribution and abundance of adults in benthic assemblages.

Propagule supply and early post-settlement mortality markedly influence both the strength of interactions among established individuals and overall patterns of distribution and abundance on rocky shores. Abundance of established individuals is often directly proportional to the density of settlement and consequently, and strength of adult interactions depends on variation of settlement. In contrast, if settlement is high enough to consistently saturate the system, then local populations tend to be driven by strong interactions among adults regardless of settlement variation. In some cases, heavy early postsettlement mortality can lead to low densities of adults despite an abundance of settlers, and this has been shown for several seaweeds and many invertebrate species. The causes of variation in propagule supply can be classified into two broad categories – oceanographic transport or regional offshore production. Although invertebrate larvae and some macroalgal spores are motile, their movements are most directly important at small spatial scales near the substrate just prior to settlement. By and large, propagules of benthic species are transported at the mercy of currents and other oceanic transport phenomena. For instance, coastal upwelling results in a net offshore transport of propagules and leads to a reduction in settlement along a shoreline. This commonly occurs with invertebrate species that have long residence times in the plankton. In contrast, seaweeds, which have very short planktonic stages, often dominate intertidal sites within regions characterized by seasonal or permanent upwelling (**Figure 5**).

Regional offshore production influences the supply of larvae to a coastal habitat in two ways. First, phytoplankton production in nearby waters offshore affects the abundance of planktotrophic larvae that feed for several weeks in the plankton potentially leading to greater larval supply in areas with greater phytoplankton production. Second and in opposition, increased production in offshore can generate increased resources and habitat for the associated pelagic community that preys upon larvae and thus leads to a reduced larval supply.

Figure 5   The intertidal zone near Antofagasta in northern Chile, a region with upwelling and abundant seaweeds. Photo by P. S. Petraitis.

Unresolved Problems and Future Directions

Marine ecologists have been remarkably successful in advancing our knowledge of how strong local interactions affect the composition of communities, yet it is not yet clear how the results of small-scale experiments can be scaled up into broad scale generalizations. This is one of the major challenges of rocky intertidal ecology since practical, everyday concerns of management, commercial harvesting, biodiversity, and restoration demand answers on the scale of square kilometers of habitat, not square meters of experimental site. One current approach has been to use teams of researchers undertake identical small-scale experiments over a broad geographical region (e.g., EuroRock in Great Britain and Europe) or over similar oceanographic conditions (e.g., the ongoing studies of rocky shore in upwelling systems on the Pacific Rim by PISCO). Another approach has been the integration of 'real time' physical, chemical, biological data from in situ and remote sensors (e.g., satellites that can reveal near shore temperature and primary productivity) with experimental studies on community dynamics.

Neither approach solves the difficulties of working with large mobile consumers such as mammals, whose importance is under appreciated because of the difficulties inherent with studying mammals. Even the rat (*Rattus norvegicus*) – the most widely recorded introduced intertidal mammal with the broadest documented intertidal diet – likely remains underreported as a rocky intertidal consumer from many coastal locations where it is known to be established. It is likely that rocky intertidal organisms supply terrestrial consumers significant amounts of energy, yet there are few data on intertidal–terrestrial linkages and how intertidal shores serve as important subsidies for terrestrial habitats.

It is also unclear if detailed information from one area can be informative about another area. For example, rocky intertidal shores on both sides of the Atlantic Ocean look surprisingly alike with not only the same species of plants and animals present but also similarities in their abundances and distributions. The similarity is so striking that a good marine ecologist, knowing little more than the direction of the prevailing swells, can list the 20 most common species on any 100 m stretch of shoreline. The average beachcomber could not tell if he or she were in Brittany, Ireland, Nova Scotia, or Maine. The causes of this similarity are not well understood. Rocky shores in Europe and North America may look similar because of strong biological interactions maintain species in balance or because of historical accident, and these opposing views are endpoints on a continuum but represent one of the major intellectual debates in ecology today.

Finally ecosystems are not static, and rocky intertidal systems, which lie at a land–sea boundary, will be doubly affected by climate change as both oceanic conditions such as storm frequency and surge extent, and terrestrial conditions, such as air temperatures, are altered. Such changes could affect local communities by altering the disturbance dynamics and changing the geographic limits of intertidal species.

See also: Saline and Soda Lakes; Salt Marshes.

Further Reading

Connell JH (1961) The influence of interspecific competition and other factors on the distribution of the barnacle *Chthamalus stellatus*. *Ecology* 42: 710–723.

Denny MW (1988) *Biology and Mechanics of the Wave-Swept Environment*. Princeton, NJ: Princeton University Press.

Graham LE and Wilcox LW (2000) *Algae*. Upper Saddle River, NJ: Prentice-Hall.

Horn MH, Martin KLM, and Chotkowski MA (eds.) (1999) *Intertidal Fishes: Life in Two Worlds*. San Diego, CA: Academic Press.

Koehl MAR and Rosenfeld AW (2006) *Wave-Swept Shore: The Rigors of Life on a Rocky Coast*. Berkeley, CA: University of California Press.

Levinton JS (2001) *Marine Biology*. New York: Oxford University Press.

Lewis JR (1964) *The Ecology of Rocky Shores*. London: English Universities Press.

Little C and Kitching JA (1996) *The Biology of Rocky Shores*. New York: Oxford University Press.

Moore PG and Seed R (eds.) (1986) *The Ecology of Rocky Coasts*. New York: Columbia University Press.

Ricketts EF, Calvin J, and Hedgpeth JW (1992) *Between Pacific Tides*, 5th edn., revised by Phillips DW. Stanford, CA: Stanford University Press.

Southward AJ (1958) The zonation of plants and animals on rocky sea shores. *Biological Reviews of the Cambridge Philosophical Society* 33: 137–177.

Stephenson TA and Stephenson A (1972) *Life between Tidemarks on Rocky Shores*. San Fransisco, CA: W. H. Freeman.

Underwood AJ (1979) The ecology of intertidal gastropods. *Advances in Marine Biology* 16: 111–210.

Underwood AJ and Chapman MG (eds.) (1996) *Coastal Marine Ecology of Temperate Australia*. Sydney: University of New South Wales Press.

Underwood AJ and Keough MJ (2001) Supply side ecology: The nature and consequences of variations in recruitment of intertidal organisms. In: Bertness MD, Gaines SD, and Hay ME (eds.) *Marine Community Ecology*, pp. 183–200. Sunderland, MA: Sinauer Associates.

Saline and Soda Lakes

J M Melack, University of California, Santa Barbara, Santa Barbara, CA, USA

Introduction
Geographic Aspects
Environmental and Biological Characteristics

Examples of Ecological Processes
Economic Aspects
Further Reading

Introduction

Saline lakes occur on all continents. Lying in hydrologically closed basins where evaporation exceeds local precipitation, their size and salinity varies markedly and they are particularly susceptible to climatic variations and water diversions. Aquatic biota from microbes to invertebrates to fish and birds frequent these environments and can attain spectacular numbers. While modern scientific techniques are increasingly being applied to a few saline lakes, many are in remote locations and require exploratory sampling as a first step, often with surprising findings. For example, a trans-Saharan expedition discovered isolated villagers eating cakes of an alga called *Spirulina* that has led to an aquaculture industry.

Since 1979 a series of eight international symposia on inland saline lakes have served to strengthen and expand the scope of scientific understanding and foster a worldwide cadre of researchers. While distinctive because of their chemical conditions and biota, all ecological processes occur in saline lakes and they provide an excellent system in which to observationally and experimentally examine these processes. A treatise by Ted Hammer and synthetic reviews by several others offer comprehensive information about these diverse and fascinating environments. This is especially important because inland saline waters are threatened in many regions by diversion of their inflows and economic development.

Geographic Aspects

Saline lakes are widespread globally and occur predominately in dry areas, regions that occupy about 30% of the world's landmass. The volume of water in saline lakes is about 80% as large as that in freshwater lakes. Though about 70% of the total volume of saline water is held in the Caspian Sea, it is worth noting that about 40% of the freshwater in lakes is held in Lake Baikal and the Laurentian Great Lakes (see Freshwater Lakes). Further, many of the world's largest lakes are saline and include Great Salt Lake (USA), Lake Shala (Ethiopia), Lake Van (Turkey), the Dead Sea, Qinghai and Lop Nor (China), Nan Tso (Tibetan Plateau, China), Balkhash (Russia), Urmia (Iran), Issyk-kul (Kyrgystan), the Aral Sea, Mar Chiquita (Argentina), and Lake Eyre (Australia) and Salar of Uyuni (Bolivia) (these two lakes vary greatly in size, as is typical of many shallow playas).

Environmental and Biological Characteristics

Lakes with salinities above $3 \, g \, l^{-1}$ are usually considered saline, though this value is somewhat arbitrary. Salinity is defined as the sum of total ions by weight and usually includes the major cations (sodium, potassium, calcium, and magnesium) and anions (bicarbonate plus carbonate, chloride, and sulfate). Natural waters can attain salinities of several hundred grams per liter and vary considerably in their chemical composition. The ionic composition of saline lakes depends on the ionic ratios in the inflows and extent of evaporative concentration. As the saturation of specific salts is exceeded, they precipitate and can lead to the formation of large evaporite deposits. Typically, calcium and magnesium carbonates are the first minerals to precipitate. If sufficient calcium remains in solution, calcium sulfate often precipitates next. In the most concentrated waters, chloride is the dominant anion and sodium is usually the dominant cation; Great Salt Lake in Utah (USA) is such an example. In rare cases, other combinations of ions can occur in

Figure 1 Lake Mahega, Uganda.

highly concentrated waters such as the sodium–magnesium–chloride waters of the Dead Sea, the sodium–chloride–sulfate brine in Lake Mahega, Uganda (**Figure 1**), or the exceptional calcium chloride brine in Don Juan Pond (Antarctica). Lakes of intermediate salinities include the sodium carbonate or soda lakes of eastern Africa and the triple salt waters (sodium carbonate–chloride–sulfate) of Mono Lake, California, USA.

A considerable diversity of halophilic microorganisms with representatives from the three domains of life, the Archaea, Bacteria, and Eukarya, inhabit saline lakes. Only recently have modern molecular techniques, such as gene sequencing, been applied to natural communities of microbes and much remains to be learned. At especially high salt concentrations, microbes lack grazers and can attain very high abundances that can color saline lakes bright red or orange. Only a very few metabolic processes have not been observed at high salinities and these include halophilic methanogenic bacteria able to use acetate or hydrogen plus carbon dioxide and halophilic nitrifying bacteria.

As salinity increases in inland waters, biodiversity tends to decrease, but in the mid-range of salinities other factors cause considerable variation in species diversity. The strongest relationship between species richness of plants, algae, and animals occurs, generally, below a salinity of about $10 \, g \, l^{-1}$. An investigation by William D. Williams, an Australian professor who pioneered studies of saline lakes, found that species richness of macroinvertebrates in Australian lakes highly correlated with salinity over a salinity from 0.3 to $343 \, g \, l^{-1}$ but nonsignificant over intermediate ranges of salinity. Many taxa had broad tolerances of salinity at the intermediate values. Instead, a variety of other factors, including dissolved oxygen concentrations, ionic composition, pH, and biological interactions, appear to influence species richness and composition.

Examples of Ecological Processes

The very wide range of environmental conditions and geographic distribution of saline lakes results in a large variety of biological communities with differing species diversity and ecological interactions. Moreover, few saline lakes have been examined sufficiently with a combination of field observations and measurements of important processes, experiments, and models. Hence, three examples of ecological processes in saline lakes that are reasonably well studied and that span a wide range of physicochemical and biological conditions are presented in this article.

Eastern African Soda Lakes

Saline lakes rich in bicarbonate and carbonate, usually called soda lakes, are widespread in eastern Africa and are among the world's most productive, natural ecosystems. A conspicuous feature of these lakes is often the presence of enormous numbers of lesser flamingos (*Phoeniconaias minor*) (**Figure 2**) grazing on thick suspensions of the phytoplankter, *Arthrospira fusiformis* (previously called *Spirulina platensis*), but species diversity is low. Heterotrophic bacteria attain very high numbers, but have not been characterized with molecular methods. Phytoplankton and benthic algae include several species of green algae, diatoms, and cyanobacteria. Of the few species of aquatic invertebrates, protozoa are the most diverse with 21 species reported from Lake Nakuru (Kenya). Consumers in Lake Nakuru at salinities of around $20 \, \text{g} \, \text{l}^{-1}$ include one species of fish, *Sarotherodon alcalicus grahami* (introduced from springs near Lake Magadi, a neighboring salt pan), one copepod (*Paradiaptomus africanus*), and two rotifers (*Brachionus dimidatus and B. plicatilis*), and several aquatic insects including corixids, a notonectid, and chironomids. Modest changes in the salinity and in the vertical distribution of salinity can have major impacts on trophic structure and nutritional status of these lakes.

Figure 2 Flamingos (Lake Bogoria, Kenya).

Figure 3 Lake Elmenteita, Kenya.

Biological communities in shallow, tropical saline lakes are susceptible to slight variations in water balances and salinities. For example, intensive sampling during a period of low rainfall and abrupt increase in salinity in Lake Elmenteita (Kenya) (**Figure 3**) and Lake Nakuru (Kenya) revealed a precipitous drop in the abundance of phytoplankton and major shift in the zooplankton. As species of phytoplanktons, such as *Arthrospira fusiformis*, were replaced by much smaller phytoplanktons, the abundance of lesser flamingos decreased markedly.

Scattered across eastern Africa are numerous saline lakes inside volcanic craters. Several of these lakes have been studied in Ethiopia, Kenya, Uganda, and Tanzania. One common feature in the saline, crater lakes of eastern Africa is persistent chemical stratification, that is, they are meromictic, which has significant biological consequences. For example, Lake Sonachi (Kenya), a meromictic crater lake, had much lower algal biomass and rates of photosynthesis than the neighboring soda lakes that mixed more often. Moreover, studies of phosphorus uptake indicated that the lake was deficient in phosphorus, although a large reservoir of phosphorus was trapped below the chemocline.

Mono Lake

One of the most thoroughly studied saline lakes is Mono Lake, which lies on the western edge of the North American Great Basin just east of the Sierra Nevada. With recent salinities in the range from 70 to $90 \, \text{g} \, \text{l}^{-1}$, a pH of about 10 and very high concentrations of bicarbonate and carbonate, it is an alkaline, saline lake. As is often typical of saline lakes, Mono Lake is productive: rapidly growing algae support a simple food web that includes very abundant brine shrimp, *Artemia monica*, and an alkali fly, *Ephydra hians*, which in turn feed thousands of birds. No fish occur in the lake. The lake is a major breeding site for the California gull (*Larus californicus*), and is a critical stop-over for migrating phalaropes (*Phalaropus* spp.) and eared grebes (*Podiceps nigricollis*). The streams that flow into Mono Lake from the Sierra Nevada are a plentiful source of freshwater that were tapped by the City of Los Angeles by a complex diversion scheme initially implemented in 1941. Largely as a consequence of this interbasin transfer of water, the lake's level had fallen about 14 m and its salinity doubled from 1941 to 1982. Laboratory experiments indicated that further increases in salinity were likely to have profound impacts on the ecology as photosynthesis was found to decline about 10% for each 10% increase in salinity, and survival and reproduction of the brine shrimp was found to be increasingly impaired to the point where cyst hatching would cease if salinities were to increase by about 50% from their 1980 values. If diversions by Los Angeles were to have continued unabated, this salinity would have been reached within several decades. The end result in the mid-1990s of almost two decades of litigation and environmental assessment was modifications to the water rights of the City of Los Angeles, which led to higher lake levels. In contrast to the dismal conditions at a number of saline lakes, such as the Aral Sea, and continuing declines in level at other lakes, such as Walker Lake (Nevada), the resolution of the contest at Mono Lake is a good example of how scientific expertise can contribute in a positive way to solutions of environmental problems.

As was observed in eastern African soda lakes, climatic variations as well as diversions have significant influences on Mono Lake. In the early 1980s, California experienced substantially above-average snow and rainfall resulting in a large rise in lake level and chemical stratification that blocked the complete vertical mixing that usually occurred during the winter. Ammonium, which would have been replenished in the upper lake, accumulated in the deep water, but remained very low in the euphotic zone. Since Mono Lake is a nitrogen-limited lake, phytoplankton abundance and productivity declined. The combination of resumed diversions and drought conditions led to sufficient evaporative concentration to weaken the chemical stratification and permit wind and cooling to turn over the lake in the late 1980s, entrain ammonium-rich water, and restore higher algal biomass and productivity. After a series of years with winter mixing and average productivity in the early 1990s, diversions were curtailed in the mid-1990s, as ordered by the revised water rights agreement, and California experienced above average precipitation. Mono Lake became meromictic again with subsequent reductions in productivity. Multiyear records of annual primary productivity by phytoplankton have conspicuous differences as a function of meromictic or monomictic conditions. During meromixis, the development of persistent anoxia below the chemocline alters other chemical conditions with biological consequences. Methane and dissolved sulfide accumulate, and bacterial

communities adapted to metabolize reduced forms of elements become active.

Artemia monica is the only macrozooplankter in Mono Lake. Each year a first generation hatches from overwintering cysts, matures, and produces a second generation via release of live nauplii. A small third generation sometimes occurs, but very few animals survive through the winter. Besides exerting strong grazing pressure on the phytoplankton, *Artemia* regenerate ammonium that supports algal growth. *Artemia* are an important food for large numbers of gulls breeding at the lake in the spring and for as many as one million grebes in the autumn. Some life-history characteristics of *Artemia* are indicative of differences in algal abundance and primary productivity. Although large numbers of eggs are produced in all years, on average, fewer cysts and live nauplii are produced during meromictic years, and maturation of the first generation can be slowed and fecundity and body size reduced as compared to nonmeromictic years.

Changes in the *Artemia* populations translate to influences on the birds feeding at the lake. The fledging rate per pair of California gulls reflects their clutch size and prefledging survival, both of which should be influenced by the adult food supply. In fact, fledging success was low immediately following the onset of meromixis and remained low during the subsequent 3 years of meromixis in the 1990s.

Dead Sea

Lying about 400 m below sea level in a rift valley along the Israel–Jordan border, the surface of the Dead Sea is the lowest of any lake, and it is one of the saltiest with a current salinity of around $340 \, \mathrm{g \, l^{-1}}$. Diversions of the Jordan River, the main inflow, resulted in a 20 m decline in lake level over the last century and an increase in salinity. One consequence of the evaporative concentration of the upper waters was the termination of meromixis that had persisted for several hundred years. With the exception of a few years, the lake now mixes completely each year.

At the time of the pioneering microbiological studies by Benjamin Elazari-Volcani in the 1930s and 1940s, the lake's salinity was about $260 \, \mathrm{g \, l^{-1}}$. Using enrichments and microscopy he was able to describe a variety of halophilic and halotolerant microbes as well as the phytoflagellate, *Dunleilla viridis*, several cyanobacteria, diatoms, green algae, and a ciliate. Subsequent application of modern molecular techniques has considerably expanded the number of microbes, but the higher salinities have eliminated some organisms noted earlier.

During times when the whole lake reaches salinities of around $340 \, \mathrm{g \, l^{-1}}$, bacterial densities are low and algae are absent. However, in response to periods with large amounts of rainfall and runoff, the upper waters can be diluted to as low as $250 \, \mathrm{g \, l^{-1}}$, and blooms of *Dunaliella* and red Archaea develop. The abrupt decline of the bacterial blooms cannot be attributed to protozoan grazing, since these organisms no longer occur, and may be caused by bacteriophages, as viruses have been identified in the lake.

Economic Aspects

The salts precipitated from saline waters are a rich source of chemicals used in a variety of industrial processes and are mined from salt lakes. In coastal areas with high evaporation rates, a series of salterns allow progressive concentration of solutes and the production of useful salts. In a few saline lakes with strong chemical stratification, transparent surface waters and a turbid layer within the chemocline, high temperatures have been recorded in the turbid layer. These features have guided the construction of artificial, so-called solar ponds, with similar characteristics, for power production and heating purposes.

A common feature of tropical African soda lakes is high concentrations of nearly unialgal populations of the cyanobacteria, *Arthrospira fusiformis*, which support huge numbers of lesser flamingos and are used as a protein-rich food by people in Chad. These observations, laboratory studies, and development of mass culture methods have led to *Arthrospira*, often marketed as *Spirulina*, becoming a widely used food supplement. Other species of algae found in saline waters are commercially exploited because of their high glycerol or β-carotene content (e.g., *Dunaliella*). Additional applications include the production of salt-resistant enzymes and the use of organic osmolytes to protect enzymes.

Artemia are an important food for aquaculture of some fish and other organisms. Typically, cysts are harvested from lakeshores and maintained dry until needed, when they are readily hatched by submerging in saline water. Occasionally, such as at Mono Lake, adult *Artemia* are collected, frozen, and shipped to aquaculture facilities.

The impressive numbers of birds that frequent saline waters and the striking scenery has led to tourism as an increasingly important aspect of their economic value. World famous examples include Lake Nakuru, with it shoreline fringed by pink flamingos, Mono Lake with its peculiar tufa towers and thousands of waterfowl, and the Dead Sea with its historical significance and highly buoyant water. Some less-saline lakes, such as Pyramid Lakes, harbor fish (e.g., Lahonton cutthroat trout, *Oncorhynchus clarki henshawi*) that support recreational fishery.

See also: Freshwater Lakes.

Further Reading

Eugster HP and Hardie LA (1978) Saline lakes. In: Lerman A (ed.) *Lakes: Chemistry, Geology, Physics*, pp. 237–293. New York: Springer.
Hammer UT (1986) *Saline Lake Ecosystems of the World*. Dordrecht: Dr. W. Junk Publishers.
Melack JM (1983) Large, deep salt lakes: A comparative limnological analysis. *Hydrobiologia* 105: 223–230.
Melack JM (2002) Ecological dynamics in saline lakes. *Verhandlungen Internationale Vereinigung Limnologie* 28: 29–40.
Melack JM, Jellison R, and Herbst D (eds.) *Developments in Hydrobiology 162: Saline Lakes*. Dordrecht: Kluwer.
Oren A (ed.) (1999) *Microbiology and Biogeochemistry of Hypersaline Environments*. New York: CRC Press.
Vareschi E and Jacobs J (1985) The ecology of Lake Nakuru. VI. Synopsis of production and energy flow. *Oecologia* 65: 412–424.
Williams WD (1996) The largest, highest and lowest lakes in the world: Saline lakes. *Verhandlungen Internationale Vereinigung Limnologie* 26: 61–79.

Salt Marshes

J B Zedler, C L Bonin, D J Larkin, and A Varty, University of Wisconsin, Madison, WI, USA

Physiography	Ecosystem Services
Extent	Challenges for Salt Marsh Conservation
Habitat Diversity	Research Value
Salt Marsh Plants	Restoration
Salt Marsh Animals	Further Reading
Ecology	

Physiography

Salt marshes are saline (typically at or above seawater, $>34\,\mathrm{g\,l^{-1}}$) ecosystems with characteristic geomorphology (sedimentary environments, fine soil texture, and relatively flat topography), herbaceous vegetation, and diverse invertebrates and birds. They occur along shores in estuaries, lagoons, forelands (open areas), and barrier islands in marine environments, and in shallow inland sinks where salts accumulate. They are not found where waves, currents, or streamflow create strong erosive forces. Salt (which stresses most species) severely limits the pool of plant species that can colonize saline sediments, and wetness typically confines the vegetation to herbaceous species, although some species are long-lived 'subshrubs'. Given a near-surface water table, most shrubs and trees cannot establish their extensive root systems.

Plants of tidal marshes are usually able to colonize sediment above mean high water during neap tides (MHWN = average higher high-tide level during lower-amplitude neap tides, which alternate with the broader-amplitude spring tides). Sediment stabilization by halophytes initiates salt marsh formation. Plants not only slow water flow and allow sediments to settle out, but also their roots help hold sediments in place. Gradual accretion around plant shoots can further elevate the shoreline, allowing development of a marsh plain and transition to upland. This process can reverse, with tides eroding accumulated sediments. When sedimentation is outweighed by erosion, salt marshes retreat.

The overriding physiochemical influence is salt, which comes from marine waters, from exposed or uplifted marine sediments, or from evaporation of low-salinity water in arid-region sinks. Salt marshes along coasts typically have tidal influence (**Figure 1**), although many nontidal lagoons have saline shores that support salt marsh vegetation. Salt marshes in inland settings occur in shallow sinks (e.g., around the Great Salt Lake, Utah, USA). The salts that contribute to salinity are primarily those of four cations (sodium, potassium, magnesium, calcium) and three anions (carbonates, sulfates, and chlorides); the relative proportions differ widely among soils of inland salt marshes, but sodium chloride is the predominant salt of seawater.

Tidal regimes differ around the globe, but most tidal marshes experience two daily high tides of slightly different magnitude, while some have the same high and low tides from day to day. Levels alternate weekly as neap and spring amplitudes, with the amplitudes readily predicted given gravitational forces between the Earth, the Moon, and the Sun (astronomic tides). Forces vary in relation to global position and coastal morphology; in southern California, mean astronomic tidal range is 3 m, while in the Bay of Fundy it is 16 m. The influence of seasonal low- and high-pressure systems on water-level oscillations (atmospheric tides) also vary greatly. For example, in Western Australia's Swan River Estuary, atmospheric tides outweigh astronomic tides. In the Gulf of Mexico, astronomic tides are minimal because of limited seawater connection with the Atlantic Ocean. Water levels within the Gulf vary only a few centimeters except during storms and seiches.

In tidal systems, marsh vegetation generally ranges from MHWN to the highest astronomic tide. Depending on tidal amplitude and the slope of the shore, salt marshes can be very narrow or kilometers wide. Strong wave action limits the lower salt marsh boundary, but a sheltered area can extend the lower boundary below MHWN.

Animal diversity is high, especially among the benthic and epibenthic invertebrates and the arthropods in the soil or plant canopies. Species that complete their life cycles within salt marshes either tolerate changing salinity and inundation regimes or avoid them by moving elsewhere or reducing contact. Globally, salt marshes are known to support large populations of migratory birds in addition to resident birds, insects, spiders, snails, crabs, and fin and shellfish. Indeed, foraging is the most visible activity in salt marshes.

Figure 1 A tidal marsh in San Quintin Bay seen from the air. Image by the Pacific Estuarine Research Lab.

Extent

Salt marsh area is not well inventoried. The global extent of pan, brackish, and saline wetlands is approximately 435 000 km^2, or 0.3% of the total surface area and 5% of total wetland area. In USA, the 48 conterminous states have about 1.7 Mha of salt marshes, out of a total of 42 Mha of wetlands.

While broadly distributed, salt marshes are most common in temperate and higher latitudes where the temperature of the warmest month is >0 °C. Closer to the equator, where the mean temperatures of the coldest months are >20 °C, salt marshes are generally replaced by mangroves. Salt marshes sometimes occur

Habitat Diversity

Habitats within the salt marsh vary with elevation, microtopography, and proximity to land or deeper water. In southern California, the high marsh, marsh plain, and cordgrass (*Spartina foliosa*) habitat tend to follow elevation contours, although cordgrass is often restricted to low elevations adjacent to bay and channel margins. Other habitats are related to minor variations in topography, which impound fresh or tidal water. For example, back-levee depressions, tidal pools, and salt pans occur where drainage is somewhat impaired. Salt marshes along the Atlantic Coast of USA are very extensive, with *S. alterniflora* creating a monotype except for a narrow transition at the inland boundary where succulent halophytes or salt pans are found.

Tidal creeks provide diverse habitats for plants and animals. Banks are often full of crab burrows, and creek bottoms harbor burrowing invertebrates and fishes. They also serve as conduits for fish, fish larve, phyto- and zooplankton, plant propagules, sediments, and dissolved materials, which move between the salt marsh and subtidal channels.

Adjacent habitats can include small, unvegetated salt pans that dry and develop a salt crust, especially during neap tides. Salt pans occur where salt concentrations exceed tolerance of halophytes. During heavy rains or high tides, water fills the pan, creating temporary habitat for aquatic algae and animals and permanent habitat for the species that survive the dry spells *in situ* as resting stages. More extensive salt pans are sometimes called salt flats. Other nearby habitats usually include mudflats (where inundation levels exceed tolerance of halophytes), brackish marsh (where salinities are low enough for brackish plants to outcompete halophytes), sandy or cobble beaches (where wave force excludes herbaceous vegetation), sand dunes (where soils are too coarse and dry for salt marsh plants), and river channels (where freshwater enters the estuary and is not sufficiently saline).

Salt Marsh Plants

Salt-tolerant plants (halophytes) include herbaceous forbs, graminoids, and dwarf or subshrubs. Many of the forbs are succulent (e.g., *Sarcocornia* and *Salicornia* spp.). Graminoids often dominate Arctic salt marshes, while subshrubs dominate salt marshes in Mediterranean and subtropical climates. Many salt marshes support monotypic stands of cordgrass (*Spartina* spp.) (**Table 1**).

Floristic diversity of salt marshes is low because few species are adapted to saline soil. Members of the family Chenopodiaceae comprise a large proportion of the flora (e.g., species of *Arthrocnemum*, *Atriplex*, *Chenopodium*, *Salicornia*, *Sarcocornia*, and *Suaeda*). In contrast to the flowering plants, salt marsh algae are diverse in both species and functional groups (green macroalgae, cyanobacteria, diatoms, and flagellates).

NaCl is a dual stressor, as it challenges osmotic regulation and sodium is toxic to enzyme systems. Salt marsh halophytes cope with salt by excluding entry into roots, sequestering salts intracellularly (leading to succulence), and excreting salt via glands, usually on leaf

---

Table 1 Representative species of global salt marshes based on a summary by Paul Adam

Arctic Puccinellia phryganodes dominates the lower elevations

Boreal Triglochin maritima and *Salicornia europea* are widespread. Brackish conditions have extensive cover of *Carex* spp.

Temperate
    Europe: *Puccinellia maritima* dominated lower elevations historically (but *Spartina anglica* often replaces it). *Juncus maritimus* dominates the upper marsh; *Atriplex portulacoides* is widespread
    USA:
        Atlantic Coast: *Spartina alterniflora* is extensive across seaward marsh plain; *S. patens* occurs more inland
        Gulf of Mexico: *Spartina alterniflora* and *Juncus roemerianus* dominate large areas
        Pacific Northwest: *Distichlis spicata* in more saline areas, *Carex lyngbei* in less saline areas
        California: *Spartina foliosa* along bays, *Sarcocornia pacifica* inland
    Japan: *Zoysia sinica* dominates the mid-marsh
    Australasia: *Sarcocornia quinqueflora* dominates the lower marsh, *Juncus kraussii* the upper marsh
    South Africa: *Sarcocornia* spp. are abundant in the lower marsh, *Juncus kraussii* in the upper marsh. *Spartina maritima* is sometimes present

Dry coasts vegetation tends toward subshrubs, such as *Sarcocornia*, *Suaeda*, *Limoniastrum*, and *Frankenia* species
Tropical Sporobolus virginicus and *Paspalum vaginatum* form extensive grasslands. *Batis maritima*, *Sesuvium portulacastrum*, and *Cressa cretica* are also found

surfaces. One succulent, *Batis maritima,* continually drops its older salt-laden leaves, which are then washed away by the tide. I. Mendelssohn has attributed moisture uptake from seawater to the ability of some species to synthesize prolines.

Prolonged inundation reduces the supply of oxygen to soils, causing anoxia and stressing vascular plants. In addition, abundant sulfate in seawater is reduced to sulfide in salt marsh soil, with high sulfide concentrations, which are toxic to roots.

Salt marsh vascular plants withstand brief inundation but do not tolerate prolonged submergence, as occurs when a lagoon mouth closes to tidal flushing and water levels rise after rainfall. Salt marshes in lagoons thus experience irregular episodes of dieback and regeneration in relation to ocean inlet condition.

Regular inundation benefits halophytes by importing nutrients and washing away salts. Salts that accumulate on the soil surface during daytime low tides and salts excreted by halophytes are removed by tidal efflux. Thus, soil salinities are relatively stable where tidal inundation and drainage occur frequently. Inland salt marshes, however, experience infrequent reductions in salinity during rainfall, and soils can become extremely hypersaline (e.g., >10% salt). In between irregular inundation events, halophytes and resident animals endure hypersaline drought.

Salt Marsh Animals

The salt marsh fauna includes a broad taxonomic spectrum of invertebrates, fishes, birds, and mammals, but few amphibians and reptiles. Resident fauna are adapted to the land–sea interface, while transient users benefit from the foraging, nursery, and reproductive support functions.

Salt marsh animals cope with inundation regimes that differ seasonally, monthly, daily, and hourly. Vertebrates accomplish this largely through mobility. For example, fishes exploit marsh surface foraging opportunities during high tides and then retreat to subtidal waters. Birds time their use to take advantage of either low or high tide. Residents, such as the light-footed clapper rail (*Rallus longirostris levipes*), nest during the minimum tidal amplitude. Migrants, such as curlews, move upslope at a high tide and feed during low tide during their seasonal visits. Many invertebrates move away from adverse conditions. Some beetles climb tall plants to escape rising tides. A springtail, *Anurida maritime,* has a circatidal rhythm of 12.4 h that enables it to emerge for feeding shortly after tides ebb and retreat underground prior to the next inundation. For less-mobile fauna, physiological adaptations are essential. Gastropods avoid desiccation

during low tides by sealing their shells. Some arthropods avert drowning by trapping air bubbles in their epidermal hairs during high tides.

Another challenge is fluctuating salinities, which salt marsh residents handle with exceptional osmoregulatory ability. The southern California intertidal crab species *Hemigrapsus oregonensis* and *Pachygrapsus crassipes* are able to hypo- and hyperosmoregulate when exposed to salt concentrations ranging from 50% to 150% of seawater (brackish to hypersaline). Tidal marsh fishes also have wide salinity tolerances. Cyprinodontiform tidal marsh fishes can tolerate salinities as high as 80–90 ppt. One species, *Fundulus majalis,* hatched at salinities up to 72–73 ppt. Lower salinity limits for mussels can be as low as $3 \, g \, l^{-1}$ and they can tolerate high salinities as well, with mussels able to tolerate losing up to 38% of their water content. Even birds have adaptations for dealing with salt water and saline foods; for example, the Savannah sparrow (*Passerculus sandwichensis beldingi*) has specialized glands that excrete salt through the nares.

Because salt marshes have continuously changing hydrology, small differences in elevation and topography (e.g., shallow, low order tidal creeks) influence foraging activities of fishes and birds by regulating inundation and exposure times, enhancing marsh access for fishes, and increasing edge habitat. Ephemeral pools of just centimeters in depth provide valuable bird habitat, enhance macroinvertebrate abundance and diversity, and support reproductive, nursery, and feeding support functions for fishes.

Ecology

Salt marshes are well studied relative to their limited global area. Knowledge of salt marsh ecology is strongest for vegetation, soil processes, and food webs. Conservation is an emerging issue, given threats of sea-level rise in concert with global warming.

Vegetation and Soils

In Europe, salt marsh ecology developed around floristics and phytosociology. In USA, research on the Atlantic and Gulf Coasts characterized salt marsh ecosystem functioning, especially productivity, microbial activities, outwelling of organic matter, food webs, and support of commercial fisheries, while on the Pacific Coast, studies concern the impacts of invasive species of *Spartina* and effects of extreme events on vegetation dynamics. In Canada, effects of geese damaging vegetation are a research focus. Studies of USA's inland salt marshes have contributed knowledge of waterfowl support functions and halophyte salt tolerance. In South Africa's small estuaries, *Spartina* productivity and shifts of vegetation in response to altered freshwater inflows have been explored. In Asia, widespread plantings of *S. alterniflora* have been undertaken in order to extend coastal land area, provide forage, and produce grass for human use. In general, salt marshes of Asia, Central America, and South America are poorly known.

Salt marshes develop primarily on fine sediments, but salt marsh plants can grow on sand and sometimes gravel. Older salt marshes have peaty soils, especially in cooler latitudes where decomposition is slow.

Both roots and burrowing invertebrates affect soil structure by creating macropores in soil. Invertebrates also cause bioturbation, a process whereby sediments are resuspended and potentially eroded away. This activity can be countered by algae and other microorganisms, which form biofilms on the soil surface. Biofilms cement soil particles and reduce erosion; they also add organic matter, and those that contain cyanobacteria fix nitrogen.

Salt marsh soils are often anoxic just below the surface due to high organic matter content and abundant moisture for microorganisms. This is especially so in lower intertidal areas and in impounded marshes. Tidal marsh soils are typically high in sulfur, which forms sulfides that blacken the soil, emit a distinctive rotten-egg smell, and stress many plants. Across intertidal elevation ranges, soil microorganisms, sulfides, and inundation regimes reduce species richness where inundation is most prolonged, often to a single, tolerant species.

Food Webs

Studies of salt marshes have made important advances in food web theory. Early papers focused on primary productivity measurements and attempts to explain differences in rates within and among salt marshes. The energy-subsidy model described *S. alterniflora*'s high productivity at low elevations as a function of increased rates of nutrient delivery and waste removal, due to frequent tidal inundation. It also explained how salt marshes with decreasing tidal energy across Long Island, New York, had a corresponding decrease in *S. alterniflora* productivity. R. E. Turner added the role of climate by relating higher productivity of *S. alterniflora* to warmer latitudes.

In the 1960s, E. Odum's interest in energy flow led several investigators at the University of Georgia to quantify productivity, consumption, and decomposition of various components of Sapelo Island salt marshes. J. Teal's energy-flow diagram depicted Georgia's *S. alterniflora* marsh as exporting organic matter. Although estimated by subtraction rather than measurement, detrital export

became a textbook example of how ecosystems channel and dissipate energy.

Later, advances were made in exploring the quantity and fate of detritus derived from salt marsh primary producers. J. Teal's suggestion that substantial organic matter is transported to estuarine waters supported E. Odum's 'outwelling hypothesis', that estuarine-derived foods drive coastal food webs and benefit commercial fisheries. A number of ecosystem-scale tests of outwelling ensued, and although outwelling did not prove to be universal, the research demonstrated connectivity between riverine, salt marsh, and open-water ecosystems. Also, the copious detrital organic matter provided by salt marshes was shown to be high in nutritional value once detrital particles were enriched by microorganisms, but microalgae were also shown to be an important food source. Even though their standing crop is low, high turnover rates lead to high primary productivity. In salt marshes with ample light penetration through the vascular plant canopy, microalgae can be as productive as macrophytes, and some species (notably cyanobacteria) are much richer in proteins and lipids. Algae also hold much of the labile nitrogen in salt marshes, widely thought to be the limiting factor for growth of invertebrate grazers.

Food webs are driven by both 'bottom-up' or 'top-down' processes. Evidence for bottom-up control of trophic interactions comes from experimental addition of nitrogen. Nitrogen has been shown to limit algae, vascular plants, grazers, and predatory invertebrates in nearly every salt marsh field experiment. Recently, however, P. V. Sundareshwar and colleagues showed that phosphorus can limit microbial communities in coastal salt marshes.

Despite widespread evidence for bottom-up effects, there is expanded recognition of the top-down role of consumers in regulating salt marsh food webs. Populations of lesser snow geese have increased due to agricultural grains that are left in the fields after harvest, and large flocks now cause large-scale destruction of vegetation in Arctic salt marshes due to rampant herbivory. In Atlantic salt marshes of southern USA, snail herbivory accompanies drought-induced die-back of *S. alterniflora*.

Ecosystem Services

Several ecosystem services provided by salt marshes are appreciated by society, and some protective measures are in place. The regular rise and fall of water in salt marshes, either daily with tides or seasonally with rainfall, enhances at least six valued functions:

Denitrification improves water quality. The sediments of tidal marshes are well suited to denitrification, which occurs most rapidly at oxic–anoxic interfaces. The first step, nitrification, occurs near soil–water or root–soil interfaces or along pores where oxygen enters the soil at low tide. The second step requires anoxic conditions and proceeds rapidly where moisture is sufficient for bacteria to respire and remove oxygen. In this step, nitrate is reduced to nitrogen gas in a series of microbially mediated steps. The rise and fall of tide waters ensures that oxic and anoxic conditions coexist.

Carbon sequestration slows greenhouse warming. The high net primary productivity of salt marshes creates high potential for carbon storage and the anoxic soils slow decomposition, so carbon can accumulate as peat. Large standing crops of roots, rhizomes, and litter are fractionated by a diversity of invertebrates and microorganisms and incorporated into soil. Rates are potentially highest at cooler latitudes, where decomposition is slowed by low temperatures. Sea-level rise is also a key factor; as coastal water levels become deeper, decomposition slows. Sedimentation also buries organic matter, making it less likely to decompose. With sea level rising a millimeter or more per year, on average, salt marsh vegetation can build new rooting zones above dead roots and rhizomes of past decades. Along the USA Gulf of Mexico, the ability of salt marshes to keep up with rising sea level is attributed to root and rhizome accumulation, not just sedimentation. If decomposition proceeds anaerobically to states that produce methane, however, not only is carbon storage reversed, but carbon is also released in a form that contributes more to global warming than carbon dioxide.

Fin- and shellfish have commercial value. Tidal marshes are valued for their nursery function, meaning that the young of many fishes, crabs, and shrimp make use of estuarine waters as 'rearing grounds'. In the USA, it is estimated that some 60% of commercial species spend at least part of their life cycle in estuaries. Several attributes of salt marshes contribute to the food-web-support function, including high productivity of both algae and vascular plants, detritus production and export to shallow water-feeding areas, refuge from deepwater predators, plant canopy cover as a refuge from predatory birds, warmer temperatures that can accelerate growth, and potential to escape disease-causing organisms and parasites that might have narrower salinity tolerance.

Forage is used to feed livestock. In Europe and Asia, graziers move cattle, horses, sheep, or goats onto the marsh plain during low tides. It is common to see ponies tethered to stakes in *Puccinellia*-dominated salt marshes of UK. The temporary availability (between tides) allows recovery between use and, potentially, high-quality forage and salt for livestock.

Recreational opportunities and esthetics are appreciated by people who live near or visit coastal areas. By virtue of their low-growing vegetation and locations between open water and urban areas, salt marshes attract both wildlife and people. The combination provides high value for birdwatchers, hikers, joggers, and artists. Where there is flat topography above and near the salt marsh, the needs of elderly and disabled visitors can be accommodated along with hikers, school children, and those seeking a refuge from city life. Of particular interest is the ever-changing view, as tides rise and fall along marine coasts, and as water levels change with season in inland systems. Visitor centers have been constructed near many urban salt marshes. Ecotourism then adds economic value to the local municipality as well as the larger region.

Shorelines are anchored by salt marsh vegetation. Recent damages from hurricanes and tsunamis have called attention to the protection that wetland vegetation provides to coastal lands, and especially high-cost real estate. Water flow is slowed by stems and leaves of salt marsh plants, and their roots and rhizomes bind inflowing sediments. Mucilage produced by biofilms (algae, fungi, and bacteria) can then cement particles until new plant growth anchors the substrate. The stems of vascular plants are often coated with biofilms, particularly those of tuft-forming cyanobacteria, such that the total surface area available for sediment-trapping and anchoring is greatly enhanced. Floating mats of green macroalgae (*Ulva, Enteromorpha*) also collect sediments and, when they move to the wrack line and join other debris, add to accretion at the upper marsh plain boundary.

Challenges for Salt Marsh Conservation
Habitat Loss

Estuaries, where rivers meet the sea, are not only suitable for salt marsh development but also ideal places for human habitation. The ocean–river connection is a navigational link, flat land is easy to build upon, the river provides drinking water, the salt marsh and coastal fisheries provide food, outgoing tides facilitate wastewater disposal, and seawater provides an essential preservative and universal seasoning, NaCl. Thus, many cities, such as Venice, Boston, Amsterdam, London, Buenos Aires, Washington, DC, and Los Angeles, were built on or rapidly grew to displace salt marsh ecosystems. Major ports within smaller natural bays, such as San Diego, have displaced nearly all the natural salt marsh, while others, such as San Francisco, sustain large salt marshes despite extensive conversion.

The process of converting salt marsh into nontidal land was historically called reclamation. The practice of building embankments to exclude tidal flows eliminated thousands of hectares of European salt marshes. In the Netherlands, embankments reclaimed substantial land as polders for agriculture. In USA, reclamation reduced salt marsh area by 25% between 1932 and 1954. While the trend is to halt or reverse this practice, estuaries are being dammed in Korea to create tillable fields from mudflats. In Vietnam, Mexico, and other coastal nations, salt marshes are yielding to fish and shrimp impoundments. In such cases, people who use mudflats for fishing and crabbing are displaced by farmers.

Although salt marshes are highly valued, they are increasingly threatened by human population growth. It is estimated that 75% of the global population will live within 60 km of the coast. Thus, coastal ecosystems are particularly at risk.

Eutrophication

Salt marshes are enriched when phosphorus and/or nitrogen flow into waters that ultimately flood the salt marsh. Agricultural fertilizers applied to fields throughout coastal watersheds move downslope into waters that flow toward salt marshes. Because many salt marshes are nitrogen-limited, the effect is to increase the productivity of both algae and vascular plants. Increased nitrogen loading stimulates algal growth, especially of green macroalgae, which form large mats that can smother vascular plants and benthic invertebrates. Indirect degradation occurs when microbial decomposition increases oxygen demand, causing soil hypoxia or anoxia and sulfide toxicity.

I. Valiela's long-term eutrophication experiment in a New England salt marsh indicates that nitrogen addition shifts *S. alterniflora* to *S. patens* and increases competition for light. Such altered competitive relationships are likely widespread, especially where considerable nitrogen is deposited from the air (e.g., from dairy operations in the Netherlands).

Sediment Supply

Both reduced and enhanced sediment supplies can threaten the persistence of salt marsh ecosystems. Sediment supplies are reduced when water is removed from rivers for irrigation, human consumption, and industrial use, or when overbank flooding is prevented by engineering works. Reduced sediment supply from the Mississippi River is one factor contributing to salt marsh loss in Louisiana.

Excessive sediments flow into salt marshes where the catchment has lost vegetative cover as a result of logging, farming, or development. Inflows also occur where mining operations discharge materials directly to streams. Wastes from California's gold rush are still making their way to San Francisco Bay. At a much smaller estuary, the marsh plain of Tijuana Estuary in southern California has elevated 25–35 cm since 1963 due to erosion from rapidly urbanizing canyons in nearby Tijuana, Mexico. The impacts have been losses

of microtopographic variation and local species richness.

Global Change

Increases in global mean temperature will have substantial impacts on the world's salt marshes. Sea levels rise when high-elevation glaciers and polar ice caps melt and when seawater warms and expands. The impacts of more rapidly rising sea level depend on rates of sedimentation and uplift. If sediment accretion is equal to sea-level rise, the salt marsh remains in place, but when sea-level rise exceeds sediment accretion, the salt marsh moves inland – unless bluffs or development limits salt marsh migration. As sea level rises relative to the land, salt marsh communities will experience increased inundation, such that plant and animal species should shift upslope. However, not all species will be able to disperse or migrate as rapidly as tidal conditions change. In a few cases, for example, Scandinavia, the coast is still rebounding from the pressure of former glaciers, and land is rising faster than sea level. Salt marsh is then lost at the upper end and slowly gained near the water.

Globally, mean sea level has risen 10–25 cm during the last century. Current models predict an additional 5.6–30 cm rise in sea level by 2040. In areas of rapid shifts in sedimentation or high erosion due to wind and waves, salt marshes are destabilized and threatened with compositional changes and/or loss of marsh area. Salt marshes are also threatened by subsidence; if the land settles faster than sediment or roots and rhizomes can accumulate, vegetated areas convert to open water. USA's largest area of salt marsh loss is along the Louisiana coastal plain, where subsidence, decreased sedimentation, canal dredging, levee construction, and other human disturbances eliminate more than 4300 ha yr^{-1}.

Coastal watersheds that experience increased storminess as a result of climate change will discharge water, sediments, nutrients, and contaminants more erratically than at present, with resulting impacts on salt marshes downstream.

Soil salinity might also rise with higher temperatures, increased evaporation, and increased evapotranspiration. With more rainfall and freshwater flooding, however, soil salinity might decrease. The net effect of warming on salt marsh soil salinity is difficult to predict. Increased storminess could translate into more or stronger dune washover events during high tides, and stronger ocean swells would transport seawater further inland. The toxic effect of salt on upland vegetation, coupled with persistent salt in the soil, would favor halophytes over glycophytes in an increasingly broader wetland–upland transition

Figure 2 Saltmarsh vegetation from the upland–wetland interface (foreground) to San Quintin Bay, Baja California Peninsula, Mexico. Photo by J. Zedler.

areas (**Figure 2**). This prediction is most likely for areas of low annual rainfall, such as Mediterranean-type climates.

Climate change is likely to affect species differently, potentially altering competitive relationships. Photosynthesis, transpiration, nutrient cycling, phenology, and decomposition are influenced by temperature. Salt marshes with a mixture of C3 and C4 plants might shift toward C4 plants as mean temperature climbs; however, elevated CO_2 might favor C3 species. In subtropical regions, a warming trend and sea-level rise would likely allow mangroves to move northward and displace salt marshes.

Impacts of climate change to plants and animals are difficult to estimate. European ecologists, however, have detailed information on bird use of salt marshes and can predict shifts in invertebrate foods and shorebirds given various scenarios of sea-level rise.

Invasive Species

Plant and animal species are inadvertently moved around the globe when ships take on ballast water in one port and discharge it in another; seeds of alien plants and either live animals or dormant stages are then available to colonize salt marshes. When the USA resumed trade with China, new invaders gained access to San Francisco Bay. Fred Nichols traced the arrival of a small clam, *Potamocorbula amurensis,* to 1876. Now it coats some benthos with thousands of clams/m^2.

Other alien species have been intentionally introduced. In the 1950s, the US Army Corps of Engineers experimentally introduced *S. alterniflora* onto several dredge spoil islands to stabilize the material and provide wildlife habitat. A region-wide invasion of the Pacific Northwestern USA followed several decades of 'benign' behavior. Today, the species is dominant along the lower

edge of salt marsh shorelines, where it displaces oysters and eliminates shorebird-feeding habitat.

Once a species has taken up residence, it might hybridize with native species and become more aggressive, either as the hybrid or subsequent genetic variants. Such is the case for *S. alterniflora,* which has been widely planted in Europe, China, Great Britain Australia, and New Zealand. In Great Britain, it hybridized with the native *S. maritima* to form *S. townsendii,* which then underwent chromosomal doubling to form *S. anglica. S. anglica* can grow at lower elevations than native species and vigorously colonizes mudflats. Dense clones of *S. anglica* reduce habitat for wading birds and displace native salt marsh plants.

Non-native strains of *Phragmites australis* were introduced to the USA 200 years ago, and they have since spread throughout much of North America. Today, the alien strain dominates the less-saline portions of salt marshes in the northeastern USA, where it displaces native plant species, alters soil conditions, and decreases waterfowl use. Disturbances such as ditching or dredging open salt marsh canopies and allow invasion of *P. australis,* while eutrophication, altered hydrologic regimes, and increased sedimentation favor its spread.

Invasive plant species have been linked to reduced diversity, shifts in trophic structure, habitat alteration, and changes in nutrient cycling. Invasive alien animals are equally problematic. In San Francisco Bay wetlands, alien mudsnails outcompete native ones and the Australasian isopod, *Sphaeroma quoyanum,* burrows into and destabilizes creek banks of tidal marshes, causing erosion. Marsh edge losses exceeding $100 \, \text{cm} \, \text{yr}^{-1}$ have been reported in heavily infested areas. Another invader, the green crab, *Carcinus maenus,* has altered the food web of Bodega Bay, California, by reducing densities of a native crab, two native mussels, and other invertebrates. As the green crab moves north, it will likely reduce food availability for shorebirds.

In the southeastern USA, fur farmers introduced nutria (*Myocastor coypus*) from South America in the 1930s. These rodents feed on roots of salt marsh plants. When fur clothing went out of style, nutria populations expanded and began converting large areas of marsh to mudflat and open water.

Chemical Contamination

Chemical contaminants accumulate in salt marshes that receive surface-water runoff and/or direct discharges of waste materials. Among the most toxic are halogenated hydrocarbons, which include many insecticides, herbicides, and industrial chemicals. When accumulated in the tissues of salt marsh animals a wide range of disorders can result, for example, immunosuppression, reproductive abnormalities, and cancer.

Petroleum hydrocarbons pollute harbors and remnant salt marshes following oil spills, urban runoff, and influxes of industrial effluent and municipal waste. Once they move into anoxic sediments, they can persist for decades, reducing primary production, altering benthic food webs, and accumulating in bird tissues. Polycyclic aromatic hydrocarbons have additional carcinogenic and mutagenic potential for aquatic organisms.

Heavy metals are also toxic to aquatic organisms and can impair feeding, respiration, physiological and neurological function, and reproduction, as well as promote tissue degeneration and increase rates of genetic mutation. Mercury is especially problematic because it is methylated in the anoxic soils of salt marshes and is then able to bioaccumulate in food chains.

Salt marsh plants in urban areas take up, accumulate, and release heavy metals. Judith Weis and others have found lowered benthic diversity and impaired fish behavior in contaminated sites. Fish are slower to catch prey and less able to avoid predators where heavy metals contaminate their habitat.

Research Value

Tidal marshes include an impressive array of environmental conditions within about a meter of elevational range. The compressed environmental gradient invites studies of species × abiotic factors, and over time, their contributions proceeded from community ecology to ecosystem science and, finally, integration of the two.

Community Ecology

The limited number of vascular plant species has made salt marshes very suitable for both descriptive and manipulative studies. Early researchers attributed plant species distributions to their physiological tolerance for the abiotic environment, without regard to species interactions. J. A. Silander and J. Antonovics used perturbation-response methods to determine that biotic forces also affected species distributions. Others effectively used reciprocal transplanting to examine the relative importance of abiotic conditions and interspecific competition to species distributions. For example, S. Pennings and R. Callaway revealed interspecific interactions among southern California halophytes, and S. Hacker and M. Bertness reported interspecific interactions among New England halophytes. Manipulative transplantation has shown that species distributions respond to abiotic conditions, facilitation, and competition.

The wide latitudinal range of salt marshes allowed study of community structure and function in relation to sea-level variations, for example, James Morris documented and modeled interannual variations in salinity and its effect on *S. alterniflora* growth. Such studies led to predictions of changes in response to global climate change.

Ecosystem Functioning

The monotypic nature of USA Atlantic Coast salt marshes aided early studies of vascular plant productivity and considerable literature developed around the rates of productivity and alternative methods of calculating gross and net productivity – work that transferred to grasslands and other nonwoody vegetation. Nitrogen dynamics were a later focus. The first marine system to have a nitrogen budget was Great Sippewisset Marsh in Massachusetts. The budget quantified nitrogen inputs from groundwater, precipitation, nitrogen fixation, and tidal flow, and nitrogen outputs from tidal exchange, denitrification, and buried sediments.

Integrating Structure and Function

A long controversy over the causes of height variation in *Spartina* spp. has involved USA researchers on both the Atlantic and Pacific Coasts and has linked plant and ecosystem ecology. The most convincing evidence for a genetic ('nature') component is that of D. Seliskar and J. Gallagher, who grew genotypes from Massachusetts, Georgia, and Delaware for 11 years in a common garden and documented persistent phenotypic differences. A series of papers on soil biogeochemistry explained the role of 'nurture'. Nitrogen was shown to be a key limiting factor for *S. alterniflora* plant growth because nitrate is quickly reduced to ammonia by bacteria in poorly drained areas away from creeks, where soils have lower soil redox potential. Sulfate-reducing bacteria were also implicated, because they reduce sulfate to sulfide, which impairs the growth of sensitive plant species. Increased soil redox potential and greater pore water turnover in creek-side habitat contributes to taller height forms of *S. alterniflora*. Thus, both genetics and environment influence height forms of *S. alterniflora*, an outcome of both community and ecosystem research.

Restoration

With recognition of lost ecosystem services, interest in restoring salt marshes is growing in Europe and the USA. One way that the British are combating rising

Figure 3 Tidal marsh vegetation is typically dominated by salt-tolerant grasses and succulent forbs, easily distinguished in this restored marsh at Tijuana Estuary, near San Diego, California. Photo by J. Zedler.

sea levels is via 'managed retreat', which involves breaching of embankments to restore tidal flushing to lands that were once salt marshes. In the Netherlands, tidal influence is being reinstated to various polders along the southwestern coast to restore natural processes and diverse estuarine biota to former polders.

Some of the earliest salt marsh restoration in USA has been accomplished as mitigation for damages to other sites as required by federal regulatory agencies. In North Carolina, *S. alterniflora* was being replanted in the 1970s, and the practice has expanded widely to mitigate damages due to development.

Some of the most innovative research on wetland restoration has been accomplished in salt marshes by replicating variables in restoration sites; for example, D. Seliskar and J. Gallagher showed that genotypic variation in *S. alterniflora* has implications for nearly every component of the food web (in Delaware), T. Minello and R. Zimmerman showed that channels in replanted salt marshes enhanced fish support (Galveston Bay, Texas), I. Mendelssohn and N. Kuhn showed that dredge spoil addition accelerated *S. alterniflora* recovery in subsiding wetlands (Louisiana), Cornu showed that topographic variation

across a tidal floodplain affected salmon use (Oregon), and J. Callaway, G. Sullivan, J. Zedler, and others showed that planting diverse assemblages and incising tidal creeks jumpstarted ecosystem functioning in salt marsh restoration sites (Tijuana Estuary, California) (**Figure 3**). In Spain, restoration of tidal ponds is being accomplished in replicate excavations that test the effect of size and depth on use by salt marsh animals (Doñana Marshlands).

In conclusion, salt marshes perform highly valued ecosystem services that are lost when habitats are developed and/or degraded. Further innovations will likely take place in both the practice and science of restoration, because salt marshes are highly amenable to experimentation.

Further Reading

Adam P (1990) *Saltmarsh Ecology*. Cambridge, UK: Cambridge University Press.

Adam P (2002) Saltmarshes in a time of change. *Environmental Conservation* 29: 39–61.

Allen JRL and Pye K (1992) *Saltmarshes: Morphodynamics, Conservation and Engineering Significance*. Cambridge, UK: Cambridge University Press.

Chapman VJ (1960) *Salt Marshes and Salt Deserts of the World. Plant Science Monographs*. London: Leonard Hill [Books] Limited.

Daiber FC (1982) *Animals of the Tidal Marsh*. New York, NY: Van Nostrand Reinhold Co.

Long SP and Mason CF (1983) *Saltmarsh Ecology*. Glasgow: Blackie & Sons Ltd.

Pennings SC and Bertness MD (2000) Salt marsh communities. In: Bertness MD, Gaines SD, and Hay ME (eds.) *Marine Community Ecology*, pp. 289–316. Sunderland, MD: Sinauer Associates Inc.

Pomeroy LR and Weigert RG (1981) *The Ecology of a Salt Marsh*. New York: Springer.

Reimold RJ and Queen WH (eds.) (1974) *The Ecology of Halophytes*. New York, NY: Academic Press Incorporated.

Seliskar DM, Gallagher JL, Burdick DM, and Mutz LA (2002) The regulation of ecosystem functions by ecotypic variation in the dominant plant: A *Spartina alterniflora* salt-marsh case study. *Journal of Ecology* 90: 1–11.

Threlkeld S (ed.) *Estuaries and Coasts: Journal of the Estuarine Research Foundation*. Lawrence, KS: Estuarine Research Federation.

Weinstein MP and Kreeger DA (eds.) (2000) *Concepts and Controversies in Tidal Marsh Ecology*. Boston, MA: Kluwer Academic Publishers.

Zedler JB (ed.) (2001) *Handbook for Restoring Tidal Wetlands*. New York, NY: CRC Press.

Zedler JB and Adam P (2002) Saltmarshes. In: Perrow MR and Davy AJ (eds.) *Handbook of Ecological Restoration* vol. 2: pp. 238–266. Ress, Cambridge, UK: Cambridge University Press.

Savanna

L B Hutley and S A Setterfield, Charles Darwin University, Darwin, NT, Australia

Introduction

This article examines the ecological features of one of most important tropical ecosystems, the savannas. Savannas feature the coexistence of both trees and herbaceous plants and are distinct from grasslands (absence of woody plants) and closed forests (tree dominant). Savanna ecosystems occur in over 20 countries, largely in the seasonal tropics. Much of the world's livestock occurs in savanna, underlining their social and economic importance. Approximately 20% of the world's land surface is covered with savanna vegetation, which produces almost 30% of global net primary production (NPP). With tree and herbaceous components, savanna biodiversity is high, often higher than associated dry deciduous forests. Globally, tenure of savanna lands incorporates pastoral, private use, indigenous and national parks, with the disparate management aims of grazing, mining, tourism, subsistence livelihoods, and conservation. Given their size, savannas affect global carbon, nutrient and water cycles, and, with their frequent fires, significantly influence atmospheric chemistry. Savanna ecosystems have existed for millions of years in many regions, although paradoxically, many ecologists regard savannas as an ecologically unstable mixture of trees and grasses. Savanna boundaries are dynamic in space and time and their occurrence and structure are determined by a combination of environmental factors, such as available water, nutrients, the frequency of disturbances (e.g., fire and herbivory), and stochastic weather events. This range of factors results in significant structural variation and providing an overarching and strict definition of what constitutes a savanna has been problematic. This article provides a commonly used definition, describes savanna distribution, and examines factors that influence their structure and function. Understanding the determinants of savanna functioning, resilience and stability are vital ingredients for improved management. Management of savannas is especially important, as they are under increasing development pressure, especially in tropical regions, and threats to their long-term sustainability are examined.

Definition and Occurrence

Savanna ecosystems predominantly occur in the seasonal tropics and are a unique mix of coexisting trees, shrubs, and grasses (**Figure 1**). Debate surrounds the use and definition of the term savanna, reflecting the range of tree:grass ratios found in these ecosystems. Savanna ecosystems feature a range of structures, from near treeless grasslands to woody dominant open-forest/woodlands of up to 80% woody cover. A widely used definition describes a savanna ecosystem as one consisting of a continuous or near continuous C4 grass dominated understorey, with a discontinuous woody overstorey. Woody components can be a mix of trees and shrubs of evergreen or deciduous phenology, broad or needle leafed. The grass-dominated understory can consist of a mix of species with either annual or perennial habit (often >1 m in height). Ecosystems that fit this definition have ambiguously been termed woodlands, rangelands, grasslands, wooded grasslands, shrublands, open-forests, or parklands.

Savanna formations occur on all continents of the world (**Figure 2**), with the largest extent found in the wet–dry tropical regions of Africa, South America, and Australia. Smaller areas occur in Asia, including Sri Lanka, Thailand, Vietnam, and Papua New Guinea. Savanna also occurs in India, although these tree and grass systems tend to be derived from dry deciduous forest and subhumid deciduous forest due to land-use changes and population pressure. Tropical savanna occupies an area of approximately 27.6 million km^2 including the Asian savanna regions. Tree–grass mixtures also occur in temperate regions, in North America (Florida, Texas), Mediterranean Europe, and Russia, although these temperate savannas are far smaller in extent at approximately 5 million km^2. In total, the savanna biome occupies one-fifth of the global land area and supports a large and growing population.

The existence of a dry season is a defining feature of savannas; rainfall is seasonal and ranges from 300 to 2000 mm, with a dry season lasting between 2 and 9

Figure 1 Savanna ecosystems of the world, featuring the coexistence of a discontinuous woody overstorey with a continuous herbaceous understorey. Plates (a) and (b) are of a north Australian savanna site that receives approximately 1100 mm rainfall and is dominated by evergreen trees (*Eucalyptus* sp.) and tall C4 tropical grasses (*Sarga* spp.). Canopy fullness and grass growth are significantly differently in the wet (a) and dry (b) seasons. Tower-mounted instrumentation in plate (a) is monitoring ecosystem productivity and water use over wet and dry seasons. Plate (c) African savanna of the Kalahari Gemsbok National Park, Botswana. (a, b) Photo courtesy of Joerg Melzheimer.

months of the year. There can be a single, extended dry season or several shorter dry periods. Inter-annual variation of rainfall is typically high, as is the commencement and cessation of the wet season and growing season length, making cropping in savanna lands difficult. Indeed, historical rainfall plays an important role in determining the vegetation structure of a savanna. Seasonally available moisture dramatically influences plant productivity, which in turn determines the timing of available resources for savanna animals.

Given their wide biogeographic range, savannas occur on a number of soils types, typically oxisols, ultisols, entisols, and alfisols (using US soil taxonomy).

In general, these soils are ancient and highly weathered, low in organic matter and cation exchange capacity (CEC). Oxisols occur in tropical savanna regions of South America and central and eastern African savanna and consist of highly weathered, transported, and deposited material occurring on fluvial terraces. Extensive weathering of primary minerals has occurred and they are dominated by clay minerals such as kaolinite and gibbsite which have low CEC. Also present in the soil are acidic Fe and Al sesquioxides, which limits nutrient availability, especially phosphorus. Savanna soils tend to be sands to sandy loams, deep and well drained but with low soil

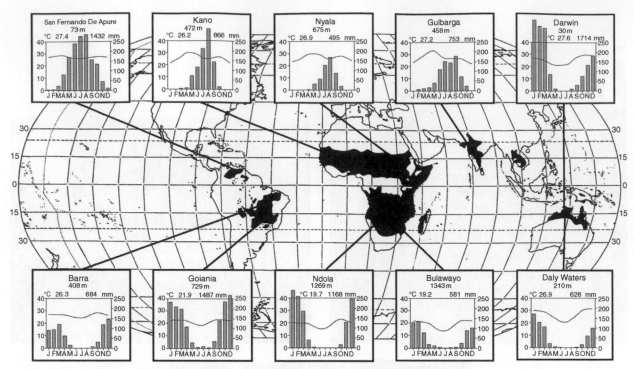

Figure 2 The distribution of the world's savannas. Temperature and monthly rainfall data for a range of savannas are also given, with highly seasonal rainfall clearly evident.

moisture-holding capacity. Entisols that occur in Australian savanna also feature the occurrence of ferruginous gravels, further reducing water- and nutrient-holding capacity. Bioturbation by earthworms and termites are critical in the cycling of nutrients through the poor soil systems. Termites essentially act as primary consumers and in savannas that lack a significant herbivore biomass (e.g., Australian and some South American savannas), they have an ecological function similar to that of herbivorous mammals.

Savannas of Australia, Africa, and South America

Tropical savanna is the predominant vegetation type across the northern quarter of Australia where rainfall is above $600 \, mm \, yr^{-1}$, an area of ~2 million km^2 (**Figures 1a, 1b,** and **2**). These savannas are open woodlands and open forests, with tree cover declining as rainfall decreases with distance from the northern coast. The overstorey flora is typically dominated by *Eucalyptus* spp., particularly *Eucalyptus tetrodonta*, *E. dichromophloia*, and *E. miniata*. *Melaleuca viridiflora*, *M. nervosa*, and *E. pruinosa* assemblages occur in the drier regions of this biome where annual rainfall $<1000 \, mm \, yr^{-1}$. The ground layer is dominated by annual and perennial grasses from the *Sarga*, *Heteropogon*, and *Schizachrium* genera. A variety of other tall grasses (>1m height) dominate the ground layer of the monsoonal savannas, which extend from Western Australia to the

Cape York Peninsula in Queensland. *Heteropogon contortus* (black speargrass) dominates the tropical savanna understory in eastern Queensland, with *Themeda triandra*, *Aristida*, *Bothriochloa*, and *Chrysopogon bladhii* becoming more dominant as rainfall declines. *Acacia*-dominated savanna communities include extensive areas of brigalow (*A. harpophylla*), lancewood (*A. shirleyi*), and gidgee (*A. cambegei* and *A. georginae*).

The neotropical savannas of South America cover more than 2 million km^2. The Brazilian *cerrado* and the Colombian and Venezuelan *llanos* are a continuous formation, interrupted by narrow gallery forests. The *cerradão* includes a range of vegetation formations from the pure or almost pure grassland of *camp limpo*, to open woodland with scattered tree cover of *campo cerrado*. These savanna can grade into denser woodland or open forests, the *cerradão*, where tree cover is greater than 50%. The dominant grasses are *Andropogon*, *Aristida*, *Paspalum*, and *Trachypogon*. The Orinoco *llanos* comprise grasslands or grasslands with scattered trees which are typically <8 m tall. Common trees include *Byrsonima* spp. *Curatella americana*, *Bowdichia virgioides*, and grasses include *Trachypogon* and *Andropogon*. Hyper-seasonally flooded savannas and esteros (savanna wetland) occur in Brazil and Bolivia. Other savanna types, such as savanna parkland and mixed woodland, occur through tropical America.

The African savannas occur across a range of soil types within a rainfall range of 200–1800 mm. One of the most extensive savanna areas is the *miombo* which covers about

2.7 million km^2 across central and southern Africa. The *miombo* is characterized by a discontinuous canopy of 10–12 m tall deciduous species of *Brachystegia*, *Isoberlinia*, and *Julbernardia*, with an herbaceous layer of tall grasses including mainly *Andropogon* species. In southern Africa, fine-leaved savannas, dominated by *Acacia* species, occur over fertile soils in low-lying, semiarid (250–650 mm yr^{-1}) areas. Broadleaved savannas, including *Burkea africana*, *Combretum* spp., and *Brachystegia*, occur on weathered, infertile soils. The northern *Sudanian* savannas have scattered deciduous trees, typically *Isoberlinia doka*, over xerophytic grasslands. These are bordered on the north by the drier *Sahelian* savannas and on the south by the wetter *Guinea*-type savannas. The arid and semiarid east African savannas are grasslands dominated by *Aristida* spp. and *Brachiara* spp., with scattered shrubs or trees (including *Acacia*, *Grewia*, and *Commiphora*), for example, the Serengeti.

Savannas occur throughout Asia, although many of these are derived from human disturbance. Savanna is fairly extensive in the Indian subcontinent, although tree clearing has increased their extent, and many areas have been converted to agriculture. The most significant and widespread savanna type in Southeast Asia is the dry dipterocarp forest, which occurs in Vietnam, Laos, Cambodia, Thailand, Burma, and a small area in India. The region receives 1000–1500 mm rain per year and is dominated by the deciduous *Dipterocarpus* species, which can grow to 20 m, over a dense grass and herb layer including *Imperata cylindrical*, *Apluda mutica*, and *Arundinaria* spp. (pygmy bamboo).

Adaptive Traits of Savanna Vegetation

Savanna plants display a suite of traits to cope with seasonal drought, low water and nutrient availability, and the impacts of regular fire and herbivory. Adaptive traits which aid in the survival of fire for woody plants include thick insulating bark, high wood moisture content, and significant resprouting capacity. Resprouting can occur via lignotubers and from other underground and stem basal tissues following the death of aerial stems. This enables recovery with minimal developmental costs. Vegetative reproduction from roots, rhizomes, or stolons is dominant in much of the savanna biome. Adaptations to low nutrient availability include root mycorrhizal associations, particularly of ectomychorrhizae. Savanna trees can rapidly translocate sequestered nutrients from the leaves to other tissues (e.g., bark) prior to leaf fall. Woody savanna plants often have thorns that restrict grazing, as well as chemical features, such as tannins making leaves less palatable. Savanna grasses also display morphological features, such as serrated edges, and chemical features, including tannins and silica bodies, to restrict grazing.

The herbaceous grass layer is dominated by grasses with a C4 photosynthetic pathway. This pathway enables high photosynthetic rates at high temperatures and irradiance and low water availability. Most savanna trees and shrubs have the C3 photosynthetic pathway that has a higher efficiency under low light when compared to the C4 pathway, a characteristic which facilitates recruitment and establishment under shaded tree canopies. The growth of savanna plants tends to occur mostly during the wet season with senescence or dormancy in the fire prone dry season, a trait that facilitates persistence in unfavorable conditions. Annual herbaceous species persist via a soil seed bank, whereas aboveground parts of perennial herbaceous species die during dry periods, with dormant, regenerative buds protected within belowground rhizomes or by cataphylls. Some annual grass species use hygroscopically active awns and pointed calluses on their seed to work them into the soil, also protecting them from fire. Perennial herbaceous species require wet season rains to produce their first green shoots as carbohydrate storage from the previous wet season is limited. Rainfall stimulates germination of annual herbaceous and grass species and the early wet season is a period of rapid growth. Most herbaceous species flower in the wet season, although in contrast, many woody species flower in the dry season.

Woody species have evolved physiological and morphological mechanisms to either tolerate (evergreen habit) or avoid (deciduous habit) prolonged periods of water stress. Deep-rooting woody plants (usually evergreen) are able to access water resources throughout the year and provides them with their full photosynthetic capacity when favorable conditions occur. Deciduous species rehydrate stems prior to onset of wet season rains, which is then followed by leaf expansion to maximize photosynthetic activity during the wet season. Deciduousness and evergreeness represent extremes of physiological adaptations to survive the seasonal savanna climate. Evergreen species invest more resources in longer lived leaves, whereas deciduous species tend to support shorter-lived leaves with high leaf photosynthetic capacity. Deciduous species need to acquire enough nutrient and photosynthate to ensure persistence and reproduction during the wet season, whereas evergreen species tend to have slower growth rates but persist throughout the seasonal cycle. Evergreeness also allows opportunistic acquisition of resources when soil nutrients are severely limiting and the cost of producing new leaves to respond to change in soil moisture is prohibitive.

Although this section has described broad seasonal growth patterns, it is important to note that the world's savanna plants include a high diversity of species and life forms, with many distinct phenological patterns. All periods of the climatic cycle is favorable to certain vegetative or flowering phenophases in at least one group of species.

Environmental Factors Determining Savanna Structure

The adaptive traits described above enable individual plant survival in seasonally variable climates, but what environmental factors operate at a landscape or regional scale that determine savanna structure? Evidence suggests that four key environmental factors are responsible: (1) plant available moisture (PAM); (2) plant available nutrients (PANs); (3) fire regime; and (4) herbivory. Herbivores include vertebrates and invertebrates and consist of both browsers consuming woody biomass and grazers consuming grasses and herbs. The overarching determinants of savanna physiognomy (relative abundance of the tree and grass layer) are climate and soil type (PAM and PAN), which determines the potential growth and survival of trees and grasses at a given site. Growth potential is moderated by disturbance agents, fire, herbivory, and stochastic events (such as cyclones). These factors act in concert to influence both competitive interactions and facilitation of tree and grass growth and determine savanna structure, floristics, and productivity (**Figure 3**). The interaction of these factors is poorly understood and their variation in space and time makes experimental testing and isolation of any single determinant difficult. Spatial heterogeneity of vegetation due to local site histories (determined by antecedent rainfall, fire history, and herbivore numbers) and an inability to quantify these factors exacerbates this difficulty.

Available Moisture and Nutrients

Savannas tend to occur in warm climates with an annual drought, with soils typically low in nutrient capital and poor water-holding capacity. Interactions between trees and grasses is dominated by competition for water and nutrients, rather than light or growing space. At a broad, continental scale, PAM is the most significant of the four ecological determinants, with increasing rainfall correlated with increased tree cover and in general, a decreased grass biomass. PAM can be quantified via a range of parameters, from simple measures such as annual rainfall or via water balance parameters (rainfall as a fraction of potential or actual evapotranspiration) or soil characteristics (water-release characteristics, soil-storage capacity). At fine spatial scales, soil physiochemical properties (PAN) have a more significant influence and the interaction with PAM is often termed the PAM/AN plane. Nutrient availability is largely a function of soil moisture and dry season nutrient uptake, and nitrogen mineralization, in particular, is limited by low levels of PAM. Significant plant growth is only possible during periods of high PAM that releases available nutrient via mineralization. Soils of semiarid savanna can have a higher intrinsic fertility when compared to highly leached soils of mesic sites, but this nutrient capital is only available for uptake during moist periods. Savanna vegetation receiving similar rainfall can exhibit contrasting structure and floristics, simply due to fine-scale changes in soil type. A good example of this interaction comes from the long-

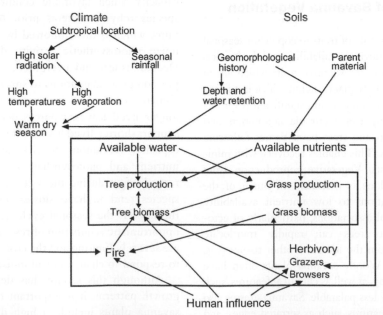

Figure 3 Interactions between the environmental determinants of savanna structure. The relative tree and grass biomass and productivity is determined by available water, nutrient, and disturbance regime (fire and herbivory). These determinants are in turn characterized by climate and soil type for any given location. Reproduced from House JI, Archer S, Breshears DD, and Scholes R (2003) Conundrums in mixed woody-herbaceous plant systems. *Journal of Biogeography* 30: 1763–1777, with permission from Blackwell Publishing.

(a)

(b)

Figure 4 Impacts of overgrazing and fire on savanna structure. Plate (a) is an overgrazed native grass paddock in semiarid savanna in north Australia (Kidman Springs Station, Victoria Rivers District, NT) at the end of the wet season of 1973. This site would be subjected to wind and water erosion, resulting in further decline in health and productivity of such sites. Exclusion of grazing and fire (Plate (b) has resulted in a complete recovery of structure and function, with return of trees and grasses stabilizing soil surfaces, increased water capture and a recovery in nutrient availability and cycling. Photos courtesy of John Ludwig, CSIRO.

term savanna research site of Nylsvley in South Africa; here, nutrient-poor, broad-leafed savanna, dominated by *Burkea africana* surround patches of nutrient-rich soil that support a very different savanna type, a fine-leafed savanna dominated by *Acacia tortilis*. Both savanna types experience the same climate, but differences in soil parent material result in higher levels of soil available N and P in the fine-leafed patches. Productivity of the fine-leafed savanna is approximately double that of the broad-leafed system and attracts a larger grazing and browsing fauna. Similarly in South American savanna, soil acidity and aluminum levels significantly affect structure and floristics independent of rainfall.

Fire

Fire is an important landscape-scale determinant that impacts all of the world's savanna. Fire is an inevitable consequence of the annual cycle of profuse herbaceous production during the wet season followed by curing of this material in the dry season, when climatic conditions are ideal for burning. Savanna fires are virtually all surface fires, consuming the highly flammable herbaceous layer. Crown fires rarely occur, as the foliage of savanna trees and shrubs are of low flammability. Human ignitions largely control fire behavior and extent in savanna, with some fires started by dry lightning strikes. Savanna fires spread rapidly through the surface fuels and high soil temperatures do not persist for longer than a few seconds to minutes. While these fires have a significant impact on aboveground plant parts, there is limited impact on savanna seed banks or belowground regenerative plant parts.

Fire has a major role in restricting tree establishment and growth, as evident from long-term fire exclusion plots (>25 years) in southern African and north Australian

savanna (**Figure 4**), which have resulted in a woody thickening. Frequent fire events can reduce tree seedling establishment and the ability of saplings to escape the flame zone via height growth. This limitation on tree establishment enables grass persistence and growth, maintaining the fuel load. The aerial stems of small seedlings and suckers are often killed during fire but the individuals are able to resprout from lignotubers or from other underground and stem basal tissues. Seedlings less than 6 months old have been observed to resprout in some species (e.g., *Eucalyptus miniata*) and frequent fire in the savannas will kill or maintain tree seedlings as a suppressed woody sprout layer until there is a sufficient fire-free period for them to escape the fire damage zone. Species can survive for at least 40 years as suppressed sprouts, during which time they develop significant lignotubers which aid in rapid growth during fire free periods.

The timing of fires in relation to reproductive phenology can constrain or promote plant reproduction. Studies on the woody species in the Brazilian *cerrado* and mesic Australian savannas have indicated that frequent fire can reduce seed production and sexual recruitment and could cause a shift in species composition, favoring vegetatively reproducing species. However, fire is also important for the sexual regeneration of some species, as burning induces flowering and fruit dehiscence in many *cerrado* species and facilitates pollination in others. Most perennial grass species are generally less affected by burning and regenerate from basal leaf sheaths protected underground. Some perennial (e.g., *Trachypogon plumosus*) and annual (e.g., *Andropogon brevifolius*) grass species decrease in abundance after a long-term absence of fire.

Prior to human occupation and use of fire in savannas, lightning would have been the dominant source of ignition and it is likely that extensive but infrequent fires would have occurred. In Australia, humans have intentionally used fire for at least 40 000 years and in Africa for potentially 1 million years or more. Large proportions of savanna regions are burnt each year for a variety of reasons: land clearing, livestock management, property protection, conservation management, and cultural purposes. In African savannas, fires burn between 25–50% of the arid 'Sudan Zone' and 60–80% of the humid 'Guinea Zone' each year. Approximately 65% of *Eucalyptus* dominated savanna woodland and 50% of savanna open forest in Kakadu National Park, northern Australia was burnt annually between 1980 and 1994. With the progression of the dry season, fire intensity increases due to fuel accumulation from curing litterfall and grass senescence resulting in an increased combustibility of fuels plus more severe fire weather (i.e., higher temperatures, stronger winds, and lower humidities). Early dry season fires (when fuel accumulation is low and curing incomplete) tend to be low-intensity, patchy, and limited in extent. Fires later in the season are of higher intensity and produce more extensive and homogeneous burning. Impacts on vegetation depend on fire intensity, distribution, and timing (fire regime) in relation to the vegetative and phenological cycles. Determining direct effects of fire on savannas is often difficult due to confounding effects of herbivory. Nevertheless, long-term burning experiments have shown that the higher-intensity, late dry-season fires are the most damaging to woody species.

Herbivory

Two common images of savannas are herbivory by large, native ungulates, particularly in Africa and the widespread grazing by domestic herds, particularly cattle. A more neglected group of savanna herbivores are the invertebrates, particularly grasshoppers, caterpillars, ants, and termites. Mammal herbivores are typically categorized as grazers, browsers, or mixed feeders, who can vary their diet depending on food availability. Mammal and insect herbivores impact on savanna structure and function via consumption of biomass, seed predation, trampling of understory, and the pushing over and killing of trees and shrubs. The importance of herbivory as a determinant varies between savanna regions, and appears to largely reflect the abundance of large herbivores present. Large herbivore diversity and abundance are much higher in Africa than in Australia, Asia, or South America. More than 40 large wild herbivore species have been described in African savanna. In contrast, only six species of megapod marsupial have been considered as large herbivorous mammals in the Australian savannas, and only three species of ungulates are regarded as native South American savanna inhabitants. Domestic animals, particularly cattle, buffalos, sheep, and goats, are now the dominant, large herbivores in most savannas.

Large herbivores can lead to changes in species composition, woody vegetation density, and soil structure. For example, grazing pressure in Africa and Australia has led to a decrease in palatable, perennial, grazing-sensitive tussock grasses, and an increase in less palatable perennial and annual grass and forb species. Changes to the soil surface can occur, including loss of crusts (important in nutrient cycling), development of scalds, compaction, increased runoff, soil erosion, and nutrient loss. In parts of Africa, woody vegetation density has sometimes been reduced by large herbivores, for example, uprooting of trees by elephants when browsing. Browsers such as giraffes can reduce woody seedling and sapling growth, thereby keeping them within a fire-sensitive heights for decades. By contrast, in many of the world's savannas the density of woody vegetation has increased at the expense of herbaceous vegetation; one of the major causes has been high rates of herbivory. A decrease in grass biomass following grazing leads to a reduction fuel and thus fire frequency and intensity, enhancing the survival of saplings and adult tress. Fire also affects herbivory as herbivores may favor postfire vegetation regrowth. Clearly, fire and herbivory have an interactive effect on savanna structure and function.

While less spectacular than large browsers and grazers, insects are often the dominant group of herbivores in savannas, especially on infertile soils supporting low mammal biomass. There is a paucity of data describing their abundance or role in these ecosystems. In a broad-leaved, low fertility savanna of southern Africa, a grasshopper biomass of $0.73\,kg\,ha^{-1}$ can consume almost $100\,kg\,ha^{-1}$ of plant material and damage an additional $36\,kg\,ha^{-1}$. This represents a loss of 16% of aboveground grass production. Grasshoppers and caterpillars can account for up to half the grass herbivory, although the rate and proportion varies substantially between years. Fertile, fine-leaved savannas are able to support a larger mammal biomass, and the proportion of herbivory resulting from insect consumption is lower when compared to infertile African sites. The impact of insect herbivores on physiognomy has not been established but they are clearly important herbivores in savannas through their impact on productivity and ecosystem properties.

Conceptual Models of Tree and Grass Coexistence

Interactions between the coexisting lifeforms in savanna communities are complex and over the last 40 years, a range of conceptual or theoretical models has been proposed to explain tree and grass mixtures. Contrasting models have all been supported by empirical evidence for particular sites, but no single model has emerged that provides a generic mechanism explaining coexistence. Models can be classified into several categories. Competition-based models feature spatial and temporal

Table 1 Conceptual models explaining the coexistence of trees and grasses in savanna ecosystems in equilibrium (tree:grass ratio relatively stable at a given site), nonequilibrium (tree : grass ratio variable) or disequilibrium (disturbance agents essential for the maintenance of tree:grass coexistence)

Competition-based	*Demographic-based*
Mechanisms of coexistence	Mechanisms of coexistence
Spatial and temporal niche separation of resource usage enables both life forms to coexist	Climatic variation and disturbance impacts on tree demography
Root-niche separation	Extremes of climate and disturbance influence tree germination and/or establishment and/or transition to mature size classes enabling coexistence
Tree and grasses exploit deep and shallow soil horizons	At low rainfall sites, tree establishment and growth occurs only in above average rainfall periods
Phenological separation	At high rainfall sites, high fuel production maintains frequent fire to limit tree dominance
Temporal differences in leaf expansion and growth, trees have exclusive access to resources at beginning and end of growing season, grasses competitive during growing season	
Balanced competition	
Trees are the superior competitor but become self-limiting for a given rainfall and unable to exclude grasses	
Competition–colonization	
Rainfall variability results in a tradeoff between tree and grass competition and colonization potential. Higher than mean rainfall	
favours tree growth, lower than mean favours grasses	Primary determinants
Primary determinants	PAM variability, PAN, fire regime, herbivory
PAM, PAN	
Secondary determinants	
Fire regime, herbivory	

separation of resource usage by trees and grasses that minimizes competition and enables the persistence of both lifeforms. Alternatively, demographic-based models have been described, where mixtures are maintained by disturbance, resulting in bottlenecks in tree recruitment and/or limitations to tree growth and grasses can persist. **Table 1** provides a summary of these models. Root-niche separation models suggest that there is a spatial separation of tree and grass root systems, with grasses exploiting upper soil horizons and trees developing deeper root systems. Trees rely on excess moisture (and nutrient) draining from surface horizons to deeper soil layers. Phenological separation models invoke differences in the timing of growth between trees and grasses. Leaf canopy development and growth in many savanna trees occurs prior to the onset of the wet season, often before grasses have germinated or initiated leaf development. As a result, trees can have exclusive access to resources at the beginning of the growing season, with grasses more competitive during the growing season proper. Given their deeper root systems, tree growth persists longer into the dry season, providing an additional period of resource acquisition at a time when grasses may be senescing. This spatial and temporal separation of resource usage is thought to minimize competition, enabling coexistence. Other competition models suggest that density of trees becomes self-limiting at a threshold of PAM and PAN and is thus unable to completely exclude grasses. These

models assume that high rainfall years favor tree growth and recruitment, with poor years favoring grasses, and high interannual variability of rainfall maintaining a relatively stable equilibrium of trees and grasses over time.

Alternatively, savannas can be viewed as meta-stable ecosystems (narrow range of stabile states) with a dynamic structure over time. Demographic-based models suggest that determinants of tree demographics and recruitment processes ultimately set the tree:grass ratios (**Table 1**). Fire, herbivory, and climatic variability are fundamental drivers of tree recruitment and growth, with high levels of disturbance resulting in demographic bottlenecks that constrain recruitment and/or growth of woody components and grass persistence results. At high rainfall sites, in the absence of disturbance, the ecosystem tends toward forest. High levels of disturbance, particularly fire, can push the ecosystem toward a more open canopy or grassland; this ecosystem trajectory is more likely at low rainfall sites.

There is observational and experimental data to support all of the above models and it is highly likely that savanna structure and function results from the interaction of all processes. In many savannas, root distribution is spatially separated with mature trees exploiting deeper soil horizons as the competitive root-niche separation model predicts. Root partitioning favors tree growth in semiarid systems where rainfall occurs during periods when grass growth is dormant; rainfall can drain to deep layers supporting tree components. By contrast, in semi-arid savanna where rainfall and growing seasons coincide, investment in deep root systems could result in tree water stress, as rainfall events tend to be sporadic and small in nature, with little deep drainage. In this case, surface roots are more effective at exploiting moisture and mineralized nutrients following these discrete events. In these savannas, tree and grass competition for water and nutrients would be intense. In mesic savanna sites, root competition between both trees and grass roots in upper soil layers is apparent, contrary to predictions of niche-separation models. Mesic savannas of north Australia (rainfall >1000 mm) are dominated by evergreen *Eucalyptus* tree species, and during the wet season these trees compete with high growth-rate annual grasses for water and nutrients in upper soil layers (0–30 cm). However, by the late dry season, tree root activity has shifted to subsoil layers (up to 5 m depth) and herbaceous species have either senesced or are physiologically dormant. These root dynamics suggest that grasses are essentially drought avoiders but are able to compete with trees during the wet season. This system serves as an example where both root-niche and phenological separation are occurring.

Tree-to-tree competition is also significant, as suggested by the strong relationship observed in most savanna regions between annual rainfall and indices of tree abundance, be it tree cover (**Figure 5**), tree basal area (area occupied by tree stems), or tree density. As PAM decreases, tree abundance declines. Competition models also fail to consider impacts of savanna determinants on different demographics of a population, such as recruitment, seedling establishment, and tree sapling growth. Root-niche or phenological

separation models largely consider impacts acting on mature individuals, whereas demographic models include impacts of climate variability and disturbance on critical life-history stages (e.g., seedling establishment and accession to fire-tolerant size classes). Demographic models assume that savanna tree dynamics are central to savanna ecosystem functioning and that savanna trees are the superior competitors under most conditions; grass persistence only occurs when determinants act to limit tree abundance. It is clear that competition, both within and between savanna life forms, occurs and that tree abundance is moderated by climate variability and disturbance. A more comprehensive model would integrate both competition and demographic theories to yield a model in which competitive effects are considered for each life-history stage.

The complexity inherent in these models is evident when savanna structure is correlated with any of the environmental determinants. **Figure 5** describes the relationship between tree cover and mean annual rainfall, in this case a surrogate for PAM. Tree cover data are shown for African and Australian savanna sites. The figure shows a large scatter of tree cover possible at any given rainfall, especially for the African sites. For African savanna, rainfall sets an upper limit on tree cover, with the relationship linear until approximately 650 mm rainfall with little increase in tree cover observed above this threshold (**Figure 5**). Points below the line represent savanna sites with a tree cover determined by PAM plus the interaction of other determinants to reduce tree cover below the maximum possible for a given rainfall. At semiarid savanna sites (<650 mm rainfall), it is likely that rainfall limits tree cover and canopy closure, permitting grass

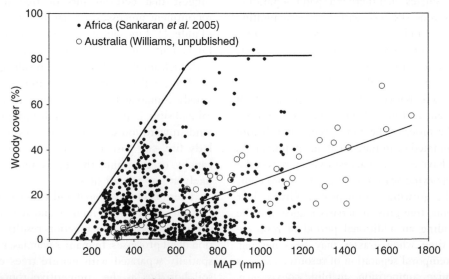

Figure 5 Relationship between mean annual rainfall (MAP) and tree cover for African and Australian savannas, with rainfall setting a maximal climate-determined woody cover. Other factors such as available nutrient, fire frequency and herbivory determine woody cover at any given site. Modified from Sankaran M, Hanan NP, Scholes RJ, *et al.* (2005) Determinants of woody cover in African savanna. *Nature* 438: 846–849 (Macmillan Publishers Ltd), with Australian tree cover data from R. J. Williams, unpublished data.

coexistence. At rainfalls >650 mm, tree canopy closure may be possible, with disturbance limiting woody dominance. For Australian savanna, there is a simpler relationship evident, with a linear increase in tree cover with annual rainfall and less scatter. Australian savannas also have a reduced tree cover (and biomass) for a given rainfall when compared to African systems (**Figure 5**). This suggests that while PAM is determining tree cover, other factors such as fire frequency or PAN are also playing a role. Australian savanna soils (PAN) may be systematically poorer than African soils or fire frequency higher, limiting tree cover and productivity.

Savanna Biomass and Productivity

Global NPP, the net production of plant biomass, is approximately $67.6 \, GtC \, yr^{-1}$ of which almost 30% occurs in savanna ecosystems ($19.9 \, Gt \, yr^{-1}$). This production occurs on 18% of the global land surface, demonstrating that savannas are relatively productive ecosystems. Mean savanna NPP has been estimated at $7.2 \, tC \, ha^{-1} \, yr^{-1}$ (**Table 2**), lower than typical values for the other major tropical ecosystem, rainforest, which ranges from 10 to $15 \, t \, C \, ha^{-1}$. Savanna NPP and biomass varies by an order of magnitude (**Table 2**), as would be expected given their geographic range and structural variation. The relative production of trees versus grasses is also highly variable, but in general, NPP of the C4 grass layer is 2–3 times that of tree NPP. Biomass stored in above- and belowground pools determines the root:shoot ratio and these data from a range of savanna sites around the world give a global mean of approximately 2 (**Table 2**). This reflects the investment in root systems and belowground storage organs, such as lignotubers, to maintain uptake of moisture and nutrient from sandy, nutrient poor savanna soils and to survive disturbance.

Table 2 Savanna biomass, soil carbon stocks and productivity

Parameter	Mean (sd)	Range
Biomass and soil stocks (t C ha⁻¹)		
Aboveground biomass	10.6 (9.0)	1.8–34
Belowground biomass	19.5 (14.9)	4.9–52
Total biomass	33.0 (22.9)	9.4–84
Root : shoot ratio	2.1 (2.0)	0.6–7.6
Soil organic carbon	174.2 (126.0)	18–373
Savanna area (M km⁻²)	27.6	
Total carbon pool (Gt C)	326	
Productivity (t C ha⁻¹ y⁻¹)		
NPP	7.2 (5.1)	1.4–22.8
NEP	0.14	

Data from Grace J, San JJ, Meir P, Miranda HS, and Montes RA (2006) Productivity and carbon fluxes of tropical savannas. *Journal of Biogeography.* 33: 387–400.

Savanna photosynthesis and growth is highly seasonal and interannual variability high. Mesic savanna may receive annual rainfall associated with rainforest ecosystems, yet productivity is significantly lower, due largely to annual drought, poor soils, and impacts of disturbance. Long-term (as opposed to annual) estimates of savanna productivity need to include loss of biomass due to fire and herbivory. Including fire and herbivory impacts on productivity estimates gives the carbon sequestration rate, which represents the net gain (sink) or loss of carbon from the ecosystem to the atmosphere. While wet season productivity can be very high in savannas, much of a wet-season's herbaceous productivity can be lost via fire or grazing. Woody biomass tends to be a less dynamic, longer-term carbon-storage pool than the herbaceous components of savanna. Savanna fire results in a significant release of greenhouse gases, including CO_2, CO, methane, nonmethane hydrocarbons, nitrous oxide, particulate matter and aerosols, equivalent to 0.5–$4.2 \, GtC \, yr^{-1}$. Fire reduces net savanna sequestration rate by about 50% and protection of savannas from fire and grazing results in an increase in woody biomass which can result in a long-term increase in stored soil carbon. Savanna sink strength in mesic Orinoco savannas in South America (~1500 mm annual rainfall) has been measured at $1 \, t \, C \, ha^{-1} \, yr^{-1}$, with this sink maintained over a 25-year period in plots with fire and grazing excluded. Similarly, the carbon sink strength of north Australian, *Eucalyptus*-dominated savannas receiving approximately the same rainfall has also been estimated at approximately $1 \, t \, C \, ha^{-1} \, yr^{-1}$, with this sink measured at sites burnt but not grazed. This carbon is likely being stored in woody biomass and soil organic carbon pools, with a small fraction being stored as black carbon (charcoal), a resilient carbon pool. Savanna soil carbon storage is by far the largest pool of carbon (**Table 2**) and soil carbon represents a longer-term storage of carbon when compared to the more dynamic vegetation components. Burning also influences nutrient dynamics via losses due to volatilization (vaporization) of lighter elements such as nitrogen and sulfur. At a global scale, savannas and tropical seasonally dry forests represent a significant source of N_2O to the atmosphere ($4.4 \, Tg \, N_2O \, yr^{-1}$). Shifts to a more frequent fire regime may result in a significant net loss of nitrogen, as savannas are in general nitrogen-poor. Many grass species are able to recover quickly after fire, with re-growth attractive to grazing animals, due to the relatively high nutrient content of the foliage.

Threats to Long-Term Sustainability

Savannas are ancient ecosystems. They are the location of human evolution, and humans are an integral component of these ecosystems. Humans have influenced the determinants of savannas for thousands of years via modification to nutrient availability from fire and clearing for agriculture. Human cultures have used fire as a vegetation management tool and introduced animal husbandry systems, changing grazing and browsing pressures and modified tree–grass competitive balances (e.g., **Figure 4**). A contemporary impact is now being experienced via climate change and its influence on rainfall distribution, temperature increases, and climate conditions conducive to fire and increased atmospheric CO_2 concentration. Human usage of the savanna biome is increasing, which can lead to degradation of vegetation and soil resources, resulting in nutrient losses and shifts in water balance and availability. Brazilian *cerradão* contains over 800 species of trees and shrubs alone; approximately 40% of the *cerradão* and *llanos* have now been cleared or altered for agricultural uses with crops such as coffee, soybeans, rice, corn, and beans. Soil management is critical given their low nutrient status, acidity and friability. Alterations in grazing pressure and fire suppression in managed savannas have also resulted in woody dominance, which ultimately reduces grazing production, severely impacting communities relying on cattle-derived incomes and reducing local biodiversity. This thickening or woody encroachment is being observed in areas subjected to extensive grazing activities in both African and Australian savannas.

Clearing for alternative land uses can also result in exotic species invasions, a problem for much of the world's savannas. African savanna, especially in South Africa, are being invaded by woody species, often *Acacia* or *Eucalyptus* species from Australia, introduced for fuel wood or timber production. Low herbivory of these species results in high growth rates and water use. The development of thickets reduces deep drainage, groundwater recharge, and streamflow, consequently affecting water supplies. In an attempt to increase the grazing potential of north Australian and South American savanna, fast-growing African grasses such as *Andropogon gayanus* have been introduced. They are more productive than native species; however, they develop far larger and more flammable fuel loads. At infested sites in north Australia, resultant fire intensity is 5 times that observed from native grass savanna and impacts on tree mortality and recruitment. This in turn will result in a demographic bottleneck, long-term loss in tree cover, and the instigation of a grass-fire cycle. Introductions of African grasses such as *Brachiaria*, *Melinis*, and *Andropogon* species have occurred in the *llanos* of Colombia and Venezuela and the *cerrado* of Brazil. These grasses are used as fodder for cattle and are displacing native species, causing a loss in biodiversity of these savannas.

Climate change will alter the distribution of rainfall, thus influencing PAM and PAN. Shifts in temperature regimes and atmospheric CO_2 concentration may also alter the relative growth rates of trees and grasses, modifying competitive balances. Trees (C3 photosynthetic pathway) can potentially utilize high CO_2 concentrations more efficiently than grasses (C4 photosynthetic pathway) due to increased carbon allocation to roots and lignotubers plus greater water use and nutrient use efficiency apparent at high atmospheric CO_2 concentrations. As CO_2 concentrations increase, physiological differences between trees (carbon-rich lifeforms) may be favored over grasses (carbon-poor) and trees may gain a competitive edge. Tree saplings may grow to fire-tolerant sizes faster, limiting the impact of fires that maintain grasses in savanna.

All of the above examples involve human impacts acting on one or more of the determinants of savanna structure and function. Clearly, increased knowledge of their interactions will provide improved understanding of savanna processes and enable better management in a rapidly changing world. Savannas may be ideal ecosystems for agro-forestry applications, rather than traditional cropping systems. Small shifts in fire regime may dramatically increase productivity; thus, savanna systems could be used for carbon sequestration and greenhouse gas mitigation schemes, providing alternative livelihoods and aiding in the maintenance of biodiversity.

See also: Mediterranean; Swamps.

Further Reading

Andersen AN, Cook GD, and Williams RJ (2003) *Fire in Tropical Savannas: The Kapalga Experiment*. New York: Springer.

Baruch Z (2005) Vegetation–environment relationships and classification of the seasonal savannas in Venezuela. *Flora* 200: 49–64.

Bond WJ, Midgley GF, and Woodward FI (2003) The importance of low atmospheric CO_2 and fire in promoting the spread of grasslands and savannas. *Global Change Biology* 9: 973–982.

du Toit JT, Rogers KH, and Bigg HC (eds.) (2003) *The Kruger Experience: Ecology and Management of Savanna Heterogeneity*. Washington, DC: Island Press.

Furley PA (1999) The nature and diversity of neotropical savanna vegetation with particular reference to the Brazilian cerrados. *Global Ecology and Biogeography* 8: 223–241.

Grace J, San JJ, Meir P, Miranda HS, and Montes RA (2006) Productivity and carbon fluxes of tropical savannas. *Journal of Biogeography* 33: 387–400.

Higgins SI, Bond WJ, and Trollope WSW (2000) Fire, resprouting and variability: A recipe for grass–tree coexistence in savanna. *Journal of Ecology* 88: 213–229.

House JI, Archer S, Breshears DD, and Scholes R (2003) Conundrums in mixed woody-herbaceous plant systems. *Journal of Biogeography* 30: 1763–1777.

Mistry J (2000) *World Savanna: Ecology and Human Use*. Harlow: Prentice-Hall.

Rossiter NA, Setterfield SA, Douglas MM, and Hutley LB (2003) Testing the grass-fire cycle: Exotic grass invasion in the tropical savannas of northern Australia. *Diversity and Distributions* 9: 169–176.

Sankaran M, Hanan NP, Scholes RJ, et al. (2005) Determinants of woody cover in African savanna. *Nature* 438: 846–849.

Scholes RJ and Archer SR (1997) Tree and grass interactions in savanna. *Annual Review of Ecology and Systematics* 28: 517–544.

Scholes RJ and Walker BH (eds.) (1993) *An African Savanna: Synthesis of the Nylsvley Study*. Cambridge: Cambridge University Press.

Solbrig OT and Young MD (eds.) (1993) *The World's Savannas: Economic Driving Forces, Ecological Constraints, and Policy Options for Sustainable Land Use*. New York: Parthenon Publishing Group.

van Langevelde F, van de Vijver CADM, Kumar L, *et al.* (2003) Effects of fire and herbivory on the stability of savanna ecosystems. *Ecology* 84: 337–350.

Williams RJ, Myers BA, Muller WJ, Duff GA, and Eamus D (1997) Leaf phenology of woody species in a north Australian tropical savanna. *Ecology* 78: 2542–2558.

Steppes and Prairies

J M Briggs, Arizona State University, Tempe, AZ, USA

A K Knapp, Colorado State University, Fort Collins, CO, USA

S L Collins, University of New Mexico, Albuquerque, NM, USA

Grasslands
Grassland Types
The Grassland Environment
Fire in Grasslands

Grazing in Grasslands
Threats to Grasslands and Restoration of Grasslands
Further Reading

Steppes and prairies (grasslands) are ecosystems that are dominated by grasses and to help understand grasslands, it is important to know something about grass morphology and growth forms. The remarkable ability of grasses to thrive in so many ecological settings and their resilience to disturbance is largely attributable to their growth form. Grasses are characterized by streamlined reduction and simplicity with tillers being the key adaptive structural element of the plant (**Figure 1**). Tillers originate from growing parts (meristems) typically just near, at, or below the surface of the soil. The meristems that produce tillers are generally well protected by their location near or beneath the soil surface. It is the location of the meristem that explains much of the resilience of grasses and thus grasslands to disturbance.

Grass leaves are narrow and generally well-supplied with fibrous supporting tissue that has thick-walled cells. These features, along with a capacity to fold or roll the leaves along the vertical plane, permit the plant to endure periods of water stress without collapse. Another feature of grass leaves is the presence of siliceous deposits and silicified cells (phytoliths). Although silica is present in many plant families, phytoliths are characteristic of grasses. Phytoliths often have distinctive forms within taxonomic groups and since they persist in soil profiles for a very long time, they can be used by paleobotanists to determine shifts in dominance from one grass form to another. Silica also makes grass forage very abrasive and it is now generally accepted that the evolution of abrasion-resistant teeth present in many modern grazing animals was an evolutionary response to tooth-wearing effects of a diet

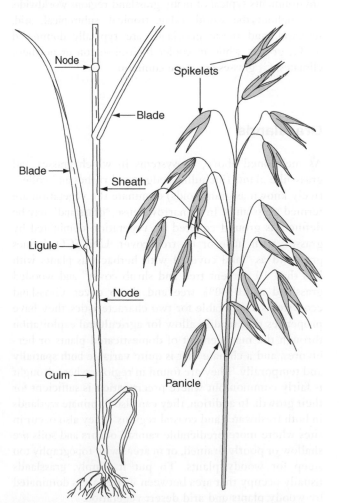

Figure 1 Common oat, *Avena sativa*, ×½. From Hubbard (1984).

high in grass. This also suggests that the grasses and their megaherbivore grazers are highly coevolved. But recent discovery of grass phytoliths in Late Cretaceous dinosaur coprolites in India suggest that grasses were already substantially differentiated and that abrasive phytoliths were present in many grasses before the explosion of grazers in the Oligocene and Miocene time periods.

Grasses show a very large variation in the way tillers are aggregated as they expand from their origin, but two general forms of grasses are recognized: bunch-forming (caespitose) and sod-forming (rhizomatous). This description captures the major features of the dominant grass species but there are some species and groups that deviate from this general pattern. The most obvious include the woody bamboos (some of which can reach tree size and for the most part are restricted to forest habitats in the tropics and subtropics).

In addition to growth form, grasses can also be roughly divided into two categories based upon their photosynthetic pathways: cool season (C_3) and warm season (C_4). C_4 photosynthesis is a variation on the typical C_3 pathway and is thought to have an advantage in high-light and -temperature environments typical of many grassland regions worldwide. Throughout the world today, tropical, subtropical, arid, semiarid, and mesic grasslands are typically dominated by C_4 grasses while in cooler high-elevation or northern climates, C_3 grasses are more common.

Grasslands

As mentioned above, ecosystems in which grasses and grass-like plants (including sedges and rushes and collectively known as graminoids) dominate the vegetation are termed grasslands. In its narrow sense, 'grassland' may be defined as ground covered by vegetation dominated by grasses, with little or no tree cover. UNESCO defines grassland as "land covered with herbaceous plants with less than 10 percent tree and shrub cover" and wooded grassland as 10–40% tree and shrub cover. Grassland ecosystems are notable for two characteristics: they have properties that readily allow for agricultural exploitation through the management of domesticated plants or herbivores, and a climate that is quite variable both spatially and temporally. They are found in regions where drought is fairly common but where precipitation is sufficient for their growth. In addition, they can also dominate wetlands in both freshwater and coastal regions. They also occur in sites where more predictable rainfall occurs and soils are shallow or poorly drained, or in areas with topography too steep for woody plants. To put it simply, grasslands usually occupy that area between wetter areas dominated by woody plants and arid desert vegetation.

Grassland biomes occur on every continent except Antarctica. It is estimated that grasslands once covered as much as 25–40% of the Earth's land surface although much of the original extent of native grassland has been plowed and converted to other grass production (corn and wheat) or other row crops such as soybeans. Indeed, grasslands are important from both agronomic and ecological perspectives. Grasslands are the basis of an extensive livestock production industry in North America and elsewhere. In addition, grasslands sequester and retain large amounts of soil carbon and thus, they are an important component of the global carbon cycle.

Indeed, because grasslands store a significant amount of carbon in their soils and they contain relatively high biodiversity, then now play a prominent role in the discussion about biofuel production. Biofuels may offer a mechanism to generate energy that releases less carbon into the atmosphere. Some energy producers recommend intensive agricultural production of corn, or other grasses such as switchgrass or elephant grass for biofuel production. However, agricultural practices have significant energy costs that may reduce the value of these fuel sources. A recent study has suggested, however, that diverse prairie communities on marginal lands are potentially 'carbon negative' because they provide significant biomass for fuel and store carbon belowground. Much additional research is needed to assess the sustainability of grasslands for biofuel production, but the prospects are certainly tantalizing to energy producers and conservationists alike.

Grassland Types

It is estimated that prior to the European settlement of North America, the largest continuous grasslands in the United States stretched across the Great Plains from the Rocky Mountains and deserts of the Southwestern states to the Mississippi river. Other extensive grasslands are, or were, found in Europe, South America, Asia, and Africa (**Figure 2**). Grasslands can be broadly categorized as temperate or tropical. Temperate grasslands have cold winters and warm to hot summers and often have deep fertile soils. Surprisingly, plant growth in temperate grasslands is often nutrient limited because much of the soil nitrogen is stored in forms unavailable for plant uptake. These nutrients, however, are made available to plants when plowing disrupts the structure of the soil. The combination of high soil fertility and relatively gentle topography made grasslands ideal candidates for conversion to crop production and thus have led to the demise of much of the grasslands across the world.

Grasslands in the Midwestern United States that receive the most rainfall (75–90 cm) are the most productive and are termed tallgrass prairies. Historically, these were most abundant in Iowa, Illinois, Minnesota, Missouri, and Kansas. The driest grasslands (25–35 cm of rainfall) and least productive

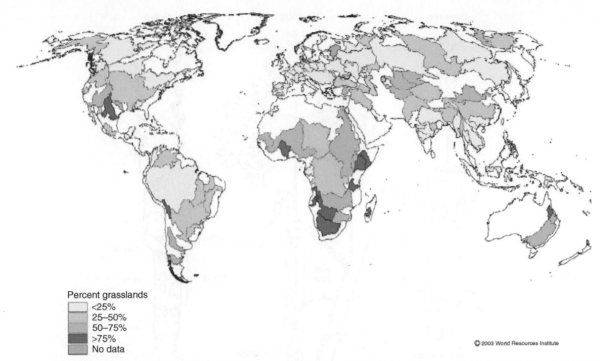

Percent grasslands
- <25%
- 25–50%
- 50–75%
- >75%
- No data

© 2003 World Resources Institute

Figure 2 Map of the grasslands of the world. World Resources Institute – PAGE, 2000. Sources: GLCCD, 1998. Loveland TR, Reed BC, Brown JF, *et al.* (1998) Development of a Global Land Cover Characteristics Database and IGBP DISCover from 1 km AVHRR Data. *International Journal of Remote Sensing* 21(6–7): 1303–1330. Available online at http://edcaac.usgs.gov/glcc/glcc.html. Global Land Cover Characteristics Database, Version 1. Olson JS (1994) Global Ecosystem Framework – Definitions, 39pp. Sioux Falls, SD: USGS EDC.

are termed shortgrass prairie or steppe. These grasslands are common in Texas, Colorado, Wyoming, and New Mexico. Grasslands intermediate between these extremes are termed mid- or mixed grass prairies. In tallgrass prairie, the grasses may grow to 3 m tall in wet years. In shortgrass prairie, grasses seldom grow beyond 25 cm in height. In all temperate grasslands, production of root biomass belowground exceeds foliage production aboveground. Worldwide, other names for temperate grasslands include steppes throughout most of Europe and Asia, veld in Africa, *puszta* in Hungary, and the pampas in South America.

Tropical grasslands are warm throughout the year but have pronounced wet and dry seasons. Tropical grassland soils are often less fertile than temperate grassland soils, perhaps due to the high amount of rainfall (50–130 cm) that occurs during the wet season and washes (or leaches) nutrients out of the soil. Most tropical grasslands have a greater density of woody shrubs and trees than temperate grasslands. Some tropical grasslands can be more productive than temperate grasslands. However, other tropical grasslands grow on soils that are quite infertile or these grasslands are periodically stressed by seasonal flooding. As a result, their productivity is reduced and may be similar to that of temperate grasslands. As noted for temperate grasslands, root production belowground far exceeds foliage production in all tropical grasslands. Other names for

tropical grasslands include velds in Africa, and the compos and llanos in South America.

Although temperate and tropical grasslands encompass the most extensive grass dominated ecosystems, grasses are present in most types of vegetation and regions of the world. Where grasses are locally dominant they may form desert (see Deserts) grassland, Mediterranean (see Mediterranean) grassland, subalpine and alpine grasslands (sometimes referred to as meadows or parks), and even coastal grassland. Most grasslands are dominated by perennial (long-lived) plants, but there are some annual grasslands in which the dominant species must reestablish each year by seed. Intensively managed, human-planted, and maintained grasslands (e.g., pastures, lawns) occur worldwide as well.

The Grassland Environment

Grassland climates can be described as wet or dry, hot or cold (typically in the same season), but on average are intermediate between the climates of deserts and forests. The climate of grasslands is best described as one of extremes. Average temperatures and yearly amounts of rainfall may not be much different from desert or forested areas, but dry periods during which the plants suffer from water stress occur in most years in both temperate and

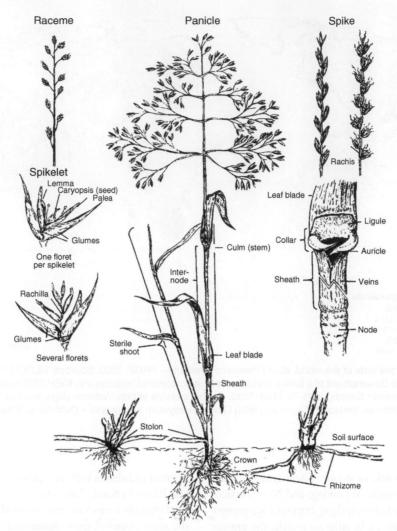

Figure 3   Structure and architecture of the grass plant. From Ohlenbusch *et al.* (1983).

tropical grasslands. An excellent example of this comes from North America, where in the area around Washington, DC (dominated by eastern deciduous forest), the annual precipitation is ~102 cm whereas at Lawrence, KS (dominated historically by tallgrass prairie), the annual precipitation is ~100 cm. But the way the rainfall is distributed is notably different. At Lawrence, KS, over 60% of the rainfall occurs in the growing season (April–September), whereas at Washington, DC, the precipitation is uniformly distributed throughout the year. The open nature of grasslands is accompanied by the presence of sustained high wind speeds. Windy conditions increase the evaporation of water from grasslands and this increases water stress in the plants and animals. Another factor that increases water stress is the high input of solar radiation in these open ecosystems. This leads to the convective uplift of moist air and results in intense summer thunderstorms. Rain falling in these intense storms may not be effectively captured by the soil and the subsequent runoff of this water into streams reduces the moisture available to grassland

plants and animals. In addition to periods of water stress within the growing season, consecutive years of extreme drought are more common in grassland than in adjacent forested areas. Such droughts may kill even mature trees, but the grasses and other grassland plants have extensive root systems and belowground buds that help them survive and grow after drought periods (**Figure 3**).

Fire in Grasslands

It is generally recognized that climate, fire, and grazing are three primary factors that are responsible for the origin, maintenance, and structure of the most extensive natural grasslands. These factors are not always independent (i.e., grazing reduces standing crop biomass which can be viewed simply as a fuel for fire, and biomass is also highly dependent upon the amount of precipitation). Historically, fires were a frequent occurrence in most large grasslands. Most grasslands are not harmed by fire, many benefit from fire, and some depend on fire for their existence. When grasses are dormant, the moisture content of the senesced foliage is low and this fine-textured fuel ignites easily and burns rapidly. The characteristic high wind speeds and lack of natural fire breaks in grasslands allow fire to cover large areas quickly. Because fire moves rapidly and much of the fuel is above the ground, temperatures peak rapidly and soil heating into the range that is biological damaging (>60 °C) occurs for only a short period of time and only at the surface or maybe a few centimeters into the soil. Thus, the important parts of the grasses (roots and buds) have excellent protection against even the most intense grass fires. Fires have been documented to be started by lightning and set intentionally by humans in both tropical and temperate grasslands. Fires are most common in grasslands with high levels of plant productivity, such as tallgrass prairies, and in these grasslands fire is important for keeping trees and adjacent forests from encroaching into grasslands. Many tree species are killed by fire, or if they are not killed, they are damaged severely because their active growing points are aboveground. Grassland plants survive and even thrive after fire because their buds are belowground where they are protected from lethal temperatures (**Figure 4**).

The response of grassland species to fire mostly depends upon the production potential of the grassland. In the more highly productive grasslands (e.g., tallgrass prairie), fire in the dormant season (usually right before the growing season) results in an increase in growth of the grasses and thus greater plant production or total biomass. This occurs because the buildup of dead biomass (detritus) from previous years inhibits growth; fire removes this layer. However, in drier grasslands, or even in years in productive grasslands when the precipitation is low, the burning of this dead plant material may cause the soil to become excessively dry due to high evaporation losses. As a result, plants become water-stressed and growth is

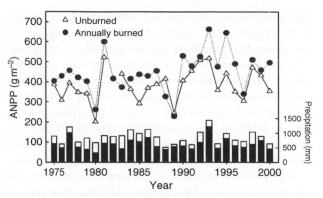

Figure 5 Long-term record (26 years) of aboveground net primary production (ANPP) at Konza Prairie Biological Field Station from unburned sites (clear triangles) and annually burned sites (solid circles). The growing season precipitation (April–September; solid bars) and annual precipitation (clear bars) is also shown.

reduced after fire, thus resulting in lower productivity. It is only with long-term data that the true impact of fires on grasslands can be determined (**Figure 5**).

So what are the mechanism(s) behind the increase in production in mesic grasslands after a fire? One of the most common misconceptions is that fire in grasslands increases productivity by increasing (releasing) the amount of nitrogen (N), a key limiting nutrient in terrestrial ecosystems. Actually, soil N decreases with burning. However, as mentioned above, the primary mechanism by which fire increases production in tallgrass prairie is through the removal of the accumulation of detritus produced in previous years. Standing dead biomass has been reported to accumulate to levels of up to $1000 \, \mathrm{g \, m^{-2}}$ in tallgrass prairie and a steady state is achieved *c*. 3–5 years after a fire. The specific effects of this blanket of dead biomass on production are numerous and manifest on individual through the ecosystem levels. This detritus may accumulate to >30 cm deep, and this nonphotosynthetic biomass shades the soil surface and emerging shoots. This reduction in light available to shoots in sites without fire occurs for up to 2 months and because soil moisture is usually high in the spring, loss of energy at this time is especially critical for primary production. In concert with reductions in light available to the grasses, the early spring temperature environment is much different between burned and unburned sites, with burned sites having a higher temperature favoring the dominant C_4 grasses. All of these factors result in less production in unburned tallgrass compared to annually burned prairie (**Figure 5**). Other evidence that fire does not increase N availability in mesic grasslands comes from N fertilization experiments. Within tallgrass prairie, in annually burned sites, N fertilizer had a strong impact on production, but in sites that have not been burned for several years, additional N did not enhance production and sites with

Figure 4 Photograph of a spring fire at the Konza Prairie Biological Field Station. The fire in the background is occurring ~2 weeks after the area in the foreground was burned. Photograph by Alan K. Knapp.

intermediate fire histories had intermediate responses to N fertilization. The results of many studies suggest that one generality regarding grasses and fire is that grasses tolerate fire extremely well and in most cases reach their maximum production in the immediate post-fire years. One qualification to this statement is that the beneficial effect of fire is not uniform across all precipitation gradients. In addition, the growth form type of the dominant grass is also very important. Highly productive grasslands on the high end of precipitation gradients show moderate to high positive response to burning whereas more arid grasslands and some bunchgrass grasslands show reduced productivity in the first few years after fire.

Most grasslands have an active growing season as well as a dormant season. Although fire can occur year-round in many grasslands, fire is most likely to occur during the dormant season and it is most rare in the middle of the growing season during normal (non drought) years. Given the fact that so many aspects of a grassland change during the yearly cycle, it seems fair to expect that a fire in different seasons would have dramatically different impacts. However, in spite of the many studies that have examined the impact of fires at different times of the year, there does not seem to be a general consensus on fire seasonality. Rather, it is probably best to say that grasslands seem somewhat sensitive to 'season of burn'. In one long-term study, it was found that the dominant grass in the tallgrass prairie (*Andropogon gerardii*) increased with burning in autumn, winter, or spring (dormant season), whereas burning in summer (growing season) resulted in an increase in many of the subdominant grasses with a reduction in *A. gerardii*.

Research indicates that community structure and ecosystem functioning in grasslands are impacted strongly by fire frequency. Plant species composition, in particular, differs dramatically between annually burned and less frequently burned sites in mesic grasslands. In tallgrass prairie, annually burned sites are dominated strongly by C4 perennial grasses. Although C4 grasses retain dominance at infrequently burned sites, C3 grasses, forbs, and woody species are considerably more abundant resulting in greater diversity and heterogeneity in unburned prairie. In fact, the flora on annually burned sites is a nested subset of that found on less frequently burned areas. Thus, the differences reflect shifts in dominance between frequently and infrequently burned sites, rather than difference in composition *per se*. Again as with response of production to fire, there appears to be a gradient of response in community structure to grassland fires. In more northern prairies of North America, burning has not been shown to strongly affect community structure. However, these northern grasslands are dominated by C3 grasses, which tend to decrease with burning, unlike the C4 grasses that dominate prairies in warmer climates. Thus, the role of competition and fire in structuring

grassland plant communities may increase along a latitudinal gradient throughout the Great Plains.

At a mesic grassland (Konza Prairie Biological Station), a clear picture of fire effects on plant community structure has emerged from the long-term (>20 years) empirical and experimental research done at the site. In the absence of large herbivores, the system is strongly driven by bottom-up forces associated with light, soil resource availability, and differential ability to compete under low-resource conditions. Although light availability increases with burning, the abundance of other critical limiting resources, N and water, declines as fire frequency increases. This is especially true in upland areas (with shallow soils) where production is likely limited by water. These changes in resource availability favor the growth and dominance of a small number of perennial C4 grasses and forbs. As dominance by these competitive species increases, general declines in plant species diversity and community heterogeneity occur.

Impact of Fire on Consumers

Direct effects

Most grassland animals are not harmed by fire, particularly if fires occur during the dormant season. Those animals living belowground are well protected, and most grassland birds and mammals are mobile enough to avoid direct contact with fire. For example, there were few differences in the kinds and abundances of ground-dwelling beetles in frequently and infrequently burned Kansas tallgrass prairie. Insects that live in and on the stems and leaves of the plants are the ones that are most affected by fire. Fire has been shown to reduce directly the abundance of caterpillars which means fewer butterflies, which are important pollinators, in frequently burned prairies. Fortunately, most natural fires are patchy in that many unburned areas remain throughout a larger burned area. These patches serve as refugia for many insect populations. Given that these animals have short generation times these refugia often allow insect populations to recover quickly following a fire.

Indirect effects

Given the distinct effects of fire frequency on plant community structure and dynamics within and among burning treatments, it seems plausible that consumers that depend on the primary producers for food and habitat structure will be indirectly affected because fire alters food availability and habitat structure. Given that fire usually homogenizes grassland plant communities, one would predict that this would hold true for consumers. However, there does not appear to be tight linkages between changes in vegetation composition and structure animal populations. Indeed, work in an Oklahoma prairie shows that more grassland birds occur in areas with patchy burns than in areas that are uniformly burned or not burned. Much more work on how fire affects habitat heterogeneity and grassland consumers communities is needed.

Grazing in Grasslands

Grazing is a form of herbivory in which most of the leaves or other plant parts (small roots and root hairs) are consumed by herbivores. Grazing, both above- and belowground, is an important process in all grasslands. The long association of grazers and grasslands has prompted the hypothesis that grasses and their megaherbivore grazers are a highly coevolved system, but, as mentioned above, there is some more recent evidence that this might not be the case. However, there is no disagreement that large grazers have been a factor in grassland ecology since their origin. The herbivory actions of many other smaller organisms including small mammals and insects may be equally important. There is no doubt that the impact of native grazers in grasslands can be extensive and work on the East African Serengeti plains estimated that 15% to >90% of the annual aboveground net primary productivity can be consumed by ungulates. However, data from small mammal exclosures suggest that small mammals can also impact grasslands as when small mammals were excluded from plots in Kenya; biomass was 40–50% higher than in adjacent plots where small mammals occurred.

Due to the ability of grasses to cope with high rates of herbivory, many former natural grasslands are now being managed for the production of domestic livestock, primarily cattle in North and South America and Africa, as well as sheep in Europe, New Zealand, and other parts of the world. Grasslands present a vast and readily exploited resource for domestic grazers. However, like many resources, grasslands can be overexploited (discussed in more detail below).

Grazing systems can be roughly divided into two main types – commercial and traditional – with the traditional type often mainly aimed at subsistence. Commercial grazing of natural grasslands is very often at a large scale and commonly involves a single species, usually beef cattle or sheep for wool production. Some of the largest areas of extensive commercial grazing developed in the nineteenth century on land which had not previously been heavily grazed by ruminants; these grazing industries were mainly developed in the Americas and Australia, and to a much less degree in southern and eastern Africa. Traditional livestock production systems vary according to climate and the overall farming systems of the area. They also use a wider range of livestock, including buffaloes, asses, goats, yaks, and camels. In traditional farming systems, livestock are often mainly kept for subsistence

and savings, and are frequently multipurpose, providing meat, milk, and manure as fuel.

Grazing aboveground by large herbivores alters grasslands in several ways. Grazers remove fuel and may lessen the frequency and intensity of fires. Most large grazers such as cattle or bison primarily consume the grasses; thus the less abundant forb species (broad-leafed, herbaceous plants) may increase in abundance and new species may invade the space that is made available. Thus, fire reduces heterogeneity in mesic grassland (a few species dominate) while grazers increase heterogeneity regardless of fire frequency. In other words, grazing decouples the impact of fire in productive grasslands (**Figure 6**). As a result, grazing increases plant species diversity in mesic grasslands. In xeric grasslands, on the other hand, grazing may lower species diversity, particularly by altering the availability of suitable microsites for forb species. These effects are strongly dependent on grazing intensity. Overgrazing may rapidly degrade grasslands to systems dominated by weedy and non-native plant species.

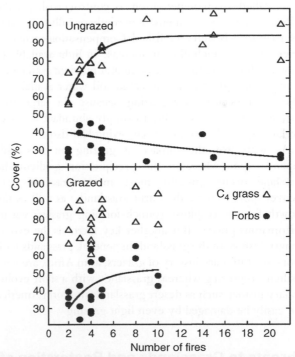

Figure 6 Aboveground biomass removal by large ungulates modulates plant community responses to fire in mesic grasslands. In ungrazed prairie (top), cover of dominant C_4 grasses increased with increasing fire frequency, while cover of forbs decreased, resulting in a loss of diversity. However, in prairie grazed by bison (bottom), the cover of forbs was positively correlated with fire frequency and the cover of grasses was unaffected, resulting in high diversity in spite of frequent fires. From Collins SL, Knapp AK, Briggs JM, *et al*. (1998) Modulation of diversity by grazing and mowing in native tallgrass prairie. *Science* 280(5364): 745–747.

Grazers may also accelerate the conversion of plant nutrients from forms that are unavailable for plant uptake to forms that can be readily used. Essential plant nutrients, such as nitrogen, are bound for long periods of time in unavailable (organic) forms in plant foliage, stems, and roots. These plant parts are slowly decomposed by microbes and the nutrients they contain are only gradually released in available (inorganic) forms. This decomposition process may take more than a year or two. Grazers consume these plant parts and excrete a portion of the nutrients they contain in plant-available forms. This happens very quickly compared to the slow decomposition process, and nutrients are excreted in high concentrations in small patches. Thus, grazers may increase the availability of potentially limiting nutrients to plants as well as alter the spatial distribution of these resources.

Some grasses and grassland plants can compensate for aboveground tissue lost to grazers by growing faster after grazing has occurred. Thus, even though 50% of the grass foliage may be consumed by bison or wildebeest, when compared to ungrazed plants at the end of the season, the grazed grasses may be only slightly smaller, the same size, or even larger than ungrazed plants. This latter phenomenon, called 'overcompensation' is controversial, yet the ability of grasses to compensate partially or fully for foliage lost to grazers is well established. Compensation occurs for several reasons, including an increase in light available to growing shoots in grazed areas, greater nutrient availability to regrowing plants, and increased soil water availability. The latter occurs after grazing because the large root system of the grasses is able to supply abundant water to a relatively small amount of regrowing leaf tissue.

As with fire, the impact of grazing on grasslands depends upon where in the precipitation gradient the grassland occurs (usually more mesic grasslands can recover more quickly than arid grasslands) as well as the growth form – cespitose (bunch-forming grasses) versus rhizomatous grasses. But another key factor is the evolutionary history of the grassland. In general, grasslands with a long evolutionary history of grazers, as in Africa, are very resilient to grazing whereas grasslands with a short evolutionary history such as desert grasslands in North America can easily be damaged by even light grazing.

Threats to Grasslands and Restoration of Grasslands

Grassland environments are key agricultural areas worldwide. In North America and elsewhere, grasslands are considered to be endangered ecosystems. For example, in US Great Plains up to 99% of native grassland ecosystems in some states have been plowed and converted to agricultural use or lost due to urbanization. Similar but less dramatic losses of mixed and shortgrass prairies have

occurred in other areas. While the loss of native grasslands due to agricultural conversion is still occurring in some places, dramatic increases in woody shrub and tree species threatens many remaining tracts of grasslands. Indeed, across the world, the last remaining native grasslands are being threatened by an increase in the abundance of native woody species from expansion of woody plant cover originating from both within the ecosystem and from adjacent ecosystems. Increased cover and abundance of woody species in grasslands and savannas have been observed worldwide with well-known examples from Australia, Africa, and South America. In North America, this phenomenon has been documented in mesic tallgrass prairies of the eastern Great Plains, subtropical grasslands and savannas of Texas, desert grasslands of the Southwest, and the upper Great Basin. Purported drivers of the increase in woody plant abundance are numerous and include changes in climate, atmospheric CO_2 concentration, nitrogen deposition, grazing pressure, and changes in disturbance regimes such as the frequency and intensity of fire. Although the drivers vary, the consequences for grassland ecosystems are strikingly consistent. In many areas, the expansion of woody species increases net primary production and carbon storage, but reduces biodiversity. The full impact of shrub encroachment on grassland environments remains to be seen.

Another threat to native grasslands is the increase of nonnative grass species. For example, in California, it is estimated that an area of approximately 7 000 000 ha (about 25% of the area of California) has been converted to grassland dominated by non-native annuals primarily of Mediterranean origin. Conversion to non-native annual vegetation was so fast, so extensive, and so complete that the original extent and species composition of native perennial grasslands is unknown. In addition, across the western United States, invasive exotic grasses are now dominant in many areas and these species have a significant impact on natural disturbance regimes. For example, the propensity for annual grasses to carry and survive fires is now a major element in the arid and semiarid areas in western North America. In the Mojave and Sonoran deserts of the American Southwest, in particular, fires are now much more common than they were historically, which may reduce the abundance of many native cactus and shrub species in these areas. This annual-grass-fire syndrome is also present in native grasslands of Australia and managers there and in North America are using growing season fire to try to reduce the number of annual plants that set seed and thus reduce the populations of exotics, usually with very mixed results.

Conservation and Restoration

Because grasslands have tremendous economic value as grazing lands and also serve as critical habitat for many plant and animal species, efforts to conserve the remaining grasslands and restore grasslands on agricultural land are underway in many states and around the world. The most obvious conservation practice is the protection and management of existing grasslands. This includes both private and public lands. Probably the largest private holder of grasslands in the world is The Nature Conservancy. The Nature Conservancy is a global organization that works in all 50 states in the United States of America, and in 27 countries, including Canada, Mexico, Australia, and countries throughout the Asia-Pacific region, the Caribbean, and the Latin America.

However, as mentioned numerous times, the factors that led to the establishment of grasslands and, in particular, the organic-rich soils derived from the dominant biota have facilitated the agricultural exploitation of grasslands. Consequently, many grasslands that were historically persistent have been converted to cropland. Thus, restoration of grasslands is also a very important conservation practice. Grassland restoration is the process of recreating grassland (including plant and animal communities, and ecosystem processes) where one existed but now is gone. Grassland restoration can include planting a new grassland where one had been broken and farmed, or it can include improving a degraded grassland (e.g., one that was never plowed but lost many plant and animal species due to prior land management practices). Restoration practices of existing grasslands may include reintroducing fires into grasslands following extended periods of fire suppression. On areas that have been moderately to heavily grazed (but not completely overgrazed), reducing the intensity of grazing may be required. In addition, mowing is also a cost-effective method of restoring grasslands. Mowing can be effective on sites that have been invaded by brush and forest, but the grasses are still present.

In areas where the grasses are completely absent (agriculture fields) or in a very degraded state, reseeding of grasses is usually necessary. There are proven techniques, complete with specialized equipment (seed drills) for restoration of grasslands, and, for the most part, it is fairly easy to get the dominant grasses established in an area. Indeed, some of the earliest examples of restoration ecology come from efforts to restore native tallgrass prairie in North America. As a result, the market for restoration of grasslands (at least in North America) has developed to the point that obtaining enough grass seed (sometimes even local native seed) is not a problem. A bigger challenge, however, in restored grasslands is increasing establishment of the nongrass species which are so critical for biodiversity. Seeds may be more difficult to obtain (especially for rarer plants), and then getting the forbs to survive and reproduce in many grassland restoration projects has been challenging. Further research is needed regarding what management techniques are important to their establishment and growth in these restored areas.

In addition to the prairie flora that is at risk, grassland animals (particularly birds and butterflies) suffer when grassland quality declines. In North America, grassland birds were historically found in vast numbers across the prairies of the western Great Plains. Today, the birds of these and other grasslands around the world have shown steeper, more consistent, and more geographically widespread declines than any other group. These losses are a direct result of the declining quantity and quality of habitat due to human activities like conversion of native prairie to agriculture, urban development, and suppression of naturally occurring fire.

See also: Agriculture Systems; Savanna.

Further Reading

Borchert JR (1950) The climate of the central North American grassland. *Annals of the Association of American Geographers* 40: 1–39.

Briggs JM, Knapp AK, Blair JM, et al. (2005) An ecosystem in transition: Woody plant expansion into mesic grassland. *BioScience* 55: 243–254.

Collins SL, Knapp AK, Briggs JM, Blair JM, and Steinauer EM (1998) Modulation of diversity by grazing and mowing in native tallgrass prairie. *Science* 280(5364): 745–747.

Collins SL and Wallace LL (1990) *Fire in North American Tallgrass Prairies*. Norman, OK: University of Oklahoma Press.

Frank DA and Inouye RS (1994) Temporal variation in actual evapotranspiration of terrestrial ecosystems: Patterns and ecological implications. *Journal of Biogeography* 21: 401–411.

French N (ed.) (1979) *Perspectives in Grassland Ecology. Results and Applications of the United States International Biosphere Programme Grassland Biome Study*. New York: Springer.

Knapp AK, Blair JM, Briggs JM, et al. (1999) The keystone role of bison in North American tallgrass prairie. *BioScience* 49: 39–50.

Knapp AK, Briggs JM, Hartnett DC, and Collins SL (1998) *Grassland Dynamics: Long-Term Ecological Research in Tallgrass Prairie*, 364pp. New York: Oxford University Press.

Loveland TR, Reed BC, Brown JF, et al. (1998) Development of a Global Land Cover Characteristics Database and IGBP DISCover from 1 km AVHRR Data. *International Journal of Remote Sensing* 21(6–7): 1303–1330.

McNaughton SJ (1985) Ecology of a grazing ecosystem: The serengeti. *Ecological Monographs* 55: 259–294.

Milchunas DG, Sala OE, and Lauenroth WK (1988) A generalized model of the effects of grazing by large herbivores on grassland community structure. *American Naturalist* 132: 87–106.

Oesterheld M, Loreti J, Semmartin M, and Paruelo JM (1999) Grazing, fire, and climate effects on primary productivity of grasslands and savannas. In: Walker LR (ed.) *Ecosystems of the World*, pp. 287–306. Amsterdam: Elsevier.

Olson JS (1994) Global Ecosystem Framework – Definitions, 39pp. Sioux Falls, SD: USGS EDC.

Prasad V, Strömberg CAE, Alimohammadian H, and Sahni A (2005) Dinosaur coprolites and the early evolution of grasses and grazers. *Science* 310: 1177–1190.

Sala OE, Parton WJ, Joyce LA, and Lauenroth WK (1988) Primary production of the central grassland region of the United States. *Ecology* 69: 40–45.

Samson F and Knopf F (1994) Prairie conservation in North America. *BioScience* 44: 418–421.

Weaver JE (1954) *North American Prairie*. Lincoln, NE: Johnsen Publishing Company.

Swamps

C Trettin, USDA, Forest Service, Charleston, SC, USA

Published by Elsevier B.V.

Introduction
General Properties of a Swamp
Ecological Functions

Restoration
Ecosystem Services and Values
Further Reading

Introduction

Swamp is a general term that is defined as "spongy land, low ground filled with water, soft wet ground" (Webster, 1983), hence its association with a wide variety of terrestrial ecosystems. Typically, a swamp is considered a forested wetland. A wetland is a type of terrestrial ecosystem that has a hydrologic regime where the soil is saturated near the surface during the growing season; the soil has hydric properties, expressing characteristics of anaerobic conditions; and the dominant vegetation is hydrophytic with adaptations for living in the wet soils. In the case of a swamp, the forest species are adapted to the wet soil conditions. Without geographic context, there is little functional information conveyed by the term swamp other than the prevalence of wetland conditions and dense forest vegetation (**Figure 1**).

The following discussion is designed to convey the common hydrologic settings, soil conditions, and vegetative communities that occur within the common usage of the term swamp. References focusing on swamp forests should be consulted for specific geographic regions.

General Properties of a Swamp
Hydrology

The hydrologic setting controls the form and function of the wetland because of the dependence on excess water to mediate biological and geochemical reactions. There are four general settings that may be used to characterize the swamp hydrology (**Figure 2**). The riverine or floodplain setting is the most commonly associated hydrogeomorphic setting for swamps. It is characterized by periodic flooding from the river or stream, and it may also receive runoff from adjoining uplands. The periodicity, and flood depth and duration are the key factors that affect the type of forest communities present in the swamp. Depressional wetlands occur where there are surface depressions which receive water from the surrounding uplands, directly from precipitation, and in certain instances, they may also intersect a shallow water table. Lacustrine and estuarine fringe wetlands

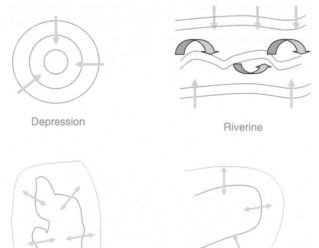

Figure 2 Influence of geomorphic position on hydrology. The arrows show the dominant direction of water flow for the four dominant types of geomorphic positions that are characteristic for swamps. After Vasander H (1996) Peatlands in Finland, 64pp. Helsinki: Finnish Peatland Society.

Figure 1 Bottomland hardwood swamp, characteristic of floodplains in the Southeastern United States.

receive their water primarily from an open-water body; runoff from adjoining uplands and precipitation also contribute to the water balance. The common hydrologic attribute of swamps in each of those settings is the presence of water above the soil surface, but the period of inundation varies widely. While it is common for swamps to have flooded conditions for periods ranging from days to months on an annual basis, it is not uncommon for there to be multiyear intervals between flood events. The factors that affect the flooding regime include timing and amount of precipitation, groundwater level, land use in the watershed contributing to the swamp, and evapotranspiration.

Soils

Swamp soils cover the full range of texture classes and degrees of organic matter accumulation (**Figure 3**). The wet mineral soils are characteristic of riverine and depressional settings. The histic mineral soils have a moderately thick accumulation of surface organic matter (<40 cm) reflecting prolonged periods of saturation and little scouring action if located in a floodplain, hence they may be found in any of the four hydrologic settings. The histosols or peat soils have a thick layer (>40 cm) of organic matter accumulation, representing the long periods of saturation on an annual basis. These soils typically occur in depressional settings and are not common in floodplains due to the periodic scouring that occurs during flood events.

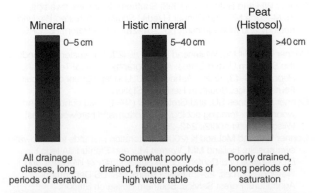

Figure 3 Types of soils common in swamp forests. The three categories reflect the amount of organic matter that has accumulated on the soil surface, which is in turn controlled by the soil drainage and hydrology. After Trettin CC, Jurgensen MF, Gale MR, and McLaughlin JA (1995) Soil carbon in northern forested wetlands: Impacts of silvicultural practices. In: McFee WW and Kelly JM (eds.) *Carbon Forms and Functions in Forest Soils*, pp. 437–461. Madison, WI: Soil Science Society of America.

Vegetation

The term swamp generally implies a forested wetland. However, due to the wide range of physical settings (see the previous two sections) and geographic locations ranging from the boreal to the tropical climatic zones on each continent, there are no consistent characteristics or attributes beyond the occurrence of hydrophytic vegetation. Accordingly, swamps may be dominated by either conifers or angiosperms, but a common situation would be a mixture of species and communities reflecting relatively minor differences in a microsite. For example, while a floodplain forest may be broadly characterized as a bottomland hardwood swamp, it contains a mosaic of vegetative communities which reflect small differences in hydrology and soils.

Ecological Functions

The ecological functions of swamps are significant, because of their prevalence and the wide range of conditions that they occupy. The following overview highlights some of the major ecosystem functions that are provided from swamp wetlands; specifics for a particular type of swamp are available from the regional references.

Hydrology

Hydrologic functions that are mediated by swamp wetlands depend on the hydrogeomorphic setting. Riverine swamps provide temporary storage for floodwaters, thereby reducing the peak flow to downstream areas. This function is physically based, with little interaction with the type of forest vegetation. However, changes in land use, especially conversion to agriculture, in the floodplain, may reduce the water storage potential, resulting in enhanced downstream conveyance of flow. The flood storage function also serves to sustain stream flow, as the waters slowly drain from the area. Swamps occurring in a depressional setting may be a source of groundwater recharge, where accumulated surface water slowly infiltrates through the subsurface sediments. In estuarine and lacustrine settings, swamps occurring at the land–water margin are important for the stability of the shoreline.

Water quality

The effects of a swamp on water quality depend on the hydrogeomorphic setting. The riverine swamp affects water quality in two primary ways – by physical and biogeochemical reactions. Sediment removal is an important function of the riverine swamps; this is a process where sediment in the floodwaters settles out onto the floodplain surface. The deposited sediment provides nutrients to the swamp vegetation and it represents the removal of a contaminant from the floodwater. Floodplains with dense understory vegetation can be more effective than open forest settings in filtering sediment from the floodwaters.

The floodplain and riparian zone swamps may also remove chemical constituents from the water, particularly nitrogen and phosphorus. As a result of the anaerobic soil conditions, nitrate nitrogen, which is a common pollutant in surface and shallow-subsurface runoff, can be converted to nitrogen gas, thereby removing it from the water. The removal of phosphorus compounds typically involves reactions associated with the sediments.

Habitat

Swamps are important for the diversity of habitat conditions that they provide. At the large scale, swamps comprise part of the mosaic of land types, yielding wet, vegetative conditions among uplands. At smaller scales, within a swamp, there are a multitude of habitat conditions that are largely dictated by elevation relative to the mean high water level.

Terrestrial

The terrestrial habitats provided by swamps are diverse due to variations in vegetative composition and structure, which are largely regulated by the hydrologic conditions of the site. The habitat also changes through the development of the forest. In early successional stages, the vegetation is typically a dense combination of shrubs and trees; then, as the trees gain dominance, the shrub layers die back yielding a less dense understory. Correspondingly, the habitat conditions for amphibians, birds, reptiles, and mammals change as the stand evolves. The swamp forests are particularly important habitat for birds, especially migratory song birds.

Aquatic

Swamps also provide important aquatic habitat for fish, birds, and amphibians. Organic matter produced in the swamp is an important energy source for aquatic organisms, including those living in water bodies within the swamp and also larger receiving bodies such as lakes, rivers, and oceans. In floodplains, the floating debris and logs provide physical structures that are an important component of the aquatic habitat.

Restoration

In many areas, swamps have been converted into agricultural use, through the use of drainage systems and clearing of the forest vegetation. The merits of restoring the converted wetlands back to swamp forests include the reestablishment of flood water storage, in the case of floodplains, and the development of wildlife habitat. The restoration of swamp forests is complicated by the myriad of soil and hydrologic conditions that one may encounter, and the effects of past management practices which necessitate the restoration may also exacerbate the situation. However, with proper consideration of the hydrologic setting and matching species to the soil and water regimes, functional restoration is feasible. The typical sequence of restoring swamp forests is to reestablish the wetland hydrology by blocking drainage ditches, and planting appropriate tree and understory species.

Ecosystem Services and Values

Swamps provide both direct and indirect values to society. Direct values include raw materials, such as timber and food stocks. Indirect values include floodwater storage, water supply, water quality, recreation, esthetics, wildlife diversity, and biodiversity. The valuation will depend on inherent characteristics of the resource that are largely constrained by the biogeographic zone and location within a watershed, societal norms, and economic conditions.

Further Reading

Barton C, Nelson EA, Kolka RK, *et al.* (2000) Restoration of a severely impacted riparian wetland system – The Pen Branch Project. *Ecological Engineering* 15: S3–S15.

Burke MK, Lockaby BG, and Conner WH (1999) Aboveground production and nutrient circulation along a flooding gradient in a South Carolina Coastal Plain forest. *Canadian Journal of Forest Research* 29: 1402–1418.

Conner WH and Buford MJ (1998) Southern deepwater swamps. In: Messina MG and Conner H (eds.) *Southern Forested Wetlands Ecology and Management*, pp. 261–287. Boca Raton, FL: CRC Press.

Conner WH, Hill HL, Whitehead EM, *et al.* (2001) Forested wetlands of the Southern United States: A bibliography. *General Technical Report SRS-43*, 133pp. Asheville, NC: US Department of Agriculture, Forest Service, Southern Research Station.

Conner RN, Jones SD, and Gretchen D (1994) Snag condition and woodpecker foraging ecology in a bottomland hardwood forest. *Wilson Bulletin* 106(2): 242–257.

Conner WH and McLeod K (2000) Restoration methods for deepwater swamps. In: Holland MM, Warren ML, and Stanturf JA (eds.) *Proceedings of a Conference on Sustainability of Wetlands and Water Resources*, 23–25 May. Oxford, MS: US Department of Agriculture, Forest Service, Southern Research Station.

de Groot R, Stuip M, Finlayson M, and Davidson N (2006) Valuing wetlands: Guidance for valuing the benefits derived from wetland ecosystem services. *Ramsar Technical Report No. 3, CBD Technical Series No. 27*, Convention on Biological Diversity. Gland, Switzerland: Ramsar Convention Secretariat. http://www.cbd.int/doc/publications/cbd-ts-27.pdf (accessed November 2007).

Messina MG and Conner WH (eds.) (1998) *Southern Forested Wetlands Ecology and Management*, 347pp. Boca Raton, FL: CRC Press.

Mitch WJ and Gosselink JG (2000) *Wetlands*, 920pp. New York: Wiley.

National Wetlands Working Group (NWWG) (1988) *Wetlands of Canada. Ecological Land Classification Series, No. 24*, 452pp. Ottawa: Sustainable Development Branch, Environment Canada.

Stanturf JA, Gardiner ES, Outcalt K, Conner WH, and Guldin JM (2004) Restoration of southern ecosystems. In: *General Technical Report SRS–75*, pp. 123–11. Asheville, NC: US Department of Agriculture, Forest Service, Southern Research Station.

Trettin CC, Jurgensen MF, Gale MR, and McLaughlin JA (1995) Soil carbon in northern forested wetlands: Impacts of silvicultural practices. In: McFee WW and Kelly JM (eds.) *Carbon Forms and Functions in Forest Soils*, pp. 437–461. Madison, WI: Soil Science Society of America.

Vasander H (1996) *Peatlands in Finland*, 64pp. Helsinki: Finnish Peatland Society.

Webster N (1983) *Unabridged Dictionary*, 2nd edn. Cleveland, OH: Dorset and Baber.

Relevant Websites

http://www.aswm.org – Association of State Wetland Managers.

http://www.ncl.ac.uk – Mangrove Swamps WWW Sites, Newcastle University.

http://www.ramsar.org – Ramsar Convention on Wetlands.

http://www.sws.org – Society of Wetland Scientists.

http://www.epa.gov – Wetlands at US Environmental Protection Agency.

http://www.wetlands.org – Wetlands International.

http://www.panda.org – World Wildlife Fund.

Temperate Forest

W S Currie and K M Bergen, University of Michigan, Ann Arbor, MI, USA

Introduction
Physiography, Climate, and the Temperate Forest Biome
Disturbance and Forest Structure
Ecological Communities and Succession

Water and Energy Flow, Nutrient Cycling, and Carbon Balance
Temperate Forest Land Cover
Further Reading

Introduction

The temperate forest biome is characterized by a distinct seasonality that includes a long growing season together with a cold winter season in which much of the vegetation may be dormant. The strong seasonality drives physiological events to occur at regular annual intervals for plant species. These include bud break, flowering, and foliar and shoot extension. As the growing season ends, marked by dropping temperatures and shortening photoperiod (day length), trees and shrubs undergo seasonal physiological changes that include the senescence and abscission of foliage (although in evergreen species some foliage is also retained) and the setting of buds for the next growing season. Because of the cold winters, the dominant woody vegetation is characterized by freeze-hardy species. During the winter season, the air temperature drops below freezing and soils are frozen or cold and wet, impeding decomposition of plant litter and promoting the accumulation of an organic layer on the soil surface.

The temperate forest is distributed over portions of five regions of the globe: North America, South America, Europe, Asia, and Australia–New Zealand (**Figure 1**). Within this biome, distinct biogeographic units are recognized, particularly the mixed-deciduous temperate forest (the largest in terms of area), the mixed-evergreen

temperate forest (sometimes called subtropical evergreen), and the temperate rainforest. Major taxa include pines (*Pinus* spp.), maples (*Acer* spp.), beeches (*Fagus* spp., *Nothofagus* spp.), and oaks (*Quercus* spp.) in the mixed-deciduous and mixed-evergreen temperate forests; spruces (*Picea* spp.), Douglas-fir (*Pseudotsuga menziesii*), and redwoods (*Sequoia sempervirens, Sequoiadendron giganteum*) in the Northern Hemisphere temperate rainforests; and southern beeches (*Nothofagus* spp.) and eucalyptus (*Eucalyptus* spp.) in Southern Hemisphere temperate rainforests.

Within a continent, forests in the temperate biome grade into subdivisions based on latitude, elevation, and large-scale patterns of precipitation. In North America, for example, the predominant natural vegetation in the eastern United States and the southern reaches of eastern Canada is mixed-deciduous temperate forest. This forest grades to the south through broad-leaved-coniferous mixtures to the mixed-evergreen forest along the Atlantic coastal plain (**Figure 2**). Temperate rainforests in North America are found in the coastal Pacific Northwest where marine climates together with orographic lifting produce high rainfall. In South America temperate forests are found in Chile and parts of Patagonia. In Europe, within the temperate forest biome the mixed-deciduous forests dominate in the western

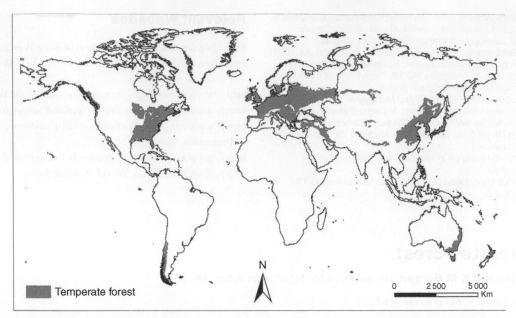

Figure 1 The distribution of temperate forests of the world. Map data: Olson DM, Dinerstein E, Wikramanayake ED *et al.* (2001) Terrestrial ecoregions of the world: A new map of life on earth, *BioScience* 51: 933–938. Map prepared by the Environmental Spatial Analysis Laboratory, University of Michigan, USA, 2006.

Figure 2 The edge of a mixed coniferous–deciduous forest in southeastern Maine, USA. Photo by W. S. Currie.

continent, Great Britain, southern Eastern Europe, and southern European Russia. In Near-East Asia, the temperate forest occurs in Turkey and Iran and a narrow band is found in Central Asia as a transition between the boreal forest to the north (see Boreal Forest) and steppe to the south. The temperate forests in East Asia occur predominantly in northern and central China, but also over most of Japan, Korea, and of the southern tip of Siberia. Temperate forests, including rainforests, are also found in parts of New Zealand, the southeast coast of Australia, and Tasmania.

Temperate forests are distinguished from boreal forests by having a 4–6-month (140–200 days) frost-free growing season with on average at least 4 months at 10 °C or above and mean annual temperatures from 5 °C to 20 °C. At higher latitudes, the temperate forest transitions to the boreal forest (see Boreal Forest), a biome of evergreen cold-tolerant forests with much shorter growing seasons. The latter are also found in middle latitudes as montane forests at high elevations and are often closer, floristically and functionally, to boreal than to temperate forests.

The occurrence of frost (at 0 °C or colder) differentiates the extratropical (including temperate) from tropical regions (see Tropical Rainforest). Moisture also distinguishes temperate regions from drier forested regions, such as chaparral (see Chaparral) and wetter forested regions such as tropical rainforests (see Tropical Rainforest). In temperate regions, precipitation exceeds potential evaporation and water is available at approximately 50–200 cm yr^{-1}. Precipitation in most temperate regions is fairly evenly distributed throughout the year in contrast to the tropics where there are typically pronounced wet and dry seasons.

Physiography, Climate, and the Temperate Forest Biome

Climatic and Physiographic Controls on the Distribution of Temperate Forests

The geographic distributions of the different vegetation biomes of the world are dependent on the physical environment and climate in the form of light, temperature, and moisture. In middle latitudes (30°–60° N and S), these controls result in a temperate forest biome within each hemisphere that is discontinuous, separated by the oceans and the tropics, and by moisture and physiographic barriers. The present-day distribution of temperate forests derives not only from present climatic controls but also from paleoclimates and past connections among the continents. Climates during the Pleistocene (c. 1.8 million to 10 000 years ago) set the stage for the present-day distribution. During glacial maxima, ice sheets covered large parts of Europe and North and South America as well as isolated areas in East Asia. In North and South America, plants migrated to unglaciated refugia and re-migrated, as glaciers receded, to their present-day distributions. Evidence suggests that many genera of forest trees that remain in North America and unglaciated East Asia were extirpated from Europe because the east–west running Alpine range blocked migrations to refugia during Pleistocene glaciations. Similarly important were continental connections between North America (the Nearctic), East Asia, and Europe (the Palearctic) at different points in geologic history. As a result, floristic differences are relatively small across the Holarctic, which spans from the west coast of North America to the east coast of Asia and includes the majority of the world's temperate forests.

Temperate forests occur across a wide range of local physiographic landforms, from rocky slopes to rolling plains and river floodplains, although generally under non-extreme physiographic conditions. Trees that occupy slopes or well-drained substrates with low organic matter such as sandy outwash plains (e.g., pines and some oaks) are adapted for drier (xeric) sites low in nutrients. Trees

adapted for moderate (mesic) sites are found on plains, glacial moraines, or low hills with greater stocks of soil organic matter. Nutrient- and moisture-demanding broad-leaved species, for example, maples and beeches, thrive in mesic landscapes. Trees occupying river floodplains, wetlands, or bogs have environments that can be very moist to wet (wet-mesic to hydric). These soils are relatively rich in organic matter but trees in these landscapes must be adapted to withstand flooding, including long periods with wet, anoxic soils with low nutrient availability.

Climatic and Physiographic Subdivisions

Given the great geographic extent of temperate forests it is not surprising that regional differences are observed. Systematic classifications of ecoregions and climates describe subdivisions within the biome (**Table 1**). The extensive temperate mixed-deciduous forest occurs primarily in Bailey's warm continental division (210), hot continental division (220), and marine division (240); these are Köppen–Trewartha classes Dcb, Dca, Do, and Cf. The warm continental division has snowy cold winters, while the hot continental division has warmer, wetter summers and milder winters. In the marine division (240) winters are mild, summers relatively cool, and precipitation occurs most of the year.

The temperate mixed-evergreen forests occur primarily in Bailey's temperate and rainy subtropical division (230) which is most analogous to the Köppen–Trewartha mid- and lower-latitude Cf (humid

Table 1 Temperate forest biome types and corresponding geographic regions, Bailey ecoregions, and Köppen–Trewartha climate classes

Temperate forest type	Geographic region	Bailey ecoregion	Köppen–Trewartha[a] climate class
Temperate mixed-deciduous forest	• Eastern North America • Asia • Europe • South America • Australia/New Zealand	Humid temperate domain (200) • Warm continental division (210) • Hot continental division (220) • Marine division (240)	Dcb: Temperate continental, cool summer Dca: Temperate continental, warm summer Do: Temperate oceanic Cf: Humid subtropical
Temperate mixed-evergreen forest	• Southeast North America • Asia • South America • Australia/New Zealand	Humid temperate domain (200) • Subtropical division (230)	Cf: Humid subtropical
Temperate rainforest	• Northwestern North America • South America • Southeast Australia/New Zealand	Humid temperate domain (200) • Marine division (240)	Cf: Humid subtropical Do: Temperate oceanic

[a]Dc: Temperate continental: 4–7 months above 10 °C, coldest month below 0 °C; Cf: Humid subtropical: 8 months 10 °C, coldest month below 18 °C, no dry season; Do: Temperate oceanic: 4–7 months above 10 °C, coldest month above 0 °C.

subtropical) class (**Table 1**). These climates have no dry season, with even the driest months having at least 30 mm of rain, and have hot summers with the average temperature of warmest month greater than 22 °C.

Temperate rainforest conditions largely occur where ocean moisture is abundant and prevented from moving inland by mountain ranges. These conditions occur in particular continental placements within Bailey's marine division (240) and Köppen–Trewartha *Do* class in higher latitudes and within Bailey's subtropical division (230) and Köppen–Trewartha *Do* and *Cf* classes in lower latitudes (**Table 1**).

Disturbance and Forest Structure

Major disturbances occur naturally in temperate forests, although particular locations vary in the types, frequencies, and severities of disturbance. Major natural disturbances include fires, windthrow during severe storms, ice storms, flooding, disease, and irruptions of defoliating or wood-boring insects. The array of natural disturbances that occur at a particular location constitutes its disturbance regime, a strong force in shaping forest structure and composition. Smaller-scale disturbances also shape forests over long time periods in the absence of a major disturbance. These include the production of forest gaps from the mortality of one to a few large trees. In some cases, idiosyncratic combinations of processes may produce repeated disturbance. An example is 'fir waves' that occur only in Japan and the northeastern US. In these waves of mortality that pass through the forest repeatedly, a fungal pathogen weakens the roots in mature trees while wind gusts cause the weakened roots to break as they rub against sharp gravel in the rocky soil. Because of the repetitive nature of natural disturbances and the long lifetimes of temperate forest trees, trees are often adapted (through what is termed 'vital attributes') to withstand particular disturbances or to regenerate following disturbance. Some examples are trees that re-sprout from stumps following fire or from branches following windthrow, cones that require fire to open, and seeds that germinate best on exposed soil.

Human activities have substantially altered the disturbance regimes in many temperate forests. The large-scale harvesting of trees for timber, whether cutting selected sizes or species of trees or cutting all of the trees in a stand, are relatively new forms of disturbance that now affect forest structure and community composition throughout much of the temperate biome. Human activities also cause large-scale chronic disturbances, including polluted rainfall (e.g., acid rain) that causes soil acidification and nitrogen enrichment over large regions of the US, Western Europe, and increasingly in eastern Asia. Still another category of human-induced disturbance is in the introduction of invasive species. In the eastern US, the introduction of a fungal pathogen in the early twentieth century caused the chestnut blight, essentially eradicating one of the dominant trees (the American chestnut, *Castanea dentata*) from a large region.

Structural Layers of Vegetation

Disturbances in temperate forests vary not only in their type and frequency but also in their intensity or severity, the latter gauged by the percentage of vegetation mortality. A major disturbance that causes widespread or near total mortality of trees in a forest stand, followed by the development of a new (secondary) forest stand, is known as a stand-initiating event. Following such an event, but mediated by the occurrence and severities of subsequent disturbances, the vertical structure of a forest stand tends to grow more complex over time. More favorable site conditions such as organic-rich, fertile soils and ample moisture also promote structural complexity. With full development, the vertical structure includes a canopy overstory, understory, a shrub layer, and an herbaceous layer. In achieving such development the forest passes through several stages. These include a stand initiation stage in which seedlings and saplings dominate and new species may continue to arrive; a stem exclusion stage in which the canopy closes, shading out shorter individuals; an understory re-initiation stage in which shade-tolerant species grow as seedlings and saplings; and finally an old-growth or steady-state stage. In the old-growth stage, the overstory typically includes both canopy dominants and subdominants (the latter with crowns only partially in sunlight) together with understory and shrub layers made up of mature, shade-tolerant individuals. Old-growth stands can be identified through a few key characteristics, including a distribution of age and size classes of trees, the absence of saw-cut stumps, and the presence of decaying logs the size of overstory trees.

The understory in a structurally complex temperate forest stand comprises trees and shrubs that spend their entire life cycle there as well as young or suppressed individuals of potential canopy-dominant species. Understory-tolerant species are those that can survive in, or even require, the shade of a forest canopy (e.g., sugar maple, *Acer saccharum*). In old-growth stands or those not recently disturbed it is common to see shade-tolerant species in both the understory and overstory because the overstory trees are those that regenerated in the shade of the canopy. Some temperate forests have a dense layer of understory shrubs, for example *Kalmia* spp., *Rhododendron* spp., and *Vaccinium* spp. (blueberry). The herbaceous layer of a temperate forest commonly contains mosses, lichens, vines, and forbs. Many shrubs and herbs are adapted to low-light environments or grow before canopy leaf extension in the spring or after overstory leaf abscission in fall; in summer only about 10% of full sunlight reaches the herbaceous layer, but this figure can rise to 70% in deciduous stands in winter. Shrubs and herbs that require more light grow in well-lighted gaps or extend their crowns into openings. Vines grow into forest canopies to access light and may be plentiful following a disturbance that kills canopy trees but leaves the dead trees standing.

Soils and Woody Debris

Soils provide a physical rooting medium, the capacity to store and release water, and the capacity to store and release nutrients for growing trees. The soils of the temperate forest regions occur in five orders of the system of soil taxonomy, namely Spodosols, Alfisols, Ultisols, Entisols, and Inceptisols. They range from somewhat infertile (Spodosols) to quite fertile (Alfisols). Spodosols are characterized by a heavily leached surface mineral horizon and a deeper accumulation of Al and Fe-rich organomineral complexes. Spodosols form under coniferous or mixed forests in relatively cool regions with substantial hydrologic leaching, particularly at the northern borders of the biome in the Northern Hemisphere. Further toward the subtropical in cooler areas of eastern North America, Europe, and parts of Asia and Australia, Alfisols form, characterized by organic-rich mineral soil horizons throughout the soil profile, moderate leaching and high fertility. Ultisols, the oldest and most highly weathered soils in temperate zones, are located in the unglaciated and warmer portions of the biome, including southern North America, Asia, Australia, and New Zealand. Because of their advanced age and weathering, these can be deep soils with relatively poor fertility. Inceptisols and Entisols, the youngest soils characterized by less weathering and poor horizon development, are widely distributed in temperate forests. In particular, these form in areas where glaciers left behind new parent material either as till or outwash.

A characteristic that distinguishes temperate from tropical forest soils is the much larger stores of soil organic matter typically present in temperate soils. In temperate regions, litter in various stages of decomposition from fresh litter to humified matter often accumulates atop the mineral soil, forming the forest floor. This organic layer is key in retaining water, retaining and releasing nutrients, and providing animal habitat. It varies in thickness from a few centimeter to tens of centimeters, depending on the age of the stand, the soil pH, the inherent decomposability of the species of litter, the amount of rainfall, and the presence or absence of earthworms.

An additional important category of organic detritus found in many temperate forests is coarse woody debris. This includes standing dead trees and downed, decomposing logs. Rotting woody debris provides a rooting medium, a habitat for soil fauna, a substrate for the saprotrophic flow of energy to the food web, and a means for returning nutrient elements to soils, as well as important structural material for forest streams. Logs undergo a wide range of decay rates, from relatively rapid (a few years) where logs are small and wetting–drying cycles are rapid, to very slow (lasting to a century) where logs are large and the environment is wet and cool. In harvested or managed forests, coarse woody debris may be absent because logs are removed for timber. In unmanaged temperate forests, the long time periods needed for large logs to be produced and decomposed produces a U-shaped curve in the mass of woody debris over time (**Figure** 3). After a stand-initiating disturbance, woody debris from the previous stand accumulates rapidly and then decays slowly. A lag time of several decades typically exists before woody debris from the new stand begins to accumulate. If the new stand remains even-aged, a second peak may occur as the new stand passes through the stem exclusion stage of development and widespread mortality occurs in smaller trees that compete unsuccessfully for light after the canopy has closed.

Ecological Communities and Succession
Vegetation Communities

Temperate forest vegetation communities span the range from single-species stands to mixed-species stands as well as the range from even-aged to all-aged stands. Which type of community is present at any point in space and time depends on the site physiography, soil, and climate together with its disturbance history. Species such as pines,

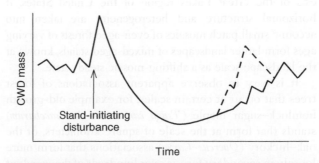

Figure 3 Dynamics in the mass of coarse woody debris (CWD) before, during, and after a major stand-initiating disturbance in a temperate forest. The solid line represents a U-shaped curve in CWD mass over time. The dashed line represents a secondary peak that may occur if the newly initiated stand remains even-aged and undergoes a self-thinning stage.

Figure 4 The canopy of an even-aged red pine plantation, aged about 75 years, in Massachusetts, USA. Photo by W. S. Currie.

eucalyptus, cottonwoods (*Populus* spp.), and others may form natural single-species, even-aged stands (**Figure 4**). Pioneer species such as aspens (*Populus* spp.) and some pines may initially form even-aged monocultures which eventually diversify in composition and vertical structure as growth, self-thinning, or succession proceeds. Long-lived hardwoods and other conifers also form stands where, increasingly with forest age, great diversity exists in tree ages and sizes. An example of the latter are hemlock–northern hardwood forests of the Great Lakes region of the United States. If horizontal structure and heterogeneity are taken into account, small patch mosaics of even-aged forests of varying ages form larger landscapes of mixed-aged stands, known at the landscape scale as a shifting-mosaic steady state.

It is easy to observe apparent associations of forest trees that occur at certain scales, for example old-growth hemlock–sugar maple (*Tsuga canadensis–Acer saccharum*) stands that form at the scale of square kilometers, or the oak–hickory (*Quercus–Carya*) associations that form more loosely in secondary forests over hundreds of thousands of square kilometers. However, a longstanding debate concerns whether forest communities represent organized associations or simply continuously varying associations as tree species respond individualistically to environmental gradients.

Temperate forest tree species form apparent associations with one another and with the abiotic environment not only across space but also over time at a particular location. A key organizing principle in understanding such temporal associations is the concept of succession, or the replacement of one dominant species or set of dominant species with another, over time, on a particular soil. Primary succession refers to the replacement of species over time occurring in the first forest stand to grow on a newly exposed soil, for example, following the retreat of a glacier. Secondary succession refers to species replacement over time following a major disturbance such as massive windthrow, mortality, or forest harvest. Early-successional species, termed pioneers, are those that are able to fix nitrogen from the atmosphere (see the section titled Nutrient cycling) or those that grow rapidly under high-light conditions but cannot tolerate shade. Late-successional forest species are typically those that can tolerate low-light or low-nutrient conditions as understory trees, while continuing to grow over long time periods, eventually reaching the overstory. Forest ecologists have long sought general principles of succession – for example, the identification of a deterministic sequence leading to a particular stable endpoint or 'climax' vegetation community in a particular climate and physiographic landform. Current understanding, however, emphasizes that while certain successional mechanisms exist, the particular sequences and possible endpoints of succession at a particular location are typically numerous, ultimately depending on a complex interplay among competition, species arrival, regeneration, disturbance regimes, and species' modification of the environment.

Temperate Forest Fauna

Faunal biodiversity in temperate forests is not as great as that in tropical forests (see Tropical Rainforest), but is greater than in boreal forests (see Boreal Forest). Because temperate forests are highly seasonal in their climate and cycles of vegetation physiology and production, faunal life cycles, ecology, and populations are often tied strongly to the seasons. Animal habitats within temperate forests are numerous and heterogeneous, including soils, the forest floor, woody debris, woody stems, and the layers of vegetation canopies. Although some animals depend on particular tree species, many are more dependent on certain aspects of forest structure.

In the temperate forest, the greatest concentration of fauna is on and just below the forest floor, in the litter, humus, and soil. Animals not only inhabit these strata, but through their activities drive soil carbon and nutrient cycling. Also within these strata are gradients of moisture, temperature, gases, and organic matter. Soil microhabitats are pore spaces, water film on soil particles, plant remains, the rhizosphere, and tunnels and burrows. Together, soil fauna and saprophytic flora contribute to the decomposition of organic matter. While most decomposition and nutrient release take place in the warm, humid summers, coinciding with the growing season of the vegetation, microorganisms and invertebrates can remain active below the insulating winter snowpack. Some animals occupy the litter in summer and move to the mineral soil in winter.

Because of the moist soil conditions, many temperate forest floors are home to reptiles (turtles and lizards) and amphibians (toads, frogs, newts, and salamanders). In the mixed-deciduous temperate forests there are over 230 species of reptiles and amphibians. These animals live on the forest floor close to streams, depressions, or lakes where there is available moisture. Lizards are found in moist woods and also in disturbed areas. Turtles live in or near bodies of water and toads and frogs are widespread, needing only shallow water. Temperate forest streams and rivers can support abundant fish populations, particularly under less-disturbed conditions and in coastal temperate rainforests.

Mammal populations in temperate forests tend to be comprised of scattered individuals or groups, and their habitat ranges from the forest floor to the canopy layers. Examples of small mammals are squirrels, rabbits, mice, chipmunks, skunks, and bats. Very large mammals are the exception and in temperate forests may include bear, mountain lions, deer, and other ungulates such as moose and elk. These mammals depend on the herb and shrub layers of the forest in addition to the litter and woody debris for food and habitat. Edge areas form transition zone habitats; for example, deer and other large animals usually live near the edges of forest openings with the trees providing shelter while edible ground vegetation is available in the openings all year.

Trunks are also habitats for spiders, beetles, and slugs. Birds are especially versatile across habitat structures; they are found on the forest floor and in several of the vegetation layers depending on nesting and foraging preferences. Types of birds that breed in mixed-deciduous forests include bark foragers (woodpeckers, flickers), canopy gleaners and pursuers (chickadee, vireo, flycatchers), ground species (thrushes, ovenbirds), and warblers. Deciduous forest are also breeding habitat for larger avian species including turkeys, vultures, owls, and hawks. In addition, moths, butterflies, and other flying insects feed and reproduce in the canopy, the understory, and the forest floor.

Water and Energy Flow, Nutrient Cycling, and Carbon Balance

Water, Evapotranspiration, and Energy

Water enters temperate forests as rainfall, snowfall, fog, and the direct condensation of water vapor onto plant or soil surfaces. Some water, amounting to less than 10% of rainfall under most conditions, is lost immediately to the atmosphere through evaporation. Depending on the season, water drips from the forest canopy to enter soils or accumulates as snow until a mid-winter thaw or spring snowmelt. Entering the soil, water is stored, taken up by plant roots, or moves to groundwater or surface water. Water taken up by plants moves upward through the xylem and exits as water vapor through leaf stomates in the process of transpiration. Typically, through the combined processes of evaporation and transpiration, less than half of the annual precipitation is passed directly back to the atmosphere as water vapor. Somewhat more than half of the annual precipitation passes through the rooting zone of the soil to enter groundwater or surface water such as streams and lakes.

Evapotranspiration, or evaporation and transpiration taken together, makes a large contribution to the ecosystem energy budget and to the regulation of temperature. In the conversion of liquid water to gas, evapotranspiration carries away large amounts of heat as latent heat. This cooling effect combines with other terms in the energy budget of a forest canopy to regulate the temperature of leaves and of the forest as a whole. Other major terms in the energy budget include the absorption or reflection of short-wave (sunlight) and long-wave radiation (from sunlight and from the atmosphere), the emission of long-wave radiation, and the gain or loss of sensible heat from the atmosphere. On a typical summer day the vegetation canopy absorbs energy in short-wave radiation from the Sun and dissipates the energy as sensible and latent heat to the atmosphere, heating the troposphere from below. On warm days with strong sunlight, the ability of forest tree canopies to dissipate heat allows the trees to maintain leaf temperatures closer to the photosynthetic optimum while also minimizing plant respiration. The opening and closing of stomates, governing transpiration, is under plant physiological control and is an important aspect of plant adaptation to life in a particular environment. During prolonged periods of drought, when trees are less able to use water to cool the canopy and maintain leaf turgor, foliar wilting and tissue damage can occur. Some temperate forest trees can be unexpectedly drought-deciduous, dropping their foliage during a late summer drought.

The photosynthetic conversion of light energy to stored chemical energy is a minor term in the physical energy budget of a forest, amounting to no more than 2% of the energy in sunlight. At the same time, this energy conversion represents the largest term in the ecological energy budget of a forest. The energy stored in photosynthate drives the life processes of all of the plants and animals in the ecosystem. A large portion of this energy is consumed by the vegetation itself through plant respiration, supplying energy for growth, metabolism, and reproduction. Another large flow of energy enters the food web through herbivory; herbivores eat seeds, fruits, and living plant tissues. The consumption of living leaves by insects, while normally minor, can grow during insect irruptions to encompass virtually the entire forest canopy over large areas. Similarly, the consumption of living leaves by forest ungulates including deer and moose are typically small energy fluxes at the ecosystem scale (although the browsing of seedlings and saplings can have a strong impact on forest regeneration and the future composition of the vegetation community). The chief means of energy flow to the faunal food web is through the saprotrophic pathway. Fungi and bacteria (often called soil flora) decompose dead and senesced plant material including leaves, roots, and woody debris. The soil flora is grazed upon by soil microfauna, which are in turn preyed upon by other fauna including arthropods, amphibians, and birds.

Nutrient Cycling and Carbon Balance

To achieve the high levels of productivity typical of temperate forests, trees require ample and reliable supplies of nutrient elements. Those required in the largest supplies include N, P, K, Ca, Mg, S, and Mn. Trees acquire most of their nutrients through root uptake from soils, which store nutrients in soil solution, on the surfaces

of mineral grains, on the surfaces of organic matter, and in decomposing organic matter itself. A forest ecosystem receives inputs of nutrients from the atmosphere and from mineral weathering, experiences losses of nutrients via leaching (the water-driven movement of elements out of the rooting zone, ultimately to streams), and cycles nutrients internally (**Figure 5**). A key internal cycle is the plant–soil cycle in which an element such as calcium (Ca) is taken up by plant roots, used nutritionally by the tree, returned to the soil in foliar litterfall, and returned to the pool of soil-available nutrients during decomposition of the litter. Temperate forests are characterized by the fact that, for most nutrient elements required for plant growth, the internal cycling is greater than the ecosystem inputs and losses of these elements.

While most of the required nutrient elements can be released through mineral weathering, a notable exception is nitrogen (N). Temperate forests rely on inputs of N from the atmosphere. Combined with the fact that trees have a high demand for N, and with the fact that N is strongly retained in unavailable forms in soil organic matter, this makes N the most limiting nutrient for plant growth in most temperate forests. Trees have a high demand for N because photosynthesis and plant metabolism require enzymes, which are made of N-rich amino acids. Amino acids are also one of the primary needs of

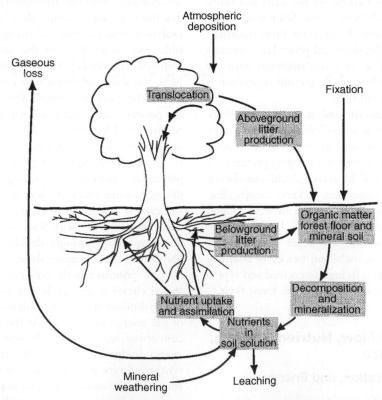

Figure 5 Schematic diagram of generalized nutrient cycling in temperate forests. Shaded terms represent nutrient cycling fluxes within the system, while unshaded terms represent ecosystem inputs or losses. From Barnes BV, Zak DR, Denton SR, and Spurr SH (1998) *Forest Ecology*, 4th edn. New York, USA: Wiley.

herbivores that consume plant tissues, including defoliating insects, deer that browse saplings, and beavers that girdle trees by eating the cambium around the tree base. Given the high demand for N by forest trees, it is somewhat ironic that trees in temperate forests are surrounded by two large, potential sources of N that are limited in availability because of the chemical form of the N. The first is N_2 gas, which is the primary constituent of the atmosphere. Most forest trees cannot access gaseous N_2, although a few exceptions include red alder (*Alnus rubra*) and black locust (*Robinia pseudoacacia*) in the USA, which access atmospheric N_2 through the process of N fixation (**Figure 5**). In this process, symbiotic bacteria living in root nodules fix the N_2 into plant-available forms. The second large pool of poorly available N occurs in humus and soil organic matter, made up of partially decayed and humified plant and microbial detritus. Typically, large accumulations of N are bound in this material in large, polyfunctional macromolecules that form during litter decomposition. Temperate forests are characterized by the combined facts that (1) cold, wet winters impede microbial decomposition and allow these pools of organic matter to accumulate, and (2) warm, humid summers promote decomposition by fungi, causing these soil organic matter pools to turn over and release nutrients at slow but continuing rates. Nutrient release during decomposition is termed mineralization because N is converted from organic to the inorganic forms of nitrate (NO_3) and ammonium (NH_4) which are easily taken up and used by plants (**Figure 5**).

Carbon is the primary elemental constituent of both forest vegetation and the organic matter in forest soils. Carbon (C) is not considered a nutrient element *per se* because a C atom passes through a forest once, in a single direction, closely linked to the flow of energy; unlike nutrients, carbon does not cycle between plants and soils repeatedly. Forests are highly open systems with respect to carbon, exchanging large quantities of CO_2 with the atmosphere. The carbon balance of a temperate forest arises from the interplay among processes controlling forest sources and sinks of atmospheric CO_2. Photosynthesis, or primary production, converts atmospheric CO_2 to reduced organic compounds, storing energy and C in the forest. Autotrophic respiration, the conversion of organic compounds to CO_2 by plants, provides energy for plant metabolism. Heterotrophic respiration, the conversion of organics to CO_2 by herbivores, microorganisms, soil fauna, and other animals in the food web, releases energy for animal life processes. Fire, the rapid oxidation of organics, also releases CO_2 to the atmosphere. Depending on the balance among these processes, temperate forests can either store or release large quantities of carbon. The primary storage pools include growing trees (particularly the woody stems), the forest floor, standing and downed woody debris

Figure 6 An old-growth sugar maple–birch–hemlock forest showing a large piece of downed woody debris. This forest is located in northern Michigan, USA. Photo by W. S. Currie.

(**Figure 6**), and soil organic matter. The transfer of carbon among these pools is linked to forest disturbance and stand dynamics including aggradation and succession. The flows of carbon into and out of the ecosystem are closely coupled to the availability of water, flows of energy, and the cycling of nutrients.

Temperate Forest Land Cover

Historical Land Cover and Land-Cover Change

Temperate forests in all regions of the globe have been significantly altered by human activities for thousands of years. Their moderate climates, fertile soils, and vegetation productivity have been favorable to human settlement and clearing for agriculture, as well as direct use of trees themselves for lumber and fuels. Agricultural and settlement activities have included development of urban areas, widespread grain and other crop (e.g., corn, vegetables) cultivation, livestock grazing, gathering of mulch, and alteration of natural water drainage. Under these historical pressures, it is estimated that only 1–2% of the original temperate forest remains as never-harvested remnants scattered around the globe. The vast majority of temperate forest land cover is in secondary forest responding to human harvest or other human-induced disturbance.

The longest histories of substantial forest clearing have been in Asia and Europe. In China clearing for agriculture probably began some 5000 years ago, where the Chinese civilization is believed to have begun around the Huang He (Yellow River). The primary sociopolitical factor contributing to deforestation of China over the centuries has probably been the focus on an agriculture-based economy. At present, there is negligible large-scale reforestation in temperate China and significant soil erosion problems hampering reforestation.

Forest clearing for agriculture in Europe began over 5000 years ago starting in present-day Turkey and Greece and moving northwest through Middle Europe to Northern Europe. Forests of Britain were substantially cleared for agriculture and grazing. Woodlands regained some area in the Middle Ages; however, even remaining European temperate forests were degraded, being used for fuelwood, woodland pasture, and later for charcoal. Coppice practices promoted species that re-sprouted more quickly than beech – including maples and oaks, and this activity altered the natural floristic composition. Tall trees in Britain and Western Europe were removed for shipbuilding. Manorial estates provided some of the few refuges for natural forests. Reforestation in recent centuries in Europe began subsequent to reduction in the use of woodlands for pasture and fuel; reforestation has also occurred through the introduction of planted managed forests and scientific forestry. However, spruce, pine, and larch have been widely planted on areas previously occupied by once deciduous temperate forests.

North American indigenous populations cleared or burned small areas for some agriculture, but land-cover change in North American temperate forests began at large scales in the late sixteenth century with the European settlement. Eastern North American was rapidly cleared as the population moved westward in the nineteenth century. By the start of the twentieth century only a small amount of the original North American temperate forest remained. When the richer soils of the topographically level Midwest and Great Plains were found to be more productive for agriculture than those of eastern North America, eastern farms were abandoned and natural forests began to re-grow. At present, secondary forests are regrowing in the eastern and central United States.

In the Near East the temperate forest occurs in a narrow belt including in Turkey and Iran. This area probably served as a plant refugium during the Ice Ages and the floristic composition is more diverse than that in Europe. Some forests have been exploited for coppice, timber, or grazing and others transformed into agriculture and fruit-tree plantations. Beech forests are the most significant of the present-day broad-leaved forests in the region. In the small area of temperate deciduous forest in South America, forests have been moderately altered since the arrival of the Spaniards in the sixteenth century; the further south one goes the more recently the vegetation has been undisturbed and wooded areas remain. Australia first saw introduction of European agricultural practices only approximately 150 years ago.

Present-Day Land Cover and Rates of Change

The global temperate forest continues to be changed by a combination of long-term effects of historical land-cover change and by present-day change agents. Present-day drivers of land-cover change in temperate forests include accelerated population growth, continued industrialization, and changes in agricultural practices. These are expressed on the landscape as continued clearing for settlement and agriculture in some regions, abandonment of agriculture and reforestation in other regions, and widespread alteration in landscape spatial structure and biodiversity.

While rates of tropical deforestation increased between 50% and 90% in the 1980s, the area of temperate forests has remained constant or increased in the last 50 years in the form of new second-growth forests. In some areas, in eastern North America and parts of Northern Europe, farming is less economically viable than in other parts of the temperate region, leading to reforestation in these areas. Preservation in the form of parks has expanded by active conservation efforts worldwide. Managed forestry has maintained existing temperate forest lands by re-planting after harvest, and sustainable forestry practices are receiving increasing attention.

While the temperate forests may have stabilized or increased in terms of total area, most regions continue to experience other alterations manifested in the landscape spatial patterns and forest biodiversity. Today the temperate forest biome is a mosaic of settlements, patches of forest, and agriculture. Large expanses of unbroken forests from past centuries have been replaced by considerable landscape-scale heterogeneity and fragmentation. Temperate forest communities have changed compositionally, as disturbance regimes have shifted from natural to a combination of natural and human-caused, producing different patterns of regeneration and succession. While some recently established nature preserves have a natural forest structure, reduced biodiversity characterizes many temperate managed and secondary forests. Considerable present-day challenges lie in understanding and addressing the impacts of land-use change and other aspects of global environmental change in the temperate biome on forest biodiversity and forest ecology.

See also: Boreal Forest; Chaparral; Tropical Rainforest.

Further Reading

Bailey RG (1998) *Ecoregions: The Ecosystem Geography of the Oceans and Continents*. New York: Springer.
Barbour MG and Billings WD (2000) *North American Terrestrial Vegetation*. Cambridge, UK: Cambridge University Press.
Barnes BV, Zak DR, Denton SR, and Spurr SH (1998) *Forest Ecology*. New York: Wiley.

Currie WS, Yanai RD, Piatek KB, Prescott CE, and Goodale CL (2003) Processes affecting carbon storage in the forest floor and in downed woody debris. In: Kimble JM, Heath LS, Birdsey RA, and Lal R (eds.) *The Potential for U.S. Forests to Sequester Carbon and Mitigate the Greenhouse Effect*, pp. 135–157. Boca Raton, FL: Lewis Publishers.

Frelich LE (2002) *Forest Dynamics and Disturbance Regimes. Studies from Temperate Evergreen-Deciduous Forests. Cambridge Studies in Ecology*. Cambridge, UK: Cambridge University Press.

Lajtha K (2000) Ecosystem nutrient balance and dynamics. In: Sala O, Jackson RB, Mooney H, and Howarth RW (eds.) *Methods in Ecosystem Science*, pp. 249–264. New York: Springer.

Olson DM, Dinerstein E, Wikramanayake ED, *et al.* (2001) Terrestrial ecoregions of the world: A new map of life on earth. *BioScience* 51: 933–938.

Rohrig E and Ulrich B (1991) *Temperate Deciduous Forests*. Amsterdam: Elsevier.

Temporary Waters

E A Colburn, Harvard University, Petersham, MA, USA

Overview	Ecosystem Ecology
Introducing Temporary Waters	Applied Ecology
The Ecology of Temporary Waters	Further Reading

Overview

What Are Temporary Waters, and Why Are They of Interest Ecologically?

Temporary waters are shallow lakes, ponds, pools, rivers, streams, seeps, wetlands, depressions, and microhabitats that contain water for a limited period of time and are otherwise dry. They occur across the globe, on all continents and oceanic islands, at all latitudes, and in all biomes, wherever water can collect long enough for allow aquatic life to develop.

Numerous and widespread, many temporary waters are small and easily studied. Their communities are diverse, with much among-site variation (i.e., high β diversity), and differ from those in permanent waters, contributing to regional (γ) biodiversity. Endemic species are often present. Organisms survive through species-specific behavioral, physiological, and life-history adaptations. Community composition and structure change in response to environmental variations. Temporary waters are highly productive and their food webs are relatively simple. For all of these reasons, temporary waters lend themselves to surveys and experimental manipulations designed to test hypotheses about biological adaptation, population regulation, evolutionary processes, community composition and structure, and ecosystem functioning.

In many parts of the world, most temporary waters have been lost. The conservation and restoration of vulnerable temporary waters is a major thrust of applied ecology. Also important are applications of ecological understanding to the control of disease vectors, especially pathogen-transmitting mosquitoes, from temporary water habitats.

What Is Covered in This Article?

This article is divided into two sections. The first introduces temporary waters – definitions, important variables, types, geographic distributions, and terminology. The second section examines the ecology of temporary waters, with an overview of the biota and their adaptations, and summaries of some key questions in organismal and community ecology, ecosystem ecology, and applied ecology.

Introducing Temporary Waters

Definition

In temporary waters, aquatic habitat is present for non-continuous lengths of time, in contrast to permanent water bodies, which are always flooded except under unusual conditions such as extreme droughts. This discontinuity in the availability of water is the defining characteristic of temporary waters.

For this article, temporary waters include temporary inland salt waters, whose chemistry and biota are allied to fresh waters and not to marine ones, but they do not include coastal areas flooded by ocean tides. Also excluded from this discussion are subterranean waters.

Important Variables

Apart from periodic drying, there are no hard and fast rules about the characteristics of temporary waters. Classification may be useful, provided it contributes to understanding. The important considerations governing how to classify temporary waters in a given situation should be: what is the purpose of classification, and what are the desired outcomes in terms of distinguishing different types of temporary water bodies? Researchers have developed many approaches to classifying temporary waters using the descriptive variables listed below.

Geography

Regional location (e.g., Ontario, Malay Archipelago), latitude (e.g., tropical, Arctic), or climate (e.g., humid, arid) may contribute to similarities among temporary waters.

Biome

Temporary waters occur in all terrestrial biomes, even the wettest. Regardless of their location globally, habitats within a particular biome, with similar hydrologic characteristics on similar substrates, often are much alike.

Water body type

Temporary waters may be lotic (flowing) or lentic (still). There are several major categories, and many unique regional names (**Tables 1–3**). Some categories overlap; for example, pools formed after thunderstorms on exposed rocks on coastal Scandinavian islands are both rainpools and rockpools.

Substrate

Substrate (e.g., rock, organic debris, sand, clay, limestone, mud, basalt, wood) influences hydrology, water chemistry, and temperature and is an important habitat variable in its own right (e.g., for seed germination or shelter for burrowing animals).

Size

Some classifications distinguish microhabitats, mesohabitats, and macrohabitats.

Table 1 Major types of temporary waters found throughout the world

Rockpools or rock pools – Accumulations of rainwater or floodwater in depressions on exposed bedrock or boulders

Rainpools or rain pools – Accumulations of rainwater on any substrate

Seasonal woodland pools – Fill annually, usually as a result of winter or spring rains, and from melting snow in northern areas, and dry later in the year

Grassland pools – Temporary ponds in grassland environments

Marsh pools – Temporary ponds that occur within larger grass, sedge, or rush-dominated wetlands and remain flooded after most of the wetland has drawn down

Swamp pools – Depressions within larger wooded wetlands that remain flooded after the surface of the swamp has dried

Floodplains – Land areas that are inundated seasonally by high waters spilling over the banks of rivers and streams

Floodplain pools – Low areas in floodplains that remain flooded after floodwaters have withdrawn and left most of the floodplain dry

Springs, seeps, and spring seeps – Sources of water derived from groundwater or from subsurface flow reaching the land surface after heavy rains. Springs are expressions of the groundwater table and tend to be relatively permanent; seeps may be more transitory. Both vary in output with rainfall over the source area, and both may provide seasonal or continuous sources of water. Flow from springs and seeps may extend from the source as marshes, pools, or streams that may contain water during cool or wet seasons and become dry during periods of high temperature and/or low precipitation

Intermittent headwater streams – The smallest tributaries at the head of stream systems, often seasonal in their flow, containing water during the wet and/or cooler months and becoming dry during the hot/dry months

Arid-land rivers, intermittent rivers, or ephemeral rivers – Flowing waters that occur in regions where the groundwater table is far below the surface and where annual potential evapotranspiration is greater than precipitation. They typically flow only during the rainy season, when runoff travels over the land and is carried downstream; some only carry storm runoff, but others may have extended flow maintained by seasonal groundwater discharges. During the dry season, there may be water below the surface and in isolated pools within the channel, and there may be brief spates of flow following cloudbursts

Dry lakes or playas – Shallow water bodies in arid regions, especially in closed basins, where water collects from large areas. Due to the arid conditions, the water usually evaporates rapidly. A long history of flooding and drying leads to accumulations of salts in these basins, and dry lakes are typically saline. Many dry lakes occupy basins that contained large freshwater lakes earlier in geological history. Deposits of salts and sediments left behind when the lakes dried may be tens or hundreds of meters deep beneath the lake beds, and they contribute to saline conditions in the playa

Sinkholes or sink holes – Depressions created in calcareous bedrock by the gradual dissolution of the rock by water. They range in diameter and depth from meters to kilometers. Sinkholes that contain water are fed by groundwater, precipitation, and/or streamflow and include both permanent and temporary waters

Snowmelt pools, icemelt pools, and meltwater pools – Formed by the seasonal melting of ice and snow in the Arctic and Antarctic, along the margins of icefields and mountain glaciers, and in areas that receive snowfall

Meltwater streams – Flowing waters that develop seasonally as glaciers, icefields, and winter snows melt; they often flow during the day and stop flowing at night as low temperatures inhibit melting

Plant-associated microhabitats or natural containers (phytotelmata) – Microhabitats formed where plants produce small depressions in which water can collect (see **Table 2**)

Artificial containers – Any human-made concavity where water can collect, including gutters, birdbaths, tires, empty cans, tractor ruts, canoes, split and discarded coconuts, and other water-holding depressions

Table 2 Examples of phytotelmata and other natural containers that provide temporary aquatic habitats for mosquito larvae and other organisms

Ant nests
Insect-bored bamboo, bamboo stumps
Fungal cap concavities
Log holes
Buttress-root slits
Eggshells
Flower bracts
Fruits
Horns
Leaf axils
Fallen leaves
Nuts
Modified leaves of pitcher plants and analogs
Pods
Reeds
Rockholes, potholes
Mollusk shells
Skulls and other skeletal remains
Stumps and trunk cavities
Treeholes

Derived from Index in Laird M (1988) *The Biology of Larval Mosquito Habitats*. Boston: Academic Press.

Hydrology

Hydrologic variables are the most important factors influencing aquatic life in temporary waters.

Water sources

Water sources include groundwater, runoff, precipitation, snowmelt, streamflow, and floodwater.

Flood timing

Flood timing encompasses both season and predictability. Vernal, estival, autumnal, and hibernal (or brumal) refer respectively to filling in spring, summer, fall, or winter. Intermittent systems flood predictably at annual (seasonal) to multiyear intervals. Waters that flood unpredictably are ephemeral if they fill several times a year, and episodic if they fill just once or twice a decade.

Seasonality and predictability of flooding influence the biota. Predictable filling of Mediterranean vernal pools by rainfall during the winter growing season facilitates plant growth and has contributed to the development of an

Table 3 Some terms used to describe temporary waters around the world

Avens	France: depressions hollowed out in limestone
Baias	South America: temporary lakes
Billabongs	Australia: pools that are left behind in floodplains as large, seasonal rivers recede after flooding
Bogs	Worldwide: freshwater peatlands with acidic water chemistry; usually with limited connection to other surface waters, often fed exclusively by rainfall
Buffalo wallows	North America: created by buffalo (*Bison bison*) rubbing their bodies on the ground, these shallow excavations on the prairies fill seasonally with water
California vernal pools	Western North America: seasonally flooded pools in Mediterranean scrub of western North America, especially California, and characterized especially by rich plant communities with large numbers of endemic species
Carolina bays	North America: round or oval depressions of uncertain origin in the coastal plain of the Southeastern United States, often supporting endemic plant communities and temporary-pond fauna
Corixos	South America: temporary-water bodies in floodplains, especially in the Pantanal region
Dambos	Southern Africa: shallow, treeless, seasonally inundated wetlands at heads of drainage networks
Dismals	North America: swamps or marshes in the Mid-Atlantic region of Virginia, Delaware, and the Carolinas
Doline	Western Balkan states/Dinaric Alps: depressions and sinkholes in limestone
Fens	Worldwide: freshwater peatlands with alkaline water chemistry
Gator holes	North America: excavations made by alligators (*Alligator mississippiensis*) in the Florida Everglades; they remain flooded when waters recede and serve as refugia for aquatic animals during droughts
Gnammas	Western Australia: temporary waters formed on granitic outcrops
Heaths	Great Britain: freshwater peatlands with acidic water chemistry
Mires	Northern Europe: freshwater peatlands with acidic water chemistry
Kettles, kettle holes	Worldwide, in areas affected by continental glaciation in the past: largely circular depressions formed by the melting of blocks of ice calved off of retreating continental glaciers and buried in morainal debris
Moors	Great Britain: freshwater peatlands with acidic water chemistry located on hilltops
Mosses	Scotland: raised bogs, i.e., freshwater peatlands with acidic water chemistry located on hilltops or above the groundwater table
Muskegs	North America: freshwater peatlands with acidic water chemistry (Algonquin)
Oshanas	Namibia and Angola: linearly linked shallow pans that are filled by floodwater and precipitation
Pakahi	New Zealand: shallow, groundwater-flooded areas with acid soils, inappropriate for cultivation (Maori)

(Continued)

Table 3 (Continued)

Pans, panes, pannes	Worldwide: shallow temporary waters that flood periodically from rainfall in arid regions; also refers to temporary pools that form in salt marshes from monthly flooding by spring tides
Phytotelmata	Worldwide: a technical term describing temporary waters associated with plants, in axils of leaves or branches, modified pitchers and similar structures, nuts
Plunge pools	Worldwide: deep holes that form in bedrock at the base of waterfalls through the action of water over time, and that retain water for a period of time after the stream has dried up
Pocosins	North America: upland-coastal floodplains or groundwater-flooded seasonal wetlands in the South Atlantic United States
Potholes, pot holes	Worldwide: rockpools in or along streambanks and streambeds, created by the action of water and rock scouring out round depressions into boulders or bedrock. Potholes may be a few centimeters to more than a meter in diameter
Prairie potholes	North America: in the Great Plains, largely circular depressions formed as blocks of ice left by departing continental glaciers were covered by morainal debris and then melted
Ramblas	Spain: temporary streams that usually flow only after rainstorms
Sabkhas, seabkhas	Arabian Gulf: saline lakes
Salinas	South America: saline lakes
Sinkholes	Worldwide: depressions in limestone, formed by the solution of surface rock or by the collapse of underground caverns or caves collapse where the subsurface has been dissolved by gradual solution in water. Sinkholes may be dry on the bottom, intermittently flooded, or contain water continuously
Sinking creeks	North America: flowing streams that disappear from the surface into one of the many cracks or sinkholes in limestone regions, or into the ground in arid areas
Sloughs	Worldwide: the term has a variety of meanings. In Great Britain it refers to muddy and shallow waters. In North America it is used to refer to prairie potholes, temporary ponds, oxbow wetlands, permanent ponds, deepwater areas in the Everglades, brackish marshes on the west coast, seasonally flowing depressions in forests, freshwater wetlands in the Great Plains. Some use the term to refer to areas where water is not stagnant but rather, flowing slowly; others specifically define sloughs as areas with stagnant water
Swallow holes	Great Britain: sinkholes in limestone, especially deep holes through which water funnels underground
Takyrs	Turkmenistan: pans in the desert
Tenajas	North America: rockpools, usually in temporary stream channels, that remain flooded for several months after the stream dries; some develop plant communities similar to those in vernal pools
Turloughs	Ireland: temporary waters formed in limestone, filled primarily by groundwater although may sometimes fill from precipitation; usually fill in fall and dry in spring or early summer
Vasante	South America: temporary streambeds connecting lakes in the Pantanal region during the rainy season
Vernal pools	North America: temporary woodland pools that fill in spring and dry in summer; applied more broadly to all seasonal woodland pools that reach maximum depth and volume in spring. Worldwide, the term applies to any temporary pools that fill in spring. The term 'California vernal pools' is used to represent a class of Mediterranean biome temporary ponds characterized primarily by their endemic plant communities
Vleis	Southern Africa: seasonally inundated wetlands in southern Africa, typically flooded by rivers at high water
Whale wallows	Eastern North America: seasonal woodland ponds along the Delaware coast in the United States

endemic flora. When the pools dry, high summer temperatures prevent the establishment of terrestrial vegetation.

Flood duration, or hydroperiod

Across most categories of temporary waters, there is a continuum of flood duration: days, weeks, months, or years. Ephemeral waters are flooded for hours, days, or weeks. Intermittent refers to flood durations of several months. Semipermanent or near-permanent waters dry only occasionally, during major droughts. Within a water body, the hydroperiod varies across filling cycles, depending on weather, with some waters being more stable than others (**Figure 1**).

Typically, with increasing hydroperiod, the potential aquatic community becomes richer, and the adaptations of the flora and fauna become less extreme. Waters with shorter hydroperiods have fewer total species but more that are unique to temporary habitats.

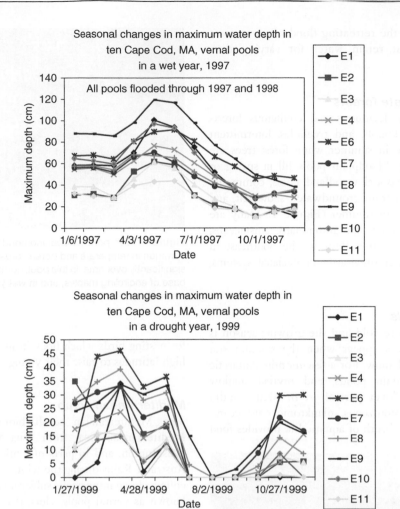

Seasonal changes in maximum water depth in
ten Cape Cod, MA, vernal pools
in a wet year, 1997

Seasonal changes in maximum water depth in
ten Cape Cod, MA, vernal pools
in a drought year, 1999

Figure 1   Water depths differ within and between years in ten temporary ponds clustered together on Cape Cod, MA, USA.

Chemistry

Important chemical characteristics include salinity (fresh, $<3 \, g \, l^{-1}$ salts; brackish, $3-35 \, g \, l^{-1}$; saline, $\geq 35 \, g \, l^{-1}$), major ions (e.g., sulfate- vs. chloride-dominated desert waters), color (e.g., clear vs. stained dark with organic acids), pH, and dissolved oxygen.

Distribution of Temporary Waters

Most types of temporary waters occur widely across the world's biomes, from the poles to the equator. Their numbers and varieties vary with annual precipitation, temperatures, and local geology and geography. They are most common in arid or cold areas where liquid water is unable to persist for long periods of time.

Tropical rainforests

Tropical rainforests, although well watered, contain many temporary waters. Cavities in bromeliads and other epiphytes retain rainwater (**Figure 2**) where decaying organic materials support microorganisms,

Figure 2   Bromeliads and other plants serve as natural containers for rainwater and provide microhabitats for microorganisms, mosquito larvae, and some tropical amphibians.

insects, and amphibians. Rainpools on the forest floor fill, dry within days, and support distinct communities. Lowlands of great tropical river systems, including the Amazon and the Paraná–Paraguay, are inundated during the rainy season, and the retreating floodwaters create a mosaic of ponds that retain water for varying time periods.

Boreal and temperate forests

Temporary waters in deciduous and coniferous forests include rainpools, rockpools, and treeholes. Intermittent headwater streams dry in summer when forest trees are transpiring (**Figure 3**). Floodplain pools fill in spring or after major storms. Seasonal woodland pools, commonly called vernal pools, fill from groundwater, snowmelt, and spring rainfall and dry in summer (**Figure 4**). Many are important breeding habitats for amphibians, crustaceans, and aquatic insects. Carolina Bays in the Southeastern United States, and other previously unglaciated systems, support endemic plants.

Tundra and icefields

Where temperatures are cold and the growing season is short, temporary waters appear when the summer sun melts glaciers, ice, and snow. For a few months, Antarctic rockpools, high-mountain ponds, and myriad shallow water bodies perched over permafrost in Arctic tundra teem with bacteria, protozoans, planktonic crustaceans, and insect larvae. This broth of aquatic life provides food

Figure 4 Temporary woodland ponds, or vernal pools, are common in temperate and boreal forests. Water levels vary significantly over time. In this pool, normal high water reaches the base of encircling maples, and in wet years it is more than a meter deep.

for nesting birds which flock in hundreds of thousands to high latitudes to raise their young.

Mediterranean scrub

The Mediterranean scrub biome occurs along the Mediterranean Sea; from Baja California to eastern Washington; and in parts of Chile, southern Africa, and Australia. Rains during the winter–spring growing season collect above impervious substrates, forming water bodies known as vernal pools, vleis, pans, Mediterranean temporary pools, and gnammas. They support endemic floras, including *Isoëtes* spp.; endemic faunas, including fairy shrimp and other crustaceans; and cosmopolitan temporary-pool plants and animals. On other substrates, temporary pools are less predictable and lack endemics. Most rivers in this biome flow only during the wet season, although isolated pools retain water for part of the dry season. Treeholes and other natural containers provide microhabitats after rains.

Deserts

In deserts, extreme aridity, high temperatures, salinity, and isolation of waters are especially stressful for aquatic life. Brief rainstorms create ephemeral pools on rocks and other surfaces. Extended rains collect water from large areas to fill closed basins, forming shallow, usually saline lakes that leave extensive deposits of encrusting salts upon drying (**Figure 5**). Many rivers and streams flow seasonally, especially in wet winters, or flash-flood unpredictably after storms, leaving behind pools of varying permanence (**Figure 6**). Permanent springs overflow during winter, creating seasonal streams, marshes, and thickets. Salts accumulate in the soil along the edges of desert waters, and temporary water bodies are generally brackish or saline.

Figure 3 Intermittent headwater streams drain up to 80% of the landscape in temperate forests and support distinctive communities of aquatic invertebrates and stream salamanders.

Figure 5 A salt crust left by evaporating water overlies dry lakes, or playas, in many desert basins.

Grasslands

Temporary waters in grasslands include pools, marshes, floodplains, and seasonal rivers and streams. Rich assemblages of plants, invertebrates, and amphibians occur in these waters and are critical for bird populations in the prairie pothole region of North America (**Figure 7**); the Eurasian steppes; the Indus, Ganges, Assam, Sylhet, and lower Mekong river plains in Asia; the southern African veldt; and the Pampas, Campos, and Pantanal regions of South America.

The Ecology of Temporary Waters

The Biota

All major groups of freshwater organisms occur in temporary waters. Many families and genera are found in similar habitats throughout the world, and there are some cosmopolitan species.

Hundreds of species of prokaryotes, including photobacteria and bacterial decomposers, proctotists, including green algae and diatoms, and protozoans, including ciliates, flagellates, and sarcodines, have been identified from temporary waters. Plants include mosses and liverworts, ferns, grasses, sedges, rushes, spike rushes, and other taxa typical of local wetlands. Microinvertebrates include rotifers, tardigrades, and gastrotrichs. Arthropods, especially water mites, crustaceans, and insects, dominate the macroinvertebrates. Many microcrustacean species, especially ostracodes and copepods, swim in the water, feed in the sediments, or cling to surfaces. Branchiopod

Figure 6 (a) Seasonal rivers in arid regions flow seasonally. (b) When flow ceases, the pools that remain persist for varying lengths of time.

Figure 7 Prairie potholes dot the landscape of the upper Great Plains in North America, provide habitat for aquatic life, and support breeding waterfowl. Reproduced from Sloan CE (1972) Ground-Water Hydrology of Prairie Potholes in North Dakota. *USGS Professional Paper 585-C.* Reston, VA: US Geological Survey.

crustaceans, particularly Notostraca (tadpole shrimps) (**Figure 8**), Anostraca (fairy shrimps), Conchostraca (clam shrimps), and some Anomola (daphnias and other water fleas), are largely restricted to temporary waters. All aquatic insect orders include temporary-water species, with the largest number in the Diptera, or true flies.

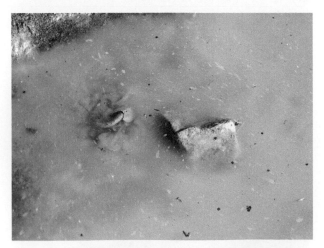

Figure 8 Notostracan crustaceans, known commonly as tadpole shrimp (left of center), are temporary-water specialists. Diapausing eggs lie for months, years, or decades in sediments of desert playas, rockpools, and woodland ponds. Hatching upon flooding, the animals are voracious predators and scavengers and their presence restricts the distributions of other temporary pool animals.

Water mites selectively feed on insects and crustaceans. Platyhelminthes (flatworms and flukes), Annelida (segmented worms and leeches), Nematoda (roundworms), Nematomorpha (gordian worms), and Mollusca (snails and bivalves) are also represented. Annual tropical killifishes (Cyprinodontiformes) and African and South American lungfishes (Lepidosireniformes) survive periodic drying, and fully aquatic fishes move seasonally into floodplains and intermittent headwater streams to feed and breed. Most anuran amphibian species, including true frogs, treefrogs, and toads, and some salamanders, preferentially breed in temporary habitats. For many of the world's birds, temporary waters are critical food sources during breeding or migration. Reptiles and mammals feed and hydrate in these seasonal waters.

Autecology: Organisms and Populations

Temporary waters lend themselves to thousands of ecological questions about adaptations, population regulation, and evolutionary pathways linking sibling species and cosmopolitan taxa.

Adaptations to drying

Inhabitants of temporary waters are distinguished by their ability to survive periodic drying. Adaptations include diapause, quiescence, and active avoidance. Rapid responses to flooding, fast growth, and flexibility in initiating the drying response maximize organisms' habitat use.

Diapause

Diapause involves suspended development. Hormonally controlled, and initiated and terminated by specific environmental cues, diapause is the most common and most effective drought-survival mechanism. It can allow survival over years – even decades – of continuous drying.

The rapid appearance of living organisms when water fills formerly dry puddles, containers, and floodplains is not, as formerly believed, spontaneous generation, or life miraculously developed from nothing. Instead, much of the life in newly flooded areas emerges from cysts, spores, seeds, or eggs diapausing on the dry substrate.

Found from bacteria to fishes in temporary waters, diapause is common in organisms with limited dispersal. Typically, the organism is replaced by a small, highly desiccation-resistant structure that awaits rehydration in the sediment. The substrate reservoir of diapausing microbes, plants, and animals is termed a seed bank, egg bank, or propagule bank. Diapause also occurs in larval and adult stages. Reproductive diapause is seen in some insects, and certain flatworms and annelids enter diapause after encysting in mucus.

Other dormancy

Other responses to drying involve decreased activity and lowered oxygen consumption. Some bdelloid rotifers, tardigrades, and nematodes survive complete dehydration to revive when flooded. Perennial plants may lose their leaves and die back to subsurface roots, tubers, or rhizomes when water levels recede. African and South American lungfish on drying floodplains encase in mud, breathe air, and more than halve their metabolism. Many mollusks burrow into sediments and estivate. Some insect pupae become dormant, delaying adult emergence. Dormancy is generally less effective than diapause for surviving extended drying or unpredictable flooding.

Avoidance

Anatomical, behavioral, or physiological adaptations can help organisms avoid drying. Some plants extend long roots deep into groundwater (**Figure 9**). Crayfish excavate burrows that remain flooded after surface drying.

Figure 9 Along the banks of ephemeral rivers and seasonal waterbodies in arid regions, deep-rooted trees and shrubs such as these tamarisks (*Tamarix* spp.) tap the groundwater and can influence drying of the waters at the surface. A gradient of increasing salinity tolerance is seen in plants radiating outward from the water source.

Animals in intermittent streams move downward into areas of high moisture or subsurface flow. Some insects and fish migrate between permanent and temporary waters. Amphibians and some insects have aquatic larvae and terrestrial adults.

Physiological ecology

Life in temporary waters may require biochemical modifications and major physiological adaptations. Many endemic plants from temporary waters use C_4 or crassulacean acid metabolism (CAM) photosynthesis, biochemical pathways that use water more efficiently than the C_3 photosynthesis of most plants. Species along salinity gradients show increasing osmoregulatory specializations.

Temporary waters have large local thermal gradients and over time may be subfreezing or above 40 °C. Biological processes vary with temperature, typically doubling with each 10 °C increase. Most species grow within narrow temperature ranges, and thermal cues regulate many life-cycle events. Enzymes need to function over temperature ranges found in temporary water bodies during organisms' life cycles. For example, inhabitants of Antarctic rockpools are active in cold water, and they diapause or secrete antifreeze substances to avoid consequences of subfreezing temperatures; their physiology differs markedly from relatives in temperate or desert pools.

Most freshwater plants and animals cannot regulate internal ionic concentrations in salt water. Inhabitants of many desert waters have impermeable body surfaces, salt-exporting cells, modified life histories, and well-developed drought-resisting adaptations (**Figure 10**).

Figure 10 The stick-like cases of salt-tolerant caddisfly larvae (Insecta: Trichoptera: Limnephilidae: *Limnephilus assimilis*) litter the bottom of Salt Creek in Death Valley, CA, in winter. The ability to regulate hemolymph osmotic and ionic concentrations in brackish waters, rapid growth, adult reproductive diapause during the hot summer months, and the presence of fewer predators than in low-salinity waters contribute to the species' persistence in this temporary desert stream.

Many are active in winter, when salinities and temperatures are low. Energetic costs of osmotic and ionic regulation must be compensated for by benefits, such as abundant food or reduced predation. Distributions in desert waters reflect species' physiological tolerances, with no plants and few highly adapted animals in hypersaline pools, greater diversity at low salinities, and low richness in highly unpredictable, ephemeral, freshwater rainpools.

Populations

Bet hedging

Desiccation-resistant seeds of annual plants from temporary waters germinate when chemicals in the seed coat are washed away. Seeds from the same parent have different levels of resistance, ensuring that not all germinate at once, and that some remain in the sediment seed bank. Similarly, crustaceans and some insects form an egg bank comparable to the seed bank of plants; some of the eggs hatch upon flooding, others hatch another time. African and South American annual killifish deposit diapausing eggs into the egg bank in the sediment of floodplain pools, and these eggs, too, hatch differentially upon flooding. This strategy of spreading risk is predicted from game theory and is termed bet hedging. The new field of resurrection ecology uses egg and seed banks to establish new communities in restoration projects, and obtains insights into evolutionary processes by growing individuals from samples collected a century or more ago and comparing them with modern individuals.

Life-history strategies

Numerous studies address short-term controls on population growth and survival in temporary waters. What are appropriate responses to flooding and drying, when they occur at different times from one year to the next, or to salinity and temperature? Theories of *r*- and *K*-selection predict that some species produce many, small progeny, raising the odds that some will survive. Others produce fewer, but larger progeny with more reserves to support them over adverse conditions. Examples tending toward both extremes can be found in temporary waters. Environmental conditions and evolutionary history shape species' responses, and many cues stimulate the initiation and termination of life cycles (**Tables 4** and **5**).

Short, irregular hydroperiods should favor *r*-selected life histories including rapid hatching/germination upon flooding, fast growth, and timely entry of many propagules into a drought-resistant state. Under longer, predictable hydroperiods, *k*-selection should produce slower growth, larger sizes, and longer life spans. In rockpools worldwide, with hydroperiods from hours to weeks, algae, insects, and crustaceans in the most ephemeral pools complete development in less than 24 h. Life spans are longer in longer-duration pools, and in those with

Table 4 Some cues stimulating the termination of diapause in temporary waters

Hydration
Hydration plus temperature
Hydration plus chemical cues
Hydration plus chemical and thermal cues
Hydration after drying (a minimum period of drying may be required)
Hydration plus chemical cues after drying
Hydration plus chemical and thermal cues after drying
Hydration after drying and low temperatures or freezing (a minimum period of drying and exposure to cold temperatures may be required)
Hydration plus chemical cues after drying and low temperatures or freezing
Hydration plus chemical and thermal cues after drying and low temperatures or freezing
Photoperiod in combination with one or more of the above

Table 5 Some cues initiating life stages adapted to drying in temporary waters[a]

Developmental stage (obligate diapause/dormancy/transformation once development reaches a critical threshold, regardless of habitat favorability)
Developmental stage plus other cues (diapause/dormancy/transformation initiated facultatively after development reaches a critical threshold, only after habitat becomes unfavorable)
Water temperature
Photoperiod
Chemical cues (pH, dissolved oxygen, chemical signals from predators or competitors, salinity, nutrients, other)
Drawdown-associated cues (chemical concentrations, crowding, depth)

[a]Note that if drying occurs before necessary developmental thresholds have been reached, cues cannot initiate drought-resistant stages, and organisms may die without completing their life cycles.

predictable flood regimes. Temperate *Eubranchipus* fairy shrimp and some aedine mosquitoes have one generation per year; they hatch, grow, mature, mate, deposit diapausing eggs, and die. In some fingernail clams, only young individuals resist drying, and they enter obligatory diapause as soon as they are born. Similar patterns are seen in many species.

Life-history tradeoffs, such as the ability to grow while water remains, potentially allow production of more offspring, or development to a larger size, which may enhance survival and reproductive fitness. A longer developmental period may also mean higher intraspecific densities and competition as habitat shrinks, and it increases the risk of being stranded if drying occurs rapidly. *Haematococcus pluvialis*, a photosynthetic flagellate related to *Volvox* from rockpools worldwide, is typical – diapausing spores develop rapidly when flooded, the organisms grow and reproduce, and upon pool drying, the motile cells form aplanospores that withstand drying and high temperatures. The diapausing spores form in less than a day, at any stage in the life cycle, providing *Haematococcus* with flexibility in the face of variable habitat duration. Widely different taxa grow rapidly to a minimum size threshold, after which they can reproduce and grow through multiple generations (e.g., *Daphnia* spp., snails, some bivalves), or become larger (e.g., amphibians, insects), as long as water is present, or until other cues initiate diapause, dormancy, or transformation.

Complex life histories

Many species' life cycles are complex. Post-hatching populations of the cosmopolitan water flea *Daphnia pulex* are all-female and reproduce parthenogenetically while conditions are favorable. When drying threatens, males are produced, and fertilized eggs develop into diapausing, drought-resisting ephippia that lie in the substrate until the next flooding event and temperature cues stimulate hatching. Similar alternating generations occur in some rotifers.

Diving beetles (*Agabus* spp.) have a 2-year life cycle. They hatch from eggs in temporary pools and, when mature, fly to permanent waters to overwinter, returning to pools to breed the following spring. The eggs they leave hatch the following year, before the next wave of adults arrives. Some water-mite larvae that parasitize *Agabus* and other migratory insects are transported by their hosts from temporary waters in fall and back in spring; they then pass through two predatory life stages before laying eggs that hatch into new parasitic larvae.

Dispersal, population maintenance, and evolutionary ecology

How nonmotile organisms disperse has long fascinated biologists and has implications for community composition and stability. Mechanisms include transport by wind; bird feet, feathers, and digestive tracts; water; humans; and insects. Many temporary-water populations are units of metapopulations; they undergo periodic local extinctions and are recolonized from other waters or provide colonizers for other sites. Genetic analysis and modeling help determine the extent of genetic mixing needed to maintain populations or allow divergence.

Endemic species, especially in crustaceans and some plant taxa, are widespread in temporary waters. New species of copepods, anostracans, and other crustaceans are still being identified from all over the world and provide exciting opportunities to understand evolutionary processes.

Community Ecology

Community studies include questions about local and regional biodiversity; community composition and structure in relation to environmental and biological variables and disturbances; patterns of colonization and extinction;

predator–prey, host–parasite, and competitive interactions between species; and food webs.

Comparable temporary waters differ in their biota. Distributions of species, and thus community composition, shift along gradients of size, hydroperiod, predictability, and salinity, with richness increasing with decreasing stress. Community composition may change between years, and it can also vary seasonally, with a succession of new hatches and migrants entering waters over time. The presence of potential community members as unhatched propagules in the sediment complicates assessments of community composition and structure.

Community theory

The theory of island biogeography postulates that species richness in isolated habitats is regulated by local extinction and colonization and should vary with habitat size and proximity to potential sources of colonizers. The intermediate disturbance hypothesis predicts high richness in communities subject to a moderate degree of disturbance or stress; according to this model, high stress leads to mortality in all but fast-growing individuals, and under low stress, inter- and intraspecific interactions such as competition and predation determine community structure. Other models look at resource and habitat partitioning/niche diversification, temporal offsets in life histories, and other mechanisms controlling community composition and structure. Studies of amphibians, plants, invertebrates, and algae in temperate woodland pools, Mediterranean temporary pools, Negev and Namibian desert pools, Scandinavian rockpools, Arctic snowmelt pools, and other areas show complex relationships between community composition and habitat variables such as size, hydroperiod, frequency of flooding, hydrologic predictability, distance from other waters, and salinity. The data suggest that community richness is related to both degrees of disturbance and the predictability of disturbance. Isolation is also important, with greater richness in waters that are connected to larger bodies (e.g., in floodplains) but also fewer taxa specifically adapted to temporary habitats. Species pools in individual water bodies are poor in comparison to the regional set of species (**Table 6**), and experimental assemblages comprised of larger subsets of available species function differently than the smaller natural communities.

Interspecific interactions

Food web manipulations allow examination of relationships among species and show interesting relationships. For instance, algae grow better when grazed by tadpoles than alone. Some potential competitors avoid conflict by preferentially choosing waters with different hydrologic or other characteristics when the other species are present. The survival outcomes for some species of amphibians and insects when they co-occur with competitors depend on which species becomes established first.

Table 6 Regional species pools (β diversity) are greater than local species pools (α diversity), as illustrated by numbers of non-dipteran macroinvertebrates found in early spring from nine adjacent temporary pools on Cape Cod, Massachusetts, USA

Water body	Number of taxa
Pool 1	34
Pool 2	22
Pool 3	24
Pool 4	37
Pool 5	12
Pool 6	38
Pool 7	48
Pool 8	28
Pool 9	22
Total species	89

Modified from figure 2 in Colburn EA (2004) *Vernal Pools: Natural History and Conservation.* Blacksburg, VA: McDonald and Woodward.

Certain aquatic insects, crustaceans, and vertebrates can survive in pools with long flood durations, but they are typically found only at the more ephemeral end of the flooding continuum. They are excluded from the longer hydroperiod pools by predators such as amphibian larvae, tadpole shrimp, and water bugs. The ovipositing females of some species explicitly avoid pools with predators. For example, vulnerable species of mosquitoes avoid laying eggs in pools containing predatory backswimmers, whereas predation-resistant midge larvae do not; American toads (*Bufo americanus*) avoid temporary pools with omnivorous wood frog tadpoles (*Rana sylvatica*).

Ecosystem Ecology

There are many questions about temporary waters as ecosystems. How do tiny, intermittently flooded water bodies produce huge numbers of insects, amphibians, and other organisms? How do nutrients, carbon, and energy flow within temporary waters, and between them and adjacent terrestrial landscapes?

Some temporary waters are among the most productive ecosystems known. In some temporary habitats, photosynthesis by microscopic producers is the base of the food web. For many, from microhabitats in plant leaves to large woodland and floodplain pools, decomposing detritus is the primary energy source. Much remains to be learned about the sources and fluxes of energy and nutrients in temporary-water ecosystems.

Applied Ecology

Vector Control

Mosquito-borne diseases including encephalitis, yellow fever, West Nile fever, and, especially, malaria affect millions of people and are a major focus of world health agencies. Most mosquitoes breed in temporary waters. Their populations have expanded following human alterations of natural habitats, the creation of flooded areas by equipment and land-use change, and the dispersal of water-retaining containers. The effective long-term control of disease vectors requires understanding of the ecology of the pest animals and of their habitats.

Ecological Engineering and Conservation

Bird and amphibian populations and unique aquatic species depend on temporary waters, and the overall contributions of these systems to biodiversity are still being explored. Losses of these habitats are severe (e.g., loss estimates for California vernal pools exceed 90%), and remaining sites face draining, filling, excavation, pollution, water abstraction, invasive species, and climate change. In many regions, seasonal rivers and pools are important water sources; elsewhere, temporary waters provide the only arable areas. Many have been dammed, or converted for rice culture and other crops. Hydraulics, hydrology, surface–groundwater interactions, and biology affect management of these systems for human use and conservation, habitat restoration, and habitat creation.

See also: Freshwater Lakes; Freshwater Marshes; Saline and Soda Lakes.

Further Reading

Batzer DP, Rader RB, and Wissinger SA (eds.) (1999) *Invertebrates in Freshwater Wetlands of North America: Ecology and Management.* New York: Wiley.

Belk DA and Cole GA (1975) Adaptational biology of desert temporary-pond inhabitants. In: Hadley NF (ed.) *Environmental Physiology of Desert Organisms*, pp. 207–226. Stroudsburg, PA: Dowden, Hutchinson and Ross, Inc.

Caceres CE (1997) Dormancy in invertebrates. *Invertebrate Biology* 116(4): 371–383.

Calhoun AJK and DeMaynadier P (eds.) (2007) *Science and Conservation of Vernal Pools in Northeastern North America.* New York: CRC Press.

Colburn EA (2004) *Vernal Pools: Natural History and Conservation.* Blacksburg, VA: McDonald and Woodward.

Eriksen C and Belk D (1999) *Fairy Shrimps of California's Pools, Puddles, and Playas.* Eureka, CA: Mad River Press.

Fryer G (1996) Diapause, a potent force in the evolution of fresh-water crustaceans. *Hydrobiologia* 320: 1–14.

Hartland-Rowe R (1972) The limnology of temporary waters and the ecology of Euphyllopoda. In: Clark RB and Wooton EF (eds.) *Essays in Hydrobiology*, pp. 15–31. Exeter, UK: University of Exeter.

Laird M (1988) *The Biology of Larval Mosquito Habitats.* Boston: Academic Press.

Simovich M and Hathaway S (1997) Diversified bet-hedging as a reproductive strategy of some ephemeral pool anostracans (Branchiopoda). *Journal of Crustacean Biology* 16(3): 448–452.

Sloan CE (1972) Ground-Water Hydrology of Prairie Potholes in North Dakota. *USGS Professional Paper 585-C.* Reston, VA: US Geological Survey.

Wiggins GB, Mackay RJ, and Smith IM (1980) Evolutionary and ecological strategies of animals in annual temporary pools. *Archiv für Hydrobiologie (Supplement)* 38: 97–206.

Williams DD (1987) *The Ecology of Temporary Waters*. Portland, OR: Timber Press.

Williams DD (2006) *The Biology of Temporary Waters*. London: Oxford University Press.

Witham CW, Bauder ET, Belk D, Ferren WR, Jr., and Ornduff R (eds.) (1998) *Ecology, Conservation, and Management of Vernal Pool Ecosystems – Proceedings from a 1996 Conference*. Sacramento, CA: California Native Plant Society.

Zedler PH (1987) *The Ecology of Southern California Vernal Pools. Biological Report 85(7.11)*. Washington, DC: US Fish and Wildlife Service.

Tropical Rainforest

R B Waide, University of New Mexico, Albuquerque, NM, USA

Introduction
Definitions
Distribution
Climate and Soils

Forest Structure
Biodiversity
Conservation Issues
Further Reading

Introduction

Of all of the Earth's ecosystems, tropical rainforests exist on the extremes of temperature, rainfall, biodiversity, and structural complexity. Tropical rainforests exist only where high year-round temperatures are found in conjunction with moderate to high rainfall, which limits both their latitudinal and elevational distribution. Compared to other ecosystems, tropical rainforests have high numbers of plant and animal species and often show great specificity in their biological relationships. This high taxonomic diversity contributes to high functional diversity, which results in a complex forest structure comprising many different life forms and sizes of plants. This diversity and structural complexity makes tropical rainforests one of the most interesting and complex ecosystems on Earth, and, as such, tropical rainforests have captivated the imaginations of scientists and the public alike.

Definitions

The common use of the term 'tropical rainforest' varies among regions of the globe depending on the ecological context. There is general agreement that tropical rainforests are tall, dense evergreen forests existing in wet and warm places, but since there is a degree of subjectivity in these terms, the name 'tropical rainforest' is applied to forests on different continents that may be quite different structurally. Moreover, since the characteristics that define tropical rainforest grade with latitude and

elevation, the boundary between tropical rainforest and other forest types is by necessity arbitrary.

The classical definition of tropical rainforest focuses on features of the vegetation: evergreen, hygrophilus, tall, and rich in lianas and epiphytes. Additional characteristics of tropical rainforest include the dominance of woody plants (**Figure 1**), principally trees; high species richness; sparse undergrowth; relatively slender trunks compared to trees of temperate forests; straight boles without branches except near the top; buttresses (**Figure 2**); large, dark green leaves with entire margins; the occurrence of flowers on the trunk or branches; and inconspicuous green or white flowers.

Alternative definitions of tropical rainforest focus on characteristics of the forest community and its environment, including the proportion of deciduous trees in the canopy, the elevation of the forest, and the length and severity of the dry season. Some local classifications of forest type incorporate floristic information, but such classifications require detailed knowledge of the flora as well as trained experts to implement them.

Several schemes attempt to classify vegetation types based on climatic conditions including temperature, rainfall, length of dry periods, and evapotranspiration. Such classification systems avoid the subjectivity inherent in definitions that depend on relative terms, but by necessity are oversimplifications of the factors controlling the distribution of tropical rainforests. The Köppen classification uses the average annual precipitation, average monthly precipitation, and average monthly temperature to divide the globe into six major climate regions and their subregions. Under this system, the tropical rain climate has no month whose mean

Figure 1 Tall, relatively slender trees with straight boles are characteristic of tropical forest on Barro Colorado Island, Panama. Credit: Nicholas V. K. Brokaw.

Figure 2 Broad buttresses are found on many rainforest trees, such as this specimen from Providencia, Antioquia, Colombia. Credit: Robert B. Waide.

temperature is less than 18 °C and the mean rainfall of the driest month is >60 mm. In the Holdridge classification, rainforests are defined as areas where the ratio of potential evapotranspiration to rainfall is low, with tropical lowland rainforests occurring where the mean annual temperature exceeds 24 °C. Shifts between forests classes are determined by changes in rainfall and temperature related to elevation and latitude. These classification systems work well in areas where plant formations are strongly controlled by climate, such as Central America and northern South America, but are less useful where edaphic factors or other environmental factors are major controlling factors, as in the lower Amazonian region in Brazil.

The rest of this article focuses on lowland, evergreen tropical forests occurring in hot, wet conditions. Tropical forests at higher elevations or in areas with a pronounced dry season are covered elsewhere.

Distribution

Tropical rainforests exist wherever conditions are appropriate, but are mostly confined to a broad belt around the equator. The latitudinal distribution of tropical rainforests is limited by the distribution of freezing temperatures, which tropical plants are unable to withstand. A circumglobal belt of dry conditions also limits the distribution of tropical rainforest and, except in rare cases, prevents a continuous transition between tropical and temperate forests. Some gaps occur in the equatorial band of tropical rainforests, such as in eastern Africa, where prevailing conditions are too dry for tropical rainforest to develop. Moreover, in some areas on the eastern margins of continents, conditions suitable for tropical rainforest exist outside of the tropics. In all areas suitable for the development of tropical rainforest, human actions limit the present distribution of forests.

Large areas of tropical rainforest exist on continents and large islands that straddle the equator. Roughly half of the tropical rainforests on the planet occur in three areas in tropical America. The largest of these forest areas (somewhat over 3 million km^2) occupies the drainage basins of the Amazon and Orinoco Rivers in northern and central Brazil and surrounding countries. A narrow strip of tropical rainforest runs along the Atlantic coast of Brazil from 7° to 28° S (from Recife nearly to São Paulo), but less than 5% of this forest remains in its original condition. A third block of forest occupies southern Mexico, Central America, and the area of northern South America west of the Andes. Many Caribbean islands also have small areas of tropical rainforest.

In Africa, another large block of tropical rainforest occupies the basin of the Congo River in the Democratic Republic of the Congo, the Republic of the Congo, Gabon, and Cameroon. Previously, part of this forest extended into Nigeria. Belts of tropical rainforest also extended along the coast of West Africa and the eastern part of the island of Madagascar, but little remains of these forests but isolated patches.

A third large area of tropical rainforest existed on the Malay Peninsula and the islands of Borneo, Sumatra, and Java. Sulawesi, the Philippines, and many of the smaller islands in Indonesia also have substantial areas of rainforest, but the condition of the remaining forest varies widely from island to island. Rainforests also occupied parts of mainland Southeast Asia where rainfall was sufficiently high. Isolated patches of tropical rainforest occur in the area of the Western Ghats in India and on the island of Sri Lanka. Most of the island of New Guinea supports tropical rainforest, and there is also a small area of rainforest in NE Australia. Patches of tropical rainforest also occur on some of the Pacific Islands (Solomons, New Hebrides, Fiji, Samoa, New Caledonia).

Climate and Soils

Tropical rainforests are found under a surprisingly wide range of climatic conditions. Annual rainfall is generally high in tropical rainforest compared to other ecosystems, but can range from 1700 to 10 000 mm. Many rainforests experience 1–4 dry months a year, when the rainfall is less than the water lost through evaporation and transpiration. These annual dry periods exert a strong effect on the phenology of biotic processes such as flowering and fruiting. In some tropical rainforests, rainfall is uniformly high throughout the year, and no annual dry periods exist. In these forests, dry periods may occur at multiyear intervals and trigger strongly synchronized biotic responses including mass flowering, increased animal reproduction, and migration. In some parts of the world, these multiyear cycles are related to periodic El Niño–Southern Oscillation (ENSO) events. While strong ENSO events result in more severe dry periods in many tropical rainforests, the strongest biological consequences seem to occur in forests that do not normally experience a dry season, especially in areas of Indonesia and Malaysia.

Mean annual temperatures in tropical rainforests generally fall in the range between 24 and 28 °C near the equator, but a consistent characteristic is the absence of a cool season. In general, diurnal temperature differences (6–10 °C) exceed monthly differences. The amount of solar radiation is higher in the topics than in temperate zones, but tropical rainforests generally have lower available solar radiation than drier tropical forests because of the greater amounts of water vapor and increased cloudiness in more humid climates. As a result, plant growth in closed-canopy tropical rainforests is often light limited.

The environments of tropical rainforests are characterized by high relative humidity during the daytime and generally saturated conditions at night. However, because much of the rainfall in tropical rainforests occurs in intense events, even months with high rainfall can have periods of a few days when little or no rain falls, saturation deficits increase, and plants wilt. The dry periods can be exacerbated by winds; evaporation rates are higher in the trade wind zone than in equatorial forests where average wind speeds are less.

Tropical cyclones can have severe effects on tropical rainforests. In general, areas within 10° latitude of the equator are not subject to tropical cyclones, but tropical rainforests in the Caribbean, Madagascar, northeastern Australia, many oceanic islands, and parts of Central America and Southeast Asia are affected by these storms. The strongest tropical cyclones can have severe but, in most cases, temporary effects on forest structure and composition. Tree mortality can be high as a result of one of these storms but the forest recovers quickly through regeneration, new growth, and refoliation of damaged trees. In those areas of tropical rainforest subject to recurrent tropical cyclones, forest structure and the biological traits of forest species may be affected by the frequency and intensity of storms.

The soils underlying tropical rainforests can have important effects on plant distribution and primary productivity. The complex interactions between soil characteristics (e.g., soil texture, age, drainage characteristics, nutrients) and the considerable topographic and geographic variation in these characteristics make it difficult to determine the importance of specific soil properties. Most areas supporting tropical rainforests have very old soils that are highly leached and weathered and as a result acidic and infertile. Such soils have low levels of the nutrients necessary for plant growth and high levels of toxic aluminum and thus are unsuitable for most forms of permanent agriculture. However, these soils can sustain high-diversity, high-biomass tropical rainforests because plants of these forests recycle nutrients efficiently. Some tropical rainforests occur on relatively fertile volcanic or floodplain soils and can sustain permanent agriculture.

In areas of the Amazon with low local relief, soil properties can have a strong effect on plant communities and therefore overall biodiversity. Small changes in topography and the depth of sand overlying the clay subsoil can cause large changes in the plant community.

Forest Structure

Tropical rainforests support a more diverse set of organisms than other kinds of forests. The number of different life forms, or synusiae, is greater in tropical forests that in temperate forests. A synusia is a group of organisms whose members are ecologically equivalent. When applied to plants, the term reflects an aggregation of species with similar life form and function. Autotrophic plants (e.g., those that photosynthesize) include those that do not need mechanical support (i.e., trees, shrubs, and herbs) and those that do (i.e., climber, epiphytes, and

Figure 3 Bromeliads and other epiphytes cover branches of trees in La Planada Reserve, Colombia. Reproduced by permission of Art Wolfe/Photo Researchers, Inc.

Figure 4 Canopy of lowland tropical rainforest of La Selva Biological Station, Costa Rica. Photographed from a light plane flying 200 feet above the canopy. Reproduced by permission of Gregory G. Dimijian, MD/Photo Researchers, Inc.

hemi-epiphytes; **Figure 3**). Heterotrophic plants include saprophytes and parasites.

The structure of a tropical rainforest arises from each synusia's methods for obtaining resources for survival and growth: water, nutrients, and sunlight. In some forests, photosynthetic, self-supporting plants seem to form distinct strata depending on their size. Such stratification is by no means a uniform characteristic of tropical rainforests. Photosynthetic plants that are not self-supporting use other plants as a platform for growth. Climbers (lianes) have roots in the ground but use other plants to support their elongated stems. Epiphytes depend on their host plants for support only, although a specialized group of this synusia (the mistletoes) obtains both support and water and dissolved substances from the support tree. Hemi-epiphytes initially live as epiphytes on supporting plants but eventually send roots down to the ground. Saprophytic and parasitic plants obtain required energy and nutrients from other living or dead plants, and therefore do not require light for growth or reproduction.

Background mortality of individual trees from natural causes is a major cause of spatial heterogeneity in tropical rainforests. Gaps caused by dead or fallen trees change the structure and the environmental characteristics of the forest. However, tropical rainforests are also dynamic ecosystems subject to a large number of natural disturbances including storms, lightning strikes, landslides, and the effects of animals, all of which can produce gaps in the forest canopy.

The canopy is an important structural element of tropical rainforests because the height and degree of closure of the canopy plays an important role in determining conditions in the understory (**Figure 4**). Moreover, the lack of easy access to forest canopies means that their importance as a source of biodiversity and an influence on ecosystem processes has probably been underestimated. Forest canopies have important roles in the regulation of nutrient cycling and in the storage of carbon. Large pools of nutrients exist in live and dead components of the canopy, and

decomposition of organic matter in the canopy influences access to these nutrient pools. The forest canopy serves to filter air- and waterborne nutrients and to provide a site for nitrogen fixation. Canopy-dwelling organisms are efficient at acquiring and storing nutrients, thus providing a buffer for pulsed nutrient releases. Forest canopies are rich in species of plants and animals that are independent of the forest floor. Moreover, canopy trees and their epiphytes provide important sources of food for birds, mammals, and insects that occupy other strata.

Biodiversity

Understanding of the biodiversity of tropical rainforests is still being refined. New species of all taxonomic groups are found every year, and knowledge of the diversity of some taxa, especially insects, is rudimentary. Tropical rainforests are extremely rich in species of all taxa compared to other terrestrial ecosystems. For example, the tropical rainforests of the world have an estimated 175 000 species of plants, which constitutes about two-thirds of the global total. Considerable variation in diversity occurs among tropical rainforests around the world, with the largest number of tree species (>250 species per hectare) occurring in Amazonia and Malaysia, followed by the islands of New Guinea and Madagascar, and then Africa. The largest areas of tropical rainforest (Neotropics, Africa) have the greatest number of primate species. Similar comparisons for other taxa are difficult because of the lack of data.

Conservation Issues

Because of their global significance with regard to carbon storage and the maintenance of biodiversity, conservation of tropical rainforests is an important and hotly debated topic. Solution of conservation issues is made more difficult because most areas of tropical rainforest occur in countries that are trying to increase the standard of living of their people. Partly because of this controversy, adequate data to judge the loss of tropical forests are difficult to come by. However, it is clear that tropical forests, including tropical rainforests, are disappearing at an increasing rate. The percent of the original forest habitat that has been lost exceeds 90% for some countries (Ghana, Bangladesh, Philippines). Estimates suggest that very little tropical rainforest will remain by the year 2050. The ultimate causes of forest loss include increasing populations in countries with tropical rainforest, extreme poverty, and the lack of effective government protection for forests. Proximate causes of forest loss include logging, clearcutting for agriculture, loss of ecosystem integrity because of forest fragmentation, and hunting (**Figure 5**). Hunting of large animals may have insidious effects on forest structure, as the populations of prey species may explode when released from predation. Increased populations of small mammals, for example, may have severe effects on other organisms, leading to the breakdown of whole ecosystems over time.

Because the issues facing tropical rainforests vary considerably from one place to another, generic conservation solutions are not practical. However, the major elements of a conservation strategy for tropical rainforests will include the creation of reserves to protect biodiversity, the regulation of exploitative use of tropical rainforest products, the engagement of traditional societies, the development of sustainable use strategies that will address the issue of poverty, and an increased effort by developed countries to form partnerships with developing countries.

Figure 5 Rainforest has been cleared for timber and agriculture in this subsistence farm in Providencia, Antioquia, Colombia. Credit: Robert B. Waide.

Further Reading

Denslow JS and Padoch C (eds.) (1988) *People of the Tropical Rain Forest*. Berkeley, CA: University of California Press.
Gentry AH (ed.) (1990) *Four Neotropical Rainforests*. New Haven, CT: Yale University Press.
Golley FB (ed.) (1989) *Tropical Rain Forest Ecosystems*. New York, NY: Elsevier.
Primack R and Corlett R (2005) *Tropical Rain Forests: An Ecological and Biogeographical Comparison*. Oxford: Blackwell Science.
Richards PW (1996) *The Tropical Rain Forest: An Ecological Study*. Cambridge: Cambridge University Press.
Sutton SL, Whitmore TC, and Chadwick AC (eds.) (1983) *Tropical Rain Forest: Ecology and Management*. Oxford, UK: Blackwell Scientific Publications.
Terborgh J (1992) *Diversity and the Tropical Rain Forest*. New York: Scientific American Library.
Whitmore TC (1998) *An Introduction to Tropical Rain Forests*. New York, NY: Oxford University Press.

Tundra

R Harmsen, Queen's University, Kingston, ON, Canada

Introduction
The Periglacial Environment
Landscape and Species Diversity
Vegetation and Succession

Ecosystem Structure and Function
Special Adaptations to Tundra Conditions
Global Warming and Other Anthropogenic Effects
Further Reading

Introduction

Tundra ecosystems are widely distributed over all continents. Tundra is characterized by climatic stress consisting of low temperatures, strong winds, low precipitation, frost

action, and long periods of shortage of liquid water caused by freezing and/or drought. These stresses combine to create what is called the periglacial environment, which is defined by repeated effects of freezing and thawing on soils and water bodies. Tundra comes in many different kinds. The two main categories are arctic and alpine tundra. Each of these can be divided into subcategories or can be seen as gradients from a richly vegetated tundra with tall shrubbery adjacent to the 'tree line' through categories with less vegetation to barren areas with only a minimum of vegetation adjacent to ice-fields and permanently frozen polar or alpine areas (see Boreal Forest and Alpine Forest). Included in the tundra biome are tundra ponds, lakes, streams, marshes, and other wetlands.

Since tundra is found at the cold limit of life forms on Earth, climatic changes of the past have had major effects on tundra ecosystems and the plant and animal species of these systems. With each Pleistocene ice age, big areas of arctic tundra were eradicated, while others shifted southwards, as entirely new areas of forest or prairie became tundra, only to be reversed with the subsequent interglacials. Similar changes would have occurred in mountainous regions. These major changes resulted in the extinction of species and in the disruption of coevolved, interactive plant, and animal assemblages. These changes in tundra communities persist today, resulting in low species diversity and the scarcity of complex food-chains. During the current interglacial, many areas that were pushed down by the weight of the ice were first flooded by the rise in sea level, but have subsequently, in part at least, re-bounded and developed into tundra. Some parts of Beringia (eastern Siberia, northern Alaska, and into the Yukon and Banks Island) were not glaciated, and retained a far northern tundra during the last glacial period. Species of plants and animals that now form arctic tundra communities survived the ice ages either south of the glaciated land area or in unglaciated refuges such as Beringia.

The Periglacial Environment

Periglacial conditions are the result of current or geologically recent frost and ice formations on a landscape. Glaciers affect landscapes in major ways, which can have long-lasting effects on geomorphology, drainage systems, and soils. But even temperature regimes that cause frequent freeze–thaw cycles – for example, annually in the high arctic and daily on high tropical mountains – affects not only plants and animals directly, but also have indirect effects on soils and water, which results in specific types of erosion and the formation of characteristic landscapes. Furthermore, the usual presence of permafrost under tundra ecosystems is of critical importance, in that it forms a permanently impenetrable floor, preventing biological penetration and vertical movement of water and nutrients.

The freeze–thaw cycle causes expansion and contraction of soils and water, while the gradual freezing of wet soils will also cause a nonrandom redistribution of water into ice lenses and ice wedges. These processes can result in frost heave and long-term vertical and horizontal movements of soils, debris, and even large rocks, creating typical landscape features (such as polygons), frost mounds (such as palsas and pingos), slope solifluction, and others. These land forms in turn affect the vegetation and all other life forms.

The permafrost is ubiquitous in the arctic tundra, but less frequently found in alpine tundra sites, as alpine landscapes are more diverse and the summers are warmer. It is not found in any but the highest tropical mountains. During spring, the thawing of the soil starts from the surface down, gradually releasing the vegetation from the grip of the frozen soil. There is usually an overlap between snow melt and the thawing of the soil, especially in undulating landscapes. All melt water must run off, accumulate in low areas, or evaporate, as no vertical movement of water is possible due to the impervious permafrost. This can cause erosion, affecting plants and small animals. During the summer, thawing of the permafrost continues till autumn, when the surface may already start to refreeze. During later autumn and early winter, the frost will penetrate deeper into the soil from the surface, as it also comes up from the main body of the permafrost. This process can cause considerable expansion and result in frost heave and can cause much damage to root systems and animal burrows. In many areas, tundra soils are low in nutrients, because the permafrost prevents vertical movement of soil water.

Some lakes (e.g., kettle lakes and moraine lakes) have their origins in major ice formations dating from the latest ice age, while others are recent formations. Tundra lakes and ponds are severely affected by annual freezing, especially those lakes that freeze each winter right to the bottom and beyond to the permafrost. Freezing of lake ice causes expansion and results in the shoreline with its vegetation being elevated above the surrounding lowlands. Lake sediments are often high in inorganic matter from spring runoff, but low in organic matter due to low productivity reflecting low nutrient levels. Frost action and wind effects on ice tend to disturb lake sediments in shallow lakes.

Landscape and Species Diversity

Many parts of the arctic tundra are flat, especially in areas adjacent to the sea. These areas are often covered with ponds and shallow lakes, separated by marshes and connected by meandering streams and rivers. These areas can accumulate peat and develop into fens. Along the seashore these habitats tend to merge into salt marshes, brackish lagoons, and beach ridges. On higher ground, with hills and rock outcrops, the landscape diversity is much greater, especially since north and south facing slopes have very different microclimates, and hence, very different biological communities. Here one can also find deep lakes and fast-flowing rivers. Both erosion and the underlying rock type will also add to ecosystem diversity. In mountainous areas the arctic tundra merges into an alpine version.

The diversity of alpine tundra worldwide is enormous, as it is found on all continents and in many climatic zones. Snow accumulation during winter, combined with slope, wind, and summer climate affect the length of the growing season of alpine tundra ecosystems. Tropical alpine tundra occurs only at very high altitudes, with unique climates varying from desert to some of the wettest conditions on Earth (**Figure 1**). It should also be noted that many of the alpine tundra zones are isolated from other such zones by hundreds or even thousands of kilometers, so that they have undergone independent evolution of their flora and fauna. Especially geologically old high mountains contain many endemic species derived from

local forest or savannah species. For instance, the Southern Alps of New Zealand have over 600 species of alpine plants, very few of which are found elsewhere on Earth.

Roughly 5% of the Earth's surface is covered with arctic vegetation and 3% with alpine vegetation. The alpine tundra worldwide, as well as per hectare for most alpine systems, has a much higher biodiversity than the arctic lowland tundra. Species richness declines with altitude on mountains and with latitude in the arctic, and also is dependent on local climatic conditions, nutrient availability, etc.

Vegetation and Succession

Whether one climbs a mountain and crosses the timberline, or travels northwards in the arctic, and crossing the tree line, one enters the low tundra, which is characterized by shrubs. A combination of low temperatures, shallow soils, and strong winds prevents tree growth, but a tight shrub cover manages to thrive under such conditions. On each mountainous area on Earth shrub tundras can be found, which are superficially quite similar to other isolated alpine shrub tundra communities; even many of the individual species have a remarkably similar appearance. However, mostly unrelated species form such shrub communities in different parts of the world. For instance, most of the species of the shrub vegetation on East Africa's Mount Kenya, New Guinea's Mount Wilhelm, and Pico Mucuñuque of the Venezuelan Andes belong to different families. This is a good example of convergent evolution acting on divergent taxa, causing adaptation to a specific environment. The shrub zone in the Canadian arctic has a more impoverished vegetation than the shrub zones on tropical mountains. It is dominated by several species of willow and birch, and a smattering of other species (**Figure 2**). Again, the arctic

Figure 1 Mount Kenya, tropical Africa. High tropical alpine tundra. In the foreground a boulder moraine with lichens, mosses, scattered tussock grasses, and a few rosettes of the large *Seneciodendron keniensis*. In the middleground a sparse stand of the yet larger *Seneciodendron keniodendron*. The genus *Seneciodendron* is endemic to the east-central African mountains. In the background the Tyndall Glacier. Photo by W. C. Mahaney.

Figure 2 Hudson Bay Lowlands, Northern Manitoba, Canada 60° N. Low arctic willow (*Salix* spp.) and graminoid tundra. Note the radio-collar on the polar bear.

tundra in Greenland, Scandinavia, or Siberia also looks very similar, but in this case the species are all close relatives or even the same circumpolar species on the different continents. Another difference is that on tropical mountains there are a lot of shrub species that are not found below the tree line, whereas in the arctic many of the shrub species are also found south of the tree line. These differences are the result of the different effects of the ice ages, which on mountains merely caused the vertical movement of more or less entire plant communities up and down alpine valleys and slopes, while in the arctic, changes in the climate can cause north–south displacements of the conditions suitable for shrub tundra of over hundreds of kilometers.

The more typical graminoid, forb, and moss tundra found higher up the mountains and further north in the arctic is adapted to extreme cold, long periods of temperatures permanently below freezing (and permanent darkness in the arctic) and strong winds. It is the strong winds blowing ice crystals which abrade any vegetation above snow level, combined with desiccation that makes tree and tall shrub growth impossible in high arctic and alpine tundra. Especially in arctic deserts, where snow cover is low, vegetation remains very low to the ground (**Figure 3**). For instance on Banks Island at 70° N, arctic willow (*Salix arctica*) grows horizontally along the ground, forming matted areas of intertwining branches that form catkins and leaves in summer. One such willow can live and grow for decades. All grasses, sedges, and forbs die back in autumn and survive the winter as belowground root masses, or as ground hugging rosettes.

One advantage of being a plant in a dense, low to the ground plant community is that on cool, sunny summer days radiant heat from the 24 h solar radiation is trapped within the air between the plants, keeping temperatures high enough for growth and maturation of seed. There are very few annual plants in the high arctic tundra, because the season is not long enough to germinate, grow, and reproduce. A few very small species, such as *Koenigia islandica* and *Montia lamprosperma*, maintain an annual life strategy. Uniquely a few species of semiparasitic members of the Scrophulariaceae, such as *Euphrasia arctica*, do so as well. These species have a distinct early season advantage being able to grow very rapidly by gaining nutrients and photosynthate from neighboring perennials.

The frequent disturbances due to the freeze–thaw cycles often lead to local eradication of vegetation. This creates openings for reinvasion and subsequent succession. One of the most interesting examples of this is the result of solifluction of soil clumps on south facing slopes in the arctic. Soil clumps with vegetation surrounded by clefts get heated by the sun on the downslope side, causing them to thaw out and slump downwards, burying the lowest vegetation, while at the same time exposing a small strip of upslope bare soil (**Figure 4**). It takes up to 30 years for the clump to make one entire downhill rotation. On each clump, one can see a successional sequence of plant maturity, species composition, and diversity, as the oldest community gets buried and an opening appears at the top end for reinvasion. Succession on a larger scale occurs after slope collapses, frost mounting, stream erosion, mud deposits after flooding, etc.

Ecosystem Structure and Function

Very few species remain active within arctic tundra ecosystems during the winter. Only most mammals such as the muskox (*Ovibos moschatus*), the reindeer (*Rangifer*

Figure 3 Banks Island, North West Territories, Canada 70° N. Upland high arctic tundra, also described as arctic desert. The vegetation is dominated by mountain avens (*Dryas integrifolia*) and various species of arctic vetch (*Oxytropis* spp. and *Astragalus* spp.) and scattered clumps of small graminoids. In the background is the Thomsen River valley with sedge meadow tundra and tundra ponds.

Figure 4 Banks Island, North West Territories, Canada 70° N. Two types of high alpine tundra. In the foreground a wet graminoid tundra fed by snowmelt water. On the opposite slope a sparsely vegetated dry tundra showing solifluction. The muskoxen feed primarily on the graminoid slope, but will venture onto drier tundra types to feed on high nitrogen species such as arctic vetch.

tarandus), the arctic hare (*Lepus arcticus*), lemmings, and the wolf (*Canis lupus arctos*) remain fully active. A few birds, for example, raven (*Corvus corax*) and rock ptarmigan (*Lagopus mutus*), manage as well. During the autumn and early winter months, soil microbial metabolic activity continues down to at least $-12\,^{\circ}\mathrm{C}$. The vast majority of organisms that spend the winter on the tundra do so in some form of dormancy. Alpine tundra, being much more diverse, and much of it having periods of daylight throughout the year, varies greatly in the degree of winter activity of the fauna. The brief summer on the tundra is enormously productive, and provides food for a wide variety of organisms. The vegetation starts to bloom and grow as soon as the snow starts to melt. At that time of year the sun hardly sets if at all and temperatures rise quickly. Dormant overwintering insect larvae start to feed and eggs eclose to add innumerable larvae in snow melt ponds, in the soil, and on the new vegetation. The ecosystem seems to burst into active life. High availability of edible vegetation, exploding insect, bird and rodent populations, and young birds lasts till just before freeze-up in autumn (**Figure 4**).

Many bird species migrate annually from more southerly wintering sites to the tundra to breed, taking advantage of, and adding to, the burst of summer productivity. Some of these species arrive in extremely large numbers. Most of these birds are insectivorous or feed on pond crustaceans, some such as loons and grebes are pisciverous, falcons and hawks are predators, and geese are herbivorous. Especially the colonially nesting geese can have major destructive effects on the vegetation, which in turn can affect many other species.

In some tundra ecosystems some small mammals, especially two species of lemmings, show extreme oscillations in population density, making them keystone species in the tundra ecosystem. For instance, on Banks Island in northern Canada both the collared lemming (*Dicrostonyx torquatus*) and the brown lemming (*Lemmus sibiricus*) undergo sharp population oscillations with a 3–5-year period. At peak populations the lemmings are all over the place, whereas the year after it is hard to find a single lemming. During the outbreak phase, several predatory birds, including snowy owls (*Nyctea scandinaca*), rough-legged hawks (*Buteo lagopus*), and jaegers (*Stercorarius* spp), migrate long distances and concentrate in the regions with high lemming populations. They lay large clutches and raise many young, only to disperse to other areas when the lemming population collapses (**Figure 5**). Mammalian predators are not as able to respond by migration. Arctic foxes (*Alopex lagopus*) and ermines (*Mustella erminea*) are the main mammalian predators; they also take advantage of lemming outbreak with large litters. However, this leaves relatively dense populations of these predators after the collapse of the lemming population. This has a major feed-forward effect in that the half-starved predators exert a

Figure 5 Nest of snowy owl (*Nyctea scandiana*) with six eggs and one hatchling. Snowy owls start incubating as soon as their first egg is laid, so that the young are hatched sequentially. Note the seven dead lemmings surrounding the nest, intended as food for the hatchlings. Later that summer, the lemming population crashed. Only the two eldest hatchlings survived to fledge, the others were eaten by the older ones.

strong negative effect on other less-favored prey species, mostly birds, from small passerines to ducklings and even goslings. Only after the predator population has collapsed can the lemming population start to grow again.

The ultimate cause of the collapse of the lemming population is not the predation pressure, but the exhaustion of quality vegetation and a delay in nutrient cycling. However, once the lemming population has collapsed, the subsequently declining predator population can drive the lemming population further down to its minimum. The vegetation, litter layer, and soils are strongly affected by the lemming cycles. This is shown by the enormous difference between the tundra in northern Canada and central Greenland, as in Greenland there are no lemmings, much more accumulated litter, differences in relative abundance of plant species, and far fewer predators. Exclosure experiments in Canadian tundra have similar results.

Special Adaptations to Tundra Conditions

Many species have evolved special adaptations to the rigorous, but often predictable conditions of the tundra. This article presents four cases of such adaptations as examples of this phenomenon: the muskox, two species of arctic bumblebee, an alpine lobelia, and two congeneric alpine beetle species.

The muskox of Banks Island in Canada's Northwest Territories

The muskox (*Ovibos moschatus*) is a surviving species of the Pleistocene megafauna; it survived the ice age both in Beringia and south of the ice sheet in what is now southern Canada and the northern United States. It has a very long adaptive history in arctic conditions, which shows in a number of very effective adaptations to extreme cold. Besides the obvious anatomical features such as the extremely effective insulating wool under the shaggy guard hair and the front hooves that are perfectly shaped to scratch the hard arctic snow to expose vegetation, this animal has a set of integrated physiological and behavioral traits making up a unique reproductive strategy. A muskox cow responds to her nutritional condition in autumn by not going into heat when in poor condition, and only going into heat early in the rutting season when in excellent condition. This means that cows in poor condition, which would not have been able to survive the winter and produce a calf the next spring, will live and have another chance at reproduction the next year. The cows that do get pregnant, when faced with a bad winter, will either abort their fetus or abandon the calf after birth. Since most calves are born well before snow melt and the reappearance of new fodder, the cows have to be in good shape to not only carry the calf to birth, but also lactate for several weeks. However, only calves born early in the year have a good chance of gaining enough weight and reserve fat to survive their first winter.

Integrated with this strategy are some significant traits. At birth, the calf weight over cow weight ratio is one of the lowest among ungulates, making abortion or abandonment a relatively minor cost for the cow, which can then cut lactation. Once the calf is born and the cow is lactating, she licks the calf when it urinates and swallows the urine. The urea of the urine is rebuilt into protein by the cow's gut flora and will eventually be available for milk production. This is important because storage of protein over the winter is difficult, and late winter forage is scarce and low in protein. As soon as new forage is available during snow melt, the cows graze selectively on high protein vegetation, such as willow catkins and sprouting rosettes of arctic vetch (*Oxytropis* spp.). In far northern parts of their range, muskox cows live long lives, but only reproduce every second or third year and still lose some of their calves.

Two Species of Bumble Bee from the Canadian Arctic

The author has a personal recollection of working in early July on the tundra on northern Banks Island when in the middle of a snow squall a bumblebee flew by. This seemingly incongruous event is explained by the fact that the common large bumblebee (*Bombus polaris*) has an unusually well insulated thorax, which allows it to keep its flight muscles at approximately 30 °C even when the ambient drops to the freezing point. What is even more special about this species is that the queen keeps her abdomen also near 30 °C, which presumably allows its eggs to develop faster. However, early in the season the queen also warms her eggs and larvae in the nest by inserting her abdomen into the middle of the nest and producing heat by vibrating her flight muscles and circulating the heat to her abdomen. The queen, after overwintering, builds the nest, often in an abandoned lemming burrow, using bits of dead vegetation and muskox wool. There she raises one brood of workers before switching to start raising reproductives for the next year. The other species of bumblebee (*B. hyperboreus*) found on Banks Island is an obligatory brood parasite of *B. polaris*. The queens of this species lay only eggs for reproductives, and lay them in the nest of their host species. This strategy is obviously adapted to the very short summer season in the high arctic tundra, but it also depends on the presence of *B. polaris*. The ratio of the densities of the two species is stabilized by frequency-dependent selection.

Flightless Beetles of the Genus *Parasystatus* on Mount Kenya

In the tussock grass alpine tundra of Mount Kenya between 3200 and 4000 m, there are six described species and at least one undescribed species of the genus *Parasystatus*. These large beetles must be adapted to the diurnal extremes of the climate, which has been described as summer each day and winter each night. Two of these species, *P. elongates* and one undescribed species, have been studied in some detail as to their adaptation to the nightly frost of that zone. *P. elongates* spends its entire larval and pupal development inside a tussock of the grass *Festuca abyssinica*, where it is not affected by the nightly frost. As an adult beetle, it is active by day, shielded from the intense solar radiation by inflated elytra and a shiny, reflective outer cuticle. At night, the beetle hides under vegetation to avoid the worst of the frost; it has an ineffectively high supercooling point, but an effective freeze tolerance. The other species of the same genus is active well into the night, and protects itself with a much lower supercooling point, but is freezing sensitive. (Cooling a liquid to below its freezing point without phase transition; here pertaining to the avoidance of ice formation due to the presence of antifreeze substances and/or the absence of crystallization nuclei.) These two different physiological adaptations to nightly frost within one genus indicate that the two species have independently invaded the alpine tundra, rather than having arisen through speciation in the alpine zone. Being flightless – a typical adaptation to mountain top ecosystems – also rules out invasion from another mountain

The Giant Lobelia and Its Insect Commensals on Kilimanjaro

Between 3000 and 4000 m on the slopes of Mount Kilimanjaro, the giant lobelia (*Lobelia deckenii*) also has to face the stress of nightly frost, which can be severe due to parts of the Kilimanjaro alpine tundra being relatively dry. The plant has evolved into a ball-shaped rosette consisting of a fleshy center surrounded by concave spiky leaves, which are arranged in such a manner as to trap rainwater. A single plant can contain, trapped in its rosette, a compartmentalized mass of several liters of water. This volume is large enough to prevent it from freezing right to the middle in any one night. Indeed, the center of the plant where the growing tip is located maintains a very even temperature throughout the diurnal cycle. Not surprisingly, this water mass of the lobelia plants with its relatively even temperature has become the breeding environment for a few species of insects with aquatic larvae, the most abundant of which is a chironomid midge. The water in the lobelias also contains microorganisms, which feed on decomposing debris and are in turn the food for the insect larvae.

Global Warming and Other Anthropogenic Effects

Extensive research in arctic and alpine regions including ice core analysis, paleolimnology, palynology, and geomorphology has provided a detailed picture of the climatic history of these regions. This allows us to conclude that, as well as the major changes at the end of the last ice age, frequent climate oscillations have subsequently occurred that caused major changes in tundra ecosystems. Furthermore, there have been times when tundra types existed that are no longer extant. The species complexes that now exist consist of species that have been sufficiently flexible and/or dispersible to have survived the climatic and landscape oscillations of the past. However, this does not necessarily bode well for the future of tundra ecosystems and species, as anthropogenic changes are certain to be increasingly imposed on the Earth. Already, the most likely reason for the extinction of most of the Pleistocene arctic megafauna is a combination of climate change and human hunting. The disappearance of the large herbivores at that time caused a major switch in plant dominance on tundra ecosystems from graminoids to mosses, with concomitant changes in long-term soil and peat formation. We must expect similar major changes in the coming century, associated with at least some extinctions. Climate change will be severe and direct human effects will also increase. Already, several species are declining due to pollution and over-hunting. Some of the most at risk tundras (and associated endemic tundra species) will be isolated alpine tundra systems on relatively low mountains, where climatic warming will cause the entire system to be replaced with forest.

See also: Alpine Ecosystems and the High-Elevation Treeline; Alpine Forest; Boreal Forest; Cycling and Cycling Indices; Freshwater Lakes; Polar Terrestrial Ecology; Steppes and Prairies.

Further Reading

Chapin FS and Körner C (eds.) (1995) *Arctic and Alpine Biodiversity. Patterns, Causes and Ecosystem Consequences*, 332pp. Berlin: Springer.
Coe MJ (1967) *The Ecology of the Alpine Zone of Mount Kenya*, 136pp. The Hague: Junk.
Craeford RMM (ed.) (1997) *Disturbance and Recovery in Arctic Lands*, 621pp. Dordrecht: Kluwer Academic.
French HM and Williams P (2007) *The Periglacial Environment*, 478pp. Toronto: Wiley.
Goulson D (2003) *Bumblebees: Their Behavior and Ecology*, 235pp. Oxford: Oxford University Press.
Jones HG, Pomeroy JW, Walker DA, and Hoham RW (eds.) (2001) *Snow Ecology: An Interdisciplanary Examination of Snow-Covered Ecosystems*, 378pp. Cambridge University Press.
Laws RM (ed.) (1984) *Antarctic Ecology, vol. 1*, 344pp. London: Academic Press.
Mahaney WC (ed.) (1989) *Quaternary and Environmental Research on East African Mountains*, 483pp. Rotterdam: Balkema.
Pienitz R, Douglas MSV, and Smol JP (eds.) (2004) *Long-Term Environmental Change in Arctic and Antarctic Lakes*, 562pp. Dordrecht: Springer.
Rosswall T and Heal OW (eds.) (1975) *Ecological Bulletin, Vol. 20: Structure and Function of Tundra Ecosystems*, 450pp. Stockholm: Swedish Natural Science Research Council.
Wielgolaski FE (ed.) (1997) *Ecosystems of the World 3: Polar and Alpine Tundra*, 920pp. Amsterdam: Elsevier.

Upwelling Ecosystems

T R Anderson and M I Lucas, National Oceanography Centre, Southampton, UK

Introduction

Throughout the world's oceans, phytoplankton community structure and rates of primary production are determined by the interplay between available light and nutrient supply (NO_3^-, Si, PO_4^{2-}, dissolved Fe) as well as by grazing. Winds blowing over the ocean create a surface mixed layer, the depth of which is of great importance for production by phytoplankton. If mixing is vigorous, as is often the case at high latitudes, then nutrients are plentiful but plankton circulating within a mixed layer that may be hundreds of meters deep are exposed to low average light intensities. In contrast, mixing is inhibited in warm stratified waters such as those of the vast subtropical gyres that cover 40% of the surface ocean, in which case light is plentiful and limitation is instead by nutrients. The unique physical circulation of upwelling systems leads to conditions that, to varying degrees, provide both light and nutrients together in quantities that considerably exceed rate-limiting requirements for sustaining maximal growth rates of phytoplankton. As a result, upwelling ecosystems are among the most productive in the ocean.

Upwelling Circulation

The Coriolis effect, whereby the Earth's rotation causes moving bodies at its surface to be deflected, means that wind-driven ocean currents turn right in the Northern Hemisphere, and left in the Southern Hemisphere. The result is horizontal flow at the ocean surface in the so-called Ekman layer, typically tens of meters deep. Upwelling occurs in areas where this flow diverges, the Ekman flow or divergence, so that water displaced at the surface must be replaced by deeper water from beneath. Depending on the nature of this divergence, two major types of upwelling systems can be distinguished.

First, coastal upwelling systems occur where the Ekman layer is directed offshore resulting in flow divergence near the coast. Such systems tend to occur on the eastern boundary of ocean basins, major examples being the Canary, Benguela, Humboldt (Peru), and California Current systems (**Figure 1**). Offshore Ekman flow in eastern boundary current (EBC) systems is driven by local equatorward winds associated with the pressure gradient between the quasi-stationary atmospheric high-pressure systems over the subtropical oceans relative to adjacent continental low-pressure atmospheric systems. Seasonal north–south progressions of these high-pressure systems (poleward in spring, summer) cause increased upwelling and nutrient supply that, along with increased day length and light, drive latitudinal shifts in phytoplankton biomass and productivity. The other major coastal upwelling system is the Somali Current, driven by seasonal monsoon winds of the Arabian Sea. Coastal upwelling is often enhanced by topographical features such as capes or canyons where local upwelling cells form.

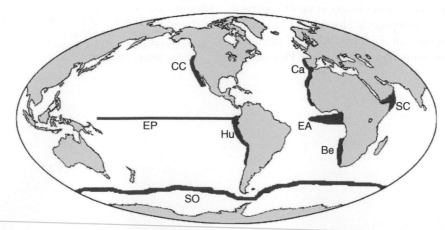

Figure 1   Global map of major upwelling systems. Be, Benguela; Ca, Canary; CC, California Current; EA, Equatorial Atlantic; EP, Equatorial Pacific; Hu, Humboldt; SC, Somali Current; SO, Southern Ocean.

Second, upwelling occurs in the open ocean, being most marked where easterly trade winds give rise to Ekman divergence north and south of the equator. The resulting area of equatorial upwelling in the Pacific is vast, extending westwards from the coast of South America to beyond the international date line. A smaller belt of upwelling occurs in the equatorial Atlantic. In the Southern Ocean, the Antarctic Circumpolar Current contains another zonal upwelling region, most vigorous between 50° and 60°S, driven by northerly Ekman flow that is generated by the strongest prevailing westerly winds in the 40–50° latitudes.

General Characteristics

Nutrients are present in high (e.g., NO_3^-, 35; Si, 30–60; PO_4^{2-}, 1–2 μmol l^{-1}) concentrations in subsurface waters of the global oceans. Upwelling brings them to the surface, fertilizing the resident phytoplankton assemblage. Stratification of the surface mixed layer between the strongest upwelling pulses provides favorable light conditions for algae to grow and take up the nutrients at their disposal. Resulting rates of primary production are often among the highest seen in marine systems. Coastal upwelling systems, for example, occupy just 0.5% of the ocean surface area, yet contribute to 2% of global marine primary production. Supporting an abundance of higher trophic orders such as fish, birds, seals, and whales, they also contain some of the world's major fin-fisheries.

Intermittence is a key feature of upwelling systems. Upwelling intensity is seasonally episodic in systems such as the Canary Current, Benguela, and Somali systems, whereas in others such as the Humboldt and Southern Ocean, upwelling is semicontinuous all year round. In all systems, wind strength varies on shorter timescales of days to weeks, leading to periods of strong and weak upwelling, or times when upwelling ceases altogether. Organisms must be able to tolerate these changes in upwelling intensity and the resulting impact on nutrient supply and spatiotemporal variation in food resources, as well as variations that may occur from year to year and on longer timescales. In addition, they face the prospect of either themselves, or their reproductive products, being swept away in the Ekman layer toward less-favorable habitats. A key feature of these organisms, including phytoplankton, zooplankton, and fish, is that their life histories and behavior are specifically geared toward maintaining populations in the regions of the upwelling centers.

Understanding the ecosystem structure and functioning of upwelling systems, and in particular how they are influenced by climate variability, is increasingly recognized as essential to the management of sustainable fishery resources.

Primary Production and Lower Trophic Levels

Primary Production

The EBC systems provide a highly favorable combination of light and nutrient supply for primary production by virtue of a strongly shoaling pycnocline (the density gradient that signifies the base of the mixed layer) toward the coast and a relatively shallow (<500 m) shelf environment. A near-surface pycnocline has the dual effect of facilitating the injection of nutrients into surface waters and maintaining phytoplankton in a shallow (usually <50 m) well-lit euphotic layer environment. Shallow shelf sediments augment the nutrient concentration of deeper upwelling source waters. As the four major EBC systems lie predominantly in mid-latitudes (40° N/S to 10° N/S), insolation rates are seasonally high, providing both the light and necessary surface warming and stratification to optimally drive photosynthetic carbon fixation supported by a nutrient-replete environment. Taken together, they have a combined productivity estimated to be $\sim$1 Gt C yr^{-1}. Chlorophyll concentrations typically exceed 2 mg m^{-3} but can reach up to 50 mg m^{-3} locally where intense dinoflagellate blooms are present. Highest production rates are consistently found in the Humboldt system (2–6 g C m^{-2} d^{-1}) due to a higher average irradiance and a less-fluctuating nutrient environment associated with relatively consistent upwelling.

A key feature of primary production in upwelling ecosystems is that it is fuelled by large amounts of NO_3^- that is 'new' (i.e., allochthonous) to the euphotic zone. Phytoplankton production based on nitrate uptake is therefore termed 'new production'. Nitrate arises almost entirely from remineralization of organic matter below the pycnocline, notably from dead phytoplankton and other material that had earlier 'rained' down from surface waters. In contrast, 'regenerated production' is based on nitrogen (NH_4^+, urea, and dissolved organic nitrogen) excreted by organisms within (i.e., autochthonous) the euphotic zone. The relative importance of new production is often stated by expressing it as a fraction of total phytoplankton production (i.e., the sum of new and regenerated production), this fraction being known as the f-ratio. Values for the f-ratio are usually high ($\sim$0.5–0.7) in most EBC upwelling ecosystems, although in seasonally pulsed upwelling systems such as the Southern Benguela, the yearly averaged f-ratio is lower ($\sim$0.3). High rates of new production make available carbon and nutrients for transfer to higher trophic levels and are the fundamental reason why EBC systems can sustain productive fisheries. They also provide the potential for large downward fluxes of sinking particles from the euphotic zone in the event that phytoplankton are inefficiently grazed. In contrast, systems based on regenerated production gradually run downhill unless there are new inputs of nitrogen because nutrients are never recycled with 100% efficiency.

Phytoplankton Community Structure

A characteristic succession is seen in the composition of phytoplankton communities of coastal upwelling ecosystems. This succession is driven by changes in the nutrient and light environment, linked closely with upwelling frequency and the three-dimensional (3D) circulation of water as it flows away from the upwelling centers. Large individual (20–200 μm), colonial, and chain-forming diatoms of up to 500 μm in length proliferate as newly upwelled water arrives and stabilizes in the sunlit surface layer. Analogous to weeds, these algae grow quickly because of intrinsically fast growth rates and an ability to take up nutrients rapidly, provided that concentrations remain sufficiently high. Cell division rates of $2-4\,d^{-1}$ quickly result in population growth that outstrips zooplankton herbivory, leading to extensive diatom-dominated blooms. The accumulating chlorophyll biomass can exceed $6\,mg\,m^{-3}$ in just a few days, high enough to be easily visible in ocean color satellite imagery (**Figure 2**).

Most diatom species within EBC systems are adapted to avoid lateral dispersal in surface currents away from the upwelling centers. Many species trigger a resting stage in their life cycle in response to diminishing nutrient availability. Spores are formed that sink rapidly and become entrained into deeper shelf-edge waters and surface sediments. Sinking of vegetative cells, or chain-formation that increases sinking rates, provide alternative strategies to counteract dispersal. Spores and physiologically inactive diatoms remain within the sediment–water interface layer and await the next upwelling event that will entrain them back into near-shore and nutrient-rich sunlit surface waters, so initiating another bloom event. The construction of silica tests and spiny frustules from dissolved silicate, along with their large size and ability to form chains and colonies, offer initial protection from herbivorous zooplankton such as copepods. The efficacy of grazers is further weakened in pulsed upwelling systems that do not settle into steady state. As initial exponential growth rates of diatoms are much faster (hours, days) than those of herbivorous copepod consumers (weeks), episodic and short-term upwelling events (days) produce a 'mismatch' between phytoplankton and zooplankton, the former breaking free from top-down grazer control that might otherwise prevent blooms from occurring. Nevertheless, the grazing that does occur permits carbon to be efficiently transferred to pelagic fish in a short two-step food chain (diatoms ⇒ mesozooplankton ⇒ pelagic fish).

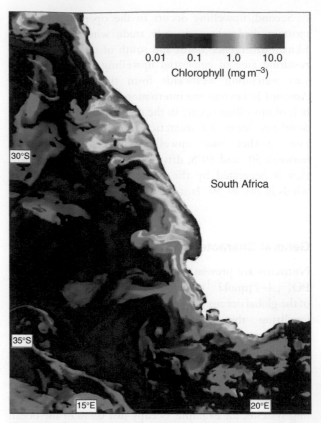

Figure 2   Satellite image of chlorophyll biomass in the Southern Benguela region. Note high chlorophyll concentrations inshore where upwelling of nutrient rich water is strongest, and also offshore filaments in the chlorophyll signal which weakens offshore as nutrients become exhausted. Courtesy Stewart Bernard, Univ. Cape Town.

A shift in phytoplankton community structure occurs as nutrients become depleted in the well-stratified surface waters downstream of coastal upwelling centers. Diatoms give way to smaller cells such as nanoplanktonic phytoflagellates (2–20 μm) as well as other smaller picoautotrophs (<2 μm) that do not require Si and are better able to scavenge nutrients at low concentrations because of their high surface area to volume ratio. As nutrients in the surface layers are scarce, these small phytoplankton primarily occupy the thermocline where nutrients diffuse slowly from below. A deep chlorophyll maximum (DCM) develops that involves a delicate tradeoff between maximizing nutrient availability, but having sufficient light in a near light-limited environment. Microzooplankton grazers keep the numbers and biomass of these small phytoplankton in check (<0.5–1 μg chl $a\,l^{-1}$). Within this 'microbial loop', particulate organic nitrogen (PON) is efficiently recycled via microzooplankton and bacteria into NH_4^+ and urea to support further phytoplankton growth. As nitrogen remineralization is usually balanced by the rapid uptake of such regenerated nitrogen by phytoplankton, concentrations of these nutrients remain low (<0.5–1 $\mu mol\,l^{-1}$).

Paradoxically therefore, grazing pressure is essential to support further algal growth. Carbon is only inefficiently transferred along an extended food chain (pico-, nanoplankton $\Rightarrow$ microzooplankton $\Rightarrow$ mesozooplankton $\Rightarrow$ fish) because, at each step, something approaching 90% of the transferred carbon is lost through respiration.

Oxygen Depletion

Dead and decaying material is a feature of all ecosystems, those of upwelling areas being no exception. Detrital particles are produced in abundance, either as senescent phytoplankton or as zooplankton fecal material. A 'rain' of sinking particulate organic material ('marine snow') is exported from the euphotic zone and decomposed either in mid-water by heterotrophic bacteria, or by both benthic organisms and bacteria on the seafloor. Large quantities of oxygen are consumed, creating an oxygen minimum zone (OMZ) in the sediments and the overlying water. The resulting oxygen concentrations of $<0.5\ ml\ l^{-1}$ are inhospitable to many animals, pelagic and benthic alike.

Coastal upwelling systems are particularly susceptible to hypoxic ($<0.5\ ml\ l^{-1}$) or anoxic (near-zero O_2) events because of the high rates of diatom-dominated phytoplankton productivity that quite suddenly become nutrient-limited and therefore senescent. The oxycline may often extend to near the surface ($<50\ m$). Most zooplankton that can actively migrate often do so in order to maintain their position in the oxygenated waters near the surface. Others, such as *Eucalanus inermis* in the Humboldt system, are able to withstand low oxygen concentrations, and indeed are known to congregate in the OMZ, perhaps exploiting it as a refuge from predation. In similar fashion, juvenile hake (*Merlucius capensis* and *M. paradoxus*) in the northern Benguela system off Namibia are known to exploit the OMZ as a refuge from their cannabilistic parents!

Oxygen consumption in upwelling areas leads to extensive seafloor habitats subject to permanent hypoxia. Diversity is low, but those animals that tolerate low-oxygen conditions are abundant. Calcareous foraminiferans, nematodes, and annelids utilize the influx of organic material from above, but rely on anaerobic metabolism to do so. Chemoautotrophic bacteria use NO_3^- and sulfur as terminal electron acceptors instead of O_2, first stripping the anoxic water column of NO_3^- (denitrification) before specific anaerobic sulfur and sulfate-reducing bacteria release sulfurous and foul-smelling H_2S into the water column creating so-called 'black-tides' and 'sulfur eruptions'. However, it appears that some nitrogen losses previously ascribed to denitrification should instead be linked to the process of ammonium oxidation 'annamox', first described in Dutch sewerage works.

Offshore hypoxia can have profound effects on the near-shore intertidal environment. When gentle upwelling begins, low-oxygen water is driven into the near-shore and intertidal zone, killing all before it except those most resistant to hypoxia. In the Benguela system, crayfish 'walkouts' by animals fleeing from low-O_2 water can leave thousands of tons of crayfish stranded on the beaches in sheltered embayments.

Zooplankton

Great variety is seen in the zooplankton of upwelling ecosystems. Smallest are the microzooplankton, including ciliates, heterotrophic dinoflagellates, and flagellates (typical size 2–5 μm) that efficiently graze the smallest phytoplankton. Reproducing by cell division, their high growth rates (e.g., $1.0\ d^{-1}$) are similar to those of their algal prey such that their grazing is sufficient to prevent small phytoplankton cells from blooming. Microzooplankton do not fit the general paradigm that organisms eat prey items significantly smaller than themselves. Instead, they have evolved various specialized feeding mechanisms including direct engulfment, tube feeding in which a feeding tube, the peduncle, pierces prey which then has its insides sucked out, and pallium feeding, where a feeding veil envelops and digests prey *in situ*. Prey items as large as, or larger than, the microzooplankter's own body size can be consumed using these adaptations. It has been suggested, for example, that heterotrophic dinoflagellates are able to compete with copepods for diatom prey, although the extent to which this competition operates in marine ecosystems is as yet poorly known.

It is the larger mesozooplankton ($\sim$0.2–2 mm), notably copepods and to a lesser extent euphausiids, that are the major grazers of diatoms and which form the main trophic link with fish and other higher trophic levels. Phytoplankton are captured by filter-feeding or by selective particle capture (raptorial feeding) based on size and/or palatability. *Calanus* is the dominant copepod genus in upwelling ecosystems. Although not reproducing as fast as microzooplankton, it may achieve as many as ten generations per year, each with its own life cycle of eggs, nauplii, copepodites, and adults. When food conditions are favorable, fecundity is high and egg production is rapid. Upon hatching, the planktivorous juvenile stages are swept along in the Ekman layer and will starve if they do not encounter suitably dense patches of appropriately sized food particles. For adults, survival is enhanced by the storage of energy reserves in the form of lipids. Given their size, copepods are unable to maintain their position within upwelling systems by swimming against lateral advection that is often offshore. This problem is overcome by diel vertical migration. Offshore surface Ekman flow is balanced by deeper shoreward flow onto the shelf. By migrating into this deeper layer by day, copepods utilize the natural circulation pattern to maintain their inshore position where food resources are richest.

Euphausiids are also a significant component of the zooplankton community in upwelling ecosystems, for example, *Euphasia lucens* in the Benguela system. They are much larger (1–2 cm) than copepods and have a longer life span of about a year. This longevity, along with an omnivorous diet, means that euphausiids are better able to cope with the fluctuating food conditions of upwelling ecosystems than are copepods. Nevertheless, physical transport away from upwelling centers remains a problem, and these animals also employ diel vertical migration into the subsurface countercurrent to maintain their position in the flow field. Their larger size makes euphausiids a key prey item for larger zooplankton consumers, including baleen whales that are often temporary residents of upwelling systems.

Open-Ocean Upwelling Systems

The general principles and characteristics that govern productivity in EBC regions apply also to the major open-ocean upwelling systems (Equatorial Pacific, Equatorial Atlantic, and Southern Ocean). There are nevertheless key differences, notably that upwelling strength tends to be lower and there is no influence of the seabed (e.g., in supplying iron) on euphotic zone processes, now that it is 3000–4000 m beneath the ocean surface.

The Equatorial Pacific is a vast upwelling system, as well as being a good example of a so-called high-nutrient low-chlorophyll (HNLC) ecosystem. Phytoplankton biomass is generally low and relatively constant ($\sim$0.2–0.4 mg chl a m^{-3}) which, along with low productivity of $\sim$0.1–0.5 g C m^{-2} d^{-1}, occurs despite the presence of sufficient macronutrients (NO_3^-, Si, PO_4^{2-}) and light. Iron, however, is in short supply. This micronutrient is needed by phytoplankton to harvest light using their photosynthetic machinery (photosystems I and II), as well as by the enzymes nitrate and nitrite reductase to reduce NO_3^- within cells to NH_4^+. Without a sedimentary source, aeolian supply is the primary source of Fe to the open ocean. However, most aeolian dust supply is from the Saharan desert, far-distant from the Equatorial Pacific. The resulting shortage of iron impacts most severely on large cells, notably diatoms, because of their inability to compete with smaller phytoplankton at low nutrient concentrations. In the western basin, phytoplankton biomass is dominated by small solitary picoplanktonic cells (0.2–2 µm) within a DCM comprising prochlorophytes, *Synechococcus*, and small eukaryotes. These cells utilize what little iron supply there is from the waters upwelled from below, starving the surface ocean of this element. Diatoms are more abundant ($\sim$6%) to the east of $\sim$140 °W where deep nutrient-rich upwelling outcrops at

the surface but, nevertheless, the overall biomass is still picoplankton dominated. Grazing by microzooplankton keeps phytoplankton stocks in check, but small natural enhancements of iron that occur in the Equatorial Pacific in response to the passage of tropical instability waves promote transient increases in primary production.

The upwelling region at the Antarctic Polar Front is another HNLC system with low Fe concentrations. Chlorophyll biomass is typically $\sim$0.5 mg m^{-3} in the austral summer with a productivity of $\sim$0.5–1 g C m^{-2} d^{-1}. Unlike the Equatorial Pacific, however, winter gales drive deep mixing that entrains nutrients, including Fe, into surface waters. This is sufficient to initiate short diatom-dominated blooms in the early spring (September, October) as the light environment improves. Iron limitation throughout the rest of the year opens the way for a more typical HNLC community of pico- and nanoplanktonic phytoflagellates that are microzooplankton controlled. Populations of the prymnesiophyte *Phaeocystis antarctica* may also develop, an organism that exists both as solitary cells and mucilaginous colonies, and which is the main producer of volatile organic sulfur (dimethyl sulphide, DMS) in the region.

The Equatorial Atlantic is unique among open-ocean upwelling systems both in terms of its hydrography and because of its close proximity to aeolian dust sources from the Sahara meaning that productivity is far less Fe-limited than in other open ocean systems. Between June and January, water flowing from the Amazon basin floods eastwards (the North Equatorial Counter Current) across the northern margin of the equatorial upwelling region. Stripped of nutrients as it crosses the Amazonian shelf, this fresher, buoyant water forms a layer $\sim$40 m deep that caps the nutrient-rich water below and also limits light penetration. A DCM forms at the juncture of these two water types, the low light intensities being particularly well exploited by the cyanobacterium *Prochlorococcus*. The nutrient-depleted waters above are home to a separate community in which nitrogen fixers such as *Trichodesmium* utilize atmospheric nitrogen as a nutrient source. The aeolian flux of dust plays an important role since nitrogen fixers have a particularly high requirement for Fe.

Between February and May, the Amazonian outflow diverts northwards toward the Caribbean. Phytoplankton now find themselves $\sim$40 m closer to the surface in a higher light environment and productivity increases but, as the rate of upwelling is weak, the upward flux of nutrients is insufficient to support diatom blooms except in the eastern basin near the African coast.

In addition to the major upwelling systems described above, the upper ocean contains numerous mesoscale eddies - whirling current systems analogous to weather systems in the atmosphere but only about a tenth of the size (tens instead of hundreds of kilometers across).

Produced through the conversion of potential energy to kinetic energy as part of the ocean's annual energy cycle, both cyclonic and anticyclonic eddies (depending on the vertical structure of the water column) can result in the localized doming of isopycnals (constant density surfaces) and upwelling of nutrient-rich waters into the euphotic zone as they form. Eddies themselves then decay as they release their potential energy over periods of weeks to months, vertical motions both upward and downward occurring on their periphery during this time. Primary production is generally stimulated through nutrient enrichment. As in other upwelling systems, regions of higher nutrients and shallower mixed layer depth associated with eddies tend to promote the growth of larger phytoplankton cells such as diatoms, their concentrations typically being higher within eddies than in surrounding waters. Ubiquitous in nature, eddies provide a significant vertical transport mechanism for nutrients throughout much of the world's oceans.

Fish and Higher Trophic Levels

The EBC upwelling ecosystems of the world support major commercial fisheries based on the shoals of sardine, anchovy, and mackerel that thrive on the abundance of phytoplankton and zooplankton food. In the Humboldt

system alone, for example, catches have been around 12 million tons in peak years, although this decreases by >50% during unfavorable conditions. Indeed, stocks of different fish species have been highly variable over the years, suggesting a remarkable responsiveness in ecosystem structure to changing conditions. Understanding the links between fish, lower and higher trophic levels, and environment is essential to ensuring the sustainable management of these important fish resources.

Small Pelagic Fish

The food chain of upwelling systems embraces phytoplankton and zooplankton at its base, linking to small pelagic fish which are in turn consumed by higher predators such as piscivorous fish, birds, and seals (**Figure 3**). A curious aspect of this trophic network is that there are many species at low (phytoplankton, zooplankton) and at high trophic levels, but only a few species of small pelagic fish in between. Indeed, the fish biomass of coastal upwelling systems is typically dominated by either a single species of sardine (*Sardinops*) or a single species of anchovy (*Engraulis*) at any one time.

Although food resources are generally favorable, the 3D circulation makes upwelling systems a hazardous environment for fish. Losses of eggs and juvenile stages may occur due to offshore transport or because of starvation when being carried by currents from the spawning to nursery areas. Spawning grounds are therefore often strategically

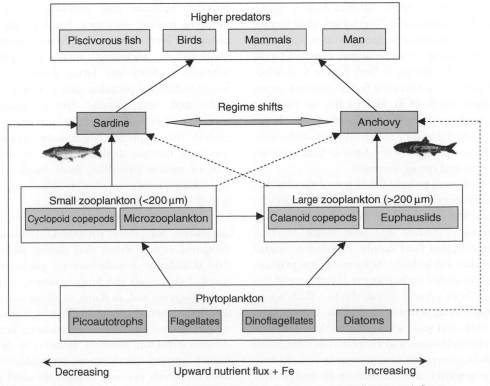

Figure 3 Idealized flow diagram for an upwelling ecosystem food web. Dashed arrows indicate weak flows.

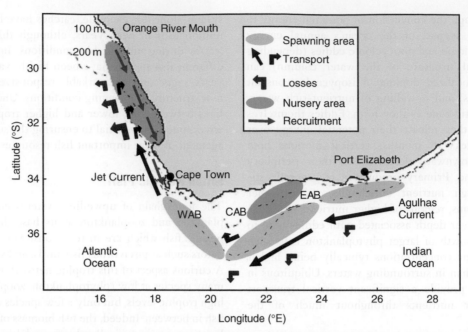

Figure 4 Map of the Southern Benguela off South Africa showing the locations of small pelagic fish spawning and nursery grounds and transport and loss processes that impact on eggs and larvae. WAB, CAB, and EAB indicate the Western, Central, and Eastern Agulhas Banks, respectively. Redrawn from Lehodey P, Alheit J, Barange M, *et al*. (2006) Climate variability, fish and fisheries. *Journal of Climate* 19: 5009–5030. © Copyright 2006 American Meteorological Society (AMS).

positioned in quieter areas surrounding the upwelling centers such as downstream of capes in sheltered embayments. The result is a complex network of spawning grounds, transport pathways, and migration patterns, a typical example being the Benguela system (**Figure 4**). Anchovy spawn on the Western Agulhas Bank in spring and summer (with a maximum in November), while sardine have a longer spawning season in the same area, peaking in both October and March. Once fertilized, eggs and larvae drift northwards in the Benguela 'Jet'. Larvae feed selectively on small particles and juvenile fish recruitment occurs at several locations north of St. Helena Bay on the West Coast. There, anchovy recruits feed primarily on larger zooplankton (copepods) as they slowly migrate southwards, returning as 1-year-old adults to the Agulhas Bank to spawn in the following austral spring/summer.

The survival of small pelagic fish is thus determined to a large degree by direct physical factors such as circulation patterns and the intensity and duration of upwelling that simultaneously control egg and larval survival, recruitment success, and food supply. Population control is therefore neither exclusively 'bottom-up' via primary producers nor 'top-down' by higher predators. Instead it is from the 'waist', both up and down, the so-called 'wasp-waist' hypothesis. Small pelagic fish provide higher tropic levels such as birds and seals with food, while at the same time keeping phytoplankton and zooplankton numbers in check. As a result, ecosystem functioning as a whole may be remarkably sensitive to fluctuations in pelagic fish numbers. Direct environmental forcing or commercial

fishery exploitation of 'wasp-waist' populations may cause disruption to these ecosystems by undermining the stability of the entire food web.

Bottom-up and top-down controls of small pelagic fish populations are nevertheless by no means unimportant. Sardines and anchovy, for example, have different feeding strategies. Sardines are mostly indiscriminate filter feeders on phytoplankton and smaller zooplankton, including small cyclopoid copepods, whereas anchovy use biting behavior to selectively ingest individual particles such as larger (~2 mm) copepods and euphausiids. Strong upwelling should therefore favor anchovy by promoting the diatom growth that supports larger zooplankton. In contrast, the more nutrient-depleted waters present during periods of weaker upwelling favor smaller phytoplankton and consequently smaller zooplankton that are preferred by filter-feeding sardines.

Variability in small pelagic fish populations has consequences for their predators. Evidence from the Benguela region shows that during periods of pelagic fish abundance populations of piscivorous fish (e.g., snoek, hake), seals and birds (gannets, cormorants) generally increase and, in doing so, begin to exert a stronger top-down control on the small pelagic fish. This in turn relaxes anchovy predation pressure on copepods and so mesozooplankton numbers recover, in turn exerting a higher grazing pressure on phytoplankton. Not only does top-down predator control on small pelagics equal or exceed that by commercial fishermen, it can substantially shape community structure right down to the level of primary producers.

Fish Production

High primary production fuelled by new nutrients undoubtedly contributes to the prodigious fish production of coastal upwelling systems. Back in 1969, John Ryther proposed that high fish yield should be expected where phytoplankton cells are large, or exist as colonies or chains, thereby leading to only one or two trophic links from primary producers to fish. The greater number of trophic links stemming from smaller phytoplankton cells should lead instead to greater respiration losses and recycling of organic matter.

Fish catches vary markedly between EBC systems, typical values being 0.05%, 0.09%, and 0.16% of primary production for the Canary, Benguela, and Humboldt Current systems respectively. Much of this variation may be due to differences in upwelling intensity and frequency that impact on lower trophic levels and fish recruitment, although intensity of fishing may also play a part. The strongly seasonal and pulsed nature of upwelling in the Canary and Benguela systems, for example, leads to temporal mismatches between primary producers and copepods, depressing their fecundity and therefore population size, so reducing the food supply for fish. In contrast the Humboldt system experiences less variation in the intensity of upwelling, leading to tighter coupling between phytoplankton and copepods, greater zooplankton production, and ultimately higher fish production. The low fish catch per unit primary production of the Canary system relative to that of the Benguela is due to its narrow continental shelf (20 km; the Benguela's is ~85 km), such that proportionately more primary production may be advected offshore away from the main zones of fish production.

The impact of environment on the reproductive success of fish in upwelling ecosystems can be thought of in terms of a fundamental triad of processes: enrichment (nutrient supply for primary production), concentration processes (convergence, water-column stability), and retention (within favorable habitat). Acting at the base of the food chain, nutrient enrichment via upwelled water fertilizes primary producers, although excessive wind may deepen the surface mixed layer, leading to light limitation. Upwelling also stimulates small-scale turbulence, increasing encounter rates between both zooplankton and fish larvae and their prey. On a larger scale, convergence promotes food particle aggregation, but divergent upwelling flow will tend to dissipate particles offshore. Based on these pros and cons, an optimal level of upwelling intensity can be defined, the 'optimal environmental window' (OEW), that maximizes fish yield (**Figure 5**). When on the left side of the OEW (too little wind) upwelling is weak and

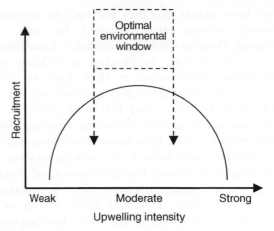

Figure 5 Optimal environmental window for fish recruitment. Reproduced from Cury P and Roy C (1989) Optimal environmental window and pelagic fish recruitment success in upwelling areas. *Canadian Journal of Fisheries and Aquatic Sciences* 46: 670–680.

primary production, and hence also food for fish, is restricted by insufficient nutrient supply. On the other hand too much upwelling (right side of the OEW) leads to dispersal of organisms away from upwelling centers and provokes light limitation in phytoplankton as they are mixed deeper into the water column because stratification is not established.

Higher Trophic Levels

The abundance of zooplankton and small pelagic fish in coastal upwelling systems provides food for a range of higher trophic levels including piscivorous fish, seabirds, pinnipeds, and cetaceans. Predatory fish such as horse mackerel and deep-water hake are themselves important fishery resources, the latter being caught by mid- and deep-water trawling. Another economically viable product of upwelling systems, particularly in the Humboldt and Benguela, is the production of bird droppings, guano, which is prized as a fertilizer because of its high N and P content. The so-called 'guano birds' such as the guanay cormorant, Peruvian booby, Chilean pelican, Cape cormorant, and Cape gannet are part of resident seabird populations that breed along the coast and on adjacent islands feeding on small pelagic fish such as anchovies and sardines. Both the Benguela and Humboldt systems also support populations of small-sized penguins. The African Penguin (*Spheniscus demersus*) extends from central Namibia to Algoa Bay on the south coast of South Africa. Its population has dwindled from more than one million in 1900 to about 200 000 now, feeding mainly on a diet of pelagic schooling fish (anchovy, sardine, redeye). Reasons for the declining population are ecologically complex, but include competition with commercial fisheries for food, habitat degradation because of removal of guano from islands that they burrow into for nesting, pollution (oiling), and predation by seals. The Humboldt Penguin (*Spheniscus humboldti*) breeds mainly from 5 to 33 °S along the Peruvian and Chilean coast, with another small colony at 42 °S. Like the African Penguin, it also feeds on small pelagic fish, and its population has declined to around 30 000 for similar reasons.

Because of the episodic nature of upwelling, higher predators must endure large seasonal or interannual fluctuations in prey availability, for example, in response to El Niño events (see below). Resident seabird and pinniped populations are particularly susceptible. In catastrophic instances, starvation may cause adult seabirds to die, although more often food scarcity affects breeding success by decreasing the proportion of adults breeding and the growth rate of the hatchlings. In similar fashion, there is often a high incidence of seal pup mortality during food shortages, the adult females being unable to provide enough milk for their survival. Seabirds and pinnipeds are able to employ various strategies to compensate when food resources are in short supply including increasing the time and/or distance spent foraging for offspring, delaying reproduction until such time that food becomes available, or ceasing breeding efforts altogether. Others may target alternative food resources, such as squid, or migrate to other regions within the system where food resources are more plentiful.

Many migratory species are attracted to the high productivity of upwelling systems. Blue whales, for example, feed on dense swarms of euphausiids that exist in the California Current system. Many birds that nest elsewhere also benefit from the abundance of food in upwelling areas. The California Current system, for example, is visited by sooty shearwaters, which breed off South America, and red-necked phalaropes which nest in the Arctic. Arctic terns migrate to the Southern Hemisphere in austral winter, feeding in the Benguela and Southern Ocean upwelling systems.

The catch of pelagic anchovy and sardine by the predators described above is generally considered to surpass that by commercial purse-seine fishermen, even in the Humboldt which is heavily exploited. Seals in particular are unpopular competitors, not least because they cause damage to nets and generally interfere with fishing operations. Nevertheless, the potential consequences of overfishing in upwelling ecosystems should not be underestimated. Over the last few decades the fishing industry has progressively concentrated on species at relatively low trophic levels, with emphasis on small pelagic species such as sardine and anchovy with decreased catches of predatory fish such as hake and horse mackerel. An apparent consequence of this 'fishing down marine food webs' has been a decline in pelagic fish and a proliferation of jellyfish that occupy the vacant niche, the two utilizing the same food resources. Jellyfish biomass in the northern

Benguela off Namibia, for example, is now thought to exceed that of commercially important fish stocks. Once established, this regime shift may be difficult to reverse because jellyfish are predatory on fish eggs and juveniles.

Climatic Forcing

Changes have occurred in the dominant fish species of upwelling ecosystems, from year to year, over decades and indeed centuries. Worldwide, populations of anchovy and sardine have exhibited 'flip-flops' in which one species is replaced by the other. Fishing is one factor that may influence these changes. Comparison of the fish catches of different upwelling systems, however, reveals a remarkable synchronicity in their behavior (**Figure 6**) suggesting a climate linkage via global 'teleconnections'.

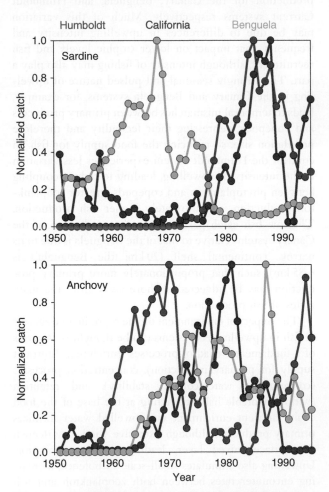

Figure 6 Comparison of sardine and anchovy catches in the Humboldt, California, and Benguela systems. Data are normalized to maximum catch (million tons): sardine: 5.62, Humboldt; 0.29, California; 1.51, Benguela; anchovy: 12.9, Humboldt; 0.32, California; 0.97, Benguela. Reproduced from Schwartzlose RA, Alheit J, Bakun A, *et al.* (1999) Worldwide large-scale fluctuations of sardine and anchovy populations. *South African Journal of Marine Science* 21: 289-347.

The implication is that fish populations are driven primarily by natural climate variability and its influence on ecosystem structure and recruitment success. Short-term events such as El Niño cause calamitous declines in fish stocks that lead to hardship for wildlife such as bird and seal populations, and of course fishermen. Superimposed on this short-term variability are longer-term trends that occur in response to factors such as climate change. Understanding these variations and their causes is crucial to the maintenance of sustainable fisheries in upwelling areas.

El Niño

The El Niño Southern Oscillation (ENSO) is the most important example of a relatively short-term impact of climatic forcing on upwelling ecosystems, with a typical periodicity of 3–5 years. In normal (La Niña) years, easterly trade winds blow across the surface of the equatorial Pacific from Peru/Chile to Indonesia creating the general divergent open ocean upwelling that occurs across the eastern half of the equatorial Pacific. This process sets up a surface temperature gradient of <20 °C in the east to >30 °C in the west, resulting in a shallow thermocline (~20 m) in the east, but a much deeper one (~80 m) in the west. At its eastern end, along-shore winds off Peru and Chile drive the coastal upwelling of the Humboldt Current System. El Niño occurs when the easterly trade winds lose intensity, allowing warm water from Indonesia and eastern Australia to flood eastwards across the Pacific, 'capping' the deeper nutrient-rich waters that lie below (**Figure 7**). The coastal winds that drive upwelling in the Humboldt System possess insufficient energy to erode and mix this stratified surface layer.

Upwelling continues during El Niño but the water arriving at the surface is depleted in nutrients with drastic consequences for marine life. Diatom blooms are suppressed with a dramatic shift to a community structure dominated by the small cells of the microbial loop. Fish stocks collapse in response to the low availability of food, the starvation of adults and/or larvae leading to recruitment failure. High mortality rates are also seen in top predators. Major ENSO events occurred in 1972 and 1976, as well as in later years, economic disaster following in their wake. For example, revenue losses to Chile and Peru resulting from decimated fish stocks were about $8 billion for the 1997–98 ENSO event.

Further afield, the effects of ENSO events are felt throughout the Southern Hemisphere, and indeed the globe. The warm waters of El Niño in turn affect atmospheric circulation, the resulting teleconnections instigating changes in other upwelling systems. For example, so-called 'Benguela Niño' events occur about 6 months after the onset of activities in the Pacific. During these events, the Angola-Benguela front moves southwards by several hundred kilometres, bringing low-oxygen warm water into the Namibian upwelling region that results in a southward displacement of pelagic fish stocks.

Long-Term Climate Variability

Having high fecundity, small pelagic fish are able to recover from events such as El Niño within a year or two. Yet the observed anchovy–sardine flip-flops persist over many years suggesting that climatic factors operating on longer timescales play an important role in structuring coastal upwelling ecosystems and fish stocks (**Figure 6**).

(a)

(b)

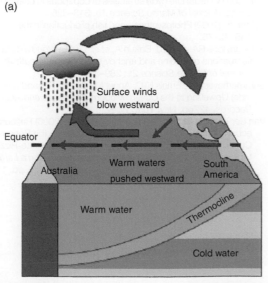
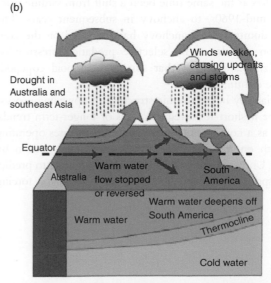

Figure 7 Atmospheric and ocean circulation patterns associated with La Niña (a) and El Niño (b). Reproduce from SEPM Photo CD-5, *Oceanography Series* (edited by Peter A. Scholle), with permission from Society for Sedimentary Geology (SEPM).

Anchovy were the dominant fish species in the Humboldt Current System until the mid-1970s. Catches peaked at about 12.9 million tons in 1970, but were followed by a severe decline that may have been precipitated by the major El Niño of 1972. Recovery of the anchovy stock did not occur until the mid-1980s, sardine being dominant during the interim period. The variability of fish catches appears to have followed cycles of around 55–65 years over the last century. Analysis of atmospheric circulation patterns (e.g., the 'atmospheric circulation index', ACI) reveals that the dominant direction of air masses has also changed on similar timescales. Fish scales preserved in anoxic and undisturbed shelf sediments off California and off Namibia reveal 50–70 year cycles of anchovy and sardine abundance that are linked to changes in sea surface temperature over the last 1600 years. In the Humboldt system, regime shifts appear to correlate with lasting periods of warm or cold temperature anomalies related to the approach and retreat of warm subtropical water toward the coasts of Peru and Chile. Sardine are favored during periods of warm water intrusion (1970–85) whereas the anchovy fishery prospers during periods when temperatures remain relatively cool (1950–70, 1985 to present). Resolving the underlying, probably basin-scale, physical processes that lead to such patterns, along with teleconnections linking different upwelling systems, remains a priority for scientific investigation.

Changes in local atmospheric pressure gradients with global warming might be expected to increase upwelling frequency and intensity, with accompanying changes in the structure and function of ecosystems. Nutrient concentrations have for example increased in the Benguela region over recent decades, suggesting an increase in upwelling. There has at the same time been a shift from sardine prior to the mid-1960s, to anchovy in subsequent years. The recent dominance of anchovy has impacted on the zooplankton population by selective predation pressure on larger zooplankton, so that smaller cyclopoid copepods become dominant.

Variability in the trophic structure of upwelling ecosystems, be it short-term regime shifts or longer-term trends, occurs as a consequence of a range of processes operating via both environment and man's direct intervention by fishing. Understanding these interactions in order to predict the response of upwelling ecosystems to climatic forcing and fishing strategies involves unravelling a multiplicity of factors that affect primary production, zooplankton and fish recruitment, a challenging task for the scientific community.

Further Reading

Alheit P and Niquen M (2004) Regime shifts in the Humboldt Current ecosystem. *Progress in Oceanography* 60: 201–222.

Bakun A (1990) Global climate change and intensification of coastal ocean upwelling. *Science* 247: 198–201.

Barange M and Harris R, eds. (2003) *Marine Ecosystems and Global Change. IGBP Science no. 5*, 32pp. Stockholm: IGBP.

Croll DA, Marinovic B, Benson S, *et al.* (2004) From wind to whales: Trophic links in a coastal upwelling system. *Marine Ecology Progress Series* 289: 117–130.

Cury P and Roy C (1989) Optimal environmental window and pelagic fish recruitment success in upwelling areas. *Canadian Journal of Fisheries and Aquatic Sciences* 46: 670–680.

Cury P, Bakun A, Crawford RJM, *et al.* (2000) Small pelagics in upwelling systems: Patterns of interaction and structural changes in 'waspwaist' ecosystems. *ICES Journal of Marine Science* 57: 603–618.

Cury P and Shannon L (2004) Regime shifts in upwelling ecosystems: Observed changes and possible mechanisms in the northern and southern Benguela. *Progress in Oceanography* 60: 223–243.

Hare CE, DiTullio GR, Trick CG, *et al.* (2005) Phytoplankton community structure changes following simulated upwelled iron inputs in the Peru upwelling region. *Aquatic Microbial Ecology* 38: 269–282.

Lehodey P, Alheit J, Barange M, *et al.* (2006) Climate variability, fish and fisheries. *Journal of Climate* 19: 5009–5030.

Lynam CP, Gibbons MJ, Axelsen BE, *et al.* (2006) Jellyfish overtake fish in a heavily fished ecosystem. *Current Biology* 16: R492–R493.

Mann KH and Lazier JRN (2006) *Dynamics of Marine Ecosystems. Biological–Physical Interactions in the Ocean.* Oxford, UK: Blackwell.

Moloney CL, Jarre A, Arancibia H, *et al.* (2005) Comparing the Benguela and Humboldt marine upwelling ecosystems with indicators derived from inter-calibrated models. *ICES Journal of Marine Science* 62: 493–502.

Murray JW, Barber RT, Roman MR, Bacon MP, and Feely RA (1994) Physical and biological controls on carbon cycling in the Equatorial Pacific. *Science* 266: 58–65.

Payne AIL, Brink KH, Mann KH, and Hilborn R (1992) Benguela trophic functioning. *South African Journal of Marine Science* 12: 1–1108.

Peterson W (1998) Life cycle strategies of copepods in coastal upwelling zones. *Journal of Marine Science* 15: 313–326.

Ryther JH (1969) Photosynthesis and fish production in the sea. *Science* 166: 72–76.

Schwartzlose RA, Alheit J, Bakun A, *et al.* (1999) Worldwide large-scale fluctuations of sardine and anchovy populations. *South African Journal of Marine Science* 21: 289–347.

Summerhayes CP, Emeis K-C, Angel MV, Smith RL, and Zeitschel B (eds) Upwelling in the Ocean: *Modern Processes and Ancient Records*, 422pp. New York: Wiley.

Van der Lingen CD, Shannon LJ, Cury P, *et al.* (2006) Resource and ecosystem variability, including regime shifts, in the Benguela Current System. In: Shannon V, Hempel G, Malanotte-Rizzoli P, Moloney C, and Woods J (eds.) *Benguela: Predicting a Large Marine Ecosystem*, Large Marine Ecosystem Series, vol. 14, pp. 147–184. Amsterdam: Elsevier.

Urban Systems

T Elmqvist, Stockholm University, Stockholm, Sweden

C Alfsen, UNESCO, New York, NY, USA

J Colding, Royal Swedish Academy of Sciences, Stockholm, Sweden

Introduction
Urbanization and Plant and Animal Communities
Urban Habitats and Gradient Analyses
Urban Systems and Ecosystem Services

Urban Restoration
Urban Landscapes as Arenas for Adaptive Management
Linking Humans and Nature in the Urban Landscape
Further Reading

Introduction

Urbanization is a global multidimensional process that manifests itself through rapidly changing human population densities and changing land cover. The growth of cities is due to a combination of four forces: natural growth, rural to urban migration, massive migration due to extreme events, and redefinitions of administrative boundaries. Half of the world's population today lives in urban areas, a proportion expected to increase by 2/3 within 50 years. Today, over 300 cities have a population of more than 10^6 and 19 megacities exceed 10^7. As urbanization is accelerating, the growth of cities forms large urban landscapes, particularly in developing countries. (Urban landscape is here defined as an area with human agglomerations with >50% of the surface built, surrounded by other areas with 30–50% built, and overall a population density of >10 ind. ha^{-1}.) For example, during the last 20 years in China, clusters of cities have emerged forming at least five mega-urban landscapes. These large and densely urbanized regions have each between 9 and 43 large cities located in close proximity and a population ranging from 27 to 75 million people. This rapid urbanization represents both a challenge and an opportunity to ensure basic human welfare and a viable global environment. The opportunity lies in that urban landscapes also are the very places where knowledge, innovations, human and financial resources for finding solutions to global environmental problems are likely to be found.

Since urbanization is a process operating at multiple scales, factors influencing environmental change in urban landscapes often originate far beyond city, regional, or even national boundaries. Fluctuation in global trade, civil unrest in other countries, health pandemics, natural disasters, and possibly climate change and political decisions are all factors driving social–ecological transformations of the urban landscape.

Mismatches between spatial and temporal scales of ecological process on the one hand, and social scales of monitoring and decision making on the other have not only limited our understanding of ecological processes in urban landscapes, they have also limited the integration of urban ecological knowledge into urban planning. In ecology there is now a growing understanding that human processes and cultures are fundamental for sustainable management of ecosystems, and in urban planning it is becoming more and more evident that urban management needs to operate at an ecosystem scale rather than within the traditional boundaries of the city.

Although studies of ecological patterns and processes in urban areas have shown a rapid increase during the last decade, there are still significant research gaps that constrain our general understanding of the effects of urbanization processes. The vast majority of studies so far have been short term (one to two seasons), conducted in cities in Northern Europe or the US, have lacked experimental approaches, focused on either birds or plants, while other taxa are rarely represented and have only included portions of a rural–urban gradient. Most significantly, we nearly completely lack studies in rapidly growing urban landscapes in tropical developing countries that are rich in biodiversity and are just beginning to address the complexity of human settlements in the tropics.

Of further significance is that urban landscapes provide important large-scale probing experiments of the effects of global change on ecosystems, since, for example, significant warming and increased nitrogen deposition already are prevalent and because they provide extreme, visible, and measurable examples of human domination of ecosystem processes. Urban landscapes may be viewed as numerous large-scale experiments producing novel types of plant and animal communities and novel types of interactions among species, and as such deserve the full attention of not only evolutionary biologists and ecologists but also of students of social–ecological interactions.

Urbanization and Plant and Animal Communities

Urbanization is today viewed to endanger more species and to be more geographically ubiquitous than any other human activity (**Figure 1**). For example, urban sprawl is rapidly transforming critical habitats of global biodiversity value, for example, in the Atlantic Forest Region of Brazil, the Cape of South Africa, and coastal Central America. Urbanization is also viewed as a driving force for increased homogenization of fauna and flora. In the urban core in Northern Hemisphere cities, a similar set of species is recorded that is, often cosmopolitan plant and animal species tolerant of anthropogenic impacts. For example, the composition of communities of wildlife species found in cities across the US is remarkably similar despite large variation in climate and geographical features. A common pattern among cities is that they often show a high turnover of species with losses of native species and gains of nonnatives over time. For example, it is documented that New York has lost 578 native plant species while it gained 411, and Adelaide lost 89 while it gained 613 new plant species over a period of 166 years. Although cities may be species rich, frequently having higher species diversity than surrounding natural habitats, this is often due to a high influx of nonnative species and formation of new communities of plants and animals. A trend of increasing nonnative species from the suburbs to the urban core is well documented for plants, birds,

Figure 1 Cape Town, South Africa with more than 3 million residents is located in the Cape Floristic Region, an area with the highest density of plant species in the world with more than 9600 plants species of which 70% are endemic. Through initiatives like Working for Wetlands and Cape Flats Nature, successful efforts are taken to address the large challenge of conserving precious biodiversity in fragmented natural habitats in an urban setting where poverty is widespread. These initiatives focus on building bridges between people and nature and demonstrating benefits from conservation for the surrounding communities, particularly areas where incomes are low and living conditions are poor, and encouraging local leadership for conservation action.

mammals, and insects. For example, in Berlin the proportion of novel species increased from 28% in the outer suburbs to 50% in the built-up center of the city. In New York, the abundance and biomass of earthworms increased tenfold when comparing rural and urban forests, mainly due to increased numbers of introduced species in urban areas. Over broad geographical scales urbanization seems to have an effect of convergence in species composition with loss of native species and invasion of exotics. Nevertheless, a remarkable amount of native species diversity is known to exist in and around large cities, such as Singapore, Rio de Janeiro, Calcutta, New Dehli, and Stockholm.

Interestingly, the number of plant species in urban areas often correlates with the human population size. Species number often increases with log number of human inhabitants, and that relationship is stronger than the correlation with city area. The age of the city also affects species richness; large, older cities have more plant species than large, younger cities. Also of interest is that diversity may correlate with measures of economic wealth. For example, in Phoenix, USA, measures of plant and avian diversity in urban neighborhoods and parks show a significant positive correlation with measures of median family income levels.

In general, urban landscapes present novel ecological conditions, such as rapid rate of change, chronic disturbances, and complex interactions between patterns and processes. Organisms that have survived in urbanized areas have been able to do so for at least two reasons: (1) they evolved rapidly and adjusted genetically or (2) they were largely preadapted to this environment and required little or no genetic adjustment. There are several documented cases of rapid evolution in urban areas, involving, for example, tolerance to toxic substances and heavy metals in plants, such as lead tolerance in urban roadside *Plantago lanceolata*. Among insects there are many cases of rapid evolution in urban areas, notable example being the famous case of industrial melanism among Lepidoptera in UK, a phenomenon also documented from areas in USA, Canada, and elsewhere in Europe. Also of interest is that specific urban and rural races have been identified within well-studied *Drosophila* species.

In **Table 1** we have summarized some of the effects of urbanization including abiotic and biotic changes. Human activities may cause increased deposition of nutrients such as nitrogen and phosphorus and emission of toxic chemicals which influence urban soil processes. Decomposition rates in urban soils are often negatively affected by pollution and toxic chemicals, but positively affected by increased soil temperature. Decomposition rates may therefore often be higher in urban than in rural soils. However, urban litter tends to have higher C:N ratios and therefore also tends to be more recalcitrant

Table 1 Ecological effects of urbanization

Physical and chemical environment	Population and community characteristics	Ecosystem structure and function
Air pollution increases	Altered reproductive rates	Altered disturbance regimes
Hydrological changes	Genetic drift, changes in selection	Altered succession
Local climate change	Social and behavioral changes	Altered decomposition rates
Soil changes	High species turn over, increase of exotic species	Altered nutrient retention
Water changes	Loss of K-species and gain of r-species	Habitat fragmentation
	Increased dominance of generalist species	Changes in trophic structure, domination of omnivores

Modified from McDonnell MJ and Pickett STA (1990) Ecosystem structure and function along urban–rural gradients: An unexploited opportunity for ecology. *Ecology* 71: 1232–1237.

than rural litter. Urbanization affects in complex ways both directly and indirectly C-pools and N-transformation rates and contrasting dynamics in urban and rural soils is an area where much more research is needed.

Biotic changes influencing ecosystem functioning are listed in **Table 1**. There are a number of reasons why new human-imposed scales for ecological processes are found within urban areas. First, compared with ecosystems in rural areas, urban systems are highly patchy and the spatial patch structure is characterized by a high point-to-point variation and degree of isolation between patches. Second, disturbances such as fire and flooding are suppressed in urban areas, and human-induced disturbances are more prevalent as well as intense human management of urban habitats. Third, because of the 'heat-island' effect, that is, higher mean temperatures in cities than in the surroundings, cities in temperate climates have significantly longer vegetation growth periods. Fourth, ecological successions are altered, suppressed, or truncated in urban green areas, and the diversity and structure of communities of plants and animals may show fundamental differences from those of nonurban areas. In general, with increasing urbanization there is a trend toward dominance of generalist species with high reproductive capacity and short generation times.

Urban Habitats and Gradient Analyses

Urban habitats are extremely diverse and examples include parks, cemeteries, vacant lots, streams and lakes, gardens and yards, campus areas, golf courses, bridges, air ports, and landfills (see Landfills). These habitats are highly dynamic, influenced by both biophysical and ecological drivers on the one hand and social and economic drivers on the other. Urban landscapes often represent cases of extreme habitat fragmentation. Habitat patches in the urban core are more or less strongly isolated from each other by a matrix of built environment making

dispersal risky and difficult at least for poorly dispersing organisms. There are numerous studies analyzing effects of isolation of urban habitats and, for example, in urban gardens in UK, the best predictor of species richness of ground arthropods was found to be the proportion of green areas within a 1 km radius of the sampling site. Analyses of the distribution of plant species in vegetation fragments in Birmingham, UK showed a positive correlation between the density of patches available to a species and the proportion of these patches that was occupied. For many plant species the rate of occupancy increased with site age, area, and similarity of adjacent habitats. Similarly, for urban amphibian assemblages in Melbourne, Australia, an increase in species richness was associated with pond size and a decrease with increasing isolation. Habitat quality also influenced species composition. The importance of isolation is likely to increase over time and, for example, in Boston an isolated urban park lost 25% of its plant diversity over 100 years. To what extent greenways and corridors increase connectivity and contribute to maintain viable populations in urban green areas is poorly understood. But greenways may, in multiple ways, provide a chain of different habitats permeating the urban environment and be of benefit for many organisms. Apart from preventing local extinction and facilitating re-colonization, increased habitat connectivity is important to maintaining vital biological interactions, for example, plant–pollinator interactions and plant–seed disperser interactions. Although most of the studies in urban landscapes address the continuous loss of green areas due to urban growth and expansion, this is not the case in all cities. For example, in Shanghai the proportion of green areas has increased in parallel with urban expansion and the total area expanded from less than $9 \, km^2$ in 1975 to more than $250 \, km^2$ in 2005.

Gradient analyses have a long tradition in ecology and go back to the pioneering work by Whittaker in the late 1960s. Gradient analyses have also been a rather common way of disentangling the complexity of urban habitats and have been used to investigate how urbanization changes ecological patterns and processes across landscapes, for example, in invertebrate, plant, and bird community composition, leaf litter decomposition and nutrient cycling, and the structure of landscape elements. Almost all the gradients that have been used for urban studies have been one dimensional in the sense that they only describe physical features of the gradient such as proportion of impervious surface, while the characteristics of the human population occupying a particular portion of the landscape often have been neglected. Because urbanization is an exceedingly complex amalgamation of factors, by using only a single axis the interpretation of the underlying processes has often been severely constrained. It has been suggested that a more comprehensive gradient analysis should include not only physical geography, demography, rates of ecological processes, and energy, but also history of land use, socioeconomic analyses, and patterns of management.

Variation in species densities across an urban gradient suggests that some individual species disappear with urbanization, whereas other species invade in response to the environmental changes associated with development. At least for birds, species richness has often been found to peak at intermediate levels of urbanization and decrease with either more or less development. Some species are classified as urban avoiders with their highest densities at the most natural sites, whereas many species seem to be able to adapt to suburban environments, with densities peaking at intermediate levels of development. Some species are urban exploiters whose highest densities are found at the urban core (see **Figure 2**).

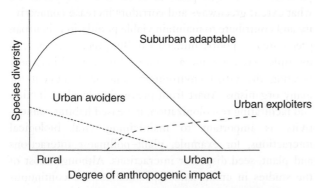

Figure 2 Plants and animals may respond differently to increasing human impact. Urban avoiders are large-bodied species or species linked to late successional stages. These species might be very sensitive and show a decline already at moderate human impacts. Suburban adaptable speceis may, to various degrees, utilize human modifications of the landscape; the majority of plant and animal species likely belong to this group. Urban exploiters directly benefit from human presence for food, reproduction or protection, and may often be cosmopolitan, generalist species. Terminology after Blair RB (1996) Land use and avian species diversity along an urban gradient. *Ecological Applications* 6(2): 506–519.

A multitude of factors are likely to influence this pattern of extinction and colonization, of which changes in predation rates have been suggested to be among the most important. Predation on artificial nests has often been found to be higher in urban parks than in neighboring woodlands and the abundance of predators such as corvids, rats, and house mice are often more in urban parks compared to the rural end of the gradient. However, there are also studies showing no correlation or a declining predation pressure along the urban–rural gradient. Observed patterns of extinction and invasion in urban landscapes may also be linked to gaps in the spectrum of body masses exhibited in the community and there are documented cases that body mass patterns are correlated with invasion and extinction in other human-transformed ecosystems.

In cities, ownership and management of urban habitats is extremely diverse and complex. In addition to land managed by government, municipalities, churches, and foundations, there is also land managed by local user groups that often covers substantial tracts. For example, domestic gardens cover 23% of the land area of Sheffield, and as much as 27% of the city of Leicester. Lands appropriated for allotment areas, domestic gardens, and golf courses were found to cover nearly 18% of the total green space of greater Stockholm, Sweden, representing well over twice the area covered by nature-protected areas. While the numbers of ecological studies of urban green areas are limited, there is evidence that different types of urban green areas used for purposes other than biodiversity conservation play an important role in sustaining urban biodiversity. For example, in many cities of Asia, educational institutions sometimes harbor the largest and last remaining green areas in extensively urban developed settings. These campus areas can be extremely significant for biodiversity. A good example is the university campuses of Pune city, India, which harbor up to half the plant, bird, and butterfly species of the region despite the fact that these campuses only cover some 5% of the land area. Another illustrative case is the Musahi Institute of Technology in Yokohama, Japan, where a former community-managed forest now restored for student education, has revived interest in reducing the loss of biodiversity-rich forests in semiurban areas of Japan. Ecological studies also show that domestic gardens sometimes hold a rich flora of plants, including rare and threatened ones. Thompson and colleagues found that the private gardens in Sheffield, UK, contained twice as many plant species as any other habitat assessed. These gardens also supported surprisingly high numbers of invertebrates and this regardless of whether garden plants were native or alien. Even such a controversial land use as golf courses can contribute with important biodiversity functions in cities when courses are wisely managed and well designed. For example, golf courses contribute to sustain urban woodlands in many cities of Japan, and in some larger cities of Sweden they may harbor significant populations of species of both amphibians and macro-invertebrates, which are declining in rural areas.

Also, smaller habitat parcels in urban settings can provide high-quality habitats. One illustrative example of this is allotment areas, common in many city-regions in developed countries. For example, while allotment areas only cover some 0.3% of the land in greater Stockholm, they tend to be extremely biodiversity rich. In Stockholm city one allotment garden was found to contain 447 different plant species in an area of 400 m^2.

Urban Systems and Ecosystem Services

The concept of ecosystem services has proved to be useful in describing human benefits from urban ecosystems. For example, urban vegetation may significantly reduce air pollution, mitigate the urban heat island effect, reduce noise, and enhance recreational and cultural values, of importance for urban citizen's well-being (**Table 2**).

The scale of importance for generation of these services is often much larger than a city, for example, for reduction of air pollutants and water regulation, while for some, such as recreational and educational services, generation often occur within city boundaries. In most cases, these services tend to be overlooked by urban planners and decision makers, despite the fact that the potential of generation of ecosystem services can be quite substantial. In a study made within Stockholm County, it was assessed that this region's ecosystems potentially could accumulate about 41% of the CO_2 generated by traffic and about 17% of total anthropogenic CO_2. In the Chicago region, trees were found to remove some 5500 t of air pollutants per

Table 2 Examples of services generated by urban ecosystems

Ecosystem services

Supporting services
Soil formation
Nutrient recycling

Provisioning services
Freshwater
Food, fiber, and fuel
Genetic resources

Regulating services
Air quality
Local climate regulation
Water purification and waste treatment
Biological control
Pollination

Cultural services
Esthetic and recreational
Educational

year, providing a substantial improvement in air quality. It was also found that the present value of long-term benefits from the trees of the Chicago region was more than twice the present value of costs related to planting. Moreover, wetlands in urban settings can substantially lower the amount of money spent on sewage treatment costs and in many cities large-scale experiments are taking place where wetlands are being used to treat sewage water. It has been estimated that up to as much as 96% of the nitrogen and 97% of the phosphorus can be retained in wetlands through the assimilation of wetland plants and animals. Green spaces of cities also provide ample opportunity for recreation. In a study on the response of persons put under stress, it was shown that when subjects of the experiment were exposed to natural environments the stress level decreased, whereas during exposure to built-up urban environments the stress levels remained high or even increased.

A major challenge in urban areas is how to sustain the capacity to generate ecosystem services. This capacity is mainly but not exclusively related to the diversity of 'functional groups' of species in a system, like organisms that pollinate, graze, predate, fix nitrogen, spread seeds, decompose, generate soils, modify water flows, open up patches for reorganization, and contribute to the colonization of such patches. In urban areas, such functional groups may be substantially reduced in size or show changes in the composition due to high species turnover, both of which may increase vulnerability in maintaining ecosystem services. To what extent exotic species contribute to reduce or enhance the flow of ecosystem services is virtually unknown for any urban area. But, since introduced species make up a large proportion of the urban biota, it is important to know not only to what extent introduced species are detrimental, but also to what degree some of the introduced species may enhance local diversity and maintain important functional roles. For urban ecology to significantly contribute to improving management of urban habitats and the maintenance of ecosystem services, the following research questions are particularly urgent to address:

- To what extent are urban ecosystems sinks for many animal and plant species and what are the effects of species loss on ecosystem functions?
- To what extent do novel species play important functional roles in urban ecosystems by replacing role of extinct native species and enhancing ecosystem functions and services?
- What is the importance of source–sink dynamics and matrix permeability for maintenance of urban biodiversity and important ecosystem services?
- How do we develop management systems that match the spatial and temporal scales of ecological processes?

Urban Restoration

Designed, replicated urban restoration experiments could substantially advance our knowledge and understanding of processes of importance for generating urban ecosystem services, for example, through better understanding of population and community responses to disturbances, patterns of self-organization and succession, assembly rules, and through identifying components that contribute to resilience or vulnerability. Urban restoration also represents an interesting opportunity for ecologists to work in partnership with landscape architects, urban designers, and architects and help in designing urban environments based on ecological knowledge but merged with the functional and esthetic design of urban space.

The majority of urban restoration projects deal with transforming brown areas (abandoned industrial lots or air fields, landfills, etc.) to functioning green areas, such as Fresh Kills landfill on Staten Island, New York, or the Olympic Park in Beijing. Other large-scale restoration projects involve, for example, substantial wetland restoration such as Kristianstad, Sweden, and New Orleans, USA. In New Orleans, coastal wetlands have eroded substantially during the last 50 years and restoring wetlands is viewed as one important measure to reduce vulnerability to hurricanes. Important lessons of urban restoration projects so far are that restoring ecological functions in urban areas is possible but time consuming and that there are often significant effects on many ecosystem services even after one or two seasons. Although the costs are initially high, these could be offset by increases in property values and increased investments in development in areas surrounding the restoration site.

Urban Landscapes as Arenas for Adaptive Management

In cities that experience rapid social and environmental transformation, it is critical to develop a capacity to respond to potential surprises. One important aspect of such capacity building is to facilitate for a wider integration of local people and interest groups in the use and management of urban green areas. There are several reasons for a wider integration of local people in urban ecosystem management. First, governments cannot entirely rest on protected area management to safeguard the native flora and fauna found in city-regions. As cities expand, there will be an increased lack of natural lands for the establishment of protected areas. Studies also show that many urban nature reserves are unable to sustain native species in the long run. In addition, protected

area management is financially costly to most local governments. In London, for example, parts of the protected green belt have been severely degraded due to lack of money and this has resulted in urban residents avoiding these areas for various activities. Second, much of the flora and fauna depend on well-functioning habitats provided by privately owned lands. In the US, for example, almost two-thirds of all the endangered and threatened species depend on private lands for their continued existence. Also, urban homeowners with gardens have been engaged to support declining pollinator populations in Great Britain through the deliberate planting of certain nectar providing plants in their yards. They have also helped sustain urban frog populations during their period of main rural declines through a massive establishment of garden ponds. Homeowners in Britain are also involved in programs for monitoring trends in the population status of birds. Third, a number of international treaties that have been signed by national governments around the world, including local Agenda 21, the Convention of Biological Diversity, and the Malawi-principles, strive toward a decentralization of biodiversity management down to local people. Recently, the Millennium Ecosystem Assessment (MA 2005) concluded that a wider cooperation among people within different sectors in society is necessary for more efficient land use that contributes to the support of ecosystem services.

One approach increasingly used to achieve collaborative partnership in urban ecosystem management is 'adaptive co-management'. The approach rests on the notion of the sharing of resource management responsibility and authority between users of ecosystems and government agencies. This typically involves local people and interest groups, scientists and local authorities, with the potential to promote information exchange to effectively deal with and respond to change and issues that often transcend locality. Adaptive co-management emphasizes 'learning-by-doing' in ecosystem management, where management objectives are treated as 'experiments' from which people can learn by testing and evaluating different management policies. This form of ecosystem management avoids set prescriptions of management that may be superimposed on a particular place, situation, or context. Such designs have the potential to lower overall costs of management, most notably costs incurred for describing and monitoring the ecosystem, designing regulations, coordinating users and enforcing regulations, and depend on the self-interest of participants. Co-management arenas could, for example, also serve as platforms for designed experiments and urban restoration as discussed above and improve ecological functions and designs in cities.

Linking Humans and Nature in the Urban Landscape

Urban landscapes are not only ecological experiments but also long-term experiments in social, economic, and cultural transformations shaped by cultures, property rights, and access rights. Since cities are places where knowledge, human and financial resources are concentrated, rapid urban transformations can likely be more readily monitored and observed than similar processes in more rural areas. Studies of transformations in urban landscapes may therefore well provide the ground for a better understanding of socio-economic drivers of changes also in other ecosystems. After decades of mutual neglect and artificial divide between nature on the one hand, and cities with their respective urban processes on the other hand, the conservation community has started to shift its perception to include cities as a component of natural landscapes. Just as it is now increasingly recognized that in protected nature reserves, conservation will not be successful as long as it is at the expense of human aspirations, urban planners increasingly acknowledge that functioning natural systems such as watersheds, mangroves, and wetlands are indispensable for reducing vulnerabilities to natural disasters and building long-term resilience.

In New Orleans for example, it has been argued that population growth and urban economic growth is necessary for meeting the costs of building a viable defense against the grave environmental problems of massive coastal erosion. In the New York Metropolitan region, sustainable management of the Catskills, the land around the upland water reservoirs supplying New York City with drinking water, has been chosen as an important complement to building water treatment plants.

The urban landscape provides a public space for the cross-fertilization of minds and various disciplines, enabling a new perspective on man in nature, one that could place human well-being at the core, break the artificial and largely culturally biased divide between the pristine and the human-dominated ecosystems, and contribute to the creation of a new language, with signs, concepts, words, tools, and institutions that would gather rather than divide, broker conflicts rather than create them, and establish responsible environmental stewardship at the heart of public interest.

See also: Landfills; Riparian Wetlands.

Further Reading

Adams CC (1935) The relation of general ecology to human ecology. *Ecology* 16: 316–335.

Adams CE, Lindsey KJ, and Ash SJ (2006) *Urban Wildlife Management*. Boca Raton: CRC Press, Taylor and Francis.

Alfsen-Norodom C (2004) Urban biosphere and society: Partnership of cities. *Annals of New York Academy of Sciences* 1023: 1–9.

Blair RB (1996) Land use and avian species diversity along an urban gradient. *Ecological Applications* 6(2): 506–519.

Colding J, Lundberg J, and Folke C (2006) Incorporating green-area user groups in urban ecosystem management. *Ambio* 35(5): 237–244.

Collins JP, Kinzig A, Grimm NB, et al. (2000) A new urban ecology. *American Scientist* 88: 416–425.

Felson AJ and Pickett STA (2005) Designed experiments: New approaches to studying urban ecosystems. *Frontiers in Ecology and the Environment* 10: 549–556.

Kinzig AP, Warren P, Martin C, Hope D, and Katti M (2005) The effects of human socioeconomic status and cultural characteristics on urban patterns of biodiversity. *Ecology and Society* 10(1): 23. http://www.ecologyandsociety.org/vol10/iss1/art23 (accessed December 2007).

McDonnell MJ and Pickett STA (1990) Ecosystem structure and function along urban–rural gradients: An unexploited opportunity for ecology. *Ecology* 71: 1232–1237.

McDonnell MJ and Pickett STA (1993) *Humans as Components of Ecosystems: Subtle Human Effects and the Ecology of Populated Areas*, 363pp. New York: Springer.

McGranahan G, Marcotullio P, Bai X, et al. (2005) Urban systems. In: Scholes R and Ash N (eds.) *Ecosystems and Human Well-being: Current State and Trends*, ch. 27, pp. 795–825. Washington, DC: Island Press. http://www.maweb.org/documents/document.296.aspx.pdf (accessed December 2007).

Millennium Ecosystem Assessment (2005) *Ecosystems and Human Well-being: Synthesis*. Washington, DC: Island Press.

Pickett STA, Cadenasso MI, Grove JM, et al. (2001) Urban ecological systems: Linking terrestrial ecological, physical and socioeconomic components of metropolitan areas. *Annual Review of Ecology and Systematics* 31: 127–157.

Sukopp H, Numata M, and Huber A (1995) *Urban Ecology as the Basis of Urban Planning*. The Hague: SPB Academic Publishing.

Turner WR, Nakamura T, and Dinetti M (2004) Global urbanization and the separation of humans from nature. *Bioscience* 54: 585–590.

Wind Shelterbelts

J-J Zhu, Institute of Applied Ecology, CAS, Shenyang, People's Republic of China

Introduction
Interactions between Wind and Trees
Shelterbelt Structures
Determination of Optical Porosity

Wind Profiles near Shelterbelts
Establishment and Management of Shelterbelts for
   Wind Protection
Further Reading

Introduction

The quantity of falling solar energy and the proportion that is absorbed by either the atmosphere or the surface varies greatly from one place to another, which results in regional variations in temperature both for Earth's surface and atmosphere. The variations of temperature cause a difference in atmospheric pressures, which leads to the movement of air from high-pressure area to low-pressure area, that is, the wind. Wind moves horizontally, vertically, and turbulently. It is affected by the conditions of the surfaces it encounters. The surface wind influences the habitats for wildlife, the growth of crop and livestock, soil erosion, snow distribution, sand-blowing, etc., and causes extreme damages when it is very strong.

Shelterbelt is always called windbreak (some authors distinguish the usage between the two terms based on their objectives; here no distinction is adopted), which can be defined as the barrier used to reduce wind speed. It usually consists of trees and shrubs, or even perennial or annual crops, wooden fences or other materials (**Figures 1** and **2**). Shelterbelts when they are reasonably designed can provide large areas of reduced wind speed because they increase the roughness. The areas of reduced wind speed, especially in windy regions, are generally called sheltered zones, which are very useful

Figure 2 Shelterbelt provides wind prevention in farmland.

for wildlife, agriculture, and for the people suffering from the severe climates. In fact, there are many ecological functions such as protecting against erosion, improving crop production, filtering air and water, ameliorating climatic extremes, improving ecological environmental qualities, reducing and ameliorating potential conflicts that may arise, fulfilled by the shelterbelts through altering wind behavior.

However, shelterbelt structural factors such as the height, density, orientation, length, width, continuity, cross-section shape, and the pattern of tree arrangements in windbreaks all influence shelter effectiveness. Therefore, how to manipulate shelterbelt structures through management practices to meet different objectives is one of the key scientific problems in windbreak studies. Obviously, shelterbelts involve comprehensively complex systems of establishment and management to satisfy multiple objectives.

Interactions between Wind and Trees

Wind Parameters

There are four parameters of wind that are measured: (1) wind direction (the direction from which the wind blows); (2) wind speed (measured from mechanical

Figure 1 Farmland shelterbelt.

Figure 3 Wind measurement with a three-dimensional ultrasonic anemometer.

anemometers, m s^{-1}) (**Figure 3**); (3) wind gusts and squalls (gust, a sudden significant increase of wind speed, the peak wind speed must reach at least 8.0 m s^{-1}, and the variation between peaks and lulls is at least 5.1 m s^{-1}, the duration is usually less than 20 s; squall, a sudden onset of strong winds with speeds increasing to at least 8.0 m s^{-1} and sustained at 11.2 m s^{-1} or more for at least 1 min); and (4) shifts (the changes of wind directions).

Ecological Effects of Wind

Wind has a wide variety of ecological effects, and plays an important role in the development of agriculture and forestry. For example, wind can transport water vapor, heat energy, pollen grains, spores, and seeds of plants, generate static electricity, and affect evaporation and transpiration, etc. In contrast, gale (wind speed from 14.3 to 28.3 m s^{-1}), can erode soil, do damage to farm, livestock, and to trees or forests.

Effects of Trees on Wind

Vegetations, especially trees, exert significant effects on winds close to the ground through altering wind behaviors. This is because the roughness of trees can provide surface friction, which makes the wind speed decrease, and the turbulence increase. Trees can be considered as a barrier on the land surface that changes the wind patterns. When wind approaches the trees, a portion of the airflow passes through the trees due to the porosity of leaves and branches. The remaining airflow is forced up and over the trees. When the wind moves through the trees, the surface area of leaves, branches, and the bark of stems provides a very frictional surface, which can reduce the wind speed effectively. Pressure on the ground increases as the wind approaches the trees and reaches a maximum at the windward edge of the trees. Meanwhile,

the pressure decreases as the wind passes through the trees, and reaches a minimum just to the lee side of the trees. The pressure gradually increases to the lee side, returning to the original pressure condition beyond a certain distance.

Tree Shelterbelts/Windbreaks

When the principles of the ecological effects of vegetations on wind are applied in practice, the shelterbelts are formed. It is reported that the origin of shelterbelts was in the middle of 1400 s when the Scottish Parliament urged the planting of tree belts to protect agricultural production. The primary purpose of a tree shelterbelt is to create a barrier that reduces the wind speed for preventing soil particles blowing away from fields. Shelterbelts are commonly established with one or more closely spaced rows, containing one or more tree species. Because a shelterbelt increases the surface roughness and thus reduces the wind speed when it is properly designed on the farmland or other areas that need protection, it can provide large areas with reduced wind speed for both agriculture, and people and animals. The main effect of a tree shelterbelt is to provide shelter. The effects of shelterbelts on wind (benefits) are more dependent on their structures, which human beings can manage or control.

Effects/Benefits of Shelterbelts

Effects of shelterbelts on microclimates. Shelterbelt can reduce soil evaporation rates when leaf area index is low. This may improve the soil water availability for the crop later in the season and thus improve seasonal water-use efficiency. The increase in diffuse radiation near a shelterbelt can increase light transmission into plant canopies. This may be a mechanism that could enhance photosynthetic activity of crops. Shelterbelts have thermal radiation effects too as a result of the sky view effect, where the cold night sky is replaced by warmer trees. A shelterbelt can reduce thermal radiation losses from the surrounding crop out to a certain distance from the shelterbelt. This may reduce frost incidence. Shelterbelts can lower soil and canopy temperatures through shading, which may limit growth of crops. However, we should also note that shading may decrease evaporation rates close to a shelterbelt and, therefore, reduce moisture losses from bare soil prior to germination and emergence of crops. Therefore, shelterbelts can provide improved growing conditions in the sheltered areas.

Effects of shelterbelts on improving crop yield and quality. Due to higher soil moisture, daytime temperatures, humidity, and nighttime carbon dioxide levels, as well as lower evaporation and nighttime air temperature in sheltered areas, the yield, quality, and maturity of crops are improved compared with unsheltered crops, particularly, in locations where snow accumulation or wind damage often occur.

Effects of shelterbelts on fauna and vegetation. Shelterbelts can provide wildlife habitats. Generally, macrofauna is lower in the field as compared to the shelterbelt and ecotones. The soil organic matter content, as well as microbial and faunal biomass, decrease gradually from the shelterbelt toward the field center. The shelterbelts influence the biomass, density, and composition of many soil and aboveground fauna taxa and individual size of animals occurring in bordering fields. In addition, the transport of pests, pollen and pathogens, etc., all rely on wind. Thus, providing shelter will modify the pathways for this around-the-crop environment or farmland.

A well-designed tree shelterbelt can increase the biological diversity that potentially may introduce natural predators to prey on pests and so reduce the need for pesticides.

Effects of shelterbelts on erosion control and others. Because of reduced wind velocity, erosion can be controlled by shelterbelts, especially on exposed sandy or dry soils. Well-designed shelterbelts can provide for uniform snow deposition/distribution on farmland across fields. Single-row tree shelterbelts are the most effective and permanent barriers for uniform snow distribution. Stockyards and around-farm buildings can also be protected through control of wind by shelterbelts. Additionally, esthetics and values in the landscape, tree-based resources such as timber, nuts, fiber or biomass, etc., are the benefits of shelterbelts.

Shelterbelt Structures

Shelterbelt structure can be considered as the distribution patterns of tree stems, branches, and leaves (tree elements) in a shelterbelt stand, which is determined by tree species, stem density, composition and the arrangement pattern of the trees in the shelterbelt, DBH (stem diameter at breast height, 1.3 m from the ground), tree height (H), tree age, etc. Additionally, the shelterbelt structure is also affected by its orientation, length, width, cross-sectional shape, continuity, and uniformity.

Internal Structural Characteristics

Porosity and density. The most commonly applied descriptor of the internal structure of shelterbelts is porosity. Shelterbelt porosity is defined as the ratio or percentage of pore space to the space occupied by tree elements. The optimum aerodynamic porosity (β_a) is usually considered to be 0.35–0.45. Porosity affects the turbulence level in and around the shelterbelt. As porosity increases, more wind passes through the shelterbelt, that is, less wind-speed reduction, but the protected distance increases and

turbulence generation decreases. In contrast, as porosity decreases, less wind passes through the shelterbelt and wind-speed reduction is greater, but the extent of the protected distance decreases with more turbulence generation.

Shelterbelt density (β_d) is the ratio of the solid portion of the shelterbelt to the total volume of the shelterbelt. It has the same meaning as the term of porosity, that is, the ratio of the open portion of the shelterbelt to the total volume. The two terms are complementary.

Optical porosity and optical density. Although shelterbelt porosity or density is important in describing the shelterbelt structure, unfortunately, it is nearly impossible to physically measure the aerodynamic porosity of plants because of the three-dimensional nature of the pores through which the wind flows. Therefore, much effort has been directed toward finding an alternative measurement. Optical porosity (β), a two-dimensional measure of porosity, which is defined as a simple ratio of perforated area to total area on the vertical section of a shelterbelt, has been employed as a descriptor of shelterbelt structure. It has proved to be a promising alternative to aerodynamic porosity, especially for narrow shelterbelts. Generally, optical porosity is not equivalent to aerodynamic porosity since it does not take into account the three-dimensional nature of the pores, but for a narrow artificial windbreak, β is close to β_a.

Optical density (β_r) has the reciprocal or complementary implication as optical porosity. It is defined as the ratio of projected solid area to the total side view area of a shelterbelt.

External Structural Characteristics

Height. Shelterbelt height (H) is the most important factor determining the extent of the protected zone. Both theory and empirical measurements have shown that the protected distance is proportional to shelterbelt height. The shelterbelt height varies according to tree species, site conditions, and management levels, and keeps increasing before the maturity of the shelterbelt. The shelterbelt height can be described by the maximum height of individual trees, the average height to the tops of the taller trees or as the height averaged over randomly located points along the length of the shelterbelt. In order to facilitate comparisons of shelter effectiveness between shelterbelts, the protected distance from the shelterbelt is usually expressed in times of the shelterbelt height.

Orientation. Shelterbelt orientation is another important factor to shelterbelt structure. The maximum shelter effectiveness is generally obtained when the shelterbelt is oriented perpendicular to the problem wind. If the direction of the problem wind becomes oblique to the windbreak, the size of protected area decreases.

When the wind becomes parallel to the windbreak, wind speed may increase under some conditions.

Length. The length of a shelterbelt varies greatly from area to area. The influence of the shelterbelt length on shelter effectiveness lies on the ends of the shelterbelt. Wind speeds at the end of a windbreak are generally greater than those at the open area. This is caused by the flow around the ends of the shelterbelt, where the wind seeks the path with least resistance (**Figure 4**).

Width. The width of a shelterbelt (W) affects the shelter effectiveness by influencing the porosity or optical porosity of the shelterbelt. Generally, shelterbelt can be widened by adding rows, but additional width of shelterbelt has minimal influence on shelter response until the ratio of width to height (W/H) of a shelterbelt is around 5 (**Table 1**). However, shelter effectiveness decreases with increasing width when the ratio of W/H is greater than 5 (**Figure 5**). Numerical simulation results suggest that the complex interactions between the shelterbelt width and the internal shelterbelt structure may be more important than previously believed; and the simulated data indicate that as the width increases, the location of maximum reduction of wind speed moves closer to the shelterbelt.

Cross-sectional shape. The cross-sectional shape of a shelterbelt is the external profile of the cross section, which is described by the geometry of its boundary. The windbreak shapes are formed by the layout of the trees, shrubs, the composition of tree species and their arrangement patterns within the windbreak. The aerodynamics of cross-sectional shape of shelterbelts indicates that the cross-sectional shapes affect the magnitude and extent of wind-speed reduction in the protected zone. There are many shelterbelt cross-sectional shapes; the most common shapes include rectangle, triangle (windward and leeward), gable roof/pitched roof (symmetry and asymmetry), and notch (**Figure 6**). The observations in the field model windbreaks indicated that the windbreaks with rectangle cross-sectional shape exhibit better shelter effectiveness (**Figure 7**). It is reported that windbreak cross-sectional shape has more effective influence than horizontal distribution on shelter effectiveness.

Continuity and uniformity. The continuity and uniformity of shelterbelts influence shelter efficiency. Any gaps or separations in a shelterbelt will concentrate wind flow, which creates a zone on the leeward side of the gap or separation where the wind speed exceeds that on the open field (**Figure 4**). Therefore, lanes or other openings through a windbreak should be avoided.

Aerodynamic Parameters of Shelterbelt

Apparent porosity. Apparent porosity (β_0) is defined as eqn [1], which is an estimate of porosity based on the minimum relative wind speed. The minimum relative wind speed is considered as a function of β_0:

$$\beta_0 = -0.2 + \sqrt{0.04 + 3.2 U_{r-min}} \qquad [1]$$

where U_{r-min} is minimum relative wind speed (m s^{-1}).

Based on the measurement of the minimum wind speed on the leeside of a windbreak, β_0 can be obtained. β_0 is generally used as a standard for comparison of the characteristics of diverse shelterbelts rather than a subjective verbal characterization.

Penetration coefficient. Penetration coefficient (α_0) is also called permeability or aerodynamic porosity. The mean value of α_0 is defined under the conditions of an infinitely long shelterbelt on flat ground with the z-axis upward, x-axis horizontally perpendicular, and y-axis parallel to the shelterbelt as

$$\alpha_0 = \frac{\dfrac{1}{H}\displaystyle\int_{z_0}^{H} U(0,z)\,\mathrm{d}z}{\dfrac{1}{H}\displaystyle\int_{z_0}^{H} U_0(z)\,\mathrm{d}z} = \frac{\displaystyle\int_{z_0}^{H} U(0,z)\,\mathrm{d}z}{\displaystyle\int_{z_0}^{H} U_0(z)\,\mathrm{d}z} \qquad [2]$$

where z is height above the ground (m), z_0 is roughness length (m), $U(0, z)$ and $U_0(z)$ are horizontal wind speeds at the leeward of the shelterbelt and at the same height in the open area (m s^{-1}), respectively, and H is the shelterbelt height (m).

Resistance coefficient. The resistance coefficient (R_c) is defined as the ratio of the difference in pressure across the shelterbelt to the product of the shelterbelt height and the dynamic pressure of the approaching flow. The aerodynamic properties of a porous shelterbelt are more directly affected by the resistance coefficient.

Drag coefficient. Drag coefficient (C_d) is defined as the ratio of the drag force per unit area on the shelterbelt to the dynamic pressure of the approaching flow. In principle, the shelterbelt exerts a drag force on the wind field, causing a net loss of momentum in the incompressible airflow and thus producing the sheltering effect. A common physical way to express the aerodynamic effect of a shelterbelt is in terms of its resistance to the flow, or in terms of dimensionless form such as a drag coefficient

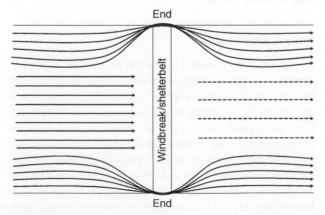

Figure 4 Flowing pattern when wind passes through and around the ends of the shelterbelt.

Table 1 Relationships between windbreak width and relative wind velocity (%)[a]

Width W (m)	Height H (m)	W/H	Rows	Optical porosity β	Relative wind speed at various distances from the shelterbelt (%)							
					−5 H	0 H	5 H	10 H	15 H	20 H	25 H	Total
54	12	4.5	25	0.320		44	44	62	72	89	89	62
5	10	0.5	3	0.335	51	75	33	52	54	64	62	55

[a]Wind direction is perpendicular to the shelterbelt.

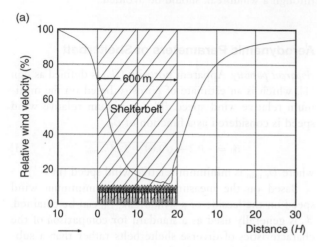

(a)

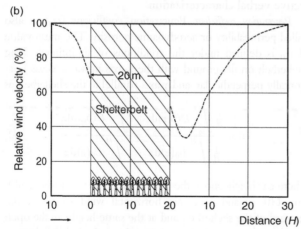

(b)

Figure 5 Comparison of protected distances between two types of shelterbelts: (a) wide shelterbelt ($H = 28$ m, $W = 600$ m) and (b) relatively narrow shelterbelt ($H = 28$ m, $W = 20$ m).

$$C_d = \frac{2D}{\rho U_c^2 H} \qquad [3]$$

where D is drag force (N), ρ is air density ($kg\ m^{-3}$), U_c is reference wind velocity ($m\ s^{-1}$), which is usually replaced by U_H, the wind speed at H.

Determination of Optical Porosity

Generally, when dealing with artificial windbreaks such as salt fences, stubble barriers, and other windbreaks consisting of nonliving materials, the calculation of β is usually a simple mathematical exercise. But when dealing with living shelterbelts, the estimation of β is much more complicated because the openings of the living shelterbelt are irregular in shape and in distribution. The method of digital image processing for measuring β has been applied. The basic principle is to partition the pores and the tree elements using computer system through digital images. The steps of the processes for determining β are as follows.

Photographing

In order to get an image with higher resolution, monochrome photographs should be taken during calm days without too strong light to avoid excessive reflection from the tree elements. Photographs should be taken as close as possible to the shelterbelt for ensuring that the smallest pores or tree elements are resolved by the computer system. In practice, a mark at a certain height (almost the same as the eye level of the photographer) is fixed on one tree of the front row before taking the photograph. Then, the photographer focuses the mark perpendicular to the shelterbelt and takes one photograph. The photographer moves 20 m left or right from the position and another two photographs are taken. The distance between the photographer and the shelterbelt usually changes according to the type of camera, for example, in a Ricoh KR-10, $f = 50$ mm with a zoom of 52 mm, the distance between the photographer and the windbreak can be as long as about 100 m.

Image Processing

In the laboratory, digital photographs (the photographic negatives should be digitized) of shelterbelts can be divided into two parts in the computer system, that is, the crown and the trunk, according to the composition of shelterbelts. Gray tones produced by light reflected from tree elements are treated the same as transmitted light of the same intensity. The light intensity of each pixel could have a value ranging between 0 (black) and 255 (white). It was possible to assess each of the pixels in the photograph by selecting a threshold value for the intensity of the transmitted light between 0 and 255. Each pixel above or below the threshold value was automatically assigned a binary value of false or

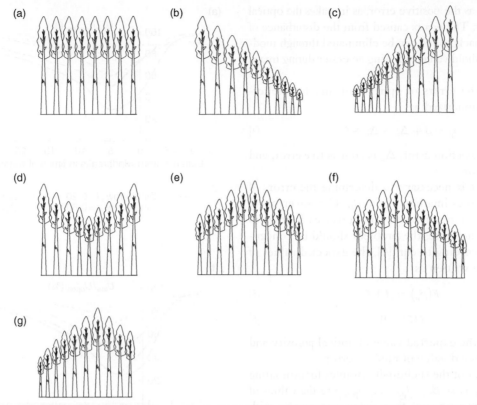

Figure 6 Sketches of various cross-sectional shapes for shelterbelts: (a) rectangle, (b) windward triangle, (c) leeward triangle, (d) notch, (e) gable roof/pitched roof (symmetry), and (f, g) gable roof/pitched roof (asymmetry).

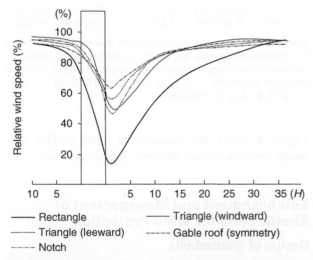

Figure 7 The shelter effectiveness of various cross-sectional shapes for model windbreaks (optical porosity = 0.60, rows = 7, height = 1.85 m).

true, respectively. A preliminary threshold value is selected and the entire image is digitized. The digitized image formed by pixels being either true or false can be displayed. If the pores or tree elements are omitted, the process can be repeated by replacing different threshold values until the representative image is obtained. Once the representative image is achieved, the software can provide the total

number of pixels of the image and the tree elements, respectively. Then, the optical porosity of a shelterbelt can be obtained.

Error Analysis of Optical Porosity

As the optical porosity of a shelterbelt is estimated from photographs through image processing, the formation of shelterbelt image is the centricity of the true shelterbelt. Particularly, if the shelterbelt is composed of more than two rows, the projective error and the contractive error from the camera must be produced. Therefore, some errors exist between the optical porosity (β_p) estimated from images and the optical porosity (β) of true shelterbelts. For these reasons, it is necessary and useful if the errors can be deleted or limited during the image processing.

The errors of β_p can be attributed to the following factors: (1) the characteristics of the formation of shelterbelt image, depending on whether it can reflect the tree elements truly or not; (2) the disturbance of the windbreak background; and (3) random errors, including the errors produced by the camera, situation of the sample shelterbelts, and the measuring processes. As the shelterbelt image is a centric projection, both projection and contractive errors should exist between true tree shelterbelt and the imaged shelterbelt. The projection error is defined as the negative error, as it makes the optical porosity larger, and contractive error is defined as the positive error, as it makes the optical porosity smaller. The errors caused from the disturbance of the windbreak background can be eliminated through modulating the threshold value and using an eraser during image processing.

Therefore, the errors of optical porosity for shelterbelts can be summarized as

$$\beta_p = \beta + \Delta_p + \Delta_c + \xi \qquad [4]$$

where Δ_p is projection error, Δ_c is contractive error, and ξ is random error.

Obviously, it is necessary to determine the errors of Δ_p, Δ_c, and ξ for estimating β from β_p. The two errors of projection and contractive can counteract each other to some extent. Therefore, the problem should be concentrated on ξ. According to eqn [4], the expected values of β_p and ξ can be written as

$$E(\beta_p) = \beta + \xi \qquad [5]$$

$$E(\xi) = 0 \qquad [6]$$

where $E(\beta_p)$ is the expected value of optical porosity and $E(\xi)$ is the expected value of random error.

If the number of the shelterbelt samples for estimating optical porosity is m, β_{p1}, β_{p2}, ..., β_{pm}, are the values of optical porosity estimated from image processing with random errors. The estimation of $E(\beta_p)$ and ξ for a shelterbelt can be obtained as

$$E(\beta_p) = \overline{\beta}_p = \frac{1}{m}\sum_{i=1}^{m}\beta_{pi} \qquad [7]$$

$$\xi = \sum_{i=1}^{m}\beta_{pi} - \overline{\beta}_p \qquad [8]$$

where $\overline{\beta}_p$ is the mean value of optical porosity estimated from the image without deleting errors.

Wind Profiles near Shelterbelts

Wind profiles near the shelterbelts have been intensively studied because of their importance related to shelter effectiveness. The wind profiles near the shelterbelts are influenced by many factors from both shelterbelt structures and climatic conditions. Although shelterbelt length, cross-sectional shape, and width may influence shelter effectiveness at leeward, tree height and shelterbelt porosity are more important. Obviously, shelter effectiveness is proportional to tree height if other factors are all the same. Therefore, porosity is considered as one of the most important key shelterbelt structural indices in determining the effectiveness in wind reduction. Generally, the observations of wind reduction were carried out in both field and wind tunnel experiments.

(a)
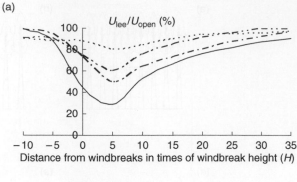

Single row $\beta = 0.20$ ...... Single row $\beta = 0.40$
Single row $\beta = 0.60$ –·–·– Single row $\beta = 0.80$

(b)
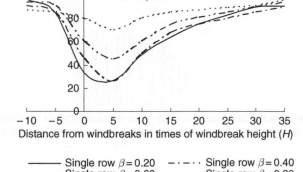

Single row $\beta = 0.20$ –·–·· Single row $\beta = 0.40$
Single row $\beta = 0.60$ ...... Single row $\beta = 0.80$

Figure 8 Wind profiles near the model shelterbelt in relation to the composition of rows and the optical porosity, U_{lee} and U_{open} are wind speeds in open field and the leeward area (m s^{-1}). (a) one row, optical porosity $\beta = 0.20$, 0.40, 0.60, and 0.80; (b) two rows, $\beta = 0.15$, 0.26, 0.47, and 0.69.

Figure 8 shows the examples of wind profiles near model shelterbelts with different optical porosities.

Establishment and Management of Shelterbelts for Wind Protection

Design of Shelterbelts

The designing of shelterbelts is determined by the objectives of shelterbelt establishment. The goal of any shelterbelt is to provide favorable microclimate conditions to landowners; these conditions are obtained by altering wind patterns directly or indirectly. Therefore, there are general principles that apply to the majority of situations.

Shelterbelt orientation. Field shelterbelts should be oriented perpendicular to the prevailing winds in order to maximize the protected zone leeward. This orientation minimizes the number of shelterbelts needed to protect a given area. The orientation can vary within a certain

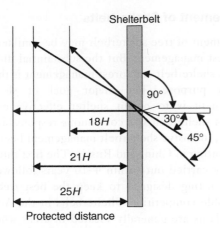

Figure 9 Effects of angles of shelterbelt to wind direction on effectively protected distance (*H*).

range according to the situations of the mechanical equipment and land use because the protected zone is not reduced greatly when the angle between the shelterbelt and the direction of prevailing winds is less than a certain degree (e.g., 30°) (**Figure 9**).

Intervals between shelterbelts. The protected zone of a shelterbelt is limited, but the requirements for protection usually exceed these limited influences. Therefore, a system of properly oriented shelterbelts should be established. The number of shelterbelts that are required to provide protection for a given area is directly related to the average height of the tallest trees in the shelterbelt. Intervals between adjacent shelterbelts are determined by tree height, structure of the shelterbelt, and problem wind velocity. Typically, the interval or distance between adjacent shelterbelts should range from 10*H* to 25*H*,

depending on the extent of protection desired and the size of the field.

Tree species selection and composition. Many factors such as local climate and soil conditions, wind firmness, features of tree species (height, crown spread, competitiveness), compatibility with the crops in the farmland and pest problem, etc., determine the selection of tree and shrub species for a shelterbelt. Species adaptability is the most critical of all the selection factors. The most desirable tree species for a variety of shelterbelt uses should be relatively free of diseases and pest problem, with a narrow crown and deep roots, a certain potential height and long-lived. Generally, native tree species are usually a good choice.

Species combination. Most shelterbelts consist of one tree species, but mixed shelterbelts can make use of the site conditions efficiently, improve the stability and pest and disease resistance, and provide good shelterbelt structures if the mixed tree species are combined reasonably. However, there are few mixed tree shelterbelts in practice because of the difficulties in planting and managing. **Table 2** lists the sample results of mixed shelterbelts. The mixed tree species include *Ulmus pumila*, *Populus xiaozhuanica*, *Salix matsudana*, *Pinus sylvestris* var. *mongolica*, and shrubs *Amorpha fruticosa*, *Lespedeza bicolor*. Five mixing patterns, that is, mixing between trees in one row, row–mixing (dissymmetry and symmetry), mixing among segments, and mixing among shelterbelts, are formed.

Shelterbelt structure/optical porosity. According to the results obtained on the basis of wind reduction experiments, an optimum optical porosity exists. Shelterbelts with low or high optical porosity are generally ineffective for wind protection. There is considerable variation in the

Table 2 Arrangements of mixing patterns for shelterbelts[a]

Mixed patterns	Arrangement types (4 rows)	Stem density
Mixing between trees in one row	1. *Salix – Populus – Salix – Populus*	2.0 m × 2.0 m
	2. *Ulmus – Populus – Ulmus – Populus*	2.0 m × 2.0 m
Mixing in rows (dissymmetry)	3. *Salix – Populus – Salix – Populus*	2.0 m × 2.0 m
	4. *Ulmus – Populus – Ulmus – Populus*	2.0 m × 2.0 m
Mixing in rows (symmetry)	5. *Salix – Salix – Pinus – Pinus*	2.0 m × 2.0 m
	6. *Pinus – Pinus – Populus – Populus*	2.0 m × 2.0 m
	7. *Ulmus – Ulmus – Populus – Populus*	2.0 m × 2.0 m
	8. *Ulmus – Populus – Populus – Ulmus*	2.0 m × 2.0 m
	9. *Salix – Populus – Populus – Salix*	2.0 m × 2.0 m
	10. *Populus – Populus – Salix – Salix*	2.0 m × 2.0 m
Mixing among segments	11. *Populus* (500 m) – *Pinus* (500 m) – *Ulmus* (500 m)	2.5 m × 2.5 m
Mixing among shelterbelts	12. *Populus* (500 m), *Pinus* (500 m)	2.0 m × 2.0 m

[a]All the shelterbelts were composed of 4 rows with shrubs. The shrubs were planted aside the shelterbelts with density of 1 m × 1 m. 10 years later (in 2002), the mixed shelterbelts of type (5) and type (6) failed because of the growth difference between *Pinus* and the deciduous species. The other shelterbelts such as type (11) and type (12), which mixed with *Pinus*, *Populus*, and *Ulmus*, succeeded. Among the mixed shelterbelts of deciduous tree species, type (8) and type (9) exhibited better patterns because the relatively fast-growing species (*Populus*) were planted in the inner rows, and the relatively slow growing species (*Salix* and *Ulmus*) were planted in the outer rows. This arrangement made full use of the edge effect of the shelterbelts, i.e., the tree species growing more slowly with age should be planted in outside rows and those growing faster should be planted in the inner rows for mixed shelterbelts.

results of optimum optical porosity. The variation might be caused by the differences in shelterbelt structures, the effects of thermal instability in the field, the type of instruments used, and the method used to determine the optical porosity. Despite these differences, most studies have suggested that better sheltering is obtained by shelterbelts with porosities ranging from 0.20 to 0.50. The characteristics of foliages and branches of individual tree or shrub are important in determining the shelterbelt porosity or optical porosity. Therefore, porosity or optical porosity can be modified by changing tree species or the spacing within and between tree rows; however, their effects on shelterbelt porosity are not well studied.

Spacing within and between rows. Spacing within and between tree rows varies with regions, tree species, desired density or optical porosity, and the number of rows. Generally, within-the-row spacing is as follows: shrubs, 1 m; trees, from 2.0 to 3.0 m according to the row number of the shelterbelts.

Planting arrangements. The planting arrangements of trees and shrubs in a shelterbelt determine the shelterbelt structure, and further influence the shelter efficiency. Generally, there are three types of planting arrangements (i.e., rectangle, triangle, and random). For maximizing the benefits of shelterbelts, the planting arrangements of multirow shelterbelts should be in triangle patterns. This is because the triangle planting arrangement of trees in a shelterbelt is favorable to the shelterbelt structure associated with shelter efficiency.

Length and width. The influence of the shelterbelt length on shelter effectiveness lies on the ends of the shelterbelt (**Figure 4**). For this reason, the length of a shelterbelt should be at least 10 times as long as its mature height for minimizing such an influence. The shelterbelt width influences the shelter effectiveness through changing the porosity of the shelterbelt. In wide or multirow shelterbelts, however, tree competition may weaken the stability of the shelterbelt stand; in order to solve this problem, mixed shelterbelts as described in **Table 2** (i.e., type 8 and type 9) are recommended. The problem of the single- or two-row shelterbelt is the potential damage of continuity when the shelterbelt loses some of trees. Therefore, the shelterbelt width should be as narrow as possible under the condition of high preserved rate of trees, that is, the smallest proportion of land devoted to tree shelterbelts and the largest complete protection for the remaining land.

Competition zone. The most common negative comments concerning tree shelterbelts are related to the impact of competition between trees and the adjacent crops, especially under conditions of limited moisture. The extent of competition varies greatly with crop species, tree species, geographic location, and weather conditions. Some types of competition can be reduced by root pruning, that is, cutting of the lateral tree roots extending into crop field.

Management of Shelterbelts

Management of tree shelterbelt may be similar to general forest management. But the substantial difference between shelterbelt and forest management is the management purpose. The major goal of shelterbelt management is to obtain shelter effectiveness. The care of a shelterbelt is a continuous responsibility, for example, intensive shelterbelt management begins soon after planting in China and Russia. The first thinning is generally carried out within 4–10 years followed by a release cutting designed to keep the best trees from undesirable competition. Shelterbelts in the American Great Plains are generally located in areas where climatic and soil conditions are not conducive to the natural occurrence and regeneration of trees and shrubs. The management of shelterbelt is to keep all grass and weeds out of shelterbelt until crown closure (by the time when the shade is too dense for weed to grow). The success of a shelterbelt establishment depends not only on the initial design, species selection, and site preparation, but also on the subsequent care and the management level. There are numerous silvicultural techniques that can be used to maintain a shelterbelt well beyond the life expectancy of the original tree planting. Generally, shelterbelt systems can provide more sheltering benefits than an individual shelterbelt. The spatial patterns and the future development of shelterbelt systems can be more easily and clearly exhibited in the landscape level. Thus, landscape consideration can provide more important references for the management of shelterbelts system.

Further Reading

Brandle JR, Hodges L, and Wight B (2000) Windbreak practices. In: Garrett HE, Rietveld WJ, and Fisher RF (eds.) *North American Agroforestry: An Integrated Science and Practice*, pp. 79–118. Madison: American Society of Agronomy.

Caborn JM (1965) *Shelterbelts and Microclimate*. London: Faber and Faber.

Cao XS (1983) *Shelterbelt for Farmland*. Beijing: Chinese Forestry Press, (in Chinese).

Ennos AR (1997) Wind as an ecological factor. *Trends in Ecology and Evolution* 12: 108–111.

Everham EM (1995) A comparison of methods for quantifying catastrophic wind damage to forest. In: Coutts MP and Grace J (eds.) *Wind and Trees*, pp. 340–357. Cambridge: Cambridge University Press.

Heisler GM and DeWalle DR (1988) Effects of windbreak structure on wind flow. *Agricultural Ecosystems and Environment* 22/23: 41–69.

Jiang FQ, Zhu JJ, Zeng DH, et al. (2003) *Management for Protective Plantation Forests*. Beijing: China Forestry Publishing House.

Kenney WA (1987) A method for estimating windbreak porosity using digitized photographic silhouettes. *Agricultural and Forest Meteorology* 39: 91–94.

Loeffler AE, Gordon AM, and Gillespie TJ (1992) Optical porosity and windspeed reduction by coniferous windbreaks in Southern Ontario. *Agroforestry Systems* 17: 119–133.

Peltola H, Kellomaki S, Kolstrom T, et al. (2000) Wind and other abiotic risks to forests. *Forest Ecology and Management* 135: 1–2.

Ruck B, Kottmeier C, Matteck C, Quine C, and Wilhelm G (2003) Preface. In: Ruck B, Kottmeier C, Matteck C, Quine C, and Wilhelm G (eds.) *Proceedings of the International Conference Wind Effects on Trees*, pp. iii. Karlsruhe, Germany: Lab Building, Environment Aerodynamics, Institute of Hydrology, University of Karlsruhe.

Zhou XH (1999) *On the Three Dimensional Aerodynamic Structure of Shelterbelts*. PhD dissertation, Graduate College at the University of Nebraska.

Zhou XH, Brandle JR, Takle ES, and Mize CW (2002) Estimation of the three-dimensional aerodynamic structure of a green ash shelterbelt. *Agricultural and Forest Meteorology* 111: 93–108.

Zhu JJ, Gonda Y, Matsuzaki T, and Yamamoto M (2002) Salt distribution in response to optical stratification porosity and relative windspeed in a coastal forest in Niigata, Japan. *Agroforestry Systems* 56(1): 73–85.

Zhu JJ, Matsuzaki T, and Jiang FQ (2004) *Wind on Tree Windbreaks*. Beijing: China Forestry Publishing House.

INDEX

NOTES:

Cross-reference terms in italics are general cross-references, or refer to subentry terms within the main entry (the main entry is not repeated to save space). Readers are also advised to refer to the end of each article for additional cross-references - not all of these cross-references have been included in the index cross-references.

The index is arranged in set-out style with a maximum of three levels of heading. Major discussion of a subject is indicated by bold page numbers. Page numbers suffixed by *t* and *f* refer to Tables and Figures respectively. *vs.* indicates a comparison.

This index is in letter-by -letter order, whereby hyphens and spaces within index headings are ignored in the alphabetization. Prefixes and terms in parentheses are excluded from the initial alphabetization.

hydrologic cycle
  schematic diagram, 365f
landfills, 305
salt marsh ecosystems, 391
savanna ecosystems, 398f
temperate forest biomes, 423
tropical ecosystems, 439–440
United States, 261
Evechinus chloroticus, 21, 22f
Everglades, Florida (USA)
  biodiversity, 279f
  Florida Everglades mesocosm project
    air/water movement, 284f
    algal biomass, 293f
    ecosystem subunits, 285f
    fish distribution, 295f
    general discussion, 290
    invertebrate abundance, 294f
    physico-chemical parameters, 292t
    plan diagram, 291f
    plant communities, 291f
    structural characteristics, 283f
    vertical/longitudinal section, 292f
    water supply, 284f
  lagoon ecosystems, 296–297
  water availability, 277
evergreen forests
  alpine ecosystems, 159t
  physiographic regions, 417, 419t
  savanna ecosystems, 397
evolution
  self-organization, 105
  desert environments, 230
exergy, 4–5, **128–139**, 5t
  anthropogenic effects, 136, 136t
  calculation techniques, 132
  definition, 128, 128f
  dissipative structure, 132
  eco-exergy
    basic concepts, 130, 130f
    ecological systems, 34
    exergy calculations, 132, 134t
    fundamental ecosystem
      theory, 34, 36f
    living organisms, 132, 134t
  ecological systems
    irreversible processes, 34
  emergence/emergent properties, 95t
  fundamental ecosystem theory, 34, 36f
  information theory, 131, 131f, 135
  loss calculations, 136, 136t
  nitrogen-phosphorus ratio plot, 139f
  organic matter, 132, 134t, 137, 138t
  thermodynamic hypothesis
    basic concepts, 137
    Le Chatelier's Principle, 137
existence, 122, 124t
exorheic environments, 218t
exotic species
  forest plantations, 267
  freshwater lakes, 273
  savanna ecosystems, 404
  See also invasive species
exploitative competition
  network environ analysis, 80t
  research areas, 83, 83f
  extinction models
    desert environments, 239, 240f
    grasses, 240
    invasive species, 240
    predators, 239
  extinction rates
    urban environments, 464
  Mediterranean ecosystems, 323
  tundra ecosystems, 449
extreme environments
  coral reefs, 205, 205f
  desert environments, 226, 230

desert stream ecosystems, 216, 218t
tundra ecosystems, 444, 447
Exuma Cays, Bahamas, 212f

F

Fabaceae
  Mediterranean ecosystems, 320t
facilitation models, 245
Fagus spp.
  Mediterranean ecosystems, 321t
  temperate forest biomes, 417
Fagus sylvatica, 158
fairness, *See* cooperation
fairy shrimp, 433, 436
falcons
  tundra ecosystems, 447
farm forest plantations, 264
farming systems
  bee pollination, 28
    desert environments, 237
    historical background, 237
  shelterbelts/windbreaks, 469
fauna, *See* animals
Faviidae, 288t
feces
  coral reefs, 206
feedback
  emergence/emergent properties, 95t
  food webs, 101, 102f
  fundamental ecosystem theory, 34, 36t
  orientation theory, 124, 124t, 125f
  self-organization, 101, 102f, 104
  simple systems, 107–108
Fennoscandia, 181
fens
  characteristics, 175, 275, 333,
    334, 429t
  development factors, 332f
  largest wetlands, 281
  patterned fens, 335, 336f
  poor fens, 335
  rich fens, 335, 335f
  water availability, 277
ferns
  Mediterranean ecosystems, 327
  temporary water bodies, 433
Ferrissia spp., 371
fertilization
  forest plantations, 269
  indirect interactions, 87
Festuca abyssinica, 448
Festuca rubra, 243f
Festuca spp., 88
Ficus spp., 259
Fiji, 441
filamentous organisms, 168f, 177
Fimbrisitylis bahiensis, 244
fine particulate organic matter (FPOM), 349,
  352f, 354f, 356, 358f, 368
fingernail clams, 371, 372f, 436
Finland
  boreal forests, 181
  peatlands, 331
  polar ecosystems, 339–342
Finn, JT, 53–54
Finn's cycling index, 52
fire-dependent regeneration, 195–196
Firehole River, Yellowstone National Park
  (USA), 357f
first law of thermodynamics, *See* exergy;
  thermodynamic laws
fir trees
  boreal forest ecosystems, 181
fir waves, 420
fish/fisheries
  African floodplain ecosystems, 259–260
    mangrove forests, 313–314
    saline/soda lakes, 384

Australian floodplain ecosystems, 261
competition studies, 83
coral reefs
  competition, 203
  importance, 203, 203f, 207
  overfishing, 205, 205f, 210f,
    211, 211f
  positive interactions, 205
  predators, 204
  recruitment effects, 206
desert stream ecosystems, 217–218, 229
eco-exergy losses, 136
ecosystem dynamics, 21, 22f, 23f
estuarine ecosystems, 249, 252
exergy calculations, 134t
floodplain ecosystems
  Africa, 259–260
  Australia, 261
  marshes, 276
  South America, 262–263
Florida Everglades mesocosm
  project, 295f
food webs
  upwelling ecosystems, 455, 455f
  freshwater ecosystems, 373
  indirect interactions, 85
  intertidal environments, 375, 377
  lake ecosystems
    saline/soda lakes, 381, 382, 384
  mangrove forests, 310, 312
  marshes, 278
  Mediterranean ecosystems, 322, 322t
  overfishing
    coral reefs, 205, 205f, 210f,
      211, 211f
    freshwater lakes, 274
    indirect interactions, 85
  recreational fishing, 384
  recruitment
    optimal environmental window (OEW),
      457, 457f
    upwelling ecosystems, 457, 457f
  riparian systems, 344
  salinity
    saline/soda lakes, 381, 382, 384
    salinity tolerances, 373
  salt marsh ecosystems, 386, 387, 389
  South American floodplain ecosystems,
    262–263
  stable isotope analysis, 21
  stream ecosystems
    characteristics, 373
    desert environments, 217–218, 229
  temporary water bodies, 433, 434
  trophic transfer efficiency, 19
  upwelling ecosystems, 451
    climate change effects, 458, 458f
    fish production, 457, 457f
    food webs, 455, 455f
    general discussion, 455
    small pelagic fish, 455, 456f
    trophic levels, 457
Fissurellidae, 288t
fitness variance
  implications
    fragmented populations, 105, 105f
flagellates
  agroecosystems, 148
  estuarine ecosystems, 249
  indirect interactions, 86
  temporary water bodies, 436–437
  upwelling ecosystems, 453, 455f
flagged branching effects, 156, 157f,
  161f, 162
flamingoes
  desert environments, 225f
  marshes, 278, 279f
  saline/soda lakes, 381f, 382, 384

Printed and bound by CPI Group (UK) Ltd, Croydon, CR0 4YY

03/10/2024

01040328-0020